Prof. PAL G. BERGAN

The Norwegian Institute
of Technology
Division of Structural
Mechanics
Trondheim, Norway

Prof. K. J. BATHE

Department of Mechanical
Engineering
MIT
Cambridge, MA, USA

Prof. W. WUNDERLICH

Institut für Konstruktiven Ingenieurbau
Ruhr-Universität Bochum
Bochum, Federal Republic of Germany

ISBN-13 :978-3-642-82706-8 e-ISBN-13 :978-3-642-82704-4
DOI : 10.1007 / 978-3-642-82704-4

Libary of Congres Cataloging in Publication Data.
Main entry under title:
Finite element methods for nonlinear problems.
Papers presented at the Europe-US Symposium on Finite Element Methods for Nonlinear Problems,
Norwegian Institute of Technology, Trondheim, Aug. 12–16, 1985.
1. Finite element method -- Congresses. 2. Nonlinear theories -- Congresses.
I. Bergan, Pål G. II. Bathe, Klaus-Jürgen. III. Wunderlich, W. (Walter)
IV. Europe-US Symposium on Finite Element Methods for Nonlinear Problems
(1985: Norwegian Institute of Technology)
TA347.F5F59 1986 620'.001'515353 85-30453

Softcover reprint of the hardcover 1st edition 1986

2161/3020 543210

Finite Element Methods for Nonlinear Problems

Proceedings of the Europe-US Symposium
The Norwegian Institute of Technology, Trondheim
Norway, August 12–16, 1985

Edited by
P. G. Bergan, K. J. Bathe, W. Wunderlich

With 317 Figures

Springer-Verlag
Berlin Heidelberg New York Tokyo

ALLE ZEIT WACH 1842

Preface

This book contains a collection of papers presented at the Europe-US Symposium on Finite Element Methods for Nonlinear Problems. The symposium was held at the Norwegian Institute of Technology, Trondheim, Norway during August 12 to 16, 1985.

The finite element method has during recent years gained a position as the most important discipline in computational mechanics. The basis for this method was laid out about two decades ago, and linear finite element techniques are today well established and well understood. Much work is still being done in order to make these linear methods more efficient and reliable. However, a substantial part of the current research efforts in the finite element field is focused on developing the nonlinear capabilities of the method. This task is highly challenging and demanding, both from a theoretical and practical point of view.

It was in this spirit that the Europe-US Symposium on Finite Element Methods for Nonlinear Problems was organized. The meeting may be seen as the continuation of the US-Germany Symposium on Finite Element Methods held in 1976 at MIT, Cambridge, USA and the Europe-US Workshop on Nonlinear Finite Element Analysis in Structural Mechanics held in 1980 at the Ruhr-Universität, Bochum, West-Germany.

Invitations for the symposium were sent to some of the most prominent research workers in the field in Europe and USA. The response was extremely positive, and 44 papers were presented during the week of the symposium. These papers represent a cross-section of the major research efforts in nonlinear finite element analysis at present. Some key words for these activities are: large strains and large rotations, improved material modelling, solution of nonlinear equations, stability and collapse of plates and shells, contact problems, nonlinear dynamics, fluid problems, efficient formulation of algorithms and computer implementations. The symposium also contained two workshop sessions with short presentations of special topics and open discussions. Written material from these sessions is not available because of their informal character.

It is hoped that the present proceedings will be of good use to researchers and practicing engineers working with finite element problems.

The editors, who served as the international organizing commitee, would like to thank all contributors who have made the symposium and this book possible. Thanks are also extended to T. Jørgensen, P.K. Larsen. S. Remseth and T. Søreide, who, in addition to the first editor, served on the local organizing commitee. The symposium would not have been possible without the substantial economical support given by Det norske Veritas, Norsk Hydro, Norwegian Contractors, Saga Petroleum, Statoil and The Norwegian Institute of Technology. All this help is greatly appreciated. Thanks are also extended to Springer-Verlag for their speedy publication of these proceedings.

October 1985

P.G. Bergan
K.J. Bathe
W. Wunderlich

Contents

Part I:
Fundamental Problems

On the Accuracy and Efficiency of Numerical Algorithms for Geometrically Nonlinear Structural Analysis

K. MATTIASSON, A. BENGTSSON, A. SAMUELSSON
Chalmers University of Technology, Gothenburg, Sweden

Summary

Procedures for the numerical analysis of geometrically non-linear problems, with special reference to beam, plate, and shell type structures, are considered. Various formulations, such as the Total and Updated Lagrangian and the Co-rotational formulations, and various numerical solution procedures are investigated from the view-point of accuracy and numerical efficiency.

1. Introduction

In the present paper some aspects of the formulation and solution of geometrically nonlinear problems will be treated. The study is concerned with small strain/large rotation problems, typically encountered in slender, flexible structures such as beams, plates, and shells.

The first point to be discussed is the choice of 'best' kinematic formulation for this kind of problems. The formulations to be treated are the Total Lagrangian (TL), Updated Lagrangian (UL), and Co-rotational (CR) formulations. These are identical from a continuum mechanics point of view, the only difference being the choice of reference configuration. However, in discretized form, applied to a structural element based on some approximating theory, the performances of the various formulations are usually quite different.

The merits and drawbacks of the various formulations in application to the above mentioned type of problems are discussed. The conclusions drawn from this theoretical discussion are verified in a couple of numerical examples. In these examples only beam elements and large deflection beam problems are

Finite Element Methods for Nonlinear Problems
Europe-US Symposium, Trondheim, Norway 1985
Editors: Bergan, Bathe, Wunderlich

studied. It is our opinion that the study of simpler structures can yield valuable insight into the more complex shell problem. At least one can be sure of that a formulation which is inefficient in beam analysis will also be inefficient in shell analysis.

The second point to be treated herein is the choice of suitable solution procedure for the present type of problems. In later years a number of solution strategies capable of tracing the complete deformation path over and beyond limit points have been presented. The desire is to have a solution procedure that can handle all types of problems and is efficient in all situations. In the present study we have tried to answer the question: how efficient are some of these procedures in application to geometrically nonlinear problems not exhibiting limit points? Comparisons are made with the conventional true and modified Newton solution methods. Numerical results are again given for a nonlinear beam problem.

2. Equilibrium. The principle of virtual displacements

The equilibrium of a body is readily expressed by means of the principle of virtual displacements. Using the *second Piola-Kirchhoff stress tensor* $\tilde{S}$ and the *Lagrangian strain tensor* $\tilde{E}$ as a suitable pair of energy conjugate stress and strain measures, the virtual work in a Lagrangian description of motion can be expressed as

$$\int_{V_R} \tilde{S}:\delta\tilde{E}\, dV_R = \int_{S_R} \tilde{t}_R\cdot\delta\tilde{u}\, dS_R + \int_{V_R} \tilde{f}_R\cdot\delta\tilde{u}\, dV_R \tag{1}$$

V_R and S_R are the volume and surface area, respectively, in an arbitrary known reference configuration. The vectors $\tilde{t}_R$ and $\tilde{f}_R$ are body forces per unit surface area and volume, respectively, in the reference state.

Let dS and ds denote the length of a line element in reference and current configuration, respectively. The Lagrangian strain tensor $\tilde{E}$ and its Cartesian components E_{ij} are then defined by

$$ds^2 - dS^2 = 2d\tilde{X}\cdot\tilde{E}\cdot d\tilde{X} = 2E_{ij}dX_i dX_j \tag{2}$$

For a rigid-body motion the difference $ds^2 - dS^2$ is equal to zero. The components of $\underset{\sim}{E}$ must in that case be equal to zero, and $\underset{\sim}{E}$ is thus an *objective* tensor.

The Lagrangian strain tensor can be expressed in terms of the *deformation gradient tensor* $\underset{\sim}{F}$ as

$$\underset{\sim}{E} = \frac{1}{2}(\underset{\sim}{F}^T \cdot \underset{\sim}{F} - \underset{\sim}{I}), \quad E_{ij} = \frac{1}{2}\left(\frac{\partial x_k}{\partial X_i}\frac{\partial x_k}{\partial X_j} - \delta_{ij}\right) \tag{3}$$

The second Piola-Kirchhoff stress tensor $\underset{\sim}{S}$ can also be shown to be an objective tensor. It is related to the true, Cauchy stress tensor $\underset{\sim}{\sigma}$ by

$$\underset{\sim}{S} = \frac{\rho_R}{\rho}(\underset{\sim}{F}^{-1}) \cdot \underset{\sim}{\sigma} \cdot (\underset{\sim}{F}^{-1})^T, \quad S_{ij} = \frac{\rho_R}{\rho}\frac{\partial X_i}{\partial x_k}\sigma_{kl}\frac{\partial X_j}{\partial x_l} \tag{4}$$

where ρ and ρ_R are mass densities in current and reference configuration, respectively.

3. Interpretation of stress and strain tensors in the case of small strains/large rotations

Noting the relation between the reference position vector $\underset{\sim}{X}$, the current position vector $\underset{\sim}{x}$, and the displacement vector $\underset{\sim}{u}$

$$\underset{\sim}{x} = \underset{\sim}{X} + \underset{\sim}{u}, \quad x_i = X_i + u_i \tag{5}$$

we can express the Lagrangian strain components E_{ij} in terms of displacement gradients as

$$E_{ij} = \frac{1}{2}(u_{i,j} + u_{j,i}) + \frac{1}{2}(u_{k,i}u_{k,j}) = \varepsilon_{ij} + \eta_{ij} \tag{6}$$

where comma denotes differentiation with respect to reference coordinates, and ε_{ij} is the *linear Lagrangian* or *engineering strain tensor*.

Consider now a small part of a continuum in a fixed Cartesian coordinate system X_i, i = 1, 2, 3, with base vectors $\hat{\underset{\sim}{I}}_i$. Apply a *finite rotation*, defined by the orthogonal rotation tensor $\underset{\sim}{R}$, so that a vector $d\underset{\sim}{X}$ turns into a new vector $d\underset{\sim}{X}^*$, where $d\underset{\sim}{X}^* = \underset{\sim}{R} \cdot d\underset{\sim}{X}$ and $|d\underset{\sim}{X}| = |d\underset{\sim}{X}^*| = dS$. Let, furthermore, X_i^*, i = = 1, 2, 3, be another Cartesian coordinate system with base vectors $\hat{\underset{\sim}{I}}{}_i^*$, which is initially coinciding with X_i, but rotating with the body.

Now, give the body an *infinitesimal deformation* so that the vector $d\underset{\sim}{X}^*$ turns into $d\underset{\sim}{x}$. The vector $d\underset{\sim}{x}$ has the component forms $d\underset{\sim}{x} = dx_i \hat{\underset{\sim}{I}}_i = dx_i^* \hat{\underset{\sim}{I}}{}_i^*$. We, furthermore, have the relation $d\underset{\sim}{X} + d\underset{\sim}{u} = d\underset{\sim}{X}^* + d\underset{\sim}{u}^*$, where $d\underset{\sim}{u}$ is the relative displacement vector between two neighbouring particles.

Let $|d\underset{\sim}{x}| = ds$. The difference $ds^2 - dS^2$ can thus, in view of Eq. (2), be written

$$\begin{aligned} ds^2 - dS^2 &= 2d\underset{\sim}{X}\cdot\underset{\sim}{E}\cdot d\underset{\sim}{X} = 2E_{ij}dX_i dX_j = \\ &= 2d\underset{\sim}{X}^*\cdot\underset{\sim}{E}^*\cdot d\underset{\sim}{X}^* = 2E_{ij}^* dX_i^* dX_j^* \end{aligned} \tag{7}$$

It should be emphasized that the displacement gradient du_i/dX_j is of finite magnitude, while $du_i^*/dX_j^* << 1$. This implies that the quadratic part of the strain tensor E_{ij}^* can be neglected, and we can write $E_{ij}^* = \varepsilon_{ij}^*$. Noting also that $dX_i = dX_i^*$, we find $E_{ij} = E_{ij}^*$, and finally

$$E_{ij} = \varepsilon_{ij}^* \tag{8}$$

This result can be stated in words as follow: *In case of small strains, but large rotations, the Lagrangian strain tensor components are equal to the engineering strain components in a system rigid-body rotating with the particle.*

According to the *polar decomposition theorem* the deformation gradient tensor $\underset{\sim}{F}$ can be multiplicatively decomposed into a *rotation tensor* $\underset{\sim}{R}$ and a *stretch tensor* $\underset{\sim}{U}$ as

$$\underset{\sim}{F} = \underset{\sim}{R}\cdot\underset{\sim}{U} \tag{9}$$

For *infinitesimal strain* the stretch tensor can be expressed as

$$\underset{\sim}{U} \approx \underset{\sim}{I} + \varepsilon\underset{\sim}{A} \tag{10}$$

where ε is a small number and $\underset{\sim}{A}:\underset{\sim}{A} \approx \underset{\sim}{I}:\underset{\sim}{I}$.

The deformation gradient can thus be approximated by

$$\underset{\sim}{F} \approx \underset{\sim}{R} \tag{11}$$

Noting that $\rho_0/\rho \approx 1$ in case of infinitesimal strain, we can express the second Piola-Kirchhoff stress tensor as

$$\underset{\sim}{S} = (\underset{\sim}{F}^{-1})^T \cdot \underset{\sim}{\sigma} \cdot (\underset{\sim}{F}^{-1}) \approx \underset{\sim}{R} \cdot \underset{\sim}{\sigma} \cdot \underset{\sim}{R}^T \tag{12}$$

The component forms of the Cauchy stress tensor $\underset{\sim}{\sigma}$ are $\underset{\sim}{\sigma} = \sigma_{ij} \hat{\underset{\sim}{I}}_i \hat{\underset{\sim}{I}}_j = \sigma^*_{ij} \hat{\underset{\sim}{I}}^*_i \hat{\underset{\sim}{I}}^*_j$, where $\hat{\underset{\sim}{I}}^*_i$ is a base vector in the previously introduced co-rotating Cartesian coordinate system.

The matrices of tensor components σ_{ij} and σ^*_{ij} are thus related by

$$[\sigma] = [R]^T [\sigma^*][R] \tag{13}$$

where [R] is the matrix of the components of $\underset{\sim}{R}$ in the fixed system.

Combining Eqs. (12) and (13), we finally get

$$[S] = [R]([R]^T[\sigma^*][R])[R]^T = [\sigma^*] \tag{14}$$

This result can be interpreted as: *In case of small strains the second Piola-Kirchhoff stress components are equal to the Cauchy stress components in a system rigid-body rotating with the particle.*

The important conclusion one can draw from the above discussion is that a constitutive relation, formulated for the case of infinitesimal strain *and* rotation, and relating Cauchy stress and engineering strain or rates of these quantities, can be used unaltered to relate Lagrangian strain and second Piola-Kirchhoff stress. For instance, constitutive relations given in the form

$$\underset{\sim}{\sigma} = f(\underset{\sim}{\varepsilon}) \quad \text{or} \quad \dot{\underset{\sim}{\sigma}} = \underset{\sim}{C} : \dot{\underset{\sim}{\varepsilon}} \tag{15}$$

can be rewritten as

$$\underset{\sim}{S} = f(\underset{\sim}{E}) \quad \text{or} \quad \dot{\underset{\sim}{S}} = \underset{\sim}{C} : \dot{\underset{\sim}{E}} \tag{16}$$

4. Finite element formulations for small strain/large rotation problems

In the present section the finite element equations for beams, plates, and shells will be established in a rather general form. The Euler-Bernoulli hypothesis for beams and the Kirchhoff hypothesis for plates will be assumed to hold. This implies that shear deformations are neglected and that normals to the midplane remain straight and normal. Strains and stresses perpendicular to the midplane/centre line are neglected. In the following we will assume that the xy-plane of a local coordinate system is coinciding with the midplane of the plate/shell element, or that the x-axis of the local system passes through the end-points of the beam element in the reference configuration.

The strains at the midplane/centre line in the current configuration are then found to be

$$\text{plate:}\quad E^0_{\alpha\beta} = \tfrac{1}{2}(u_{\alpha,\beta} + u_{\beta,\alpha}) + \tfrac{1}{2}u_{\gamma,\alpha}u_{\gamma,\beta} + \tfrac{1}{2}w_{,\alpha}w_{,\beta} \qquad \alpha,\beta,\gamma = 1,2 \tag{17}$$

$$\text{3D beam:}\quad E^0_{xx} = u_{,x} + \tfrac{1}{2}u^2_{,x} + \tfrac{1}{2}(v^2_{,x} + w^2_{,x}) \tag{18}$$

$$\text{2D beam:}\quad E^0_{xx} = u_{,x} + \tfrac{1}{2}u^2_{,x} + \tfrac{1}{2}w^2_{,x} \tag{19}$$

In the above strain-displacement relations the displacement components u, v, and w refer to the x-, y-, and z-directions, respectively.

The element nodal rotation and transverse displacement components are collected in a vector denoted $\{\tilde{u}_b\}$ (b as in bending), and the in-plane nodal displacement components are collected in a vector $\{\tilde{u}_m\}$ (m as in membrane). The displacement-components associated with bending and membrane action, respectively, $\{\tilde{u}_b\}$ and $\{\tilde{u}_m\}$, can be approximated as

$$\{u_b\} = [\phi_b]\{\tilde{u}_b\} \tag{20}$$

$$\{u_m\} \quad [\phi_m]\{\tilde{u}_m\} \tag{21}$$

where $[\phi_b]$ and $[\phi_m]$ are matrices with interpolation functions. The strain components in a arbitrary point in the element can be expressed in matrix form as

$$\{E\} = ([B_L^m] + \frac{1}{2}[B_{NL}^m])\{\tilde{u}_m\} + ([B_L^\kappa] + \frac{1}{2}[B_{NL}^b])\{\tilde{u}_b\} \quad (22)$$

The strain matrices $[B_L^m]$, $[B_{NL}^m]$, and $[B_{NL}^b]$ correspond to the first, second, and third term, respectively, in Eqs. (17)-(19). The matrix $[B_L^\kappa]$ gives the contribution to the strain due to curvature, which usually is approximated as second derivative of transverse displacement. The matrix is a function of the z-coordinate in case of a plate or a 2D beam, and of both y- and z-coordinates in case of a 3D beam.

The relation between strain rate and nodal velocities is given by

$$\{\dot{E}\} = ([B_L^m] + [B_{NL}^m])\{\dot{\tilde{u}}_m\} + ([B_L^\kappa] + [B_{NL}^b])\{\dot{\tilde{u}}_b\} \quad (23)$$

The equilibrium equations in discretized form can now, in view of Eqs. (1) and (23), be written

$$\{\tilde{U}_b\} = \int_{V_R} ([B_L^\kappa]^T + [B_{NL}^b]^T)\{S\}\, dV_R \quad (24a)$$

$$\{\tilde{U}_m\} = \int_{V_R} ([B_L^m]^T + [B_{NL}^m]^T)\{S\}\, dV_R \quad (24b)$$

where $\{\tilde{U}_b\}$ and $\{\tilde{U}_m\}$ are nodal force vectors, and $\{S\}$ is a vector with second Piola-Kirchhoff stress components.

Collecting the nodal forces in one vector $\{\tilde{U}\}$ and the strain matrices in one matrix $[B]$, and taking the time derivative, we obtain the equilibrium equations in rate form as

$$\{\dot{\tilde{U}}\} = \int_{V_R} ([B]^T\{\dot{S}\} + [\dot{B}]^T\{S\})\, dV_R \quad (25)$$

Introducing the constitutive relation in rate form

$$\{\dot{S}\} = [C]\{\dot{E}\} \quad (26)$$

we can rewrite Eq. (25) as

$$\{\dot{\tilde{U}}\} = [K]\{\dot{\tilde{u}}\} \quad (27)$$

where

$$[K] = [K_L] + [K_{LD}] + [K_S] \tag{28}$$

and

$$[K_L] + [K_{LD}] = \int_{V_R} [B]^T[C][B]\, dV_R \tag{29}$$

$$[K_S]\{\dot{\tilde{u}}\} = \int_{V_R} [\dot{B}]^T\{S\}\, dV_R \tag{30}$$

The tangent stiffness matrix $[K]$ can formally be divided in the linear stiffness matrix $[K_L]$, the large displacement stiffness matrix $[K_{LD}]$ which is a function of current displacements, and the initial stress stiffness matrix $[K_S]$ which is a function of current stress.

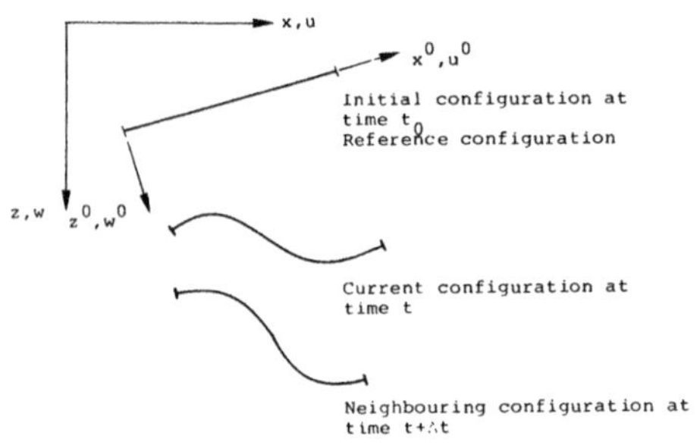

(a) TL-formulation

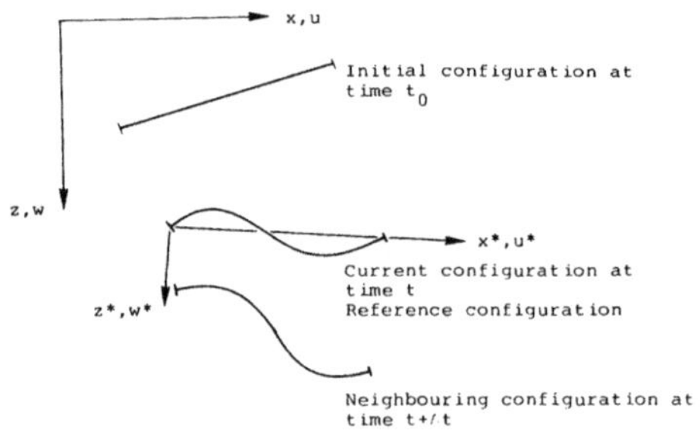

(b) UL-formulation

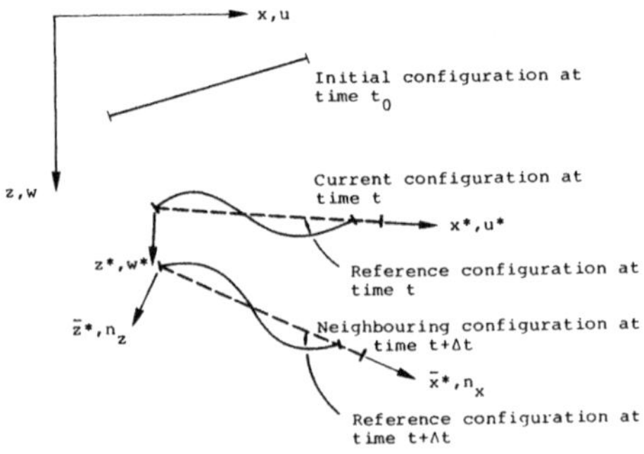

(c) CR-formulation

Fig.1. Various kinematic formulations illustrated for a beam in 2D space

4.1 The Total Lagrangian formulation

In the TL-formulation the *initial* configuration is taken as reference configuration. The TL-formulation is illustrated for a 2D beam element in Fig.1a. The tangent stiffness matrix is assumed to be established at time t, and strains, stresses, and nodal forces are calculated at time t+Δt. Displacements are measured from the initial configuration.

4.2 The Updated Lagrangian formulation

In the UL-formulation the *current* configuration is taken as reference configuration. Since the displacements are measured from the current configuration at time t, the $[K_{LD}]$ matrix will not enter the formulation. The nodal forces at time t+Δt are calculated according to Eq. (24a,b). The nonlinear strain matrices in these expressions, $[B^b_{NL}]$ and $[B^m_{NL}]$, are functions of incremental displacements during the time step Δt. For the present types of finite elements a local coordinate system is usually attached to each element and is redefined in every iterative step of the calculations. The UL-formulation is illustrated in Fig.1b.

4.3 The Co-rotational formulation

The CR-formulation is probably less well-known than the previous two ones. The theoretical bases of the formulation are explained in Mattiasson et. al. [1,2], where also reference is made to a number of works in which the CR-formulation has been used.

In the CR-approach a local Cartesian coordinate system is attached to each finite element, and *continously* translates and rotates with the element as the deformations proceed. In the TL-form of the CR-approach (CR-TL) the reference configuration is taken as the initial configuration, but translated and rotated in accordance with the motion of the co-rotating local system. In the UL-form of the CR-approach (CR-UL) the translated and rotated configuration at time t is taken as reference configuration during the time step Δt. The CR-TL-approach is illustrated in Fig.1c for a 2D beam element.

A finite element formulation based on the CR-approach and the use of so-called 'natural' variables is described in Refs. [1,2]. The natural element displacements represent the minimum number of geometric variables necessary to completely determine the deformations of the element. The tangent stiffness matrix can in the CR-formulation be written

$$[K] = [A]^T[K^*][A] \tag{31a}$$

$$[K^*] = [E]^T[k][E] + [K_N^*] \tag{31b}$$

The matrix $[K^*]$ is the global tangent stiffness matrix referred to local coordinate axes at time t, and [A] is a transformation matrix. The matrix [k] is the 'natural tangent stiffness matrix', established for the natural variables in the co-rotating system, and [E] is a transformation matrix that relates the global and natural variables. The matrix $[K_N^*]$, finally, is a kind of 'initial stress matrix' and is a function of natural nodal forces.

The global nodal forces at time $t+\Delta t$ are calculated as

$$\{\tilde{U}\} = [A]^T[E]^T\{\tilde{N}\} \tag{32a}$$

$$\{\tilde{N}\} = \int_{V_R} ([b_L]^T + [b_{NL}]^T)\{S^*\}\, dV_R \tag{32b}$$

where $[b_L]$ and $[b_{NL}]$ are 'natural strain matrices', and $\{\tilde{N}\}$ is the natural nodal force vector.

5. Approximations and their effects on the accuracy in various formulations

The object of the present section is to discuss and analyse the effects of certain approximations which are frequently made in the analysis of the present type of structures. The results of this discussion are by no means claimed to be original. These are well-known facts to many, but certainly not to all researchers in the field. Anyway, we have found it important to shed light on these matters.

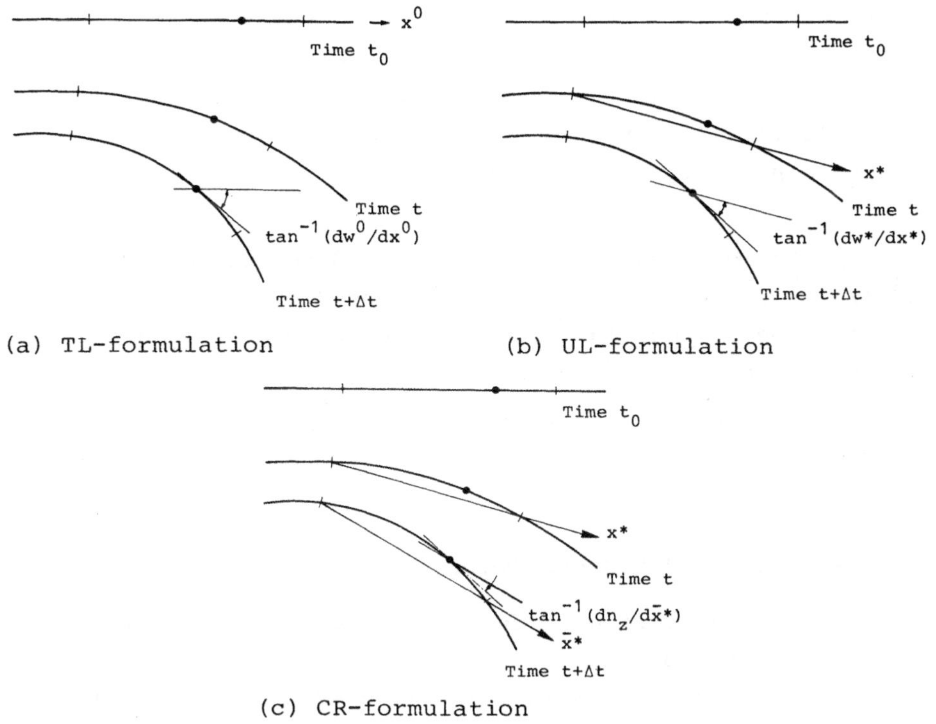

Fig.2. The angle of inclination at a point of a beam in 2D space in various formulations

5.1 Approximation of curvature

In most theories of beams, plates, and shells curvatures are approximated by second derivatives of transverse displacement. We will now discuss under what circumstances such an approximation is justified and when it is not. For simplicity we will consider a beam element in 2D space. In Fig.2 the inclination angle at time t+Δt of a beam element, measured relative various coordinate systems, is shown. The curvature κ can be expressed as

$$\kappa = -\frac{d^2w^0/dx^{0\,2}}{[1 + (dw^0/dx^0)^2]^{3/2}} = -\frac{d^2w^*/dx^{*2}}{[1 + (dw^*/dx^*)^2]^{3/2}} =$$

$$= -\frac{d^2n_z/d\bar{x}^{*2}}{[1 + (dn_z/d\bar{x}^*)^2]^{3/2}} \tag{33}$$

It is obvious that in a TL-formulation the approximation $\kappa \approx -d^2w^0/dx^{02}$ is *not* justified in problems involving large rotations, since the gradient dw^0/dx^0 may be of the order one. It is also noted that the error done by introducing such an approximation is not influenced by the element size or the step length.

In a UL-formulation usuallv $dw^*/dx^* << 1$ and the approximation $\kappa \approx -d^2w^*/dx^{*2}$ is justified. The error caused by this approximation is, however, depending on the step size and, to a certain degree, on the element size.

In a CR-formulation the approximation $\kappa \approx -d^2n_z/d\bar{x}^{*2}$ is fully justified. The error is to some extent influenced by the element size, but not by the step size.

5.2 Approximative strain-displacement relation

The in-plane displacement gradients are often considered to be so small that the squares of them can be neglected. This implies that the second term in Eqs. (17)-(19), or the matrix $[B^m_{NL}]$ in Eqs. (22)-(24), is neglected. In large displacement problems the in-plane displacement gradients are mainly due to bending. Take for instance a line element dx in the xz-plane, which is given a rigid rotation with the displacement dw at the right end. The displacement gradient du/dx can then easily be shown to be

$$\frac{du}{dx} = -(1 - \sqrt{1 - (\frac{dw}{dx})^2}) \tag{34}$$

This shows that the in-plane displacement gradient is depending on the transverse displacement gradient, or, to put it shortly, that the conditions for the present approximation to be justified are exactly the same as for the curvature approximation discussed above.

5.3 Approximate volume integration

The volume integrations can often be carried out exact in the TL- and CR-TL-formulations, since the elements in the initial state usually have a simple shape. In the UL- and CR-UL-formulations these integrations are carried out over the deformed element which have a complicated shape. This complicated integration domain is, however, in practice often approximated by the initial, simple domain.

For this reason the CR-TL-formulation is considered to be obviously superior to the CR-UL-formulation in the present kind of problems, and the latter formulation will therefore not be treated in the numerical examples. In the UL-formulation the error caused by the approximation mentioned above is strongly influenced by the element size.

6. Numerical examples

In this section the conclusions drawn in the preceding section will be verified in a couple of numerical examples. For the present purpose it is fully adequate and very illustrative to study elastic, plane beam problems. This choice also has the advantage that 'exact' solutions are available for some test problems in form of elliptic integral solutions.

The beam elements used in this study are all based on the Euler-Bernoulli beam theory, and assumed third order polynomials for both transverse and axial displacements. The nodal degrees of freedom are the rotations and the axial and transverse displacements at the two end points, plus the axial displacements at two internal nodal points.

The highest order polynomial terms in the volume integrations are of the order eight. The integrations along the length of the beam are carried out numerically by means of four Gauss points, which implies that a polynomial of the order seven is integrated exactly. The use of three Gauss points has been tested in some cases.

6.1 Example one: Cantilever beam with a transverse pointload

The Total Lagrangian formulation:

The 'exact' load-displacement curves for the cantilever beam in Fig.3a are shown in Fig.3b, together with results from a FEM-solution with eight elements and the TL-formulation. It is obvious that the TL-formulation can be expected to give reliable

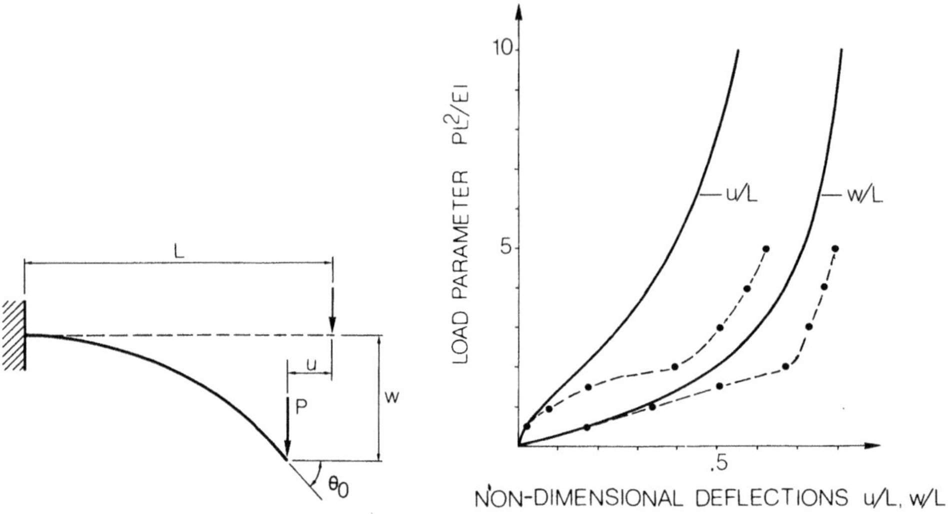

(a) (b)

Fig.3. (a) Cantilever beam with a transverse point load

(b) 'Exact' load-displacement curves (solid lines) [3, 4], and FEM-solution with eight elements and a TL-formulation (dots)

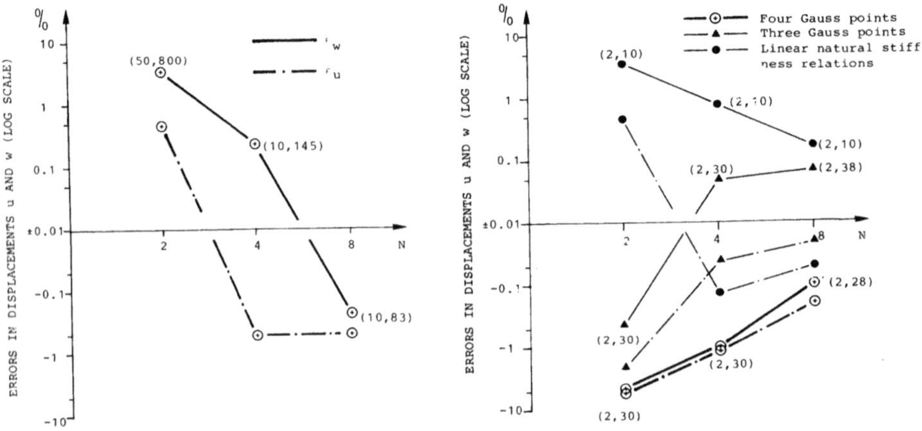

(a) UL-formulation (b) CR-formulation

Fig.4. Errors in displacements u and w for various numbers of elements at the load level $PL^2/EI = 5$

results only in case of small or moderate displacements. The errors were found to be very little depending on the number of elements. No solutions at all were obtained when the term $(du^0/dx^0)^2$ in the strain-displacement relation was neglected.

The Updated Lagrangian formulation:

The results of the UL-formulation follow closely the 'exact' load-displacement curves in Fig.3b. The errors in displacements at the load level $PL^2/EI = 5$ are shown in Fig.4a for various numbers of elements. The 'exact' displacements at this load level are according to Refs. [3,4] : $w/L = 0.71379$ and $u/L = 0.38763$. At each mark in the diagram there are two numbers. The first one indicates the minimum number of load steps necessary to obtain a converged solution within 60 iterations in each step. The second number is the total number of iterations. The rate of convergence is obviously very slow for this formulation.

It was difficult to check the influence of the number of load-steps, since the minimum number was so high. A small but noticable change in results could, however, be observed when the number of load steps was increased. The neglection of the term $(du^*/dx^*)^2$ in the strain-displacement relation did have a negligible influence on the results. The use of three instead of four Gauss points did have very small influence on the results, but seemed to improve the convergence properties somewhat.

The Co-rotational formulation:

The CR-formulation also gave very good results. The convergence properties were excellent, in contrast to what was the case for the UL-formulation. The results did not depend on the step size or the neglection of the term $(dn_z/d\bar{x}^*)^2$. The use of three Gauss points improved the performance of the element considerably. Even when a linear natural stiffness relation was used, the performance of the element improved somewhat. Especially it seemed to improve the convergence properties considerably.

6.2 Example two: Square frame with two opposite point loads

The square frame in Fig.5a was analysed with the various formulations and with various numbers of elements. In general this example seems to confirm the observations made in the previous example. With the UL-formulation no results were obtained with 16 elements. In the CR-formulation the element with three Gauss points shows a remarkable good performance. Quite acceptable results are obtained with only two elements (one quarter of the frame is analysed). 'Exact' results for the load level $PL^2/EI = 2$ are according to Refs. [3,4] : $w/L = 0.54087$ and $u/L = 0.24854$.

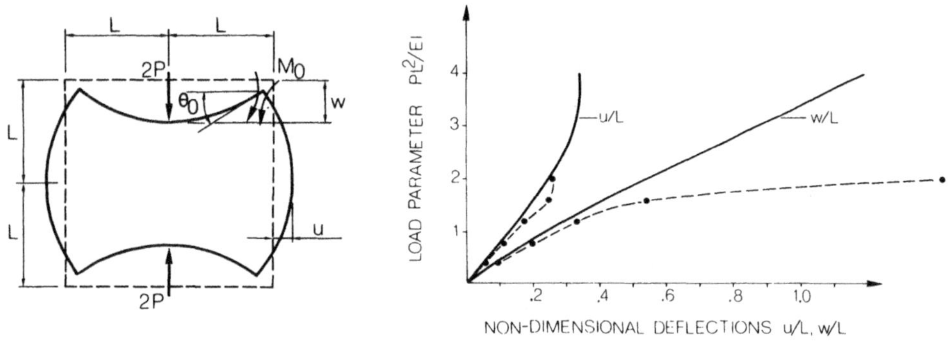

(a) (b)

Fig.5. (a) Square frame with two opposite point loads
(b) 'Exact' load-displacement curves (solid lines)[3,4], and FEM-solution with eight elements and a TL-formulation (dots). One quarter of the frame is analysed

7. Solution propagation methods

The solution of the geometrically nonlinear equations for thin elastic structures are generally complicated due to instabilities, snap-throughs and other irregularities. Mostly the solution has to be traced step-by-step from a zero pair of loads and displacements to a final pair with only load or only displacements prescribed.

Mostly, for each step only an iterative solution is possible - Newton (N) methods are resorted to. Steplength factors λ may be

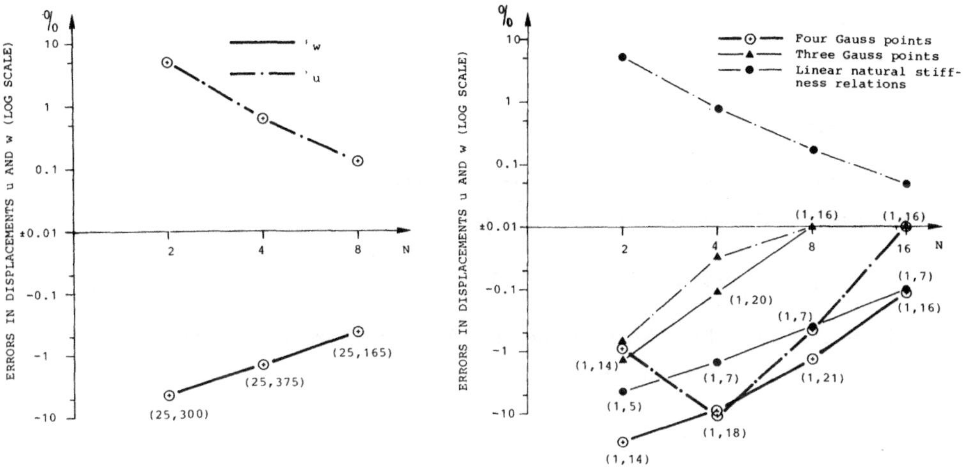

(a) UL-formulation (b) CR-formulation

Fig.6. Errors in displacements u and w for various numbers of elements at the load level $PL^2/EI = 2$

prescribed or may be controlled by a constraint equation. A few such equations have been proposed - by Riks [5], Batoz-Dhatt [6], Ramm [7], Bergan [8], Crisfield [9], [10]. It is important that the constraint equation is chosen in such a way that it does not make the iterative procedure too timeconsuming or in other ways ineffective.

Step-length control is indeed necessary to make it possible to pass limit points, but it is also of interest to know how efficient it is in other situations. For that purpose we have again calculated the displacements of a straight, elastic cantilever loaded transversally at the free end, see above. The calculations have been performed with pure Newton iterations combined with prescribed load-steps and also with two different constraint equations. A finite element discretization with 8 elements of equal length has been assumed. The transverse element displacements have been approximated as a third degree polynomial while the axial force is taken constant, see Jennings [11]. This element is known to be very efficient in combination with co-rotational formulation.

The two constraint equations investigated are the displacement vector constraint for iterations i and i+1

$$\Delta \underset{\sim}{u}_i^T \Delta \underset{\sim}{u}_i = \Delta \underset{\sim}{u}_{i+1}^T \Delta \underset{\sim}{u}_{i+1} = \Delta l^2 \tag{35}$$

giving a version of the so-called arc-length method (A-L), and the displacement component constraint

$$(u_p)_i = \text{const}, \quad (\Delta u_p)_i = 0 \tag{36}$$

where index p denotes the prescribed component, giving the incremental displacement method (I-D).

The iterations are performed according to the Batoz-Dhatt scheme [6]. After an initiation Δu_i and $\Delta\lambda_i$ are in each iteration calculated from

$$\Delta u_i = K_i^{-1} R_i + \Delta\lambda_i K_i^{-1} P \tag{37}$$

and the constraint equation expressed in $\Delta\lambda_i$. Here R_i is the residual and P a reference load. The matrix K_i may be the tangent matrix, giving a Newton method, or some approximation of that matrix. The constraint equation (36) is linear in $\Delta\lambda_i$ while (35) is quadratic in $\Delta\lambda_i$.

For the tested geometrically nonlinear problem pure Newton was found to be effective. The so-called modified Newton method with constant iteration matrix in each step required a large number of iterations.

The exact load-displacement curve, Fig. 3 , [3], was reproduced by our numerical calculations with 5 digits accuracy. The three methods differed, however, considerably in efficiency. While the Newton method without constraints could reach the load level $2PL^2/EI$ in one step with 7 iterations, the Newton method with arc-length control could reach the same level in one step with 13 iterations. The Newton method with incremental end displacement control could only reach $0.2\ PL^2/EI$ in one step and 9 iterations.

Fig. 7 , also shows the results of automatic step-lengths controls [7] according to

$$\Delta l_{i+1}^2 = \Delta l_i^2 (N_d/N_i) \tag{38}$$

where N_d is the desired number of iterations and N_i is the required number of iterations in step number i.

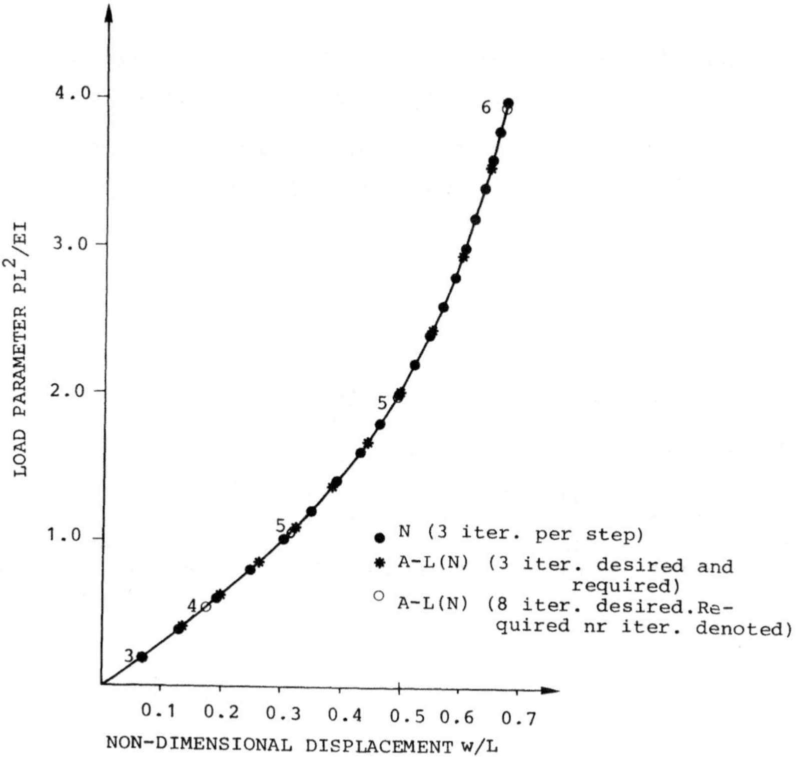

Fig.7. Exact solutions of the cantilever end deflection due to vertical load. Automatic load-step control for the A-L-methods

8. Conclusions

The conclusions from the present study can be summarized as follow:

- In small strain/large rotation problems the CR-formulation is to be preferred, since it is the most accurate formulation and has the best convergence properties. It is also independent of step size

- The squares of the in-plane displacement gradient can be neglected in the CR- and UL-formulations, but not in the TL-formulation
- The TL-formulation can only be used in problems with small or moderate displacements
- The well known Newton method is a highly competitive solution method for problems not exhibiting limit points
- The Newton method with arc-length control is an efficient solution method for problems with limit points

References

1. Mattiasson, K.: On the co-rotational finite element formulation for large deformation problems, Dr. Thesis, Department of Structural Mechanics, Chalmers University of Technology, Publ. 83:1, Göteborg, 1983.

2. Mattiasson, K. and Samuelsson, A.: Total and updated Lagrangian forms of the co-rotational finite element formulation in geometrically and materially nonlinear analysis, Proc. of the Int. Conf. on Numerical Methods for Non-Linear Problems, Vol. 2 (Eds. Taylor et. al.), Pineridge Press, Swansea, 1984, pp. 134-151.

3. Mattiasson, K.: Numerical results from large deflection beam and frame problems analysed by means of elliptic integrals, Int. J. for Num. Meth. in Eng., Vol. 17, 1981, pp. 145-153.

4. Mattiasson, K.: Numerical results from elliptic integral solutions of some elastica problems of beams and frames, Department of Structural Mechanics, Chalmers University of Technology, Publ. 79:10, Göteborg, 1979.

5. Riks, E.: An incremental approach to the solution of snapping and buckling problems, Int. J. Solids and Structs., Vol. 15, 1979, pp. 529-551.

6. Batoz, J.L. and Dhatt, G.: Incremental displacement Algorithms for nonlinear problems, Int. J. for Num. Meth. in Eng., Vol. 14, No. 8, 1979.

7. Ramm, E.: Strategies for tracing non-linear responses near limit points, Non-linear Finite Element Analysis in Structural Mechanics (Ed. Wunderlich et. al.), Springer-Verlag, New York, 1981, pp. 63-89.

8. Bergan, P.G.: Solution algorithms for nonlinear structural problems, Engineering application of the finite element method, Computas, 1979, pp. 13.1-13.39.

9. Crisfield, M.A.: Variable step-lengths for nonlinear structural analysis, Department of the Environment Department of Transport, TRRL Laboratory Report 1049, Crowthorne, 1982.

10. Crisfield, M.A.: Solution procedures for nonlinear structural problems, Recent Advances in Non-Linear Computational Mechanics (Ed. Hinton et. al.), Pineridge Press, Swansea, 1982.

11. Jennings, A.: Frame analysis including change of geometry, J. Struct. Div., ASCE 94, No. ST3, 1968, pp. 627-644.

Large Rotations in Problems of Structural Mechanics

J. F. BESSELING
Delft University of Technology, The Netherlands

Summary

The application of Euler parameters in the analysis of large rotations in spatial movements of pipelines and of links and hinges of mechanisms is discussed. The structural equations are based upon a finite element representation of the elastic line concept with the introduction of generalized deformations and generalized stresses. Experience with various aspects of the solution strategy for nonlinear equations in structural mechanics is reported.

1. Introduction

The description of large rotations in a computer analysis of structures should be such, that no singularity positions can be encountered. A description of arbitrary spatial rotation by three rotation parameters does not satisfy this condition, though they can be chosen such that for values of the rotation angles smaller than $\pi/2$ no problems occur. However the use of the four Euler parameters eliminates all singularity difficulties and has the additional advantage, that the elements of the transformation matrix are algebraic quadratic expressions in terms of these parameters instead of goniometric functions, as in the case of the use of Euler angles for instance. Quadratic and linear algebraic expressions are computationally more economic than goniometric functions.

Since the sum of the squares of the Euler parameters should remain equal to one, application of these parameters entails a subsidiary condition. However this condition is mathematically of the same form as the constitutive equations for the generalized deformations and it can therefore easily be incorporated in a finite element analysis, formulated with the aid of such generalized deformations. With the introduction of each set of Euler parameters a so-called lambda-element is defined, and the fulfillment of the subsidiary condition is enforced by giving this element a large or infinite "stiffness" against "deformation". Then of course in a numerical analysis the same problem arises, that stems from infinite or relatively very large

Finite Element Methods for Nonlinear Problems
Europe-US Symposium, Trondheim, Norway 1985
Editors: Bergan, Bathe, Wunderlich

stiffnesses against physical deformation. In both cases the solution of this problem is obtained by the introduction of lagrangian multipliers as new variables. In the case of the physical deformations these multipliers have a clear physical significance as generalized stresses: the duals of the rates of deformations.

The systemmatrix, obtained by the procedure indicated, is symmetric, but not positive-definite. With some simple precautions the usual matrix decomposition method can be used, however.

As an illustration a finite element for deep sea pipelaying analysis is discussed and characteristic link- and hinge-elements for kinematic and dynamic analysis of spatial mechanisms such as robots are given.

For the solution of the equations attention is drawn to the possibility of the application of increments of work for loading, thereby avoiding the singularities of limit points, while keeping the systemmatrix symmetric.

2. Euler parameters

The finite element representations of the elastic line concept requires the description of the displacements and of the change of orientation of triads in the nodal points. Whilst the displacements can be given in terms of elements of a linear vector space, for the change of orientation we must introduce an orthogonal transformation. We introduce the transition from a global orthogonal triad with unit vectors $\underline{e}_i$ to body fixed orthogonal triads with unit vectors $\underline{e}_i$, in each nodal point. This transition is accomplished by an orthogonal transformation of the unit vectors $\underline{e}_i$.

Such orthogonal transformations can be defined in terms of various sets of three parameters, such as Euler angles, Bryant angles and Rodrigues parameters. However based upon a description with three parameters there are always orientations for which the transformation matrix becomes singular. Then in a computerprogram without restrictions on the magnitude of the rotations special precautionary measures have to be taken. A description by the four Euler parameters avoids this complication, but introduces a subsidiary condition since the parameters are no longer independent.

By Euler's theorem any orthogonal transformation can be defined in terms of a single rotation about an unique axis, that is referred to as an orientational axis of rotation. The direction of the orientational axis is specified by a unit vector $\underline{e}_\mu$ with components $\cos\alpha$, $\cos\beta$, $\cos\gamma$. If the angle of rotation about the axis is denoted by μ, then the four Euler parameters λ_i are defined as

$$\lambda = \begin{vmatrix} \lambda_0 \\ \lambda_1 \\ \lambda_2 \\ \lambda_3 \end{vmatrix} = \begin{vmatrix} \cos\frac{1}{2}\mu \\ \sin\frac{1}{2}\mu\cos\alpha \\ \sin\frac{1}{2}\mu\cos\beta \\ \sin\frac{1}{2}\mu\cos\gamma \end{vmatrix}, \quad \lambda_0^2 + \lambda_1^2 + \lambda_2^2 + \lambda_3^2 = 1. \tag{1}$$

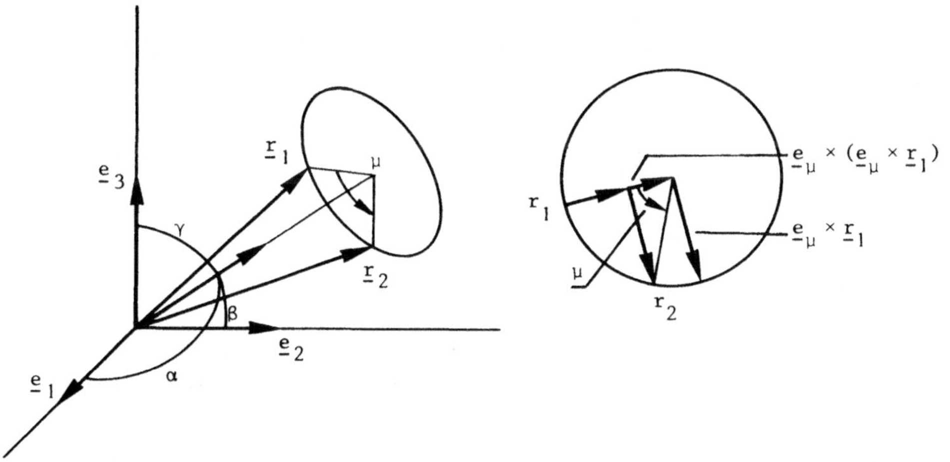

Fig. 1. Transition from vector $\underline{r}_1$ to vector $\underline{r}_2$ by rotation over angle μ about axis $\underline{e}_\mu$.

While rotating over an angle μ from the vector $\underline{r}_1$ to the vector $\underline{r}_2$ the endpoint of a vector describes a circle in a plane perpendicular to the axis of orientation $\underline{e}_\mu$. As is illustrated in Fig. 1 the vector difference of the vectors $\underline{r}_1$ and $\underline{r}_2$ is given by

$$\begin{aligned} \underline{r}_2 - \underline{r}_1 &= (1-\cos\mu)\, \underline{e}_\mu \times (\underline{e}_\mu \times \underline{r}_1) + \sin\mu\, (\underline{e}_\mu \times \underline{r}_1) \\ &= 2\sin^2\tfrac{1}{2}\mu\underline{e}_\mu \times (\underline{e}_\mu \times \underline{r}_1) + 2\cos\tfrac{1}{2}\mu\sin\tfrac{1}{2}\mu(\underline{e}_\mu \times \underline{r}_1). \end{aligned} \tag{2}$$

This result can also be expressed in terms of the Euler parameters by

$$\underline{r}_2 = \underline{r}_1 + 2\underline{p} \times (\underline{p} \times \underline{r}_1) + 2\lambda_0\, (\underline{p} \times \underline{r}_1) \tag{3}$$

with

$$\underline{p} = \lambda_1\underline{e}_1 + \lambda_2\underline{e}_2 + \lambda_3\underline{e}_3. \tag{4}$$

By taking for $\underline{r}_1$ and $\underline{r}_2$ the unit vectors $\underline{e}_i$ and $\underline{e}_{i'}$, and by taking into account condition (1) we derive

$$\begin{vmatrix} \underline{e}_{1'} \\ \underline{e}_{2'} \\ \underline{e}_{3'} \end{vmatrix} = \begin{bmatrix} \lambda_0^2+\lambda_1^2-\lambda_2^2-\lambda_3^2 & 2(\lambda_1\lambda_2+\lambda_0\lambda_3) & 2(\lambda_1\lambda_3-\lambda_0\lambda_2) \\ 2(\lambda_1\lambda_2-\lambda_0\lambda_3) & \lambda_0^2-\lambda_1^2+\lambda_2^2-\lambda_3^2 & 2(\lambda_2\lambda_3+\lambda_0\lambda_1) \\ 2(\lambda_1\lambda_3+\lambda_0\lambda_2) & 2(\lambda_2\lambda_3-\lambda_0\lambda_1) & \lambda_0^2-\lambda_1^2-\lambda_2^2+\lambda_3^2 \end{bmatrix} \begin{vmatrix} \underline{e}_1 \\ \underline{e}_2 \\ \underline{e}_3 \end{vmatrix} \tag{5}$$

We note that the coordinate transformation from body fixed to global coordinates according to

$$x^{i'}\underline{e}_{i'} = x^{i'}R^k_{i'}\underline{e}_k = x^k\underline{e}_k \Rightarrow x^k = x^{i'}R^k_{i'} \tag{6}$$

is given by the transposed matrix, which we shall denote by R^T.

$$x = R^T x'. \tag{7}$$

By differentiation of the identity $R^T R = I$ with respect to time we obtain

$$\dot{R}^T R = - R^T \dot{R} = \Omega, \ \dot{x} = \Omega x. \tag{8}$$

With each skew-symmetric matrix we may associate an axial vector. The axial vector associated with Ω is the angular velocity vector $\underline{\omega}$ and we have the following relation:

$$\underline{\omega} = \omega^1\underline{e}_1 + \omega^2\underline{e}_2 + \omega^3\underline{e}_3,$$

$$\Omega = \begin{bmatrix} 0 & -\omega^3 & \omega^2 \\ \omega^3 & 0 & -\omega^1 \\ -\omega^2 & \omega^1 & 0 \end{bmatrix}. \tag{9}$$

The relation between the components of the angular velocity vector and the rates of change of the Euler parameters is then determined by equating corresponding elements in (9).

$$\begin{vmatrix} \omega^1 \\ \omega^2 \\ \omega^3 \end{vmatrix} = 2 \begin{bmatrix} -\lambda_1 & \lambda_0 & -\lambda_3 & \lambda_2 \\ -\lambda_2 & \lambda_3 & \lambda_0 & -\lambda_1 \\ -\lambda_3 & -\lambda_2 & \lambda_1 & \lambda_0 \end{bmatrix} \begin{vmatrix} \dot{\lambda}_0 \\ \dot{\lambda}_1 \\ \dot{\lambda}_2 \\ \dot{\lambda}_3 \end{vmatrix}. \tag{10}$$

These expressions are needed if a loading of the structure is determined by external moments, which are the duals of the angular velocities. For the derivation of the equilibrium equations or of the equations of motion by the principle of virtual power the virtual angular velocities must be expressed in terms of the virtual rates of change of the Euler parameters, that are used for the description of the motion of the elastic line.
The expressions given above and other most useful expressions in terms of Euler parameters are discussed in detail in [1].

3. Pipeline element

If the description of a pipeline is based upon the elastic line concept the interface of the finite elements has shrunk into nodal points. The location of these nodal points and the orientation of the orthogonal triads, rigidly attached to each nodal point, can be given by the position vector with components $x^i = \xi^i + u^i$ and by the orthogonal transformation R (7), which transforms the components of a vector in the x^i-system into components with respect to the body fixed triad at the nodal point. In [2] six deformation parameters or generalized strains for a straight bar connecting nodal points p and q were defined.

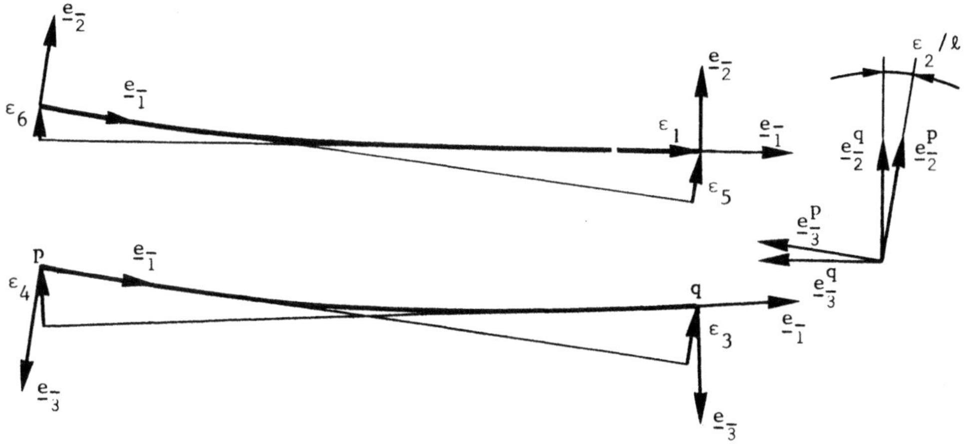

Fig. 2. Generalized strains of bar element between nodal points p and q.

The six generalized strains can be expressed as analytic functions of the $2 \times 7 = 14$ nodal parameters of the bar element: six displacement components and eight Euler parameters.

$$\varepsilon_i = D_i^e(u_k). \tag{11}$$

Since the generalized strains of Fig. 2 are defined with respect to orthogonal triads, oriented according to the bar-axis and the principal axes of its cross-section, we need the orthogonal matrix $R^{i}_{\bar{k}} = R^{i'}_{\bar{k}}$, which determines in the deformed as well as in the undeformed structure the orientation of the bar-axis and of the cross-sectional plane with reference to the local triads at the nodal points.

$$\underline{e}_{\bar{k}} = R^{i'}_{\bar{k}}\, \underline{e}_{i'} = R^{i'}_{\bar{k}}\, R^{\ell}_{i'}\, \underline{e}_{\ell}. \tag{12}$$

With

$$\ell^2 = (\xi^i_p - \xi^i_q)(\xi^i_p - \xi^i_q),$$

$$\underline{\ell}_{pq} = (\xi^i_p + u^i_p - \xi^i_q - u^i_q)\, \underline{e}_i, \tag{13}$$

the functions (11) for the generalized strains of Fig. 2 can now be written as follows

$$\varepsilon_1 = \frac{1}{2\ell}\left[\underline{\ell}_{pq} \cdot \underline{\ell}_{pq} - \ell^2\right] + \frac{1}{30\ell}\left[2\varepsilon_3^2 + \varepsilon_3\varepsilon_4 + 2\varepsilon_4^2 + 2\varepsilon_5^2 + \varepsilon_5\varepsilon_6 + 2\varepsilon_6^2\right],$$

$$\varepsilon_2 = \frac{\ell}{2}\left[\underline{e}^p_{\bar{3}} \cdot \underline{e}^q_{\bar{2}} - \underline{e}^p_{\bar{2}} \cdot \underline{e}^q_{\bar{3}}\right],$$

$$\varepsilon_3 = \underline{e}^p_{\bar{3}} \cdot \underline{\ell}_{pq},$$

$$\varepsilon_4 = -\,\underline{e}^q_{\bar{3}} \cdot \underline{\ell}_{pq},$$

$$\varepsilon_5 = -\,\underline{e}^p_{\bar{2}} \cdot \underline{\ell}_{pq},$$

$$\varepsilon_6 = \underline{e}^q_{\bar{2}} \cdot \underline{\ell}_{pq}. \tag{14}$$

If the deformations ε^e_i remain sufficiently small ($\varepsilon^e_i \ll 1$), then in the elastic range they are linearly related to a normal force σ_1, a twisting moment σ_2 and four bending moments σ_3, σ_4, σ_5, σ_6 by a symmetric matrix of elasticity coefficients.

$$\sigma^e_i = S^e_{ij}(\varepsilon^e_j - \varepsilon^{eo}_j). \tag{15}$$

Here we introduced initial deformations ε^{eo}_i to represent inelastic strains, or simply misfits in the perfect, undeformed structure ($u_k = 0$). For a prismatic bar with the $\bar{2}$- and $\bar{3}$-axis in a cross-section along the principal axes the elasticity coefficients are given by the matrix

$$S^e = \begin{bmatrix} S_1 & 0 & 0 & 0 & 0 & 0 \\ & S_2 & 0 & 0 & 0 & 0 \\ & & 4S_3 & -2S_3 & 0 & 0 \\ & & & 4S_3 & 0 & 0 \\ & \text{symm.} & & & 4S_4 & -2S_4 \\ & & & & & 4S_4 \end{bmatrix}, \quad S_1 = \frac{EA}{\ell}, \quad S_2 = \frac{S_T}{\ell^3}, \quad S_3 = \frac{EI_{\bar{2}}}{\ell^3}, \quad S_4 = \frac{EI_{\bar{3}}}{\ell^3}. \tag{16}$$

Needless to say that for a pipeline element will hold $S_3 = S_4$.
In addition to the physical deformations we must consider the condition (1), to be imposed on the Euler parameters. For each orthogonal triad we introduce a "deformation" of the so-called lambda element:

$$\varepsilon^\lambda = D^\lambda(u_k) = 1 - \lambda_0^2 - \lambda_1^2 - \lambda_2^2 - \lambda_3^2. \tag{17}$$

The deformation, which should remain equal to zero, can be kept arbitrarily small by attributing to the lambda element a "stiffness" sufficiently large against this deformation. Numerical difficulties arising from this large stiffness can be overcome in the same way as the difficulties, that are due to the large differences between longitudinal and bending stiffnesses of the bar elements in (16), as we shall discuss in section 5 of this paper.

4. Links and hinges of mechanisms

Links in mechanisms may be treated as rigid or as deformable, just as the problem requires. For instance slow motions of robots usually can be analysed with sufficient accuracy by a kinematic analysis followed by a kineto-static analysis. However with increasing speed the interaction between the dynamic characteristics of the mechanism and the control characteristics of the actuators will more and more be determined by the flexibilities of the mechanism.
For the description of the deformation of links in mechanisms we may use the generalized deformations defined in (14). If so needed we can use a number of elastic line elements to model a more complicated link or a link for which a sufficiently accurate determination of the dynamic characteristics requires the introduction of internal degrees of freedom. It will be clear that the

stiffness matrix for an element in a mechanism will in general be quite different from the matrix given in (16) for a prismatic bar element. However as long as in the nodal points, which act as interfaces between links, the action of a moment (or the absence of the action of a moment: hinge) about a body fixed axis has to be considered a proper description of the system can be obtained by the introduction of an orthogonal triad, rigidly attached to the nodal point. The Euler parameters, determining the orientation of the triad with respect to the global system, are variables in the deformation expressions of the link and of the associated lambda element, but they also serve in the specification of the hinge characteristics.
In the case of a hinge between two links we have two orthogonal triads in one nodal point, each rigidly attached to a link. A relative rotation is now defined as a deformation of a so-called hinge element. For the definition of this element we consider the one nodal point as two coinciding nodal points p and q, with one orthogonal triad attached to a link in point p and one attached to the other link in point q. The rest position of the hinge is now easiest specified by letting the triads coincide, but at least one of the unit vectors of the triads should be along the body fixed axis of rotation. Three possible deformations of such a hinge element can be defined

$$\begin{aligned} \varepsilon_1 &= \underline{e}^p_{\bar{2}} \cdot \underline{e}^q_{\bar{3}} , \\ \varepsilon_2 &= \underline{e}^p_{\bar{3}} \cdot \underline{e}^q_{\bar{1}} , \\ \varepsilon_3 &= \underline{e}^p_{\bar{1}} \cdot \underline{e}^q_{\bar{2}} . \end{aligned} \tag{18}$$

Clearly relative rotation about the $\bar{1}$-axis implies $\varepsilon_1 \neq 0$. Similarly ε_2 and ε_3 are relative rotations about the $\bar{2}$- and $\bar{3}$-axes respectively. If the axis of rotation has thus been specified, the other two deformations have to be kept equal to zero. In a numerical analysis this is accomplished by attributing to the hinge element a sufficiently large stiffness against each of the deformations in question.
Since the orientation of the triads at hinges is dictated by the hinge axis, in the case of links, that may be modelled as bar elements the triads may no longer coincide with the principal axes of the cross-section. Consequently for the use of the stiffness matrix (16) of a bar element another orthogonal transformation may be needed.

5. Equilibrium equations

For the derivation of the equilibrium equations or of the equations of motion we may turn to the principle of virtual power. In this paper we shall only consider quasi-static analysis. For the derivation of the equations of motion with the intertial terms expressed by Euler parameters the reader is referred to [1] and [3].

According to the principle of virtual power in the case of equilibrium the power of the external forces and moments shall be equal to zero for all kinematic admissible motions without deformation. If we write for the power of the external forces and moments $f_k \dot{u}_k$, then the principle of virtual power requires for equilibrium

$$f_k \dot{u}_k = 0 \quad \forall \quad \dot{u}_k \in \{\dot{u}_k \mid D_{i,k} \dot{u}_k = 0\}. \tag{19}$$

The subsidiary condition $D_{i,k} \dot{u}_k = 0$ may be taken into account by lagrangian multipliers σ_i. Hence

$$f_k \dot{u}_k = \sigma_i D_{i,k} \dot{u}_k \quad \forall \quad \dot{u}_k \tag{20}$$

leading to the equilibrium equations

$$f_k = \sigma_i D_{i,k}. \tag{21}$$

Now we have to specify the quantities f_k, which are the duals of the rates of change $\dot{u}_k$. If u_k is a component of a displacement, then f_k is simply the dual force component. However if $\dot{u}_k$ is a rate of change of a Euler parameter, then we must realize that normally the external moment will be specified, which is the dual of the angular velocity. Let the external moment in a nodal point be denoted by $\underline{f}^{\omega}$. Then the virtual power is given by

$$\underline{f}^{\omega} \cdot \underline{\omega} = f_k^{\omega} \Lambda_{k\ell} \dot{\lambda}_{\ell} = f_{\ell}^{\lambda} \dot{\lambda}_{\ell}, \tag{22}$$

where Λ is the matrix occuring in (10). It follows that

$$f_k^{\lambda} = f_{\ell}^{\omega} \Lambda_{\ell k}. \tag{23}$$

Next we turn to the lagrangian multipliers σ_i in (21), which are the duals of $\dot{\varepsilon}_i$.

For an elastic bar element we have an elastic potential

$$P = \frac{1}{2} (\varepsilon_i^e - \varepsilon_i^{eo}) S_{ij} (\varepsilon_j^e - \varepsilon_j^{eo}) \tag{24}$$

and from

$$f_k \dot{u}_k = \frac{\partial P}{\partial \varepsilon_i} \dot{\varepsilon}_i = \sigma_i \dot{\varepsilon}_i \qquad (25)$$

relation (15) follows.
In the case of a pipe element there is a complication due to the defective treatment of the effect of fluid pressure when we model the pipe by the elastic line concept. The buoyancy of a finite element has to be corrected for the lack of pressure at the ends of the element. This can be interpreted as a normal force, which is determined by the area of the cross-section and the local external and internal fluid pressures, and which has to be added to the normal force determined by the longitudinal strain of the elastic line. In a sea with dept h for a finite element at an average distance w above the ocean floor we thus have the following constitutive equation:

$$\sigma_1 = \sigma_1^{\varepsilon} + \sigma_1^{w},$$

$$\sigma_1^{\varepsilon} = S_1 D_1, \quad \sigma_1^{w} = g(\rho_{water} A^e - \rho_{fluid} A^i)\,(h-w), \qquad (26)$$

where A^e denotes the external cross-sectional area of the pipe and A^i the internal cross-sectional area, which is filled with a fluid. Thus the buoyancy of the finite element, that as an external force has to be added to the loads f_k, is corrected by taking into account an artificial stress-strain relationship. We should realize that the elastic line concept cannot cope with the effect of circumferential stresses in the pipe due to the waterpressure. Hence a lengthening by circumferential compression due to Poisson's ratio is neglected, but this is of minor importance in the determination of the deformation of the pipeline. Of course for the analysis of the strength of the pipeline the elastic line concept has to be abandoned. However the values of the resulting normal force σ_1^{ε}, the twisting moment and the bending moments from an elastic line analysis remain valid with sufficient accuracy for a local analysis of the pipe.
Finally we have to introduce artificial stiffness quantities for the use of Euler parameters, for which we enforce condition (1) by making the "stiffness" S^{λ} in

$$\sigma^{\lambda} = S^{\lambda} \varepsilon^{\lambda}, \qquad (27)$$

very large or infinite.

6. Solution strategy

The equilibrium equations (21) are highly nonlinear. For arbitrarily large displacements and rotations they can only be solved in a step by step manner. We therefore write down incremental equations, making use of the following notations and relations.

$$
\begin{aligned}
\sigma_i^B + \Delta\sigma_i^B &= S_{ij}^B D_j^B(u_k) + S_{ij}^B D_{j,k}^B \Delta u_k, \\
\sigma_i^N + \Delta\sigma_i^N &= \sigma_i^N + \Delta\sigma_i^\varepsilon - g(\rho_{water}A^e - \rho_{fluid}A^i)\Delta w_i, \\
D_i^N(u_k) + D_{i,k}^N \Delta u_k &= S_{ij}^{N-1}(\sigma_j^\varepsilon + \Delta\sigma_j^\varepsilon), \\
D_i^\lambda(u_k) + D_{i,k}^\lambda \Delta u_k &= S_{ij}^{\lambda-1}(\sigma_j^\lambda + \Delta\sigma_j^\lambda).
\end{aligned}
\tag{28}
$$

Here the superscript B refers to the bending and twisting of the finite elements, the superscript N to the normal force, the longitudinal strain and the associated stiffness, while the superscript λ denotes the quantities, defined for the lambda element. D_i^λ are the "deformations" (17), which are to be kept sufficiently close to zero value by the large "stiffness" S_{ij}^λ. The incremental equations we shall consider now read as follows:

$$
\begin{bmatrix}
D_{i,k}^B S_{ij}^B D_{j,\ell}^B + \ldots + \sigma_i^N D_{i,k\ell}^N + \sigma_i^\lambda D_{i,k\ell}^\lambda & D_{q,k}^N & D_{s,k}^\lambda \\
D_{p,\ell}^N & -S_{pq}^{N-1} & 0 \\
D_{r,\ell}^\lambda & 0 & -S_{rs}^{\lambda-1}
\end{bmatrix}
\begin{vmatrix}
\Delta u_\ell \\
\Delta\sigma_q^\varepsilon \\
\Delta\sigma_s^\lambda
\end{vmatrix}
=
$$

$$
=
\begin{vmatrix}
\Delta f_k + g(\rho_{water}A^e - \rho_{fluid}A^i)\Delta w_i D_{i,k}^N + f_k - \sigma_i D_{i,k} \\
-D_p^N + S_{pq}^{N-1}\sigma_q^\varepsilon \\
-D_r^\lambda + S_{rs}^{\lambda-1}\sigma_s^\lambda
\end{vmatrix}.
\tag{29}
$$

First we observe that the systemmatrix of these incremental equations is not positive definite. Furthermore the matrix

$$
K^B = [D_{i,k}^B S_{ij}^B D_{j,\ell}^B]
\tag{30}
$$

may have zero's on the diagonal due to the absence in S_{ij}^B of a stiffness

against elongation, and against deviation from condition (1) for the Euler parameters. In addition an infinite stiffness against elongation and against deviation from condition (1) for the Euler parameters gives permanent zero's on the diagonal. Still, if we make sure that we do not start the decomposition procedure with those rows and columns, that may have a zero on the diagonal, no problems were encountered with this decomposition.
In the analysis of pipelines we may neglect the influence of the twisting and bending terms in the geometrical stiffness, but the terms

$$\sigma_i^N D_{i,k\ell}^N + \sigma_i^\lambda D_{i,k\ell}^\lambda \tag{31}$$

are essential. In particular the omission of σ_1^w in (26) at great waterdepths will lead to a fictitious instability of the pipeline, that will spoil the analysis.
In the form we have written down the incremental equations for a pipeline under water the quantities Δw_i appear as unknown increments in the right hand side of the equations. To have brought them to the left hand side would have spoilt the symmetry of the systemmatrix. But now they have to be neglected in the computation of the increments. However their effect can be properly taken into account by letting every incremental step to be followed by a number of iterations with prescribed increments of loads and displacements equal to zero. For this purpose the right hand side of equations (29) was augmented by terms, that in a true equilibrium solution should be equal to zero. On closer inspection the equations (29) then appear to be the equations for a Newton-Raphson type of iteration procedure.
In pipe laying analysis we need still another type of iteration due to the fact that at the ocean floor we have an overcomplete set of boundary conditions: moments as well as slopes have to be zero. In a finite element analysis this problem is conveniently handled by an iterative procedure of adding or cancelling elements or parts of elements at the ocean floor end of the pipeline.
In order to discuss stepsize and iteration truncation criteria we turn to the simpler case, when in the equilibrium equations no distinction is made between the types of generalized stresses. We shall indicate external forces, which are the duals of prescribed displacements, by f_k^o and the prescribed forces by λf_k^c, λ being a load parameter. With a corresponding splitting up of $D_{i,k}$ into $D_{i,k}^o$ and $D_{i,k}^c$ the equilibrium equations read

$$\sigma_i D^c_{i,k} = \lambda f^c_k,$$

$$\sigma_i D^o_{i,m} = f^o_m. \qquad (32)$$

Incremental equations of the form

$$[D^c_{i,k} S_{ij} D^c_{j,\ell} + \sigma_i D^c_{i,k\ell}]\, \Delta u^c_\ell = \Delta\lambda f^c_k - D^c_{i,k} S_{ij} D^o_{j,\ell} \Delta u^o_\ell$$
$$+ \lambda f^c_k - \sigma_i D^c_{i,k} \qquad (33)$$

can be used for determining a linearized solution for a loading step ($\Delta\lambda \neq 0$, $\Delta u^o \neq 0$) as well as for an iteration procedure ($\Delta\lambda = 0$, $\Delta u^o = 0$) to reduce the errors to an acceptable level.

In a computer analysis we want an automatic stepsize correction procedure and a reliable truncation criterion for the iteration process. The stepsize procedure can be chosen as follows, where eps1 is the controlling quantity (eps1 < 0.01).

We start with $\Delta\lambda = 1$ and compute the values

$$\Delta u^c_k,\; D_i(u_k + \Delta u_k),\; D_{i,k}(u_k + \Delta u_k),$$

$$f^{error}_k := (\lambda + \Delta\lambda)\, f^c_k - D_i S_{ij} D^c_{j,k},$$

$$\text{delta} := \sum_k \text{abs}\,(f^{error}_k \Delta u^c_k),$$

$$\text{norm} := \sum_k \text{abs}\,(D_i S_{ij} D_{j,k} \Delta u_k). \qquad (34)$$

If delta/norm > eps1 then

$$\Delta\lambda := [\text{eps1.norm/delta}]^{\frac{1}{2}} \qquad (35)$$

else $\Delta\lambda := 1$. If $\lambda + \Delta\lambda > 1$ then $\Delta\lambda := 1 - \lambda$.

Each step with $\Delta\lambda < 1$ is followed by an iteration process for which a control parameter eps2 is introduced (eps2 < 0.01). We compute for $\Delta\lambda = 0$ and $\Delta u^o = 0$ the values of

$$\Delta u_k^c,\ D_i(u_k + \Delta u_k),\ D_{i,k}(u_k + \Delta u_k),$$

$$f_k^{error} := \lambda f_k^c - D_i S_{ij} D_{j,k}^c,$$

$$delta := \sum_k abs\ (f_k^{error}(u_k^c + \Delta u_k^c)),$$

$$norm := \sum_k abs\ (D_i S_{ij} D_{j,k}(u_k + \Delta u_k)). \tag{36}$$

until delta/norm < eps2.

We note that we look at the error in terms of work quantities. Our experience is that for equations based upon the principle of virtual power this is an effective procedure, which may have something to do with the physical basis of it.

It is wellknown that the solution of equations (32) breaks down if the load reaches a limit point or a bifurcation point in the loading path. Effective means have been developed to deal with these possibilities. We mention the arc-length method [4]. It is an inconvenience of this method that a special formulation is needed to overcome the difficulty that in the first instance it leads to a unsymmetric systemmatrix [5]. We have tried out a simple alternative method, which so far seems not to have been considered. We may as loading parameter introduce the amount of work needed to achieve the required loading and deformation of the structure. We write the equations in the form

$$\begin{bmatrix} D_{i,k} S_{ij} D_{j,\ell} + \sigma_i D_{i,k\ell} & -f_k \\ -f_\ell & 0 \end{bmatrix} \begin{vmatrix} \Delta u_\ell \\ \Delta\lambda \end{vmatrix} = \begin{vmatrix} \lambda f_k - \sigma_i D_{i,k} \\ -\Delta Work \end{vmatrix}. \tag{37}$$

Again the systemmatrix with zero's on the diagonal can with proper precautions be successfully decomposed, and a stepwise, iterative process as outlined above solves the nonlinear equations efficiently. Of course bifurcation points lead to singularity of the systemmatrix, but they can be handled in much the same fashion as in the application of the arc-length method.

7. Concluding remarks

For the nonlinear problems discussed in this paper computer programs were developed and run on a 68000- processor desk-computer system. The language Pascal permitted a well-structured set up of the programs with much flexibility in trying out various strategies for the solution. Robot mechanisms and pipe laying analysis with up to 60 finite elements were thus successfully analysed. In addition a multipurpose computer program SPACAR [6] was the product of a fruitful cooperation between the Laboratory for Production Automation and Mechanisms and the Laboratory of Engineering Mechanics of the TH-Delft.

8. References

1. Jonker, J.B.; Dynamics of Spatial Mechanisms with Flexible Links. Delft University of Technology, Lab. Eng. Mech., 1985, WTHD 171.

2. Besseling, J.F.; Derivatives of Deformation Parameters for Bar Elements and their use in Buckling and Post-buckling Analysis. Comp. Meths. Appl. Mech. Eng. 12 (1977) pp. 7-124.

3. Besseling, J.F., Jonker, J.B., Schwab, A.L.; Kinematics and Dynamics of Mechanisms. Delft Progress Report 10 nr. 2 (1985) pp. 160-172.

4. Riks, E.; An Incremental Approach to the Solution of Snapping and Buckling Problems. Int. J. Solids and Structures, Vol. 15 (1979) pp. 529-551.

5. Ramm, E.; Strategies for Tracing the Nonlinear Response near Limit Points, In: Nonlinear Fin. El. An. in Str. Mech., Eds. W. Wunderlich, E. Stein, K.J. Bathe, Springer-Verlag (1980) pp. 63-89.

6. SPACAR User Manual, Publication of the Cadom group (Delft Univ. of Techn., Dept. of Mech. Eng., 1984).

Kinematic and Dynamic Analysis of Mechanisms. A Finite Element Approach Based on Euler Parameters

M. GERADIN, G. ROBERT, P. BUCHET
University of Liège, Belgium

ABSTRACT

Euler parameters provide an efficient representation of finite rotations which arise in the kinematic description of rigid and flexible mechanical linkages. The properties of Euler parameters are best described in terms of quaternion algebra. The first part of the paper is thus devoted to quaternions, their algebraic properties and their matrix notation. In the second part, advantage is taken of the ability of Euler parameters to describe in an efficient way the kinematics of relative motion and deformation. The relationships so obtained are applied to derive a library of finite elements specialized to kinematic and dynamic analysis of mechanical linkages. The library is made of various types of joints (hinge, universal and prismatic joints) and of flexible members. The third part is devoted to problem solving. Aspects such as the choice of structural displacements, initial assembly of the mechanism, kinematic and dynamic response are considered. Finally, preliminary numerical examples are presented which demonstrate the effectiveness of the concepts on which a computer code under development is based.

I. INTRODUCTION

Kinematic and dynamic analysis of mechanisms is inherently a class of extremely nonlinear problems with respect to geometry, and no general agreement exists yet on the most appropriate numerical approach to it. Most of the litterature on the subject is still limited to semi-analytical methods which are not very well suited to the development of large-purpose codes.

In the recent past, two methods of approach have emerged. On one hand, the method of lagrangian coordinates [1] (which is largely used in robotics [2])has the merit to keep minimum the number of generalized coordinates, but it suffers from drawbacks such as complex topology description of multi-loop mechanisms [1] ,difficult generalization to flexible mechanisms [8] and high degree of nonlinearity. On the other hand, the method of cartesian coordinates (pioneered by Chace [3] , Haug [4] and others) has attractive features such as lower degree of nonlinearity and easy topological description. Coupling at joints is simply expressed as a set of

Finite Element Methods for Nonlinear Problems
Europe-US Symposium, Trondheim, Norway 1985
Editors: Bergan, Bathe, Wunderlich

kinematical constraints ; as a result, the number of unknowns is significantly larger than in the former approach.

The finite element concept [5,6] may be regarded as an interesting variant of the cartesian coordinates approach. It is conceptually easy to think of a mechanism as an assembly of either rigid or flexible member finite elements interconnected by coupling elements such as hinges, prismatic joints, Hooke joints, ... Kinematic freedoms play then the same role as rigid body modes in structures. The main difficulty of the finite element modelling of mechanisms lies in the 3-D representation of arbitrarily large displacements and rotations.

It has already been shown in an earlier paper by the first author[9] that using Euler parameters greatly simplifies the description of finite rotations.

The same idea is developed in this paper, starting from quaternion algebra. It is shown in the first part that the concept of quaternions leads in a natural way to the algebraic representation of finite rotations, and greatly simplifies the problem of representing and calculating relative rotations between elements undergoing relative motion. Euler parameters are shown to be a specific set of quaternions, and very simple relationships are derived to describe in terms of them the kinematics of relative motion at joints and the kinematics of deformation in flexible members.

From the point of view of the finite element approach to mechanism analysis, rigid bodies, members and joints are equally considered as specific finite elements as suggested earlier by Van der Werff [6] . According to its nature, a single finite element is either made of kinematic constraints which are taken into account by the augmented lagrangian method or it contributes also to the kinetic and strain energies of the system.

The problem of assembling the system of finite element equations has deserved special attention. The rotation parameters at structural level are defined as Euler parameters describing the rotation relatively to an initial configuration. In which case, the assembling of the whole system reduces simply to an identification of parameters.

The problems of kinematics and dynamics of mechanisms are then solved in much the same way as static and dynamic problems of structural mechanics. Even the problem of finding an initial feasible configuration of the mechanism that verifies member lengths and kinematic constraints is treated as an equivalent static problem in which the equilibrium configuration corresponds to vanishing of the strain energy.

Finally some preliminary numerical examples are presented which demonstrate the effectiveness of the concepts on which a computer code currently under development is based.

II. KINEMATIC DESCRIPTION OF FINITE ROTATION USING EULER PARAMETERS

2.1 ELEMENTS OF QUATERNION ALGEBRA [10]

Definition

A quaternion is defined as a 4-dimensional complex number

$$\hat{a} = a_0 + ia_1 + ja_2 + ka_3 \tag{2.1}$$

with i, j and k being imaginary unit numbers such that

$$\begin{aligned} i^2 &= j^2 = k^2 = -1 \\ jk &= -kj = i \; , \; ki = -ik = j \; , \; ij = -ji = k \end{aligned} \tag{2.2}$$

It can alternately be written in *vector notation*

$$\hat{a} = a_0 + \vec{a} \tag{2.3}$$

where a_0 and $\vec{a}$ are respectively the scalar and vector parts of the quaternion.

Multiplication rule

The multiplication rule is an immediate consequence of the definition (2.1)

$$\begin{aligned} \hat{a}\hat{b} = {} & a_0b_0 - a_1b_1 - a_2b_2 - a_3b_3 + \\ & + i(a_0b_1 + a_1b_0 + a_2b_3 - a_3b_2) + \\ & + j(a_0b_2 + a_2b_0 + a_3b_1 - a_1b_3) + \\ & + k(a_0b_3 + a_3b_0 + a_1b_2 - a_2b_1) \end{aligned} \tag{2.4}$$

It is *associative*, but *non-commutative*. In vector notation, it can be written

$$\hat{a}\hat{b} = a_0b_0 - \vec{a}\cdot\vec{b} + a_0\vec{b} + b_0\vec{a} + \vec{a}\times\vec{b} \tag{2.5}$$

Conjugate quaternion to $\hat{a}$

The conjugate quaternion to $\hat{a}$ is

$$\begin{aligned} \hat{a}^* &= a_0 - ia_1 - ja_2 - ka_3 \\ &= a_0 - \vec{a} \end{aligned} \tag{2.6}$$

It is easily verified that the conjugate of a quaternion product is such that

$$(\hat{a}\hat{b})^* = \hat{b}^* \hat{a}^* \tag{2.7}$$

Norm of a quaternion

The norm of a quaternion is calculated by

$$|\hat{a}|^2 = \hat{a}\,\hat{a}^* = a_0^2 + a_1^2 + a_2^2 + a_3^2 \tag{2.8}$$

In particular, $\hat{a}$ is a *unit quaternion* if

$$|\hat{a}| = 1 \tag{2.9}$$

Vector quaternion

A quaternion is a vector quaternion if $\hat{a} = o + \vec{a}$, in which case

$$\hat{a} + \hat{a}^* = 0 \tag{2.10}$$

2.2 REPRESENTATION OF FINITE ROTATIONS IN TERMS OF QUATERNIONS

In classical mechanics, vector quantities such as the position vector of a given point, the associated velocity vector and the angular velocity vector of a rigid body can be regarded as pure vector quaternions. The benefit of the representation of finite rotations in terms of quaternions results then from the following properties.

2.2.1 Representation of finite rotations in terms of quaternions

Given - a *unit* quaternion $\hat{e} = e_0 + \vec{e}$
- the position vector $\hat{x} = o + \vec{x}$

the finite rotation of $\vec{x}$ to a new position $\vec{y}$ may be represented by the triple quaternion product

$$\hat{y} = \hat{e}\,\hat{x}\,\hat{e}^* \tag{2.11}$$

Proof

The proof holds by observing that the result of (2.11) is a spherical motion; since

(i) $\hat{y}$ is also a vector quaternion

$$\hat{y} + \hat{y}^* = \hat{e}\,\hat{x}\,\hat{e}^* + (\hat{e}\,\hat{x}\,\hat{e}^*)^*$$
$$= \hat{e}(\hat{x} + \hat{x}^*)\,\hat{e}^* = 0$$

(ii) its norm is preserved by the transformation

$$\hat{y}\hat{y}^* = (\hat{e}\,\hat{x}\,\hat{e}^*)(\hat{e}\,\hat{x}\,\hat{e}^*)^*$$
$$= \hat{e}\,\hat{x}\,\hat{x}^*\hat{e}^* = |\hat{x}|^2$$

2.2.2 Inverse rotation

The inverse rotation is directly obtained in terms of the conjugate quaternion

$$\hat{x} = \hat{e}^* \hat{y} \hat{e} \tag{2.12}$$

2.2.3 Successive finite rotations

Let the position vector $\hat{x}$ undergo two successive rotations

$$\hat{y} = \hat{e}_1 \hat{x} \hat{e}_1^*$$

$$\hat{z} = \hat{e}_2 \hat{y} \hat{e}_2^* = (\hat{e}_2\hat{e}_1) \hat{x} (\hat{e}_2\hat{e}_1)^*$$

The resulting rotation is then given by

$$\hat{z} = \hat{e} \hat{x} \hat{e}^* \quad \text{with} \quad \hat{e} = \hat{e}_2\hat{e}_1 \tag{2.13}$$

2.3 CALCULATION OF VELOCITIES IN TERMS OF QUATERNIONS

Starting from a time-dependent spherical motion

$$\hat{y}(t) = \hat{e}(t) \hat{x} \hat{e}^*(t) \tag{2.14}$$

velocities are calculated in the transformed frame by

$$\dot{\hat{y}} = \dot{\hat{e}} \hat{e}^* \hat{y} + \hat{y} \hat{e} \dot{\hat{e}}^* \tag{2.15}$$

By noting next that the normality condition $|\hat{e}|^2 = 1$ involves

$$\dot{\hat{e}} \hat{e}^* + \hat{e} \dot{\hat{e}}^* = 0 \tag{2.16}$$

and defining the vector quaternion of angular velocities measured in the transformed frame

$$\hat{\omega} = 2 \dot{\hat{e}} \hat{e}^* \tag{2.17}$$

the velocity vector (2.15) can be written in the form

$$\dot{\hat{y}} = \frac{1}{2} (\hat{\omega} \hat{y} - \hat{y} \hat{\omega}) \tag{2.18}$$

The velocity vector can also be calculated in the initial frame by

$$\hat{v} = \hat{e}^* \dot{\hat{y}} \hat{e} \tag{2.19}$$

in which case it takes a similar form to (2.18)

$$\hat{v} = \frac{1}{2} (\hat{\omega}'\hat{x} - \hat{x} \hat{\omega}') \tag{2.20}$$

with the angular velocity vector measured in the initial frame

$$\hat{\omega}' = 2 \hat{e}^* \dot{\hat{e}} \tag{2.21}$$

2.4 MATRIX NOTATION

An arbitrary unit quaternion $\hat{e}$ may be represented in matrix form by the 4-component column matrix

$$\underline{p} = \underline{\hat{e}} = [\, e_0 \; e_1 \; e_2 \; e_3 \,]^T = \begin{bmatrix} e_0 \\ \underline{e} \end{bmatrix} \quad \text{with} \quad \underline{p}^T\underline{p} = 1 \tag{2.22}$$

in which case the quaternion product $\hat{\underline{e}} = \hat{\underline{e}}_1 \hat{\underline{e}}_2$ can be written in either form

$$\hat{\underline{e}} = \hat{\underline{e}}_1 \hat{\underline{e}}_2 = \underline{\underline{A}}_1 \hat{\underline{e}}_2 = \underline{\underline{B}}_2 \hat{\underline{e}}_1$$

or (2.23)

$$\underline{p} = \underline{\underline{A}}_1 \underline{p}_2 = \underline{\underline{B}}_2 \underline{p}_1$$

with the (4 × 4) matrices

$$\underline{\underline{A}} = \begin{bmatrix} e_0 & -\underline{e}^T \\ \underline{e} & e_0 \underline{\underline{I}} + \tilde{\underline{\underline{e}}} \end{bmatrix} \quad \text{and} \quad \underline{\underline{B}} = \begin{bmatrix} e_0 & -\underline{e}^T \\ \underline{e} & e_0 \underline{\underline{I}} - \tilde{\underline{\underline{e}}} \end{bmatrix} \tag{2.24}$$

where $\underline{\underline{I}}$ is the unit matrix, and $\tilde{\underline{\underline{e}}}$ is the skew-symmetric matrix attached to $\underline{e}$

$$\tilde{\underline{\underline{e}}} = \begin{bmatrix} 0 & -e_3 & e_2 \\ e_3 & 0 & -e_1 \\ -e_2 & e_1 & 0 \end{bmatrix} \tag{2.25}$$

2.4.1 Matrix form of finite rotations

By making repetitive use of (2.23), the rotation operation (2.11) can be recasted in matrix form

$$\begin{aligned} \hat{\underline{y}} &= \hat{\underline{e}}\, \hat{\underline{x}}\, \hat{\underline{e}}^* \\ &= \underline{\underline{A}}\, \hat{\underline{x}}\, \hat{\underline{e}}^* \\ &= \underline{\underline{A}}\, \underline{\underline{B}}^T \hat{\underline{x}} \end{aligned} \tag{2.26}$$

One finds

$$\underline{\underline{A}}\, \underline{\underline{B}}^T = \begin{bmatrix} 1 & \underline{o}^T \\ \underline{o} & \underline{\underline{R}} \end{bmatrix} \tag{2.27}$$

where the 3 × 3 submatrix $\underline{\underline{R}}$ is the classical rotation operator. Developing the product (2.27) provides the result

$$\underline{\underline{R}} = (2\, e_0^2 - 1)\, \underline{\underline{I}} + 2\, (\underline{e}\, \underline{e}^T + e_0 \tilde{\underline{\underline{e}}}) \tag{2.28}$$

which shows that the components of the quaternion $(e_0\,,\, e_1\,,\, e_2\,,\, e_3)$ are nothing other than the well known *Euler parameters*

$$e_0 = \cos \frac{\phi}{2} \qquad e_i = u_i \sin \frac{\phi}{2} \qquad i = 1,2,3 \tag{2.29}$$

Their geometric meaning is given by the figure 1. It is related to the Euler-Chasles theorem which expresses any finite rotation as a unique rotation ϕ about an arbitrary direction defined by the unit vector $\underline{u}$.

As suggested in [7] , it is convenient to extract from $\underline{\underline{A}}$ and $\underline{\underline{B}}^T$ the (3 × 4) matrices

$$\underline{\underline{E}} = [-\underline{e} \quad e_0 \underline{\underline{I}} + \tilde{\underline{\underline{e}}}\,] \quad \text{and} \quad \underline{\underline{G}} = [-\underline{e} \quad e_0 \underline{\underline{I}} - \tilde{\underline{\underline{e}}}\,] \tag{2.30}$$

in which case $\underline{\underline{R}}$ may also be expressed as a bilinear form in terms of quaternions

$$\underline{\underline{R}} = \underline{\underline{E}}\ \underline{\underline{G}}^T \tag{2.31}$$

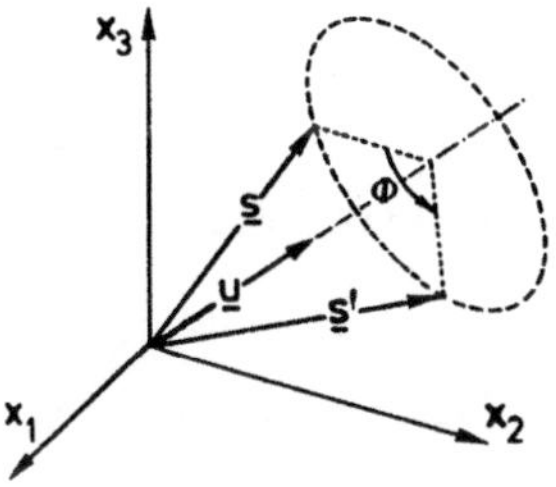

Fig.1 Finite rotation ϕ about arbitrary axis $\underline{u}$

2.4.2 Matrix expression of velocities

By making use again of (2.23) the angular velocity vectors (2.17) and (2.18) expressed in the transformed and initial frames respectively are given by

$$\underline{\hat{\omega}} = 2\ \underline{\underline{B}}^T\ \underline{\dot{\hat{e}}} \qquad \text{and} \qquad \underline{\hat{\omega}}' = 2\ \underline{\underline{A}}^T\ \underline{\dot{\hat{e}}} \tag{2.32}$$

It is straightforward to extract from (2.31) the vector part of the quaternions $\underline{\hat{\omega}}$ and $\underline{\hat{\omega}}'$, giving the important expressions

$$\underline{\omega} = 2\ \underline{\underline{E}}\ \underline{\dot{p}} \qquad \text{and} \qquad \underline{\omega}' = 2\ \underline{\underline{G}}\ \underline{\dot{p}} \tag{2.33}$$

In order to calculate the linear velocities (2.18) and (2.20), on has to further note that, as a direct consequence of (2.5)

$$\begin{aligned} \underline{\hat{a}}\ \underline{\hat{b}} - \underline{\hat{b}}\ \underline{\hat{a}} &= 0 + 2\ \underline{\underline{\tilde{a}}}\ \underline{b} \\ &= 0 - 2\ \underline{\underline{\tilde{b}}}\ \underline{a} \end{aligned} \tag{2.34}$$

Making then use of (2.34) in (2.18) provides the expression of the velocity vector in transformed coordinates

$$\underline{\dot{y}} = -2\ \underline{\underline{\tilde{y}}}\ \underline{\underline{E}}\ \underline{\dot{p}} \tag{2.35}$$

Starting from (2.19), it can similarly be expressed in terms of components in the initial frame

$$\underline{\dot{y}} = \underline{\underline{R}}\ \underline{v} = -2\ \underline{\underline{R}}\ \underline{\underline{\tilde{x}}}\ \underline{\underline{G}}\ \underline{\dot{p}} \tag{2.36}$$

2.4.3 Additional properties

The following additional relationships are straightforward to derive and may be useful in further developments

$$\begin{aligned} &\underline{\underline{E}}\ \underline{\underline{E}}^T = \underline{\underline{G}}\ \underline{\underline{G}}^T = \underline{\underline{I}} \quad , \quad \underline{\underline{E}}^T\underline{\underline{E}} = \underline{\underline{G}}^T\underline{\underline{G}} = \underline{\underline{I}} - \underline{p}\ \underline{p}^T \\ &\underline{\underline{E}}\ \underline{p} = \underline{\underline{G}}\ \underline{p} = 0 \quad , \quad \underline{\underline{\dot{E}}}\ \underline{\dot{p}} = \underline{\underline{\dot{G}}}\ \underline{\dot{p}} = 0 \\ &\underline{\underline{\dot{R}}} = 2\ \underline{\underline{E}}\ \underline{\underline{\dot{G}}}^T = 2\ \underline{\underline{\dot{E}}}\ \underline{\underline{G}}^T \end{aligned} \tag{2.37}$$

$$2\,\underline{\underline{E}}\,\dot{\underline{\underline{E}}}^T = \tilde{\underline{\omega}} \qquad , \quad 2\,\underline{\underline{G}}\,\dot{\underline{\underline{G}}}^T = \tilde{\underline{\omega}}'$$

III. RIGID BODY DYNAMICS [7]

3.1 GENERALIZED COORDINATES OF THE RIGID BODY

Let $\underline{r}$ be the position vector of the center of mass O' of the rigid body in $Ox_1x_2x_3$, and $\underline{s}'$ the constant relative position of P in $O'\xi_1\xi_2\xi_3$ (see fig. 2)

$$\underline{r} = [\,x_1 x_2 x_3\,]^T \quad , \quad \underline{s}' = [\xi_1\xi_2\xi_3\,]^T$$

The position vector of an arbitrary point P in $Ox_1x_2x_3$ is

$$\underline{r}_p = \underline{r} + \underline{s} = \underline{r} + \underline{\underline{R}}\,\underline{s}' \tag{3.1}$$

and is thus a function of the generalized coordinates

$$\underline{q}^T = [\underline{r}^T\ \underline{p}^T\,] = [\,x_1 x_2 x_3 e_0 e_1 e_2 e_3\,] \tag{3.2}$$

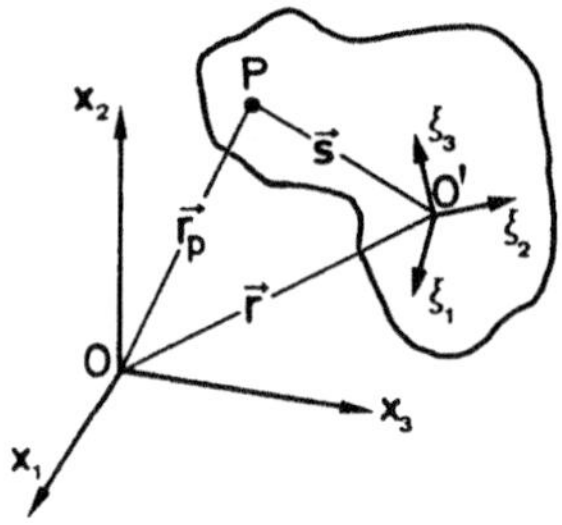

Fig. 2 Position vector of an arbitrary point P

The velocity vector is given by

$$\dot{\underline{r}}_p = \dot{\underline{r}} + \dot{\underline{s}}$$

Owing to (2.34) and (2.35), it may be written in either one of the forms

$$\dot{\underline{r}}_p = \dot{\underline{r}} - 2\,\tilde{\underline{\underline{s}}}\,\underline{\underline{E}}\,\dot{\underline{p}} \tag{3.3}$$

or

$$\dot{\underline{r}}_p = \dot{\underline{r}} - 2\,\underline{\underline{R}}\,\tilde{\underline{\underline{s}}}'\underline{\underline{G}}\,\dot{\underline{p}} \tag{3.4}$$

3.2 KINETIC ENERGY

Assuming that O' is the center of mass, the kinetic energy of the rigid body reduces to

$$T = \frac{m}{2}\,\dot{\underline{r}}^T\,\dot{\underline{r}} + 2\,\dot{\underline{p}}^T\,\underline{\underline{G}}^T\,\underline{\underline{J}}'\,\underline{\underline{G}}\,\dot{\underline{p}} = \frac{1}{2}\,\dot{\underline{q}}^T\,\underline{\underline{M}}\,\dot{\underline{q}} \tag{3.5}$$

where $\underline{J}'$ is the inertia tensor in the relative frame $O\xi_1\xi_2\xi_3$

$$\underline{\underline{J}}' = \frac{1}{2}\int_V \tilde{\underline{s}}'^T\,\tilde{\underline{s}}'\,dm = \int (\underline{s}'^T\,\underline{s}\,\underline{\underline{I}} - \underline{s}'\,\underline{s}'^T)\,dm \tag{3.6}$$

and m is the total mass

The (7 × 7) mass matrix $\underline{\underline{M}}$ is given by

$$\underline{\underline{M}} = \begin{bmatrix} \underline{\underline{N}} & \underline{0} \\ \underline{0} & 4\,\underline{\underline{G}}^T \underline{\underline{J}}' \underline{\underline{G}} \end{bmatrix} \qquad \text{with } \underline{\underline{N}} = \text{diag}(m) \tag{3.7}$$

3.3 VIRTUAL WORK AND GENERALIZED FORCES

Let $\underline{t}$ be the force applied per unit of volume, and $\underline{n} = \tilde{\underline{\underline{s}}}\ \underline{t}$ the corresponding torque about O'. The virtual work expression is

$$\delta W = \int_V \underline{t}^T \delta\underline{r}_p \, dV = \int_V \underline{t}^T (\delta\underline{r} + \delta\underline{s})\, dV \tag{3.8}$$

By analogy with (2.33) let us introduce the infinitesimal angular displacements

$$\delta\underline{\alpha} = 2\,\underline{\underline{E}}\,\delta\underline{p} \quad \text{such that } \delta\underline{s} = -2\,\tilde{\underline{\underline{s}}}\underline{\underline{E}}\ \delta\underline{p} \tag{3.9}$$

The virtual work expression (3.8) can then be written in the form

$$\delta W = \delta\underline{q}^T \underline{g} = \delta\underline{r}^T \underline{f} + \delta\underline{p}^T \underline{k} \tag{3.10}$$

with the vector of generalized forces $\underline{g}^T = [\ \underline{f}^T \ \ \underline{k}^T\ ]$
calculated by

$$\underline{f} = \int_V \underline{t}\, dV \qquad \underline{k} = 2\,\underline{\underline{E}}^T \int_V \underline{n}\, dV = 2\underline{\underline{G}}^T \int_V \underline{n}'\, dV \tag{3.11}$$

3.4 DYNAMIC EQUILIBRIUM OF THE RIGID BODY

The dynamic equilibrium equations of the rigid body are calculated from the Lagrange equations with the lagrangian function

$$L = T + \lambda\ (\underline{p}^T \underline{p} - 1)$$

where λ is a lagrangian multiplier associated with the normality constraint

$$\underline{p}^T\underline{p} = 1 \tag{3.12}$$

The resulting system of equations of motion is

$$\begin{bmatrix} \underline{\underline{N}} & \underline{0} & \underline{0} \\ \underline{0} & 4\,\underline{\underline{G}}^T \underline{\underline{J}}' \underline{\underline{G}} & 2\,\underline{p} \end{bmatrix} \begin{bmatrix} \ddot{\underline{r}} \\ \ddot{\underline{p}} \\ \lambda \end{bmatrix} = \begin{bmatrix} \underline{f} \\ \underline{g} + 8\,\dot{\underline{\underline{G}}}^T \underline{\underline{J}}' \dot{\underline{\underline{G}}} \quad \underline{p} \end{bmatrix} \tag{3.13}$$

and has to be supplemented by the constraint (3.12). It has cubic linearity in the Euler parameters and its time derivatives.

IV. KINEMATIC DESCRIPTION OF DEFORMATION AND RELATIVE MOTION

Let us consider the representation of a mechanical linkage as an assembling of either rigid or elastic members (bar and beam elements with finite or infinite stiffness) connected by coupling elements such as spherical joints,

Hooke joints and hinges. The Euler parameter formalism provides an efficient representation of relative rotation due either to kinematic freedom of to elastic deformation.

4.1 LOCAL TORSION AND BENDING

If one notes that $\underline{\omega}' = 2\,\underline{\underline{G}}\,\dot{\underline{p}}$ provides an expression of the angular velocities in local axes, it it possible to write a similar expression for infinitesimal rotations

$$\delta\,[\,\phi_1 \phi_2 \phi_3\,]^T = 2\,\underline{\underline{G}}\,\delta\underline{p} \tag{4.1}$$

Supposing then that the Euler parameters are functions of the ξ_1 variable the curvature matrix $\underline{\gamma}$ may be defined as

$$\underline{\gamma} = \frac{\partial}{\partial \xi_1}\,[\,\phi_1 \phi_2 \phi_3\,]^T = 2\,\underline{\underline{G}}\,\frac{\partial \underline{p}}{\partial \xi_1} \tag{4.2}$$

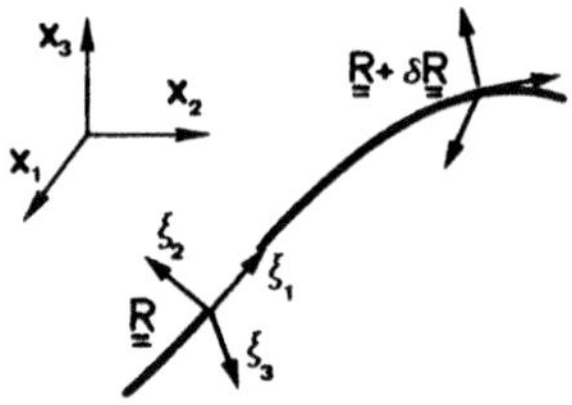

Fig.3 Description of local torsion and bending

with components γ_1, γ_2 and γ_3 which have respectively the meaning of torsion deformation about $O\,\xi_1$, bending deformation about $O\,\xi_2$ and bending deformation about $O\,\xi_3$.

Equation (4.2) shows that Euler parameters are specially well suited to describe arbitrary large rotations of beams. They can thus be used not only in the kinematic analysis of 3-D mechanisms, but also in the structural analysis of beams undergoing arbitrarily large displacements.

4.2 HINGE JOINT ELEMENT

The kinematic constraints involved in a hinge joint are most easily deduced from the composition rule (2.23).

The Euler parameters of the relative rotation

$$\underline{\underline{R}}'' = \underline{\underline{R}}^T\,\underline{\underline{R}}' \tag{4.3}$$

are given by

$$\begin{bmatrix} e_0'' \\ \underline{e}'' \end{bmatrix} = \begin{bmatrix} e_0 & \underline{e}^T \\ -\underline{e} & e_0\,\underline{\underline{I}} - \underline{\underline{\tilde{e}}} \end{bmatrix} \begin{bmatrix} e_0' \\ \underline{e}' \end{bmatrix} \tag{4.4}$$

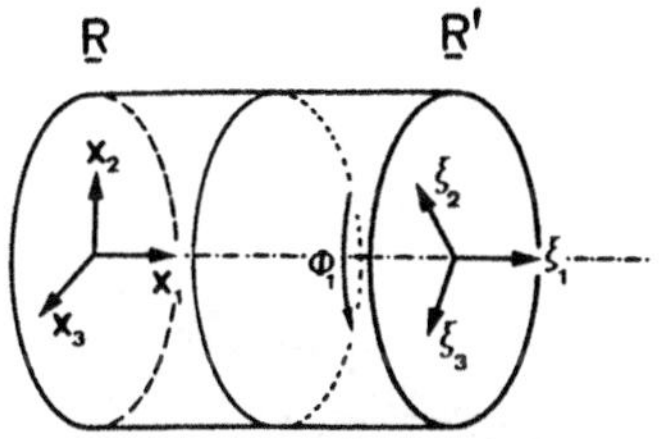

Fig.4 Hinge with relative rotation ϕ_1 *about* Ox_1 *axis*

ant the relative rotation ϕ_1 is such that

$$\phi_1 = 2\,\mathrm{ATAN2}\,(e''_1\ ,\ e''_0) \tag{4.5}$$

For a *perfect hinge*, the conditions of zero relative rotation are

$$e''_2 = e''_3 = 0 \tag{4.6}$$

while for a *flexible hinge* the transverse bending deformation are computed from

$$\phi_2 = 2(e''_2\, e''_0 + e''_3\, e''_1) \quad \text{and} \quad \phi_3 = 2(e''_3\, e''_0 - e''_2\, e''_1) \tag{4.7}$$

4.3 HOOKE JOINT ELEMENT

Similarly, the kinematical constraints in a perfect Hooke joint are easy to express in terms of the Euler parameters (4.4) of the relative rotation (4.3). The Hooke joint allows for a relative rotation which consists of two elementary rotations about orthogonal axes

$$\underline{\underline{R}}'' = \underline{\underline{R}}_2(-\beta)\ \underline{\underline{R}}_3(+\alpha) \tag{4.8}$$

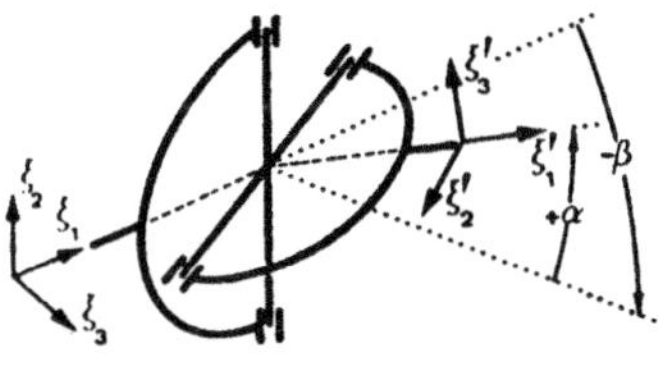

Fig.5 Hooke joint

It is easy to verify that the Euler parameters of a rotation matrix of type (4.8) have to satisfy the kinematic constraint

$$e''_2 e''_3 - e''_1 e''_0 = 0 \tag{4.9}$$

V. BEAM ELEMENT FOR KINEMATIC AND DYNAMIC ANALYSIS OF MECHANICAL LINKAGES

Let us consider a straight beam element of length ℓ_0 in underformed configuration. If small deformation is assumed, axial strain, torsion deformation and bending deformations may be conveniently measured in the local frame $\underline{\underline{R}}$ attached to its end point Q. The columns of $\underline{\underline{R}}$

$$\underline{\underline{R}} = [\, \underline{m}_1 \underline{m}_2 \underline{m}_3 \,] \tag{5.1}$$

are easily computed in terms of the Euler parameters $\underline{p}_Q$.

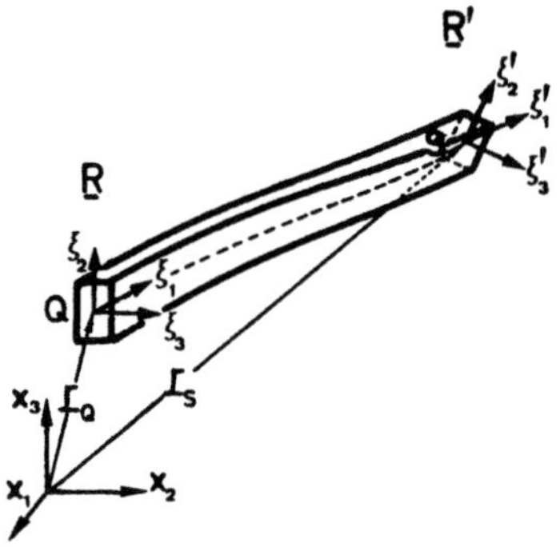

Fig.6 Beam element

Assume next that Euler parameters vary linearly on the beam length

$$\underline{p} = (1 - \frac{\xi_1}{\ell_0})\, \underline{p}_Q + \frac{\xi_1}{\ell_0}\, \underline{p}_S \tag{5.2}$$

The beam deformation is conveniently measured by

$$\underline{\varepsilon}_e^T = [\, \underline{d}^T \ \underline{\gamma}^T \,] \tag{5.3}$$

with - the local measure of curvature (eqn 4.2):

$$\underline{\gamma} = \frac{2}{\ell_0} \underline{\underline{G}} (\underline{p}_S - \underline{p}_Q) = \frac{2}{\ell_0} \underline{\underline{G}}\, \underline{p}_S \tag{5.4}$$

- the Green measure of axial strain :

$$d_1 = \frac{1}{2\,\ell_0^2} (\underline{r}_S - \underline{r}_Q)^T (\underline{r}_S - \underline{r}_Q) - \frac{1}{2} \tag{5.5}$$

- the shear deformations :

$$d_i = \frac{1}{\ell_0} \underline{m}_i^T (\underline{r}_S - \underline{r}_Q) \qquad i = 2,3 \tag{5.6}$$

Or, with the definition of element generalized coordinates

$$\underline{q}_e^T = \underline{r}_Q^T \ \underline{p}_Q^T \ \underline{r}_S^T \ \underline{p}_S^T \tag{5.7}$$

$$\underline{\varepsilon}_e = \underline{S}_e (\underline{q}_e) \tag{5.8}$$

It is straight forward to derive from eqns (5.4) to (5.6) the strain matrix $\underline{B}_e$ such that

$$\delta \underline{\varepsilon}_e = \underline{\underline{B}}_e\, \delta \underline{q}_e \tag{5.9}$$

It is found that $\underline{\underline{B}}_e$ is linear in the displacements $\underline{r}$ and quadratic in the Euler parameters $\underline{p}$.

The strain energy is a quadratic expression of the integral stiffness matrix of the element

$$W_e = \frac{1}{2} \underline{\varepsilon}_e^T \underline{\underline{H}}_e \underline{\varepsilon}_e = W(\underline{q}_e) \tag{5.10}$$

Since absolute displacements have been used in the kinematic description, the kinetic energy is easily computed by assuming a linear interpolation of displacements

$$\underline{r} = \underline{r}_Q \left(1 - \frac{\xi_1}{\ell_0}\right) + \underline{r}_S \frac{\xi_1}{\ell_0} \tag{5.11}$$

and yields to

$$T_e = \frac{\ell_0}{2} \int_0^1 \mu \dot{\underline{r}}^T \dot{\underline{r}} \, d\xi_1 = \frac{1}{2} \dot{\underline{q}}_e^T \underline{\underline{M}}_e \dot{\underline{q}}_e \tag{5.12}$$

with a mass matrix $\underline{\underline{M}}_e$ which does not depend on displacements.

In order to express the dynamic equilibrium equations of a flexible member, one has to remember that Euler parameters must satisfy the constraint $\underline{p}^T \underline{p} = 1$. The contribution of each flexible element to the lagrangian is thus

$$L_e = T_e(\dot{\underline{q}}_e) - W(\dot{\underline{q}}_e) + \lambda_Q (\underline{p}_Q^T \underline{p}_Q - 1) + \lambda_S (\underline{p}_S^T \underline{p}_S - 1)$$

where λ_Q, λ_S are lagrangian multipliers.

VI. PRISMATIC JOINT ELEMENT

The prismatic joint developed (fig. 7) allows to represent sliding motion

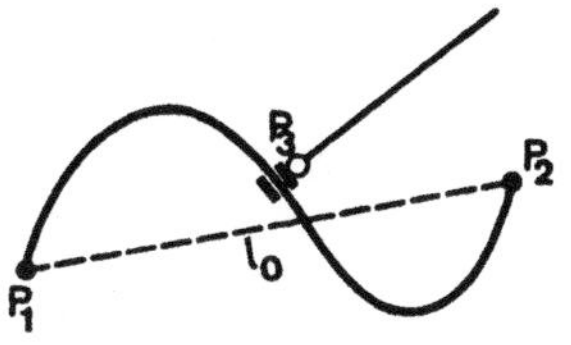

Fig.7 Sliding curvilinear motion

on a curve with the following restrictions :

- point contact,
- planar sliding motion,
- undeformable sliding curve.

A convenient way to describe the sliding curve consists to use the Bézier interpolation functions [11] as illustrated by fig. 8.

Let

$$\underline{a}_0^T = [\,0 \quad 0 \quad 0\,]$$
$$\underline{a}_1^T = [\,x_1 \quad y_1 \quad 0\,]$$
$$\underline{a}_m^T = [\,x_m \quad y_m \quad 0\,]$$

be the local coordinates of the knot points A_0 , ... A_m defining the Bézier descriptor. The local coordinates of an intermediate point on the Bézier curve of degree m is then given by

$$\underline{a}(s) = \sum_{i=0}^{m} C_m^i (1-s)^i s^{m-i} \underline{a}_i \qquad (6.1)$$

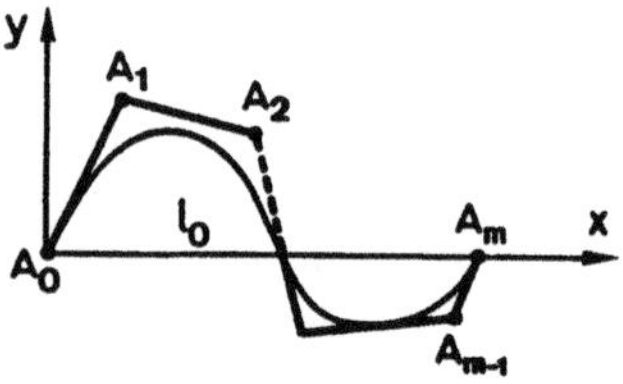

Fig. 8 Parametric description of sliding curve

where s is the current parameter. Its position in global coordinates is similarly given by

$$\underline{r}(s) = \underline{r}_1 + \underline{\underline{R}}\, \underline{a}(s) \qquad (6.2)$$

where $\underline{r}_1$ is the position vector of the origin P_1, and $\underline{\underline{R}}$ is the rotation operator at P_1. It can be shown that eqn (6.2) may still be rewritten in a form involving the coordinates of end nodes

$$\underline{r}(s) = (1 - s^m)\underline{r}_1 + s^m r_2 + \underline{g}(s) \qquad (6.3)$$

with the function

$$\underline{g}(s) = \sum_{i=1}^{m-1} C_m^i (1-s)^{m-i} s^i \underline{\underline{R}}\, \underline{a}_i \qquad (6.4)$$

In order to express the kinematics of sliding motion, two types of constraints have to be expressed

(i) sliding motion along the curve. If $\underline{r}_3$ is the position vector of P_3, this constraint is

$$\underline{r}_3 - (1 - s^m)\underline{r}_1 - s^m \underline{r}_2 - \underline{g}(s) = 0 \qquad (6.5)$$

(ii) indeformability of the sliding curve.

The longitudinal and transverse indeformability can be expressed by the conditions

$$\underline{m}_i^T (\underline{r}_2 - \underline{r}_1) = 0 \qquad i = 1,2,3 \qquad (6.6)$$

while the rigid body rotation is preserved by the condition on Euler parameters

$$\underline{p}_2 = \underline{p}_1 \qquad (6.7)$$

The constraints (6.6) depend implicitly on Euler parameters through the fact that the $\underline{m}_i$ are functions of $\underline{p}$.

VII. FINITE ELEMENT ASSEMBLING

Let us consider the rotation part of the displacement represented by fig. 9. $\underline{\underline{R}}$ represents the rotation operator of the initial configuration in the structural system $Ox_1x_2x_3$: it can be specified by its Euler parameters $\underline{r}$. Let similarly $\underline{\underline{R}}'$ be the rotation operator describing the deformed configuration at time t, with Euler parameters $\underline{p}$.
The additional rotation due to displacement may be represented in *structural axes* by defining a displacement transformation $\underline{\underline{D}}(t)$ such that

$$\underline{\underline{R}}'(t) = \underline{\underline{D}}(t) \cdot \underline{\underline{R}} \tag{7.1}$$

with the initial condition

$$\underline{\underline{D}}(o) = \underline{\underline{I}} \tag{7.2}$$

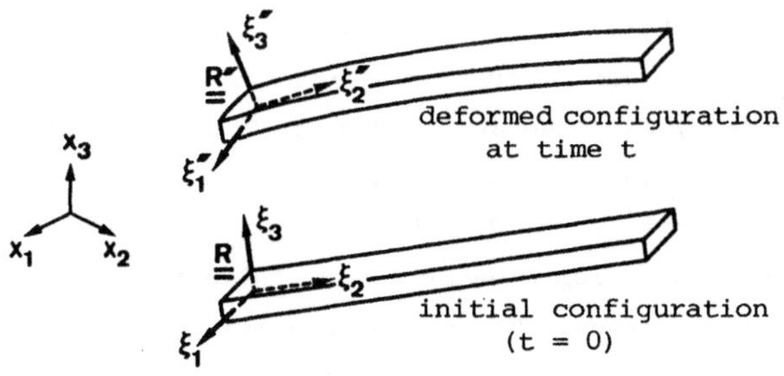

Fig. 9 Representation of rotation displacement in a given point

In order to show that the Euler parameters of $\underline{\underline{D}}(t)$ can effectively be used to express the assembling of the finite element system at structural level, let us consider a rigid connexion between two elements 1 and 2 (fig. 10).

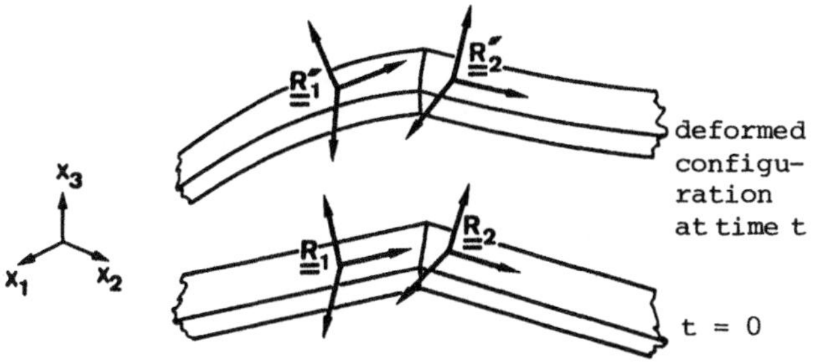

Fig. 10 Representation of rotation displacements at a rigid joint

The rotations at both ends of the connection are described in initial configuration by distinct rotation matrices $\underline{\underline{R}}_1$ and $\underline{\underline{R}}_2$.
After displacement, both rotation matrices are transformed according to (7.1) into

$$\underline{\underline{R}}'_1(t) = \underline{\underline{D}}_1(t)\ \underline{\underline{R}}_1 \tag{7.3}$$

$$\underline{\underline{R}}'_2(t) = \underline{\underline{D}}_2(t)\ \underline{\underline{R}}_2 \tag{7.4}$$

But the rigid joint hypothesis implies unchanged relative rotation

$$\underline{\underline{R}}'^T_1\ \underline{\underline{R}}'_2 = \underline{\underline{R}}^T_1\ \underline{\underline{R}}_2 \tag{7.5}$$

Owing to (7.3) and (7.4), the condition (7.5) holds if and only if the displacement matrices are identical

$$\underline{\underline{D}}_1(t) = \underline{\underline{D}}_2(t) \tag{7.6}$$

Equation (7.6) constitutes thus a convenient basis for the assembling process of rotational degrees of freedom at structural level.
Let us next denote by $\underline{d}(t)$ the Euler parameters of $\underline{\underline{D}}(t)$. Owing to the initial condition (7.2), they are such that

$$\underline{d}(o)^T = [\,1 \quad 0 \quad 0 \quad 0\,] \tag{7.7}$$

They provide a very easy representation of finite rotations in an arbitrary mechanism at a global level. The quaternion multiplication rule (2.23) allows to calculate from them the Euler parameters $\underline{p}$ of the instantaneous angular configuration

$$\underline{p} = \underline{\underline{T}}\ \underline{d} \tag{7.8}$$

where $\underline{\underline{T}}$ is the constant matrix constructed from the Euler parameters of the initial configuration

$$\underline{\underline{T}} = \underline{\underline{B}}(\underline{r}) = \begin{bmatrix} r_0 & -\underline{r}^T \\ \underline{r} & r_0\underline{\underline{I}} - \underline{\underline{\tilde{r}}} \end{bmatrix} \tag{7.9}$$

VIII. SOLVING THE PROBLEM OF KINEMATICS AND DYNAMICS

An augmented lagrangian function can be formed for the system which takes the most general form

$$T(\underline{q},\underline{\dot{q}}) - W(\underline{q}) + \underline{q}^T \underline{g}(t) - \underline{\lambda}^T \underline{\phi} - \frac{1}{2} \underline{\phi}^T \underline{C} \underline{\phi} \tag{8.1}$$

where

- $T(\underline{q},\underline{\dot{q}})$ is the total kinetic energy of the system, calculated from expressions (3.5) and (5.12). It does not depend on displacements in the absence of rigid bodies with large rotational inertia ;
- $W(\underline{q})$ is the total strain energy, gathering contributions from flexible members and possibly, flexible hinges ;
- $\underline{g}(t)$ is the set of externally applied loads.
- $\underline{\lambda}$ is a set of lagrangian multipliers conjugated to the constraints $\underline{\phi}(\underline{q}) = 0$;
- $\underline{C}$ is a diagonal matrix of positive constants defining the penalty terms.

The constraints $\underline{\phi}(\underline{q})$ can themselves be separated into three kinds of constraints :

1. a first subset $\underline{\phi}_1(\underline{q}) = 0$ expresses either kinematic constraints at joints such as (4.6) and (4.9) or indeformability conditions such as (6.6) to (6.8) ;
2. The second subset $\underline{\phi}_2(q) = 0$ expresses the normality conditions $\underline{p}^T\underline{p} = 1$ on every set of Euler parameters ;
3. the third subset arises whenever kinematically driven degrees of freedom exist in the mechanism. They take the general form $\underline{\phi}_3(\underline{q},\underline{q}_D) = 0$ where the $\underline{q}_D$ are the driving displacements. In particular, eqn (6.5) is a constraint of this type.

The functional (8.1) has to be specialized according to the nature of the problem to be solved. Five specific problems may be defined

- *assembling* the mechanism, i.e. finding an initial feasible configuration,
- calculating a *static equilibrium* configuration,
- *motion analysis* of a kinematically-driven mechanism,
- *dynamic analysis* of a *force-driven* mechanism,
- *dynamic analysis* of a *kinematically-driven* mechanism.

Starting from (8.1), all of them are formulated as specific static and dynamic equilibrium problems. Equilibrium iteration is always performed using a Newton-Raphson type technique, and time integration is best achieved using the HHT algorithm [13] .

IX. APPLICATIONS

STATIC RESPONSE OF A CANTILEVER LOADED AT ITS TIP

The problem as shown in fig. 11.a consists of a cantilever beam loaded by a force at the tip acting in a constant direction. Since an exact solution to this problem exists for an inextensible beam [11], it provides a good example to test the adequacy of the flexible member element developed to undergo arbitrarily large rotations. The numerical solution given uses 10 elements of equal length. A finite element solution to the same problem, with the same number of elements, is also available in [12].
The non-dimensional displacement-load curves of fig. 11.b were obtained up to $\frac{P\ell^2}{EI} = 16$ by applying 20 equal load increments $\Delta P = 3.10^4$ kg with an average of 4 iterations/increment. The curves match exactly the analytical solution given in [11] .

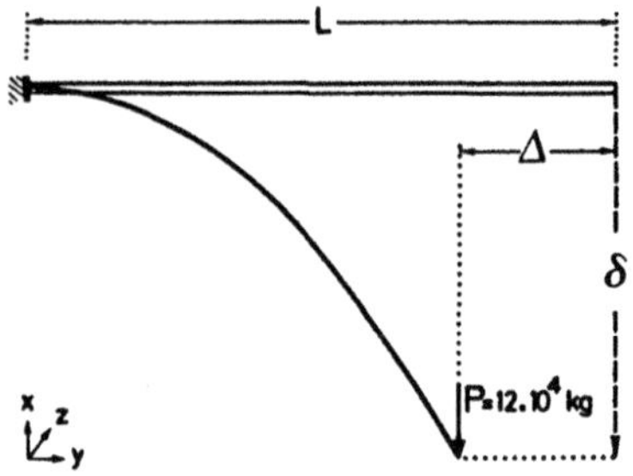

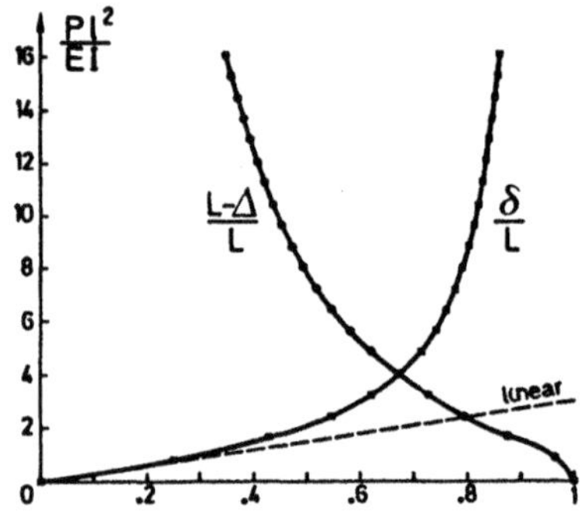

Fig.11 (a) Cantilever beam problem *(b) Displacement-load curves*

BRICARD MECHANISM [14]

The Bricard mechanism shown in fig. 12.a is an interesting example for testing the effectiveness of the formalism for modelling 3-D articulated systems. In the present analysis, it is force-driven by a torque applied on θ_1 . Flexibility effects have not been taken into account. Fig. 12.b displays the effective motion with time.

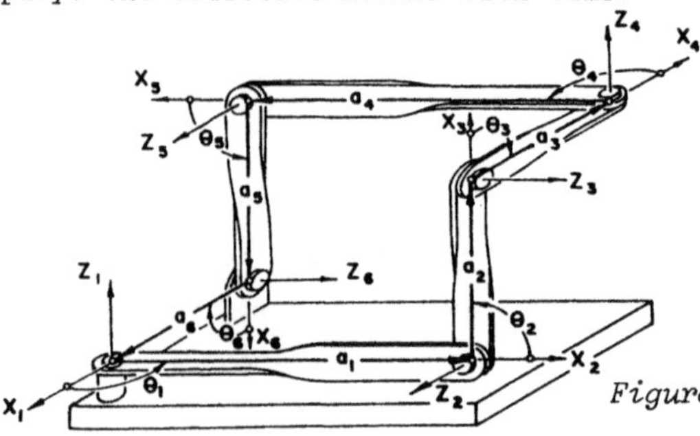

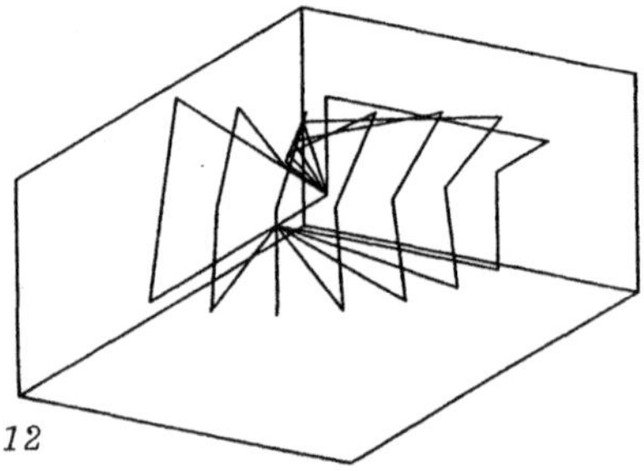

Figure 12

(a) Bricard mechanism *(b) displacement evolution with time*

The characteristics of the system were as follows :

$$E = 2.1\ 10^{11}\ N/m^2 \ , \quad \nu = 0.3 \ , \quad \rho = 7\,800\ kg/m^3 \ ,$$

$$A = 5.10^{-4}\ m^2 \ , \quad I_X = 2.10^{-7}\ m^4 \ , \quad I_Y = I_Z = 10^{-7}\ m^4 \ .$$

The lengths were respectively :

$$\ell_X = \ell_Z = 0.2\ m \ , \quad \ell_Y = 0.5\ m.$$

The torque applied at the input member is a triangular impulse with a maximum value of 1 000 N × m at time t = 0.02 sec.

X. CONCLUSIONS

A new approach based on Euler parameters has been presented to the problem of representing finite rotations which is particularly adapted to kinematic and dynamic analysis of mechanisms with flexible members using finite elements. Finite element models for flexible members and various types of joints have been developed. Only preliminary results obtained using this approach have been presented in the paper. The computational efficiency of the method will be more extensively discussed in a next paper on the same subject.

XI. REFERENCES

[1] WITTENBURG, J., *Dynamics of systems of rigid bodies*, Teubner, Stuttgart, 1977

[2] PAUL, R., *Robot manipulators. Mathematics, programming and control*, MIT Press, Cambridge, Mass., 1981

[3] CHACE, M., *Computer-aided engineering of large displacements dynamic mechanical systems*, CAE/CAD/CAM Conference, April 1983, Salt Lake City, Utah

[4] HAUG, E., WEHAGE, R., BARMAN, N., *Dynamic analysis and design of constrained mechanical systems*, Technical Report n° 50, Div. of Eng., Univ. of Iowa, Nov. 1978

[5] SONG, J., HAUG, E., *Dynamic analysis of mechanisms*, Comp. Meth. in Appl. Mech. and Eng., 24(1980), 359-381

[6] VAN DER WERFF, L., *Kinematic and dynamic analysis of mechanisms. A finite element approach*, PhD Thesis, Delft Univ. Press, 1977

[7] NIKRAVESH, P., *Spatial kinematics and dynamic analysis with Euler parameters*, NATO A.S.I. on Computer aided analysis and optimization of mechanical system dynamics, Univ. of Iowa, Aug. 1983

[8] GERADIN, M., ROBERT, G., BERNARDIN, C., *Dynamic modelling of manipulators with flexible members*, in "Advanced software in robotics"(Ed. by A. Danthine and M. Géradin), North Holland, 1984

[9] GERADIN, M., *Finite element approach to kinematic and dynamic analysis of mechanisms using Euler parameters*, in "Numerical methods for nonlinear problems II",Ed. C. Taylor et al, Pineridge Press, Swansea, U.K., 1984

[10] WEHAGE, R.A., *Quaternions and Euler parameters. A brief exposition*, in "Computer aided analysis and optimization of mechanical systems dynamics", Ed. E. Haug,Springer-Verlag,1984

[11] BISSHOPP, K.E., DRUCKER, D.C., *Large deflections of cantilever beams*, Q. Appl. Math. 3 (1945)

[12] HIBBIT, H.D., BECKER, E.B., TAYLOR, L.M., *Nonlinear analysis of some slender pipelines*, Comp. Meth. in Appl. Mech. and Eng., 17/18 (1979), 203-225

[13] HUGHES, T.J.R., *Analysis of transient algorithms with particular reference to stability problems*, in "Computational methods for transient analysis", Ed. T. Belytschko and T.J.R. Hughes, North Holland, 1983

[14] UICKER, J.J., DENAVIT, J., HARTENBERG, R.S., *An iterative method for the displacement analysis of spatial mechanisms* Jnl of Appl. Mech., Trans. ASME, 1964, 309-314.

A Discussion of Cauchy Stress Formulations for Large Strain Analysis

J. P. HALLEUX, J. DONEA
Commission of the European Communities, Ispra, Italy

Summary

Formulations based on some objective rate of the Cauchy stress have often been advocated in recent years in the development of computational procedures for solving finite deformation problems. As a matter of fact, considerable success has been obtained using such formulations, even for cases where other approaches fail to give reasonable answers. Though the Zaremba-Jaumann-Noll (ZJN) rate of the Cauchy stress still seems to be the most popular rate, Dienes recently proposed an alternative approach usually referred to as rotated Cauchy stress and originally attributed to Green and Naghdi (GN). This last method is currently considered by various authors as being the best one available. Both rates of the Cauchy stress are critically examined in the present paper on the basis of an elementary rectilinear shear test for which an analytical solution is derived. It is demonstrated that both rates are not fully satisfactory from a fundamental point of view. Furthermore, it is shown that, despite some interesting features, the GN rate does not appear to be clearly superior to the ZJN rate for practical applications. Finally, a rate-form is proposed which reproduces the analytical solution to the rectilinear shear test.

1. Introduction

Evaluation of non-linear finite element methods involves quite a number of aspects which are all of importance. These include not only performance analysis but also correctness proof or at least demonstration when no rigorous approach is available. The topic which is addressed here goes beyond strict numerical aspects and questions the fundamental theoretical descriptive concepts underlying some of the currently used large strain formulations. Computational tools are indeed unlikely to be reliable if they are not based on careful numerical modelling on the one hand, and on sound mathematical and physical principles on the other. In large strain analysis, which nowadays is becoming of engineering importance, the descriptive aspects are still matter of vehement debate. This is fortunate for researchers as can be seen from the number of papers issued in recent years on this still rather specialized subject. This situation is, how-

Finite Element Methods for Nonlinear Problems
Europe-US Symposium, Trondheim, Norway 1985
Editors: Bergan, Bathe, Wunderlich

ever less fortunate in a practical environment (for instance in the field of metal forming) where engineers are puzzled by the amount of the often very contradictory proposals and claims which are made.

It is a fact that classical small-strain elasto-plasticity and respective stress-rate formulations have gained large popularity in many of the computer codes developed in the past twenty years. This fact quickly focused attention on the extension of these concepts to the field of large strain problems, using some objective rate-type constitutive assumptions. It is well known that such an apparently natural inductive extension is not without difficulties. Nagtegaal and de Jong [1] have shown in 1981 some results of the stresses obtained in a simple finite shear problem (shear test). They found that for a bilinear elasto-plastic stress-strain relationship, the use of the most advocated stress rate, i.e. the Zaremba-Jaumann-Noll (ZJN) rate of the Cauchy stress, produced a radical difference in shear stress response depending on the assumption made for the hardening behaviour: while the isotropic hardening assumption led to a monotonically increasing shear stress, the kinematic assumption led to a periodically varying shear stress. It was quickly recognized that this problem, also occurring in purely elastic cases, lies in the use of the stress rate itself together with the assumption of a hypoelastic material of grade zero, already found irrelevant by Truesdell [2]. It was further discovered that Dienes [3] had already provided an original solution to this problem. He defined a rate of material rotation on which the rate of change of stress should be dependent, resulting in the substitution of the ZJN rate by a stress rate attributed [4] to Green and Naghdi (GN).

Following this evolution, the so-called hydro-codes such as HONDO [5] and DYNA [6] which first made use of the ZJN stress rates, are now progressively upgraded in order to include the GN rates. In these finite element explicit codes for fast non-linear dynamics, derived from an early finite difference code, HEMP [7], equilibrium is tested in the current configuration by a direct application of the principle of virtual work (i.e. expressed in terms of the Cauchy stress). Robustness, soundness and reproducibility seem to be the main characteristics of this family of codes leading to an interesting cost-performance ratio. Moreover, suitable adaptations of these codes extend their capabilities to the field of slower phenomena which require the use of implicit time-stepping algorithms such as those used for example in NIKE [8] where the Green-Naghdi stress rate seems to be firmly established.

Despite the undeniable success of the ZJN stress rate or that of the apparently even better GN rate in the usual elasto-plastic approach to finite strain problems, these both - as will be shown here on the basis of the simple elastic rectilinear shear problem - do not appear to be completely satisfactory. In this paper both rates are thus critically examined and the respective shear test solutions compared to an analytical finite elasticity reference solution. Finally, a rate form [9] delivering this reference solution is also proposed.

2. The elastic rectilinear shear problem

In a Lagrangian description, the motion of a continuum defined by a set of material points is expressed as:

$$\underline{x} = \underline{x}(\underline{X}, t) \tag{1}$$

where:

$\underline{X}$ is the position vector of a material point in the undeformed or reference configuration corresponding to time t_o (X_i are material coordinates);

$\underline{x}$ is the current (corresponding to time t) position vector of the material point which was located in $\underline{X}$ in the reference configuration at time $t = t_o$ (x_i are spatial coordinates).

Note that throughout the text, when not otherwise specified, all quantities are referred to fixed cartesian axes ($\underline{e}_1, \underline{e}_2, \underline{e}_3$).

Partial differentiation of (1) with respect to the material coordinates gives the deformation gradient $\underline{F}$ with components

$$F_{ij} = \frac{\partial x_i}{\partial X_j} \tag{2}$$

The simple test to be analysed here is defined by

$$x_i = H_{ij}(t)\, X_j \tag{3}$$

where $\underline{H}$ is the particular deformation gradient corresponding to the elastic rectilinear shear problem of interest, and has components

$$H_{ij} = \begin{bmatrix} 1 & T(t) & 0 \\ 0 & 1 & 0 \\ 0 & 0 & 1 \end{bmatrix} \tag{4}$$

Dependence on time t is unimportant here as only kinematic arguments will be raised.

While the solution which is sought is non-linear in the geometrical sense, the material is supposed to behave instantaneously in a linear elastic way when no rotations are present (i.e. the rate of true stress is directly proportional to the rate of true strain). As the bulk modulus plays no role due to the isochoric character of the given shear problem chosen, the shear modulus (μ) alone governs the stress-strain relationship.

An analytical solution is sought with the assumption that a pure shear state in a linear elastic material must be reachable by straining at constant volume, accompanied by some rigid body rotation. Despite the fact that this assumption appears to be quite natural in the way one may think of a linear elastic material, it must be understood that it implies a constitutive choice which is not necessarily equivalent to the equation of hypo-elasticity. As a matter of fact, it involves the choice of a preferential configuration and is, therefore, to be related to what is commonly referred to as finite elasticity [10] in which the deformation gradient plays an essential role. Moreover, it should be noted that this way of thinking does not follow on obviously from hyper-elastic concepts either, and should, therefore, be handled with care in eventual generalizations. It remains that it possesses sound physical meaning and - as will be seen - the solution obtained exhibits some very satisfactory features.

Consider (Fig.1) two distinct fibres PA and PB, a fibre being here defined as a collection of material points lying on a straight line. These fibres are supposed to be in their initial state. In the deformed state they lie along QA and QB, respectively. The deformation process defined by eqs.(3) and (4) is of course such that straight lines always remain straight. Suppose further that the choice is made that the fibres QA and QB are perpendicular in the deformed state (one could as well have chosen the initial state). Because RQ = RP + PQ one immediately obtains:

$$\tan\alpha - T = \tan\beta \tag{5}$$

and since PS = PQ + QS:

$$\tan\left(\frac{\pi}{2} - \alpha\right) + T = \tan(\gamma - \beta) \tag{6}$$

where α, β and γ are angles as defined in Fig.1.

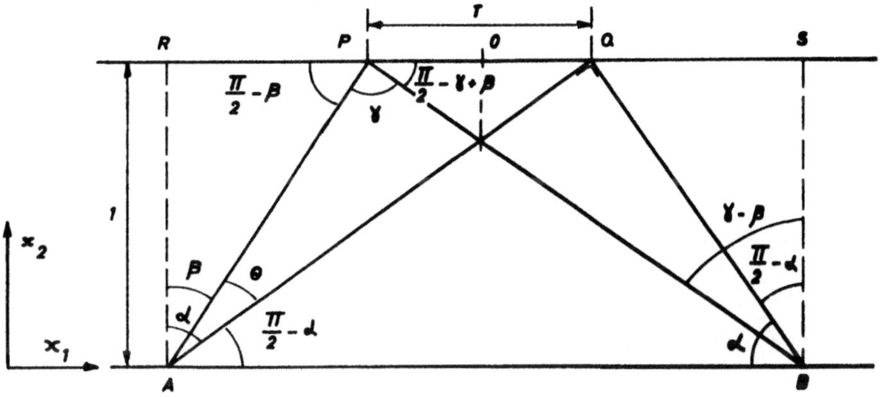

Fig. 1 - The rectilinear **shear problem.**

Now, the condition for which γ is a right angle **is sought. The fibres con-**sidered are then perpendicular both in the initial **and deformed state,** i.e. they necessarily correspond in the deformed state to the **principal** directions in a pure shear problem. It should be noted that **this does not** mean that these fibres are always perpendicular during the motion defined by T. Eq.(6) can then be written as:

$$\frac{1}{\tan\alpha} + T = \frac{1}{\tan\beta} \tag{7}$$

and replacing in eq.(5):

$$T\tan^2\alpha - T^2\tan\alpha - T = 0 \tag{8}$$

which admits two roots α_1 and α_2 which are the angles of the principal directions given by:

$$\tan\alpha = \frac{T^2 \pm \sqrt{T^4 + 4T^2}}{2T} = \frac{T}{2} \pm \sqrt{\left(\frac{T}{2}\right)^2 + 1} \tag{9}$$

It is easily verified that: $\alpha_1 = \pi/2 + \alpha_2$.

For β one finds:

$$\tan\beta = -\frac{T}{2} \pm \sqrt{\left(\frac{T}{2}\right)^2 + 1} \tag{10}$$

Due to symmetry it is seen that α and β play the same role ($\beta = \pi/2 - \alpha$), only T changes sign. This means that point 0 is exactly in the middle of PQ (and also of RS, of course).

The ratio between final length AQ (ℓ_f) and initial length AP (ℓ_i) of the principal fibre in extension is easily obtained by noting again on Fig.1 that:

$$\ell_f = \sqrt{1 + tg^2\alpha} = \frac{1}{\cos\alpha} \tag{11}$$

and

$$\ell_i = \sqrt{\frac{1}{tg^2\alpha} + 1} = \frac{1}{\sin\alpha} \tag{12}$$

The principal engineering strain (e) is given by

$$e = \frac{\ell_f - \ell_i}{\ell_i} = \frac{\sin\alpha - \cos\alpha}{\cos\alpha} \tag{13}$$

and the true strain is given by

$$\varepsilon = \ell n(1+e) = \ell n \tan\alpha \tag{14}$$

In the same way for the other principal direction it is consistently found that:

$$\varepsilon = -\ell n \tan\alpha \tag{15}$$

It is easily seen that the angle of material rotation (θ) corresponding to T is given by:

$$\theta = \frac{\pi}{2} - 2\beta \quad \text{or alternatively} \quad \theta = 2\alpha - \frac{\pi}{2} \tag{16}$$

and via its tangent

$$\tan\theta = \frac{1 - \tan^2\beta}{2\tan\beta} \tag{17}$$

and using eq.(10)

$$\theta = \tan^{-1}\frac{T}{2} \tag{18}$$

From eqs.(9), (14) and (15), and using the shear modulus (μ), the stress components are:

$$\sigma_{11} = -\sigma_{22} = \mu\sin 2\alpha\, \ell n(\tan\alpha) = \frac{2T\mu}{\sqrt{T^2+4}}\, \ell n\left(\frac{T + \sqrt{T^2+4}}{2}\right) \tag{19}$$

$$\sigma_{12} = \mu\cos 2\alpha\, \ell n(\tan\alpha) = \frac{4\mu}{\sqrt{T^2+4}}\, \ell n\left(\frac{T + \sqrt{T^2+4}}{2}\right) \tag{20}$$

or in terms of the angle of material rotation:

$$\sigma_{11} = -\sigma_{22} = 2\mu \sin\theta \ \ln\left(\frac{\sin\theta + 1}{\cos\theta}\right) \tag{21}$$

$$\sigma_{12} = 2\mu \cos\theta \ \ln\left(\frac{\sin\theta + 1}{\cos\theta}\right) \tag{22}$$

The principal stress and the angle of principal direction are plotted on Figs.2 and 3, respectively, as a reference solution (designated as REF.). It is seen that in the limit of vanishingly small strains the result coincides with the geometrically linear solution with principal directions oriented at 45°. The stress increases monotonically and behaves logarithmically for T tending to infinity while the principal directions tend to be oriented following the cartesian axes. The behaviour at very large strains appears thus to be physically sound.

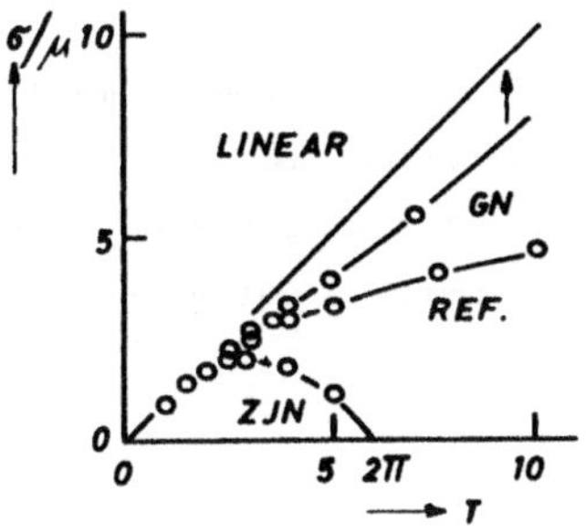

Fig. 2 - Principal stress.

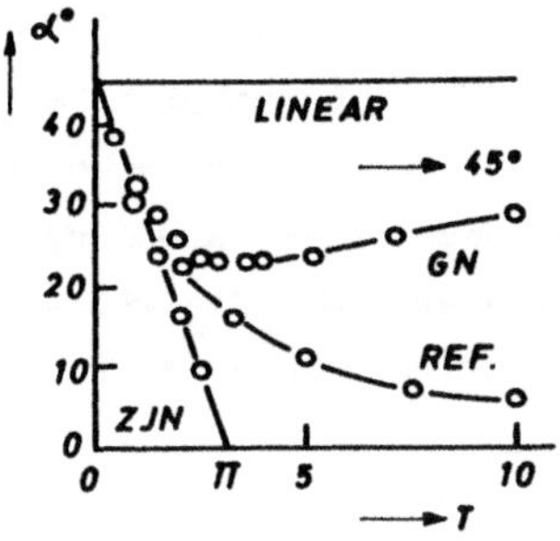

Fig. 3 - Principal direction.

3. The Zaremba-Jaumann-Noll (ZJN) rate of Cauchy stress

A hypoelastic material [2] can be defined as behaving in the following way:

$$\overset{\nabla}{\underline{\sigma}} = \underline{C}(\underline{\sigma})\underline{D} \tag{23}$$

where the rate $\overset{\nabla}{\underline{\sigma}}$ called the co-rotational or ZJN stress rate is given by:

$$\overset{\nabla}{\underline{\sigma}} = \dot{\underline{\sigma}} - \underline{W}\underline{\sigma} + \underline{\sigma}\underline{W} \tag{24}$$

It is easily verified that this stress rate is objective, i.e. preserves invariance under superposed rotation.

The velocity gradient ($\underline{L}$):

$$\underline{L} = \frac{\partial \dot{\underline{x}}}{\partial \underline{x}} = \underline{D} + \underline{W} \tag{25}$$

where the dot denotes time derivative, furnishes the rate of deformation

in its symmetric part (the stretching $\underset{\sim}{D}$) and the rate of rotation in its skew-symmetric part (the spin $\underset{\sim}{W}$).
It may be noted that the ZJN stress rate ($\overset{\nabla}{\underset{\sim}{\sigma}}$) is linear in the stretching ($\underset{\sim}{D}$) and that the material response function ($\underset{\sim}{C}$) depends on the stress ($\underset{\sim}{\sigma}$) only.
If $\underset{\sim}{C}$ is further restricted to be stress-independent, a hypoelastic material of grade zero (Jaumann type) is obtained. $\underset{\sim}{C}$ can then be viewed as a generalized Hooke matrix depending on the elastic constants K (bulk modulus) and μ (shear modulus) only. This offers a computationally tempting opportunity to include elasto-plastic material behaviour by merely replacing the elastic matrix by the elasto-plastic matrix as classically used in small strain analysis. As this depends on the stress state only, a form coherent with eq.(23) is apparently obtained. However, theoretical considerations [2] certainly oblige to handle hypoelastic materials of grade 0, or their generalization, with care.

As a matter of fact, a number of anomalies can be found when use is made of computational approaches based on this sort of material modelling. For instance, the shear test as defined by eqs.(3) and (4) and with a constitutive relation such as (23) where $\underset{\sim}{C}$ is the elastic matrix, delivers the following set of relations after application of (24):

$$\begin{aligned} \dot{\sigma}_{11} - \dot{T}\sigma_{12} &= 0 \\ \dot{\sigma}_{22} + \dot{T}\sigma_{12} &= 0 \\ \dot{\sigma}_{12} + \frac{1}{2}\dot{T}(\sigma_{11}-\sigma_{22}) &= \dot{T}\mu \end{aligned} \tag{26}$$

A periodic response is obtained

$$\sigma_{11} = -\sigma_{22} = \mu(1-\cos T) \tag{27}$$

$$\sigma_{12} = \mu \sin T \tag{28}$$

Principal value and angle are plotted in Figs.2 and 3.
This result is usually considered to be unsatisfactory from a physical point of view because of the oscillatory nature of the stresses. Among the remedies which have been proposed to this situation, the approach by Dienes [3] consisting of using a modified stress rate but still based on the stretching, has generated particular enthusiasm; this even to the point that many computer programs originally based on the ZJN rate switched to using this new rate. The desirability of this abrupt change is criti-

cally examined hereafter where it is found that the use of this new rate is less convincing than numericists would like it to be. In a further section a procedure based on the rate of natural (logarithmic) strains is derived from finite elasticity considerations and a new stress rate is obtained.

4. The Green-Naghdi (GN) rate of the Cauchy stress

The Green-Naghdi stress rate [3,4] is based on mixed considerations taken on the one hand from hypoelasticity and on the other from finite elasticity. A hypoelastic-like constitutive equation is used relating some objective stress rate ($\overset{\Delta}{\underline{\sigma}}$) to the stretching ($\underline{D}$):

$$\overset{\Delta}{\underline{\sigma}} = \underline{C}(\underline{\sigma})\underline{D} \tag{29}$$

As in hypoelasticity the rate of deformation ($\underline{D}$) is thus taken as the measure of strain rate. The stress rate ($\overset{\Delta}{\underline{\sigma}}$) is, however, not related to the rate of Cauchy stress via the spin ($\underline{W}$) as in (24) but via a rate of rotation ($\underline{\Omega}$) which is derived from the proper-orthogonal matrix ($\underline{R}$) obtained in the polar decomposition of the deformation gradient ($\underline{F}$):

$$\overset{\Delta}{\underline{\sigma}} = \dot{\underline{\sigma}} - \underline{\Omega\sigma} + \underline{\sigma\Omega} \tag{30}$$

$$\underline{\Omega} = \dot{\underline{R}}\,\underline{R}^T \tag{31}$$

$$\underline{F} = \underline{UR} = \underline{RV} \tag{32}$$

where $\underline{U}$ and $\underline{V}$ are the left and right stretch tensors, respectively.

Among the general justifications given by Dienes [3] to sustain this approach, the following two summarizing points are interesting for this discussion:

- physical insight in the global process of deformation and rotation is provided by the polar decomposition of the deformation gradient; the material rotation matrix ($\underline{R}$) and the stretch tensor ($\underline{U}$ or $\underline{V}$) should thus play a central role in the analysis of deforming bodies;
- while it appears in hypoelasticity to be possible to pass from some objective stress rate to another by adequate modification of the material response function ($\underline{C}$), it remains that "when a definite material law is specified, the stress-strain relations are different when the various definitions of stress rates are used". In particular it is convenient at least for numericists to look at elasticity in a rate form,

i.e. with an objective stress rate depending on the stretching ($\underline{D}$) and on a response function ($\underline{C}$) explicitly containing the usual elastic coefficients. This preference is to be attributed to the ease with which this rate form apparently permits generalization to elasto-visco-plastic material behaviour.

Though the first argument is pertaining to finite elasticity rather than to hypoelasticity, it is suggested by Dienes to relate the rate of change of the logarithm of the left stretch tensor ($\underline{U}$) directly to the stretching ($\underline{D}$) and thereby apparently to justify some link with the second argument. Unfortunately, Dienes' proof is limited to the case where no rotation is present. It will be shown in the next section that the rate of change of the stretch tensor, and thereby the strain-rate measure to be used in the formulation is also dependent on the rotation, if present. Furthermore, the objective stress rate ($\overset{\Delta}{\underline{\sigma}}$) obtained via the rate of material rotation ($\underline{\Omega}$) for which he finds an explicit expression depending exclusively on spin, stretching and stretch, lets him conclude that "this makes it possible to determine the rate of material rotation when the velocity field is known". This last point appears to be misleading if not handled with care. While it is obvious that spin and stretching depend only on the velocity and the current configuration, this is not the case for the stretch and not even for its rate as shown hereafter. These quantities (like the material rotation itself) always imply the choice of some preferential initial configuration. Therefore, it seems preferable to view the rate of material rotation ($\underline{\Omega}$) as such, i.e. strictly and only depending on the evolution of the proper orthogonal matrix ($\underline{R}$) obtained by the polar decomposition theorem.

Finally, the objective stress rate to be used in order to reproduce a consistent finite strain solution - as shown in the next section - is not the rate of material rotation ($\underline{\Omega}$) but a linear combination of it with the spin ($\underline{W}$).

Anyway, the set of eqs.(29-31) can be chosen to define some sort of generalized hypoelastic [11] behaviour. To complete the analysis attempted here of the Green-Naghdi stress rate approach, it remains to check how this affects the solution of the test case which was chosen. Thus, turning again to the shear test, the following solution is obtained [3] in terms of the angle of material rotation (θ) as given in (18).

$$\sigma_{11} = -\sigma_{22} = 4\mu(\cos 2\theta \ \ell n \cos\theta + \theta \sin 2\theta - \sin^2\theta) \tag{33}$$

$$\sigma_{12} = 2\mu \cos 2\theta(2\theta - 2\tan 2\theta \ \ell n \ \cos\theta - \tan\theta) \tag{34}$$

Principal value and angle are again found in Figs.2 and 3.
While at first sight the GN solution certainly looks better than the ZJN solution, it does not appear to present a very sound behaviour for large values of T (θ approaching $\pi/2$). Despite the presence of logarithms the limiting behaviour at infinity is identical to a purely linear solution, i.e. principal directions at $\pm$ 45° and linear dependence on T. In the authors' opinion a consistent behaviour should yield exactly the opposite, i.e. principal directions at 0 and 90° and, of course, true logarithmic dependence on T. As a matter of fact, rapid deviation from the analytical solution (section 2) is observed as soon as T exceeds 1, i.e. when ZJN solutions start to exhibit their oscillating character.
GN solutions are, therefore, not felt as being exceptionally remarkable and hardly can justify a large superiority over ZJN solutions.

5. Rate of natural strains

In order to reproduce finite elasticity solutions via a rate form, it is proposed to derive an expression for the rate of change of natural (logarithmic) strains. This may be achieved by expressing the left stretch tensor ($\underline{U}$), obtained (see eq.(32)) in the polar decomposition of the deformation gradient, in principal form:

$$\underline{U} = \underline{S}^T \underline{\underline{Q}} \underline{S} \tag{35}$$

where $\underline{S}$ is a proper orthogonal matrix formed with the direction cosines of the principal directions and $\underline{Q}$, consisting of the principal values of the stretch tensor ($\lambda_{(n)} \geq 0$, n = 1,2,3), is purely diagonal.
The principal values of the natural strain tensor are then defined by

$$\varepsilon_{(n)} = \ell n \ (\lambda_{(n)}), \quad n = 1,2,3 \tag{36}$$

and the cartesian strain components are obtained via

$$\underline{\varepsilon} = \underline{S}^T \ \ell n \underline{Q} \ \underline{S} \tag{37}$$

which differentiated with respect to time furnishes the rate of change of natural strains:

$$\dot{\underline{\varepsilon}} = \underline{\underline{A\varepsilon}} - \underline{\underline{\varepsilon A}} + \underline{S}^T \dot{\underline{Q}} \underline{Q}^{-1} \underline{S} \tag{38}$$

where

$$\underline{A} = \dot{\underline{S}}^T \underline{S} \tag{39}$$

Restricting now to two dimensions $\underline{S}$ and $\underline{A}$ are simply given by

$$\underline{S} = \begin{pmatrix} \cos\alpha & \sin\alpha \\ -\sin\alpha & \cos\alpha \end{pmatrix} \qquad \underline{A} = \begin{pmatrix} 0 & -\dot{\alpha} \\ \dot{\alpha} & 0 \end{pmatrix} \tag{40}$$

where α is the angle of the principal stretch direction.

It is then possible - as shown in [9] - to express the rate of rotation ($\underline{A}$) in the form of a linear combination between the spin ($\underline{W}$) and the rate of material rotation ($\underline{\Omega}$):

$$\underline{A} = \underline{W} + \frac{\lambda_{(1)}^2 + \lambda_{(2)}^2}{(\lambda_{(1)} - \lambda_{(2)})^2} (\underline{\Omega} - \underline{W}) \tag{41}$$

To complete the determination of the rate of natural strains, it remains to evaluate the contribution from the last term in (38) which accounts for the rate of change of the principal stretches. One finally obtains [9]:

$$\dot{\underline{\varepsilon}} = \underline{A}\underline{\varepsilon} - \underline{\varepsilon}\underline{A} + \underline{D} - \left\{\frac{1}{2}(D_{22} - D_{11})\sin 2\alpha + D_{12}\cos 2\alpha\right\} \begin{bmatrix} -\sin 2\alpha & \cos 2\alpha \\ \cos 2\alpha & \sin 2\alpha \end{bmatrix} \tag{42}$$

In order to interpret this form it is more useful to define a rotated strain tensor $\underline{\varepsilon}'$ in its principal form

$$\underline{\varepsilon}' = \underline{S}\,\underline{\varepsilon}\,\underline{S}^T \tag{43}$$

where $\underline{\varepsilon}'$ is thus diagonal and is from (37) $\ell n\, \underline{Q}$.
Then

$$\underline{S}^T \dot{\underline{\varepsilon}}' \underline{S} = \underline{S}^T \dot{\underline{S}}\,\underline{\varepsilon} + \dot{\underline{\varepsilon}} + \underline{\varepsilon}\,\dot{\underline{S}}^T \underline{S} \tag{44}$$

and replacing in eq.(42) and expressing in principal form:

$$\dot{\underline{\varepsilon}}' = \begin{pmatrix} \dot{\varepsilon}_{(1)} & 0 \\ 0 & \dot{\varepsilon}_{(2)} \end{pmatrix} = \underline{SDS}^T - \left\{\frac{1}{2}(D_{22} - D_{11})\sin 2\alpha + D_{12}\cos 2\alpha\right\} \begin{bmatrix} 0 & 1 \\ 1 & 0 \end{bmatrix} \tag{45}$$

The corresponding rate of rotated stress ($\underline{\sigma}'$) is

$$\dot{\underline{\sigma}}' = \underline{C}(\underline{\sigma})\dot{\underline{\varepsilon}}' \tag{46}$$

while the Cauchy stress is given at any time by

$$\underline{\sigma} = \underline{S}^T \underline{\sigma}' \underline{S} \tag{47}$$

When compared to Dienes' approach (section 4), it is seen that $\underline{S}$ (or $\underline{A}$)

plays the role of $\underline{R}$ (or Ω) and that the constitutive relation in cartesian axis makes use of a modified stretching ($\underline{D}'$) which reads:

$$\underline{D}' = \underline{D} - \frac{1}{2}\left\{D_{22}-D_{11})\sin 2\alpha + D_{12}\cos 2\alpha\right\}\begin{bmatrix} -\sin 2\alpha & \cos 2\alpha \\ \cos 2\alpha & \sin 2\alpha \end{bmatrix} \tag{48}$$

As can immediately be seen from eq.(45), this modification forces the strain rate to have its principal directions along the axes of principal stretch.

The straightforward possibility thereby offered to work exclusively in principal directions is certainly computationally attractive. The validity of this very peculiar property in relation to the apparent generality of eq.(46) remains, however, to be assessed.

As expected from its construction, the application of the above procedure to the shear test delivers a result which perfectly reproduces the analytical solution derived in section 2, and reproduced on Figs.2 and 3.

6. Conclusions

Two commonly used objective rates of the Cauchy stress have been critically examined on the basis of the elastic rectilinear shear problem: the classical ZJN rate and the GN rate as proposed by Dienes [3]. They are found to be unsatisfactory for very large deformations. As a matter of fact, both rates deliver good results up to about T = 1, corresponding to 100% shear strain in a purely linear solution. Afterwards, when T exceeds 1, they rapidly diverge from the analytical solution chosen here as a reference solution. A rate form is also proposed which perfectly reproduces this analytical solution of the shear test up to infinite strains. In the light of these findings, it is felt that use of stress rates still requires attention and that a claim of superiority of a certain rate over another, besides being probably illusory, is also hazardous because it tends to hide the importance of the underlying constitutive assumptions.

References

1. Nagtegaal, J.C.; de Jong J.E.: Some aspects of nonisotropic work hardening in finite strain plasticity, in: Lee, E.H.; Mallett R.L. (eds.) Plasticity of Metals at Finite Strain: Theory, Experiment and Computation, Stanford U. and Rensselaer P.I. (1981) 65-102.

2. Truesdell, C.; Noll, W.: The nonlinear field theories of mechanics, in Flügge S. (ed.) Handbuch der Physik III/3, Springer Verlag, 1965.

3. Dienes, J.K.: On the analysis of rotation and stress rate in deforming bodies. Acta Mechanica 32 (1979) 217-232.

4. Johnson, G.C.; Bammann, D.J.: A discussion of stress rates in finite deformation problems, submitted for publication in Int. J. Solids and Structures.

5. Key, S.W.; Krieg, R.D.: On the numerical implementation of inelastic time dependent and time independent finite strain constitutive equations in structural mechanics. Comp. Meths. in Appl. Mech. and Engng. 33 (1982) 439-452.

6. Goudreau, G.L.; Hallquist, J.O.: Recent developments in large scale finite element lagrangian hydrocode technology. Comp. Meths. in Appl. Mech. and Engng. 33 (1982) 689-723.

7. Wilkins, M.L.: Calculation of elasto-plastic flow, University of California, Lawrence Livermore National Laboratory, Rep. UCRL-7322, Rev.1, 1969.

8. Hallquist, J.O.: NIKE 2D - an implicit, finite deformation, finite element code for analysing the static and dynamic response of two-dimensional solids. Univ. of Calif., Lawrence Livermore National Laboratory, Rep. UCRL-52678, 1979.

9. Donea, J.: The rate of natural strains and its use in large-deformation analysis, submitted for publication in Engineering Computations.

10. Gurtin, M.E.: An introduction to continuum mechanics, in Bellman R. (ed.) Mathematics in Science and Engineering 158, Academic Press, 1981.

11. Green, A.E.; McInnis, B.C.: Generalized hypoelasticity, Proc. Roy. Soc. Edinburgh A57, Part III (1967) 220-230.

Part II: Nonlinear Materials

Distributed Cracking and Nonlocal Continuum

Z. P. BAŽANT
Northwestern University, Evanston, USA

Summary

Finite element analysis of failure of certain types of concrete structures requires modeling of distributed cracking. Propagation of failure zones with distributed cracking must be treated from the viewpoint of fracture mechanics if the calculation should be objective and properly converge as the mesh is refined. After a brief review of the crack band model, the lecture expounds a novel approach to strain-softening — a nonlocal formulation, in which stress at a point depends on the entire strain field from a neighborhood of the point. Application of the theory is demonstrated by finite element analysis of waves in strain-softening materials. Comparisons with other strain-softening fracture models due to Hillerborg et al. and Willam et al. are also made, and stability of distributed cracking is analyzed.

Introduction

Heterogeneous brittle materials such as concretes or rocks fail by progressive fracturing distributed over a finite-size zone within the material. In the continuum approximation, this zone is characterized by strain-softening, i.e., a stress-strain relation in which the maximum principal stress decreases at increasing strain. The purpose of the present lecture is to present an overview of some recent results obtained at Northwestern University in the use of strain-softening material models for the description of fracture propagation in concrete structures.

Crack Band Model

Due to the distributed nature of microcracking, as well as the fact that the path of a final crack is generally not smooth but highly tortuous, it is not unrealistic to model cracking by means of stress-strain relations, introducing strain-softening in which the major principal tensile stress is reduced to zero (Fig. 1). This approach is particularly convenient for finite element analysis, since it necessitates merely an adjustment in the incremental stiffness matrix of the finite element.

Finite Element Methods for Nonlinear Problems
Europe-US Symposium, Trondheim, Norway 1985
Editors: Bergan, Bathe, Wunderlich

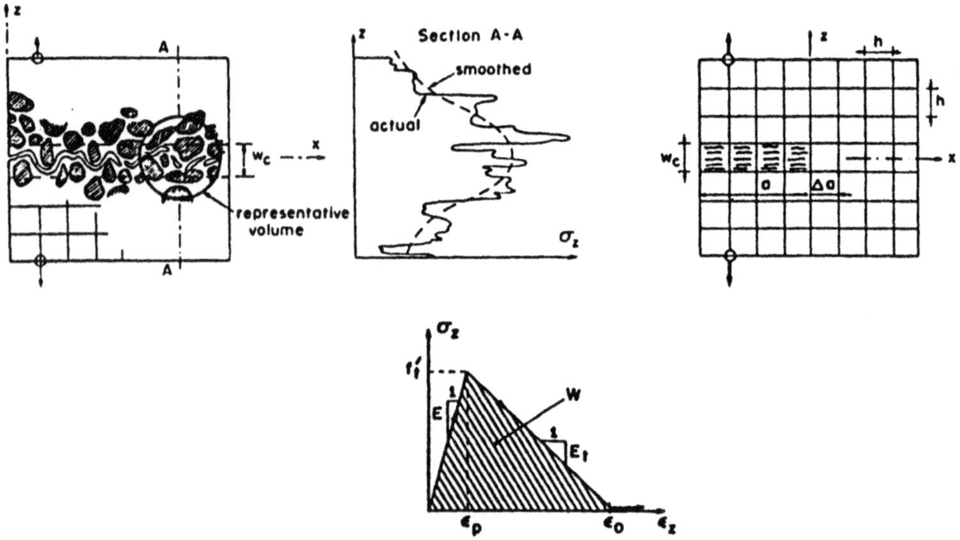

Fig. 1 - Fracture in a Heterogeneous Aggregate Material, Statistical Scatter of Stress, Crack Band Model, and Strain-Softening Stress-Strain Diagram.

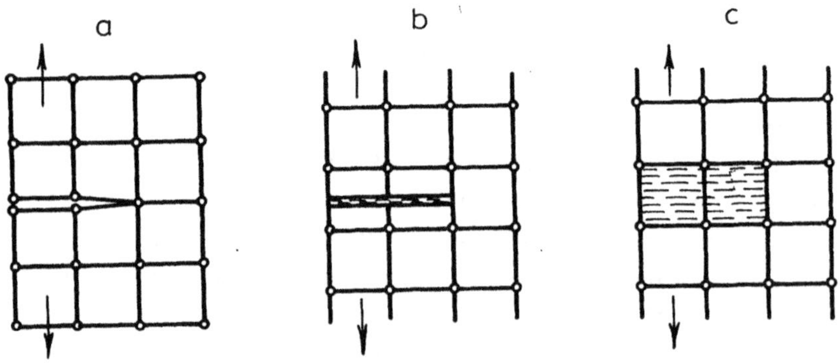

Fig. 2 - Softening Fracture Models: (a) Fictitious Crack Model of Hillerborg et al., (b) Composite Damage Model of Willam et al., anc (c) Crack Band Model of Bazant et al.

In the form of sudden cracking, this approach was introduced in 1967 by Rashid [27,19]. For strain-softening analysis of fracture this approach was developed in Ref. 4-5, in which the strain-softening triaxial stress-strain relation was introduced in the form

$$\underset{\sim}{\varepsilon} = \underset{\sim}{D}\underset{\sim}{\sigma} + \underset{\sim}{\xi}. \tag{1}$$

Here $\underset{\sim}{\varepsilon}$ and $\underset{\sim}{\sigma}$ are the column matrices of the components of strain and stress, $\underset{\sim}{D}$ is the 6×6 matrix of elastic constants, and $\underset{\sim}{\xi} = (\xi_{11}, \xi_{22}, \xi_{33}, 0, 0, 0)^T$, where superscript T denotes a transpose and the numerical subscripts refer to cartesian coordinates x_i $(i = 1,2,3)$; ξ is the column matrix of additional smeared-out strains due to cracking. The normal stresses are assumed to be uniquely related to their associated cracking strains (Fig. 1), i.e.,

$$\sigma_{11} = C(\xi_{11})\xi_{11}, \quad \sigma_{22} = C(\xi_{22})\xi_{22}, \quad \sigma_{33} = C(\xi_{33})\xi_{33} \tag{2}$$

in which C is the secant modulus which reduces to zero at very large cracking strain and may be calibrated from direct tensile test data [21-24,27,28]. Different algebraic relations are used for unloading. In this formulation it is assumed that cracking can occur only in three mutually orthogonal directions which are known in advance and do not rotate against the material once the cracking starts.

In the classical formulation of finite element analysis of concrete structures, as employed in the current large computer programs, a strain-softening stress-strain relation, mostly one that exhibits a sudden stress drop, is used indiscriminately for an arbitrarily chosen finite element size. It has been demonstrated, however, that this approach is unobjective with regard to the analyst's choice of the element size. It can yield greatly different results for different mesh sizes and converges to a physically meaningless solution as the mesh size is refined to zero [3,7-9]. For failures of the brittle type, the load causing further fracture extension converges to zero as the mesh size tends to zero. So does the total energy consumed by failure of the structure, which is an unrealistic result.

It has been shown that these incorrect features can be eliminated by using an energy criterion rather than the stress-strain relation as the primary condition for the extension of the cracked band of finite elements. Considering a crack band of a single-element width, consisting of finite elements with a uniform strain distribution across the band, the fracture energy, i.e., the energy consumed per unit extension of the crack band (and per unit thickness), may be expressed as

$$G_f = w_c \int \sigma_{33}\, d\varepsilon_{33} = \frac{w_c}{2} f_t'^2 \left(\frac{1}{E_0} - \frac{1}{E_t}\right) \qquad (3)$$

in which w_c is the width of the crack band front, which must be considered as a material property [3], σ_{33} is the normal stress across the central plane of the crack band, f_t' is the direct tensile strength of the material, E_0 is the initial elastic Young's modulus, and E_t is the mean downward slope of the strain-softening segment of the stress-strain diagram, which is negative.

The crack band finite element model yields results which agree with all basic experimental data from fracture testing of concrete [5] as well as rock [32], both the maximum load data and the R-curves.

Numerical Implementation and Comparison with Strain-Softening Fracture Models of Hillerborg, Willam and Others

The crack band model [5,32] for which the basic idea of a fixed crack band front width was proposed in 1974 on the basis of stability analysis [3], has three independent material parameters: fracture energy G_f, tensile strength f_t', and crack band front width w_c. The value of the strain-softening modulus E_t then follows from Eq. 3. Although the optimum fit of the entire data set was obtained when $w_c \simeq 3\,d_a$ for concrete [5] and $w_c \simeq 5\,d_a$ for rock [32], it was found, however, that within the relatively broad range $d_a \leq w_c \leq 10\,d_a$ (where d_a = maximum size of aggregate or grain) the effect of w_c on the results is weak, and almost equally good fits of the test data can be obtained with various values of w_c within this range. Another reason for choosing the value of $w_c = 3\,d_a$ for concrete was the fact that the corresponding E_t value represented well the mean slope of the descending branch for some direct tensile tests in which the specimen seemed to be in an approximately homogeneous state of strain.

Therefore, as far as the analysis of localized fracture is concerned, only two independent parameters of the crack band model, namely f_t' and G_f, are important, same as for the fictitious crack model of Hillerborg et al. [11,21,22] and the composite damage model of Willam, Sture et al. [33,34]. Although these two models have not yet been systematically compared with the basic fracture test data from the literature, they should be capable of representing these data equally well as the crack band model since they are essentially equivalent to it. Each of them is characterized, too, by two independent material parameters G_f and f_t to be found from fracture test

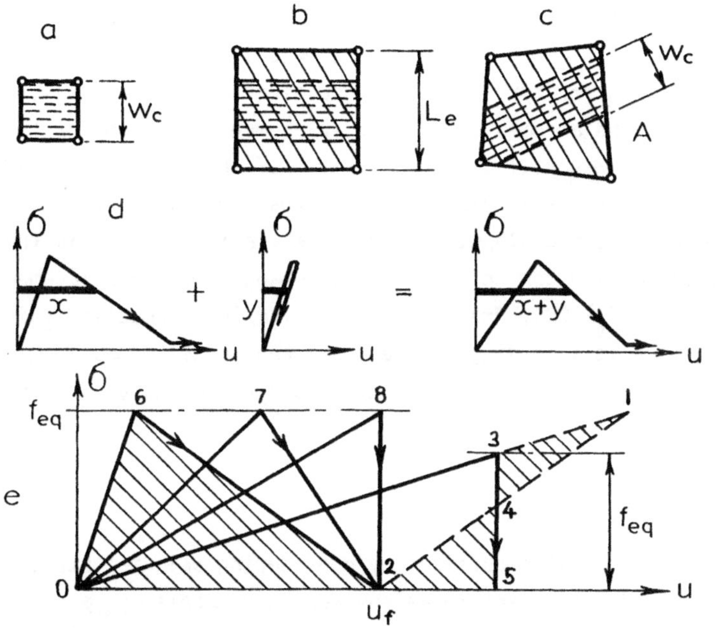

Fig. 3 - Crack Band Model when Element Size is Increased (a-c), Determination of Overall Stress-Strain Relation for Finite Element (d) and Resulting Stress-Displacement Relations for Elements of Various Sizes.

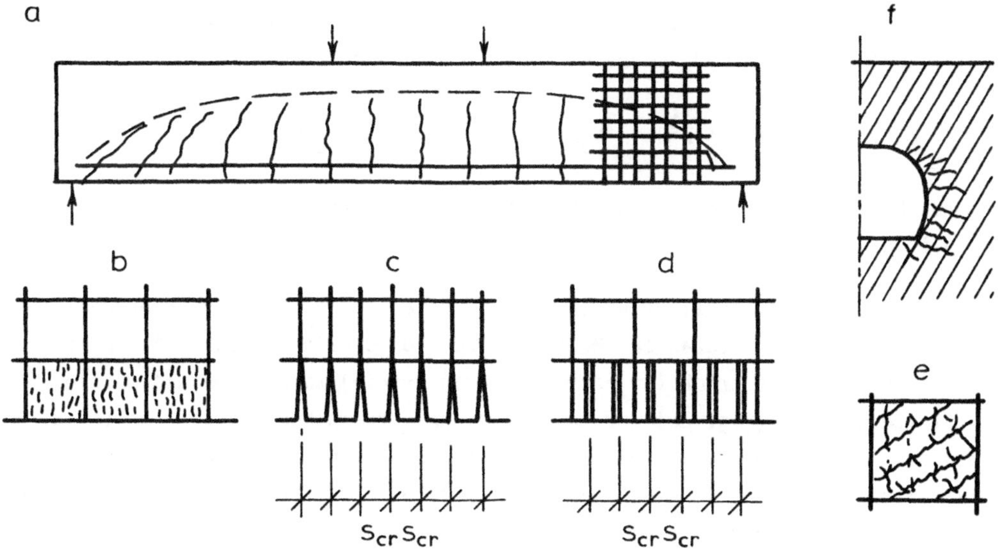

Fig. 4 - Nonlocalized Distributed Cracking (a,f) and Its Treatment by Crack Band Model (b), Fictitious Crack Model (c) and Composite Damage Model (d).

data. Hillerborg's model [11,21,22], which represents an adaptation to concrete of the formulations of Barrenblatt, Dugdale, Knauss, Kfouri, Rice and others for ductile fracture of metals, considers a line crack with a sharp tip (Fig. 2a) and assumes a stress-displacement relation for this crack to be the basic material characteristic. If this relation is approximated as linear, the relative displacement across the line crack is $\delta = (\sigma - f_t')/C_f$ where C_f is the slope; $C_f < 0$. The fracture energy is then represented by the area under this strain-softening stress-displacement diagram, i.e., $G_f = -f_t'^2/2C_f$, from which $C_f = -f_t'^2/2G_f$. Equating this expression for G_f to that according to the crack band model we have $G_f = -f_t'^2/2C_f = w_c(E^{-1} + E_t^{-1})f_t'^2/2$, from which we obtain the equivalence condition

$$\frac{1}{C_f} = -w_c\left(\frac{1}{E} - \frac{1}{E_t}\right) . \tag{4}$$

The same relation may be also obtained by considering uniaxial stress and equating the stress displacement relations for both models for a bar element of length $L_e \geq w_c$. For the crack band model, segment w_c of the length L_e exhibits strain-softening, while the rest of the bar element undergoes elastic unloading, as may be concluded from stability analysis [3]. For Hillerborg's model, only one line crack is allowed within the bar length L_e, according to stability conditions. Thus, the declining portion of the σ-δ relationship is described according to these two models as

$$\delta(\sigma) = w_c\left(\frac{f_t' - \sigma}{-E_t} + \frac{f_t'}{E}\right) + (L_e - w_c)\,\frac{\sigma}{E} = \frac{f_t' - \sigma}{-C_t} + L_e\,\frac{\sigma}{E} \tag{5}$$

($E_t < 0$, $C_t < 0$), and it may be checked that this equation is satisfied for any σ if Eq. 4 holds.

The composite damage model of Willam, Sture et al. [33,34], which represents an adaption of Pietruszczak and Mro'z's model for strain-softening in shear [35], assumes a line crack (or narrow cracking strip) to be embedded within the finite element (Fig. 2b) and the additional strain due to cracking to be added to that due to elastic deformation of the element. For uniaxial stress in a finite element of length L_e in a bar, one obtains for this element the same stress-displacement relationship as Eq. 5, which again confirms the equivalence of this model.

We may thus conclude that the crack band model, the fictitious crack model of Hillerborg et al., and the composite damage model of Willam et al. are equivalent for uniaxial stress, except that for the crack band model no length $L < w_c$ may be considered; but neither other models can be applied for this case because the continuum description is no longer admissible if $L < w_c$. For problems with localized fracture and arbitrary geometry, these three models are not exactly equivalent, but they are nearly equivalent, with differences that are probably too small to be clearly detected from test results. This is because, according to St. Venant's principle, only the total additional relative displacement due to cracking at the fracture front matters for the energy release rate from the entire body.

It has been claimed [33,34] that the composite damage model and the fictitious crack model differ from the crack band model by having only two rather than three independent material constants, G_f and f_t'. However, this claim is unjustified because w_c, the third independent constant in the crack band model, has almost no effect in fitting fracture test data, as already pointed out, and can be only determined if the softening modulus E_t is specified independently, which was done in Ref. 3 on the basis of direct tensile tests.

It has been also claimed that the composite damage model and the fictitious crack model permit the use of finite elements of any size, L_e, whereas the crack band model is restricted to square finite elements of size $L_e = w_c$. This is not true, however. The crack band model permits finite elements of sizes $L_e > w_c$ if the average stress strain relation for the cracked finite element is modified (as shown in Refs. 7, 36,37), and this modification turns out to yield the same stress displacement relation as the other two models. In general this modification may be formulated as follows.

Consider an arbitrary finite element of area A which is not too elongated in any direction and has one integration point (a triangle or a quadrilateral); Fig. 3c. The overall stress displacement relation for the entire element is obtained by adding the strain on a crack band of width w_c to the strain in the finite element (Fig. 3d). This yields for the mean strain tensor in the finite element the general rule

$$\underset{\sim}{\varepsilon} = \frac{w_c}{\sqrt{A}} \underset{\sim}{\varepsilon}_{soft} + \left(1 - \frac{w_c}{\sqrt{A}}\right) \underset{\sim}{\varepsilon}_{el} \tag{6}$$

where the element is approximately treated the same as a square element of side $\sqrt{A}$, $\underset{\sim}{\varepsilon}_{soft}$ is the strain according to the strain-softening constitutive relation for the crack band and $\underset{\sim}{\varepsilon}_{el}$ is the elastic strain at the same stress. It should be noted, though, that the use of an element size L_e that is too large may cause excessive numerical error.

Equation 6 assures that the energy dissipation at full cracking, as obtained from the stress-displacement relation for the entire element, is the same as the energy dissipation for the crack band width w_c. This may be illustrated by the $\delta(\sigma)$ expression for uniaxial stress in Eq. 5. For various element lengths L_e, these relations give the curves in Fig. 3e. While the peak stress point in the stress-displacement relation moves to the right as the element length L_e increases, the end point with $\sigma = 0$ remains fixed because the area under this diagram, representing the fracture energy, must remain constant. The declining portion of the $\delta(\sigma)$ diagram has the inverse slope $d\delta/d\sigma = (w_c/E_t) + (L_e - w_c)/E$; it becomes steeper as L_e increases (Fig. 3e) and becomes vertical when $L_e = L_{cr}$,

$$L_{cr} = w_c\left(1 + \frac{E}{-E_t}\right) = \frac{2G_f E}{f_t'^2} \qquad \left(\frac{d\sigma}{d\delta} \to \infty\right) \tag{7}$$

For $L_e > L_{cr}$, the post-peak portion of the $\delta(\sigma)$ diagram has a positive slope. The post-peak states are then unstable under displacement control. This means that a static analysis of strain-softening problems is impossible. The response of the finite element system after the peak is *dynamic* and the inertia forces must be taken into account.

However, if only the final static state after strain-softening is of interest, strain-softening problems can be analyzed statically even for $L > L_{cr}$ provided that the post-peak response of positive slope is replaced by an equivalent vertical stress drop (Fig. 3e). This drop represents a dynamic motion which begins and ends with a zero kinetic energy. Consider the vertical stress drop $\overline{35}$ in Fig. 3e. The potential energy represented by area $\overline{134}$ accelerates the motion along segment $\overline{34}$, and the potential energy represented by area $\overline{245}$ decelerates the motion along segment $\overline{45}$. If these two potential energies cancel each other, i.e., if the areas $\overline{134}$ and $\overline{245}$ are the same, then the motion which begins with an equilibrium state at point 3 will end with an equilibrium state (zero kinetic energy) at point 5. On the other hand, if the areas 134 and 245 were unequal, the final state at point 5 would not be an equilibrium state and a static analysis would then be incorrect.

We may conclude that, for the purpose of static analysis, in which some finite elements are larger than L_{cr}, the reverse post-peak behavior with positive slope can be replaced by a vertical stress drop for which the areas $\overline{134}$ and $\overline{245}$ in Fig. 3e are equal. This means that the area $\overline{035}$ under the entire diagram with the vertical stress drop must be equal to the area $\overline{012}$ under the actual stress-displacement diagram for the finite element of length L_e, which in turn equals the fracture energy G_f. Thus, the strength limit at the beginning of the equivalent vertical stress drop, which has been called the equivalent strength f'_{eq} [7,36,29], must satisfy the condition $f'_{eq}\ \delta_0/2 = G_f$ where $\delta_0 = L_e f'_{eq}/E$. This yields for the equivalent strength at $L > L_{cr}$ the expression

$$f'_{eq} = \sqrt{\frac{2G_f E}{L_e}} = f'_t \sqrt{w_c\left(1 - \frac{E}{E_t}\right)}\ \frac{1}{\sqrt{L_e}} \tag{8}$$

This is the same expression as that deduced before in a different manner [7,36,29]. We see that, for $L_e > L_{cr}$, the strength limit used in the calculation must be increased in inverse proportion to $\sqrt{L_e}$.

When the vertical stress drop is used, and the mesh is not too crude, the results of analysis are very close to those of linear elastic fracture mechanics. The essential nonlinear characteristics of the response are caused by strain-softening, and so the strain-softening behavior cannot be analyzed with large elements that require a vertical stress drop, unless the structure is so large that linear elastic fracture analysis is applicable.

In the preceding we considered only the situations where the front of distributed cracking is known to localize to its minimum admissible width (w_c in the crack band model, zero in Hillerborg's and Willam's models). When typical fracture specimens are analyzed by finite elements, we know from testing experience that such complete localization will occur. Fracture test specimens are of course designed to obtain just that (localized fracture). In general situations, though, it is not known whether complete localization of the cracking front will occur, and in many practical situations it will not. For example, large, nonlocalized zones of distributed cracking may develop if they are stabilized by compression zones ahead of the cracking front, by reinforcement, or by boundary restraint (Fig. 4). Whether or not complete localization occurs must be decided on the basis of stability analysis.

The crack band model is ideally suited to handle also the situations of nonlocalized cracking. No additional material parameters are needed for this purpose because a stress-strain relation is used.

On the other hand, further extensions are needed for Hillerborg's fictitious crack model and Willam-Sture's composite damage model in order to handle nonlocalized cracking. An array of line cracks must be introduced for these models, either as a system of interelement line cracks or as a system of cracks with two or more cracks (or cracking strips) embedded in each finite element (see Fig. 3). These line cracks cannot have arbitrary spacing s_{cr}, or else the analysis would be unobjective; any desired results could be obtained depending on the analyst's choice of s_{cr}. Thus we again see that the previous claims that these models with line cracks have one independent material parameter less than the crack band model (i.e., two vs. three) are unjustified. These models with line cracks require a characteristic length such as spacing s_{cr} as the *third* material parameter in order to be capable of handling any general situations other than fracture specimens, which are designed so that the cracking is certain to localize.

It may be concluded that no strain-softening fracture model can be objective with regard to mesh size unless it involves some characteristic lengths as a material property.

Hillerborg's and Willam-Sture's models with line cracks do not seem very suitable for handling the situations where cracking orientation varies over the space (Fig. 4f). These models require proper crack spacing s_{cr} to be objective, but what is the proper spacing if the cracks are nonparallel? This must be expected in general situations. The question how to handle such situations apparently has not yet been addressed, and would have to be addressed if models should be implemented in a general finite element code. A strain-softening constitutive relation appears to be the only practically manageable way to deal with nonlocalized cracking whose orientation is spatially variable (Fig. 4f).

Other behaviors which have been modeled with the crack band approach but appear difficult for the models based on line cracks are: 1) cracking in more than one direction at a point and 2) cracking process where the prevailing crack direction at one point changes its orientation during the loading process.

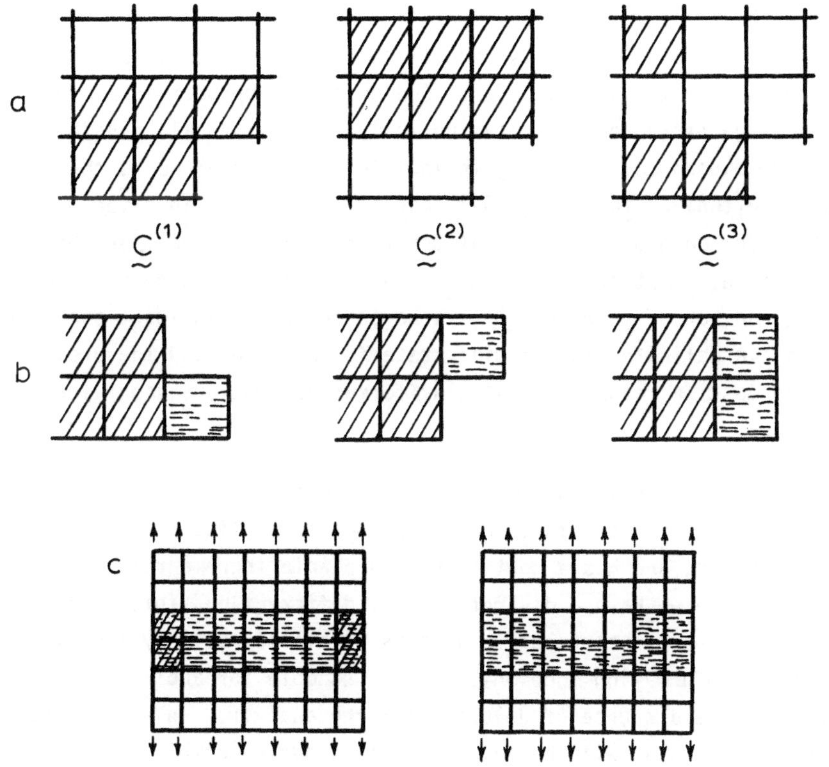

Fig. 5 - (a) Various Possible Unloading Zones for which Stiffness Matrix $\underset{\sim}{C}$ Needs to Be Determined, and (b,c) Various Possible Extensions of Crack Band.

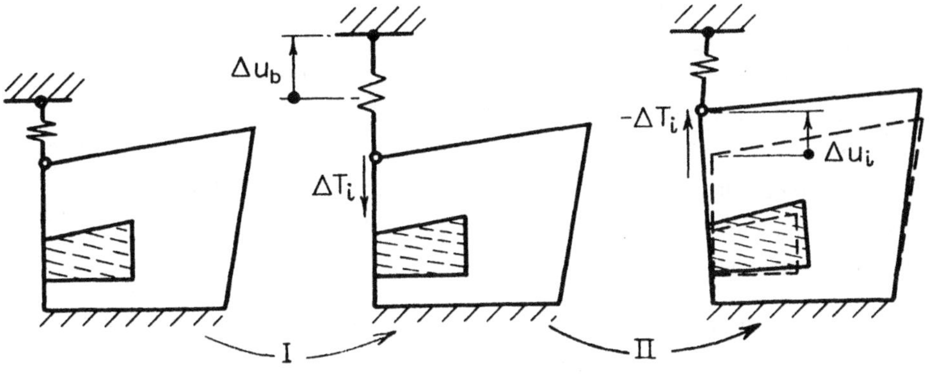

Fig. 6 - Decomposition of Extension of Crack Band into Two Stages.

Stability of Equilibrium Path

Let us now consider the stability conditions for strain localization to be used in general finite element codes. One possible approach is as follows: 1) At each stage of incremental loading analysis consider all possible combinations of loading and unloading in individual finite elements that are in a strain-softening state (Fig. 5a), assuming no change in given loads, and 2) check for each of these situations ($i = 1,2,\ldots,n$) whether the incremental stiffness matrix $\underset{\sim}{C}^{(i)}$ is positive definite. In general, there is a very large number of matrices $\underset{\sim}{C}^{(i)}$ to be checked. If at least one of them is not positive definite, the initial state is unstable, and the previous loading step must be reanalyzed using different assumptions about loading and unloading.

Note that the situation is fundamentally different from that in analyzing instability of elastic structures due to geometric nonlinearity. In those classical problems, one has a continuously variable incremental stiffness matrix, while here we have a discrete set of different possible stiffness matrices. Thus the structure which appears stable may be only locally stable for one deformation mode while being unstable for another deformation mode. This may be a source of delusion for the analyst. Unless certain deformation increments, with certain loading-unloading combinations, are checked, the structure appears to be stable, the finite element program runs, the iterations converge, etc., leaving an unsuspecting analyst happy that he has obtained the solution, although he may in fact have missed detecting an instability.

The checking of all possible matrices $\underset{\sim}{C}^{(i)}$ may overtax the computer's capability. There is another possibility of checking for instabilities which seems less demanding. Instead of the increments taking place at no change in loads, one may consider the loading step at a given change of loads. Again one may consider different combinations of loading and unloading in the individual strain-softening elements, and then calculate for each of them the corresponding increments of displacements, $\Delta\underset{\sim}{u}$, strains, $\Delta\underset{\sim}{\varepsilon}$, and stresses, $\Delta\underset{\sim}{\sigma}$, for the equilibrium path at given increments of loads and boundary displacements.

It is now important to note that there may exist more than one equilibrium path, i.e., the equilibrium path may bifurcate due to various possible combinations of loading and unloading. For example, in the analysis of the

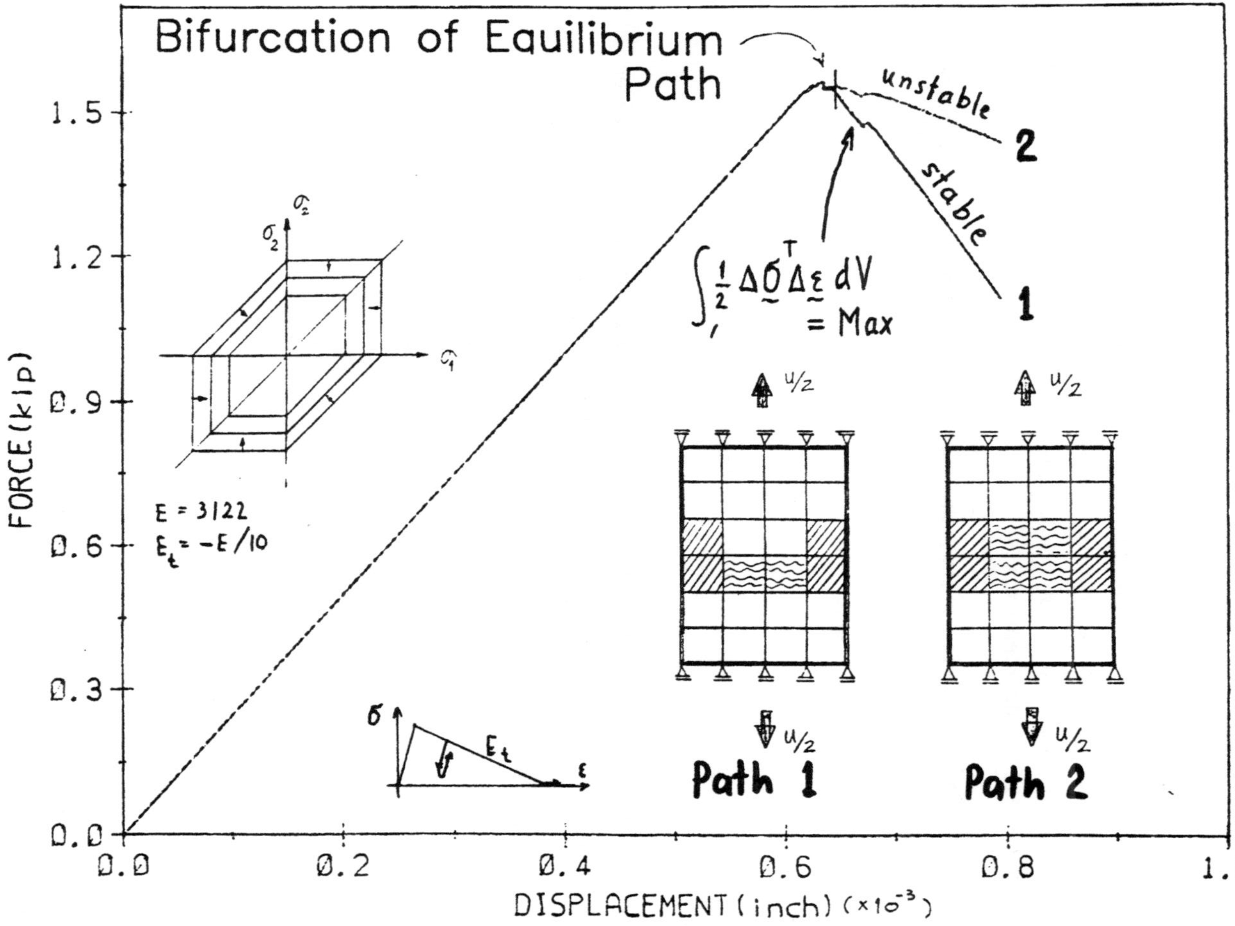

Fig. 7 - Bifurcation of Equilibrium Loading Path in a Strain-Softening Panel.

symmetric panel in Fig. 4c, in which the shaded finite elements are assumed to have a slightly smaller strength, it is found that both a symmetric solution with a crack band two elements wide and a nonsymmetric solution with a crack band one element wide represent equilibrium paths and both are locally stable (as found by F.B. Lin and Z.P. Bažant in 1985). Now, which one is the correct solution? This type of question may be in general answered by the following theorem:

Among all possible equilibrium paths, the stable one (correct one) maximizes the second order energy release:

$$\delta^2 U = \int_V \frac{1}{2} \Delta\underset{\sim}{\varepsilon}^T \Delta\underset{\sim}{\sigma}\, dV = \text{Max} \tag{9}$$

where the integral is taken over the entire volume V of the structure, and $\Delta\underset{\sim}{\varepsilon}$, $\Delta\underset{\sim}{\sigma}$ are the column matrices of the increments of strain and stress components.

Equation 9 may be proven by imagining all loads to be applied through added fictitious elastic springs, whose spring constants are denoted as C_i ($i = 1,2,\ldots$). If C_i is very small, it is nearly as if the load were prescribed (controlled), and if C_i is very large, it is nearly as if the displacement were prescribed. The loading increment may be decomposed into 2 stages (Fig. 6). In stage I, the deformation of the body is fixed (frozen), while the ends of the springs C_i are displaced by Δu_i to effect the prescribed changes of loads or displacements. The reactions produced in the springs are denoted as ΔT_i. In stage II, the body is released (i.e., unfrozen, let to deform freely to achieve a new equilibrium state). This amounts to applying opposite loads $-\Delta T_i$ on the body, and a new equilibrium state, in which the displacements of spring attachment points are Δu_i, is calculated. The change of the free energy of the structure is $\Delta F = \Delta F_1 + \Delta F_2$, where ΔF_1 and ΔF_2 correspond to stages I and II. ΔF_1 is due solely to the fictitious added springs and is obviously the same for all possible equilibrium loading paths of the body for given changes of loads. Stage II is a change of state of an isolated system, and so the stable path is that for which ΔF_2 is minimum. Therefore, $\Delta F_2 = \Sigma_i \frac{1}{2}(-\Delta T_i)\Delta u_i$ = Min. According to the principle of virtual work, $\int_V \Delta\underset{\sim}{\sigma}^T \Delta\underset{\sim}{\varepsilon}\, dV + \Sigma_i(-\Delta T_i)\Delta u_i = 0$, and so $\Delta F_2 = -\int_V \Delta\underset{\sim}{\sigma}^T \Delta\underset{\sim}{\varepsilon}\, dV$ = Min. This proves Eq. 9.

In the example in Fig. 7, finite element calculations [38] show that $\delta^2 U$

in Eq. 9 is larger for the nonsymmetric equilibrium path in which the crack band front is one rather than two elements wide.

Nonlocal Continuum Approach

In the crack band model, the finite element that undergoes cracking (strain-softening) cannot be smaller than the length w_c. This is disconcerting from the mathematical viewpoint; the mesh cannot be refined to zero, so there exists no limiting case of a continuum that the finite element model is supposed to approximate. Furthermore, to determine accurately the energy dissipated due to cracking, it is necessary to have the possibility of refining the mesh size below w_c. To make arbitrary mesh refinements possible, it appears necessary to abandon the concept of a local continuum and introduce a nonlocal continuum.

From the works of Kröner and others [12-16], it is known that in a statistically heterogeneous medium which is not in a macroscopically homogeneous state of strain, the averaged (smoothed) stress at a certain point depends not only on the gradient of the averaged displacements at that same point (local properties), but also on the averaged displacements within a certain characteristic finite neighborhood of that point. The properties of such a medium cannot, therefore, be said to be local, and the medium is called nonlocal.

The nonlocal displacement gradient may be defined by the relation

$$D_i u_j(\underset{\sim}{x}) = \frac{1}{V} \int_{V(\underset{\sim}{x})} \frac{\partial u_j(x')}{\partial x_i} \, dV' = \frac{1}{V} \int_{S(\underset{\sim}{x})} u_j(\underset{\sim}{x}') \, n_i(\underset{\sim}{x}') \, dS' \qquad (10)$$

in which u_j are the cartesian displacement components ($j = 1,2,3$), $\underset{\sim}{x}$ is the coordinate vector of the given point characterized by cartesian coordinates x_j, $V(\underset{\sim}{x})$ is the characteristic volume of the material centered at point $\underset{\sim}{x}$ (Fig. 8), $S(\underset{\sim}{x})$ is the surface of this volume, $n_j(\underset{\sim}{x}')$ is the unit normal of this surface at point $\underset{\sim}{x}'$, and D_i is the gradient averaging operator. The surface integral in Eq. 10 follows from the volume integral by application of the Gauss integral theorem. More generally, a weighting function can be introduced in Eq. 10. Using the gradient averaging operator, the mean strains may be defined as

$$\bar{\varepsilon}_{ij} = \frac{1}{2} (D_i u_j + D_j u_i). \qquad (11)$$

In previous works dealing with nonlocal continua it has been generally assumed that the continuum equation of motion may be written as

$$\frac{\partial}{\partial x_j} \bar{C}_{ijkm}(\bar{\varepsilon})\ D_m u_k = \rho \ddot{u}_i \tag{12}$$

in which C_{ijkm} are secant elastic moduli which, in general, depend on the mean strain, ρ is the mass density, and superior dots refer to time derivatives. It is found, however, that Eq.3 is incapable of describing a strain-softening continuum. It always leads to unstable response as soon as strain-softening begins. The difficulty has been traced to the asymmetry of these equations due to the combination of partial derivatives $\partial/\partial x_j$ with the gradient averaging operator D_m. This nonsymmetry gives rise to nonsymmetric finite element matrices even if C_{ijkm} are constant, i.e., if the medium is elastic. Such a nonsymmetry is certainly an objectional property.

For this reason, a systematic derivation of the continuum equation of motion on the basis of Eq. 10 has been attempted, using the calculus of variations. It has been found [2,25] that the proper form of the continuum equation of motion is

$$(1 - \gamma)D_j \bar{C}_{ijkm}(\bar{\varepsilon})D_m u_k + \gamma \frac{\partial}{\partial x_j} C_{ijkm}(\varepsilon) \frac{\partial}{\partial x_m} u_k = \rho \ddot{u}_i \tag{13}$$

in which γ is an empirical coefficient between 0 and 1, and C_{ijkm} are the local secant moduli. In contrast to Eq. 12, each term of the last equation has a symmetric structure and averaging is carried out twice (in D_m and in D_j). Consequently, discretization by finite elements leads to symmetric stiffness matrices if the elastic moduli $\bar{C}_{ijkm}$ and C_{ijkm} are symmetric.

Another essential difference from the classical nonlocal continuum formulations is the use of a local term (term with γ), which corresponds to a local continuum overlaid over the nonlocal continuum. Stability requires that $\gamma > 0$, i.e., the local term cannot be omitted [1]. In theory any positive value of γ assures stability, but in practice values around $\gamma = 0.1$ need to be used to suppress noise which would be excessive for smaller γ.

Equation 10 can also be written in the form

$$(1 - \gamma)D_j \sigma_{ij} + \gamma\ \tau_{ij,j} = \rho\ \bar{u}_i \tag{14}$$

in which

$$\sigma_{ij} = \bar{C}_{ijkm}(\bar{\underset{\sim}{\varepsilon}})\ \varepsilon_{km} = \bar{C}_{ijkm}(\bar{\underset{\sim}{\varepsilon}}) D_m u_k \tag{15}$$

$$\tau_{ij} = C_{ijkm}\ \varepsilon_{km} = C_{ijkm} \frac{\partial u_k}{\partial x_m}\ . \tag{16}$$

τ_{ij} are the usual, local stresses, and σ_{ij} are the stresses characterizing the stress state in the entire representative volume of the material, called the broad-range stresses [1,2]. From Eq. (14) we see that averaging must be carried out not only for strains, but also for the stresses obtained from the strains.

When the continuum defined by Eq. 13 is discretized by finite elements the size of which is smaller than the size ℓ of the representative volume, one obtains a system of overlapping (or imbricated) finite elements visualized in Fig. 9. Therefore, the present type of nonlocal continuum has been called imbricate. The finite elements keep a constant size ℓ as the mesh is refined, and the number of finite elements crossing a given point is inversely proportional to the mesh size, while the cross section of these elements diminishes so that all imbricated elements have the same total cross section for any mesh size. It can also be shown that the limiting case of the finite difference equations describing such an imbricated system of finite elements is the differential equation in Eq. 10 [1]. If the finite element size h is larger than the characteristic length ℓ, then the finite element model of the imbricate continuum becomes identical to that for the classical local continuum.

To assure convergence and stability, the local stress-strain relations (Eq. 16) may not exhibit strain-softening, or else unstable response and spurious sensitivity to mesh size, along with incorrect convergence, may be obtained. The strain-softening properties must be described solely by the broad-range stress-strain relation in Eq. 15.

Examples of Strain-softening Finite Element Solutions

To demonstrate that the concept of imbricate nonlocal continuum yields convergent finite element solutions for problems with a finite-size strain-softening zone, we will now describe two examples of one-dimensional waves. As the first example, Fig. 8 reproduces some of the results of explicit dynamic finite element calculations from Ref. 13, in which wave propagation in a strain-softening bar of length ℓ was analyzed. The characteristic

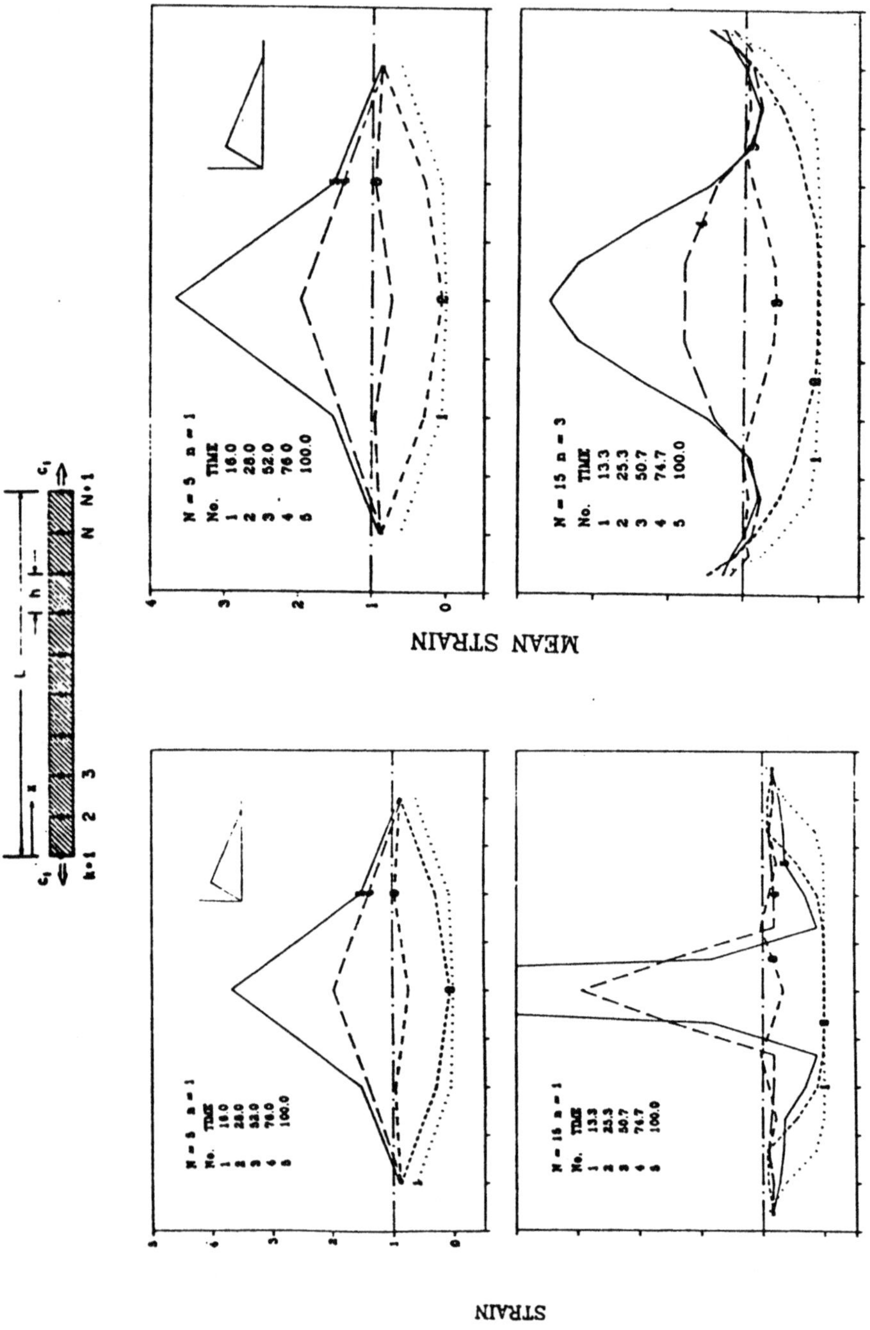
N = 5 n = 1
No. TIME
1 16.0
2 28.0
3 52.0
4 76.0
5 100.0
N = 5 n = 1
No. TIME
1 16.0
2 28.0
3 52.0
4 76.0
5 100.0
N = 15 n = 3
No. TIME
1 13.3
2 25.3
3 50.7
4 74.7
5 100.0
N = 15 n = 1
No. TIME
1 13.3
2 25.3
3 50.7
4 74.7
5 100.0
MEAN STRAIN
STRAIN
N N+1
k=1 2 3
L
h

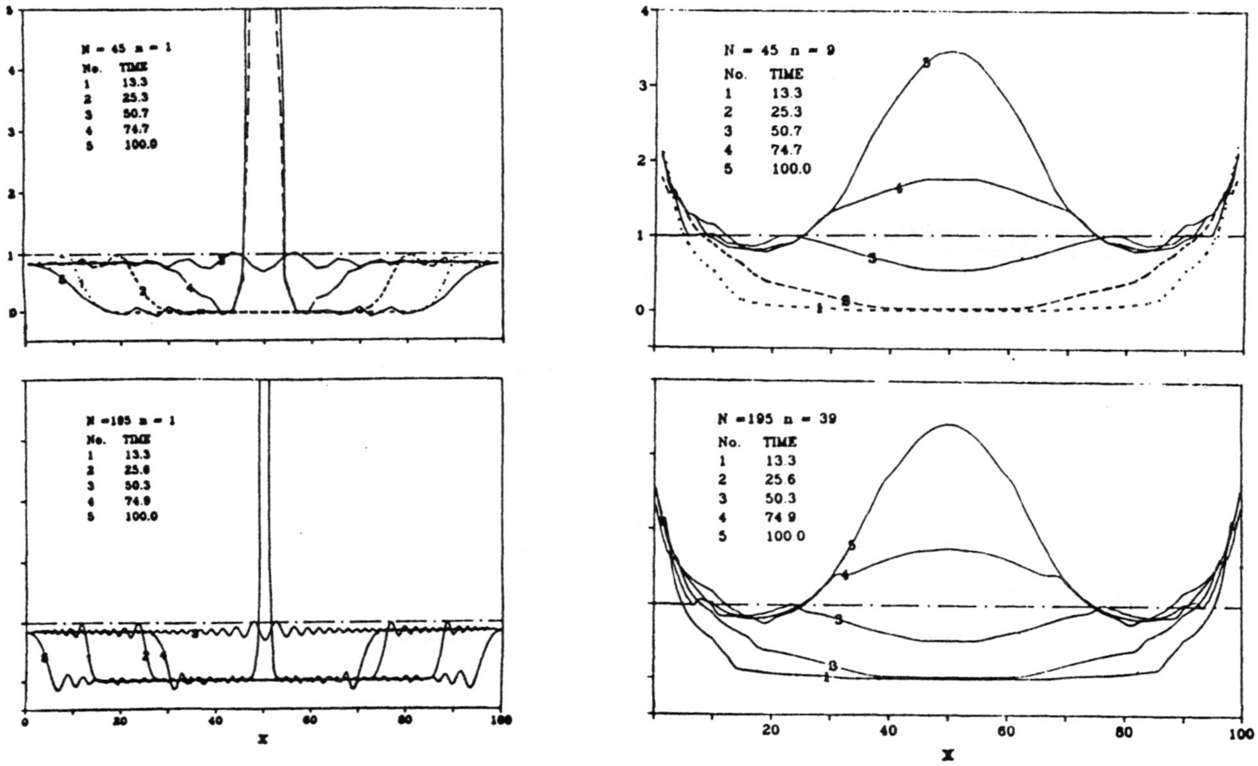

Fig. 8 - Strain Profiles at Various Times Obtained for Various Finite Element Subdivisions (N = number of elements).

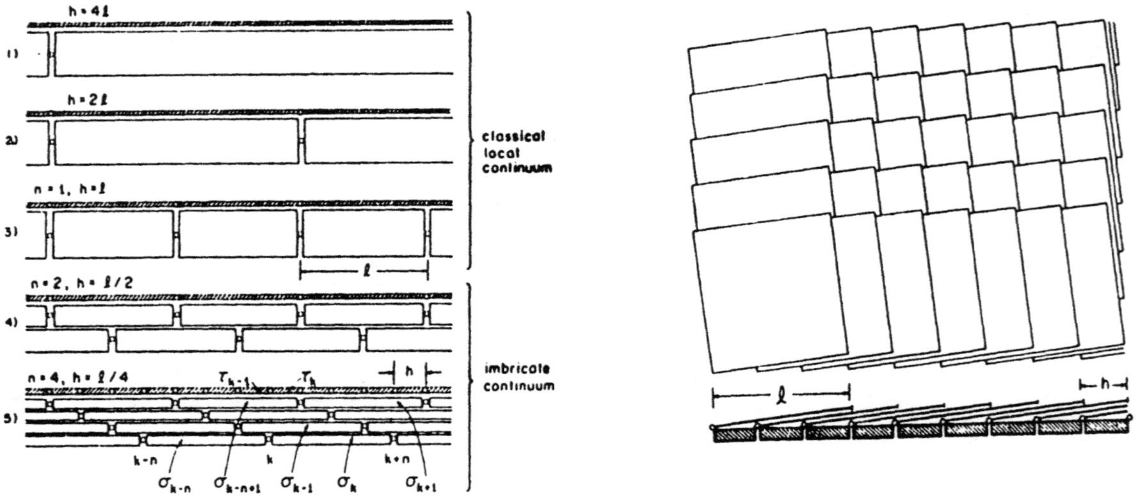

Fig. 9 - Illustration of Mesh Refinement for a One-Dimensional Imbricate Continuum (left) and of Element Arrangement for a Two-Dimensional Imbricate Continuum (right) (the elements are slightly rotated for the purpose of illustration).

length ℓ for nonlocal solution is assumed to be 1/5 of bar length L. Both ends of the bar are subjected to a constant outward velocity v_0 beginning at time t = 0. This loading produces step waves of strain propagating inward. When these waves meet at midlength, the strain suddenly increases and strain-softening ensues.

If this problem is analyzed with the usual finite element method for local continuum, it is found that strain-softening always localizes into a single finite element. Thus, the width of the strain-softening zone reduces to zero as the element mesh is refined. As a consequence, the energy W consumed by failure decreases with decreasing mesh size and approaches zero as the mesh size tends to zero (Fig. 10). Moreover, the finite element model of local continuum exhibits a discontinuous dependence of response on the prescribed end velocities as well as on the slope E_t of the strain-softening branch [30]. The solution, however, converges to a unique exact solution [30], although this solution is unrealistic from the physical point of view.

By contrast, for the present imbricate continuum, the solution of wave propagation in the strain-softening bar exhibits correct convergence, with a strain-softening zone of a finite size in the limit (Fig. 8, right column). Also, the energy consumed by failure in the bar converges to a finite value, as shown in Fig. 10.

As another example, Fig. 11 shows the refinements of the finite element mesh for a two-dimensional panel which was solved [39] by the imbricated element approach. The cross-hatched elements at the sides of the panel were assumed to have tensile strength that is 5% smaller than the strength of all other elements. This causes that strain-softening bands start propagating inwards from the edges. The same triaxial strain-softening constitutive relation as in Ref. 5 was used. Uniform displacement u is imposed at top and bottom sides of the panel and is incremented in small steps. The results obtained for the subsequent mesh refinements [39] are shown in Fig. 12. We see that for the imbricate nonlocal model the solution converges quite well in terms of the load-deflection diagram, dissipated energy and strain profile. By contrast, the local (usual) finite element solutions are very different for the three meshes and, if they converge, they do so to physically unrealistic results.

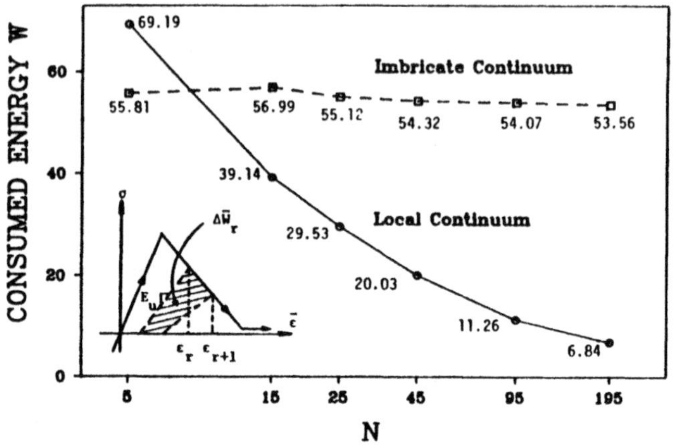

Fig. 10 - Energy Consumed by Failure Caused by Wave Propagation in a One-Dimensional bar.

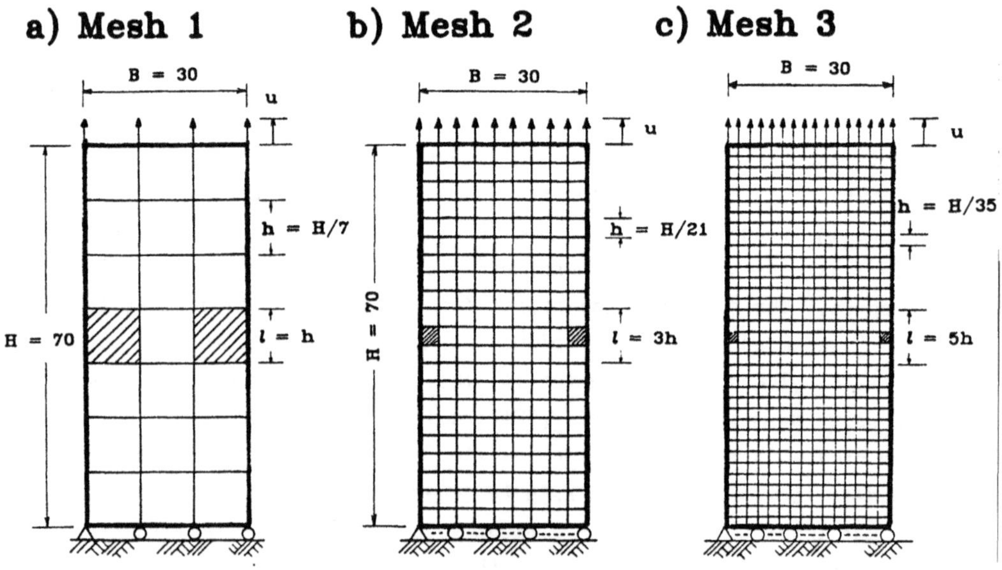

Fig. 11 - Mesh Refinement for a Rectangular Strain-Softening Panel With Weaker Elements at the Sides.

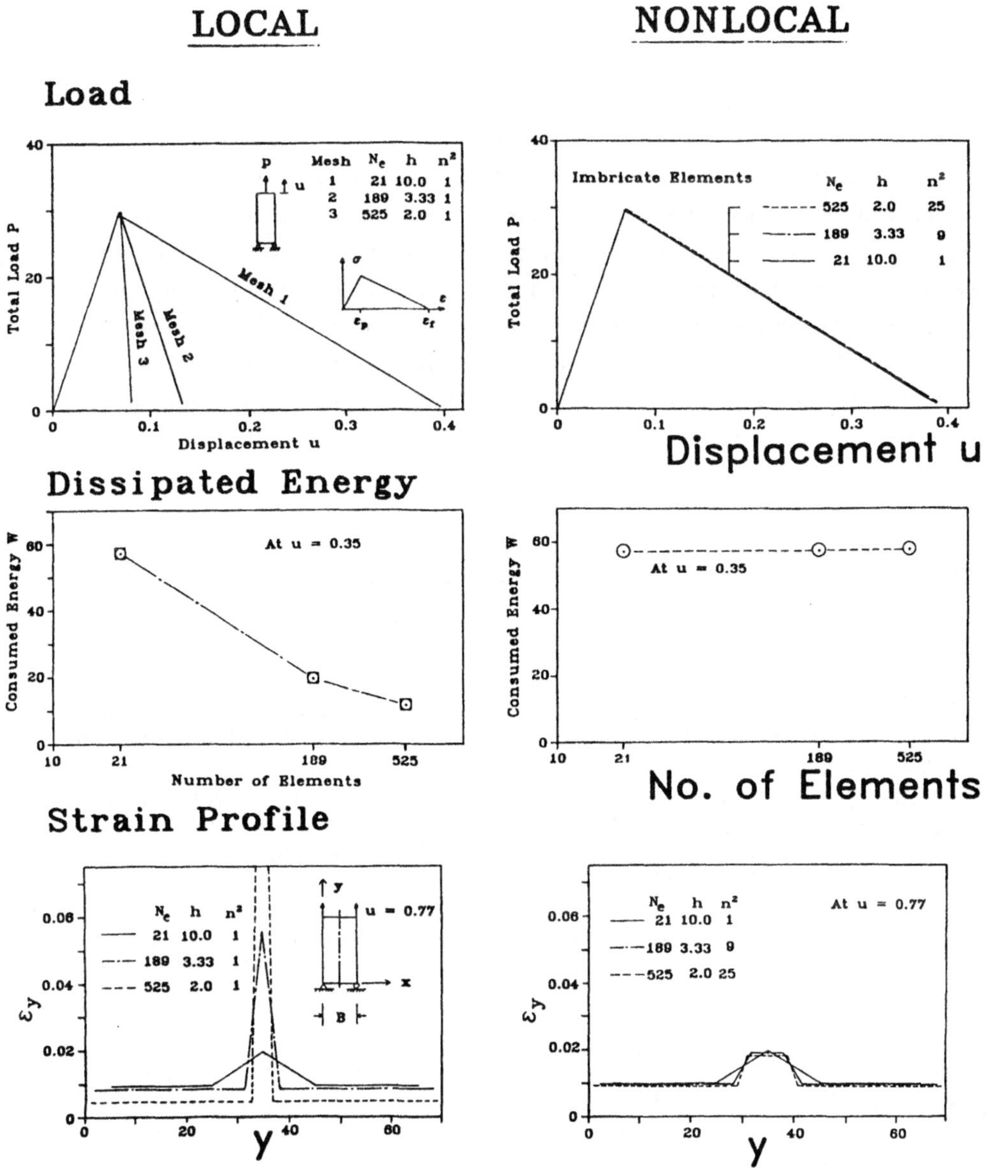

Fig. 12 - Convergence of Finite Element Results for Panel in Fig. 11 [47].

Conclusion

The concept of imbricate nonlocal continuum, along with strain-softening stress-strain relations, allows a mathematically consistent and realistic description of progressive distributed cracking of the concrete structures. When fine resolution inside the strain-softening zone is not needed, this approach reduces to the crack band theory. Comparison with Hillerborg's fictitious crack model and Willam-Sture's composite crack model shows that these models are essentially equivalent for fully localized cracking but not for nonlocalized cracking, for which they need to be generalized by adding a characteristic length of the material. Stability checks need to be carried out to decide where cracking will localize. Finite element solutions demonstrate that stable strain-softening regions of finite size can be modeled.

Acknowledgment. - Partial financial support under AFOSR No. 83-0009 to Northwestern University is gratefully acknowledged.

References

1. Bažant, Z.P., Chang, T.P., Belytschko, T.B.: Continuum Theory for Strain Softening, Report No. 83-11/428c, Center for Concrete and Geomaterials, Northwestern University, Evanston, Illinois, Nov. 1983; also Journ. Eng. Mech. ASCE, Vol. III, Dec. 1984, 1666-1692.

2. Bažant, Z.P.: Imbricate Continuum and Its Variational Derivation, Report No. 83-11/428i, Center for Concrete and Geomaterials, Northwestern University, Evanston, Illinois, Nov. 1983; also Journ. Eng. Mech. ASCE, Vol. III, Dec. 1984, 1693-1712.

3. Bažant, Z.P.: Instability, Ductility and Size Effect in Strain-Softening Concrete; Journal of the Engineering Mechanics Division ASCE, Vol. 102, Apr. 1976, pp. 331-344; Discussions, Vol. 103, pp. 357-358, 775-777, Vol. 104, pp. 501-502 (presented at the University of Colorado, Boulder, in June 1974).

4. Bažant, Z.P.: Crack Band Model for Fracture of Geomaterials, Proc. 4th Intern. Conf. on Numerical Methods in Geomechanics, held in Edmonton, Alberta, Canada, June 1982, ed. by Z. Eisenstein, Vol. 3, 1137-1152.

5. Bažant, Z.P., Oh, B.H.: Crack Band Theory for Fracture of Concrete, Materials and Structures (RILEM, Paris), Vol. 16 (1983) 155-177.

6. Bažant, Z.P., Oh, B.H.: Microplane Model for Fracture Analysis of Concrete Structures, Proc. Symp. on the Interaction of Nonnuclear Munitions with Structures, U.S. Air Force Academy, Colorado Springs, May, 1983, 49-55.

7. Bažant, Z.P., Cedolin, L.: Blunt Crack Band Propagation in Finite Element Analysis, Journal of the Engineering Mechanics Division, ASCE, Vol. 105, No. EM2, April 1979, 207-315.

8. Bažant, Z.P., Cedolin, L.: Fracture Mechanics of Reinforced Concrete, Journal of the Engineering Mechanics Division, ASCE, Vol. 106, No. EM6,

Proc. Paper 15917, December 1980, 1287-1306; with Discussion and Closure in Vol. 108, 1982, EM., 464-471.

9. Bažant, Z.P., Cedolin, L.: Finite Element Modeling of Crack Band Propagation, Journal of Structural Engineering, ASCE, Vol. 109, No. ST2, Feb. 1983, 69-92.

10. Belytschko, T.B., Bažant, Z.P., Hyun, V.V., Chang, T.P.: "Strain-Softening Materials and Finite Element Solutions," submitted to Computers and Structures.

11. Hillerborg, A., Modéer, M., Petersson, P.E.: Analysis of Crack Formation and Crack Growth in Concrete by Means of Fracture Mechanics and Finite Elements, Cement and Concrete Research, Vol. 6, 1976, 773-782.

12. Kröner, E.: Elasticity Theory of Materials with Long-Range Cohesive Forces, Int. J. Solids Structures, Vol. 3, 1967, 731-742.

13. Kröner, E.: Interrelations between Various Branches of Continuum Mechanics, Mechanics of Generalized Continua, ed. by E. Kröner, Springer-Verlag, 1968, 330-340.

14. Krumhansl, J.A.: Some Considerations of the Relation between Solid State Physics and Generalized Continuum Mechanics, Mechanics of Generalized Continua, ed. by E. Kröner, Springer-Verlag, 1968, 298-311.

15. Kunin, I.A.: The Theory of Elastic Media with Microstructure and the Theory of Dislocations, Mechanics of Generalized Continua, ed. by E. Kröner, Springer-Verlag, 1968, 321-328.

16. Levin, V.M.: The Relation between Mathematical Expectation of Stress and Strain Tensors in Elastic Microheterogeneous Media: Prikladnaya Matematika i Mekhanika, Vol. 35, 1971, 694-701 (in Russian).

17. Eringen, A.C., Edelen, D.G.B.: On Nonlocal Elasticity, Int. J. Engng. Science, Vol. 10, 1972, 233-248.

18. Eringen, A.C., Ari, N.: Nonlocal Stress Field at Griffith Crack, Cryst. Latt. Def. and Amorph. Mat., Vol. 10, 1983, 33-38.

19. ASCE State-of-the-Art Report on Finite Element Analysis of Reinforced Concrete, prepared by a Task Committee chaired by A. Nilson, Am. Soc. of Civil Engrs., New York, 1982.

20. Wittmann, F.H.; (Editor): Fracture Mechanics of Concrete, Elsevier, Netherlands, 1983.

21. Petersson, E.: Fracture Energy of Concrete: Method of Determination, Cement and Concrete Research, Vol. 10, 1980, 78-89, and Fracture Energy of Concrete: Practical Performance and Experimental Results, Cement and Concrete Research, Vol. 10, 1980, 91-101.

22. Petersson, P.C.: Crack Growth and Development of Fracture Zones in Plain Concrete and Similar Materials, Doctoral Dissertation, Lund Institute of Technology, Lund, Sweden, Dec. 1981.

23. Evans, R.H., Marathe, M.S.: Microcracking and Stress-Strain Curves for Concrete in Tension, Materials and Structures (RILEM, Paris), No. 1, Jan.-Feb., 1968, 61-64.

24. Heilmann, H.G., Hildsdorf, H.K., Finsterwalder, K.: Festigkeit und Verformung von Beton unter Zugspanungen: Deutscher Ausschuss für Stahlbeton, Heft 203, W. Ernst & Sohn, West Berlin, 1969.

25. Bažant, Z.P.: Fracture in Concrete and Reinforced Concrete, Chapter 13 in Mechanics of Geomaterials: Rocks, Concretes, Soils, John Wiley and

Sons, London, 1985 (Proc. of IUTAM W. Prager Symposium held at Northwestern University in Sept. 1983).

26. Rashid, Y.R.: Analysis of Prestressed Concrete Pressure Vessels, Nuclear Engng. and Design, Vol. 7, No. 4, April 1968, 334-344.

27. Reinhardt, H.W., Cornelissen, E.: Post-Peak Cyclic Behavior of Concrete in Uniaxial Tension and Alternating Tensile and Compressive Loading, Cement and Concrete Research, Vol. 14, 1984, 263-270.

28. Rüsch, H., Hilsdorf, H.: Deformation Characteristics of Concrete under Axial Tension, Vorunterschungen, Bericht Nr. 44, Munich, May 1963.

29. Bažant, Z.P.: Mechanics of Fracture and Progressive Cracking in Concrete Structures, Chap. 1 in Fracture Mechanics Applied to Concrete Structures, ed. by G.C. Sih, Martinus Nijhoff Publishers B.V., The Hague, Netherlands, 1985.

30. Bažant, Z.P., Belytschko, T.B.: Wave Propagation in a Strain-Softening Bar: Exact Solution, J. of Engineering Mechanics ASCE, Vol. 111, No. 3, March 1985.

31. Achenbach, J.D.: Wave Propagation in Elastic Solids, North Holland Publ. Co., Amsterdam, 1973.

32. Bažant, Z.P., Oh, B.H.: Rock Fracture via Strain-Softening Finite Elements, J. of Engng. Mechanics ASCE, Vol. 110, July 1984, 1015-1035.

33. Willam, K.J., Bicanic, N., Sture, S.: Constitutive and Computational Aspects of Strain-Softening and Localization in Solids, Proc. Sym. on Constitutive Equations: Micro, Macro and Computational Aspects, K.J. Willam, Ed., held at ASME Winter Annual Meeting, New Orleans, Dec. 1984, Am. Soc. of Mech. Engrs., New York, 1984.

34. Willam, K.J., Hurlbut, B., Sture, S.: Experimental, Constitutive and Computational Aspects of Concrete Failure, Preprints, U.S.-Japan Seminar on Finite Element Analysis of Reinforced Concretes, Tokyo, May 1985, 149-172.

35. Pietruszczak, S., Mróz, Z.: Finite Element Analysis of Deformation of Strain-Softening Materials, Intern. J. of Num. Meth. in Engng., Vol. 17, 1981, 327-334.

36. Bažant, Z.P., Cedolin, L.: Fracture Mechanics of Reinforced Concrete, J. of the Engineering Mechanics Division, ASCE, Vol. 106, 1980, 1287-1306.

37. Bažant, Z.P., Cedolin, L.: Finite Element Modeling of Crack Band Propagation, J. of Structural Engineering ASCE, Vol. 109, 1983, 69-92.

38. Lin, F.B.: Private Communication on Doctoral Dissertation in Preparation, Northwestern University, May 1985.

39. Chang, T.P.: Finite Element Analysis of Strain-Softening Failures, Doctoral Dissertation, Northwestern University, in preparation, May 1985.

On the Numerical Simulation of Tensile Fracture

L. G. NILSSON, M. OLDENBURG
Luleå University of Technology, Luleå, Sweden

Summary

A continuum approach to tensile fracturing is developed. It is related to fracture mechanics concepts. The effects of specimen size on fracture are included in the model. The constitutive equations are derived in a manner similar to the derivation of the theory of plasticity.

The behaviour of the numerical results obtained from the material model when different element types and different schemes of numerical integration are used is investigated. The objectivity of the model with respect to element size is demonstrated. A numerical analysis of a mixed-mode fracturing beam has been performed and compared with experiment. Using a combined plastic-fracture model, both the correct crack path, and the post-peak behaviour could be predicted.

Introduction

The structural analysis of materials such as ceramics, concrete and rock requires a knowledge of how cracks are formed and how they propagate. Under tensile stresses, fracture is dominated by the formation of distributed microcracks, which, if the stress intensity increases, eventually coalesce into macrocracks.

Within the framework of finite elements, two computational approaches have been used for the fracture analysis. In the discrete crack approach, individual cracks are simulated by relaxing the element nodal connectivity, i.e. creating new boundaries. Thus cracks are constrained to propagate along element boundaries. In the smeared crack approach, the post failure behaviour is modelled in terms of stress-strain relations in which a continuous displacement field is maintained.

In recent years, it has been widely recognized that the size effect of the fracturing process is well described in terms of nonlinear fracture mech-

Finite Element Methods for Nonlinear Problems
Europe-US Symposium, Trondheim, Norway 1985
Editors: Bergan, Bathe, Wunderlich

anics. By using the discrete crack approach, Hillerborg [1] and his colleagues have completed pioneering work in the development of the so-called fictitious crack model (FCM). The parameters of this model, which govern the structural stiffness degradation, are the tensile strength, the fracture energy and the shape of the stress-crack width relation. The discrete fracture approach, however, is badly suited for large scale finite element analysis, due to the exhaustive book-keeping required to keep track of all cracks. Considerable effort has therefore been expended in the development of the smeared crack approach to account for size effects. The blunt crack band theory by Bazant and Oh [2] accounts for these effects by introducing the width of the band over which microcracks are distributed as an additional parameter. In work by ourselves, Nilsson and Oldenburg [3], a smeared crack approach based on the FCM was presented. By using a plastic fracturing theory and energy considerations, we were able to show that the basic three parameters of the FCM were sufficient, and that the crack width can be obtained from the model.

In this paper some computational aspects in connection with the smeared crack approach are discussed. Examples of these are the choice of element type and spatial integration rule and the consideration of plastic deformations in the compressive and tensile-compressive stress regions.

Constitutive model

In the past considerable research has been devoted to the nonlinear behaviour of heterogeneous materials. Considering compressive states of stress, we believe that the constitutive modelling has reached a stage where reasonable accuracy may be obtained. For tensile stress states, however, the modelling of fracture processes is still rather primitive. Since the nonlinear response is often dominated by the formation of cracks, considerable research must be directed towards studying this problem. Therefore, in this paper we only discuss the fracturing part of the material model, and only mention briefly some problems regarding material failure in combined states of stress.

The fracture model

With the fictitious crack model Hillerborg et al [1] introduced a model suited for the numerical analysis of crack propagation in heterogeneous materials. The fictitious crack model is an extension of the Barenblatt's cohesive force model, but the forces are due to macroscopic effects from microcracking in the fracture process zone. The numerical implementation by Hillerborg et al [1] uses a discrete model, where finite element nodal forces along the fictitious cracks follow an assumed stress-crackwidth relation.

To treat general problems, where the onset and the direction of the crack progagation is unknown, the cohesive force model must be generalized to a continuum model. In this, the effect of the localized damage is smeared out in an objective manner, based on fracture mechanics concepts and on an energy balance principle.

Assuming that the total strain rate has elastic and fracture strain components, we have

$$\dot{\underset{\sim}{\varepsilon}} = \dot{\underset{\sim}{\varepsilon}}^{e} + \dot{\underset{\sim}{\varepsilon}}^{c} \qquad (1)$$

where $\dot{\underset{\sim}{\varepsilon}}^{e}$ is the elastic component and $\dot{\underset{\sim}{\varepsilon}}^{c}$ is the fracture component of the strain rate tensor. The contributions to the fracturing strain rate tensor $\dot{\underset{\sim}{\varepsilon}}^{c}$ are non-zero whenever a fracture criterion $F^{c} = 0$ is satisfied.

The tensile fracture creates a crack perpendicular to the direction of the maximum principal stress. The existence of the crack and the direction of the crack plane is memorized for all future time. The tensile strength across the crack plane will then depend on the crack width w. A compressive stress across the crack plane can exist only after crack closure (w = 0) has occured. Secondary cracks may develop across pre-existing crack planes so long as a tensile strength persists in any principal direction. In the present development of the model, however, secondary crack planes can only appear perpendicular to the existing crack planes. Thus, the maximum number of crack planes is three. This assumption is acceptable if the strains increase approximately proportionally within the fracture zone.

The tensile fracturing of the material is governed by the maximum stress criterion

$$F^{c}(\underset{\sim}{\sigma}, H_r) = \sigma_i - \sigma_f(w_i, H_r) = 0 \; ; \qquad i = 1, 2, 3 \qquad (2)$$

where σ_i is the principal stress, σ_f is the tensile strength, w_i is the crack width in direction i, and H_r is a hardening function, which may depend on the strain-rate, see Nilsson and Oldenburg [3]. This hardening is neglected in the present study.

Based on experimental results from Petersson [4], we suggest an exponential function for the stress-crackwidth relation. By using the fracture energy relation

$$G_f = \int_0^\infty \sigma(w)\,dw \tag{3}$$

from Rice [5], we find

$$\sigma_f = f_t e^{-(f_t w_i / G_f)} \tag{4}$$

where f_t is the initial tensile strength.

We also assume that the stresses are caused by the elastic components of the strain. Since the crack is assumed to open in the principal stress direction, we obtain analogously to an associated flow rule,

$$\dot{\underset{\sim}{\sigma}} = \underset{\sim}{S}^e : (\dot{\underset{\sim}{\varepsilon}} - \dot{\underset{\sim}{\varepsilon}}^c) \tag{5}$$

$$\dot{\underset{\sim}{\varepsilon}}^c = \dot{\lambda} \frac{\partial F^c}{\partial \underset{\sim}{\sigma}} \tag{6}$$

where $\underset{\sim}{S}^e$ is the elastic stiffness tensor and $\dot{\lambda}$ is a scalar loading parameter.

The relations (5) and (6) together with the consistency relation $F^c = 0$ yield the stress-strain relation for materials with softening tensile fracture

$$\dot{\underset{\sim}{\sigma}} = \underset{\sim}{S}^c : \dot{\underset{\sim}{\varepsilon}} \tag{7}$$

where

$$\underset{\sim}{S}^c = \underset{\sim}{S}^e - \frac{1}{D_i^c} \left(\underset{\sim}{S}^e : \frac{\partial F^c}{\partial \underset{\sim}{\sigma}}\right)\left(\frac{\partial F^c}{\partial \underset{\sim}{\sigma}} : \underset{\sim}{S}^e\right) \tag{8}$$

and

$$D_i^c = \frac{E(1-\nu)}{(1+\nu)(1-2\nu)} + \frac{\partial \sigma_f}{\partial w_i} \frac{\partial w_i}{\partial \varepsilon_i^c} \tag{9}$$

The derivation of the fracture stiffness tensor $\underset{\sim}{S}^{cc}$ and $\underset{\sim}{S}^{ccc}$ when two and three crack planes have been formed follows an identical procedure.

To relate the crack opening w to the fracture strain tensor $\underset{\sim}{\varepsilon}^c$, a control volume V_e enclosing a crack plane of area A_c is considered. For this volume the energy rate equation

$$\int_{A_c} \sigma_i \dot{w}_i dA = \int_{V_e} \underset{\sim}{\sigma} : \underset{\sim}{\dot{\varepsilon}}^c dV \tag{10}$$

must hold. In the subsequent loading, the principal stresses may rotate, and Eq. (10) must include frictional forces on the crack plane. Thus,

$$\int_{A_c} \sigma_c^{t} \dot{\delta}_c dA = \int_{V_e} \underset{\sim}{\sigma} : \underset{\sim}{\dot{\varepsilon}}^c dV \tag{11}$$

where $\sigma_c = [\sigma_{nn} \; \sigma_{nt}]$ is the vector of normal and shear stresses on the crack plane and $\delta_c = [w \; \delta]$ is the vector of crack opening and relative sliding, respectively.

By using Eq. (10), and assuming constant conditions within the control volume, we find

$$\frac{\partial w_i}{\partial \varepsilon_i^c} = \frac{V_e}{A_c} \tag{12}$$

The quantity V_e/A_c may be interpreted as a characteristic length. In finite element analyses we use the element volume as control volume V_e.

From the expressions (9), (3) and (12) we derive the relation

$$D_i^c = \frac{E(1-\nu)}{(1+\nu)(1-2\nu)} - \frac{V_e}{A_c} \frac{\sigma_f f_t}{G} \tag{13}$$

where we note that material instability occurs when $D_i^C = 0$, i.e. when

$$\left(\frac{V_e}{A_c}\right)_{critical} = \frac{E(1-\nu)}{(1+\nu)(1-2\nu)} \cdot \frac{G_f}{\sigma_f f_t} \tag{14}$$

For uni-axial stress conditions, we obtain the well-known expression for perfect brittle failure

$$\ell_{critical} = \left(\frac{V_e}{A_c}\right)_{critical} = \frac{EG}{f_t^2} \tag{15}$$

For given elastic constants, the stress-strain relation within the control volume will depend on the characteristic length V_e/A_c, the initial tensile strength and the fracture energy. Fig. 1 shows stress-strain relations for different values of the characteristic length and a given set of material parameters.

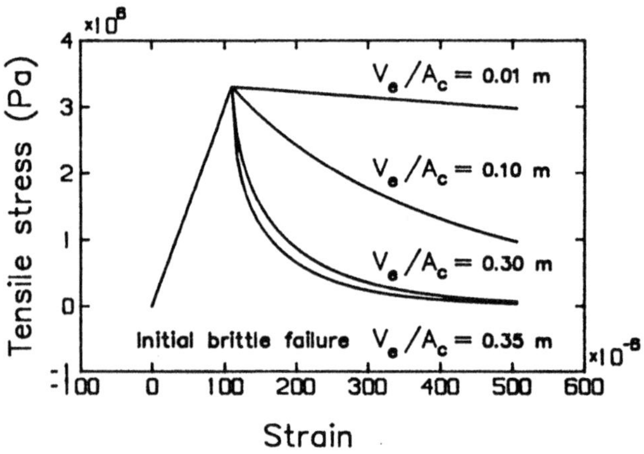

Fig. 1. Influence of the characteristic length V_e/A_c on the uniaxial tensile stress-strain curve. In the diagram, the following material parameters are assumed: f_t = 3.3 MPa, E = 30 GPa, ν = 0.2, G_f = 127 N/m.

Computational aspects

Local stress calculations

During the analysis the stress increment due to fracture is calculated from Eq. (7). The three principal stress directions are checked for penetration of the fracture criterion (2), and the stresses are adjusted to satisfy the consistency condition $\dot{F}^c = 0$.

Three different states of the material in our constitutive model are possible. The first is the uncracked state, the second is the cracked state during equilibrium iterations when the damage can both increase and decrease. Finally, in the cracked equilibrium state the damage is irreversible and can only increase. These three states are illustrated in Fig. 2, where the possible stress paths from each state are plotted in uni-axial stress-strain diagrams of the material model.

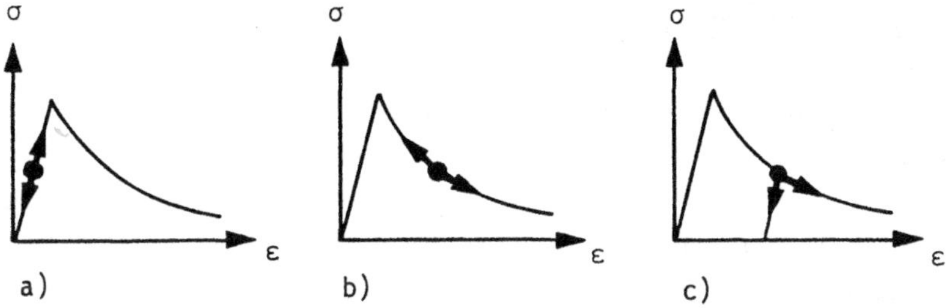

Fig. 2. Three different stress states of the material model. a) Elastic state. b) Cracked state during equilibrium iterations. c) Cracked state at equilibrium.

The maximum crack width obtained after convergence represents the damage of the material. During equilibrium iterations, cracks may form and disappear before equilibrium is satisfied. No spurious cracks will therefore be formed due to the iterative solution process.

Rotation of the stress direction

In the smeared crack approach, the discrete crack is modelled as a change in the constitutive matrix by constraints on the stresses in one or many directions. If these constraints are fixed and the load direction changes, stiffnesses in directions parallel to the crack will influence the stiffnesses in the load direction.

To overcome this difficulty, the constraint on the stress field due to fracture can be released. The concept proposed is to perform the strain and crack width calculations under the assumption that the crack remains in the initial crack direction. However, the reference system of the stress constraints resulting from the fracture model, is rotated with the principal strain direction. Hence, the fully developed crack will not be permitted to transfer load in any direction. This concept is similar to "the rotating crack model" by Milford and Schnobrich [6].

A more refined fracture model may be obtained by consideration of multiple cracks in any direction. Whether a new fracture forms or a relative motion on an existing crack plane takes place may depend on which process demands activation of the smallest quantity of energy.

Element types for fracture analyses

We confine our discussion to two-dimensional elements for plane and axisymmetric conditions.

If e i g h t - n o d e i s o p a r a m e t r i c e l e m e n t s (or higher order elements) are used, we evaluate the crack length at each discrete point used for numerical integrations. The linear strain-field approximation implies a spurious continuity when fracture occurs in one point of the element. Thus, high order elements should not be used to model localized phenomena like fracture.

In general, four points are required to integrate the f o u r - n o d e i s o p a r a m e t r i c e l e m e n t. Due to the bilinear interpolation functions the stress states in the different integration points are strongly coupled. Thus, the dissipated fracture energy cannot be calculated correctly, and the post-fracture stress-strain relations will be erroneous.

One method by which this difficulty may be circumvented, is to use a mixed numerical integration scheme. The total stress state is divided into one elastic and one inelastic component. The inelastic component is the difference between the total stresses and the elastic stresses obtained from the total strains. This component of the stress state is integrated at one point, while the elastic component is integrated in four integration points.

To avoid spurious bending effects in fractured elements, a change to one integration point also of the elastic components is often necessary. Due to the stabilizing effect of the surrounding uncracked elements the solution will be unaffected by zero-energy modes in the fractured element, However, the sudden loss of bending stiffness of the element at fracture may influence the results obtained.

To overcome the problems mentioned above, we have found it most convenient and economical to use a one point integration of the total stiffness at the expence of eliminating possible hourglass modes. We have used the Kosloff and Frazier [7] approach to this problem.

Two types of crack length calculations have been tested in connection with the use of the four-node element. The crack length has been calculated either as the square-root of the element area, or alternatively a more exact algorithm has been used. When the refined algorithm was used, an improvement in the performance of the crack model was observed only when the element mesh contained distorted elements.

The t h r e e - n o d e e l e m e n t has constant stress and strain fields and needs only one integration point. The major drawback of this element is, however, its excessive stiffness.

Example

Earlier versions of our constitutive model have been used for analysis of wave propagation in concrete structures, Nilsson and Oldenburg [3], and for analysis of rock blasting, Digby et al [13]. In this study we use the refined constitutive model for the analysis of a quasi-statically loaded concrete beam. Further examples can be found in Oldenburg [14].

Experiments on the mixed-mode fracture of a notched concrete beam performed by Arrea and Ingraffea [8] have been widely used to evaluate the validity of different fracture models in structural analyses. The experiments have been simulated by the use of smeared fracture models by Glemberg [9] and Rots et al [10], and by use of discrete fracture models by Arrea and Ingraffea [8] and Gylltoft et al [11], among others. The specimen geometry and test set-up are shown in Fig. 3.

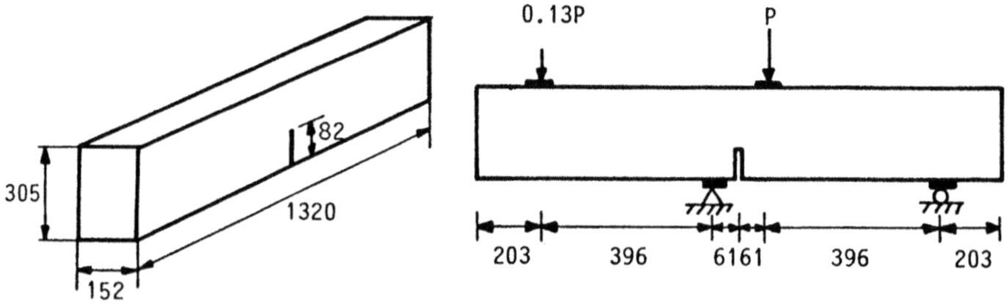

Fig. 3. Test set-up and specimen geometry for the mixed-mode fracture experiment.

The crack paths observed in the experiments are sketched roughly in Fig. 4. The loading was displacement controlled by the Crack Mouth Sliding Displacement (CMSD) shown in Fig. 4.

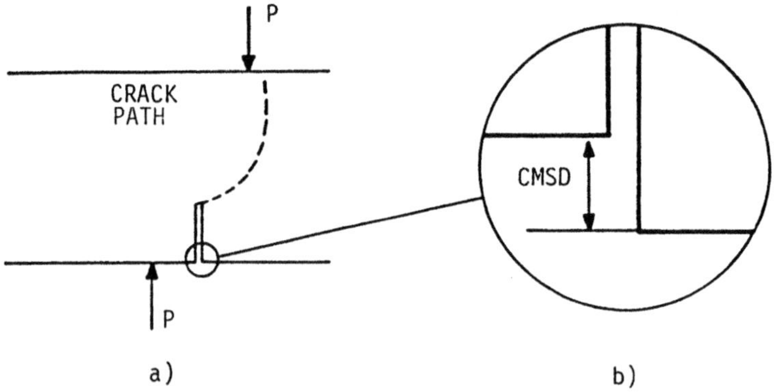

Fig. 4. a) Typical crack paths observed from the test.
b) Definition of Crack Mouth Sliding Displacement (CMSD).

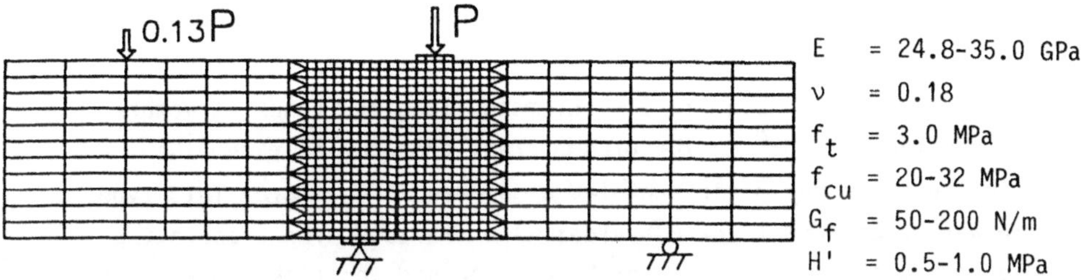

Fig. 5. Finite element mesh for analysis of the mixed-mode fracture test.

Analyses using the fracture model with four-node elements, one-point integration, and allowing the fracture constitutive properties to rotate with the principal strain direction were performed. Fig. 5 shows one choice of element mesh and material data used in the analyses.

When a linear-elastic/fracturing constitutive model was used, we were not able to reproduce the behaviour of the tested beams. The structure was "too stiff", and the crack paths differed from those observed experimentally. The crack paths predicted by the analyses were essentially those shown in Fig. 6a.

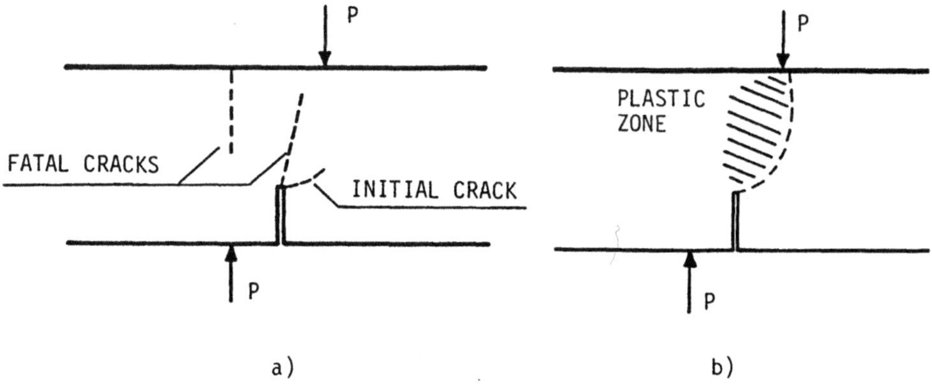

Fig. 6. a) Crack paths obtained from linear-elastic/fracturing analyses
b) Zone where irregular fracturing occurs.

A number of plausible explanations may be advanced in connection with the discrepancies mentioned above. One is that the fracture model is excessively sensitive to shear, i.e. in each element along the crack, fracture occurs at approximately 45^{0} to the global crack propagation path. The crack above the notch may be interpreted as a Mode II fracture. This type of failure was obtained for low G_f-values. The fracture type propagating from above and to the left of the notch was obtained for higher G_f-values. With high G_f-values in the analysis, the structure can sustain higher loads, and an unstable crack will propagate from the top of the beam due to bending. Another possible explanation for the stiff and unstable behaviour of the beam, predicted by the analyses, is that the fracture was the only energy consuming process.

In tests by Gylltoft, Ohlsson and Oldenburg [12], the experiments of Arrea and Ingraffea [8] were reproduced with monotonic loading. The energy consumption in the tests was more than six times larger than that expected from crack formation alone, if realistic values of the fracture energy are assumed. This implies that in addition to tensile crack formation, there are other sources of energy dissipation. One possible explanation is that the energy dissipates from shear deformation in the main crack. This was contradicted by an elastic solution of the problem, where no shear stresses could be found in the direction of the real crack path. However, shear deformation can still occur during the fracture process. Another explanation is that extensive, irregular cracking occurs in a region shown in Fig. 6b.

To model ductile behaviour, i.e. plasticity or irregular fracture interacting with localized fracture, the fracture model was combined with a plastic material model. The Drucker-Prager yield criterion with isotropic hardening and an associated flow rule were assumed.

By using the refined fracture model, we found results which approach the test results. In Fig. 7 the crack pattern and deformation modes are shown.

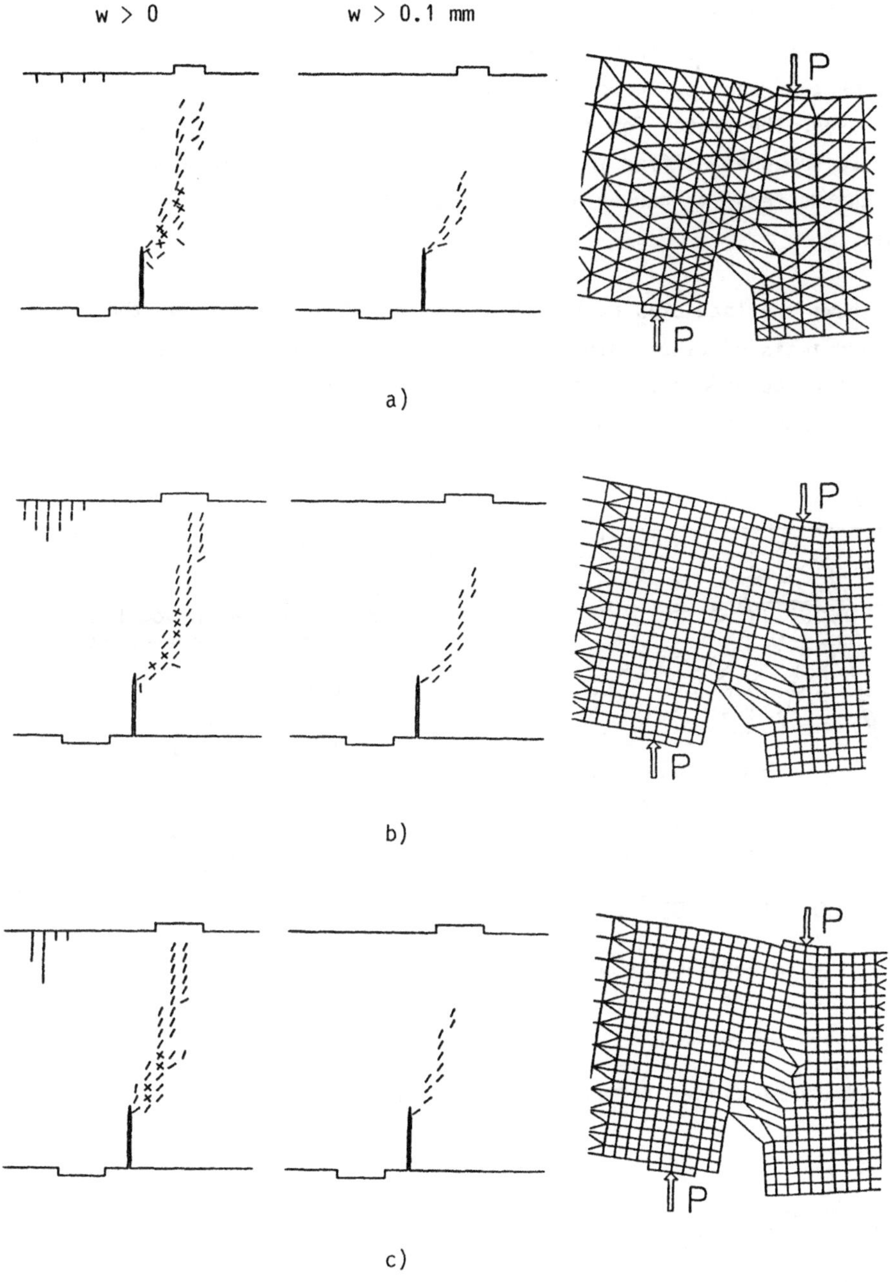

Fig. 7. Crack patterns and magnified deformations from analysis with a) three-node elements, b) four-node elements with mixed integration, and c) four-node elements with one-point integration

To obtain crack propagation characteristics with three-node elements, similar to those obtained from the analysis with four-node elements and the experimental results, the plastic hardening parameter H' = 0.5 GPa, had to be used. The reason for this could be due to that the three-node elements are somewhat stiffer than the four-node isoparametric elements, and that the decrease in hardening modulus compensated this increase in stiffness.

Figure 8 shows the load-CMSD curves obtained from our analysis and the analyses and tests of other authors. We note that the present model predicts the descending part of the curve better than the other models, although the CMSD obtained is still too large.

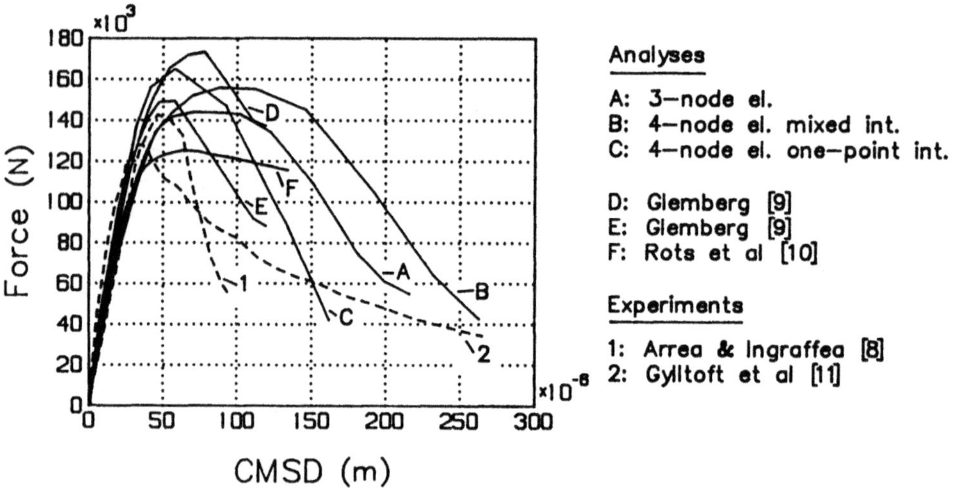

Fig. 8. Load-CMSD curves from analyses and experiments.

References

1. Hillerborg, A.; Modeer, M; Petersson, P-E: Analysis of crack formation and crack growth in concrete by means of fracture mechanics and finite element analysis. Cement and Concrete Research 6 (1976) 773-782.

2. Bazant, Z.P.; Oh, B.H.: Concrete fracture via stress-strain relations. Rep. 81-10/665c, Dept. of Civil Engineering, Northwestern University, Evanstone (1981).

3. Nilsson, L.; Oldenburg, M.: Nonlinear wave propagation in plastic fracturing materials. IUTAM Symposium, Nonlinear Waves, Berlin Heidelberg: Springer (1983) 209-217.

4. Petersson, P-E.: Crack growth and development of fracture zones in plain concrete and similar materials. Rep. TVBM-1006, Div. of Building Materials, Lund Institute of Technology, Lund (1981).

5. Rice, J.R.: Mathematical analysis in the mechanics of fracture. In Fracture - an advanced treatise (ed. Liebiwitz, H.), Vol. II, New York: Academic Press (1968) 191-311.

6. Milford, R.V.; Schnobrich, W.C.: Numerical model for cracked reinforced concrete. Proc. of the int. conf. on computer aided analysis and design of concrete structures (eds. Damjanic, F. et al), Swansea: Pineridge Press (1984) 71-84.

7. Kosloff, D.; Frazier, G.A.: Treatment of hourglass patterns in low order finite element codes. Numerical and Analytical Methods in Geomechanics 2 (1978) 57-72.

8. Arrea, M.; Ingraffea, A.R.: Mixed-mode crack propagation in mortar and concrete. Rep. No. 81-13, Dept. of Structural Engineering, Cornell University, New York 1982.

9. Glemberg, R.: Dynamic analysis of concrete structures, Publ. 84:1, Dept. of Structural Mechanics, Chalmers University of Technology, Göteborg 1984.

10. Rots, J.G.; Nanta, P.; Kusters, G.M.A.; Blaauwendraad, J.: Smeared crack approach and fracture localization in concrete, HERON 30, No. 1, Delft 1985.

11. Gylltoft, K.; Ohlsson, U.: Finite element analysis of mixed mode fracture using a discrete crack approach. Rep. in preparation, Div. of Structural Engineering, Luleå University of Technology, Luleå 1985.

12. Gylltoft, K.; Ohlsson, U.; Oldenburg, M.: Mixed mode fracture tests of notched concrete beams under monotonic loading. Rep. in preparation, Div. of Structural Engineering, Luleå University of Technology, Luleå 1985.

13. Digby, P.J.; Nilsson, L.; Oldenburg, M.: Finite element simulation of time dependent fracture and fragmentation processes in rock blasting. Numerical and Analytical Methods in Geomechanics, Accepted for publication.

14. Oldenburg, M.: Finite element analysis of tensile fracturing structures. Rep. 1985:010L, Luleå University of Technology, Luleå 1985.

Stability and Uniqueness of Strain-Softening Computations

K. J. WILLAM, E. PRAMONO, S. STURE
University of Colorado at Boulder, USA

Summary

The paper addresses computational issues of strain-softening calculations in solids. The main focus is directed towards the numerical simulation of the localization of softening within structural components with simultaneous unloading of the elastic neighborhood. The spatial aspect of the localization process requires special provisions of the incremental/iterative solution strategy in order to survive on one hand local material instabilities and capture on the other hand bifurcation of the solution path within load increments which are truly finite. The performance of several solution algorithms will be assessed with two simple model problems which illustrate the strong interaction between the numerical integration of strain-softening constitutive relations and the spatial localization of contained softening within elastic unloading zones.

Introduction

Strain-softening computations encounter puzzling phenomena during the numerical simulation of distributed damage and localized fracture which transcend previous experience with the inelastic analysis of "stable" materials. In fact these numerical difficulties are exacerbated by the loss of "normality" and the dominance of shear dilatancy under mixed mode fracture which results in predominant straining orthogonal to the loading direction. On one side the failure analysis of brittle solids and structures is very susceptible to the formulation of strain-softening constitutive relations which are equivalent in the sense of strength, stiffness as well as fracture energy to the actual discontinuous failure process. On the other side, the numerical implementation of strain-softening models is far from successful if accepted and conventional numerical tools for strain-hardening plasticity are simply extended into the softening regime. In fact it is the localization phenomenon which leads to simultaneous loading and unloading in adjacent elements which has to be monitored properly. Therefore many numerical algorithms which perform very well in the case of weak continuous nonlinearities are not successful at all when a hard discontinuity takes place within a contiguous material in a finite load interval which draws the demarcation between the localized strain-softening regime and the elastically unloading surrounding.

Finite Element Methods for Nonlinear Problems
Europe-US Symposium, Trondheim, Norway 1985
Editors: Bergan, Bathe, Wunderlich

The intent of this paper is to clarify the computational issues of strain-softening calculations. To this end the composite fracture model for strain-softening will be reviewed together with local and global uniqueness conditions for non-associated hardening/softening solids in order to delineate guidelines for successful strain-softening computations. Subsequently three incremental/iterative solution methods will be scrutinized with regard to their ability to capture strain localization. Two simple model problems will be adopted in order to illustrate the respective performance of initial load and tangential stiffness methods when applied to strain-softening elastic-plastic computations. In conclusion the constant arc length method will be examined for stabilizing the post-critical response calculations beyond the range of direct displacement control.

Composite Fracture Model for Strain-Softening

Strain-softening or rather the degradation of strength is normally attributed to the change of internal microstructure of the material. Two philosophies have recently emerged for the interpretation of the post-critical response regimes of stroke-controlled experiments.

(i) The progressive fracture [1] and continuous damage theories [2] start from the notion of distributed microdefects which coalesce continuously during progressive failure. The simultaneous decrease of strength and stiffness essentially results in a secant-type material formulation where no permanent deformations accompany the failure process during loading-unloading cycles. This concept corresponds to a parallel arrangement of elastic-perfectly brittle elements which fail progressively under increased deformation.

(ii) The blunt crack approach [3] and the composite fracture model [4] on the other hand start from the notion of a constant energy release rate which accompanies the progressive failure process in the form of a contained crack band. The simultaneous decrease of strength and the increase in irreversible fracture deformations essentially result in a tangent-type material formulation where no degradation of stiffness takes place as long as the elastic recovery process remains uncoupled from the progressive failure process. This concept corresponds to a series arrangement of intact elastic and strain-softening failure zones very much along the line of strain-softening plasticity which is based on the additive decomposition of deformation rates into elastic and plastic components.

In view of recent post-critical experiments on concrete [5] the composite fracture model will be adopted which collapses to previous strain-softening formulations of elastic-plastic solids [6,7] when the fracture energy release concept is abandoned. For simplicity we will focus on tensile softening, i.e. the composite fracture formulation of an equivalent strain-softening continuum which maintains constant fracture energy release rate during Mode I type cracking. For the extension to general three-dimensional conditions with Mode II and mixed mode failure the reader is referred to Ref. [5].

In the case of tensile cracking, fracture initiation is defined by the maximum tensile stress criterion

$$F(\boldsymbol{\sigma},\mathbf{q}) = \sigma_n - \sigma_t = 0; \quad \text{where } \sigma_n = \sigma_1 \text{ and } \sigma_1 \geq \sigma_2 \geq \sigma_3$$

$$\text{and } \sigma_t = \sigma_t(\mathbf{q}) \tag{1}$$

For initially isotropic conditions the normal tensile stress σ_n in an active crack coincides with the major principal stress, i.e. $\sigma_n = \sigma_1$. The tensile strength σ_t describes the degradation of bond stresses in terms of the internal variables $\mathbf{q}$ which characterize the change of the internal microstructure during softening. In the case of single localized macrodefects the fracture energy release rate G_f^I will provide the missing link between the macroscopic measure of strength and the internal fracture state. The consistency condition

$$\dot{F} = \frac{\partial F^t}{\partial \boldsymbol{\sigma}} \dot{\boldsymbol{\sigma}} + \frac{\partial F}{\partial \mathbf{q}} \dot{\mathbf{q}} = \mathbf{n}^t \mathbf{E}(\dot{\boldsymbol{\epsilon}} - \dot{\boldsymbol{\epsilon}}_f) + \mathbf{p}^t \dot{\mathbf{q}} = 0 \tag{2}$$

leads to the following strain-based loading conditions for progressive cracking

$$\text{for} \quad \mathbf{n}^t \mathbf{E}\, \dot{\boldsymbol{\epsilon}} \geq 0 \quad \text{then } \dot{\boldsymbol{\epsilon}}_f \neq 0 \tag{3}$$

$$\text{and } \dot{\mathbf{q}} \neq 0$$

Fig. 1 illustrates the progressive degradation of tensile strength σ_t when the elastic trial stress rate $\dot{\boldsymbol{\sigma}}_t = \mathbf{E}\dot{\boldsymbol{\epsilon}}$ has a positive component in the direction of the gradient $\mathbf{n} = \partial F/\partial \boldsymbol{\sigma}$. The constitutive behavior is characterized by the following evolution laws:

$$\text{Elastic:} \quad \dot{\boldsymbol{\sigma}} = \mathbf{E}(\dot{\boldsymbol{\epsilon}} - \dot{\boldsymbol{\epsilon}}_f)$$

$$\text{Fracture:} \quad \dot{\boldsymbol{\epsilon}}_f = \dot{\lambda}_f \mathbf{m} \quad \text{where } \mathbf{m} = \frac{\partial Q}{\partial \boldsymbol{\sigma}} \tag{4}$$

$$\text{Internal State:} \quad \dot{\mathbf{q}} = \mathbf{G}\, \dot{\boldsymbol{\epsilon}}_f$$

When the elastic response behavior is not affected by the progressive fracturing process the elasticity tensor remains invariant, $\mathbf{E}$=const, and

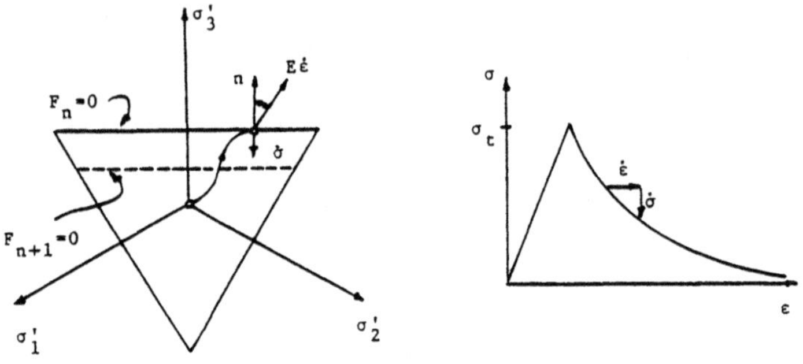

FIG. 1 STRAIN - SOFTENING MODEL FOR TENSION

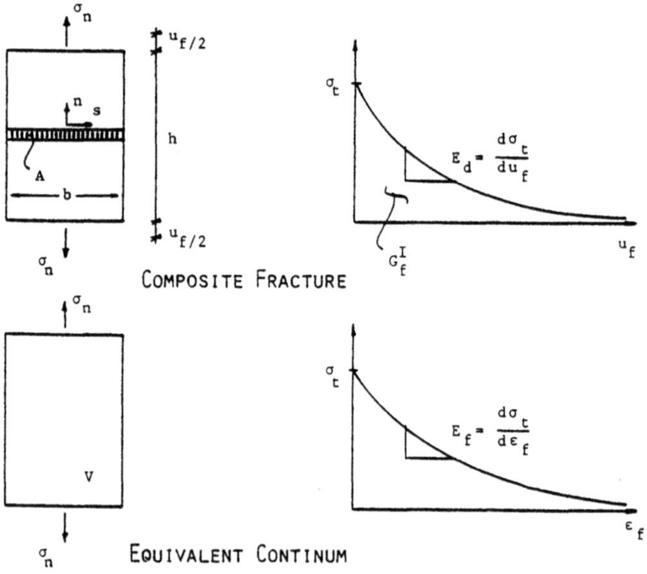

FIG. 2 COMPOSITE FRACTURE MODEL FOR TENSILE CRACKING

does not depend on the internal variable q. In this simple case the elastic stiffness and the degradation of strength during the progressive fracture process are decoupled entirely which is partly supported by the experimental evidence reported in Ref. [5].

The evolution of fracture strains $\boldsymbol{\epsilon}_f$ is characterized by the fracture rule $\dot{\boldsymbol{\epsilon}}_f=\dot{\lambda}_f\mathbf{m}$ which defines direction and magnitude of the macroscopic strain-softening response via the gradient of the fracture potential $Q=Q(\boldsymbol{\sigma})$ and the fracture parameter λ_f. In the case of Mode I type cracking in the direction of the major principal stress the associated fracture rule with $\mathbf{m} = \mathbf{n}$ degenerates to the unit vector in three space

$$\mathbf{m} = \mathbf{n} = \begin{bmatrix} \frac{\partial Q}{\partial \sigma_1} \\ \frac{\partial Q}{\partial \sigma_2} \\ \frac{\partial Q}{\partial \sigma_3} \end{bmatrix} = \begin{bmatrix} 1 \\ 0 \\ 0 \end{bmatrix} \tag{5}$$

The general setting for the internal state variable $\mathbf{q}$ follows the early proposal by Kröner [8] who introduced the material tensor $\mathbf{G}$ in order to map changes of the microstructure, i.e. the moments of dislocation pile-ups in metals onto the macrostructure. In our case of tensile cracking we will resort to this generalization of the traditional definitions of strain/and work hardening in order to incorporate the concept of a constant fracture energy release rate.

To this end let us consider the elementary volume V=hbt which is composed of an intact elastic portion and a contained crack band of the area A=bt where t denotes the thickness. Fig. 2 illustrates the composite fracture element and the equivalent continuum element which dissipate the same fracture energy when the tensile strength diminishes to zero residual level during increased straining. In the post-critical response regime the tensile softening localizes into a single tensile crack band, the formation of which requires the following supply of separation energy

$$G_f^I = \int_0^{u_r} \sigma_t \, du_f \tag{6}$$

Note that this expression is a first approximation for the extraction of the critical fracture energy release rate from direct tension experiments, whereby the inelastic dissipation mechanisms which accompany the coalescence of distributed microdefects into a single macrodefect have been neglected. If we adopt the basic postulate of fragmentation theory where the surface area generated during fracture is proportional to the energy expended [9]

$$G_f = -\frac{\partial \Pi}{\partial A} = \text{const.} \tag{7}$$

then strain-softening may be monitored directly in terms of the raw experimental data involving the nominal tensile strength σ_t and the overall extension u_f. Because of the localized character of the tensile crack and the elastic unloading process of the remaining region, the overall extension u_f is really the crack opening displacement in the final stage of separation.

The question remains how can this relative displacement measure be converted into an equivalent strain in order to be amenable to strain rather than displacement-softening computations? To achieve this we resort to the notion of an equivalent continuum element which dissipates the same amount of fracture energy in the σ_t-ε_f space as does the composite fracture model in the σ_t-u_f space. An elementary energy consideration leads to the following relationship between the softening modulus E_f and the fracture modulus E_d

$$E_f = E_d \frac{V}{A} \simeq E_d h \qquad \text{where } E_d = \frac{d\sigma_t}{du_f} = \frac{d^2 G_f^I}{du_f^2} \tag{8}$$

Clearly in the case of the direct tension test the gage length V/A=h corresponds to the height of the specimen and thus to the crack spacing which converts the overall extension into a nominal strain, $du_f = h d\varepsilon_f$. Therefore, spatial discretization with a finite element mesh size of that particular height would yield the correct fracture energy, while a finer mesh resolution would yield far more brittle response predictions, if the fracture modulus E_f were not adjusted to facilitate the corresponding element height. On the other hand if an elemental philosophy were adopted in which h corresponds to the height or rather the governing element size in the direction of the major principal stress then we have to ascertain that localized cracking is contained to a single element at a time. In this sense the fictitious crack approach by Hillerborg et al. [10] and the blunt crack approach by Bazant et al. [3] solve the spatial problem by following a particular fracture topology similar to the discrete fracture model of Ingraffea and Saouma [11]. In contradiction the current composite fracture model does not involve remeshing but requires a solution strategy for the numerical integration of the strain-softening relations which captures the strain localization within a single element in order to monitor the proper fracture energy release rate.

For the construction of a general fracture-based strain-softening theory we have to map the crack opening displacement du_f and the equivalent tensile fracture strain $d\varepsilon_f$ into three dimensions. To this end let us return to the general formulation outlined in Eqs. 1-5. Introducing the evolution laws in Eq. 4 into the consistency condition in Eq. 2, we obtain the missing expression for the fracture parameter

$$\dot{\lambda}_f = \frac{\langle \mathbf{n}^t \mathbf{E}\, \dot{\boldsymbol{\varepsilon}} \rangle}{E_f + \mathbf{n}^t \mathbf{E}\, \mathbf{m}} \qquad \text{where } E_f = -\mathbf{p}^t \mathbf{G}\, \mathbf{m} \tag{9}$$

If we resort to a strain definition of the internal state variable $d\mathbf{q} = d\boldsymbol{\varepsilon}_f$ with $\mathbf{G}=\mathbf{I}$ then the metric of tensile deformation is defined by the equivalent tensile fracture strain

$$\dot{\varepsilon}_f = \langle \dot{\boldsymbol{\varepsilon}}_f^t \, \dot{\boldsymbol{\varepsilon}}_f \rangle^{1/2} = \sqrt{\langle \dot{\varepsilon}_1 \rangle_f^2 + \langle \dot{\varepsilon}_2 \rangle_f^2 + \langle \dot{\varepsilon}_3 \rangle_f^2} \tag{10}$$

which extracts from the strain tensor only the positive tensile components. The fracture modulus E_f in Eq. 9 characterizes the strain-softening behavior of the equivalent continuum as follows

$$E_f = -\frac{\partial F}{\partial \boldsymbol{\varepsilon}_f}^t \frac{\partial Q}{\partial \boldsymbol{\sigma}} = \left(\frac{\partial \sigma_t}{\partial u_f} \frac{\partial u_f}{\partial \varepsilon_f} \frac{\partial \varepsilon_f}{\partial \boldsymbol{\varepsilon}_f}\right)^t \mathbf{m} \tag{11}$$

$$\text{where} \quad \frac{\partial \sigma_t}{\partial u_f} = E_d \quad \text{and} \quad \frac{\partial u_f}{\partial \varepsilon_f} = \frac{V}{A}$$

For uniaxial tension this expression reduces to the relationship in Eq. 8, i.e. $E_f = E_d h$, where h defines the characteristic dimension of the elementary volume in the direction of major tensile straining. Note that this geometric quantity lends itself to a more fundamental micromechanical interpretation since $h = V/A$ is nothing but the reciprocal value of the microcrack density, i.e. the crack surface per unit volume within the solid.

Combining the evolution laws for the elastic and inelastic fracture behavior we arrive at the tangential constitutive relation to be compared with non-associated elastic-strain softening formulations

$$\dot{\boldsymbol{\sigma}} = \mathbf{E}_T \dot{\boldsymbol{\varepsilon}} \qquad \text{where } \mathbf{E}_T = \mathbf{E} - \frac{\mathbf{E}\, \mathbf{m}\, \mathbf{n}^t \mathbf{E}}{E_f + \mathbf{n}^t \mathbf{E} \mathbf{m}} \tag{12}$$

The principal difference is the definition of the strain-softening modulus E_f in Eq. 11 which monitors directly the fracture energy release rate via E_d

$$E_f = E_d \frac{V}{A} (\frac{\partial \varepsilon_f}{\partial \boldsymbol{\varepsilon}_f})^t \mathbf{m} \qquad \text{where} \qquad E_d = \frac{d^2 G_f^I}{d u_f^2} \tag{13}$$

On the other hand the fracture-based formulation reproduces the traditional strain-softening models which describe the degradation of strength directly in strain rather than displacement space.

Uniqueness

Let us consider for a moment local and global uniqueness conditions of the elastic plastic strain-softening solution. In the past considerable discussion has revolved around the well-posedness of strain-softening computations with repeated reference to Hadamard [12]. The argument relates to wave propagation in elastic solids whereby the hyperbolic nature of the characteristic equation and the concomitant speed of wave propagation is limited by the positive definiteness of the tangential material law, i.e.

$$\boldsymbol{x}^t \mathbb{E}_T \boldsymbol{x} > c\, \boldsymbol{x}^t \boldsymbol{x} \qquad \text{with } c > 0 \tag{14}$$

The local stability postulate of Drucker [13] leads to a similar conclusion, since the condition

$$\dot{\boldsymbol{\sigma}}^t \dot{\boldsymbol{\varepsilon}} > 0 \tag{15}$$

is synonymous with the positive definiteness of E_T according to Eq. (14). Clearly this stringent requirement leads to severe restrictions of the material model to positive behavior with unlimited ductility. Therefore, if we intend to capture the degradation of strength by strain-softening material models we have to resort to contiguous uniqueness arguments which are considerably less restrictive. In this context we recall the original work of Melan [14] who studied uniqueness of strain-hardening elastic-plastic solutions. The classical approach revolves around the solution of an auxiliary problem in which we examine the condition under which two different solutions may exist at the time t_{n+1} when a unique solution exists at the time t_n. For quasistatic conditions the proof of uniqueness of the solution of the initial value problem results in the question whether the following relation can vanish for $\dot{\boldsymbol{\sigma}}_1 \neq \dot{\boldsymbol{\sigma}}_2$ and/or $\dot{\boldsymbol{\varepsilon}}_1 \neq \dot{\boldsymbol{\varepsilon}}_2$

$$\int_V (\dot{\boldsymbol{\sigma}}_2 - \dot{\boldsymbol{\sigma}}_1)^t (\dot{\boldsymbol{\varepsilon}}_2 - \dot{\boldsymbol{\varepsilon}}_1) dV = \int_V \Delta\dot{\boldsymbol{\sigma}}^t \Delta\dot{\boldsymbol{\varepsilon}}\, dV = 0 \tag{16}$$

Local uniqueness is therefore assured whenever the governing tangential material law is positive definite i.e. when

$$\mathbf{x}^T \mathbf{E}_T \mathbf{x} > 0 \qquad \text{where} \qquad \mathbf{x} = \Delta\dot{\boldsymbol{\varepsilon}} = (\dot{\boldsymbol{\varepsilon}}_2 - \dot{\boldsymbol{\varepsilon}}_1) \tag{17}$$

In the case of the non-associated strain-softening law in Eq. 12 this constraint leads to the following limiting inequality examined originally by Mroz [15]

$$E_f + \mathbf{n}^t \mathbf{E}\, \mathbf{m} > \frac{(\mathbf{x}^t \mathbf{E}\, \mathbf{m})\, (\mathbf{n}^t \mathbf{E}\, \mathbf{x})}{\mathbf{x}^t \mathbf{E}\, \mathbf{x}} \tag{18}$$

This leads in the worst case to the "hard" restriction on the strain-softening modulus [16] that

$$2E_f > (\mathbf{m}^t \mathbf{E}\, \mathbf{m})^{1/2} (\mathbf{n}^t \mathbf{E}\, \mathbf{n})^{1/2} - \mathbf{n}^t \mathbf{E}\, \mathbf{m} \tag{19}$$

Clearly, for associated behavior with $\mathbf{n} = \mathbf{m}$ we recover the classic statement of Melan [14] that uniqueness is guaranteed for strain-hardening behavior with $E_f > 0$.

On the other hand, when the strain rate satisfies either orthogonality condition

$$\mathbf{x}^t\, \mathbf{E}\, \mathbf{m} = 0 \tag{20}$$

$$\text{or} \qquad \mathbf{x}^t\, \mathbf{E}\, \mathbf{n} = 0$$

then we recover the far less restrictive uniqueness condition for the strain-softening modulus E_f

$$- E_f < \mathbf{n}^t \mathbf{E}\, \mathbf{m} \tag{21}$$

Therefore in the most fortuitous case of orthogonality between the elastic stress rate $\mathbf{E}\Delta\dot{\boldsymbol{\varepsilon}}$ and the gradients $\mathbf{n}, \mathbf{m}$ uniqueness is assured as long as the softening modulus $-E_f$ is smaller than $\mathbf{n}^t\, \mathbf{E}\, \mathbf{m}$, see also Ref. [7]. In the case of uniaxial tension this condition implies that the softening slope must be less than the elastic modulus, i.e.

$$- E_f < E \tag{22}$$

As a result the choice of softening modulus is quite restrictive. Recalling that the fracture based softening formulation introduces with h a characteristic dimension into the definition of $E_f = E_d h$ such that the restriction in Eq. 22 results in a following limitation of the mesh size h

$$h < h_c = \frac{E}{-E_d} \tag{23}$$

This last expression clearly illustrates the interrelationship between the fracture-based strain-softening formulation and the element size in which fracture energy is released. In fact the ratio between the elastic and the fracture moduli governs the maximum mesh size h_c for which weak uniqueness is ascertained.

In this context Valanis [17] demonstrated very recently that rate dependent constitutive relations for strain-softening viscoelastic and viscoplastic solids satisfy the local uniqueness condition as long as the elastic component remains positive.

Since the pivotal uniqueness condition in Eq. 16 involves the volume integral over the entire domain, let us examine the implication related to global uniqueness, see also Ref. [6]. To achieve this we subdivide the domain into a zone $V-V_f$ in which the positive material operator assures local uniqueness in the strong sense and into a zone V_f in which the non-positive material operator may lead to the loss of uniqueness.

$$\int_{V-V_f} \Delta\dot{\boldsymbol{\sigma}}^t \, \Delta\dot{\boldsymbol{\epsilon}} \, dv + \int_{V_f} \Delta\dot{\boldsymbol{\sigma}}^t \, \Delta\dot{\boldsymbol{\epsilon}} \, dv = 0 \tag{24}$$

Clearly, the concomitant finite element discretization results in the question, if the tangent stiffness remains positive definite, i.e.

$$\Delta\dot{\mathbf{r}}^t \, \mathbb{K}_T \, \Delta\dot{\mathbf{r}} > 0 \quad \text{where } \mathbb{K}_T = \mathbb{K} - \mathbb{K}_f$$

$$\text{with } \mathbb{K} = \int_V \mathbb{B}^t \, \mathbb{E} \, \mathbb{B} \, dv \tag{25}$$

$$\text{and } \mathbb{K}_f = \int_{V_f} \mathbb{B}^t \, \mathbb{E}_f \, \mathbb{B} \, dv$$

In the case of contained strain-softening computations the splitting of the tangential stiffness operator into a positive elastic component $\mathbb{K}$ and into an non-positive contribution $\mathbb{K}_f$ leads to the question: Under which circumstances will the softening material operator contaminate the overall stiffness and lead to a loss of positiveness of $\mathbb{K}_T$? Clearly, there is no general answer to this problem, because the topology and boundary conditions determine the extent of the localized strain-softening region V_f for a given load history. As a guideline one might speculate that sharp localization of the softening zone will enhance the confinement and will thus retard the loss of positiveness. Therefore, the numerical integration of the differential constitutive relations has to capture the spatial localization of strain-softening as well as contiguous elastic unloading of the neighborhood.

In conclusion let us consider global stability which corresponds to the local statement in Eq. 15. In this case the change of stress leads to the following exchange of internal and external energy

$$\frac{1}{2}\int_{V-V_f} \dot{\sigma}^t \dot{\varepsilon}\, d\sigma + \frac{1}{2}\int_{V_f} \dot{\sigma}^t \dot{\varepsilon}\, dV = \frac{1}{2}\int_{S_\sigma} \dot{p}_s^t \dot{u}\, dS + \frac{1}{2}\int_{S_u} \dot{p}^t \dot{u}_s\, ds \tag{26}$$

Adsorption Release Positive Positive/Negative

where body forces are neglected and a small displacement gradient theory is presumed. The softening in region V_f results in a release of energy which is monitored by the fracture-based softening law in Eq. 12. On the other hand the rate of external work by the prescribed surface tractions p_s is always positive, while that of prescribed surface displacements u_s can be positive or negative. This clearly demonstrates the stabilizing influence of displacement control which provides the computational tool for tracing the post-critical response behavior. On the finite element level Eq. 26 reduces to the incremental equilibrium relation

$$(\mathbb{K} - \mathbb{K}_f)\, \dot{\mathbb{r}} = \dot{\mathbb{R}}_p + \dot{\mathbb{R}}_u \tag{27}$$

If we transform the non-positive softening stiffness $\mathbb{K}_f$ into an equivalent force contribution on the right hand side we obtain

$$\mathbb{K}\, \dot{\mathbb{r}} = \dot{\mathbb{R}}_f + \dot{\mathbb{R}}_p + \dot{\mathbb{R}}_u \tag{28}$$

where $\dot{\mathbb{R}}_f$ denotes the crack driving force defined as

$$\dot{\mathbb{R}}_f = \mathbb{K}_f \dot{\mathbb{r}} = \int_{V_f} \mathbb{B}^t \mathbb{E}\, \dot{\varepsilon}_f\, dV \tag{29}$$

Note that the fracture strain rate $\dot{\varepsilon}_f$ monitors crack extension via the fracture energy release rate embedded in Eqs. 9 and 11. We recognize that $\dot{\sigma}_f = \mathbb{E}\, \dot{\varepsilon}_f$ in expression Eq. 29 corresponds to the crack extension stress in fracture mechanics which monitors the separation of nodes between contiguous elements when a discrete fracture approach is adopted [10,11]. The principal difference between this scheme and the discrete fracture approach is that equivalent strain-softening takes place within an element rather than at the interface between adjacent elements. This fracture energy equivalence is assured as long as strain-softening is localized in a single element and does not spill over into contiguous elements.

Model Problems

In conclusion we examine the stability and uniqueness of the numerical solution with the two elementary model problems shown in Fig. 3. The series arrangement of two axial force members of different cross-section was selected in order to investigate the performance of three different incremental/iterative solution methods for the numerical integration of the strain-softening relations. Of particular interest was the question whether the computational strategies were able to capture the localization of fracture in the weaker element, and if a unique solution could be obtained independently of the size of the load increment. Of further interest was the overall computational efficiency and the limiting conditions for convergence in the light of the theoretical considerations delineated in the previous two sections. Particular care was exercised to implement the three algorithms in their full generality and not to bias the solution effort by taking advantage of the simple structural configuration for which hand calculations suffice to determine the reference solution.

The odd geometry and material parameters were chosen to fit the direct tension experiments on cylindrical concrete specimens reported in Ref. [5]. A linear softening model was adopted to monitor the fracture energy release rate G_f = 0.322 lb/in. which was required to separate the tension specimen into two competent parts. For the height h = 2.125 in. of each bar element the equivalent strain-softening modulus E_f is obtained as follows:

$$E_f = E_d h = \frac{1}{2}\frac{\sigma_t^2}{G_f^I} h = 0.530 \times 10^6 \text{ psi} \tag{30}$$

with σ_t = 400 psi

and E = 2.8 x 10^6 psi

Figure 4 illustrates the element deformations at three stages under the prescribed end displacement $r_p>0$. The results clearly illustrate that beyond peak softening leads to a rapidly increasing extension of the weak element while the stronger elements contract due to elastic unloading. Therefore severe strain localization takes place during monotonically increasing loading of the structure under displacement control which results in a discontinuous strain distribution along the axis, whereby the deformation rates in the softening and the elastically unloading element have opposite directions.

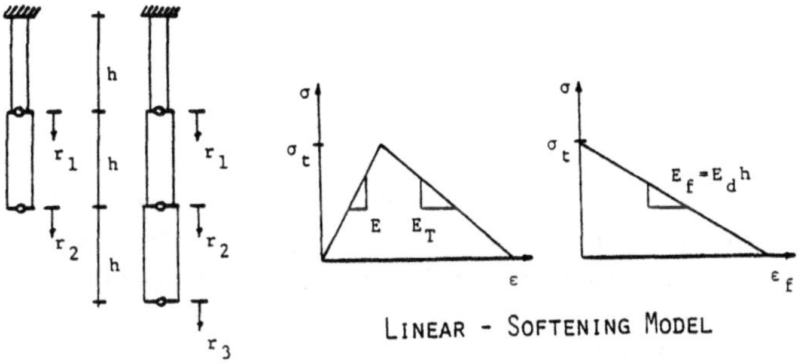

FIG. 3 LAY - OUT OF TENSILE MODEL PROBLEMS

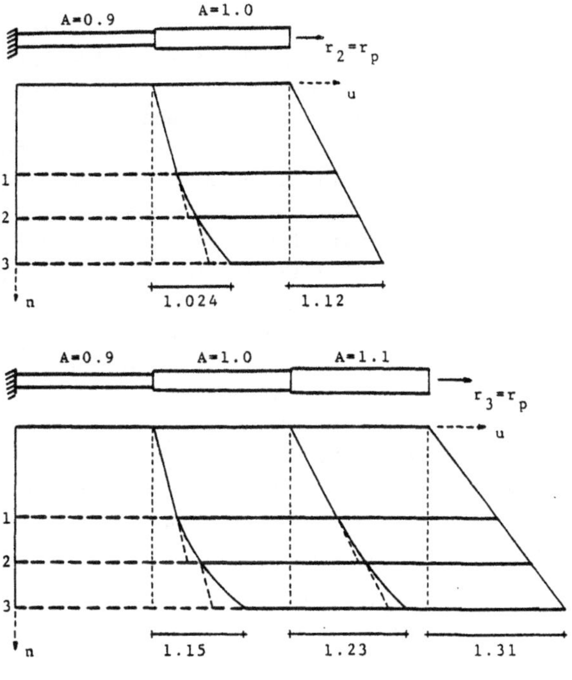

FIG. 4 LOCALIZATION OF STRAIN - SOFTENING REGIME

Figure 5 shows the representative behavior of the two bar structure when the continuation process is resolved into six relatively fine increments. The iterative refinement within each increment demonstrates rapid convergence of the strain-softening and elastic unloading deformations in the two elements. The accompanying relaxation of "real" stresses is plotted when an elastic predictor is used along the line of the initial load iteration method which dates back to the original work in Refs. [18,19]. The algorithm is summarized in the Appendix which corresponds to the elastic plastic solution strategy developed in Ref. [20]. We observe that the initial elastic predictor step at the beginning of each increment is iteratively refined until the constitutive constraint and equilibrium are satisfied simultaneously. It is intriguing that the associated inelastic flow corrections lead to a stress relaxation which is particularly pronounced in the stronger element B subjected to elastic unloading.

In the previous study the incrementation was chosen in such a way that the initial predictor step was causing strain-softening in the weaker element only. In this case all three algorithms summarized in the Appendix captured the localization of strain-softening in element A, whereby the tangent modulus-return scheme was most efficient because of its rapid convergence within two iterations. However, this conclusion only holds for the refined staging strategy in which incipient softening is traced for one element at a time. Therefore in a complex structure the continuation process would have to be subdivided into extremely fine increments in order to capture separate localization events in each softening element. This procedure corresponds to the original proposals of elastic-plastic computations at the end of the 1960's in which the load incrementation was adapted to incipient yielding of the most critical element within the mesh. In computational practice we have to resort to much bolder staging strategies which are independent of the different localization points in the continuation process. In fact the determination of the next localization point is in general quite costly particularly when arc-control is used. Therefore a more pragmatic approach resorts to the following strategy for a prescribed load increment :

(i) determine from the elastic predictor step the most critical trial stress state within the structure

(ii) monitor the structural response due to softening within this critical element, while preventing all other elements from strain localization.

This strategy was certainly successful when the present model problems were analyzed with large load increments, which might otherwise have resulted in the simultaneous softening of all elements. Because of the extensive search and sorting effort this approach is equally costly since it involves continuous exchange between structure and element oriented

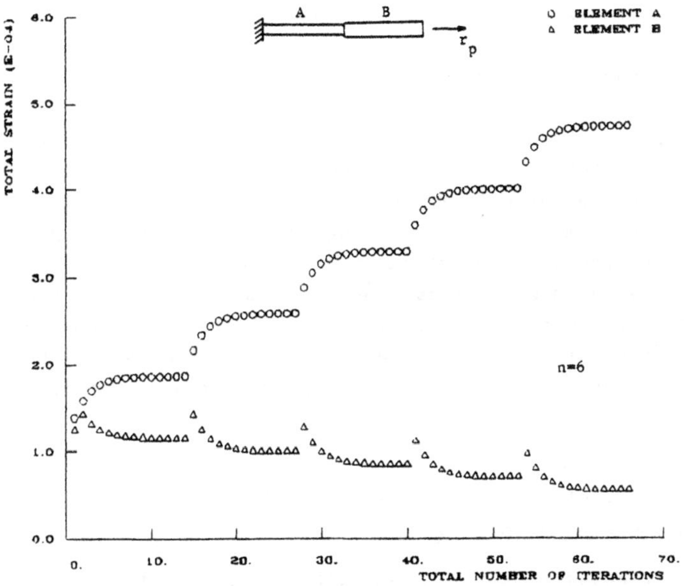

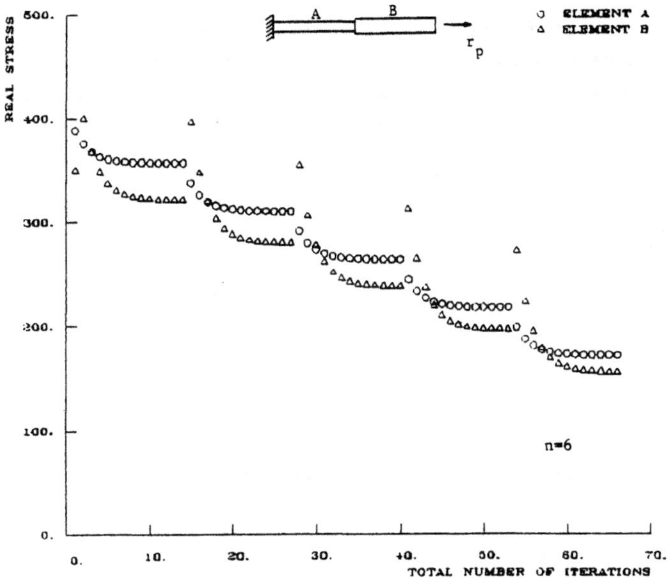

FIG. 5 RESPONSE BEHAVIOUR FOR FINE STAGING

oriented data within large scale finite element programs. Therefore it is highly desirable that the solution algorithm capture localization automatically without resorting to the artifacts above.

To this end we return to our two model problems and examine the performance of the three solution algorithms summarized in the Appendix. The load is applied in two large increments; the first is purely elastic while the second leads to simultaneous softening of all elements. Fig. 6 depicts the localization of deformations within the structure as predicted by the initial load method. Observe the snap back and the elastic recovery of the stronger elements within the second iteration of the second increment. The initial load method captures the strain localization in spite of the fact that the initial elastic predictor step mobilizes strain-softening in all elements. In fact this snap back in the stronger elements is captured by two essential ingredients of the algorithm

(i) the initial elastic stiffness is utilized throughout the analysis along the line of Eq. 28.

(ii) during elastic-inelastic transition the penetration point is located through the contact parameter, ξ.

In fact it is the latter argument which results in elastic snap back of the stronger elements because of predominant softening in the weak element. In the case of the two bar model problem the initial load method determines the increment of the free displacement degree of freedom Δr_1 according to the following iteration scheme

$$(k_A + k_B)\ \Delta r_1^{i+1} = \Delta R_f^i + \Delta R_u \tag{31}$$

where k_A and k_B denote the elastic stiffnesses of elements A and B. The crack driving force due to strain-softening in the two elements is given by

$$\Delta R_f^i = (\Delta\lambda_f^i\ AE)_A + (\Delta\lambda_f^i\ AE)_B \tag{32}$$

where the fracture parameter for uniaxial tension is

$$\Delta\lambda_f^i = (1 - \xi^i)\ \frac{E}{E_f + E}\ \Delta\varepsilon = 1.232\ (1 - \xi^i)\ \Delta\varepsilon \tag{33}$$

The contact parameter ξ locates the penetration point for elastic-inelastic transition according to linear interpolation

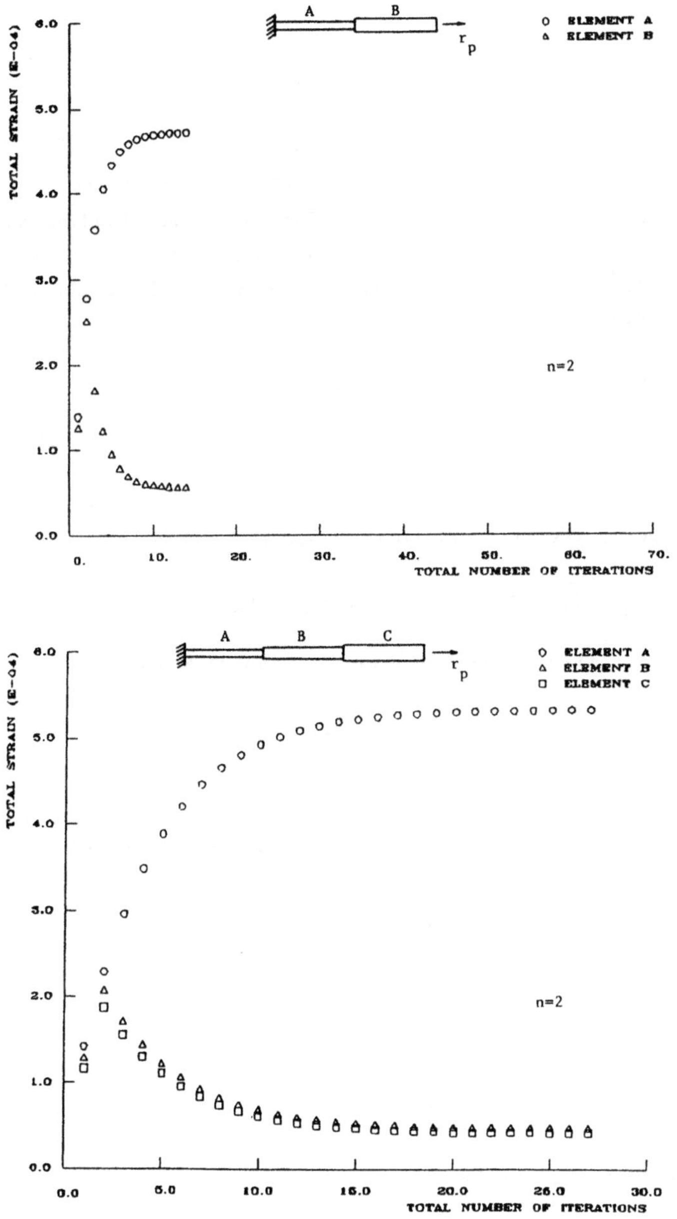

FIG. 6 STRAIN LOCALIZATION FOR COARSE STAGING

$$\xi^i = \frac{\sigma_y - \sigma_n}{\sigma^i_{n+1} - \sigma_n} \qquad \text{where } 0 < \xi \leq 1 \quad \text{for } \sigma^i_{n+1} > \sigma_y \tag{34}$$

σ_y denotes the initial value of the tensile strength $\sigma_y = \sigma_t(\varepsilon^n_f)$ and σ_n and σ^i_{n+1} are the initial stress and the current value of the trial stress state. We observe that $\xi = 0$ for continuous loading and reduces to $\xi=1$ if $\sigma^i_{n+1} \leq \sigma_y$. The strain increment $\Delta\varepsilon$ in Eq. 33 can be expressed in terms of the displacement increment which yields for the crack extension force

$$\Delta R^i_f = 1.232\ k_A(1 - \xi^i_A)\Delta r^i_1 - 1.232\ k_B(1 - \xi^i_B)(\Delta r_p - \Delta r^i_1) \tag{35}$$

where $\Delta r_2 = \Delta r_p$ denotes the prescribed displacement increment. In the case when the weak element has arrived at the elastic limit $\xi^\circ_A = 0$ and when the sufficiently small load increment is applied such that $\xi^\circ_B = 1$ then the iterative corrector equation (Eq. 31) reduces to

$$(k_A + k_B)\ \Delta r^{i+1}_1 = 1.232\ k_A\ \Delta r^i_1 + k_B\ \Delta r_p \tag{36}$$

Convergence of the initial load iteration is assured since

$$\frac{1.232\ k_A}{k_A + k_B} = 0.58 < 1 \tag{37}$$

We observe that strain-softening results in overrelaxation of the fracture parameter $\Delta\lambda_f$ in Eq. 33 because of the magnification factor $E/(E_f + E) = 1.232 > 1$. Therefore we could ask the question what is the limiting softening modulus E^ℓ_f for which convergence breaks down when

$$\frac{E}{E^\ell_f + E}\ \frac{k_A}{k_A + k_B} = 1 \tag{38}$$

The convergence limit is reached when $E_f = E^\ell_f = -1423$ ksi. For the fracture energy release rate of G^I_f=0.322 lb/in. this limitation results in the maximum element height of

$$h_c = 5.47 \text{ in} \tag{39}$$

below which convergence can be assured. Note that this global convergence limitation is considerably more restrictive than the weak uniqueness condition in Eq. 22 which was based only on the local material consideration

$$-E_f < E = 2800 \text{ ksi} \tag{40}$$

Clearly these elementary convergence observations are complicated by the general transition problem which has to be considered for large load increments, when both elements extend into the softening regime. However, from the higher stress state in the weak element A we conclude that $\xi_A < \xi_B$ which signals predominant softening in element A and results in the snap-back of the stronger element B into the elastic recovery regime according to Eq. 35.

Let us now examine the performance of the modified tangent modulus return scheme summarized in the Appendix. This particular strategy was developed by Vermeer [16] for the solution of non-associated shear dilatancy problems and it perfomed best among the numerous variations of tangential methods which were recently assessed in Ref. [21]. This strategy essentially consists of three parts an elastic predictor step and for plastic loading a subsequent plastic corrector step which is combined with the out-of-balance equilibrium step depicted in the flow chart of the Appendix. Unfortunately, many recent publications focus on a particular aspect of the algorithm such as the plastic corrector step and leave the important details of initial staging and synthesis with the out-of-balance equilibrium correction to the reader. In any case the plastic correction involves in general the solution of elastic-plastic transition problem for the contact point denoted by the subscript "c" and the tangential computation of the intermediate state "I" which is subsequently returned to the current yield surface by the plastic flow corrector. The essential step is in this case the calculation of $\Delta\lambda^I$ from the incremental format of the consistency condition using a Taylor expansion about the intermediate configuration and assuming that the total strain increment during flow correction remains constant. This yields for the plastic parameter the following expression

$$\Delta\lambda^I = \frac{< F\,(\boldsymbol{\sigma}^I,\ \sigma_y) >}{E_f + \mathbf{m}^t\,\mathbf{E}\,\mathbf{m}} \tag{41}$$

when the intermediate trial stress state exceeds the current yield condition. In contrast to the initial load method which returns after each iteration to the initial stage "n" the state variables are now continuously updated in the spirit of the Newton-Raphson iteration strategy.

In the case of our single degree of freedom model problem the tangent modulus return scheme results in the following iterative corrector of the incremental equilibrium relation

$$(k_A + k_B)_T \, \Delta r_1^{i+1} = R - S^i + \Delta R_f^c + \Delta R_f^I \tag{42}$$

For uniaxial conditions and linear hardening/softening laws the intermediate predictor state will remain on the yield surface, thus $\Delta R_f^I = 0$. Similiarly if we monitor the load history such that only the weaker element is subjected to incipient softening $\Delta R_f^c = 0$ and the tangent stiffness is confined to element A. Since there are no external loads applied at the free degree of freedom, Eq. 42 reduces in the case of the fine staging strategy to

$$(k_{A,T} + k_B) \, \Delta r_1^{i+1} = -(A\sigma^i)_A + (A\sigma^i)_B \tag{43}$$

$$\text{where} \quad \sigma_A^i = \sigma_A + E_T \, \Delta\varepsilon_A^i - \Delta\lambda_f^i E$$

$$\text{and} \quad \sigma_B^i = \sigma_B + E \, \Delta \, \varepsilon_B^i$$

where the righthand side represents the out-of-balance force of the free degree of freedom. For the fine staging strategy the tangent modulus return scheme captures immediately softening of the weak element A and elastic unloading of the stronger element B resulting in a convergent solution after one iteration.

However, in the case of the coarse staging strategy, where both elements mobilize softening the tangent modulus return scheme results in divergent oscillations since $(k_A + k_B)_T$ is now negative definite. As a result the tangent modulus return scheme breaks down prematurely because of its inability to snap back. Only in the case of extremely careful staging of the load increments the tangent approach captures the strain localization within the structure. Ultimately failure will occur when the tangent stiffness looses positive definiteness which leads to the same restriction of the softening modulus $E_f^\ell = -1.423$ ksi as the initial load method.

The third method, the initial modulus return scheme in the Appendix is very similar to the tangent modulus return method except for the initial stiffness which is retained throughout the iteration process. The plastic flow corrector step has been discussed extensively in Ref. [22] and corresponds essentially to the flow correction from the intermediate state, Eq. 41. The performance of this algorithm is very similar to the initial load method. For fine staging convergence is obtained in approximately the same number of iterations. For coarse staging snap back is oscillatory and not as prompt as in the initial load method. However, the same strain localization is attained in the weak element when two large load increments are used.

We conclude that the localization of strain-softening computations imposes severe restrictions on the staging strategy when tangential approaches are used. The initial stiffness methods are in this regard less restrictive and capture strain localization irrespectively of the magnitude of the load increments. All methods, however, break down when the tangent stiffness looses positive definiteness. Therefore an in-depth effort is presently on the way to explore the capabilities of the arc-length methods [23,24] to capture strain localization. Initial results for elastic softening material laws indicate very promising results when the modified constant arc length normal-projection method [23] is extended by orthogonality constraints between the initial displacement predictor and the iterative displacement corrections using the initial stiffness throughout the iteration history. In this case stable strain localization was obtained for our model problems using strain-softening moduli $E_f <$ -1.423 ksi which fully contaminated the positive definiteness of the tangent stiffness operator. These results will be detailed in a forthcoming publication.

Acknowledgement

The authors would like to thank Dr. N. Bicanic for his constructive critique. The investigation was supported by the U.S. Air Force Office of Scientific Research Contract AFOSR 82-0273 with the University of Colorado, Boulder under the liaison of Lt. Col. Hokanson.

References

1. Dougill, J.W., "On Stable Progressively Fracturing Solids", J. Appl. Math. Phys., ZAMP, Vol. 27 (1976), pp. 423-436.

2. Janson, J. and Hult, J., "Fracture Mechanics and Damage Mechanics, a Combined Approach," J. De Mécanique Appliquée, Vol. 1 (1977), pp. 69-84.

3. Bažant, Z.P. and Oh, B.H., "Crack Band Theory for Fracture of Concrete", Rilem, Materiaux and Constructions, Vol. 16 (1983), pp. 155-177.

4. Willam, K.J., Bićanić, N. and Sture, S., "Constitutive and Computational Aspects of Strain Softening and Localization in Solids", ASME/WAM 1984 Symposium on Constitutive Equations Macro and Computational Aspects, (K. Willam, ed.), ASME Vol. G00274, New York (1984), pp. 233-252.

5. Willam, K.J., Hurlbut, B. and Sture, S., "Experimental Constitutive and Computational Aspects of Concrete Failure", US-Japan Joint Seminar on Finite Element Analysis of Reinforced Concrete Structures, University of Tokyo May 21-24, 1985, Proc. Vol. 1 (1985), pp. 149-171.

6. Prevost, J.-H. and Höeg, K., "Soil Mechanics and Plasticity Analysis of Strain Softening", Geotechnique, Vol. 25 (1975), pp. 279-297.

7. Nguyen, Q.S. and Bui, H.D., "Sur les Matériaux Elastoplastiques a Ecrouissage Positif ou Négatif", J. de Mecanique, Vol. 13 (1974), pp. 321-342.

8. Kröner, E., "Plastizität und Versetzungen", Chapter 9 in Volume 2, Vorlesungen über theoretische Physik, A. Sommerfeld, Akademische Verlags gesellschaft, Leipzig (1970).

9. Von Rittinger P., "Lehrbuch der Aufbereitungskunde", Ernst und Korn, Berlin, 1867.

10. Hillerborg, A., Modéer and Peterson, P.E., "Analysis of Crack Formation and Crack Growth in Concrete by Means of Fracture Mechanics and Finite Elements", Cement and Concrete Research, Vol. 6 (1976), pp. 773-781.

11. Ingraffea, A.R. and Saouma, V., "Numerical Modeling of Discrete Crack Propagation in Reinforced and Plain Concrete", Chapter 4 in Fracture Mechanics of Concrete, M. Nijhoff Publ., Dordrecht (1985), pp. 171-225.

12. Hadamard, J.S., "Leçons sur la Propagation des Ondes", Hermann, Paris (1903).

13. Drucker D.C., "A Definition of Stable Inelastic Materials", J. Appl. Mech., Vol. 26 (1959), pp. 101-106.

14. Melan, E., "Zur Plastizität des räumlichen Kontinuums", Ing. Archiv Vol. 9, (1938), 116-126.

15. Mroz, Z., "On the Description of Anisotropic Workhardening," J. Mech. Phys. Solids, Vol. 15 (1967), pp. 163-175.

16. Vermeer, P.A., "Formulation and Analysis of Sand Deformation Problems,", Report No. 195 of the Geotechnical Laboratory, Delft University of Technology, 1978.

17. Valanis, K.C., "On the Uniqueness of Solution of the Initial Value Problem in Softening Materials", Joint ASME/ASCE Summer Mechanics Conference, June 24-26, 1985, Albuquerque, JAM, 85-APM-29 (1985), pp. 1-5.

18. Argyris, J.H. and Scharpf, D.W., "Methods of Elasto-Plastic Analysis, ISD-ISSC Symp. Stuttgart (1969), pp. 381-416; also ZAMP, Vol. 23 (1972), pp. 517-551.

19. Nayak, G.C. and Zienkiewicz, D.C., "Elasto-Plastic Stress Analysis. A Generalization for Various Constitutive Relations Including Strain-Softening", Int. J. Num. Meth. Engng., Vol. 5, (1972), pp. 113-135.

20. Balmer, H., Doltsinis, J. St. and König, M., "Elastoplastic Creep Analysis with the ASKA Program System", Comp. Meth. Appl. Mech. Eng. Vol. 3 (1974), pp. 87-104.

21. Potts, D.M. and Gens, A., "A Critical Assessment of Methods of Correcting for Drift from Yield Surface in Elasto-Plastic Finite Element Analysis", Int. J. Num. Anal. Meth. Geomech. Vol. 9 (1985), pp. 149-159.

22. Ortiz, M. and Simo, J.C., "An Analysis of New Class of Integration Algorithms for Elastoplastic Constitutive Relations", Techn. Report, Brown University, Division of Engineering, Providence (1984).

23. Ramm, E., "Strategies for Tracing Nonlinear Response Near Limit Points", Europe-US Workshop on Nonlinear Finite Element Analysis in Structural Mechanics," Proc. (Bathe, Stein, Wunderlich eds.) Springer Verlag, Berlin (1980), pp 63-89.

24. Crisfield, M.A., "A Fast Incremental/Iterative Solution Procedure that Handles Snap Through ", Computers and Structures, Vol. 13 (1981), pp. 55-62.

Appendix

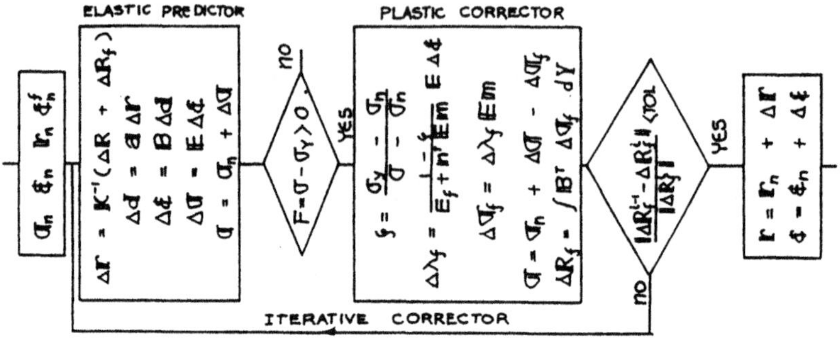

INITIAL LOAD METHOD

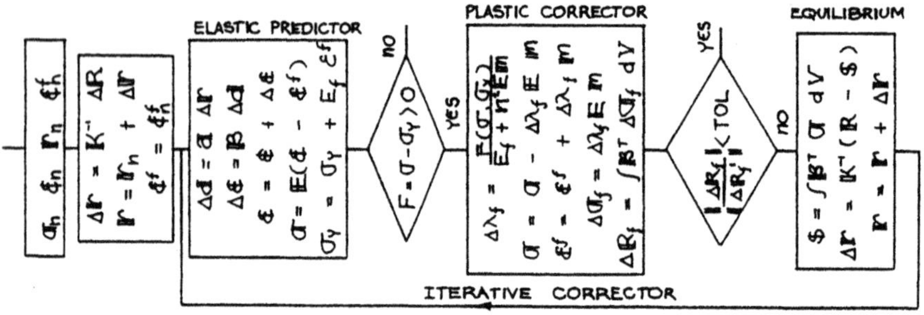

INITIAL MODULUS RETURN SCHEME

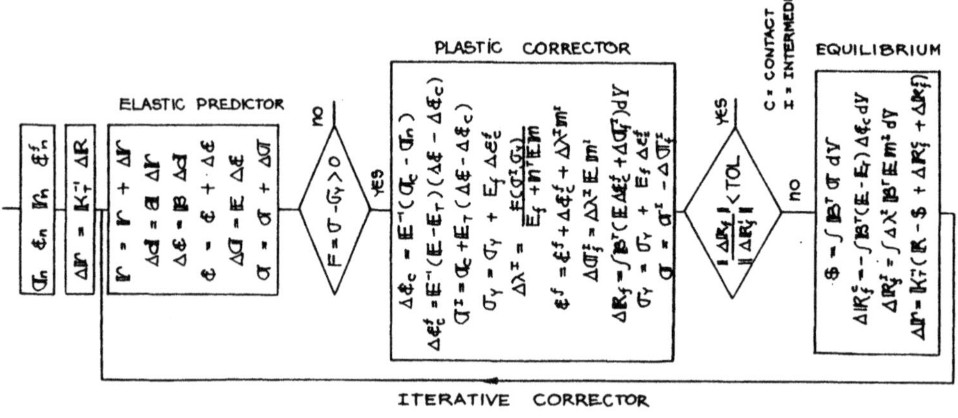

TANGENT MODULUS RETURN SCHEME

An Endochronic Approach and Other Topics in Small and Finite Deformation Computational Elasto-Plasticity

S. N. ATLURI
Georgia Institute of Technology, Atlanta, USA

Abstract

(i) A synopsis of a recently developed methodology for computational small-deformation elasto-plastic analyses of cyclically loaded structures, based on a variant of an endochronic theory, is presented. The present plasticity model is shown to provide a unified basis for the earlier "multiple-yield-surface" or "nonlinear kinematic hardening" theories of Mroz, Krieg, Dafalias and Popov, Chaboche, and others. (ii) Issues related to constitutive modeling of finite deformation elasto-plasticity, and attendant numerical implementations, are discussed. Consistent rate forms of elastic-plastic evolution equations, which, in the limit, model the hyperelastic or small-deformation elastic-plastic behaviours appropriately, are presented.

I. An Endochronic Computational Approach to Cyclic Plasticity

Without much loss of generality, we treat only deviatoric plasticity. Let $\underset{\sim}{\sigma}$ be the stress, $\underset{\sim}{\sigma}'$ its deviator; $d\underset{\sim}{\varepsilon}$ the strain-rate, $d\underset{\sim}{\varepsilon}'$ the deviatoric strain; $d\varepsilon_m$ the mean strain, $d\underset{\sim}{\varepsilon}' = (d\underset{\sim}{\varepsilon}^e)' + d\underset{\sim}{\varepsilon}^p$; the plastic strain-change $d\underset{\sim}{\varepsilon}^p$ is purely deviatoric; and thus the differential of mean-strain, denoted by $d\varepsilon_m$, is purely elastic, i.e., $d\varepsilon_m \equiv d\varepsilon_m^e$. We consider the solid to be elastically isotropic. Using additive decomposition of $d\underset{\sim}{\varepsilon}$ into $d\underset{\sim}{\varepsilon}^e$ and $d\underset{\sim}{\varepsilon}^p$, we have:

$$d\underset{\sim}{\varepsilon}^p = d\underset{\sim}{\varepsilon}' - \frac{d\underset{\sim}{\sigma}'}{2\mu} \tag{1.1a}$$

Following Valanis [1], we define endochronic (internal time) (but Newtonian time-like) parameters:[1]

$$d\zeta = \left(d\underset{\sim}{\varepsilon}^p : d\underset{\sim}{\varepsilon}^p\right)^{1/2} \; ; \quad dz = \frac{d\zeta}{f(z)} \; ; \quad f(o) = 1 \; , \quad d\zeta \geq 0 \tag{1.1b,c,d}$$

where $f(\zeta)$ is monotonically increasing.

As in Valanis [1], the stress in the elastic-plastic solid is represented through the integral:

[1] In what follows, $\underset{\sim}{A}\cdot\underset{\sim}{B} = A_{ij}B_{jk}$; and $\underset{\sim}{A}:\underset{\sim}{B} = A_{ij}B_{ij}$.

Finite Element Methods for Nonlinear Problems
Europe-US Symposium, Trondheim, Norway 1985
Editors: Bergan, Bathe, Wunderlich

$$\underset{\sim}{\sigma}' = 2\mu \int_0^z \rho(z - z') \frac{\partial \underset{\sim}{\varepsilon}^p}{dz'} \tag{1.2}$$

where μ is the initial (elastic) shear modulus, and $\rho(z)$ is a material-specific kernel. Equation (1.2) thus appears to circumvent the need for a yield surface as well as for the flow rules of classical plasticity theory. Differentiation of (1.2) leads to:

$$d\underset{\sim}{\sigma}' = 2\mu_p \{d\underset{\sim}{\varepsilon}' + \frac{\underset{\sim}{h}(z)}{\rho(o)f(\zeta)} [(d\underset{\sim}{\varepsilon}' - \frac{d\underset{\sim}{\sigma}'}{2\mu}):(d\underset{\sim}{\varepsilon}' - \frac{d\underset{\sim}{\sigma}'}{2\mu})]^{1/2}\} \tag{1.3}$$

$$\rho(o) = \rho \text{ at } z = 0; \ \mu_p = \mu[1 + \rho(o)]^{-1}; \ \underset{\sim}{h}(z) = \int_0^z \frac{\partial \rho}{\partial z}(z-z') \frac{\partial \underset{\sim}{\varepsilon}^p}{\partial z'} dz \tag{1.4}$$

While the classical loading/unloading criteria (or criteria for elastic or plastic processes) are apparently bypassed in Eq. (1.3), there are, nevertheless, prices extracted for this seeming simplicity. Some of these counterbalancing difficulties of the above endochronic approach, as compared to a classical plasticity theory, are as follows: (i) The determination of stress history (and $d\underset{\sim}{\sigma}$) for a given strain history (or $d\underset{\sim}{\varepsilon}$) at each material point becomes highly iterative in nature, as seen from (1.3); (ii) In a finite element/boundary-element/or other weak solution of the boundary value problem, the trial solution $d\underset{\sim}{\varepsilon}'$ is derived by differentiation of trial displacements $d\underset{\sim}{u}$. To determine the trial stresses $d\underset{\sim}{\sigma}'$ and yet retain a piecewise-linear-equation solution strategy, there is no recourse other than to approximate Eq. (1.3) as $d\underset{\sim}{\sigma}' = 2\mu_p d\underset{\sim}{\varepsilon}'$. Thus, the stiffness matrix at any stage of loading is essentially the linear-elastic stiffness matrix; and the elastic-plastic solution method becomes the so-called "initial-strain" method; (iii) To model the uniaxial stress-strain curve of a material that does exhibit a sharp 'knee' near the elastic limit, the kernel $\rho(z)$ has to be weakly singular at $z = 0$. These drawbacks notwithstanding, Valanis and Fan [2] have recently presented a series of papers dealing with a direct computational implementation of an iterative, initial strain method based on Eq. (1.3) and using exponential functions for the kernel $\rho(z)$ in Eq. (1.4). Details of computational times for achieving convergence of plasticity iterations of the global finite element equations, or of the iterations for stress integration, are not readily available in [2].

Watanabe and Atluri [3-6] have recently presented alternate characterizations of cyclic-plasticity constitutive relations using the essential concepts of an endochronic theory, but with the following features: (i) The notion of a yield-surface, and the demarcation in the definitions of the elastic processes and plastic processes, are retained; (ii) The stress history (or $d\underset{\sim}{\sigma}$), for a given strain history (or $d\underset{\sim}{\varepsilon}$), can be determined quite easily, as in a classical plasticity theory, by using a "generalized-midpoint-radial-return" algorithm; (iii) The finite-element formulation can be based on a "tangent-stiffness" approach, wherein the material constitutive law at each point can be chosen differently depending on whether an elastic process or a plastic process is postulated at each point during the current 'load' increment; and (iv) The present [3-6] endochronic approach provides a unified basis for other well-documented theories to model cyclic-plasticity, viz., the "multiple-yield-surface" theories of Mroz [7], Krieg [8], and Dafalias and Popov [9]; and the kinematic hardening theories of Prager, and Chaboche [12]. The starting point in the work of Watanabe and Atluri [3-4] is the representation of the kernel $\rho(z)$ in Eq. (1.2) in the form:

$$\rho(z) = \rho_o \, \delta(z) + \rho_1(z) \tag{1.5}$$

where $\delta(z)$ is a Dirac function and $\rho_1(z)$ is a non-singular function. It turns out [3-4] that the term $\rho_o \, \delta(z)$ in Eq. (1.5) leads to the notion of a yield surface; the function $f(\zeta)$ in (1.2b) leads to the notion of a yield-surface-expansion (isotropic hardening); and the function $\rho_1(z)$ in (1.5) leads to the notion of yield-surface translation (kinematic hardening). Use of (1.5) in (1.2) leads to:

$$\underset{\sim}{\sigma}' = 2\mu \, \rho_o \frac{d\underset{\sim}{\varepsilon}^p}{dz} + 2\mu \int_o^z \rho_1(z-z') \frac{d\underset{\sim}{\varepsilon}^p}{dz'} \, dz' \tag{1.6a}$$

$$\equiv \tau_y^o \frac{d\underset{\sim}{\varepsilon}^p}{dz} + \underset{\sim}{\alpha}'(z) \tag{1.6b}$$

wherein the definitions of τ_y^o and $\underset{\sim}{\alpha}'$ (the "back stress") are apparent. Eq. (1.6b) can be written as:

$$d\underset{\sim}{\varepsilon}^p = \frac{(\underset{\sim}{\sigma}' - \underset{\sim}{\alpha}')}{\tau_y^o f} \cdot d\zeta \ ; \qquad d\zeta \geq 0 \tag{1.7}$$

Of course, Eq. (1.7) is entirely reminiscent of the classical flow-rule and normality relation for plastic strain-rate using a Mises' yield criterion. However, at this point, this similarity is purely formal.

From the very definition of $d\zeta$ as in (1.1b), it follows that, <u>during plastic flow</u>,

$$\frac{d\underset{\sim}{\varepsilon}^p}{d\zeta} : \frac{d\underset{\sim}{\varepsilon}^p}{d\zeta} = 1 \ , \quad \text{i.e.,} \ (\underset{\sim}{\sigma}' - \underset{\sim}{\alpha}'):(\underset{\sim}{\sigma}' - \underset{\sim}{a}') = [\tau_y^o \ f(\zeta)]^2 \tag{1.8a,b}$$

Equation (1.8b) clearly indicates that during plastic flow, the stress point, in the deviatoric stress space, remains on a Mises-cylinder of radius $\tau_y^o \ f(\zeta)$, with the center of the surface at $\underset{\sim}{\alpha}'$.

By differentiating (1.7) with respect to ζ, one obtains the following relation which holds during plastic flow:

$$\left(\frac{d^2\underset{\sim}{\varepsilon}^p}{d\zeta^2} + \frac{d\underset{\sim}{\varepsilon}^p}{d\zeta}\frac{df}{d\zeta} \right) = \left(\frac{d\underset{\sim}{\sigma}'}{d\zeta} - \frac{d\underset{\sim}{\alpha}'}{d\zeta} \right) \frac{1}{\tau_y^o} \tag{1.9}$$

From (1.1), and the definition of $\underset{\sim}{\alpha}'$ as in (1.6), respectively, we see that:

$$\frac{d\underset{\sim}{\sigma}'}{d\zeta} = 2\mu \left(\frac{d\underset{\sim}{\varepsilon}'}{d\zeta} - \frac{d\underset{\sim}{\varepsilon}^p}{d\zeta}\right) \tag{1.10}$$

and

$$\frac{d\underset{\sim}{\alpha}'}{d\zeta} = 2\mu \left[\rho_1(o) \frac{d\underset{\sim}{\varepsilon}^p}{d\zeta} + \frac{\underset{\sim}{h}^*}{f}\right] \tag{1.11a}$$

$$\text{where } \underset{\sim}{h}^* = \int_o^z \frac{d\rho_1}{dz} (z-z') \frac{\partial \underset{\sim}{\varepsilon}^p}{\partial z'} dz' \tag{1.11b}$$

Use of (1.10) and (1.11a) in (1.9) results in:

$$d\underset{\sim}{\varepsilon}' = [1 + \rho_1(o) + \tau_y^o \frac{(df/d\zeta)}{2\mu}]d\underset{\sim}{\varepsilon}^p + \frac{\underset{\sim}{h}^* d\zeta}{f} + \frac{\tau_y^o f}{2\mu} \left(\frac{d^2\underset{\sim}{\varepsilon}^p}{d\zeta^2} \right) d\zeta \tag{1.12}$$

Also, during plastic flow, it follows from (1.8a) and (1.7) that:

$$\frac{d^2\underset{\sim}{\varepsilon}^p}{d\zeta^2} : \frac{d\underset{\sim}{\varepsilon}^p}{d\zeta} = 0 = \left(\frac{d^2\underset{\sim}{\varepsilon}^p}{d\zeta^2} \right) : \left(\frac{\underset{\sim}{\sigma}' - \underset{\sim}{\alpha}'}{\tau_y^o f} \right) \tag{1.13}$$

Taking the trace of both sides of (1.12) with $[(\underset{\sim}{\sigma}' - \underset{\sim}{\alpha}')/(\tau_y^o f)]$ [or which is also equal to $(d\underset{\sim}{\varepsilon}^p/d\zeta)$] and using (1.13), one obtains:

$$d\underset{\sim}{\varepsilon}' : \frac{(\underset{\sim}{\sigma}' - \underset{\sim}{\alpha}')}{\tau_y^o f} = \left[1 + \rho_1(o) + \frac{\tau_y^o (df/d\zeta)}{2\mu} + \frac{\underset{\sim}{h}^* : (\underset{\sim}{\sigma}' - \underset{\sim}{\alpha}')}{\tau_y^o f}\right] d\zeta \equiv C\, d\zeta \tag{1.14}$$

wherein the definition of C is apparent. Eq. (1.14) can be rewritten as:

$$d\zeta = \frac{1}{C}\left[\frac{(d\underset{\sim}{\varepsilon}'):(\underset{\sim}{\sigma}' - \underset{\sim}{\alpha}')}{\tau_y^o f} \right] \triangleq \frac{1}{C}\, d\underset{\sim}{\varepsilon}' : \underset{\sim}{N} \tag{1.15}$$

where $\underset{\sim}{N} = (\underset{\sim}{\sigma}' - \underset{\sim}{\alpha}')/\tau_y^o f$ is a unit "Normal".

Now, by definition, during a "Plastic Process", i.e. when $d\underset{\sim}{\varepsilon}^p \neq \underset{\sim}{0}$, we have $d\zeta > 0$. Thus, (1.15) clearly indicates:

(A) <u>Definition of a Plastic Process (P):</u> $d\zeta > 0$

$$\text{(P) if (i) } (\underset{\sim}{\sigma}' - \underset{\sim}{\alpha}'):(\underset{\sim}{\sigma}' - \underset{\sim}{\alpha}') = (\tau_y^o f)^2 \; ; \quad \underline{\text{and}} \quad d\underset{\sim}{\varepsilon}':\underset{\sim}{N} > 0 \tag{1.16}$$

Equation (1.15) also indicates that a "plastic process" is not possible if $\underset{\sim}{N}:d\underset{\sim}{\varepsilon}' \leq 0$. In confirmity with this, we define an "elastic process" as follows:

(B) <u>Definition of an Elastic Process (E):</u> $d\zeta = 0$

$$\text{(i) if } (\underset{\sim}{\sigma}' - \underset{\sim}{\alpha}'):(\underset{\sim}{\sigma}' - \underset{\sim}{\alpha}') < (\tau_y^o f)^2 \; ; \quad \text{or} \tag{1.17a}$$

$$\text{(ii) if } (\underset{\sim}{\sigma}' - \underset{\sim}{\alpha}'):(\underset{\sim}{\sigma}' - \underset{\sim}{\alpha}') = (\tau_y^o f)^2 \quad \underline{\text{and}} \quad \underset{\sim}{N}:d\underset{\sim}{\varepsilon}' \leq 0 \tag{1.17b}$$

It is interesting to observe that the (Elastic) and (Plastic) processes defined above, for the present endochronic theory, depend directly on whether $(\underset{\sim}{N}:d\underset{\sim}{\varepsilon}) \gtreqless 0$; while in the classical plasticity theory these processes depend, <u>ab initio</u>, on whether $(\underset{\sim}{N}:d\underset{\sim}{\sigma}) \gtreqless 0$. In computational mechanics, the central problem of plasticity is to determine $d\underset{\sim}{\sigma}$, given as $d\underset{\sim}{\varepsilon}$. In this context, the (E) and (P) criteria of (1.16) and (1.17) are more direct and more meaningful. Using (1.15) in (1.7), we obtain:

During (P):

$$d\underset{\sim}{\varepsilon}^p = \frac{1}{C}\,\underset{\sim}{N}(\underset{\sim}{N}:d\underset{\sim}{\varepsilon}') \equiv \frac{1}{C}\,\underset{\sim}{N}(\underset{\sim}{N}:d\underset{\sim}{\varepsilon}) \tag{1.18}$$

since $\underset{\sim}{N}$ is deviatoric. Recall that $\underset{\sim}{\alpha}'$ [see (1.6)] and $\underset{\sim}{h}^*$ [see (1.11b)], and through them, the coefficient C [see (1.14)] depend on the kernel $\rho_1(z)$.

A convenient choice for the kernel $\rho_1(z)$ is:

$$\rho_1(z) = \sum_i \rho_{1i}\,\exp(-\,\beta_i z) \tag{1.19}$$

such that, from (1.6) it follows that:

$$\underset{\sim}{\alpha}' = \sum_i 2\mu_o \int_o^z \left\{\rho_{1i}\,\exp[-\,\beta_i(z-z')]\,\frac{d\underset{\sim}{\varepsilon}^p}{dz'}\,dz'\right\} \equiv \sum_i \underset{\sim}{\alpha}'^{(i)} \tag{1.20a}$$

and

$$\underset{\sim}{h}^* = \sum_i \left(\frac{-\beta_i}{2\mu_o}\right)\underset{\sim}{\alpha}'^{(i)} \tag{1.20b}$$

with $\underset{\sim}{\alpha}'^{(i)}$ being defined in an apparent fashion. From (1.20) it follows:

$$d\underset{\sim}{\alpha}' = 2\mu_o\,\rho_1(o)\,d\underset{\sim}{\varepsilon}^p - \left\{\sum_i \frac{\beta_i\underset{\sim}{\alpha}'^{(i)}}{f}\right\}\left(d\underset{\sim}{\varepsilon}^p:d\underset{\sim}{\varepsilon}^p\right)^{1/2} \tag{1.21}$$

Thus, the evolution equation for $\underset{\sim}{\alpha}'$ is nonlinear in $d\underset{\sim}{\varepsilon}^p$ and thus is similar to a nonlinear-kinematic-hardening relation [12]. It has been discussed in detail by Watanabe and Atluri [4] that the present theory, with the translation of the yield surface as in (1.20), and the expansion of the yield surface as specified by:

$$f = (1 + \gamma\zeta) \qquad [\text{linear}]\,, \tag{1.22a}$$

or

$$f = a + (1 - a)\exp(-\psi\zeta) \qquad [\text{saturated}]\,, \tag{1.22b}$$

where γ and ψ are constants; and $\zeta = \int d\zeta$; includes the multiple-yield-surface theories of Mroz [7], Krieg [8], and Dafalias and Popov [9] as special cases.

Based on (1.18), the stress-strain relation in the present theory may be written as:

$$d\underset{\sim}{\sigma}' = 2\mu_o\left[d\underset{\sim}{\varepsilon}' - \Gamma\,\frac{1}{C}\,\underset{\sim}{N}\,(\underset{\sim}{N}:d\underset{\sim}{\varepsilon}')\right] \tag{1.23a}$$

$$(d\underset{\sim}{\sigma}:\underset{\sim}{I}) = (2\mu + 3\lambda)(d\underset{\sim}{\varepsilon}:\underset{\sim}{I})$$

where $\Gamma = 1$ in (P) and $\Gamma = 0$ in (E).

By assuming $\Gamma = 1$ or 0 appropriately, one may proceed to develop a tangent-stiffness finite-element method in the usual fashion. If the stress $\underset{\sim}{\sigma}_n$ at state C_n, in an incremental solution, is known, the incremental stresses $\Delta\underset{\sim}{\sigma}$ corresponding to the trial-solutions $\Delta\underset{\sim}{\varepsilon}$ for incremental strains are determined in the usual fashion. We assume that $\underset{\sim}{\sigma}_n$ is on the yield surface and further assume that the process had been plastic; i.e., $\|(\underset{\sim}{\sigma}'_n + 2\mu\Delta\underset{\sim}{\varepsilon}') - \underset{\sim}{\alpha}'_n\| > (\tau_y^o f_n)$. Then, for any θ such that $0 < \theta < 1$, the algorithm for determining the actual stress-increment $\Delta\underset{\sim}{\sigma}$ in the plastic process proceeds as follows:

$$\underset{\sim}{N}_\theta = \frac{(\underset{\sim}{\sigma}'_n + 2\mu\theta\Delta\underset{\sim}{\varepsilon}') - \underset{\sim}{\alpha}'_n}{\|(\underset{\sim}{\sigma}'_n + 2\mu\theta\Delta\underset{\sim}{\varepsilon}') - \underset{\sim}{\alpha}'_n\|} \tag{1.24a}$$

$$\Delta\underset{\sim}{\varepsilon}^p = \frac{1}{C_n}(\underset{\sim}{N}_\theta:\Delta\underset{\sim}{\varepsilon})\underset{\sim}{N}_\theta \;; \tag{1.24b}$$

$$\left\{\text{where } C_n = \left[1 + \rho_1(o) + \frac{\tau_y^o(df/d\zeta)}{2\mu_o} + \frac{\underset{\sim}{h}^*:(\underset{\sim}{\sigma}' - \underset{\sim}{\alpha}')}{\tau_y^o f}\right]_n\right\}$$

$$\Delta\underset{\sim}{\sigma}' = 2\mu\left[\Delta\underset{\sim}{\varepsilon}' - \frac{1}{C_n}(\underset{\sim}{N}_\theta:\Delta\underset{\sim}{\varepsilon})\underset{\sim}{N}_\theta\right] \tag{1.24c}$$

$$\Delta\underset{\sim}{\sigma}:\underset{\sim}{I} = (2\mu + 3\lambda)(\Delta\underset{\sim}{\varepsilon}:\underset{\sim}{I}) \tag{1.24d}$$

$$\Delta\zeta = (\Delta\underset{\sim}{\varepsilon}^p:\Delta\underset{\sim}{\varepsilon}^p)^{1/2} \;; \quad f_{n+1} = f_n + \Delta f \tag{1.24e}$$

$$\alpha'_{n+1} = \underset{\sim}{\alpha}_n + \left\{2\mu\,\rho_1(o)\Delta\underset{\sim}{\varepsilon}^p - \left(\sum_i \frac{\beta_i \alpha'^{(i)}_n}{f_n}\right)\Delta\zeta\right\} \tag{1.24f}$$

Of course, several variants of the above algorithm, such as sub-incremental ones, are possible. The above tangent-stiffness finite-element, and generalized mid-point-radial-return-stress-integration, algorithm has been used by Watanabe and Atluri [5,6] to solve several problems of cyclic plasticity and non-proportional biaxial loading. It has been found that the present models capture the experimentally observed phenomena of cyclic hardening, cross-hardening, ratcheting, etc.

Because of the superior predictive capabilities of the present model and the fact that it is no more difficult to implement than the usual (classical) plasticity models, it may be a candidate for further exploitation in general purpose computational programs.

II. Towards a Consistent Finite Deformation Plasticity Theory and Attendant Computations:

In dealing with finite deformation problems, it has been (and to a large extent still is) customary in the computational mechanics literature to postulate constitutive theories in the form of linear relations between an objective rate (usually the Jaumann rate) of stress or an objective rate of an internal variable (such as the back stress in a kinematic hardening plasticity theory) on the one hand, and (the objective) velocity strain (or the symmetric part of velocity gradient) on the other. In this section, we reexamine such approaches and point out several alternative paths towards rational and consistent theories and attendant computational algorithms for finite deformation elasto-plasticity.

First we examine hyperelasticity, and then elastoplasticity, with the aim that a finite-deformation elastoplasticity theory should remain valid in the limits of hyperelasticity as well as of small-deformation elastoplasticity.

II.1 Hyperelasticity

We consider here an isotropic hyperelastic solid wherein: $\underset{\sim}{F}$ is the deformation gradient with polar decomposition $\underset{\sim}{F} = \underset{\sim}{R}\cdot U_e = \underset{\sim}{V}_e\cdot\underset{\sim}{R}$, where $\underset{\sim}{U}_e$ and $\underset{\sim}{V}_e$ are the elastic stretches ($\underset{\sim}{U}_e \equiv \underset{\sim}{U}$ and $\underset{\sim}{V}_e \equiv \underset{\sim}{V}$ in this case); $\underset{\sim}{R}$ the rigid rotation; W_{oe} the elastic-strain-energy density per unit initial volume; $\underset{\sim}{\sigma}$ the Kirchhoff stress ($= J\underset{\sim}{\tau}$ where $J = \det \underset{\sim}{F}$, and $\underset{\sim}{\tau}$ the Cauchy stress). For observer transformations denoted by an orthogonal rotation $\underset{\sim}{Q}$, it is well known that the various quantities transform as: $\underset{\sim}{F} \to \underset{\sim}{Q}\cdot\underset{\sim}{F}$; $\underset{\sim}{R} \to \underset{\sim}{Q}\cdot\underset{\sim}{R}$; $\underset{\sim}{U}_e \to \underset{\sim}{U}_e$; $\underset{\sim}{V} \to \underset{\sim}{Q}\cdot\underset{\sim}{V}_e\cdot\underset{\sim}{Q}^t$; and $\underset{\sim}{\sigma} \to \underset{\sim}{Q}\cdot\underset{\sim}{\sigma}\cdot\underset{\sim}{Q}^t$. Based on these oberver-frame related transformations, one may write an objective constitutive relation for an isotropic hyperelastic solid in any number of alternative forms as follows:

$$\underset{\sim}{\sigma} = \underset{\sim}{R}\cdot J\underset{\sim}{\Gamma}(\underset{\sim}{U}_e)\cdot\underset{\sim}{R}^t = \underset{\sim}{F}\cdot\underset{\sim}{S}(\underset{\sim}{U}_e)\cdot\underset{\sim}{F}^t = \underset{\sim}{F}^{-t}\cdot\underset{\sim}{C}(\underset{\sim}{U}_e)\cdot\underset{\sim}{F}^{-1} = \underset{\sim}{F}^{-t}\cdot\underset{\sim}{M}(\underset{\sim}{U}_e)\cdot\underset{\sim}{F}^t =$$

$$= \underset{\sim}{F}\cdot\underset{\sim}{r}(\underset{\sim}{U}_e)\cdot\underset{\sim}{R}^t = \underset{\sim}{V}\cdot\underset{\sim}{T}(\underset{\sim}{V}_e) = \underset{\sim}{\sigma}(\underset{\sim}{V}_e) \tag{2.1}$$

where $\underset{\sim}{\Gamma}$ is the "rotated" stress, $\underset{\sim}{S}$ is the 2nd Piola-Kirchhoff stress, $\underset{\sim}{C}$ the "convected" stress, $\underset{\sim}{M}$ is another "induced" convected stress, $\underset{\sim}{r}$ the "Biot-Lure-Jaumann" stress. The physical interpretations of these tensors, as well as that of $\underset{\sim}{T}$, are given in Atluri [13]. The tensors $\underset{\sim}{\Gamma}$, $\underset{\sim}{S}$, $\underset{\sim}{C}$, $\underset{\sim}{M}$, and $\underset{\sim}{r}$ are functions of $\underset{\sim}{U}_e$ alone and, hence, are observer invariant, while $\underset{\sim}{T}$ is a function of $\underset{\sim}{V}_e$ and thus transforms as $\underset{\sim}{T} \rightarrow \underset{\sim}{Q}\cdot\underset{\sim}{T}\cdot\underset{\sim}{Q}^t$. From (2.1) it is easy to note the following equalities:

$$J\underset{\sim}{\Gamma}(\underset{\sim}{U}_e) = \underset{\sim}{U}_e\cdot S(\underset{\sim}{U}_e)\cdot\underset{\sim}{U}_e = \underset{\sim}{U}_e^{-1}\cdot\underset{\sim}{C}(\underset{\sim}{U}_e)\cdot\underset{\sim}{U}_e^{-1} = \underset{\sim}{U}_e^{-1}\cdot\underset{\sim}{M}(\underset{\sim}{U}_e)\cdot\underset{\sim}{U}_e = \underset{\sim}{U}_e\cdot\underset{\sim}{r}(\underset{\sim}{U}_e) \tag{2.2}$$

As shown in detail in Atluri [13], for isotropic elastic solids, the following relations exist between the various stresses and the elastic-strain-energy density W_{oe}:

$$\Gamma(\underset{\sim}{U}_e) = \frac{\partial W_{oe}}{\partial \ell n \underset{\sim}{U}_e} \;;\quad \underset{\sim}{S}(\underset{\sim}{U}_e) = \frac{\partial W_{oe}}{\partial(\underset{\sim}{U}_e^2)} \;;\quad \underset{\sim}{C}(\underset{\sim}{U}_e) = \frac{\partial W_{oe}}{\partial(\underset{\sim}{U}_e^{-2})} \;;\quad \underset{\sim}{r}(\underset{\sim}{U}_e) = \frac{\partial W_{oe}}{\partial \underset{\sim}{U}_e} \tag{2.3}$$

and $$\underset{\sim}{T}(\underset{\sim}{V}_e) = \frac{\partial W_{oe}}{\partial \underset{\sim}{V}_e} \;;\quad \underset{\sim}{\sigma}(\underset{\sim}{V}_e) = \frac{\partial W_{oe}}{\partial \ell n \underset{\sim}{V}_e} \tag{2.4}$$

where $\ell n\,(\_)$ denotes the natural logarithm of $(\_)$. For initially unstressed solids, we have the restrictions:

$$\underset{\sim}{\Gamma} = \underset{\sim}{S} = \underset{\sim}{C} = \underset{\sim}{r} = \underset{\sim}{0} \text{ when } \underset{\sim}{U}_e = \underset{\sim}{I} \;;\text{ and } \underset{\sim}{T} = \underset{\sim}{\sigma} = 0 \text{ when } \underset{\sim}{V}_e = \underset{\sim}{I}\,. \tag{2.5}$$

Based on (2.4) and (2.5), one may define a resticted class of <u>"semi-linear" isotropic</u> hyperelastic solid through the relations:

$$\underset{\sim}{\Gamma} = 2\mu\, \ell n \underset{\sim}{U}_e + \lambda[(\ell n \underset{\sim}{U}_e):\underset{\sim}{I}]\underset{\sim}{I} \tag{2.5a}$$

$$\underset{\sim}{S} = 2\mu(\underset{\sim}{U}_e^2 - \underset{\sim}{I}) + \lambda[(\underset{\sim}{U}_e^2 - \underset{\sim}{I}):\underset{\sim}{I}]\underset{\sim}{I} \tag{2.5b}$$

$$\underset{\sim}{C} = 2\mu(\underset{\sim}{U}_e^{-2} - \underset{\sim}{I}) + \lambda[(\underset{\sim}{U}_e^{-2} - \underset{\sim}{I}):\underset{\sim}{I}]\underset{\sim}{I} \tag{2.5c}$$

$$\underset{\sim}{r} = 2\mu(\underset{\sim}{U}_e - \underset{\sim}{I}) + \lambda[(\underset{\sim}{U}_e - \underset{\sim}{I}):\underset{\sim}{I}]\underset{\sim}{I} \tag{2.5d}$$

$$\underset{\sim}{T} = 2\mu(\underset{\sim}{V}_e - \underset{\sim}{I}) + \lambda[(\underset{\sim}{V}_e - \underset{\sim}{I}):\underset{\sim}{I}]\underset{\sim}{I} \tag{2.5e}$$

$$\underset{\sim}{\sigma} = 2\mu\, \ln \underset{\sim}{V}_e + \lambda[(\ln \underset{\sim}{V}_e):\underset{\sim}{I}]\underset{\sim}{I} \tag{2.5f}$$

wherein it is possible that the coefficients μ and λ may have <u>different</u> numerical values in each of Eqs. (2.5a-f).

Let $\underset{\sim}{L} = \dot{\underset{\sim}{F}}\cdot\underset{\sim}{F}^{-1}$ be the velocity gradient, $\underset{\sim}{D} = (\underset{\sim}{L})_s$ [$= \frac{1}{2}(\underset{\sim}{L} + \underset{\sim}{L}^t)$] be the velocity strain [$(\,_\sim)_s$ and $(\,_\sim)_a$ denote the symmetric and antisymmetric parts, respectively, of $(\,_\sim)$]; $\underset{\sim}{W} = (\underset{\sim}{L})_a$ [$= \frac{1}{2}(\underset{\sim}{L} - \underset{\sim}{L}^t)$] be the spin; and $\underset{\sim}{\Omega} = \dot{\underset{\sim}{R}}\cdot\underset{\sim}{R}^t$. Now, by differentiating Eqs (2.1) with respect to time, one easily obtains a set of objective rate relations:

$$\dot{\underset{\sim}{\sigma}} - \underset{\sim}{\Omega}\cdot\underset{\sim}{\sigma} + \underset{\sim}{\sigma}\cdot\underset{\sim}{\Omega} \equiv \dot{\underset{\sim}{\sigma}}_G = \underset{\sim}{R}\cdot\dot{\underset{\sim}{\Gamma}}\cdot\underset{\sim}{R}^t \tag{2.6a}$$

$$\dot{\underset{\sim}{\sigma}} - \underset{\sim}{W}\cdot\underset{\sim}{\sigma} + \underset{\sim}{\sigma}\cdot\underset{\sim}{W} \equiv \dot{\underset{\sim}{\sigma}}_J = \underset{\sim}{R}\cdot\{\dot{\underset{\sim}{\Gamma}} - \tfrac{1}{2}(\dot{\underset{\sim}{U}}_e\cdot\underset{\sim}{U}_e^{-1} - \underset{\sim}{U}_e^{-1}\cdot\dot{\underset{\sim}{U}}_e)\cdot\underset{\sim}{\Gamma}$$

$$- \underset{\sim}{\Gamma}\cdot\tfrac{1}{2}\cdot(\underset{\sim}{U}_e^{-1}\cdot\dot{\underset{\sim}{U}}_e - \dot{\underset{\sim}{U}}_e\cdot\underset{\sim}{U}_e^{-1})\}\cdot\underset{\sim}{R}^t \equiv \underset{\sim}{R}\cdot\{\dot{\underset{\sim}{\Gamma}} - \bar{\underset{\sim}{W}}\cdot\underset{\sim}{\Gamma} + \underset{\sim}{\Gamma}\cdot\bar{\underset{\sim}{W}}\}\cdot\underset{\sim}{R}^t \tag{2.6b}$$

$$\dot{\underset{\sim}{\sigma}} - \underset{\sim}{L}\cdot\underset{\sim}{\sigma} - \underset{\sim}{\sigma}\cdot\underset{\sim}{L}^t \equiv \dot{\underset{\sim}{\sigma}}_T = \underset{\sim}{F}\cdot\dot{\underset{\sim}{S}}\cdot\underset{\sim}{F}^t;\quad \dot{\underset{\sim}{\sigma}} + \underset{\sim}{L}^t\cdot\underset{\sim}{\sigma} + \underset{\sim}{\sigma}\cdot\underset{\sim}{L} = \dot{\underset{\sim}{\sigma}}_R = \underset{\sim}{F}^{-t}\cdot\dot{\underset{\sim}{C}}\cdot\underset{\sim}{F}^{-1} \tag{2.6c,d}$$

$$\dot{\underset{\sim}{\sigma}} + \underset{\sim}{L}^t\cdot\underset{\sim}{\sigma} - \underset{\sim}{\sigma}\cdot\underset{\sim}{L}^t = \underset{\sim}{F}^{-t}\cdot\dot{\underset{\sim}{M}}\cdot\underset{\sim}{F}^t;\quad \dot{\underset{\sim}{\sigma}} - \underset{\sim}{L}\cdot\underset{\sim}{\sigma} + \underset{\sim}{\sigma}\cdot\underset{\sim}{\Omega} = \dot{\underset{\sim}{\sigma}}_A = \underset{\sim}{F}\cdot\dot{\underset{\sim}{r}}\cdot\underset{\sim}{R}^t \tag{2.6e,f}$$

$$\dot{\underset{\sim}{\sigma}} = \dot{\underset{\sim}{V}}_e\cdot\underset{\sim}{T}(\underset{\sim}{V}_e) + \underset{\sim}{V}_e\cdot\dot{\underset{\sim}{T}}(\underset{\sim}{V}_e)\;;\quad \dot{\underset{\sim}{\sigma}} = 2\mu\frac{d}{dt}(\ln\underset{\sim}{V}_e) + \lambda[\frac{d}{dt}(\ln\underset{\sim}{V}_e):\underset{\sim}{I}]\underset{\sim}{I} \tag{2.7}$$

In (2.6b), $\bar{\underset{\sim}{W}}$ is the observer-invariant form of spin:

$$\bar{\underset{\sim}{W}} = \underset{\sim}{R}^t\cdot\underset{\sim}{W}\cdot\underset{\sim}{R} - \underset{\sim}{R}^t\cdot\dot{\underset{\sim}{R}} = \underset{\sim}{R}^t\cdot(\underset{\sim}{W} - \underset{\sim}{\Omega})\cdot\underset{\sim}{R} \tag{2.8}$$

Also, $\dot{\underset{\sim}{\sigma}}_G$ is the Green-Naghdi rate, $\dot{\underset{\sim}{\sigma}}_J$ the Jaumann rate, $\dot{\underset{\sim}{\sigma}}_T$ the Truesdell rate, $\dot{\underset{\sim}{\sigma}}_R$ the Rivlin rate, etc. However, using (2.2) and (2.1), <u>all</u> the rate equations (2.6a-f) may be written in a unified form as:

$$\underset{\sim}{R}^t\cdot\dot{\underset{\sim}{\sigma}}_G\cdot\underset{\sim}{R} \equiv \dot{\underset{\sim}{\Gamma}} \equiv \frac{d}{dt}(\underset{\sim}{U}_e\cdot\underset{\sim}{S}\cdot\underset{\sim}{U}_e) = \frac{d}{dt}(\underset{\sim}{U}_e^{-1}\cdot\underset{\sim}{C}\cdot\underset{\sim}{U}_e^{-1}) \equiv \frac{d}{dt}(\underset{\sim}{U}_e^{-1}\cdot\underset{\sim}{M}\cdot\underset{\sim}{U}_e) \equiv \frac{d}{dt}(\underset{\sim}{U}_e\cdot\underset{\sim}{r}) \tag{2.10}$$

where $\underset{\sim}{R}^t\cdot\dot{\underset{\sim}{\sigma}}_G\cdot\underset{\sim}{R}$ may be considered as the Green-Naghdi rate in the <u>rotated</u> frame. If the eigendirections of $\underset{\sim}{U}_e$ are represented by the matrix α and those of $\underset{\sim}{V}_e$ are represented by $\underset{\sim}{\beta}$, we have (see Atluri [13]):

$$\frac{d}{dt}(\ln \underset{\sim}{U}_e) = \underset{\sim}{\alpha}^t \cdot \frac{d}{dt}(\ln \underset{\sim}{\lambda}) \cdot \underset{\sim}{\alpha} + \dot{\underset{\sim}{\alpha}}^t \cdot \ln \underset{\sim}{\lambda} \cdot \underset{\sim}{\alpha} + \underset{\sim}{\alpha}^t \cdot \ln \underset{\sim}{\lambda} \cdot \dot{\underset{\sim}{\alpha}} \tag{2.11a}$$

$$\frac{d}{dt}(\ln \underset{\sim}{V}_e) = \underset{\sim}{\beta}^t \cdot \frac{d}{dt}(\ln \underset{\sim}{\mu}) \cdot \underset{\sim}{\beta} + \dot{\underset{\sim}{\beta}}^t \cdot \ln \underset{\sim}{\mu} \cdot \underset{\sim}{\beta} + \underset{\sim}{\beta}^t \cdot \ln \underset{\sim}{\mu} \cdot \dot{\underset{\sim}{\beta}} \tag{2.11b}$$

where $\underset{\sim}{\lambda}$ and $\underset{\sim}{\mu}$ are diagonal matrices consisting of the eigenvalues of $\underset{\sim}{U}_e$ and $\underset{\sim}{V}_e$ respectively, and $\frac{d}{dt} \ln \underset{\sim}{\lambda} = \underset{\sim}{\lambda}^{-1} \cdot \dot{\underset{\sim}{\lambda}} = \dot{\underset{\sim}{\lambda}} \cdot \underset{\sim}{\lambda}^{-1}$. Thus when (2.11a) and (2.11b) are used in (2.5a) and (2.5f), respectively, the expressions for $\dot{\underset{\sim}{\Gamma}}$ and $\dot{\underset{\sim}{\sigma}}$ become rather complicated. On the other hand, when the elastic stretches are <u>small</u>, i.e., $\underset{\sim}{U}_e = \underset{\sim}{I} + \underset{\sim}{\varepsilon}_U$ and $\underset{\sim}{V}_e = \underset{\sim}{I} + \underset{\sim}{\varepsilon}_V$ and where each of the components $(\varepsilon_{ij})_U$ and $(\varepsilon_{ij})_V$ is << 1, then

$$\ln \underset{\sim}{U}_e = \underset{\sim}{\varepsilon}_U - \frac{\underset{\sim}{\varepsilon}_U^2}{2} + \frac{\underset{\sim}{\varepsilon}_U^3}{3} + \ldots \; ; \quad \ln \underset{\sim}{V}_e = \underset{\sim}{\varepsilon}_V - \frac{\underset{\sim}{\varepsilon}_V^2}{2} + \frac{\underset{\sim}{\varepsilon}_V^3}{3} - \ldots \tag{2.12a,b}$$

Thus, for moderately <u>small</u> elastic stretches, (2.5a), (2.5f), and (2.12) lead to:

$$\dot{\underset{\sim}{\Gamma}} = 2\mu[\dot{\underset{\sim}{\varepsilon}}_U - \tfrac{1}{2}(\dot{\underset{\sim}{\varepsilon}}_U \cdot \underset{\sim}{\varepsilon}_U + \underset{\sim}{\varepsilon}_U \cdot \dot{\underset{\sim}{\varepsilon}}_U) + \tfrac{1}{3}(\dot{\underset{\sim}{\varepsilon}}_U \cdot \underset{\sim}{\varepsilon}_U^2 + \underset{\sim}{\varepsilon}_U^2 \cdot \dot{\underset{\sim}{\varepsilon}}_U + \underset{\sim}{\varepsilon}_U \cdot \dot{\underset{\sim}{\varepsilon}}_U \cdot \underset{\sim}{\varepsilon}_U)]$$

$$+ \lambda[\{\dot{\underset{\sim}{\varepsilon}}_U - \tfrac{1}{2}(\dot{\underset{\sim}{\varepsilon}}_U \cdot \underset{\sim}{\varepsilon}_U + \underset{\sim}{\varepsilon}_U \cdot \dot{\underset{\sim}{\varepsilon}}_U) + \ldots\} : \underset{\sim}{I}]\underset{\sim}{I} \tag{2.13a}$$

and a similar relation for $\dot{\underset{\sim}{\sigma}}$ with $\dot{\underset{\sim}{\varepsilon}}_U$ being replaced by $\dot{\underset{\sim}{\varepsilon}}_V$. For arbitrary $\underset{\sim}{U}_e$ and $\underset{\sim}{V}_e$, it appears more algebraically convenient to work with relations of the type (2.5b), (2.5d), and (2.5e). Considering, for instance, (2.5d) and (2.5e), one has from (2.10):

$$\underset{\sim}{R}^t \cdot \dot{\underset{\sim}{\sigma}}_G \cdot \underset{\sim}{R} = 2\mu(\dot{\underset{\sim}{U}}_e \cdot \underset{\sim}{U}_e + \underset{\sim}{U}_e \cdot \dot{\underset{\sim}{U}}_e) + \lambda[\dot{\underset{\sim}{U}}_e(\underset{\sim}{U}_e : \underset{\sim}{I}) + \underset{\sim}{U}_e(\dot{\underset{\sim}{U}}_e : \underset{\sim}{I})] - (2\mu + 3\lambda)\dot{\underset{\sim}{U}}_e \tag{2.14}$$

or, alternatively,

$$\dot{\underset{\sim}{\sigma}} = 2\mu(\dot{\underset{\sim}{V}}_e \cdot \underset{\sim}{V}_e + \underset{\sim}{V}_e \cdot \dot{\underset{\sim}{V}}_e) + \lambda[\dot{\underset{\sim}{V}}_e(\underset{\sim}{V}_e : \underset{\sim}{I}) + \underset{\sim}{V}_e(\dot{\underset{\sim}{V}}_e : \underset{\sim}{I})] - (2\mu + 3\lambda)\dot{\underset{\sim}{V}}_e \tag{2.15}$$

Now, consider the polar decomposition, $\underset{\sim}{F} = \underset{\sim}{R} \cdot \underset{\sim}{U}_e = \underset{\sim}{V}_e \cdot \underset{\sim}{R}$. From this we have $\underset{\sim}{L} = \dot{\underset{\sim}{F}} \cdot \underset{\sim}{F}^{-1} = \underset{\sim}{\Omega} + \underset{\sim}{R} \cdot \dot{\underset{\sim}{U}}_e \cdot \underset{\sim}{U}_e^{-1} \cdot \underset{\sim}{R}^t = \dot{\underset{\sim}{V}}_e \cdot \underset{\sim}{V}_e^{-1} + \underset{\sim}{V}_e \cdot \underset{\sim}{\Omega} \cdot \underset{\sim}{V}_e^{-1}$.

$$\text{Thus, } \underset{\sim}{D} = \underset{\sim}{R} \cdot (\dot{\underset{\sim}{U}}_e \cdot \underset{\sim}{U}_e^{-1})_s \cdot \underset{\sim}{R}^t = (\dot{\underset{\sim}{V}}_e \cdot \underset{\sim}{V}_e^{-1})_s + (\underset{\sim}{V}_e \cdot \underset{\sim}{\Omega} \cdot \underset{\sim}{V}_e^{-1})_s \tag{2.16}$$

$$\underset{\sim}{W} = \underset{\sim}{\Omega} + \underset{\sim}{R} \cdot (\dot{\underset{\sim}{U}}_e \cdot \underset{\sim}{U}_e^{-1})_a \cdot \underset{\sim}{R}^t = (\dot{\underset{\sim}{V}}_e \cdot \underset{\sim}{V}_e^{-1})_a + (\underset{\sim}{V}_e \cdot \underset{\sim}{\Omega} \cdot \underset{\sim}{V}_e^{-1})_a \tag{2.17}$$

By defining coordinate-invariant velocity-strain $\bar{\underset{\sim}{D}}$, and spin $\bar{\underset{\sim}{W}}$, such that:

$\bar{\underset{\sim}{W}} = R^t \cdot (\underset{\sim}{W} - \underset{\sim}{\Omega}) \cdot \underset{\sim}{R}$ and $\bar{\underset{\sim}{D}} = \underset{\sim}{R}^t \cdot \underset{\sim}{D} \cdot \underset{\sim}{R}$, it is easy to derive from (2.16 and 2.17) that:

$$\dot{\underset{\sim}{U}}_e = (\bar{\underset{\sim}{D}} + \bar{\underset{\sim}{W}}) \cdot \underset{\sim}{U}_e \equiv \underset{\sim}{U}_e \cdot (\bar{\underset{\sim}{D}} - \bar{\underset{\sim}{W}}) \tag{2.18}$$

and $$\dot{\underset{\sim}{V}}_e = (\underset{\sim}{D} + \underset{\sim}{W}) \cdot \underset{\sim}{V}_e - \underset{\sim}{V}_e \cdot \underset{\sim}{\Omega} \equiv \underset{\sim}{V}_e \cdot (\underset{\sim}{D} - \underset{\sim}{W}) + \underset{\sim}{\Omega} \cdot \underset{\sim}{V}_e \tag{2.19}$$

Use of (2.18) and (2.19) in (2.14) and (2.15) results in consistent forms of 'objective' rate-constitutive relations for hyperelastic solids. The resultant rate equations clearly indicate the fallacy in the current practice wherein a finite deformation (even hyper-elastic) stress-strain law is simply expressed as a linear relation between an objective rate of stress (usually $\overset{\circ}{\underset{\sim}{\sigma}}_J$) and $\underset{\sim}{D}$. For instance, it is currently common practice to write,

$$(\overset{\circ}{\underset{\sim}{\sigma}}_J) \text{ or } (\overset{\circ}{\underset{\sim}{\sigma}}_G) \equiv 2\lambda \underset{\sim}{D} + \lambda(\underset{\sim}{D}:\underset{\sim}{I})\underset{\sim}{I} \tag{2.20}$$

Equations (2.14 and 2.18) and (2.15 and 2.19) clearly show that (2.20) is not valid. In fact, as is well known, Eq. (2.20) with $\overset{\circ}{\underset{\sim}{\sigma}}_J$ leads to oscillatory shear stresses in a finite deformation simple shear test [14].

While the above hyperelastic rate relations have their own intrinsic reasons for being, our objective here is to use them as guides to postulate consistent rate-type finite-deformation elasto-plastic relations. This is pursued next.

II.2 Rate-Type Finite-Deformation Elasto-Plastic Stress-Strain Relations

The primary criteria we shall demand these relations to satisfy are: (1) The finite elasto-plastic rate relations should, in the limit, be valid in the cases of hyperelasticity as well as small-deformation elasto-plasticity; (2) They should obey the so-called Prager's [15] criterion. This criterion implies that when the objective stress rate vanishes [by virtue of the vanishing of the right-hand side of the equation with the said stress rate on the left-hand side], then the second invariant of the Kirchhoff stress ($\underset{\sim}{\sigma}$) should remain constant. Otherwise, in a J_2-flow theory of plasticity, spurious plastic flow may result; (3) The classical plasticity concepts, such as Drucker's postulates regarding plastic work and plastic normality, should hold.

We will assume that the solid is elastically isotropic. The main theme adopted now is that: the stress in a finitely deformed elasto-plastic solid may be derived from an elastic-strain-energy density function W_{oe}, such that:

$$\underset{\sim}{\Gamma}(\underset{\sim}{U}_e) = \frac{\partial W_{oe}}{\partial \ell n \underset{\sim}{U}_e} \; ; \quad \underset{\sim}{S}(\underset{\sim}{U}_e) = \frac{\partial W_{oe}}{\partial (\underset{\sim}{U}_e^2)} \; ; \quad \underset{\sim}{C}(\underset{\sim}{U}_e) = \frac{\partial W_{oe}}{\partial (\underset{\sim}{U}_e^{-2})} \; ;$$

$$\underset{\sim}{r}(\underset{\sim}{U}_e) = \frac{\partial W_{oe}}{\partial \underset{\sim}{U}_e} \; ; \quad \underset{\sim}{T}(\underset{\sim}{V}_e) = \frac{\partial W_{oe}}{\partial \underset{\sim}{V}_e} \; ; \quad \underset{\sim}{\sigma}(\underset{\sim}{V}_e) = \frac{\partial W_{oe}}{\partial \ell n \underset{\sim}{V}_e} \qquad (2.21)$$

i.e., in a manner entirely analogous to Eqs. (2.3 and 2.4). Thus, in a finitely deformed elastic-plastic solid, the 'objective' relations for the stress rate are still given by equations of the type Eqs. (2.14 and 2.18) or Eqs. (2.15 and 2.19), provided now $\dot{\underset{\sim}{U}}_e$ and $\dot{\underset{\sim}{V}}_e$ are appropriately defined in terms of the kinematics of an elastic-plastic deformation. As to this, Lee [16] has originally suggested:

$$\underset{\sim}{F} = \underset{\sim}{F}_e \cdot \underset{\sim}{F}_p \qquad (2.22)$$

with subscripts e and p denoting 'elastic' and 'plastic', respectively. However, it is well known that $\underset{\sim}{F}_e$ in (2.22) is not unique, but can be determined to only within a rigid rotation. Fardshisheh and Onat [17] have suggested that this non-uniqueness can be remedied by requiring $\underset{\sim}{F}_e$ to be a pure, symmetric stretch tensor, $\underset{\sim}{V}_e$. Thus,

$$\underset{\sim}{F} = \underset{\sim}{V}_e \cdot \underset{\sim}{F}_p \qquad (2.23)$$

By requiring $\underset{\sim}{V}_e$ to be similar to that in an otherwise hyperelastic solid, we see that under observer transformations, the following results hold: $\underset{\sim}{F} \rightarrow \underset{\sim}{Q} \cdot \underset{\sim}{F}$; $\underset{\sim}{V}_e \rightarrow \underset{\sim}{Q} \cdot \underset{\sim}{V}_e \cdot \underset{\sim}{Q}^t$; and $\underset{\sim}{F}_p \rightarrow \underset{\sim}{Q} \cdot \underset{\sim}{F}_p$. Note that in (2.23), the rigid-rotation of a material element is not explicitly determined. With the apparent loss of some generality, however, we study here the decompositions:

$$\underset{\sim}{F} = \underset{\sim}{R} \cdot \underset{\sim}{U}_e \cdot \underset{\sim}{U}_p \equiv \underset{\sim}{V}_e \cdot \underset{\sim}{V}_p \cdot \underset{\sim}{R} \qquad (2.24)$$

where $\underset{\sim}{R}$ is the rigid rotation, $\underset{\sim}{U}_e$ and $\underset{\sim}{V}_e$ are 'elastic' stretches, and $\underset{\sim}{U}_p$ and $\underset{\sim}{V}_p$ are the 'plastic' stretch tensors. We requre $\underset{\sim}{U}_e$, $\underset{\sim}{U}_p$, $\underset{\sim}{V}_e$, and $\underset{\sim}{V}_p$ to be symmetric. Under observer transformations it then follows that: $\underset{\sim}{U}_e \rightarrow \underset{\sim}{U}_e$;

$\underset{\sim}{U}_p \rightarrow \underset{\sim}{U}_p$; $\underset{\sim}{V}_e \rightarrow \underset{\sim}{Q} \cdot \underset{\sim}{V}_e \cdot \underset{\sim}{Q}^t$; $\underset{\sim}{V}_p \rightarrow \underset{\sim}{Q} \cdot \underset{\sim}{V}_p \cdot \underset{\sim}{Q}^t$; and also $\underset{\sim}{V}_e = \underset{\sim}{R} \cdot \underset{\sim}{U}_e \cdot \underset{\sim}{R}^t$; and $\underset{\sim}{V}_p = \underset{\sim}{R} \cdot \underset{\sim}{U}_p \cdot \underset{\sim}{R}^t$. Under the decomposition (2.23), we have:

$$\underset{\sim}{L} = \dot{\underset{\sim}{V}}_e \cdot \underset{\sim}{V}_e^{-1} + \underset{\sim}{V}_e \cdot \dot{\underset{\sim}{F}}_p \cdot \underset{\sim}{F}_p^{-1} \cdot \underset{\sim}{V}_e^{-1} \; ; \tag{2.25a}$$

$$\underset{\sim}{D} = \left(\dot{\underset{\sim}{V}}_e \cdot \underset{\sim}{V}_e^{-1}\right)_s + \left(\underset{\sim}{V}_e \cdot \dot{\underset{\sim}{F}}_p \cdot \underset{\sim}{F}_p^{-1} \cdot \underset{\sim}{V}_e^{-1}\right)_s \; ; \quad \underset{\sim}{W} = \left(\dot{\underset{\sim}{V}}_e \cdot \underset{\sim}{V}_e^{-1}\right)_a + \left(\underset{\sim}{V}_e \cdot \dot{\underset{\sim}{F}}_p \cdot \underset{\sim}{F}_p^{-1} \cdot \underset{\sim}{V}_e^{-1}\right)_a \tag{2.25b,c}$$

From (2.25) it easily follows that:

$$\dot{\underset{\sim}{V}}_e = (\underset{\sim}{D} + \underset{\sim}{W}) \cdot \underset{\sim}{V}_e - \underset{\sim}{V}_e \cdot \left(\dot{\underset{\sim}{F}}_p \cdot \underset{\sim}{F}_p^{-1}\right) \equiv \underset{\sim}{V}_e \cdot (\underset{\sim}{D} - \underset{\sim}{W}) - \left(\dot{\underset{\sim}{F}}_p \cdot \underset{\sim}{F}_p^{-1}\right)^t \cdot \underset{\sim}{V}_e \tag{2.26}$$

Now, the stress-power, or the rate of stress-work, in a finitely deformed elastic-plastic body (per unit initial volume) is given by:

$$\dot{W}_o = \underset{\sim}{\sigma} : \underset{\sim}{D} \equiv \dot{W}_{oe} + \dot{W}_{op} \tag{2.27a}$$

Thus, if $\underset{\sim}{D}$ is decomposed into $\underset{\sim}{D} = \underset{\sim}{D}_e + \underset{\sim}{D}_p$, we will have:

$$\dot{W}_{oe} = \underset{\sim}{\sigma} : \underset{\sim}{D}_e \; ; \quad \dot{W}_{op} = \underset{\sim}{\sigma} : \underset{\sim}{D}_p \tag{2.27b,c}$$

The well-known postulates of Drucker, concerning $\dot{W}_{op}$, may then be used to establish the normality of $\underset{\sim}{D}_p$ to the yield surface, expressed as a function of $\underset{\sim}{\sigma}$. This suggests the definitions:

$$\underset{\sim}{D}_p = \left(\underset{\sim}{V}_e \cdot \dot{\underset{\sim}{F}}_p \cdot \underset{\sim}{F}_p^{-1} \cdot \underset{\sim}{V}_e^{-1}\right)_s \; ; \quad \underset{\sim}{W}_p = \left(\underset{\sim}{V}_e \cdot \dot{\underset{\sim}{F}}_p \cdot \underset{\sim}{F}_p^{-1} \cdot \underset{\sim}{V}_e^{-1}\right)_a \tag{2.28}$$

where $\underset{\sim}{D}_p$ and $\underset{\sim}{W}_p$ are the "plastic velocity-strain" and "plastic spin", respectively. Using (2.28) in (2.26), we have:

$$\dot{\underset{\sim}{V}}_e = (\underset{\sim}{D} + \underset{\sim}{W}) \cdot \underset{\sim}{V}_e - \left(\underset{\sim}{D}_p + \underset{\sim}{W}_p\right) \cdot \underset{\sim}{V}_e \equiv \underset{\sim}{V}_e \cdot (\underset{\sim}{D} - \underset{\sim}{W}) - \underset{\sim}{V}_e \cdot \left(\underset{\sim}{D}_p - \underset{\sim}{W}_p\right) \tag{2.29}$$

$$= \left(\underset{\sim}{D}_e + \underset{\sim}{W}_e\right) \cdot \underset{\sim}{V}_e \equiv \underset{\sim}{V}_e \cdot \left(\underset{\sim}{D}_e - \underset{\sim}{W}_e\right) \tag{2.30}$$

The use of (2.29) in (2.15) will then result in an objective, elastic-plastic stress-strain rate relation for finite deformations. <u>We now need evolution equations for $\underset{\sim}{D}_p$ and $\underset{\sim}{W}_p$</u> as defined in Eq. (2.28), which may be written down, using the general isotropic function representation theorems, such as due to Wang [18]. In this connection, the work of Loret [19] in representing directly the quantity $(\dot{\underset{\sim}{F}}_p \cdot \underset{\sim}{F}_p^{-1})$ is useful and noteworthy. From this single representation, Loret [19] determines both $\underset{\sim}{D}_p$ and $\underset{\sim}{W}_p$.

However, there appears a slight inconsistency in the above approach. If the process is purely elastic (and thus the material behaviour approaches hyperelasticity) or when there is no plastic process as determined by a flow rule, it follows that $\underset{\sim}{D}_p = 0$. This necessitates $(\dot{\underset{\sim}{F}}_p \cdot \underset{\sim}{F}_p^{-1})$ to be zero, in an elastic process. Thus, in an elastic process, (2.29) reduces to:

$$\dot{\underset{\sim}{V}}_e = (\underset{\sim}{D} + \underset{\sim}{W}) \cdot \underset{\sim}{V}_e \tag{2.31}$$

which does not agree with the hyperelastic relation, (2.19), i.e., $\dot{\underset{\sim}{V}}_e = (\underset{\sim}{D} + \underset{\sim}{W}) \cdot \underset{\sim}{V}_e - \underset{\sim}{V}_e \cdot \underset{\sim}{\Omega}$.

On the other hand, using the decomposition (2.24), we have:

$$\underset{\sim}{L} = \dot{\underset{\sim}{R}} \cdot \underset{\sim}{R}^t + \underset{\sim}{R} \cdot \dot{\underset{\sim}{U}}_e \cdot \underset{\sim}{U}_e^{-1} \cdot \underset{\sim}{R}^t + \underset{\sim}{R} \cdot \underset{\sim}{U}_e \cdot \dot{\underset{\sim}{U}}_p \cdot \underset{\sim}{U}_p^{-1} \cdot \underset{\sim}{U}_e^{-1} \cdot \underset{\sim}{R}^t \tag{2.32}$$

or $$\underset{\sim}{D} = \underset{\sim}{R} \cdot (\dot{\underset{\sim}{U}}_e \cdot \underset{\sim}{U}_e^{-1})_s \cdot \underset{\sim}{R}^t + \underset{\sim}{R} \cdot (\underset{\sim}{U}_e \cdot \dot{\underset{\sim}{U}}_p \cdot \underset{\sim}{U}_p^{-1} \cdot \underset{\sim}{U}_e^{-1})_s \cdot \underset{\sim}{R}^t \tag{2.33}$$

and $$\underset{\sim}{W} = \dot{\underset{\sim}{R}} \cdot \underset{\sim}{R}^t + \underset{\sim}{R} \cdot (\dot{\underset{\sim}{U}}_e \cdot \underset{\sim}{U}_e^{-1})_a \cdot \underset{\sim}{R}^t + \underset{\sim}{R} \cdot (\underset{\sim}{U}_e \cdot \dot{\underset{\sim}{U}}_p \cdot \underset{\sim}{U}_p^{-1} \cdot \underset{\sim}{U}_e^{-1})_a \cdot \underset{\sim}{R}^t \tag{2.34}$$

Equations (2.33} and (2.34) may be rearranged as:

$$\bar{\underset{\sim}{D}} = \underset{\sim}{R}^t \cdot \underset{\sim}{D} \cdot \underset{\sim}{R} = (\dot{\underset{\sim}{U}}_e \cdot \underset{\sim}{U}_e^{-1})_s + (\underset{\sim}{U}_e \cdot \dot{\underset{\sim}{U}}_p \cdot \underset{\sim}{U}_p^{-1} \cdot \underset{\sim}{U}_e^{-1})_s \tag{2.35}$$

$$\bar{\underset{\sim}{W}} = \underset{\sim}{R}^t \cdot (\underset{\sim}{W} - \underset{\sim}{\Omega}) \cdot R = (\dot{\underset{\sim}{U}}_e \cdot \underset{\sim}{U}_e^{-1})_a + (\underset{\sim}{U}_e \cdot \dot{\underset{\sim}{U}}_p \cdot \underset{\sim}{U}_p^{-1} \cdot \underset{\sim}{U}_e^{-1})_a \tag{2.36}$$

Now, the stress-power per unit initial volume may be written as:

$$\dot{W}_o = \underset{\sim}{\sigma} : \underset{\sim}{D} \equiv (J\, \underset{\sim}{R} \cdot \underset{\sim}{\Gamma} \cdot \underset{\sim}{R}^t) : \underset{\sim}{D} \equiv J \underset{\sim}{\Gamma} : (\underset{\sim}{R}^t \cdot \underset{\sim}{D} \cdot \underset{\sim}{R}) = J \underset{\sim}{\Gamma} : \bar{\underset{\sim}{D}} \tag{2.37}$$

Thus, if $\bar{\underset{\sim}{D}}$ is decomposed into $\bar{\underset{\sim}{D}} = \bar{\underset{\sim}{D}}_e + \bar{\underset{\sim}{D}}_p$, we have $\dot{W}_{oe} = J\underset{\sim}{\Gamma} : \bar{\underset{\sim}{D}}_e$; $\dot{W}_{op} = J\underset{\sim}{\Gamma} : \bar{\underset{\sim}{D}}_p$. Now, the yield function may be expressed as:

$$f(\underset{\sim}{\Gamma}, W_{op}) = 0 \tag{2.38}$$

since the invariants of $J\underset{\sim}{\Gamma}$ are the same as those of $\underset{\sim}{\sigma}$. The Drucker-normality would then apply to $\bar{\underset{\sim}{D}}_p$ in the deviatoric space of $\underset{\sim}{\Gamma}$. Thus, we may define:

$$\bar{\underset{\sim}{D}}_p = (\underset{\sim}{U}_e \cdot \dot{\underset{\sim}{U}}_p \cdot \underset{\sim}{U}_p^{-1} \cdot \underset{\sim}{U}_e^{-1})_s \;;\; \bar{\underset{\sim}{W}}_p = (\underset{\sim}{U}_e \cdot \dot{\underset{\sim}{U}}_p \cdot \underset{\sim}{U}_p^{-1} \cdot \underset{\sim}{U}_e^{-1})_a \tag{2.39}$$

Using (2.39), (2.35), and (2.36), we have:

$$\dot{\underset{\sim}{U}}_e = (\bar{\underset{\sim}{D}} + \bar{\underset{\sim}{W}})\cdot\underset{\sim}{U}_e - (\bar{\underset{\sim}{D}}_p + \bar{\underset{\sim}{W}}_p)\cdot\underset{\sim}{U}_e \equiv \underset{\sim}{U}_e\cdot(\bar{\underset{\sim}{D}} - \bar{\underset{\sim}{W}}) - \underset{\sim}{U}_e\cdot(\bar{\underset{\sim}{D}}_p - \bar{\underset{\sim}{W}}_p) \qquad (2.40)$$

$$\equiv (\bar{\underset{\sim}{D}}_e + \bar{\underset{\sim}{W}}_e)\cdot\underset{\sim}{U}_e \equiv \underset{\sim}{U}_e\cdot(\bar{\underset{\sim}{D}}_e - \bar{\underset{\sim}{W}}_e) \qquad (2.41)$$

the use of (2.40) in (2.14) results in an objective elastic-plastic stress-strain relation that is consistant for finite deformations. We may use representation theorems [18] to write down a general expression for $(\dot{\underset{\sim}{U}}_p\cdot\underset{\sim}{U}_p^{-1})$ from which both $\bar{\underset{\sim}{D}}_p$ and $\bar{\underset{\sim}{W}}_p$ can be determined from (2.39), such that $\bar{\underset{\sim}{D}}_p$ satisfies the normality condition. We may essentially follow Loret [19].

The development in (2.40), in conjunction with (2.14), will remain consistent in the limit of hyperelasticity. Thus, when $U_p \equiv I$, $\dot{\underset{\sim}{U}}_p \equiv 0$, $\bar{\underset{\sim}{D}}_p = \underset{\sim}{0} = \bar{\underset{\sim}{W}}_p$, and thus $\dot{\underset{\sim}{U}}_e \equiv (\bar{\underset{\sim}{D}} + \bar{\underset{\sim}{W}})\cdot\underset{\sim}{U}_e$, which agrees with the hyperelastic relation, (2.18).

Also, the present suggestion of an elastic-plastic constitutive relation based on (2.40) and (2.14) does satisfy the Prager condition [15]. Thus, when $\underset{\sim}{R}^t\cdot\dot{\underset{\sim}{\sigma}}_G\cdot\underset{\sim}{R} = 0$, it is clear that since the rate of stress in a rigidly rotating system is zero, the invariants of the Kirchhoff tensor remain constant. This is not the case when other stress-rate equations as in (2.6c,d,e,f) [i.e., rates in non-rigidly-spinning systems] are used and when the right-hand sides of (2.6c,d,e, and f) are consistently determined (through time differentiation) from (2.5b,c,d), respectively. If, on the other hand, one uses ad-hoc postulations, as is commonly done currently, with only a linear relation between an objective stress rate and $\underset{\sim}{D}$, such as:

$$(\dot{\underset{\sim}{\sigma}}_T) \text{ or } (\dot{\underset{\sim}{\sigma}}_R) \equiv 2\mu[\underset{\sim}{D} - \Lambda(\underset{\sim}{N}:\underset{\sim}{D})\underset{\sim}{N}] \qquad (2.42)$$

then, $\dot{\underset{\sim}{\sigma}}_T$ or $\dot{\underset{\sim}{\sigma}}_R = 0$ implies that $\underset{\sim}{D} \equiv 0$, and then $\underset{\sim}{L} \equiv \underset{\sim}{W}$, and hence $\dot{\underset{\sim}{\sigma}}_T$ and $\dot{\underset{\sim}{\sigma}}_R$ reduce to $\dot{\underset{\sim}{\sigma}}_J$. Thus, since $\dot{\underset{\sim}{\sigma}}_J$ is a spin-based rate, when $\dot{\underset{\sim}{\sigma}}_T$ or $\dot{\underset{\sim}{\sigma}}_R$ is zero in (2.42), the invariants of $\underset{\sim}{\sigma}$ themselves remain constant, and hence Prager's condition is obeyed by the inconsistent relation (2.42).

To close the description of other plasticity features, such as kinematic hardening, we may introduce a Kirchhoff-stress-like internal variable $\underset{\sim}{\alpha}$, which we may label as the finite deformation back-stress. If

$$\overset{\bullet}{\underset{\sim}{\alpha}}_G = \overset{\bullet}{\underset{\sim}{\alpha}} - \underset{\sim}{\Omega}\cdot\underset{\sim}{\alpha} + \underset{\sim}{\alpha}\cdot\underset{\sim}{\Omega} \, , \tag{2.43}$$

one may introduce an isotropic tensor function representation

$$\overset{\bullet}{\underset{\sim}{\alpha}}_G = \underset{\sim}{f}(\underset{\sim}{\alpha}, \underset{\sim}{\sigma}, \underset{\sim}{D}) \tag{2.44}$$

Such general representations, and specific examples, were considered by Reed and Atluri [22],who attempted, successfully, to model the results of finite-torsion experiments. For instance, Reed and Atluri [21] demonstrated that an excellent agreement with the experimental results of Swift [22] for stresses and strains in a finite-torsion test, can be obtained from the simple model:

$$\overset{\bullet}{\underset{\sim}{\alpha}}{}'_J = C_1 \, \underset{\sim}{D}^p + C_2 \, \underset{\sim}{\alpha}' \tag{2.45}$$

where C_1 = constant ; $C_2 = C_2(\underset{\sim}{\alpha}' : \underset{\sim}{D}^p)$,

in conjunction with

$$\overset{\bullet}{\underset{\sim}{\sigma}}{}'_J = A_1 \, \underset{\sim}{D} + A_2 \, \underset{\sim}{\sigma}' \tag{2.46}$$

where A_1 = constant ; $A_2 = A_2(\underset{\sim}{\sigma}' : \underset{\sim}{D})$.

Another "nonlinear-kinematic-hardening" model, which is a special case of the representation in (2.44), and a generalization of (1.21), may be generated by writing:

$$\overset{\bullet}{\underset{\sim}{\alpha}}{}'_G = C \, \underset{\sim}{D}_p - \beta_i \, \underset{\sim}{\alpha}' (\underset{\sim}{D}_p : \underset{\sim}{D}_p)^{1/2} \tag{2.47}$$

The primary reason for using such models as in (2.45) and (2.47) is that simple linear models of the type

$$\overset{\bullet}{\underset{\sim}{\alpha}}{}'_J = C_1 \, \underset{\sim}{D}^p \tag{2.48}$$

can easily be shown to lead (see Atluri [14]) to oscillatory values for the components of $\underset{\sim}{\alpha}$ in cases such as finite simple shear. If this linearity is insisted upon, non-oscillatory back stress $\underset{\sim}{\alpha}$ may be generated by replacing the left-had side of (2.48) with other objective rates such as $\overset{\bullet}{\underset{\sim}{\alpha}}{}'_G$, $\overset{\bullet}{\underset{\sim}{\alpha}}{}'_T$, or $\overset{\bullet}{\underset{\sim}{\alpha}}{}'_C$, as shown by Atluri [14], or by creating other stress rates such as based on the spin of material fibers aligned with the principal directions of $\underset{\sim}{\alpha}$, as done by Lee [23]. While the use of Green-Naghdi's, Truesdell's, or Lee's, or some other's rate on the left-hand side of (2.48) as well as on the left-hand side of the evolution equation for stress [viz., objective stress rate $\equiv 2\mu \underset{\sim}{D}^e + \lambda(\underset{\sim}{D}^e : \underset{\sim}{I})\underset{\sim}{I}$] in finite-deformation, kinematic-hardening, elasto-plasticity, may lead to a non-oscillatory shear stress in a simple shear test, the physically unrealistic direct stresses produced in such a test (as well as the unrealistic axial strains produced in a finite-torsion test), by these stress rates, are discussed in detail in Reed and Atluri [21]. Also, since modeling material behaviour under cyclic loading necessitates the use of nonlinear kinematic hardening models as discussed in Section I of this paper, it may be worthwhile to circumvent the arbitrariness surrounding the left-hand side of (2.48) [while retaining only the linear term in $\underset{\sim}{D}^p$ as in (2.48)] and concentrate instead on the right-hand side of (2.48) and consider more general representations involving $\underset{\sim}{D}^p$, $\underset{\sim}{\alpha}$, $\underset{\sim}{\sigma}$ as in the right-hand sides of (2.44) and (2.47). This type of modeling the evolution of the back stress $\underset{\sim}{\alpha}$, along with the use of Eqs. (2.40) and (2.14) for representing the evolution of Kirchhoff stress $\underset{\sim}{\sigma}$, may form a consistent basis for a finiteelasto-plastic theory. This remains to be verified through systematic computational modeling of available experimental data on finite plasticity.

Acknowledgements

The results presented herein were obtained during the course of research supported by the U.S. Office of Naval Research and the U.S. National Science Foundation. This support , and the encouragement of Drs. Alan Kushner, Yapa Rajapakse, and Clifford J. Astill, are gratefully acknowledged. Thanks are expressed to Ms. Joyce Webb for her assistance in the preparation of this manuscript.

References

1. Valanis, K.C.: Fundamental consequence of a new intrinsic time measure-plasticity as a limit of endochronic theory. Arch. Mech. 31 (1980) 171.

2. Valanis, K.C. and Fan, J.: Endochronic analysis of cyclic elasto-plastic strain fields in a notched plate. Jrl. Appl. Mech. 50. 4a (1983) 789.

3. Watanabe, O. and Atluri, S.N.: Constitutive modeling of cyclic plasticity and creep, using an internal time concept. Int. Jrl. Plast. (1985) (in press).

4. Watanabe, O. and Atluri, S.N.: Internal time, general internal variable, and multi-yield-surface theories of plasticity and creep: a unification of concepts. Int. Jrl. Plast. (1985) (in press).

5. Watanabe, O. and Atluri, S.N.: A new endochronic approach to computational elastoplasticity: example of a cyclically loaded cracked plate. Jrl. of Appl. Mech. (1985) (in press).

6. Watanabe, O. and Atluri, S.N.: Endochronic approach to computational plasticity: non-proportional multiaxial cyclic loading. (in preparation).

7. Mroz, Z.: An attempt to describe the behavior of metals under cyclic loads using a more general workhardening model. Acta Mech. 7 (1969) 199.

8. Krieg, R.D.: A practical two surface plasticity theory. Jrl. Appl. Mech. 42 (1975) 641.

9. Dafalias, Y.F. and Popov, E.P.: Plastic internal variable formalism of cyclic plasticity. Jrl. Appl. Mech. 43 (1976) 645.

12. Chaboche, J.L. and Rousselier, G.: On the plastic and viscoplastic constitutive equations, part I: rules developed with internal variable concept. In Inelastic analysis and life prediction in elevated temperature design. ASME-PVP 59 (1982) 33.

13. Atluri, S.N.: Alternate stress and conjugate strain measures, and mixed variational formulations involving rigid rotations, for computational analyses of finitely deformed solids, with application to plates and shells-I theory. Comp. & Struct. 18. 1 (1983) 93.

14. Atluri, S.N.: On constitutive relations at finite strain: hypo-elasticity and elasto-plasticity with isotropic or kinematic hardening. Comp. Meth. Appl. Mech. & Engg. 43 (1984) 137.

15. Prager, W.: An elementary discussion of definitions of stress rate. Quart. Appl. Mech. 18 (1961) 403.

16. Lee, E.H.: Elastic-plastic deformations at finite strains. Jrl. Appl. Mech. 37 (1969) 1.

17. Fardshisheh, F. and Onat, E.T.: Representation of elastoplastic behaviour by means of state variables. In Problems of plasticity. A. Sawezuk (ed.) (1974) 89.

18. Wang. C.C.: A new representation theorem for isotropic functions. Parts 1 and 2. Arch. Rat. Mech. Analysis. 36 (1970) 166.

19. Loret, B.: On the effects of plastic rotation in the finite deformation of anisotropic elastic-plastic solids. Mech. of Mat. 2 (1983) 287.

21. Reed, K.W. and Atluri, S.N.: "Constitutive modeling and computational implementation for finite strain plasticity. Int. Jrl. Plast. 1. 1 (1985) 63.

22. Swift. H.W.: Length changes in metals under torsional overstrain. Engineering. 163 (1947) 253.

23. Lee, E.H., Mallett, R.L., and Wertheimer, T.B.: Stress analysis for anisotropic hardening in finite-deformation plasticity. Jrl. Appl. Mech. 50 (1983) 554.

Elastoplastic Material Laws with Linear Yield Conditions

E. ANDERHEGGEN, H. ELMER
Swiss Federal Institute of Technology, Zürich, Switzerland

Summary

Nonlinear elasto-plastic material constitutive laws can be described by sets of linear yield conditions which define polyhedric flow figures in the stress space. Two different methods based on this idea are presented and shortly discussed. It is felt that this approach can be useful specially when generalised stress components, e.g. bending moments and stretching forces in a shell element, are used.

Introduction

When modelling elasto-plastic material behaviour in nonlinear finite element analysis the following problem arises: given (1) the current strain state $\{\varepsilon\}$, (2) a finite strain increment $\{\Delta\varepsilon\}$ and (3) appropriate information both on material properties and stress-strain history at a given integration point, find (1) the stress increment $\{\Delta\sigma\}$ due to $\{\Delta\varepsilon\}$ and (2) the tangent elasticity matrix $[D_T]$ defined by

$$d\{\sigma\} = [D_T]\, d\{\varepsilon\} \tag{1}$$

which relates the infinitesimal stress increment $d\{\sigma\}$ in the strain state $\{\varepsilon\}+\{\Delta\varepsilon\}$ with the infinitesimal strain increment $d\{\varepsilon\}$ in the direction of $\{\Delta\varepsilon\}$.

According to the theory of plasticity the stresses $\{\sigma\}$ are subjected to the yield condition

$$0 \leq R - F(\{\sigma\}) \tag{2}$$

where F is a given positive-definite function of the N components σ_1 to σ_N of the stress vector $\{\sigma\}$ and R is the corresponding plastic resistance, i.e. a material constant (for simplicity, hardening shall not be

Finite Element Methods for Nonlinear Problems
Europe-US Symposium, Trondheim, Norway 1985
Editors: Bergan, Bathe, Wunderlich

considered). In the N-dimensional stress space the yield condition (2) defines the flow figure which delimits the allowable stress domain. The flow figure is always convex. In some cases it extends to infinity (e.g. in the direction of hydrostatic pressure).

A prescribed infinitesimal strain increment $d\{\varepsilon\}$ can be subdivided into an elastic part $d\{\varepsilon_e\}$ and a plastic part $d\{\varepsilon_p\}$.

$$d\{\varepsilon\} = d\{\varepsilon_e\} + d\{\varepsilon_p\} \tag{3}$$

For the elastic part Hooke's law is valid:

$$d\{\sigma\} = [D]\ d\{\varepsilon\} \qquad \text{resp.} \qquad d\{\varepsilon_e\} = [D]^{-1} d\{\sigma\} \tag{4}$$

where [D] represents the positive-definite elasticity matrix. According to the flow rule of the plasticity theory the plastic strain increment $d\{\varepsilon_p\}$ is given by the relation

$$d\varepsilon_{pn} = \frac{\delta F}{\delta \sigma_n} d\lambda \qquad \text{with} \qquad \begin{cases} d\lambda = 0 \text{ if } F(\{\sigma\}) < R \\ d\lambda \geq 0 \text{ if } F(\{\sigma\}) = R \end{cases} \tag{5}$$

valid for each of the N coefficients $d\varepsilon_{p1}$ to $d\varepsilon_{pN}$ of the plastic strain increment vector $d\{\varepsilon_p\}$. $d\lambda$ is a nonnegative plastic flow parameter.

In many practical applications $F(\{\sigma\})$ is either a single nonlinear function (e.g. according to the Von Mises yield condition) or a set of nonlinear functions of the stress components σ_1 to σ_N (e.g. according to the Tresca yield condition). In the following we suggest to describe elasto-plastic material behaviour by a set of linear yield conditions which, in the N-dimensional stress space, define a polyhedric flow figure.

This approach is not new (see references). However, we believe that it has not received the attention it deserves. In the following three sections, the fundamentals of the algorithm are presented. For clarity, many details concerning both the practical implementation of the method and the treatment of special cases are left out. The applicability of this method to different problems is commented on in the last section.

Linear yield conditions: the face method and the vertex method

In the N-dimensional stress space a polyhedric yield figure can be defined either by the linear equations of its faces or by the coordinates of its vertices. A simple flow figure with 5 faces and 5 vertices in a 2-dimensional stress space is shown below.

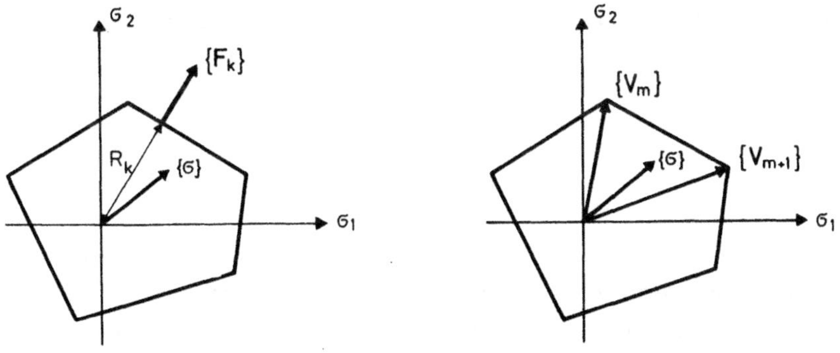

If the polyhedric flow figure is to be defined by the linear equations of its faces (this procedure will be called the "face method"), then the stresses $\{\sigma\}$ must satisfy the following K linear yield conditions

$$0 \leq S_k = R_k - \{F_k\}^T\{\sigma\} \qquad (k = 1 \text{ to } K) \tag{6}$$

Introducing the NxK face matrix [F] and the vectors $\{S\}$ and $\{R\}$ the following equation can be written

$$\{S\} = \{R\} - [F]^T\{\sigma\} \qquad \text{with} \qquad \{S\} \geq 0 \tag{7}$$

where the vectors $\{F_k\}$ building the columns of the matrix [F] are perpendicular to the faces of the flow figure and point towards the outside of the allowable stress domain. The positive material constants $R_k > 0$ of the vector $\{R\}$ are the corresponding plastic resistances. The nonnegative slack variables $S_k \geq 0$ of the vector $\{S\}$ represent the residual plastic resistances for a stress state $\{\sigma\}$.

If the polyhedric flow figure is to be defined by the coordinates of its vertices (this approach will be called the "vertex method") the stresses $\{\sigma\}$ satisfy the yield conditions if

$$\{\sigma\} = \sum_{m=1}^{M} \{V_m\} S_m \qquad \text{with} \quad \begin{cases} S_m \geq 0 \\ \Sigma S_m \leq 1 \end{cases} \tag{8}$$

Introducing the NxM vertex matrix [V], the vector {S} with M components, the summation vector {U} whose M components are all equal to unity and the nonnegative slack variable S' eq. (8) can be written as follows:

$$\{\sigma\} = [V]\,\{S\} \qquad \text{with} \quad \begin{cases} \{S\} \geq 0 \\ S' = 1 - \{U\}^T\{S\} > 0 \end{cases} \tag{9}$$

The vectors $\{V_m\}$ building the columns of the vertex matrix [V] define the M vertices of the polyhedric flow figure. The nonnegative slack variables S_1 to S_M of the vector {S} represent generalised stress parameters defining a coordinate transformation, i.e. a new stress space. In this space the yield conditions (6) are transformed to simple nonnegative conditions for the newly introduced stress parameters S_1 to S_M and for the additional slack variable S' defined in (9). (This procedure is related to the so-called decomposition method of linear programming.)

For a given stress state it is easy to see that if the inequalities (6) are satisfied, then (8) will also be satisfied and vice-versa.

Flow rules for both methods

The flow rule valid for the face method and associated with the yield condition (6) is

$$d\{\varepsilon_p\} = \sum_{k=1}^{K} \{F_k\}\, d\lambda_k = [F]\, d\{\lambda\} \qquad \text{with} \quad \begin{cases} d\lambda_k = 0 \text{ if } S_k > 0 \\ d\lambda_k \geq 0 \text{ if } S_k = 0 \end{cases} \tag{10}$$

The components λ_1 to λ_K of the vector $\{\lambda\}$ are nonnegative plastic flow parameters associated with the slack variables S_1 to S_K. From (3), (4) and (7) the following incremental relations are obtained

$$d\{\varepsilon\} = [D]^{-1}\, d\{\sigma\} + [F]\, d\{\lambda\} \tag{11}$$

$$d\{S\} = -[F]\, d\{\sigma\} \tag{12}$$

leading to

$$d\{\sigma\} = [D]\, d\{\varepsilon\} - [D]\,[F]\, d\{\lambda\} \tag{13}$$

$$d\{S\} = -[F]^T[D]\, d\{\varepsilon\} + [F]^T[D]\,[F]\, d\{\lambda\} \tag{14}$$

For the vertex method the strain parameters η_1 to η_M of the vector $\{\eta\}$ associated with the stress parameters S_1 to S_M have to be introduced. Due to the work invariance condition

$$\{\sigma\}^T\{\varepsilon\} = \{S\}^T\{\eta\} \tag{15}$$

and considering (9), the relation

$$\{\eta\} = [V]^T\{\varepsilon\} \tag{16}$$

is obtained,or, using Hooke's law and subdividing $\{\eta\}$ in an elastic part $\{\eta_e\}$ and a plastic part $\{\eta_p\}$:

$$\{\eta\} = \{\eta_e\} + \{\eta_p\} = [V]^T\{\varepsilon\} = [V]^T[D]^{-1}\{\sigma\} + \{\eta_p\} \tag{17}$$

The flow rule valid for the plastic part $d\{\eta_p\}$ associated with the yield conditions (8) is given by

$$d\eta_{pm} = -\,d\lambda_m + d\lambda' \qquad \text{with} \qquad \begin{cases} d\lambda_m = 0 & \text{if} \quad S_m > 0 \\ d\lambda_m \geq 0 & \text{if} \quad S_m = 0 \\ d\lambda' = 0 & \text{if} \quad S' > 0 \\ d\lambda' \geq 0 & \text{if} \quad S' = 0 \end{cases} \tag{18}$$

or, in matrix notation

$$d\{\eta_p\} = -\,d\{\lambda\} + \{U\}d\lambda' \tag{19}$$

The coefficients λ_1 to λ_M of the vector $\{\lambda\}$ and λ' represent nonnegative strain increment parameters similar to those introduced in (10). From (9), (15) and (17) the plastic strain increments $d\{\eta_p\}$ can be eliminated leading to

$$d\{\lambda\} = -[V]^T d\{\varepsilon\} + [V]^T[D]^{-1}[D]\, d\{S\} + \{U\}\, d\lambda' \tag{20}$$

Writing (9) in incremental form and considering (18) the following relations are obtained

$$d\{\sigma\} = \qquad [V]\, d\{S\} \tag{21}$$

$$d\{\lambda\} = -[V]^T d\{\varepsilon\} + [V]^T [D]^{-1} [V]\, d\{S\} + \{U\} d\lambda' \tag{22}$$

$$dS' = \qquad -\{U\}^T d\{S\} \tag{23}$$

The incremental relations (13), (14) and (21), (22), (23) valid for the face method and for the vertex method are given in the tableaus shown below. The algorithm needed for determining $\{\Delta\sigma\}$ for a given $\{\Delta\varepsilon\}$ as well as the tangent stiffness matrix valid for the strain state $\{\varepsilon\}+\{\Delta\varepsilon\}$ operates on the coefficients of these tableaus. This algorithm however is exactly the same for both methods. Therefore the only difference between the two methods lies in the tableau's coefficients, which must be specified as input parameters. The dimension of the square and almost symmetric tableau matrix (N+K for the face method resp. N+K+1 for the vertex method) will in general also be different.

	$d\{\varepsilon\}$	$d\{\lambda\}$	
$d\{\sigma\}=$	$[D]$	$-[D][F]$	N
$d\{S\}=$	$-[F]^T[D]$	$[F]^T[D][F]$	K
	N	K	

	$d\{\varepsilon\}$	$d\{S\}$	$d\lambda'$	
$d\{\sigma\}=$	·	$[V]$	·	N
$d\{\lambda\}=$	$-[V]^T$	$[V]^T[D]^{-1}[V]$	$\{U\}$	M
$dS'=$	·	$-\{U\}^T$	·	1
	N	M	1	

Solution algorithm

The algorithm used for determining both the stress increment $\{\Delta\sigma\}$ due to a strain increment $\{\Delta\varepsilon\}$ and the tangent elasticity matrix $[D_T]$ valid for the strain state $\{\varepsilon\}+\{\Delta\varepsilon\}$ operates on the coefficients of the tableaus given above. The basic step of this algorithm is the exchange step also used for inverting a matrix or in linear programming. With an exchange step a dependent variable S_i (or λ_i) is "exchanged" with the corresponding independent variable λ_i (resp. S_i). Thus the dependent variable becomes independent and vice-versa. The coefficients of the tableau matrix change their values according to simple well known rules. After a number of such exchange steps, the original tableaus of both the face method and the vertex method may look as shown below.

	$d\{\varepsilon\}$	$d\{S_I\}$	$d\{\lambda_I\}$
$d\{\sigma\}$ =	$[D_T]$	$[A_1]^T$	$[A_2]^T$
$d\{\lambda_D\}$=	$[A_1]$	$[A_3]$	$[A_4]$
$d\{S_D\}$=	$-[A_2]$	$[A_4]^T$	$[A_5]$

In this tableau the dS- and dλ-Variables build two vectors $d\{S_I\}$ and $d\{\lambda_I\}$ of "independent" variables and two vectors $d\{S_D\}$ and $d\{\lambda_D\}$ of "dependent" variables. If dS_i is dependent then $d\lambda_i$ is independent and vice-versa.

The matrices $[A_1]$ to $[A_5]$ appearing in the tableau shall not be derived here in detail. It should only be noted that $[D_T]$, $[A_3]$ and $[A_5]$ are symmetric matrices, so that the whole tableau matrix remains almost symmetric. More important is the fact that the coefficients of the matrices $[D_T]$, $[A_1]$ and $[A_2]$ appearing in the tableau now describe the infinitesimal stress-strain relations valid in the current state. The stress increment $d\{\sigma\}$ for a given strain increment $d\{\varepsilon\}$ can be evaluated using the tangent elasticity matrix $[D_T]$:

$$d\{\sigma\} = [D_T]\, d\{\varepsilon\} \tag{24}$$

The dependent plastic flow parameters $d\{\lambda_D\}$ corresponding to the linear yield conditions which are active in the current state, i.e. whose slack variables vanish ($\{S_I\}=0$) and do not change with $d\{\varepsilon\}$ ($d\{S_I\}=0$), are computed from

$$d\{\lambda_D\} = [A_1]\ d\{\varepsilon\} \geq 0 \tag{25}$$

Because the λ-variables may not become negative, the tableau is only valid for the strain increment $d\{\varepsilon\}$ if the above condition is satisfied. Finally the increments $d\{S_D\}$ of the slack variables $\{S_D\}$ (which are dependent on $\{\sigma\}$ and $d\{\varepsilon\}$ because the corresponding yield conditions are inactive) are given by

$$d\{S_D\} = -\ [A_2]\ d\{\varepsilon\} \tag{26}$$

Based on these preliminary considerations the algorithm used for determining $\{\Delta\sigma\}$ and $[D_T]$ for a given strain increment $\{\Delta\varepsilon\}$ can be described as follows:

1. Initialise: $\{\Delta\sigma\} := 0$; $c := 0$;
2. Multiply each row of $[A_1]$ (if any) with $\{\Delta\varepsilon\}$. If the result is negative the corresponding dependent λ_i-variable is to be exchanged with the independent S_i-variable. Unloading has occured. The material is "more elastic" than it was before. The number of rows in the matrix $[A_1]$ is reduced by one whereas it is increased by one in matrix $[A_2]$. The rank of the tangent elasticity matrix $[D_T]$ is also increased by one. This step is to be repeated until the product of each row of $[A_1]$ (if any are left) with $\{\Delta\varepsilon\}$ is positive. The tableau is then valid for a strain increase in the direction $\{\Delta\varepsilon\}$.
3. Compute the increments $\{\Delta S_D\}$ due to a strain increment $q\{\Delta\varepsilon\}$ with $q=1-c$ from

$$\{\Delta S_D\} = q\ [A_2]\ \{\Delta\varepsilon\} \leq \{S_D\} \tag{27}$$

4. If one or more ΔS_{Di}-increments are larger than the corresponding S_{Di}-variables meaning that due to $q\{\Delta\varepsilon\}$ additional yield conditions are becoming active then:
   a) reduce the parameter q to $q:=\text{Min}_i(S_{Di}/\Delta S_{Di})$ with $\Delta S_{Di} > 0$.
   b) update the slack variables $\{S_D\}$:

   $$\{S_D\} := \{S_D\} - q\ [A_2]\ \{\Delta\varepsilon\} \qquad \text{(thus } S_{Di} \text{ will become zero)}$$

   c) update the stress increment $\{\Delta\sigma\}$:

   $$\{\Delta\sigma\} := \{\Delta\sigma\} + q\ [D_T]\ \{\Delta\varepsilon\}$$

d) exchange the dependent dS_i variable with the independent $d\lambda_i$ variable. This means that the i-th yield condition becomes active. The material is "more plastic" than it was before. The number of rows is increased by one in the matrix $[A_1]$ and decreased by one in matrix $[A_2]$. The rank of the tangent elasticity matrix $[D_T]$ is also decreased by one.

e) update c: c:=c+q. If c < 1 repeat from 2 else goto 5)

else:

a) update the slack variables $\{S_D\}$:

$$\{S_D\} := \{S_D\} - q\ [A_2]\ \{\Delta\varepsilon\}$$

b) update the stress increment $\{\Delta\sigma\}$:

$$\{\Delta\sigma\} := \{\Delta\sigma\} + q\ [D_T]\ \{\Delta\varepsilon\}$$

5) End: The tangent elasticity matrix $[D_T]$ valid for the strain state $\{\varepsilon\}+\{\Delta\varepsilon\}$ and for a strain increment in the direction of $\{\Delta\varepsilon\}$ appears in the upper left corner of the tableau matrix.

To understand the algorithm it is useful to consider the initial tableaus shown above. With $[D_T]=[D]$ the initial tableau of the face method describes a fully elastic state. No yield condition is active. By increasing the strains in the direction of $\{\Delta\varepsilon\}$, some yield conditions may become active. If the stresses $\{\sigma\}$ reach a vertex of the yield figure N yield conditions are active and $[D_T]$ becomes zero.

With $[D_T]=0$ the initial tableau of the vertex method describes a fully plastic but instable state. Before starting the first step, N plastic flow parameters λ_i are exchanged with the corresponding stress parameters S_i. Now the material is fully elastic with $[D_T] = [D]$. If the strains are increased, the condition S'=0 will become active, meaning that exactly one linear yield condition is satisfied by the partially plastified material. By further increasing the strains, up to N-1 additional yield conditions ($S_m=0$) will become active, meaning that the stresses $\{\sigma\}$ have reached a vertex of the polyhedric flow figure. Then the material is fully plastic with $[D_T]=0$.

Remarks and conclusions

For reasons of clarity hardening was not considered. It is easy to show however, that different kinds of hardening can be taken into account simply by applying the solution algorithm described above to somewhat different initial tableau matrices. For the same reasons the case of a flow figure extending to infinity in a given direction was not discussed. Again this case is easy to be treated in both methods.

An important property of the linear yield condition methods is that in a general purpose program (e.g. see [1]), the vectors $\{F_k\}$ and resistances R_k of the face method or the vectors $\{V_m\}$ of the vertex method can be specified as input data, i.e. independently of any predefined yield criterium. This means greater flexiblity in the modelling of user-defined elasto-plastic material laws.

An additional important advantage is that the actual nonlinear yield conditions which have to be satisfied by the stress components do not need to be formulated explicitely. This is readily apparent for the vertex method. Let us consider a structural element like a frame or shell element made of reinforced concrete or some other composite material. The standard approach of considering each stress component in different integration points over the cross section of the bar or the thickness of the shell is not a good one for this kind of elements. Much simpler and also more appealing from an engineering point of view is the use of integral section forces (e.g. bending and twisting moments, shear and normal forces) as generalized stress parameters. This is difficult by the standard approach but rather easy by the vertex method. In fact it is only necessary to determine an adequate number of "well-chosen" but to a large extend arbitrary limit stress states within the section. The corresponding section forces are the coefficients of the vertex vectors $\{V_m\}$ needed for modelling the nonlinear elasto-plastic behaviour of the whole section. Conceivably the vertex vectors can even be determined experimentally and fed in to the analysis program without any analytical manipulation. More difficult is to determine the face vectors $\{F_k\}$ and the corresponding resistances R_k for the face method.

If the vertex vectors represent true limit stress states, the vertex method will never overestimate the load carrying capacity of the structure. In fact the linearised flow figure will automatically be inscribed in the convex nonlinear flow figure it approximates. When designing a structure for a given collapse load this is a welcome property, because underestimating the static collapse load means additional safety. This is also true for the face method, provided the linearised flow figure is inscribed in the nonlinear one, which, however, in general is more difficult to check.

Of course also standard nonlinear yield conditions like Von Mises's or Drucker-Prager's can be approximated by the linear yield condition methods. Invariance will be lost, but a significant reduction in computing time can be obtained. Concerning computing time and storage requirements, the algorithm described in section 3, as expected, has proven to be faster than corresponding algorithms based on nonlinear yield conditions which require iteration and convergence checks. With some clever programming the efficiency and the storage requirement of the linear yield conditions algorithm can be much improved. In fact the steps described in section 3 can, to a large extend, be done "by inspection", i.e. without updating explicitely the coefficients of the tableau matrix at each step. This greatly reduces both the storage requirement and the computing time.

When comparing the face method with the vertex method the following question arises: How many vertices has a polyhedric flow figure with K faces in a N-dimensional stress space? It is easy to prove that when N is larger than 2 (e.g. in the practical case of a shell element with 3 force- and 3 bending-stress components, i.e. with N=6) the number of vertices M will generally be smaller than the number of faces K. This means that a given polyhedric approximation of a nonlinear flow figure can be described by a smaller number of data, i.e. by a smaller tableau matrix, when using the vertex method than the face method. Also for this reason the vertex method, which is a little more difficult to derive and to understand, looks preferable.

Finally it should be noted that it is conceivable to extend the algorithm described above for elasto-plastic materials to other kinds of nonlinear materials, e.g. involving cracks or friction. Linear inequalities constraints and exchange steps are the tools to be used for modelling many kinds of nonlinear material laws.

References:

1. Anderheggen, E.; Bartelt, P.; Bazzi, G.; Elmer, H.; Friedrich, T., Klemenz, U.; Maag, H.; Theiler, J: FLOWERS user's manual. Inst. f. Informatics, ETH Zurich. Third editon (1985)
2. Anderheggen, E.: Nonlinear FE-analysis with direct user interaction. Preecedings of the Fourth World Congres on FEM Technology. Robinson and associates (1984)
3. Cohn M.Z.; Maier G. (eds.): Enginering plasticity by mathematical programming. Proc. of the NATO Advanced Study Institute. Unv. of Waterloo. Canada. Pergamon press (1979)

On Large Strain Elasto-Plastic and Creep Analysis

K.-J. BATHE, R. SLAVKOVIĆ, M. KOJIĆ*
Massachusetts Institute of Technology, Cambridge, USA
*ADINA Engineering, Watertown, USA

An incremental (rather than rate-type) formulation based on the logarithmic strain and a material stress is briefly discussed. Our 'effective-stress-function' algorithm is used in the computation of the stresses. The efficiency of the solution procedure is discussed, and it is employed to analyse the large strain elasto-plastic compression of a cylinder.

1. Introduction

Large strain inelastic finite element analyses are mostly conducted using the Jaumann stress rate tensor in a form of updated Lagrangian formulation [1]. Although many questions can be raised, this is a natural approach in many regards because Cauchy stresses and velocity strains — both being physical engineering quantities — are employed in the computations. However, much research has been conducted in recent years to develop alternative formulations that more accurately model inelastic deformations and that are also more effective computationally [1,2].

Our objective in this paper is to further discuss a formulation that we believe to be attractive in many regards [3]. We are using the formulation for elasto-plastic and creep analyses with strains up to about forty percent. For strains up to that magnitude the usual material descriptions are deemed to be applicable.

The formulation we present below — referred to as the updated Lagrangian Hencky (ULH) formulation — has the important attribute that it is a t o t a l formulation, in contrast to rate-type formulations. Hence, strain increments are determined from total strains at times t+Δt and t, the time step increment being Δt. Such an incremental approach has been found to be very effective considering the total and updated Lagrangian formulations discussed in refs. [4-6] for large rotation but small strain inelastic analysis, and it is natural to extend the approach to large strain solutions.

Finite Element Methods for Nonlinear Problems
Europe-US Symposium, Trondheim, Norway 1985
Editors: Bergan, Bathe, Wunderlich

We can make two important observations for the ULH formulation. Firstly, for any closed elastic strain path the final stress (and accumulated strain energy) is zero. This does not hold for conventional Jaumann stress rate based formulations and is surely a matter of concern. Secondly, using the ULH formulation very large rigid body rotations can be accommodated in a single load (or time) step. The amount of rigid body rotation that can be modeled in a single load step Δt is only limited by the need to converge in the equilibrium iterations. If the BFGS or full Newton iteration methods with line searches are employed, very large rigid body rotations can be solved in single load steps.

In the next section, we present the ULH formulation. An important ingredient of the overall solution procedure is also our 'effective-stress-function' algorithm to integrate the stresses for the given strains, and we briefly discuss this algorithm in Section 3. Finally, in Section 4 we present the solutions of some demonstrative large strain problems. These solutions illustrate the applicability and effectiveness of the ULH formulation.

2. The Updated Lagrangian Hencky Formulation

In our earlier work we presented the updated Lagrangian formulation for large deformation analysis [4,5]. Using the notation of refs. [4-6], the final incremental equation is,

$$\int_{{}^tV} \delta_t e_{ij}\, C_{ijrs}\, \Delta_t e_{rs}\, {}^t dV + \int_{{}^tV} {}^t\tau_{ij}\, \delta\Delta_t \eta_{ij}\, {}^t dV = {}^{t+\Delta t}\mathcal{R} - \int_{{}^tV} {}^t\tau_{ij}\, \delta_t e_{ij}\, {}^t dV \qquad (1)$$

where

${}^t\tau_{ij}$ = components of Cauchy stress tensor at time t,

C_{ijrs} = components of tangent constitutive tensor at time t,

$\Delta_t e_{ij}$ = components of linear part of incremental Green-Lagrange strain tensor

$\Delta_t \eta_{ij}$ = components of nonlinear part of incremental Green-Lagrange strain tensor,

$^{t+\Delta t}\mathcal{R}$ = total external virtual work at time t+Δt,

and all integrations are performed over the volume at time t, ^{t}V.

Equation (1) is obtained by linearizing the basic equilibrium equation

$$\int_{^{t+\Delta t}V} {}^{t+\Delta t}\tau_{ij}\ \delta_{t+\Delta t}\, e_{ij}\ {}^{t+\Delta t}dV = {}^{t+\Delta t}\mathcal{R} \tag{2}$$

about the configuration at time t. Here,

$$\delta_{t+\Delta t}\, e_{ij} = \frac{1}{2}\left(\frac{\partial \delta u_i}{\partial\, {}^{t+\Delta t}x_j} + \frac{\partial \delta u_j}{\partial\, {}^{t+\Delta t}x_i}\right) \tag{3}$$

and $^{t+\Delta t}x_i$, i=1,2,3 are the material point coordinates at time t+Δt.

On discretizing Eq. (1) using finite element procedures and solving the resulting finite element nodal point equilibrium equations, an increment in displacements, Δu_i, for the material particles is obtained and then

$$^{t+\Delta t}x_i \doteq {}^{t}x_i + \Delta u_i \tag{4}$$

The approximate sign in Eq. (4) is used because Eq. (1) is a linearization of the relation in Eq. (2). To obtain an accurate solution of Eq. (2), equilibrium iterations are performed (e.g., modified or full Newton-Raphson iterations, the BFGS method can be employed). In these iterations the process of calculating an incremental displacement is repeated until Eq. (2) is satisfied to sufficient accuracy. The basic problem in each such iteration is to evaluate the new stress state for the material particles corresponding to the current displacements. Let $^{t+\Delta t}x_i$ be the current material particle coordinates, those at time t being equal to $^{t}x_i$, then the Cauchy stresses corresponding to $^{t+\Delta t}x_i$ are calculated as follows:

<u>step 1</u>: Decompose the current deformation gradient,

$$^{t+\Delta t}_{\ 0}\underline{X} = {}^{t+\Delta t}_{\ 0}\underline{R}\ {}^{t+\Delta t}_{\ 0}\underline{U} \tag{5}$$

step 2: Calculate the logarithmic strains,

$$ {}^{t+\Delta t}_{0}\bar{\underline{e}} = \ell n \; {}^{t+\Delta t}_{0}\underline{U} \tag{6} $$

or

$$ {}^{t+\Delta t}_{0}\bar{\underline{e}} = {}^{t+\Delta t}_{0}\underline{\Phi} \left(\ell n \; {}^{t+\Delta t}_{0}\underline{\Lambda} \right) {}^{t+\Delta t}_{0}\underline{\Phi}^{T} \tag{7} $$

where ${}^{t+\Delta t}_{0}\underline{\Lambda}$ is a diagonal matrix storing the principal stretches (eigenvalues) of ${}^{t+\Delta t}_{0}\underline{U}$ and ${}^{t+\Delta t}_{0}\underline{\Phi}$ contains the corresponding directions (eigenvectors).

step 3: Evaluate the incremental logarithmic strains,

$$ \bar{\underline{e}} = {}^{t+\Delta t}_{0}\bar{\underline{e}} - {}^{t}_{0}\bar{\underline{e}} \tag{8} $$

step 4: Calculate total material stresses,

$$ {}^{t+\Delta t}\bar{\underline{\tau}} = {}^{t}\bar{\underline{\tau}} + \int_{t}^{t+\Delta t} \underline{C} \; d\bar{\underline{e}} \tag{9} $$

step 5: Transform ${}^{t+\Delta t}\bar{\underline{\tau}}$ to principal stretch directions,

$$ {}^{t+\Delta t}\tilde{\underline{\tau}} = {}^{t+\Delta t}_{0}\underline{\Phi}^{T} \; {}^{t+\Delta t}\bar{\underline{\tau}} \; {}^{t+\Delta t}_{0}\underline{\Phi} \tag{10} $$

step 6: Calculate Cauchy stresses ${}^{t+\Delta t}\underline{\tau}^{c}$, in principal stretch directions,

$$ {}^{t+\Delta t}\tau^{c}_{ii} = {}^{t+\Delta t}\tilde{\tau}_{ii} \tag{11} $$

$$ {}^{t+\Delta t}\tau^{c}_{ij} = {}^{t+\Delta t}\tilde{\tau}_{ij} \; \frac{2 \; {}^{t+\Delta t}\lambda_i \; {}^{t+\Delta t}\lambda_j}{({}^{t+\Delta t}\lambda_i)^2 - ({}^{t+\Delta t}\lambda_j)^2} \; \ell n \; \frac{{}^{t+\Delta t}\lambda_i}{{}^{t+\Delta t}\lambda_j} \qquad (i \neq j) \tag{12} $$

step 7: Transform ${}^{t+\Delta t}\underline{\tau}^{c}$ to obtain final Cauchy stresses,

$$ {}^{t+\Delta t}\underline{\tau} = {}^{t+\Delta t}_{o}\underline{R}_{tot}\ {}^{t+\Delta t}\underline{\tau}^{c}\ {}^{t+\Delta t}_{o}\underline{R}^{T}_{tot} \tag{13} $$

$$ {}^{t+\Delta t}_{o}\underline{R}_{tot} = {}^{t+\Delta t}_{o}\underline{R}\ {}^{t+\Delta t}_{o}\underline{\Phi} \tag{14} $$

These Cauchy stresses, defined in the stationary Cartesian coordinate system, can now directly be used as usual to evaluate the nodal point force vector corresponding to the finite element internal stress states. Hence, the equilibrium iterations and step-by-step solution can proceed in the usual manner [6].

The stress and strain measures and transformations used have been discussed earlier, see refs. [2,3,7], but an important assumption lies in the use of Eq. (9). Here we update the material stress due to the increase in the logarithmic strain, and the appropriate stress-strain matrix $\underline{C}$ must be employed. In our study we have assumed so far that the matrix $\underline{C}$ is calculated as usual [6] but employing the material stress and logarithmic strain components.

To actually carry out the steps of analysis effectively, it is necessary to have efficient procedures to calculate the polar decomposition in Eq. (5) and the eigensystem of the stretch tensor ${}^{t+\Delta t}_{o}\underline{U}$. We are using the algorithms presented in ref. [3].

3. Integration of Stresses

The finite element solution involves the integration of stresses as expressed in Eq. (9) for each finite element integration point. Considering thermo- elastic-plastic and creep problems a procedure widely employed is the α-method, which uses as the basic assumption in the stress integration [6]

$$ {}^{t+\alpha\Delta t}\underline{\tau} = (1-\alpha)\ {}^{t}\underline{\tau} + \alpha\ {}^{t+\Delta t}\underline{\tau} \tag{15} $$

where ${}^{t}\underline{\tau}$ are the stress components at time t, and $0 \le \alpha \le 1$. When α=0 the procedure corresponds to an Euler forward integration (explicit method), and when $\alpha = 0.5, 1$, the procedure corresponds to the trapezoidal and Euler backward integrations (implicit methods). It is discussed in ref. [8] that regarding stability considerations, large time steps Δt can be used when the trapezoidal or Euler backward methods are employed. This is an important consideration in the solution of creep problems (for which the response may evolve over large time spans), and in particular in the analysis of large strains, where with too severe time step size restrictions, an inordinately large computational effort may be required. Hence, the use of relatively large time steps in the response solution is much desired.

In practice, the magnitude of the time steps that can be employed in implicit integration is however restricted by two considerations. Firstly, the implicit integration requires iteration in the evaluation of the integral in Eq. (9). In the procedure of ref. [8], a form of Newton-Raphson iteration is used, which requires that the starting iteration stress vector is "sufficiently close" to the solution sought. Secondly, the selected time step size must be small enough to obtain an accurate response solution.

To increase the effectiveness of our solution procedure we have developed and researched the effective-stress-function algorithm. The details of the solution procedure will be published in a forthcoming communication and we only briefly summarize the method below.

A basic equation of thermo-plasticity is, considering isotropic hardening, and denoting tensor components by a curl ($\tilde{\ }$)

$$ {}^{t}\dot{\tilde{e}}^{P} = {}^{t}\Lambda \; {}^{t}\tilde{S} \tag{16} $$

where the ${}^{t}\dot{\tilde{e}}^{P}$ are the time derivatives of the plastic strains, the ${}^{t}\tilde{S}$ are the deviatoric stresses and ${}^{t}\Lambda$ is a scalar,

$$ {}^{t}\Lambda = \frac{3}{2\;{}^{t}\sigma_y} \left(\frac{\partial\,{}^{t}\bar{e}^{P}}{\partial\,{}^{t}\sigma_y}\;{}^{t}\dot{\sigma}_y + \frac{\partial\,{}^{t}\bar{e}^{P}}{\partial\,{}^{t}\theta}\;{}^{t}\dot{\theta} \right) \tag{17} $$

In Eq. (17) ${}^{t}\sigma_y$ is the instantaneous yield stress at time t, ${}^{t}\bar{e}^{P}$ is the effective plastic strain and ${}^{t}\theta$ is the temperature. With the yield stress defined as a function of the effective plastic strain and temperature, Eq. (17) can directly be used to evaluate ${}^{t}\Lambda$.

Considering the classical description of creep, the inelastic strains are given by

$$ {}^{t}\dot{\underset{\sim}{e}}^{C} = {}^{t}\gamma \; {}^{t}\underset{\sim}{S} \qquad ; \qquad {}^{t}\gamma = \frac{3}{2} \; \frac{{}^{t}\dot{\bar{e}}^{C}}{{}^{t}\bar{\sigma}} \tag{18) (19} $$

where ${}^{t}\bar{e}^{C}$ and ${}^{t}\bar{\sigma}$ are the effective creep strain and the effective stress, respectively.

For the numerical solution, we rewrite Eqs. (16) to (19) in the form

$$ \Delta\underset{\sim}{e}^{P} = \Delta t \; {}^{\tau}\Lambda \; {}^{\tau}\underset{\sim}{S} \qquad ; \qquad {}^{\tau}\Lambda \; \Delta t = \frac{3}{2} \; \frac{\Delta\bar{e}^{P}}{{}^{\tau}\bar{\sigma}} \tag{20) (21} $$

$$ \Delta\underset{\sim}{e}^{C} = \Delta t \; {}^{\tau}\gamma \; {}^{\tau}\underset{\sim}{S} \qquad ; \qquad {}^{\tau}\gamma \; \Delta t = \frac{3}{2} \; \frac{\Delta\bar{e}^{C}}{{}^{\tau}\bar{\sigma}} \tag{22) (23} $$

where the superscript τ denotes weighted values using the α-method; for example,

$$ {}^{\tau}\underset{\sim}{S} = (1-\alpha) \; {}^{t}\underset{\sim}{S} + \alpha \; {}^{t+\Delta t}\underset{\sim}{S} \tag{24} $$

and so on. As pointed out above, in practice, we want to be able to use $\alpha \geq 0.5$ with relatively large time steps Δt.

The deviatoric stresses at time $t+\Delta t$ are

$$ {}^{t+\Delta t}\underset{\sim}{S} = 2 \; {}^{t+\Delta t}G \left({}^{t+\Delta t}\underset{\sim}{e}^{*} - {}^{t+\Delta t}\underset{\sim}{e}^{P} - {}^{t+\Delta t}\underset{\sim}{e}^{C} \right) \tag{25} $$

where

$$ {}^{t+\Delta t}G = \frac{{}^{t+\Delta t}E}{2(1 + {}^{t+\Delta t}\nu)} \tag{26} $$

and ${}^{t+\Delta t}E$, ${}^{t+\Delta t}\nu$ are the Young's modulus and Poisson's ratio at time t+Δt (that is at the temperature ${}^{t+\Delta t}\theta$). In Eq. (25) ${}^{t+\Delta t}\underset{\sim}{e}^{*}$ stores the total deviatoric strains. Use of Eqs. (20) to (25) yields

$$
{}^{t+\Delta t}\underset{\sim}{S} = \frac{{}^{t+\Delta t}E}{1 + {}^{t+\Delta t}\nu + \alpha\Delta t\ {}^{t+\Delta t}E\ ({}^{\tau}\Lambda + {}^{\tau}\gamma)} \Bigg({}^{t+\Delta t}\underset{\sim}{e}^{*} - \Delta t(1-\alpha)\ ({}^{\tau}\Lambda + {}^{\tau}\gamma)\ {}^{t}\underset{\sim}{S} - {}^{t}\underset{\sim}{e}^{P} - {}^{t}\underset{\sim}{e}^{C} \Bigg) \tag{27}
$$

and the mean stress is obtained from

$$
{}^{t+\Delta t}\tau_m = \frac{{}^{t+\Delta t}E}{1-2\,{}^{t+\Delta t}\nu} \left({}^{t+\Delta t}e_m - {}^{t+\Delta t}e^{TH} \right) \tag{28}
$$

where ${}^{t+\Delta t}e_m$ is the mean strain and ${}^{t+\Delta t}e^{TH}$ is the thermal strain.

With the displacements ${}^{t+\Delta t}u_i$ at the material point given, it is possible to iterate on Eq. (27), using $\alpha \geq 0.5$, until the stresses for the evaluation of ${}^{\tau}\Lambda$ and ${}^{\tau}\gamma$ are very close to the stresses evaluated in Eq. (27). However, this iteration requires that the time (or load) step be small enough for convergence of the iteration. If a Newton-Raphson type of iteration is employed, experience shows that in creep problems the time step Δt may have to be unduly small.

The basic idea in the development of our effective-stress-function algorithm is to iterate on a single unknown variable — the effective stress — rather than on the six unknown stresses in Eq. (27). The governing equation is obtained by taking the scalar product of both sides of Eq. (25), which gives

$$
\left({}^{t+\Delta t}\bar{\sigma}\right)^2 = 9\left({}^{t+\Delta t}G\right)^2 \left({}^{t+\Delta t}\bar{e}^{*E}\right)^2 \tag{29}
$$

where ${}^{t+\Delta t}\bar{e}^{*E}$ denotes the elastic deviatoric effective strain at time t+Δt,

$$
{}^{t+\Delta t}\bar{e}^{*E} = \sqrt{\frac{2}{3}\ {}^{t+\Delta t}\underset{\sim}{e}^{*E} \cdot {}^{t+\Delta t}\underset{\sim}{e}^{*E}} \tag{30}
$$

Since the right-hand-side of Eq. (29) depends only on the effective stress ${}^{t+\Delta t}\bar{\sigma}$, our problem has reduced to finding the effective stress which corresponds to the zero of the effective stress function,

$$f = \left({}^{t+\Delta t}\bar{\sigma}\right)^2 - 9\left({}^{t+\Delta t}G\right)^2\left({}^{t+\Delta t}\bar{e}^{*E}\right)^2 \tag{31}$$

To solve for the zero of this function, i.e. the unknown effective stress ${}^{t+\Delta t}\bar{\sigma}$, we substitute for ${}^{t+\Delta t}\bar{e}^{*E}$ from Eq. (27) and use a simple but very stable bisection algorithm. Formally, the iteration is continued until with k=1,2,3...,ℓ and

$$f = \left({}^{t+\Delta t}\bar{\sigma}^{(k)}\right)^2 - 9\left({}^{t+\Delta t}G\right)^2\left({}^{t+\Delta t}\bar{e}^{*E(k)}\right)^2 \tag{32}$$

the condition

$$\left| f\left({}^{t+\Delta t}\bar{\sigma}^{(\ell)}\right) \right| \leq \varepsilon \tag{33}$$

is satisfied, where ε is a convergence tolerance.

Once the effective stress ${}^{t+\Delta t}\bar{\sigma}$ is known (assumed equal to ${}^{t+\Delta t}\bar{\sigma}^{(\ell)}$ where ℓ denotes the last iteration), Eq. (27) is used to calculate ${}^{t+\Delta t}\underset{\sim}{S}$ and Eq. (28) gives ${}^{t+\Delta t}\tau_m$.

The importance of the algorithm lies in that only one unknown variable, the effective stress, need be solved for in the stress point iteration, which allows very large time steps without difficulties of iteration convergence. Hence, with this algorithm the step Δt is only selected based on accuracy considerations. The algorithm can of course also be employed with sub-incrementation.

4. Some Analysis Results

The results presented below demonstrate some of the characteristics of the solution procedure. A specific question addressed is what difference in response is predicted using the ULH formulation when compared to using a conventional Jaumann formulation (ULJ).

Figure 1 shows the results obtained in the analysis of a one 4-node element large strain problem. Solutions were calculated using the conventional updated Lagrangian Jaumann formulation (see, for example, refs. [5,6]) and the ULH formulation. The figure shows that, as expected, practically identical response is predicted with the two formulations and that the 10 step solutions are very close to the 100 step solutions. The computational effort for the ULJ and ULH solutions is almost the same.

Figure 2 shows the analysis of the same element but subjected to a different loading. In this case some differences between the ULH and ULJ solutions are observed.

Figure 3 gives the results of a similar study but now the problem also includes large rigid body rotations. The difference in the response predictions in the 100 step solutions is very small and can only be plotted for the τ_{yy} stress component. However, in this case a ULJ solution with a small number of time steps requires much more computational effort than when using the ULH formulation.

Figure 4 shows the results obtained in the solution of the large creep strains of a 4-node plane strain element. We note that the stress curves are similar to those calculated in the elasto-plastic solution, see Fig. 1, but, as expected, the stress curves are now smoother.

In practice, surely, in all these analyses best accuracy for a given computational effort would be obtained using variable time step sizes.

Finally, Figure 5 gives some results obtained in the analysis of the upsetting of a cylinder. The analysis of such problem is of much interest in the metal forming industry, and this problem was also considered by Carter and Lee [9]. We used our contact algorithm to model the frictional contact conditions between the cylinder and the press [10]. Our results compare well with those of Carter and Lee. Figure 5 shows that essentially the same results for the contact stresses are obtained using the ULJ and ULH formulations.

5. Concluding Remarks

The objective in this paper was to present some recent developments regarding inelastic large strain analysis methods. We have researched and implemented a large strain formulation, which is a total strain formulation based on the Hencky strain, and for the stress integration an effective-stress-function algorithm.

The ULH formulation gives for certain analysis conditions practically the same results as the ULJ formulation, when the ULJ formulation can be employed. However, it would be valuable to further study the detailed assumptions used in the ULH stress integration and the response predicted. The ULH formulation shows much promise, because relatively large incremental steps can be used and in an elastic closed strain path the accumulated strain energy is zero. Very large incremental steps can be employed in elastic loading or unloading, even when there are large rotations. We are currently pursuing the development of a self-adaptive procedure for the automatic selection of variable time or load step sizes, which shows promise to result overall into a very effective large strain solution technique.

References

1. Lee, E.H. and Mallett, R.L. (eds.) Plasticity of Metals at Finite Strain: Theory, Experiment and Computation, Proceedings of Research Workshop, Stanford University, Palo Alto, California, July 1982.

2. Willam, K. (ed.), Constitutive Equations: Macro and Computational Aspects, ASME, AMD, December 1984.

3. Rolph III, D.W. and Bathe, K.J., "On a Large Strain Finite Element Formulation for Elasto-Plastic Analysis", contribution in ref. [2].

4. Bathe, K.J.; Ozdemir, H.; Wilson, E.L., "Static and Dynamic Geomemetric and Material Nonlinear Analysis," Technical Report UCSESM 74-4, University of California, Berkeley, February 1974.

5. Bathe, K.J.; Ramm, E.; Wilson, E.L., "Finite Element Formulations for Large Deformation Dynamic Analysis," Int. J. Num. Meth. in Eng. 9 (1975) 353-386.

6. Bathe, K.J., Finite Element Procedures in Engineering Analysis, Prentice-Hall, Inc. Englewood Cliffs, New Jersey, 1982.

7. Hill, R., "Aspects of Invariance in Solid Mechanics," in Advances in Applied Mechanics, 18, Academic Press, New York, 1978.

8. Snyder, M.D. and Bathe, K.J.,"A Solution Procedure for Thermo-Elastic-Plastic and Creep Problems", J. Nuclear Eng. and Design, 64 (1981) 49-80.

9. Carter, W., Jr. and Lee, D., "A Finite Element Analysis of Cylinder and Ring Compression and its Experimental Verification", Computers and Structures, 21, No. 1/2, (1985) 1-19.

10. Bathe, K.J., and Chaudhary, A., "A Solution Method for Planar and Axisymmetric Contact Problems", Int. J. Num. Meth. in Eng., 21 (1985) 65-88.

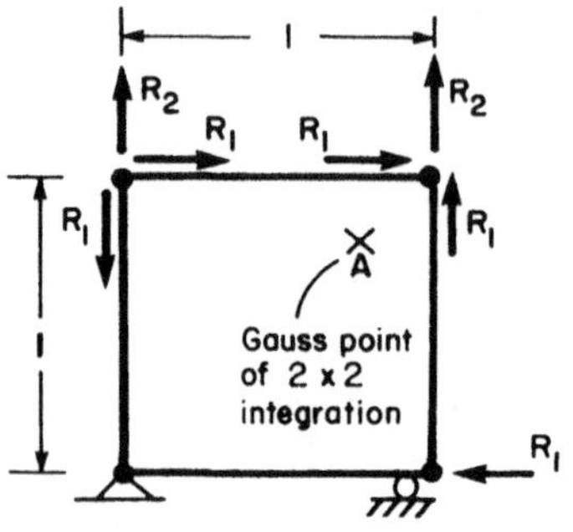

a) Element considered

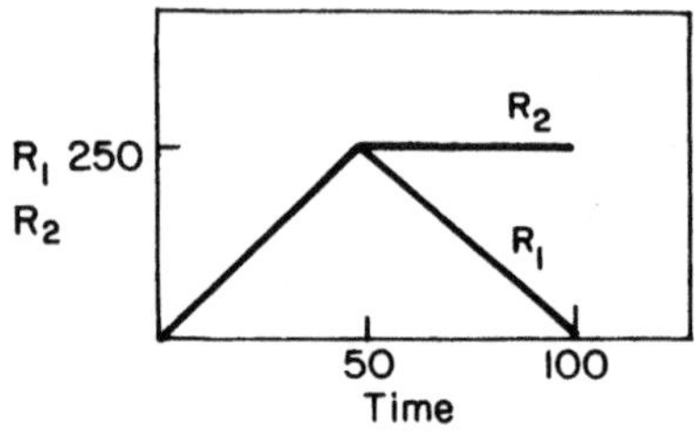

b) Load curves for 100 step solution

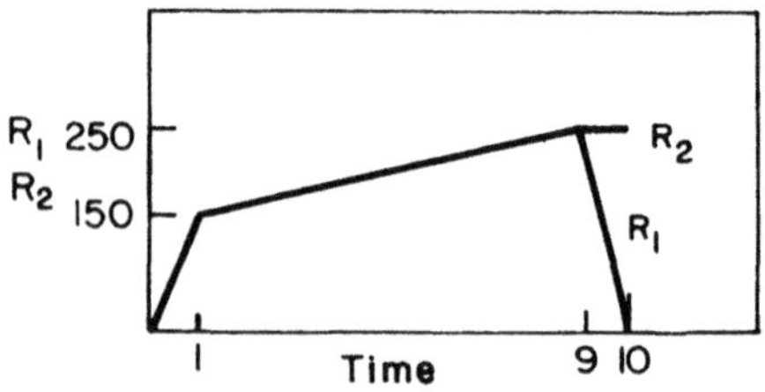

c) Load curves for 10 step solution

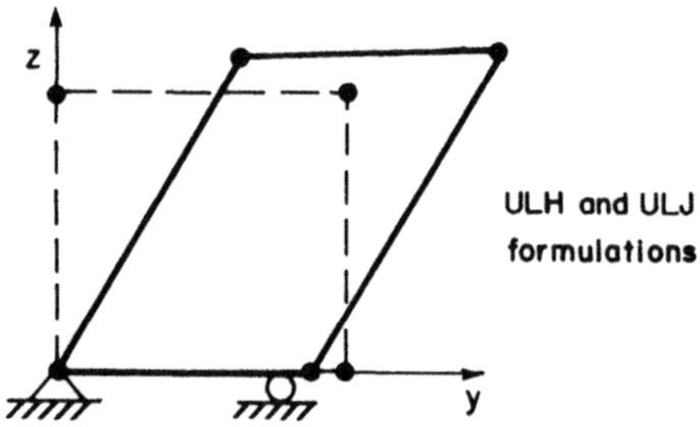

d) Deformed element at time 100

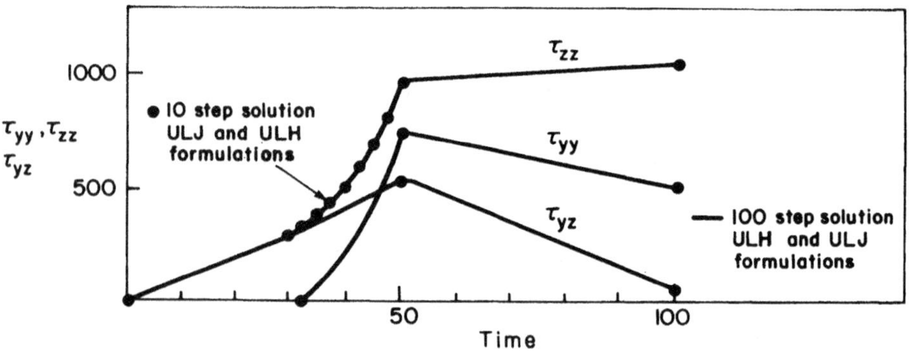

e) Predicted stresses at point A

FIGURE 1 Elastic-plastic stress response of plane strain element; $E = 10^5$, $\nu = 0.3$, $\sigma_y = 600$, $E_T = 1000$; $\Delta t=1$.

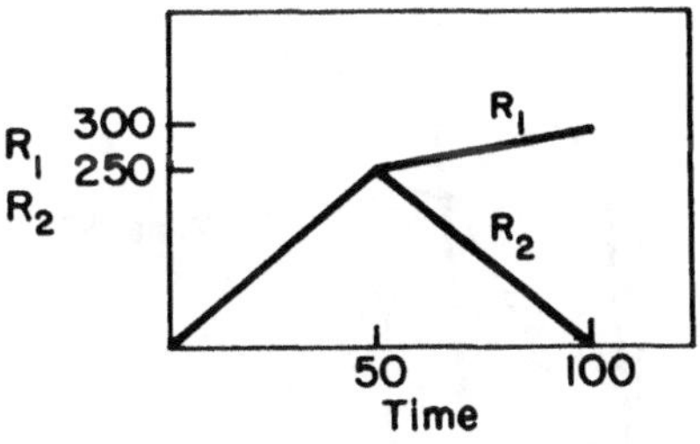

a) Loading curves

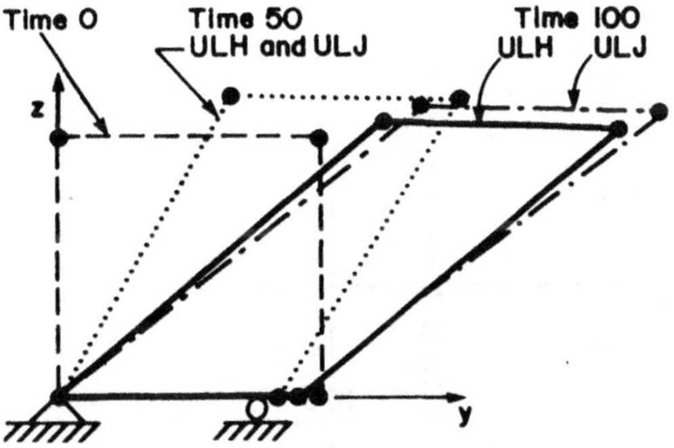

b) Deformed element

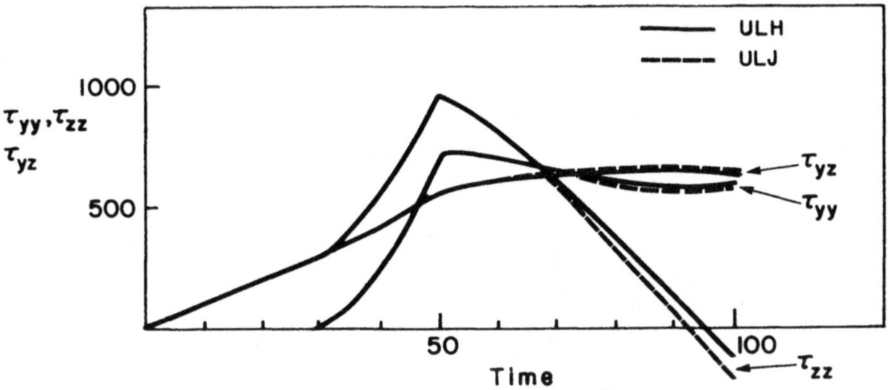

c) Predicted stresses at point A

FIGURE 2 Elastic-plastic response of plane strain element; $\Delta t=5$; for material data and boundary conditions see Figure 1.

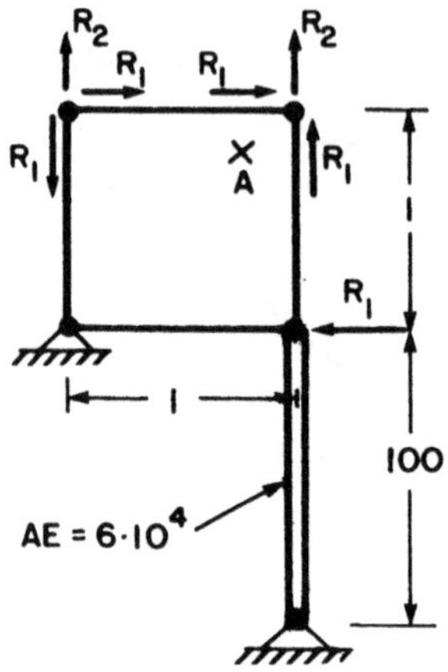

a) Problem analysed

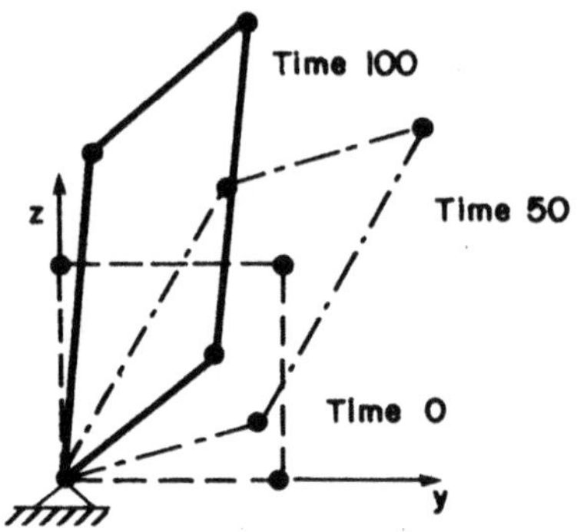

b) Deformed element ULH and ULJ formulations

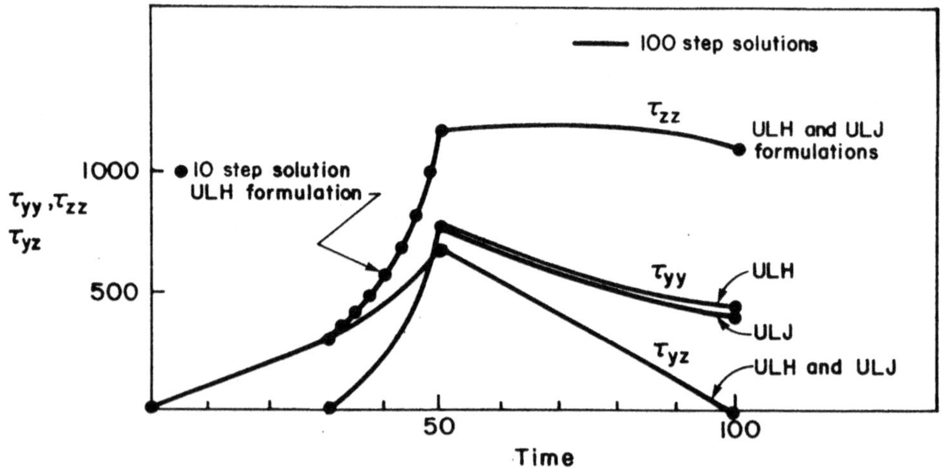

c) Predicted stresses at point A

Figure 3 Elastic-plastic stress/rotation response of one plane strain element; $\Delta t = 1$; for material data and load-time functions see Figure 1.

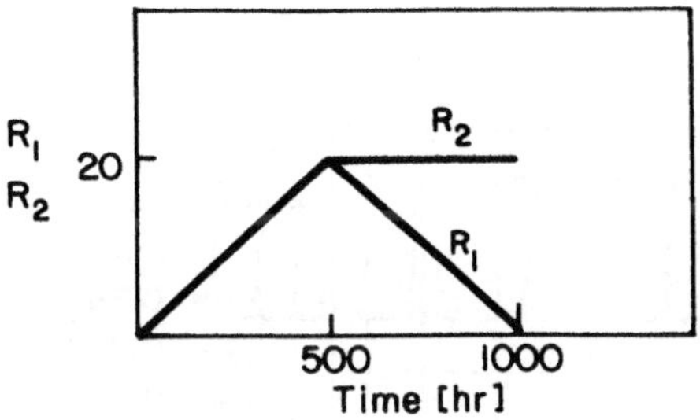

a) Specified loading on element in Figure 1

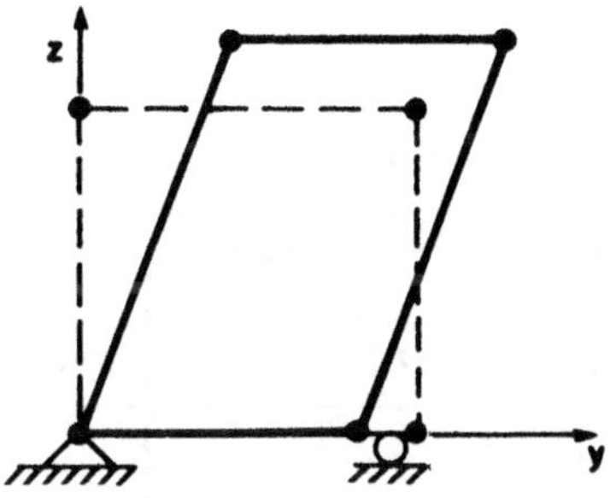

b) Deformed element at 1000 hr

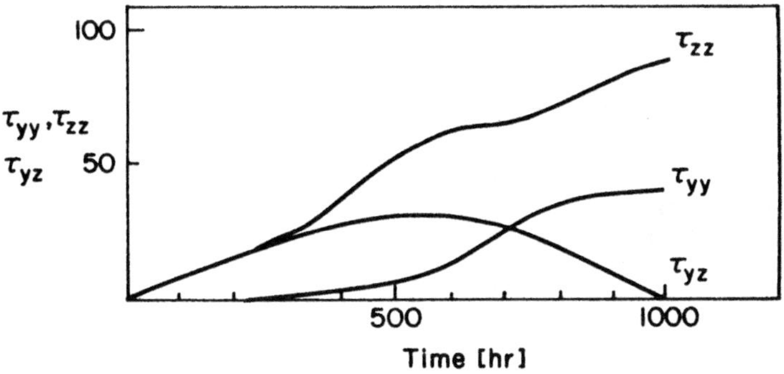

c) Predicted stresses at point A

FIGURE 4 Creep response of plane strain element; $E = 10^5$, $\nu = 0.3$, $\dot{e}^C = 0.987 \cdot 10^{-8} \cdot \bar{\sigma}^{2.66}$; $\Delta t = 10$ hr.

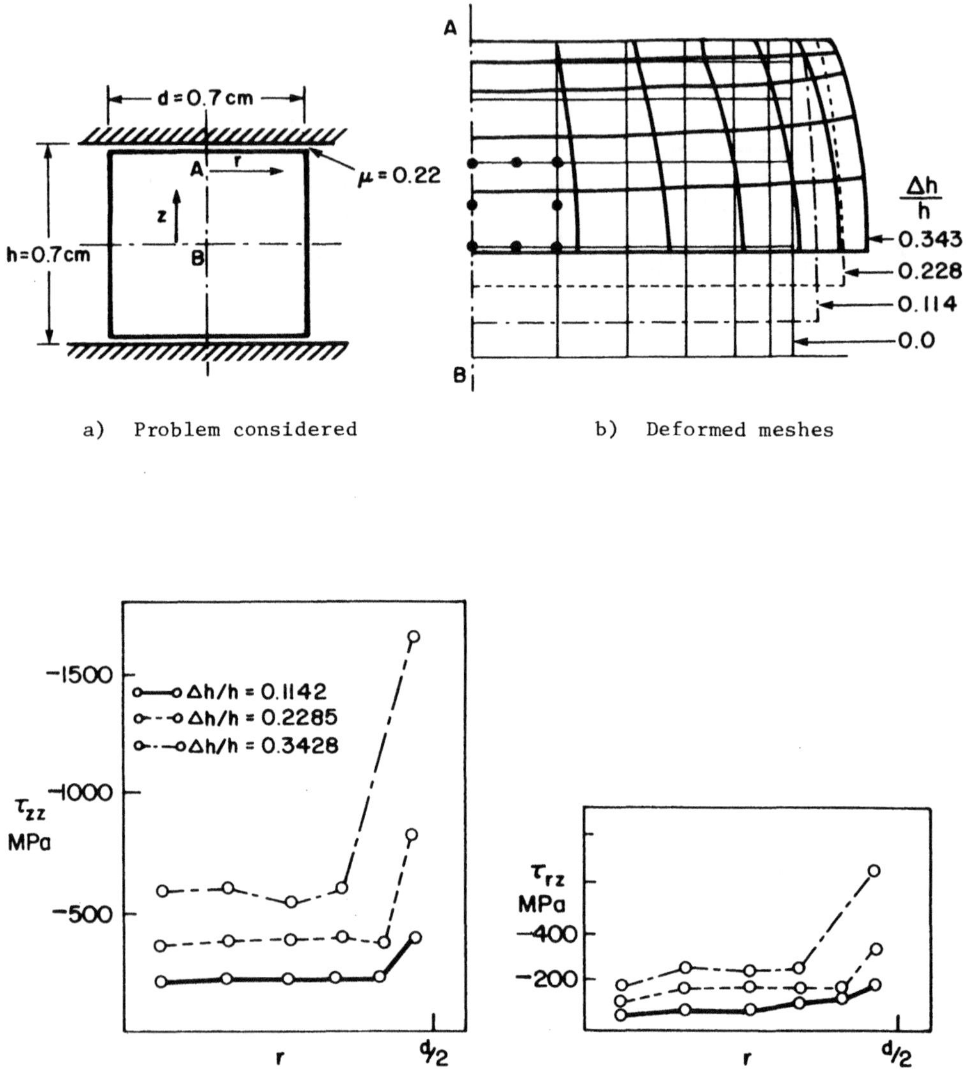

a) Problem considered

b) Deformed meshes

c) Predicted stresses at contact surface, ULJ and ULH formulations

FIGURE 5 Analysis of large strain cylinder compression; $E = 10^5$ MPa, $\nu = 0.3$, $\sigma_y = 100$ MPa, $E_T = 10^3$ MPa.

Dynamic Soil-Structure Interaction with Constitutive Modelling for Soils and Interfaces

C. S. DESAI
University of Arizona, Tucson, USA

Summary

Importance of appropriate modelling and laboratory testing for constitutive behavior of geologic materials and interfaces in dynamic soil-structure is identified. Brief description of new constitutive models developed by the author and co-workers is given. The models are implemented in a nonlinear finite element procedure, and a number of problems are solved. Details of two typical field problems are included to illustrate the applicability of the proposed models and procedure.

Introduction

The response of structures founded on geologic media, soil or rock (Fig. 1), subjected to static and dynamic loadings such as earthquakes, wave forces and blasts is influenced significantly by the properties of the structural and geologic materials and interfaces/joints. Hence, it is essential to include the effect of nonlinear behavior of geologic materials and interfaces, and also the various relative motions, translation, debonding, rebonding, experienced by the interfaces. When these effects are included, it is difficult to obtain closed-form solutions, and it becomes necessary to use numerical techniques such as the finite element method.

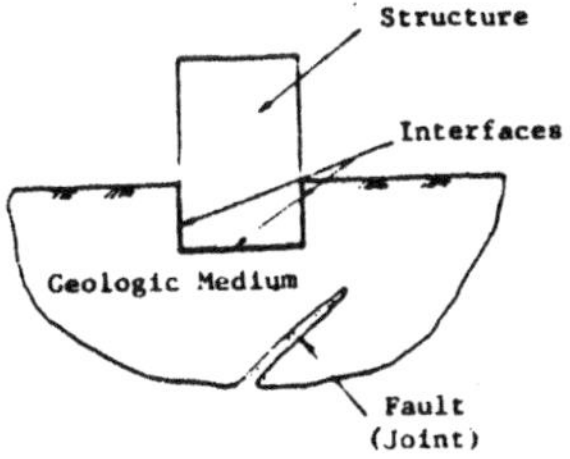

Fig.1. Typical structure-foundation system

Finite Element Methods for Nonlinear Problems
Europe-US Symposium, Trondheim, Norway 1985
Editors: Bergan, Bathe, Wunderlich

The subject of soil-structure interaction is wide in scope and a number of (recent) developments have considered the foregoing topics. It is not possible to present a detailed review of these works; such reviews are available elsewhere [1, 2]. The main objective of this paper is to present mainly a review of the recent research performed by the author and his co-workers [1, 3-9], with references to only those works that are directly relevant.

NONLINEAR BEHAVIOR

Nonlinear response of geologic materials is influenced by various factors such as state of stress, stress path, inelasticity, volume changes, type and rate of loading and inherent and induced anisotropy. Factors such as roughness, state of stress, type and rate of loading, modes of deformation (no-slip, slip, debonding, rebonding) influence the behavior of interfaces.

Geologic Materials

Nonlinear elastic and elastic-plastic hardening and softening constitutive models are proposed and used to describe behavior of geologic materials. Here, in addition to consistent mathematical formulation, it is essential to use appropriate laboratory tests to derive the constants in the constitutive models. An approach that is based on general principles, and yet leads to simple constitutive models is proposed and used for this research on static and dynamic soil-structure interaction performed by the author and co-workers [10, 15]; brief details of this approach are presented below.

Based on the frame invariance principle, a function that is used to define yield and plastic potential functions in the context of the theory of plasticity, is given by [10-15]:

$$F = F\ (J_i,\ I_i^p,\ K_j,\ a_1,\ a_2\ \dots\ a_n) \tag{1}$$

where J_i (i = 1, 2, 3) = direct invariants of stress tensor, σ_{ij}, I_i^p = direct invariants of plastic strain tensor, ε_{ij}^p, K_j (j = 1, 2, 3) = mixed or joint invariants of stress and plastic strain tensors, a_i (i = 1, 2, ..n) = (scalar) valued functions. For isotropic material and isotropic hardening, F does not involve K_j.

Various specialized cases of F in Eq. (1) can be derived for different types of behavior such as associative with isotropic hardening [12, 14], nonassociative with isotropic hardening [11, 17], isotropic hardening with strain softening [18] and anisotropic hardening [16]; inclusion of joint or mixed invariants is required for the anisotropic case.

A form of Eq. (1) that can allow derivation of various available models such as von Mises, Drucker-Prager, Mohr-Coulomb [15], critical state [19], cap [20] and recent models [21, 22], as well as the proposed model for isotropic hardening as special cases, is given by [16, 17]

$$F = J_{2D} - \left(\frac{-\alpha}{\alpha_o^{\,n-2}} J_1^{\,n} - \gamma J_1^{\,2}\right) (1 - \beta S_r)^m \tag{2}$$

where α, β, γ, m, n, are response functions, α_o is chosen to be unity (with stress units), $S_r = J_{3D}^{\,1/3}/J_2^{\,1/2}$ is stress ratio, J_{2D} and J_{3D} are second and third invariants of the deviatoric stress tensor S_{ij}. The functionals β, γ, m, and n are assumed to be constants associated with ultimate behavior, and α is growth or hardening function given by

$$\alpha = \alpha\ (\xi \quad \xi_v,\ \xi_D) \tag{3a}$$

where $\xi = \int (d\varepsilon_{ij}^p\ d\varepsilon_{ij}^p)^{1/2}$ is the trajectory of plastic strain, and ξ_v and ξ_D are its volumetric and deviatoric parts. A simple form used is

$$\alpha = \frac{a_1}{\xi^{\eta_1}} \tag{3b}$$

where a_1 and η_1 = constants found from hydrostatic and shear tests.

Typical plots of F in Eq. (2) for a soil in $J_1 - \sqrt{J_{2D}}$, octahedral and triaxial spaces are shown in Fig. 2, where $J_1 = \sigma_{ii}$ is the first invariant of the stress tensor.

Function F is adopted as yield function for defining isotropic hardening response. Then, according to the theory of plasticity with associative response:

$$d\varepsilon_{ij}^p = \lambda \frac{\partial F}{\partial \sigma_{ij}} \tag{4}$$

and from the consistency condition

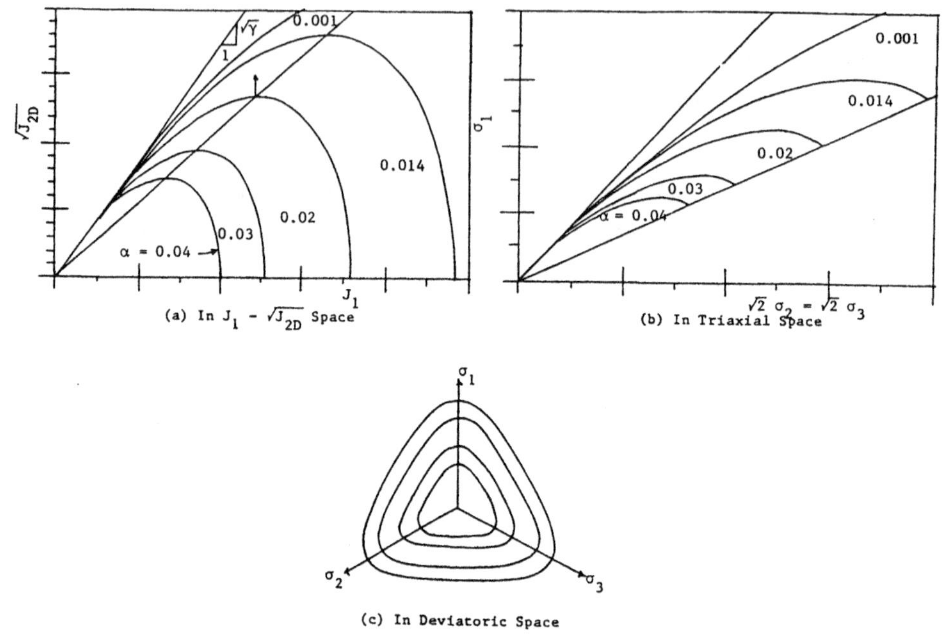

Fig.2. Plot of the yield function in various spaces; Munich sand

$$dF = 0 \tag{5}$$

the following incremental stress-strain relation can be derived:

$$\{d\sigma\} = [C^{ep}]\ \{d\varepsilon\} = ([C^e] - [C^p])\ \{d\varepsilon\} \tag{6}$$

where $\{d\sigma\}$ and $\{d\varepsilon\}$ are vectors of incremental stress and strain, respectively, and $[C^{ep}]$ = elastic-plastic constitutive matrix composed of elastic part $[C^e]$ and plasticity part $[C^p]$.

For materials that require inclusion of behavioral aspects such as nonassociativeness, softening, and induced anisotropy, it is necessary to modify the above approach and F; in this case, the form and content of Eq. (2) will change. In the proposed concept the basic idea is to consider such aspects as resulting from deviation of the response from the fundamental isotropic hardening response of (initially) isotropic material defined by F (Eq. 2). In other words, the fundamental response is defined by a set of (fixed) yield surfaces as in Fig. 2. Then factors such as nonassociativeness, softening and anisotropic hardening are introduced as perturbations or deviations from the fundamental response. For instance, nonassociative behavior is described by modifying F to $\overline{F}$ through a correction function h as

$$\overline{F} = F + h(J_i, \xi) \tag{7}$$

Then $\overline{F}$ is used as plastic potential function, Q, in the context of nonassociative plasticity [11, 15, 17]. For strain softening, a damage parameter defined in terms of plastic strains, and is superimposed on the fundamental behavior defined by F [18].

For anisotropic hardening (due to cyclic loading), an additional (potential) surface, Q, is defined, Fig. 3. This surface translates in the fixed field of yield surfaces defined by F during (cyclic) loading, unloading and reloading. The resulting model can allow for induced anisotropy due to (cyclic) loading [16].

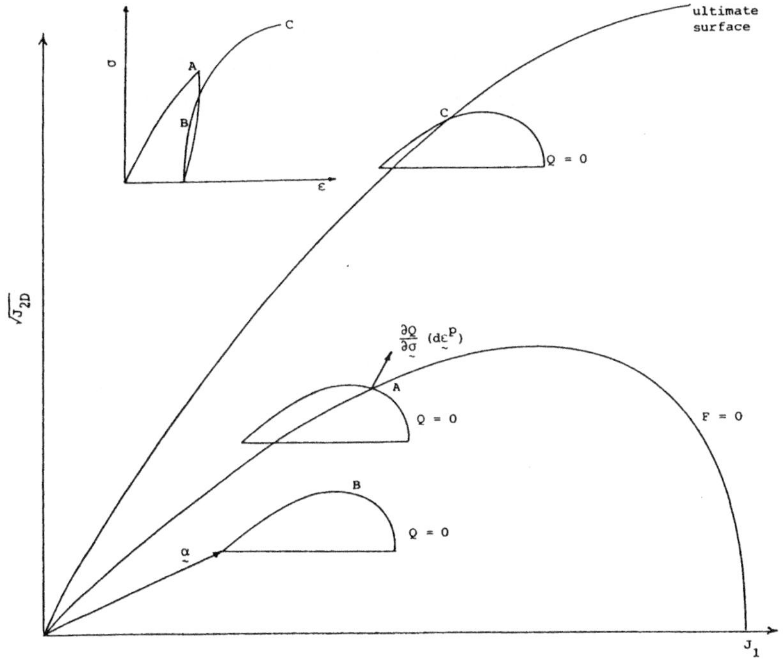

Fig.3. Anisotropic hardening

Determination of Constants

Determination of the elastic and plasticity constants is relatively straightforward. Elastic constants (E = Young's modulus and ν = Poisson's ratio) are found from unloading response. α, β, m, and γ are found from ultimate condition, which is envelope of stress states asymptotic to stress-strain curves. a_1 and η_1 are found from hydrostatic and shear tests under various stress paths. For the research herein, multiaxial tests on cubical specimens of geologic media under different stress paths are used to determine the constants.

Interface Model

A rather new and simple model called thin-layer element is proposed to simulate the interface behavior. This model can be based on nonlinear elastic or elastic-plastic concepts and can incorporate loading, unloading and reloading behavior. It allows for various modes of deformation such as no-slip, slip, debonding or separation, and rebonding, and can control interpenetration at interface [4, 6].

The thin-layer element, Fig. 4, is treated essentially like a solid element with small finite thickness. However, its constitutive behavior is defined differently. The constitutive matrix $[C_i]$ for the interface is defined through incremental stress-strain relation:

$$\{d\sigma\} = [C_i]\ \{d\varepsilon\}$$

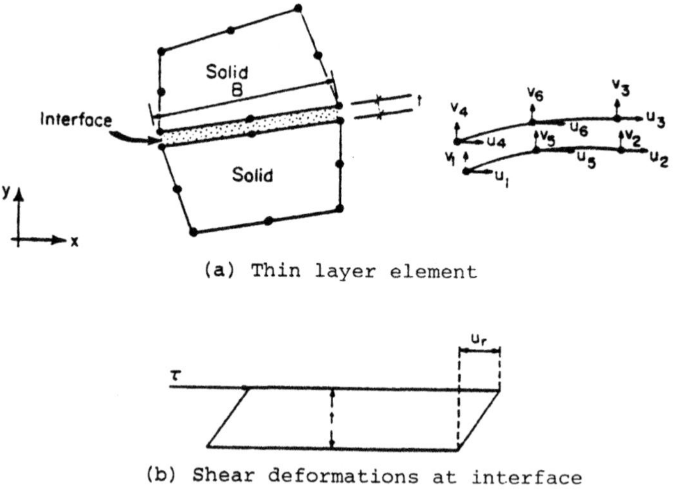

Fig.4. Interface model

where

$$[C_i] = \begin{bmatrix} [C_n] & [0] \\ \\ [0] & [C_s] \end{bmatrix} \tag{8}$$

where $[C_n]$ and $[C_s]$ are normal and shear components, respectively.

Shear Behavior - For the slip mode with compressive normal stress, the shear part of the stiffness (for nonlinear elastic model) is defined through the shear modulus, G_i, as

$$G_i \cong \frac{\tau}{u_r} t$$

where τ = shear stress, u_r = relative displacement and t = (small) thickness. It can be defined as tangent modulus from:

$$\tau = \tau \ (u_r, \sigma_n, \gamma_i, N) \tag{9}$$

where σ_n = normal stress, γ_i = initial density and N = number of loading cycles.

For nonlinear elastic model, a modified form of the Ramberg-Osgood model [8, 9] is used; Fig. 5

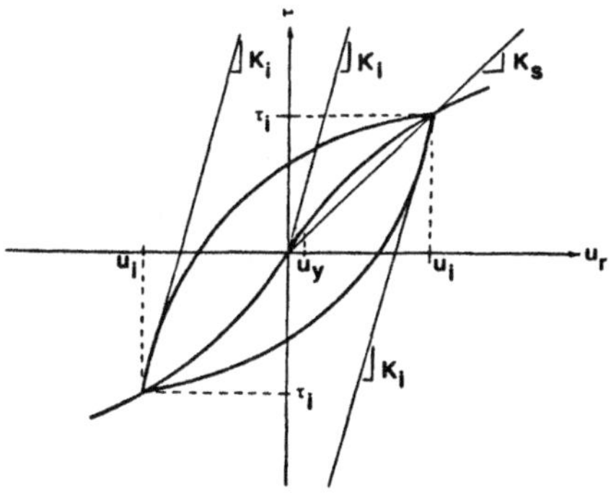

Fig.5. Ramberg-Osgood model for cyclic stress-deformation relationship

Initial Loading or Backbone Curve:

$$u_r = u_y \left(\frac{\tau}{K_i u_y}\right) \left(1 + \alpha \left|\frac{\tau}{K_i u_y}\right|^{R-1}\right) \quad (10a)$$

Unloading-Reloading Curves:

$$u_r - u_i = u_y \left(\frac{\tau - \tau_i}{K_i u_y}\right) \left(1 + \frac{2\alpha}{2^R} \left|\frac{\tau - \tau_i}{K_i u_y}\right|^{R-1}\right) \quad (10b)$$

where u_r = the relative displacement of the interface, τ = the interface shear stress, and (u_i, τ_i) = the relative displacement and shear stress, respectively, at the point of loading reversal, as shown in Fig. 5. The remaining terms in Eq. (10) are: K_i = initial shear stiffness; u_y = a reference displacement, and material parameters α and R. The tangent stiffness at the point of load reversal is assumed to be equal to K_i, as indicated in Fig. 5.

The shear modulus of an interface is usually smaller than that of the surrounding material (soil). If it is assumed that the slippage (failure) occurs in the soil, then as a simplification, its value may be adopted same as that of the soil. During slip defined by criteria such as Mohr-Coulomb, and with compressive normal stress, the value of the shear stiffness is reduced to a small value equal to about 0.01 x G_o, where G_o is the initial shear modulus.

Normal Behavior

It is relatively difficult to define the normal component $[C_n]$ for various modes of deformation. In fact, it is this normal behavior that has not been fully understood and requires careful mathematical and experimental studies. In general, the normal stiffness, C_n, is dependent on various factors:

$$C_n = C_n \left(v_r, \tau, \sigma_{ni}, \gamma_i, A_c, \sigma^{ss}, N\right) \quad (11a)$$

where v_r = (relative) normal displacement, σ_{ni} = initial stress, A_c = area of (partial contact, and σ^{ss} = stress in the surrounding elements. A form of Eq. (10) is given by

$$\sigma_n = a_o \left(e^{a_1 v_r} - 1\right) \quad (11b)$$

where a_1 and a_o are constants found from stress-strain curves for normal loading.

Determination of Constants

A series of laboratory tests are performed by using the cyclic multi-degree-of-freedom shear device [3, 7-9]. Here cyclic (direct) shear tests are performed with different values of σ_n by changing γ_i, amplitude of applied load/displacement and N to find the constants for the shear modes. For normal behavior, cyclic tests are performed with various amplitudes of σ_n with different values of γ_i and N. Attention is given to appropriate algorithms for simulation of various modes such as no-slip, slip, debonding, rebonding and control of penetration. Details of these topics are given in [7-9].

FINITE ELEMENT PROCEDURE

For dynamic analysis, the incremental form of finite element equations is

$$[m]\ \{\ddot{q}\} + [c]\ \{\Delta\dot{q}\} + (k)\ \{\Delta\dot{q}\} = \{\Delta Q(t)\} \tag{12}$$

where [m] = mass matrix, [c] = damping matrix, [k] = stiffness matrix, {q} = vector of displacements, {Q} = vector of loads, Δ denotes increment and over-dot denotes time derivative. Hysteretic damping is accounted for if nonlinear loading-unloading-reloading response through [k] is considered. Then, for this case, it may not be necessary to include the second term.

Equations (12) are solved in the time domain by using techniques such as the Newmark and Wilson schemes [23]. In the results presented subsequently, the Newmark β-method is used. However, other methods such as implicit-explicit schemes [24] are also implemented. Currently, improved time integration schemes are being developed for the nonlinear problems [25] and are implemented for the soil-structure problems. An incremental iterative technique is used for the material nonlinearity of soil and interfaces.

APPLICATION

The proposed constitutive models for soils and interfaces are implemented in a nonlinear finite element procedure which is applied for solution of a number of boundary value problems. The latter include closed-form solutions, simulation of laboratory test data, as well as (field) problems involving soil-structure interaction. Brief details of two typical problems are given below.

Pier Problem

Figures 6(a) and (b) show schematic of a pier structure founded in a non-linear soil medium (Ottawa sand), and finite element mesh. A sinusoidal motion was applied to the rigid base. Details of the material properties are given below.

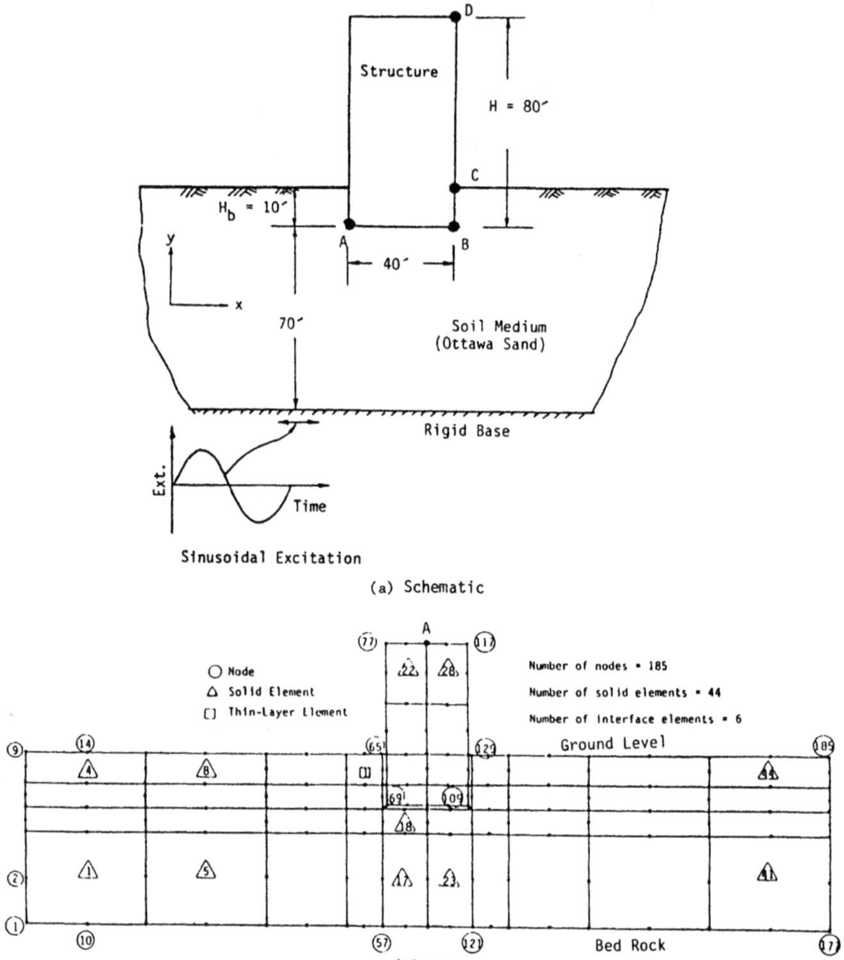

Fig.6. Pier problem

Structure (Concrete)

Young's modulus, $E_c = 3 \times 10^6$ psi (20.7×10^6 kPa)
Poisson's ratio, $\nu_c = 0.167$
Unit weight, $\gamma_c = 140$ lb/ft^3 (2.24×10^{-3} kg/cm^3)

Soil (Ottawa Sand)

Young's modulus, E_s = 38,600 psi (2.6×10^5 kPa)
Poisson's ratio, ν_s = 0.37
Unit weight, γ_s = 105 lb/ft^3 (0.0017 kg/cm^3)

Parameters for Cap Model (6, 20)

$\bar{\alpha}$ = 29.0 psi (200.0 kPa)
$\bar{\gamma}$ = 29.0 psi (200.0 kPa)
$\bar{\beta}$ = 0.01402 1/psi (0.00203 1/kPa)
$\bar{\theta}$ = 0.4
$\bar{R}$ = 2.5
$\bar{D}$ = 0.0085 1/psi (0.0012 1/kPa)
$\bar{W}$ = 0.00267
$\bar{Z}$ = 0.0

Interface

Initial modulus, E_i = 38,600 psi (2.6×10^5 kPa)
Poisson's ratio, ν_i = 0.37
Unit weight, γ_i = 105 lb/ft^3 (0.0057 kg/cm^3)
Cohesion, c_a = 0.0
Friction angle, tan δ = 0.576
$\tau = \alpha_1 + \sigma_2 u_r + \alpha_3 u_r^2$
(α_i are material constants [6])

Excitation (Sinusoidal)

Frequency, Ω = 2.0 Hz
Amplitude, A_o^* = 0.2g
(g = acceleration due to gravity)

Figures 7(a) to (d) show computed displacements, and velocities at a typical point for two cases, bonded and frictional interfaces. The bonded case assumes complete bonding between the structure and soil and no relative motion at interface, whereas the frictional case involves proposed interface model allowing for various deformation modes. It can be seen that when relative motions are allowed, the two responses are significantly different. Such differences will be dependent on factors such as location, time, frequency, loading and physical properties. However, it is evident that provision of the relative motions does influence the response of the structure-foundation system.

Field Problem

The field test program, known as SIMQUAKE, was conducted by the University of New Mexico under a research project sponsored by the Electric Power Research Institute (EPRI). Here response of model nuclear containment structures (in the field) due to strong ground motion (generated by blasts) was recorded; details of the tests and measurements are given by Vaughan and Isenberg [26, 27]. The response of a 1/8 scale model structure is analyzed here by using a plain strain finite element idealization. The schematic of the structure and the finite element mesh are shown in Fig. 8(a) and (b), and the material properties are given below.

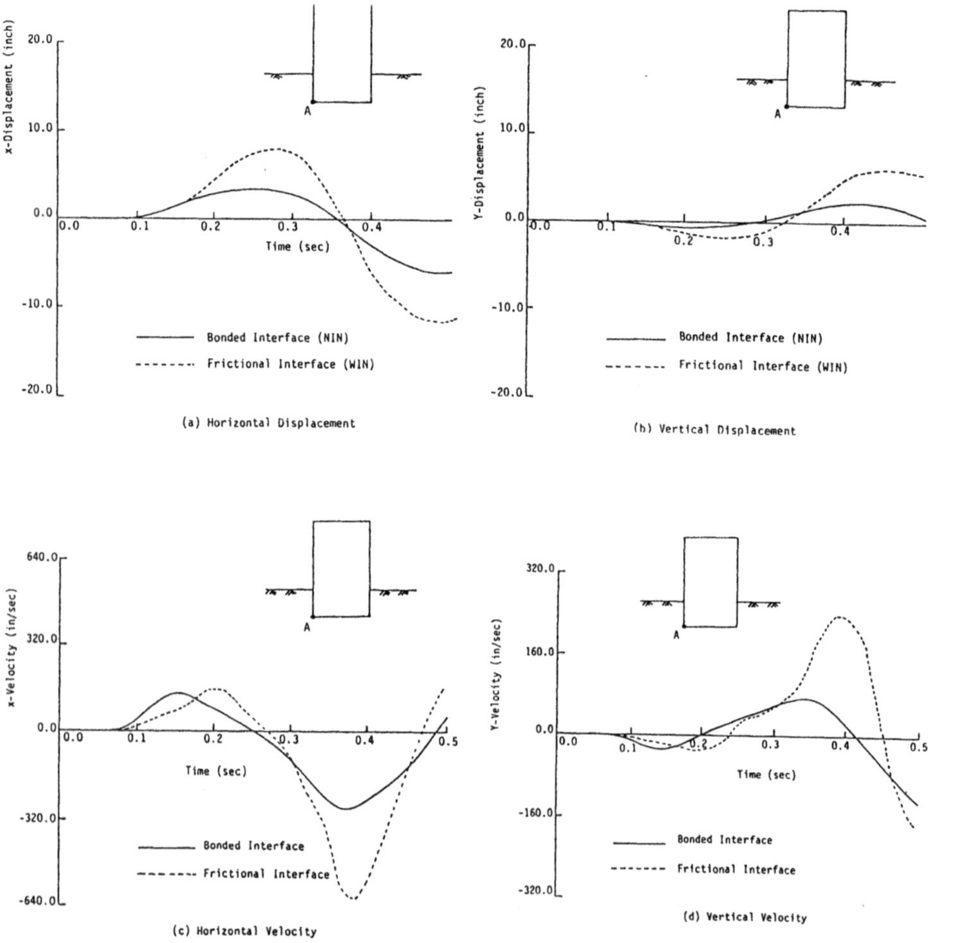

Fig.7. Comparisons for predicted behavior for bonded and frictional interfaces

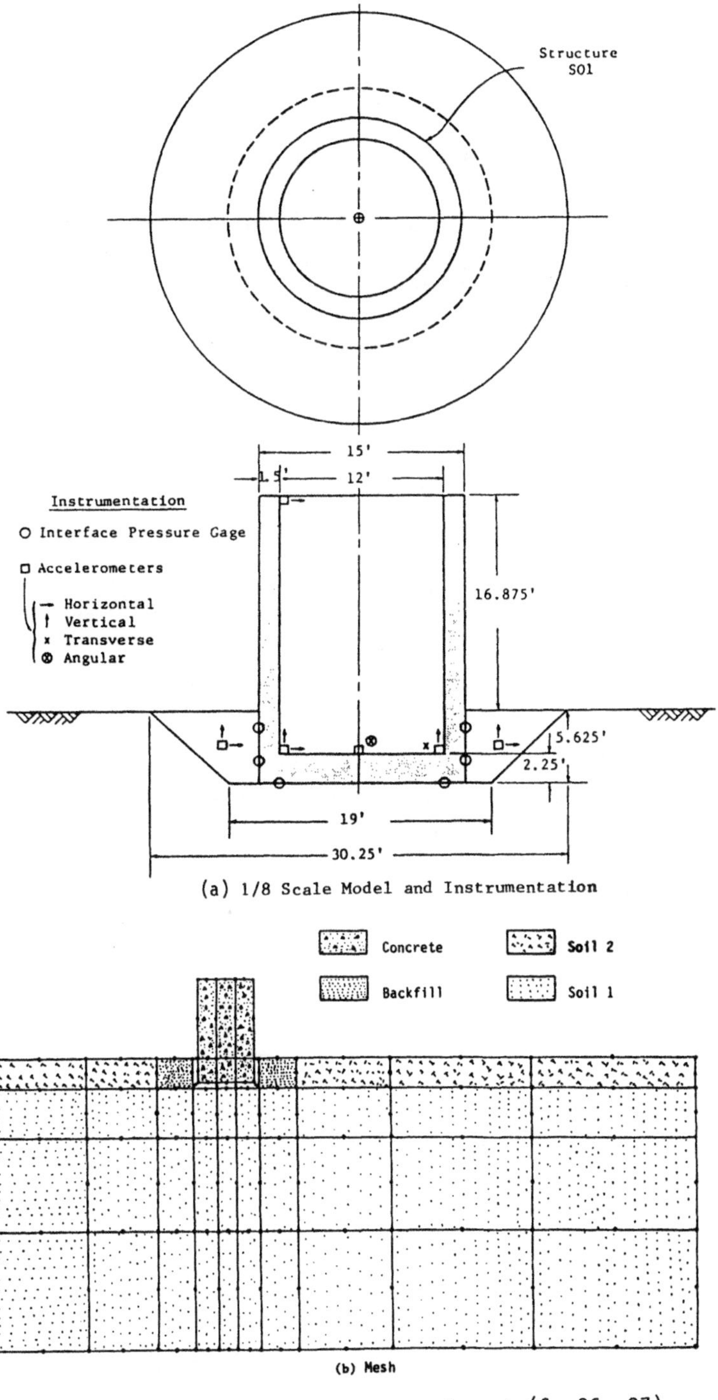

Fig.8. SIMQUAKE structure and mesh (6, 26, 27)

Cap Model Parameters for Materials for Soils and Elastic Parameters for Concrete [26, 27]

Parameters (1)	Soil 1 (2)	Soil 2 (3)	Backfill (4)	Concrete (5)
E psi (kPa)	46,435 (3.2×10^5)	22,113 (1.5×10^5)	13,100 (9.0×10^4)	4×10^6 (2.7×10^7)
ν	0.3	0.3	0.26	0.2
α psi (kPa)	470.0 (3,240.2)	470.0 (3,240.2)	470.0 (3,240.2)	
θ	0.0	0.0	0.0	
β 1/psi (1/kPa)	0.165 (0.024)	0.165 (0.024)	0.165 (0.024)	
γ psi (kPa)	390.0 (2,688.7)	390.0 (2,688.7)	390.0 (2,688.7)	
D 1/psi (1/kPa)	0.0007 (0.000102)	0.0007 (0.000102)	0.00065 (0.000094)	
W	0.06	0.06	0.06	
Z	0.0	0.0	0.0	
R	2.5	2.5	2.5	
γ_w lb/in^3 (kg/cm^3)	0.0637 (0.00176)	0.0637 (0.00176)	0.0637 (0.00176)	0.058 (0.0016)

Figure 9 shows measured and applied displacements on the boundary nodes of the mesh in Fig. 8. A cap model [20] was used for soils and the interface response was obtained from laboratory tests [6, 8, 9].

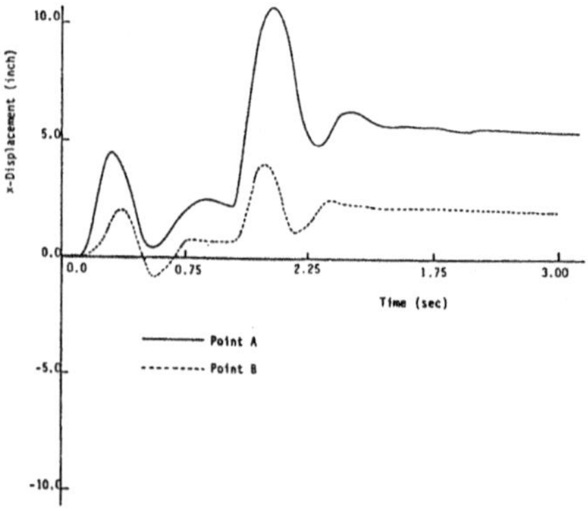

Fig.9. Input horizontal displacement-time history

Typical results for the comparison of predicted and observed velocities and and contact pressures are shown in Figs. 10 and 11, respectively.

Difference between typical computed results for frictional and bonded cases for acceleration is shown in Fig. 12. It can be seen that the comparisons are very good for the first second of the response and thereafter they are not very satisfactory. In view of the complexity of the problem, overall, the correlation is considered to be satisfactory. There is significant difference between results from the bonded and frictional cases.

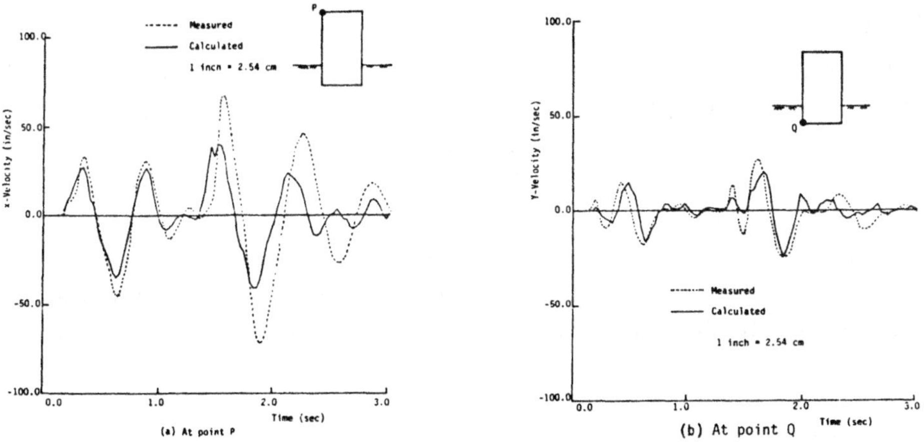

Fig.10. Comparison of computed and measured velocity-time histories

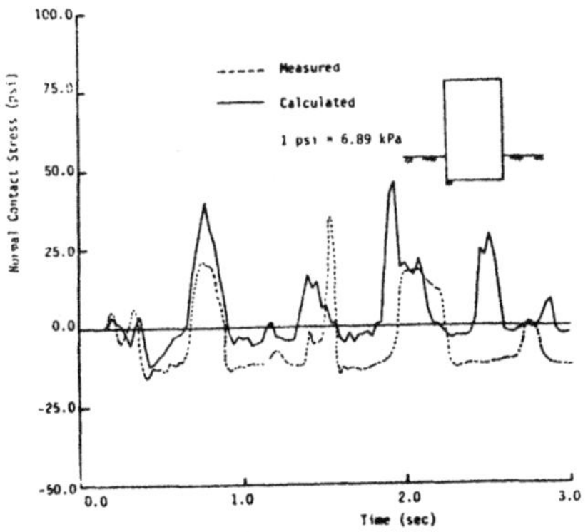

Fig.11. Comparison of calculated and measured normal contact stress-time histories beneath upstream corner of structure

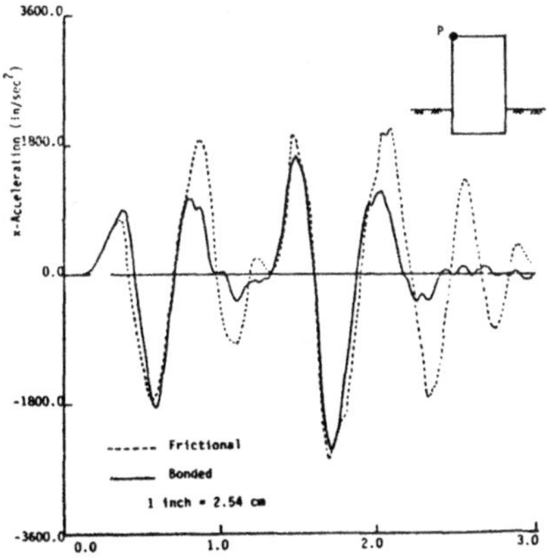

Fig.12. Influence of interface condition on horizontal acceleration-time history at top of structure in SIMQUAKE, point P

It may be noted that provision of interface motions can increase or decrease the response (displacements, velocities, accelerations) depending upon the location, time period, frequency, and other factors related to loading and physical properties.

Based on the comparisons for the field problem and the verification, it is believe that the proposed interface model can provide a satisfactory and consistent representation of the interface behavior.

Conclusions

For realistic prediction of the dynamic behavior of structure-foundation systems, it is necessary to allow for nonlinear behavior of soils and interfaces. For the latter, it is required to include various deformation modes such as no-slip, slip, debonding and rebonding with appropriate shear and normal responses. When these are incorporated in solution procedures such as nonlinear finite element schemes, it is possible to obtain improved and realistic evaluation of displacements, velocities, accelerations and contact pressures in dynamic loadings. Two typical examples are described to show this capability.

ACKNOWLEDGMENT

A part of this research was conducted under Grant No. 81-CE-94 from the National Science Foundation. Contributions of Dr. M. M. Zaman and Dr. E. C. Drumm towards the results reported in this paper are gratefully acknowledged.

References

1. Desai, C. S.: Behavior of interfaces between structural and geologic media. Int. Conf. on Recent Advances in Geotech. Earthquake Eng. and Soil Dynamics, St. Louis, MO.. 1981.

2. Wolf, P.: Dynamic soil-structure interaction. Prentice-Hall, Englewood Cliffs, N.J. 1985.

3. Desai, C. S.: A dynamic multi-degree-of-freedom shear device. Report No. 80-36, Dept. of Civil Eng., VA TECH, Blacksburg, VA, 1981.

4. Desai, C.S.; et al. Thin-layer element for interfaces and joints. Int. J. Num. Analyt. Methods in Geomech., 8, 1, 1984.

5. Zaman, M. M.; Desai, C.S.: Analysis of interface behavior in dynamic soil-structure interaction. 7th Int. Conf. on Struct. Mech. in Reactor Tech., Chicago, August 1983.

6. Zaman, M. M.; Desai, C. S.; Drumm, E. C.: Interface model for dynamic soil-structure interaction. J. Geotech. Eng. Div., ASCE, 110, 9, 1984.

7. Desai, C. S.; Zaman, M. M.: Modelling and evaluation of interface behavior in dynamic soil-structure interaction. Proc., 8th World Conference on Earthquake Eng., San Francisco, July 1984.

8. Desai, C. S.; Drumm, E. C.; Zaman, M. M.: Cyclic testing and modelling of interfaces. J. Geotech. Eng. Div., ASCE, 111, 6, June 1985.

9. Drumm, E. C.; Desai, C. S.: Determination of parameters for a model for cyclic behavior of interfaces. J. of Earthquake Eng. & Struct. Dyn., accepted for publication.

10. Desai, C. S.: A general basis for yield, failure and potential functions in plasticity. Int. J. Num. Analyt. Methods in Geomech., 4, 4, 1980.

11. Desai, C. S.; Siriwardane, H. J.: A concept of correction functions to account for nonassociative characteristics of geologic media. Int. J. Num. Analyt. Methods in Geomech., 4, 4, 1980.

12. Desai, C. S.; Faruque, M. O.: Constitutive model for (geological) materials. J. Eng. Mech. Div., ASCE, 110, 9, Sept. 1984.

13. Baker, R.; Desai, C. S.: Induced anisotropy due to plastic straining. Int. J. Num. Analyt. Meth. in Geomech. 8, 2, 1984.

14. Faruque, M. O.; Desai, C. S.: Implementation of a general constitutive model for geological materials. Int. J. Num. Analyt. Methods in Geomech., in press.

15. Desai, C. S.; Siriwardane, H. J.: Constitutive laws for engineering materials. Prentice-Hall, Englewood Cliffs, N.J. 1984.

16. Desai, C. S.; Frantziskonis, G.; Somasundaram, S.: Constitutive modelling for geological materials. Proc., 5th Int. Conf. on Num. Meth. in Geomech., Nagoya, Japan, April 1985.

17. Frantziskonis, G.; Desai, C. S.; Somasundaram, S.: Constitutive model for nonassociative behavior. J. Eng. Mech. Div., ASCE, submitted.

18. Frantziskonis, G.; Desai, C. S.: Constitutive model for strain softening. Int. J. Solids & Structures, submitted.

19. Schofield, N.; Wroth, P.: Critical state soil mechanics, McGraw-Hill, London, 1968.

20. DiMaggio, F. L.; Sandler, I.: Material model for granular soils. J. of Eng. Mech. Div., ASCE. 97, 3, 1971.

21. Lade, P.: Elastic-plastic stress-strain theory for cohesionless soil with curved yield surfaces. Int. J. Solids & Struct., 13, 1977.

22. Matsuoka, H.; Nakai, T.: Stress-deformation and strength characteristics of soil under three different principal stresses. Proc., Japan Soc. of Civil Engineers, 232, 1974.

23. Bathe, K. J.: Finite element procedures in engineering analysis. Prentice-Hall, Englewood Cliffs, N. J., 1982.

24. Belytschko, T.; Hughes, T.J.R. (Eds): Computational methods in transient analysis. Elsevier, Amsterdam, 1983.

25. Kujawski, J.; Desai, C. S.: Generalized time finite element algorithm for nonlinear dynamic problems. Engineering Computations, 1, 3, 1984.

26. Vaughan, D. K.; Isenberg, J.: Nonlinear soil-structure analysis of SIMQUAKE II. Research Project 810-2, Final Report, Electric Power Res. Inst., (EPRI), Weidlinger Associates, Menlo Park, California, 1982.

27. Vaughan, D. K.; Isenberg, J.: Nonlinear rocking response of model containment structures. Earthquake Eng. & Struct. Dyn., 11, 1983.

Part III:
Frames, Plates and Shells

Analysis Procedure for Space Frames with Material and Geometrical Nonlinearities

H. NEDERGAARD, P.T. PEDERSEN
The Technical University of Denmark, Lyngby, Denmark

Abstract

It is the purpose of the paper to present an efficient structural analysis procedure which can handle highly non-linear, complicated space frames using only one beam element to model any prismatic beam member.

Shear deformation of beam members is included but warping is neglected. In the first step of the derivation of the beam stiffness matrices the material is assumed to be linearly elastic. Using shape functions associated with the classical Timoshenko beam element the symbolic computer language FORMAC is used to derive a consistent geometrically non-linear stiffness matrix. The formulation takes into account the coupling between axial- and lateral deformations. The resulting equations are linearized in load increments and the geometry is updated after each increment.

The material non-linearities are taken into account by use of a plastic hinge method including the effect of gradual plastification of the cross-sections. The procedure is based on a yield surface formulation using generalized beam forces instead of stresses. Plastic hinges are allowed to be formed at the beam ends and at midspan.

The numerical procedure is based on Bergan's "current stiffness parameter" technique such that also structures with snap-through behaviour can be traced into the post ultimative load region.

The theory is first checked on some simple, plane structures. Finally, it is applied to an offshore jacket structure.

1. Introduction

Analysis of three dimensional frame structures based on linear elastic stiffness methods are nowadays standard engineering procedures. Such methods are reliable and easy to use. Presently under development, and finding ever wider use, are the non-linear methods taking into account geometric non-linearities and material non-linearities. The ultimate mode of failure of a

Finite Element Methods for Nonlinear Problems
Europe-US Symposium, Trondheim, Norway 1985
Editors: Bergan, Bathe, Wunderlich

particular structure in consideration depends on the mutual interaction of these two effects. Exact determination of this interaction and its effects are normally the very subject of non-linear analyses. Good progress has been made recently in this field for example with development of program systems such as FENRIS described by Arnesen and Bergan in [1] and DYCAST described by Hayduk et al in [2]. But, common to all truly non-linear programs are that they are relatively complicated to use and need experts to produce reliable results. Therefore, there is still a need for research in this area.

It is the purpose of the present paper to contribute to the development of practical analysis procedures which in an efficient way can handle highly non-linear space frames.

In all non-linear diplacement formulations the stiffness matrix must be evaluated many times. Therefore, in order to save computer time the procedure to be described here is based on a formulation where the element stiffness matrix can be calculated without numerical integration.

Also from a practical point of view the element size employed in a given problem should be as large as possible, of course still ensuring that all the important aspects of the structural problem are modelled properly. Consequently, it has been an objective to derive a procedure where any prismatic member in all cases can be modelled as one element.

The geometrical non-linearity of beam elements can be included in different ways. Use of an element stiffness matrix which depends on loads has been made by Akkoush et al [3], Argyris et al [4], Archer [5], Bagci [6], Powell [7], Saran [8] and Moan et al [9]. Common to these formulations are that they even with few elements describe the pre-buckling deformation well. But the axial shortening of the element due to large lateral displacements can only be taken into account by using many elements and then updating the geometry. Reissmann [10], Meek and Tan [11], Ueda et al [12] and Wen and Rahimzadeh [13] have described procedures which include shortening of the element due to lateral deformations. Thereby the number of elements can be reduced even for analysis in the post-buckling region. The model developed by Reissmann is based on the application of Green's strain tensor for derivation of a set of equations of equilibrium for an incremental iterative-calculation procedure. In the present procedure we shall also make use of Green's strain-tensor. However, by using the symbolic computer language FORMAC we can include important terms and still obtain analytical expressions for the incremented stiffness matrices.

Material non-linearities can be included in beam elements in an economic way by the plastic node concept which has been described by Ueda et al [14]. This theory rests on the assumption of an ideal elastic-plastic material but takes into account the effect of redistribution of generalized forces at a plastic hinge using the concept of a yield surface. In the present formulation also the gradual plastification of cross-sections has been included in the mathematical model in order to obtain a more realistic model and at the same time avoid the numerical difficulties associated with discontinuous stiffness properties.

Since the material non-linearities are modelled by a flow theory the numerical solution procedure must be based on an incremental procedure. The size of the increments is in the present procedure determined by use of the "current stiffness parameter" described by Bergan & Søreide [15] and Bergan [16].

2. Non-linear Elastic Beam Element

Figure 1 shows a prismatic beam element with six generalized displacements and associated forces and with external distributed line loading. One of the local axes, the x-axis, is taken to be joining the two end nodes, while the y- and the z-axis are obtained from the principal directions of the end sections. Since in the deformed configuration the two end sections may be rotated differently, the y- and z-axis are defined to be the average of the principal directions of the two end sections. Shear deformation will be included in the model but warping is neglected.

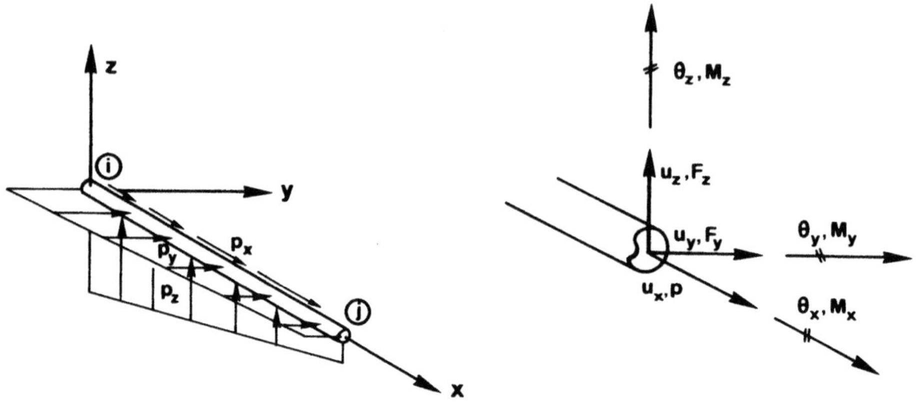

Fig.1. Prismatic beam element with loading and generalized forces and displacements.

For this element we shall in the following set up an incremental stiffness matrix with updating of nodal coordinates after each increment. Stresses and strains are derived in the local convective coordinate system which rotates and translates with the member.

In deriving the basic member force-deformation relations we use the interpolation functions associated with linear Timoshenko beam theory. That is, we assume that the lateral deflection curve is cubic in each of the two principal lateral directions and that the axial displacement and torsion varies linearly along the elastic part of the member. We shall assume that plane sections remain plane. Thereby we obtain the following strain vector

$$\bar{\varepsilon} = \begin{bmatrix} U_x' + \frac{1}{2}\left((U_x')^2 + (U_y')^2 + (U_z')^2\right) \\ U_y' - \theta_z \\ U_z' + \theta_y \\ \theta_x' \\ \theta_y' \\ \theta_z' \end{bmatrix} \qquad (1)$$

For the elastic part of the beam we shall use constitutive equations in the form

$$
\begin{aligned}
N &= EA\cdot\left(U_x' + \tfrac{1}{2}\left((U_x')^2 + (U_y')^2 + (U_z')^2\right)\right)\\
F_y &= k_y GA(U_y' - \theta_z)\\
F_z &= k_z GA(U_z' + \theta_y)\\
M_x &= GI_x\theta_x'\\
M_y &= EI_y\theta_y'\\
M_z &= EI_z\theta_z'
\end{aligned}
\tag{2}
$$

where EI_y and EI_z are the principal flexural rigidities, k_yGA and k_zGA shear stiffnesses, EA the axial rigidity and GI_x the torsional rigidity. In matrix form the constitutive equations take the form:

$$\overline{Q} = \overline{\overline{C}}\,\overline{\varepsilon} \tag{3}$$

A condition for equilibrium with an external load $\overline{p}$ is that the variation of the potential energy π equals zero:

$$\delta\pi(\overline{p}) = \int_0^1 \delta\overline{\varepsilon}^T\overline{Q}\,dx - \int_0^1 \delta\overline{U}^T\overline{p}\,dx = 0 \tag{4}$$

If we now add a load increment $\Delta\overline{p}$ we get an associated strain increment

$$\Delta\overline{\varepsilon} = \begin{bmatrix} \Delta U_x' \\ \Delta U_y' - \Delta\theta_z \\ \Delta U_z' + \Delta\theta_y \\ \Delta\theta_x \\ \Delta\theta_y \\ \Delta\theta_z \end{bmatrix} + \begin{bmatrix} U_x'\Delta U_x' + U_y'\Delta U_y' + U_z'\Delta U_z' \\ 0 \\ 0 \\ 0 \\ 0 \\ 0 \end{bmatrix} + \begin{bmatrix} \tfrac{1}{2}\Delta U_x'^2 + \tfrac{1}{2}\Delta U_y'^2 + \tfrac{1}{2}\Delta U_z'^2 \\ 0 \\ 0 \\ 0 \\ 0 \\ 0 \end{bmatrix} \tag{5}$$

Or

$$\Delta\bar{\varepsilon} = \Delta\bar{\varepsilon}_e + \Delta\bar{\varepsilon}_{g_2} + \Delta\bar{\varepsilon}_g$$

With the load increment $\Delta\bar{p}$ the potential energy of the element is changed. The new equilibrium position is found by setting the variation of this change to zero:

$$\delta\Delta\pi(\Delta\bar{p}) = \int_0^1 \delta\Delta\bar{\varepsilon}^T(\bar{Q} + \bar{Q}_f + \Delta\bar{Q})\,dx - \int_0^1 \delta\Delta\bar{U}^T(\bar{p}+\Delta\bar{p})\,dx = 0 \qquad (6)$$

Here $\bar{Q}$ denotes internal generalized forces at the previous step, $\bar{Q}_f$ those moments which arise due to the changed position of the normal force P and $\Delta\bar{Q}$ denotes the increase in generalized forces due to $\Delta\bar{p}$. For convenience we shall now introduce a strain vector $\bar{\varepsilon}_{g1}$ through the relation

$$\delta\Delta\bar{\varepsilon}_g^T\bar{Q} = \delta\bar{\varepsilon}_{g1}^T P\bar{\varepsilon}_{g1} \qquad (7)$$

Furthermore, we shall introduce a matrix $\bar{\bar{C}}_f$ through the relation

$$\bar{Q}_f = \bar{\bar{C}}_f\ \Delta\bar{U}_{i-1} \qquad (8)$$

where ΔU_{i-1} is the displacement increment from the previous step.

Analogous to Eq.(3) we have

$$\Delta\bar{Q} = \bar{\bar{C}}\ \Delta\bar{\varepsilon} \qquad (9)$$

Substituting Eqs.(4), (8) and (9) into Eq.(6) leads to

$$\delta\Delta\pi(\Delta\bar{p}) = \int_0^1 \delta\Delta\bar{\varepsilon}^T\bar{\bar{C}}\Delta\bar{\varepsilon}\,dx + \int_0^1 \delta\Delta\bar{\varepsilon}_{g1}^T P\Delta\bar{\varepsilon}_{g1}\,dx$$

$$+ \int_0^1 \delta\Delta\bar{\varepsilon}_e^T\bar{\bar{C}}_f\Delta\bar{U}_{i-1}\,dx + \int_0^1 \delta\Delta\bar{U}^T\Delta\bar{p}\,dx = 0 \qquad (10)$$

The interpolation matrix $\bar{\bar{N}}$ mentioned above is defined through

$$\Delta\bar{U} = \bar{\bar{N}}\,\Delta\bar{q}$$

By differentiation we can then obtain the following relations between strains and nodal displacements $\Delta\bar{q}$:

$$\begin{aligned}\Delta\bar{\varepsilon}_e &= \bar{\bar{B}}_e \Delta\bar{q}\\ \Delta\bar{\varepsilon}_{g1} &= \bar{\bar{B}}_{g1}\,\Delta\bar{q}\\ \Delta\bar{\varepsilon}_{g2} &= \bar{\bar{B}}_{g2}\,\Delta\bar{q}\end{aligned} \tag{11}$$

Introducing Eq.(11) into Eq.(10), performing the variation and linearizing the resulting equation yields

$$(\bar{\bar{K}}_e + \bar{\bar{K}}_{g1} + \bar{\bar{K}}_{g2})\,\Delta\bar{q} \quad = \quad \bar{F}_{outer} - \bar{F} \tag{12}$$

where

$$\bar{\bar{K}}_e \quad = \int_0^1 \bar{\bar{B}}_e^T \bar{\bar{C}}\,\bar{\bar{B}}_e\,dx$$

$$\bar{\bar{K}}_{g1} \quad = \int_0^1 \bar{\bar{B}}_{g1}^T\,P\bar{\bar{B}}_{g1}\,dx$$

$$\bar{\bar{K}}_{g2} \quad = \int_0^1 (\bar{\bar{B}}_e^T \bar{\bar{C}}\,\bar{\bar{B}}_{g2} + \bar{\bar{B}}_{g2}^T\,\bar{\bar{C}}\,\bar{\bar{B}}_e + \bar{\bar{B}}_{g2}^T\,\bar{\bar{C}}\,\bar{\bar{B}}_{g2})\,dx$$

$$\bar{F} \quad = \int_0^1 \bar{\bar{B}}_e^T \bar{\bar{C}}_f \bar{\bar{N}}\,dx\;\cdot\Delta\bar{q}_{i-1}$$

Analytical expressions for the elements in these matrices have been derived by the symbolic computer language FORMAC. However, one simplification has been introduced whereby the complexity is reduced considerably. The last term in $\bar{\bar{K}}_{g2}$ is deleted since it only contains terms which are quadratic in x-derivatives of displacements, whereas the other terms are linear in the same quantities.

In Eq.(12) $\overline{\overline{K}}_e$ is the standard element stiffness matrix for a Timoshenko beam, $\overline{\overline{K}}_{g1}$ is a matrix which depends on the normal force acting on the beam element. It contains the eigenvalues for buckling and it has been derived earlier for instance by Archer [5] and Saran [8]. The $\overline{\overline{K}}_{g2}$ matrix, see Appendix I, is believed to be new. It is a function of the deformation of the element and gives the coupling between axial and lateral deformations. It is the inclusion of this matrix which makes it possible to model each prismatic beam element as one element even in the post-buckled region. The vector $\overline{F}_{outer}$ is the consistent load vector known from linear theory whereas $\overline{F}$ denotes corrections due to the displacement of the element.

3. Material Non-linearities

As mentioned in the introduction yielding of the beam elements will be modelled by the plastic node concept introduced by Ueda and co-workers [14]. Using this approach a plastic node is inserted into one or both ends of the beam when the combination of generalized forces reaches a yield surface. According to [14] the reduction in stiffness due to plasticity can be introduced through a plastic reduction matrix defined as

$$\overline{\overline{K}}_P = -\overline{\overline{K}}_{te}\overline{\overline{G}}(\overline{\overline{G}}^T\overline{\overline{K}}_{te}\overline{\overline{G}})^{-1}\overline{\overline{G}}^T K_{te} \qquad (13)$$

where $\overline{\overline{K}}_{te}$ is the incremental elastic tangent stiffness matrix:

$$\overline{\overline{K}}_{te} = \overline{\overline{K}}_e + \overline{\overline{K}}_{g1} + \overline{\overline{K}}_{g2}$$

If we let $\Gamma_P(N, F_y, F_z, M_x, M_y, M_z) = 0$ denote the plastic yield surface in terms of generalized forces the gradients to this yield surface at the two beam ends i and j can be obtained as

$$\overline{G}_i^T = \left[\frac{\partial\Gamma_P}{\partial Q_{1i}}, \frac{\partial\Gamma_P}{\partial Q_{2i}}, \ldots\ldots\ldots, \frac{\partial\Gamma_P}{\partial Q_{6i}} \right]$$

The matrix $\overline{\overline{G}}$ is then defined as

$$\bar{\bar{G}} = \begin{bmatrix} \bar{G}_i & \bar{0} \\ \bar{0} & \bar{G}_j \end{bmatrix}$$

In the numerical examples to follow the plastic yield surface for thin-walled circular tubes have been approximated by the expression

$$\Gamma_P = \frac{\sqrt{M_y^2 + M_z^2}}{M_P} - \cos\left(\frac{1}{4rh\sigma_y}(P^2 + 3F_y^2 + 3F_z^2 + 3\left(\frac{M_x}{r}\right)^2)^{\frac{1}{2}}\right) = 0$$

where M_P is the plastic moment, h denotes the wall thickness and r the radius of the tube.

Let us also introduce an initial yield surface Γ_e inside the plastic yield surface, see Fig.2. Again for circular tubes the initial yield surface has been taken in the form:

$$\Gamma_e = \text{Max}\begin{cases} \dfrac{1}{\pi rh\sigma_y}\sqrt{\left(\dfrac{\sqrt{M_y^2+M_z^2}}{r} + \dfrac{|P|}{2}\right)^2 + 3\left(\dfrac{M_x}{2r}\right)^2} - 1 \\ \\ \dfrac{1}{\pi rh\sigma_y}\sqrt{\left(\dfrac{P}{2}\right)^2 + 3\left(\dfrac{|M_x|}{2r} + \sqrt{F_y^2+F_z^2}\right)^2} - 1 \end{cases} = 0$$

In the region between the two surfaces the degree of plastification at the two beam ends i and j will be described by the matrix

$$\bar{\bar{\eta}} = \begin{bmatrix} \eta_i & 0 \\ 0 & \eta_j \end{bmatrix} \tag{14}$$

Here η_i and η_j are determined such that they take the value one when the ends are fully plastified and the value zero when the initial yield surface has not yet been reached. Between these two limits η_i should be chosen such that the transition from elastic to fully plastic behaviour is modelled as well as possible. A suitable form is

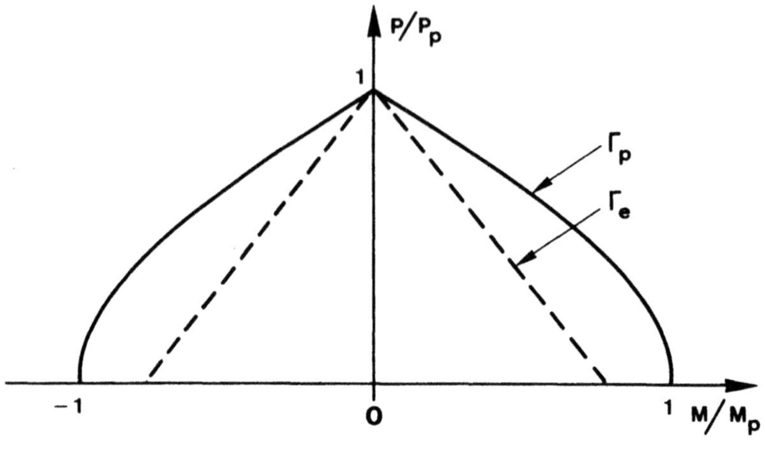

Fig.2. Initial yield Γ_e and plastic yield Γ_p surfaces for combinations of bending moments and normal forces.

$$\eta_i = \begin{cases} 0 & \text{for } \Gamma_e \leqq 0 \\ \Gamma_e^{-\alpha\Gamma_p} & \text{for } \Gamma_e > 0 \text{ and } \Gamma_p < 0 \\ 1 & \text{for } \Gamma_p \geqq 0 \end{cases}$$

It can be shown that introducing Eq.(14) into the plastic node theory leads to an expression analogous to Eq.(13) for the plasticity reduction matrix:

$$\bar{\bar{K}}_p = -\bar{\bar{K}}_{te}\bar{\bar{G}}_\eta(\bar{\bar{G}}^T\bar{\bar{K}}_{te}\bar{\bar{G}})^{-1}\bar{\bar{G}}_\eta^T\bar{\bar{K}}_{te} \qquad \text{where} \qquad \bar{\bar{G}}_\eta = \overline{\bar{\bar{G}}\eta} \tag{15}$$

Altogether the incremental tangent stiffness equation for the beam element takes the form

$$\bar{\bar{K}}_t\Delta\bar{q} = \bar{F}_{outer} - \bar{F} \tag{16}$$

where the incremental tangent stiffness matrix $\bar{\bar{K}}_t$ is determined as

$$\bar{\bar{K}}_t = \bar{\bar{K}}_{te} + \bar{\bar{K}}_p = \bar{\bar{K}}_e + \bar{\bar{K}}_{g1} + \bar{\bar{K}}_{g2} + \bar{\bar{K}}_p \qquad (17)$$

The theory described above can only introduce plastic nodes at beam ends. When it is necessary to introduce plastic nodes also in the middle of a beam element this can be done by first introducing an additional node, and then condensate the internal degrees of freedom.

4. Numerical Solution Procedure

As for linear systems the element stiffness matrices and the consistent load vectors are formed in the local coordinate systems and then transformed to a fixed global set of coordinates. Here the global stiffness matrix and load vector are assembled. The main difference between the present non-linear solution procedure and a linear formulation lies in the special element stiffness matrices, updating of the local coordinate systems and selection of load increments.

Updating of the local coordinate systems is performed as follows When the global displacements in one increment have been calculated the incremented generalized forces in the global system can be evaluated. Then summation of forces and displacements is performed. With the updated geometry new transformation matrices can be calculated whereafter the updated nodal forces and deformations are transformed to the local system and used for calculation of the next incremental stiffness matrix and load vector.

The selection of load increments characterized by a scalar parameter λ_i is performed by the "current stiffness parameter" introduced by Bergan & Søreide [15] and by Bergan [16]. According to this method the parameter

$$S_p^i = \frac{\lambda_i}{\lambda_1} \frac{(\Delta\bar{q}^1)^T \bar{F}_{ref}}{(\Delta\bar{q}^i)^T \bar{F}_{ref}} \qquad (18)$$

is an expression for the stiffness of the system. In order to avoid large deviations from the curve of equilibrium the load increments shall be small when the stiffness is small and when large changes of the stiffness occur. In the examples to follow the scaling of the load parameter λ_i is based on

$$\lambda_i = \text{Min}\left[S_P^i; \frac{S_P^i}{(1-\Delta S_P^{nom}+\Delta S_P)^{\beta}} ; \frac{\Delta\Gamma_P^{nom}}{\Delta\Gamma_P}\right] \tag{19}$$

where

$$\Delta S_P = |S_P^i - S_P^{i-1}|$$

$$\Delta\Gamma_P = |\Gamma_P^i - \Gamma_P^{i-1}|$$

and where typical values are

$$\Delta S_P^{nom} \simeq 0.05$$

$$\Delta\Gamma_P^{nom} \simeq 0.1$$

$$\beta \simeq 20$$

5. Examples and Discussion

The analysis task indicated in Fig.3 has been solved analytically and tested experimentally by Williams [17] in 1963. Since then several authors, see for example Wood & Zienkiewicz [18] and Meek & Tan [11] have used this example to test non-linear

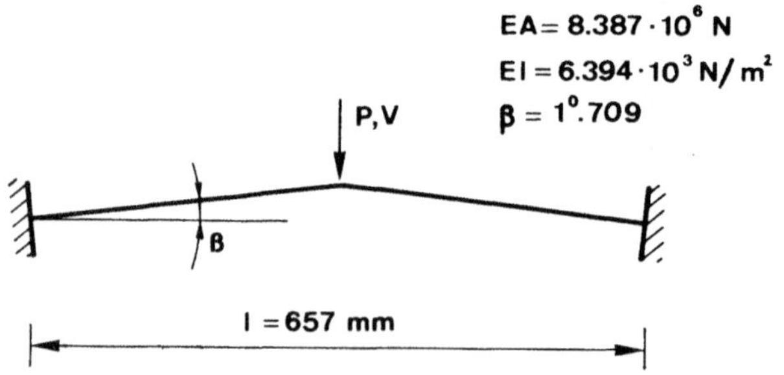

Fig.3. Williams' toggle frame.

elastic solution procedures.

The elastic solution by Williams takes into account large deformations, the influence of the axial forces on the flexural stiffness and axial shortening due to bending. The present theory has been tested by modelling each of the two beams as one element. As seen in Fig.4 the present results are in close agree-

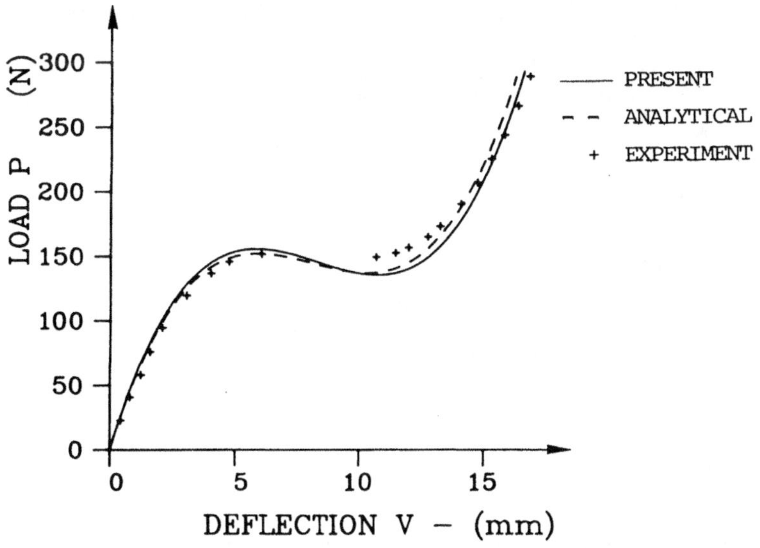

Fig.4. Load-deflection curve for toggle frame.

ment with the analytical solution by Williams.

As a second example which includes the effect of plasticity we have chosen the K-brace indicated in Fig.5. This frame has been analysed numerically using the program FENRIS, and tested experimentally by Moan et al [9]. In the FENRIS calculations each beam were divided into eight elements. Thus 16 elements were used to model half the frame. The material properties were modelled by a strain-hardening law and initial geometrical imperfections were included with amplitudes around 2 mm . In the present numerical model each beam is modelled as only one element so that the frame model consists of only two elements. An ideal elastic-plastic material model is used with a yield

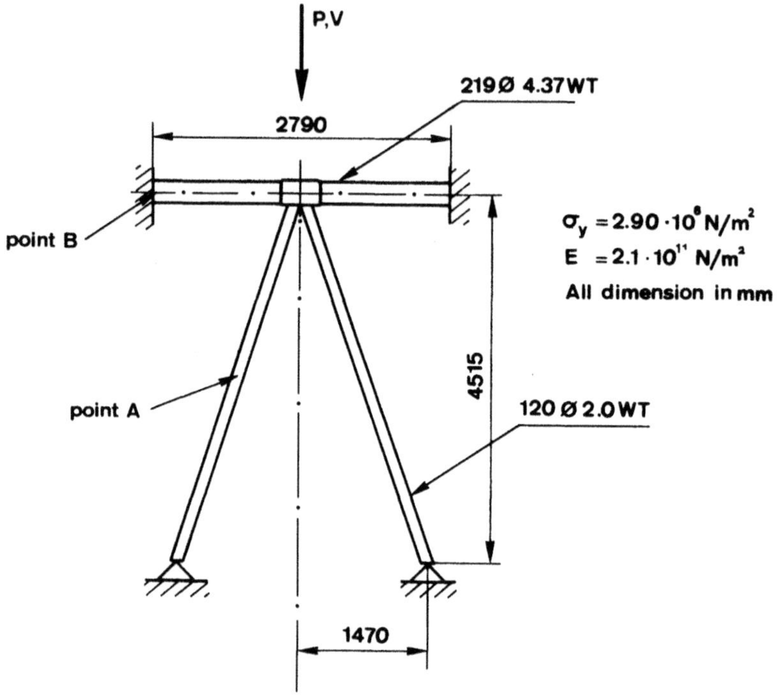

Fig.5. K-brace dimensions.

stress which is the mean between the lowest and the highest yield stress used in the FENRIS strain-hardening model. No geometrical imperfections are introduced. The results are shown in Fig.6.

The ultimate load was in the experiment determined as 0.55 MN, the numerical FENRIS model gave 0.61 MN and the present procedure gives 0.62 MN. This difference between the experimental ultimate load and the numerical results can probably be explained by the fact that local changes in cross-sections have not been taken into account in either of the numerical procedures.

The experiment shows a considerable loss in strength just after the ultimate load has been reached. This behaviour is apparently

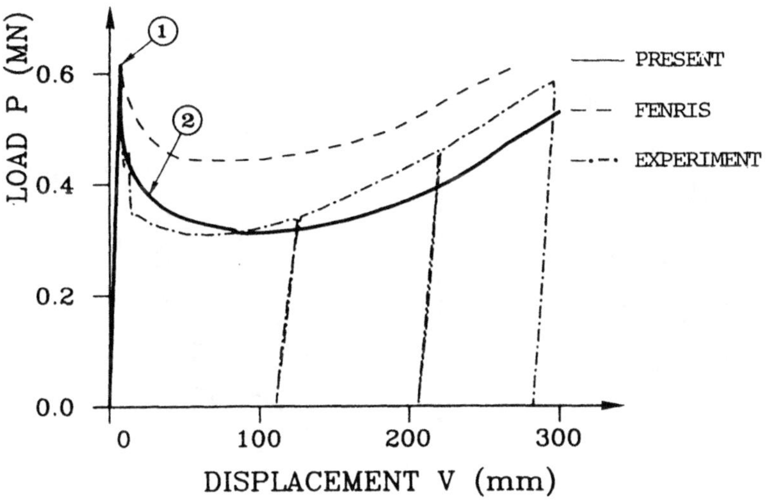

Fig.6. Load-deflection curve for K-brace.

modelled better by the present procedure than by the FENRIS program. An explanation may be that local buckling caused by global buckling of the diagonals cancels the effect of strain-hardening introduced in the FENRIS analysis.

In order to illustrate the effect of the yield surface concept Fig.7 shows the path of generalized forces at the sections A and B indicated in Fig.5. It is seen that the yield surface is followed nicely and that sharp corners on the path have been avoided by introducing the gradual plastification of cross-sections. At point A the normal force increases until elastic buckling of the diagonals occur. Thereafter a moment hinge is gradually introduced. At point B a moment hinge is first formed but as deflections increase pure axial yielding takes over.

As a last example the results from the analysis of an offshore jacket structure is included. The purpose of this example is to demonstrate the use of the procedure for a large complicated, three dimensional frame. The jacket structure in consideration is shown in Fig.8. The flexibility of the piles driven into the

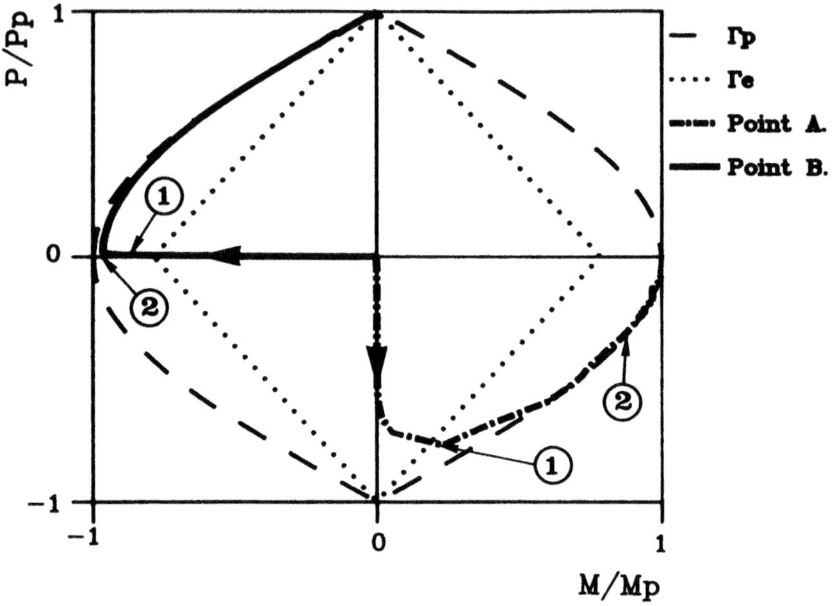

Fig.7. Node forces at points A and B for K-brace shown in Fig.5.

soil is modelled by springs at the ocean floor. The loading is a horizontal concentrated force P (a collision force).

The results of the analysis are shown in Fig.8 in the form of a load-deflection curve for the point where the external force is applied. At the point marked 1 on the load-deflection curve the horizontal strut situated directly under the load buckles plastically. At point 2 on the load-deflection curve also the diagonals in the horizontal plane of the load buckles. Thereafter plastic nodes are formed at several positions in the frame and the structure rapidly looses its strength.

6. Conclusions.

A procedure for evaluation of beam stiffness properties has been presented. The procedure seems to be relatively efficient for

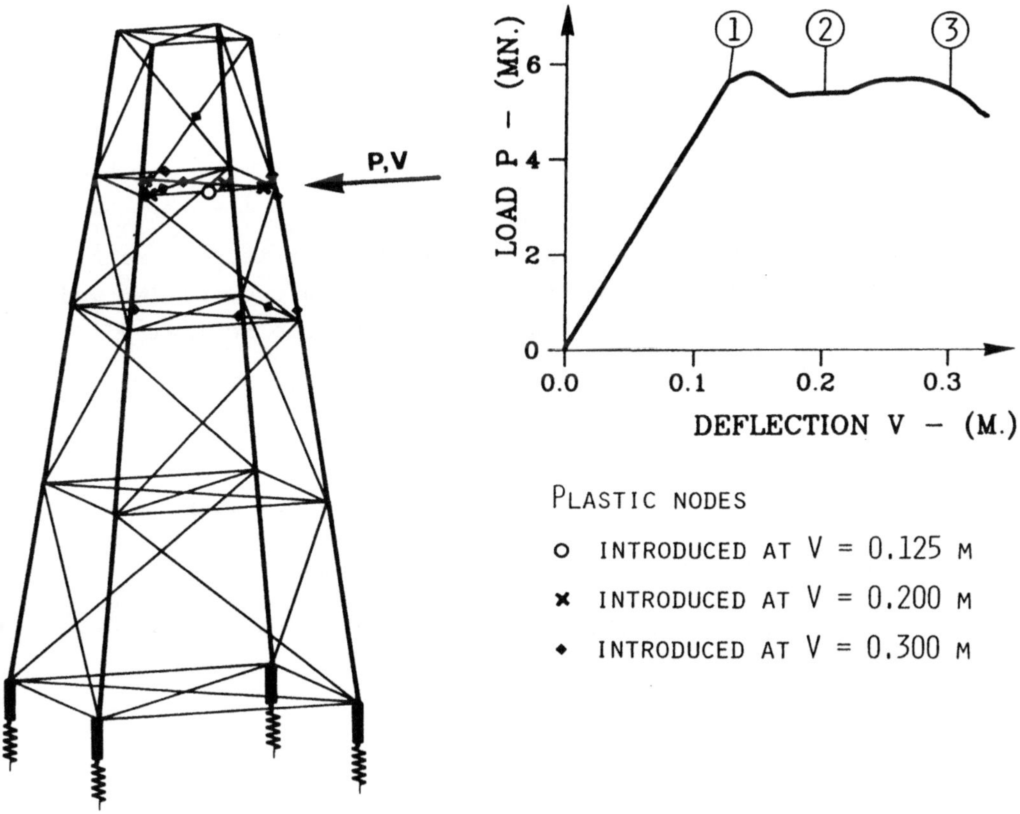

Fig.8. Steel-jacket subjected to a horizontal load P.

analysis of complicated frame structures with material as well as geometrical non-linearities. The major characteristic of the method is that each prismatic beam member always can be modelled as one element. For large structures composed of many elements this has several advantages. Firstly, it makes it easier to create the mathematical model since it is not necessary to anticipate in advance which members will behave strongly non-linear and therefore have to be modelled by many elements. Secondly, the computer time is reduced since those members which behave linearly can be introduced just as efficiently as in a normal linear program and the evaluation of stiffness properties of those members which have strong non-linear behaviour is still relatively simple.

Appendix.

Below is given those elements of the symmetric stiffness matrix $\bar{\bar{K}}_{g2}$ which are different from zero. The expressions are derived by the symbolic manipulation language FORMAC. The following notation has been used:

- D is the length of the element
- P is the acting normal force
- EA is the cross-sectional area multiplied by Young's modulus
- Q_i is the displacement of degree of freedom no. i in the local system
- DQ_i is the previous displacement increment for degree of freedom no. i in the local system

$ETY = EI_y/k_zGA \cdot D^2$ and $ETZ = EI_z/k_yGA \cdot D^2$

$ROY = 1/(1+12 \cdot ETY)$ and $ROZ = 1/(1 + 12 \cdot ETZ)$

```
KG2(1,2)=-D**(-1)*ROZ**2*EA*(Q6+Q12)*(1/10.)
KG2(1,3)=D**(-1)*ROY**2*EA*(Q5+Q11)*(1/10.)
KG2(1,5)=-ROY**2*EA*(ETY**2*Q5*12-ETY**2*Q11*12+ETY*Q5*2-ETY*
Q11*2+Q5*(2/15.)-Q11*(1/30.))
KG2(1,6)=-ROZ**2*EA*(ETZ**2*Q6*12-ETZ**2*Q12*12+ETZ*Q6*2-ETZ*
Q12*2+Q6*(2/15.)-Q12*(1/30.))
KG2(1,8)=D**(-1)*ROZ**2*EA*(Q6+Q12)*(1/10.)
KG2(1,9)=-D**(-1)*ROY**2*EA*(Q5+Q11)*(1/10.)
KG2(1,11)=ROY**2*EA*(ETY**2*Q5*12-ETY**2*Q11*12+ETY*Q5*2-ETY*
Q11*2+Q5*(1/30.)-Q11*(2/15.))
KG2(1,12)=ROZ**2*EA*(ETZ**2*Q6*12-ETZ**2*Q12*12+ETZ*Q6*2-ETZ*
Q12*2+Q6*(1/30.)-Q12*(2/15.))
KG2(2,7)=D**(-1)*ROZ**2*EA*(Q6+Q12)*(1/10.)
KG2(3,7)=-D**(-1)*ROY**2*EA*(Q5+Q11)*(1/10.)
KG2(5,7)=ROY**2*EA*(ETY**2*Q5*12-ETY**2*Q11*12+ETY*Q5*2-ETY*
Q11*2+Q5*(2/15.)-Q11*(1/30.))
KG2(6,7)=ROZ**2*EA*(ETZ**2*Q6*12-ETZ**2*Q12*12+ETZ*Q6*2-ETZ*
Q12*2+Q6*(2/15.)-Q12*(1/30.))
KG2(7,8)=-D**(-1)*ROZ**2*EA*(Q6+Q12)*(1/10.)
KG2(7,9)=D**(-1)*ROY**2*EA*(Q5+Q11)*(1/10.)
KG2(7,11)=-ROY**2*EA*(ETY**2*Q5*12-ETY**2*Q11*12+ETY*Q5*2-ETY*
Q11*2+Q5*(1/30.)-Q11*(2/15.))
KG2(7,12)=-ROZ**2*EA*(ETZ**2*Q6*12-ETZ**2*Q12*12+ETZ*Q6*2-ETZ*
Q12*2+Q6*(1/30.)-Q12*(2/15.))
```

Similarly those components of the displacement dependent load vector $\overline{F}$ which are different from zero are given below:

```
FX2(2)=-ROZ**2*D**(-2)*P*(DQ8*ETZ*2+DQ8*(1/5.)-DQ2*ETZ*2-DQ2
*(1/5.))*6
FX2(3)=-D**(-2)*ROY**2*P*(DQ9*ETY*2+DQ9*(1/5.)-DQ3*ETY*2-DQ3
*(1/5.))*6
FX2(5)=-D**(-1)*ROY**2*P*(DQ9*ETY**2*72+DQ9*ETY*6-DQ9*(1/10.
)+DQ3*ETY**2*72+DQ3*ETY*18+DQ3*(11/10.))
FX2(6)=ROZ**2*D**(-1)*P*(DQ8*ETZ**2*72+DQ8*ETZ*6-DQ8*(1/10.)
+DQ2*ETZ**2*72+DQ2*ETZ*18+DQ2*(11/10.))
FX2(8)=ROZ**2*D**(-2)*P*(DQ8*ETZ*2+DQ8*(1/5.)-DQ2*ETZ*2-DQ2*
(1/5.))*6
FX2(9)=D**(-2)*ROY**2*P*(DQ9*ETY*2+DQ9*(1/5.)-DQ3*ETY*2-DQ3*
(1/5.))*6
FX2(11)=D**(-1)*ROY**2*P*(DQ9*ETY**2*72+DQ9*ETY*18+DQ9*(11
/10.)+DQ3*ETY**2*72+DQ3*ETY*6-DQ3*(1/10.))
FX2(12)=-ROZ**2*D**(-1)*P*(DQ8*ETZ**2*72+DQ8*ETZ*18+DQ8*(
11/10.)+DQ2*ETZ**2*72+DQ2*ETZ*6-DQ2*(1/10.))
```

7. References.

1. Arnesen, A.; Bergan, P.G.: FENRIS - a general nonlinear finite element program. 4th Int. Conference on Finite Element Systems, Southampton, Finite Element Systems Handbook, 1983.

2. Hayduk, R.J.; Winter, R.; Pifko, A.B.; Fasanella, E.L.: Application of the non-linear finite element computer program DYCAST to aircraft crash analysis. Structural Crashworthiness, Butterworth & Co. Ltd., 1983, pp.283-307.

3. Akkoush, E.A.; Toridis, T.G.; Khozeimeh, K.; Huang, H.K.: Bifurcation, pre- and postbuckling analysis of frame structures. Computers & Structures, Vol.8, 1978, pp.667-678.

4. Argyris, J.H.; Hilpert, O.; Malejannakis, G.A.; Scharpf, D.W.: On the geometrical stiffness of a beam in space - a consistent V.W. approach. Computer Methods in Applied Mechanics and Engineering 20, 1979, pp.105-131.

5. Archer, J.S.: Consistent matrix formulations for structural analysis using finite-element techniques. AIAA-journal, Vol.3, No.10, 1965, pp.1910-1918.

6. Bagci, C.: Elastic stability and buckling loads of three dimensional nonuniform framed systems by finite element method using spatial line elements. Computers & Structures, Vol.10, 1979, pp.731-743.

7. Powell, G.H.: Theory of nonlinear elastic structures. Journal of the Structural Division, Proceedings of the American Society of Civil Engineers, Vol.95, 1969, pp. 2687-2701.

8. Saran, M.: On the influence of the discretization density in the nonlinear analysis of frames. Computer Methods in Applied Mechanics and Engineering 43, 1984, pp.173-180.

9. Moan, T.; Amdahl, J.; Engseth, A.G.; Granli, T.: Collapse behaviour of trusswork steel platforms. BOSS'85, Delft 1985.

10. Reissmann, C.: Grundlagen der nichtlinearen Festkörpermechanik - eine methodische Einführung. Technische Mechanik 5, Heft 1, 1984, pp.19-30.

11. Meek, J.L.; Tan, H.S.: Geometrically nonlinear analysis of space frames by an incremental iterative technique. Computer Methods in Applied Mechanics and Engineering 47, 1984, pp.261-282.

12. Ueda, Y.; Rashed, S.M.H.; Nakacho, K.: New efficient and accurate method of nonlinear analysis of offshore tubular frames (the idealized structural unit method). 3rd. Int. Offshore Mechanics and Arctic Engineering Symposium, New Orleans, Lousianna, 1984, pp.260-267.

13. Wen, R.K.; Rahimzadeh, J.: Nonlinear elastic frame analysis by finite element. Journal of Structural Engineering, Vol. 109, No.8, 1983, pp.1952-1971.

14. Ueda, Y.; Yamakawa, T.; Akamatsu, T.; Matsiushi, M.: A new theory on elastic plastic analysis of framed structures. Technology Reports of the Osaka University, Vol.19, No.875, 1969, pp.263-276.

15. Bergan, P.G.; Søreide, T.H.: Solution of large displacement and instability problems using the current stiffness parameter. Finite Elements in Nonlinear Mechanics, Tapir 1977, pp.647-670.

16. Bergan, P.G.: Automated incremental - interactive solution schemes. First Int. Conference on Numerical Methods for Nonlinear Problems, Swansea, U.K., 1980.

17. Williams, F.W.: An approach to the non-linear behaviour of the members of a rigid jointed plane framework with finite deflections. Quart. Journ. Mech. and Applied Math., Vol.17, Pt.4, 1964, pp.451-469.

18. Wood, R.D.; Zienkiewicz, O.C.: Geometrically nonlinear finite element analysis of beams, frames, arches and axisymmetric shells. Computers & Structures, Vol.7, 1977, pp.725-735.

Nonlinear Finite Element Analysis of Offshore Structures

A. ARNESEN, A. ENGSETH*, S. D. ESKESEN**
A. S. VERITEC, Høvik, Norway
*SINTEF, Trondheim, Norway
**COWICONSULT, Virum, Denmark

Summary

The present paper deals with nonlinear analyses of offshore structures. The reserve and residual strength of a simple jacket structure is analyzed and discussed. Further, a design study of a caisson platform exposed to extreme wave loading is presented. A short outline of some basic methods of analysis of the nonlinear finite element program applied is given.

1. Introduction

Design of offshore structures has been, and still is, mostly based on linear structural analysis methods. However, there has been an increasing demand for nonlinear analysis over the last few years to cope with problems that can not be handled with linear methods of analysis. Production of oil is planned for deeper water and in more hostile environments (e.g arctic sea) as new oil reservoirs are being detected. This requires often use of sophisticated methods to assess operation functions and safety aspects under extreme conditions.

Fixed platforms has been designed for water depths down to 300 and 400 m, and even deeper. The overall behaviour of such giant platforms depends on nonlinear effects associated with inelastic material behaviour in the structure and its foundation, large displacements and various interaction phenomena, which is of great importance to account for.

New production systems have emerged alternatively to the fixed platforms. Some of these systems, like compliant structures, deep water mooring lines, flexible risers, etc., basically involve large displacements or nonlinear constraints. This

Finite Element Methods for Nonlinear Problems
Europe-US Symposium, Trondheim, Norway 1985
Editors: Bergan, Bathe, Wunderlich

applies also to offshore operations like pipe laying, towing and placing of fixed concrete platforms, emergency operations, etc. Concrete structures in arctic sea represent a special challenge with extreme pressure from ice and high temperature gradients with possible nonlinear cracking behaviour.

Unexpected events occasionally occur. This has recently been confirmed by the severe subsidence of the sea bottom at the Ekofisk field in the North Sea. Assessment of the safety of the structure is of vital importance under such circumstances. This may involve elaborate nonlinear analyses like stability and collapse calculations, nonlinear dynamics, etc.

2. Nonlinear Methods of Analysis

The examples described in this paper have been analyzed by the nonlinear finite element program FENRIS /1,2/. Within the scope of this paper it is impossible to give a broad presentation of the nonlinear methods of analysis of this program, however, some techniques utilized will be described. Further details may be found in Reference /1-9/.

The formulation of motion accomodates for translations and rotations of unlimited size. Large rotations are defined in terms of transformation matrices, rotational vectors are only used incrementally. The rotation matrix is continuously updated in an incremental-iterative solution scheme in accordance with the incremental rotation which is of limited size.

The equilibrium equations is expressed in terms of a virtual work equation using Green strains and symmetric Piola-Kirchhoff stresses. Assembling contributions from all elements, the system balance equations are obtained

$$\mathbf{R}_{int} = \mathbf{R}(\lambda) - \mathbf{M}\ddot{\mathbf{r}} - \mathbf{C}\dot{\mathbf{r}} \qquad (1)$$

$\mathbf{R}_{int}$ is the internal reaction forces for the current state of internal stresses and $\mathbf{R}(\lambda)$ is the assembly of all external applied loads. $\mathbf{M}$ and $\mathbf{C}$ are the mass and the damping matrices, respectively, and **r** is the nodal displacement of the system. All applied forces are referred to a common time or load parameter λ. Any combination of different types of loading may be prescribed such as concentrated loads, conservative and non-conservative element loads, prescribed displacements, current and wave forces, buoyance forces, etc., i.e.

$$\mathbf{R}(\lambda) = p_1(\lambda)\mathbf{R}1 + p2(\lambda)\mathbf{R}2 + \ldots \qquad (2)$$

$p_1(\lambda)$, $p2(\lambda)$, etc. are prescribed load intensity functions that define the contributions of the different types of loading.

The solution of the nonlinear equations assumes a linearized incremental form of the equilibrium equations based on two configurations that are close to each other

$$\mathbf{K}_T\Delta\mathbf{r} = \Delta\mathbf{R} - \mathbf{M}\Delta\ddot{\mathbf{r}} - \mathbf{C}\Delta\dot{\mathbf{r}} \qquad (3)$$

$\mathbf{K}_T$ is the tangent stiffness matrix which is calculated on basis of current stress-strain state and deformation of the structure. Eq. 3 is solved using a "skyline profile" block solver that can handle positive definite as well as indefinite systems of equations.

The Newmark-β method is implemented for integration of the dynamic equation. Simple equilibrium corrections or equilibrium iterations at each time step is possible, and it is also possible to prescribe how often the tangent stiffness matrix should be updated during the incrementations.

Solution of nonlinear static equations is based on incremental-iterative procedures. The stiffness matrix $\mathbf{K}_T$ may be updated in different ways. Modified as well as true Newton-Raphson iterations are implicitly included in the algorithm.

The Morison equation is used for calculation of hydrodynamic inertia and drag forces. For harmonic wave condition the water particle kinematics are determined according to Airy's or Stokes fifth order theories /7/.

Two standard wave spectra are available for irregular sea states, these are the Pierson-Moskowitch spectrum and the JONSWAP spectrum /7/. Both long-crested and short-crested sea states may be simulated. A three dimensional prismatic grid is established and the time histories of surface elevation, wave particle velocity and acceleration are generated at the coordinate points of the grid. After calculating random phases, angular frequencies and wave numbers, the time history of the surface elevation, water particle velocities and accelerations are calculated by a Fast Fourier Transform. The time histories are used by the program for nonlinear time-domain dynamics.

3. Case Studies

3.1 Ultimate Capacity of 3-D Tubular Structures

The jacket type of platform is frequently used in the offshore oil production. The load carrying structural system is usually designed with redundancy, which means that failure of a single member does not necessarily result in collapse of the whole structure. Redistributuion of forces among the neighbouring members will take place and the structure will be capable of carrying increased load. Offshore platforms may be subjected to severe environmental and accidental loads which are much larger than those encountered under normal service conditions. Evaluation of the consequences of such extreme conditions by using a linear elastic analysis may be conservative provided that a sufficient force redistribution capability is possible. Such an approach may therefore lead to an uneconomical design, and it is thus important to improve our knowledge about the system behaviour and the behaviour of structural components when loading exceeds the elastic range.

The case studies in this section comprise simple 3-D X-braced subsystems of a jacket structure. Numerical results illustrating factors that may influence the strength of such systems are presented. Both damaged and undamaged systems are considered.

Two structural systems with different geometric properties are chosen for the present analyses /9/. Main dimensions and member arrangement are equal for both systems, and are shown in Fig. 1. The X-braces in the top horizontal plane, the horizontals and the legs are identical for both systems. However, the vertical X-braces have different slenderness ratios in the two systems, i.e. L/r=97 and L/r=195, where r denotes the radius of gyration. In the following these systems are denoted S1 and S2, respectively. The member dimensions and other key parameters are listed in Table 1. D and t denote diameter and thickness, while N_p and M_p denote plastic axial load and plastic moment, respectively. An elastic, perfectly plastic material law with a yield stress of 330MPa is used.

Results from five different cases are summarized in Table 2. The first three cases are for system S1 and the last two cases are for the more slender system S2. The structure is subjected to nodal point loads both in horizontal and vertical direction. The vertical loads (F_v) are 30MN and are applied prior to incrementation of the horizontal loads (F_H). The present loading is thought to represent the loading situation on a subsystem from the lower part of a real structure where no wave loading gives rise to distributed element forces. As seen from Table 2, two different types of damage are considered; 1) removal of compression member EI and 2) removal of tension member IF. The intact structures is analysed with small initial imperfections on the members, typically δ_0/L = 0.0015. Further details are given in /9/.

Fig. 2 shows the horizontal load and displacement curves for the three first load cases. It is observed that removal of the tension member causes the most serious reduction of the collapse load. The curves for load cases 2 and 3 do not start from origo, since the vertical loads F_v, applied prior to the horizontal loads, give small horizontal displacements also.

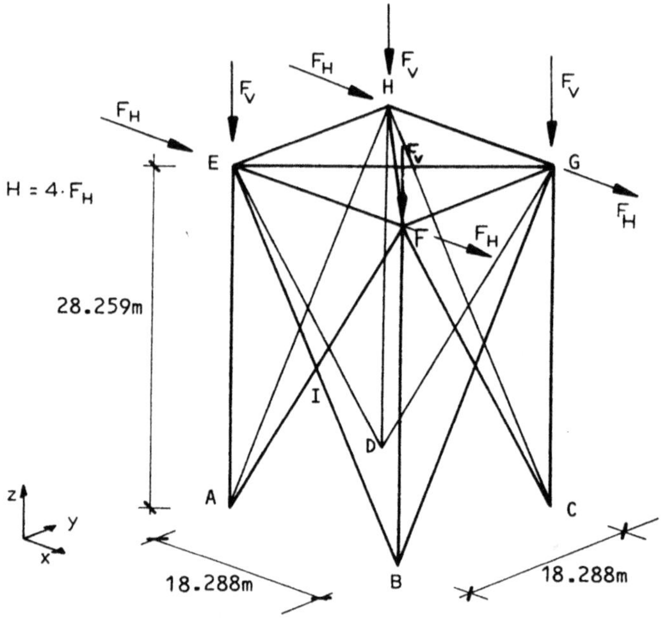

Fig. 1: 3-D subsystem of a jacket structure

Table 1: Dimensions for members in 3-D X-braced systems

MEMBER TYPE	SYSTEM	D [m]	t [m]	L [m]	N_p[MN]	M_p[MNm]	D/t	L/r
Legs	S1, S2	2.0	0.060	28.259	120.67	74.54	33	41
Horizontals	S1, S2	1.0	0.040	18.288	39.80	12.17	25	54
Horizontal X-brace	S1, S2	1.0	0.040	25.86	39.80	12.17	25	76
Vertical X-braces	S1	1.016	0.038	33.66	38.53	12.00	26.7	97
	S2	0.508	0.019	33.66	9.63	1.50	26.7	195

Table 2: Summary of load cases for 3-D-systems

SYSTEM	CASE	DESCRIPTION OF ANALYSIS
S1 (L/r=97)	1 2 3	Undamaged structure Damaged structure with compression member EI removed Damaged structure with tension member IF removed
S2 (L/r=195)	4 5	Undamaged structure Damaged structure with tension member IF removed

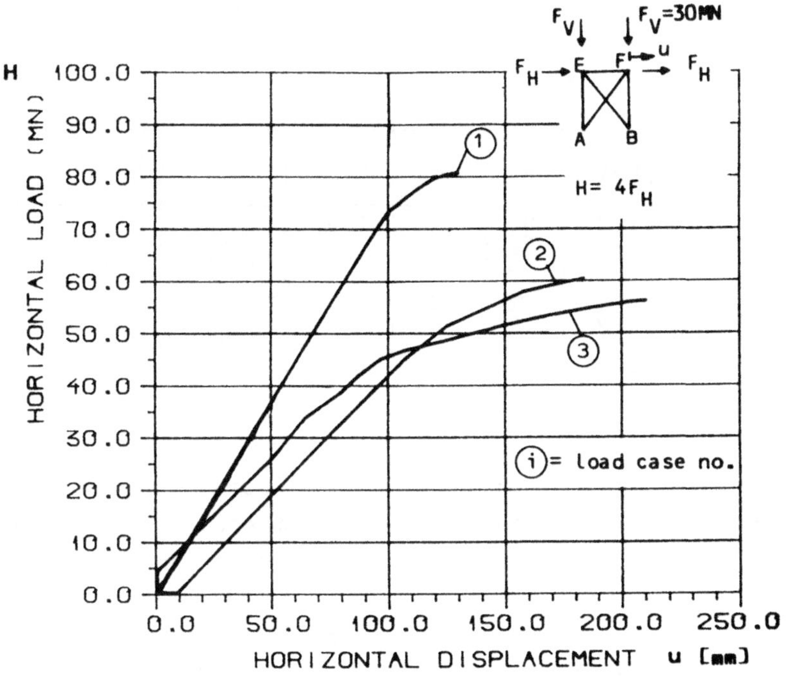

Fig. 2: Load/displacement curves for load cases 1-3 (System S1, L/r=97)

A summary of the analysis results is given in Table 3. H_{coll} denotes the total horizontal load at collapse, and H_d is the design load determined from linear analysis using the interaction equations recommended by API /9, 10/ to perform the member checks. REF and RIF are two simple measures for structural redundancy suggested by Lloyd and Clawson /11/. The Reserve Resistance Factor (REF) is defined as the ratio between the environmental load at ultimate limit state (H_{coll}) and the design environmental load (H_d). REF is a measure of the ability of the undamaged structure to sustain loads exceeding the design load. The Residual Strength Factor (RIF) is a measure of the system strength in a damaged condition, and is defined as the ratio between the collapse load for the damaged structure and the collapse load for the undamaged structure. Note that REF•RIF>1.0, indicates that a damaged structure can sustain the design load without collapsing.

Table 3: Summary of analysis results

SYSTEM	LOAD CASE	H_d[MN]	H_{coll}[MN]	REF	RIF	REF•RIF
S1 (L/r=97)	1	44.0	80.4	1.83	-	-
	2	-	60.0	-	0.75	1.37
	3	-	55.8	-	0.69	1.26
S2 (L/r=195)	4	8.0	16.9	2.11	-	
	5	-	10.6	-	0.63	1.32

The results show that relative simple structural systems may have considerable strength reserves compared to estimates obtained by simple design checks, and tolerate quite serious damages to the structural members. The results seem to indicate that structures having large slenderness ratios, possess larger reserve strength than structures having smaller slenderness ratios, however, they suffer larger strength reductions in cases involving damages to the members. This observation may be explained by the fact that slender members exhibit larger strength differences in tension and compression than of the more stocky members.

Linear analyses are often used in practice to check the residual strength of a platform for damaged condition. Such a method is used for case no. 3, taking all safety factors equal to unity /9/. This procedure gives a collapse load of 29MN compared to 55.8MN for the FENRIS-analysis (see Table 3). This indicates that use of linear elastic methods for checking the residual strength of a damaged structure may give very conservative results, and nonlinear analyses should be undertaken if excessively conservative designs are to be avoided.

The results from the analyses also show that non-structural members, i.e. members that carry practically no load in a design condition, may play an important role in the redistribution of forces for a damaged structure. An illustration is given in Fig. 3, where the axial force development in system S1 for load case no. 3 is shown. It is clearly demonstrated how surrounding members take over the load carrying function when a member reaches its capacity (which for this specific case is compression member BE). The results document the capability of the program to follow the load redistribution process when a member in the structural system fails. Such analysis may be used to identify load paths and structural members that will play a vital role in the redistribution process. A few supplementing case studies may be found in /9/.

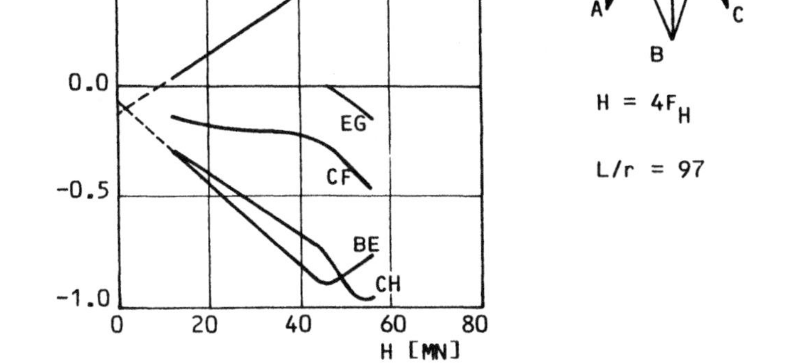

Fig. 3: Axial force development for members of System S1 for load case no. 3.

3.2 Nonlinear Design Study of a Caisson Platform

In the Danish Sector of the North Sea some of the oil fields discovered are marginal fields. To develop this type of oil fields, it is necessary to minimize the initial investments in the fields. Various alternatives of small remotely operated satellite installations have been evaluated showing an unmanned free-standing caisson as a promising candidate. A detailed design of such a platform has been performed by the Danish company Cowiconsult for Mærsk Olie og Gas A/S.

The free-standing caisson analyzed is a fixed steel platform constructed as a single tapered pile, 82m long, and driven to a depth of 28m below the mudline. The water depth is 34m. The pile diameter varies from 2.0m to 4.5m, and its wall thickness from 80mm to 100mm. The weight is approximately 700t. The platform supports a wellhead topside package, which contains a helideck and facilities for up to four wellhead completions. The caisson platform is shown in Fig. 4.

The design consideration includes extreme event analyses and fatigue analyses. The purpose of the analyses is to ensure that the caisson has sufficient structural capacity to withstand the 50-year storm condition and sufficient fatigue capacity to withstand a design life of 25 years, respectively. Only one of the design analysis, the nonlinear stochastic storm wave analysis, is reported in this paper.

The nonlinear stochastic storm wave analysis is a dynamic analysis in the time-domain which provides a means to examine nonlinear drag loading and free surface nonlinearities directly, and to examine the influence of these nonlinearities on the structural response to wave loading. The main features of the analysis are 1) Time histories of wave particle kinematics by inverse Fourier transformation of wave energy spectrum for the storm condition, 2) Time integration of the dynamic structural response, 3) Statistical interpretation of response extremes.

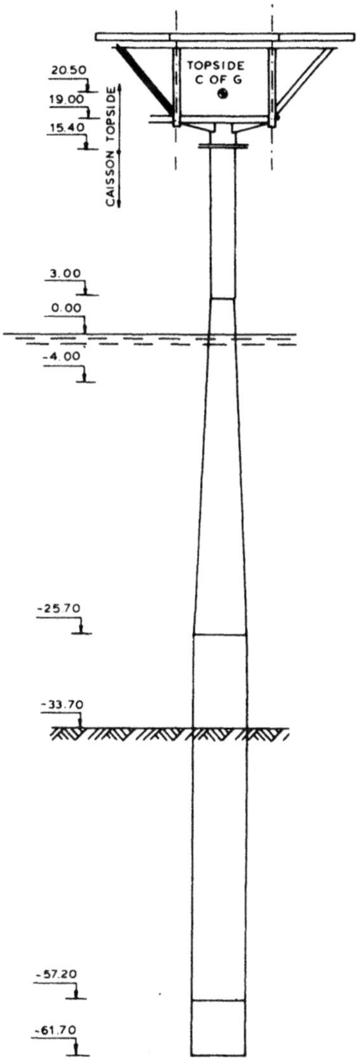

Fig. 4: Caisson platform

The analysis is performed on a 3-D space frame computer model representing the caisson, deck and foundation. The soil structure interface is modelled by applying nonlinear soil-pile springs. Fig. 5 shows the finite element model. A total of 81 beam elements are applied. The natural period for mode 1 (and 2) of the structure is found to be T_1 = 1.9s.

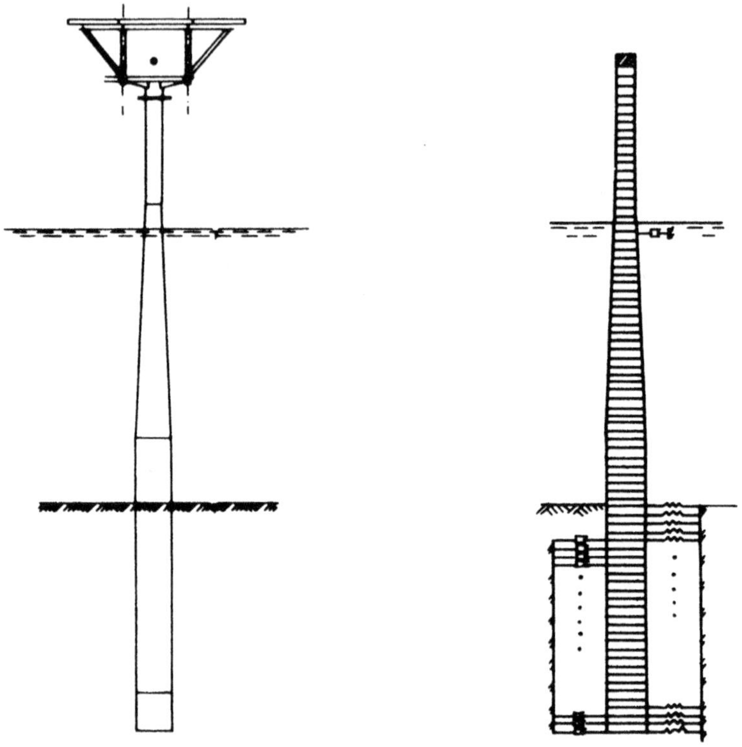

Fig. 5: Finite element model

In Fig. 6a the free vibration at deck level is shown when an impulse load of duration t = 0.2s is applied. Fig. 6b shows the shape of the first eigenmode. A total damping of 2% of critical damping is obtained.

For the extreme storm event the wave spectrum is described by the modified Pierson-Moskowitz spectrum with H_S = 9.41m and T_z = 10.3s. Fig. 7a shows the wave input spectrum, and Fig. 7b shows the surface elevation obtained by inverse Fourier transform of the frequency spectrum. Time integration is carried out using the constant average acceleration method. Each simulation has a real time duration of 24min. and a time step of 0.175s is used. Time history of stresses at lower cone-cylinder junction is shown in Fig. 8. By statistical interpretation of the time histories the most probable maximum response is determind /12/. The maximum stress obtained from wave and current load is 61MPa. Fig. 9 shows the horizontal displacement history at deck level. A maximum peak deflection of 0.23m occurs at t = 13s. The CPU-time consumption for such a simulation (8200 time steps) on a VAX-11/785 is for a linear dynamic case 290 minutes, and for a nonlinear dynamic analysis approximately 14 hours when updating the stiffness matrix at every fourth time step.

Three short-time simulations are shown in Fig. 10; Fig. 10a is for a linear dynamic case, in Fig.10b nonlinear geometric effects are accounted for, and in Fig.10c nonlinear soil behaviour is also present. The dominating response is a low frequent oscillation following the wave loading with a high frequent structural dynamic oscillation superimposed. The dynamic oscillation is greatly influenced by nonlinear effects. Pronounced changes in damping and in period are observed. The peak deflection for the nonlinear case in Fig. 10c is 0.33m, and for the linear case in Fig. 10a. 0.23m.

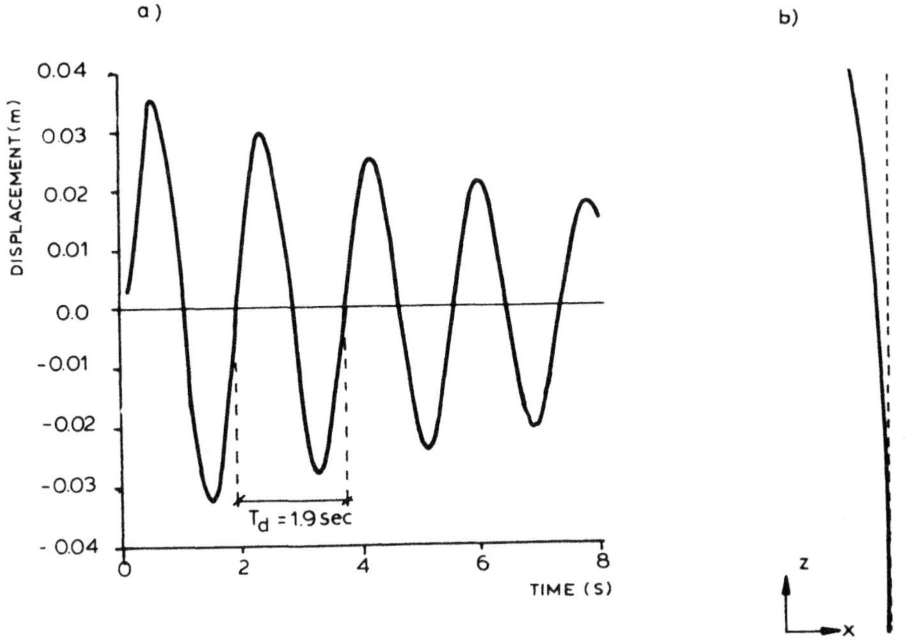

Fig. 6: a) Free vibration, b) First mode shape

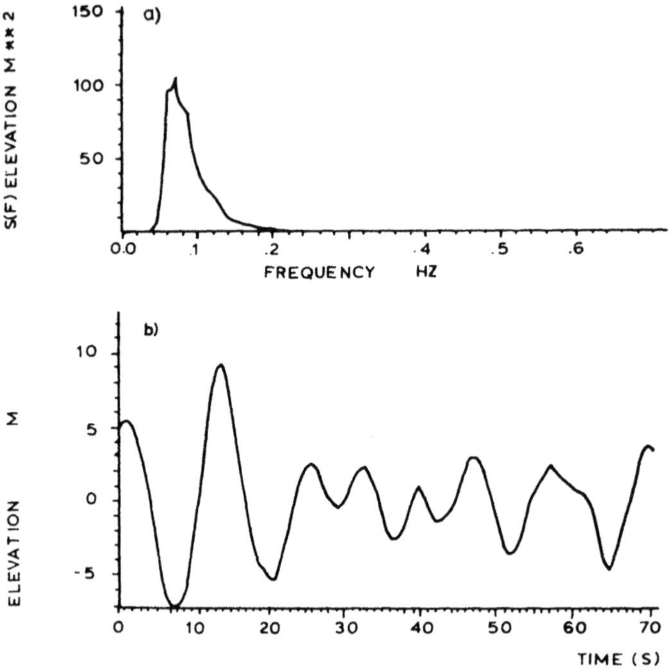

Fig. 7: a) Input wave spectrum, b) Wave surface elevation

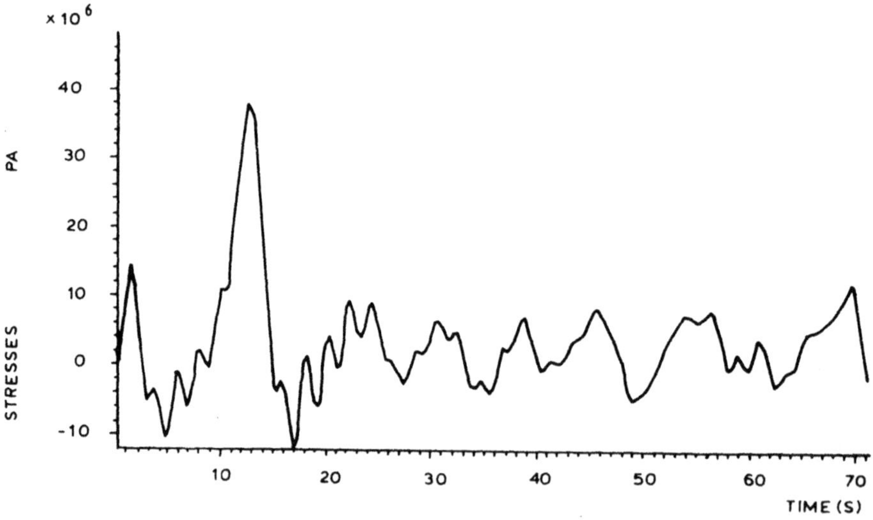

Fig. 8: History of stress at lower cone-cylinder junction

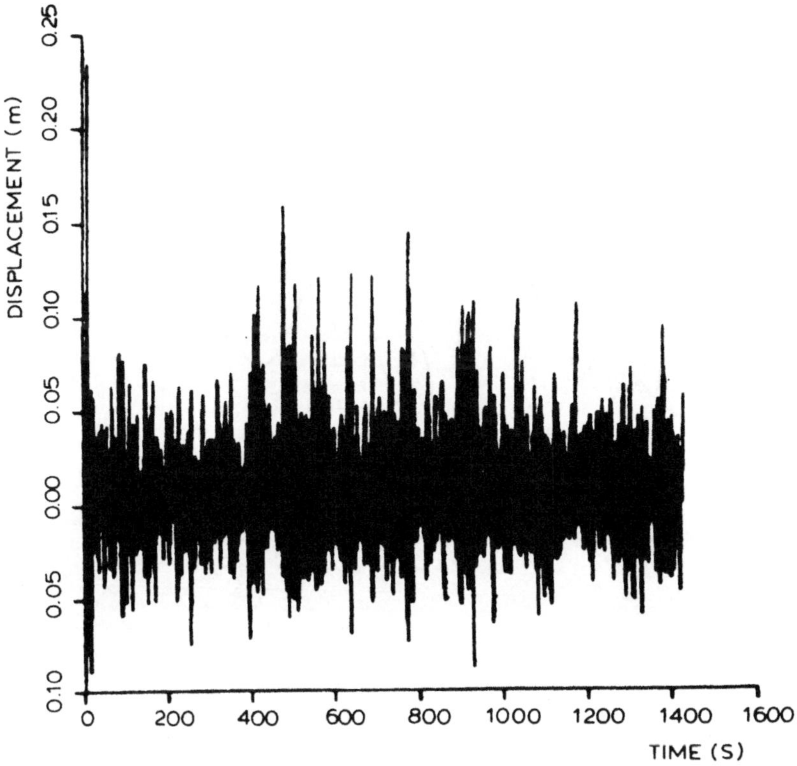

Fig. 9: Time history of horizontal displacement at deck level

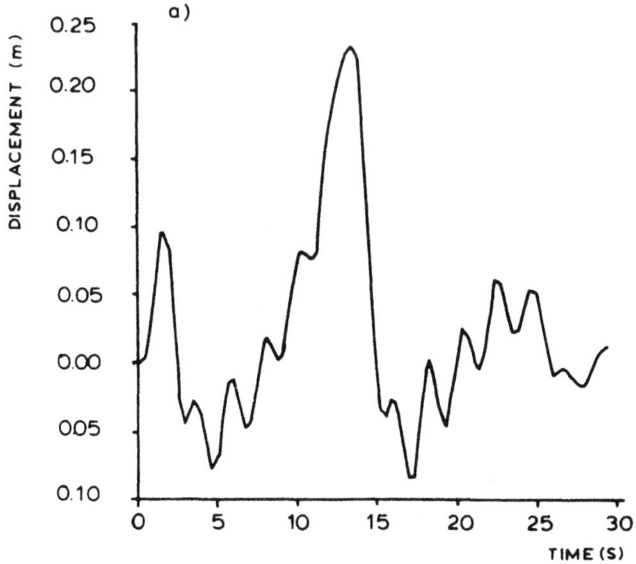

Fig. 10: Deflection at deck level: a) Linear dynamic

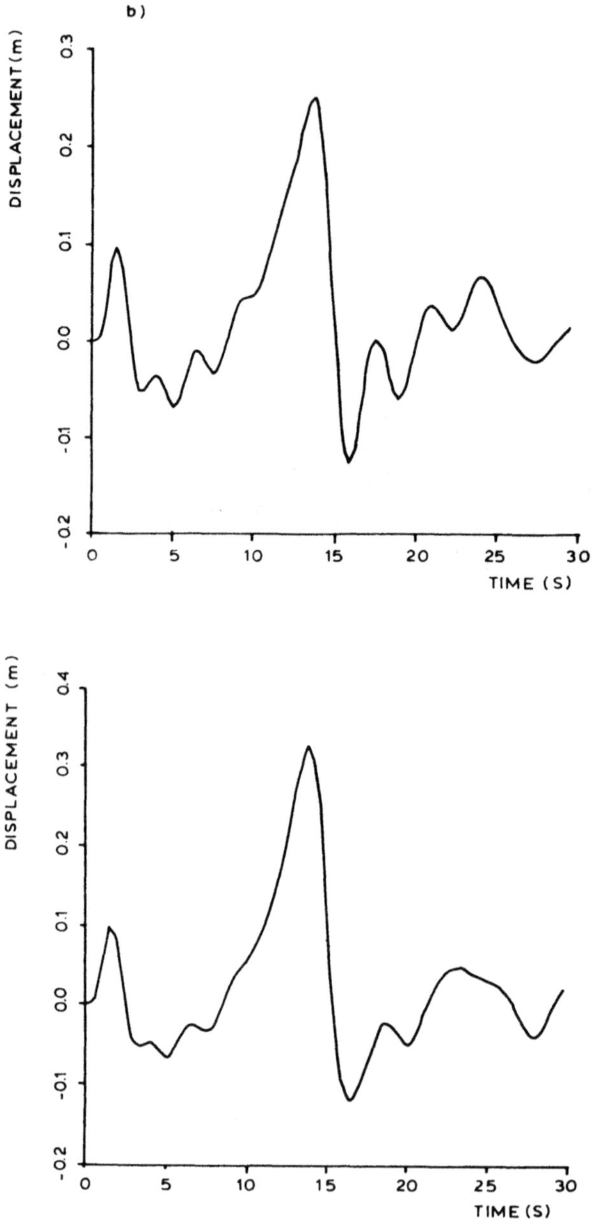

Fig. 10: (Continues) b) Nonlinear dynamic accounting for geometric effects, c) Nonlinear dynamic accounting also for nonlinear soil behaviour.

4. REFERENCES

1. FENRIS Manuals: Theory - Program Outline - Data Input, NTH, SINTEF, VERITEC, July 1985.

1. Bergan, P.G.; Arnesen, A.: "FENRIS - A General Purpose Nonlinear Finite Element Program", Proc. on 4th Int. conf. on Finite Element Systems, Southampton, July 6-8, 1983.

2. Bergan, P.G.: "Automated Incremental-Iterative Solution Methods in Structural Mechanics", In Recent Advances in Nonlinar Computational Mechanics, editors E. Hinton, D.R.J. Owen and C. Taylor, Pineridge Press, 1982.

3. Bergan, P.G.; Mollestad, E.: "An Automatic Time-Stepping Algorithmn for Dynamic Problems", Computer Methods in Applied Mechanics and Engineering, Vol. 49, 1985

4. Bergan, P.G.; Mathisen, K.M.; "Modelling of Hydrostatic Loading and Stability for Coupled Fluid - Structure Problems", In Numerical Methods in Engineering: Theory and Application, edited by J. Middelton, G.N. Pande, Balkema Press, Rotterdam 1985.

5. Mathisen, K.M.: "Doctoral Dissertation, Division of Structural Mechanics, The Norwegian Institute of Technology, to appear in 1986.

6. Hansen, H.T.: Doctoral Dissertation, Division of Structural Mechanics, The Norwegian Institute of Technology, to appear in 1986.

7. Bergan, P.G.; Nygård, M.K.: "Finite Elements with Increased Freedom in Choosing Shape Functions", Proc. on Inter. Journal of Numerical Methods in Engineering, Vol. 20, pp 643-663, 1984.

8. Bergan, P.G.; Nygård, M.K.: "Nonlinear Shell Analysis Using Free Formulation Finite Elements", Proc. of the Europe-US Symposium on "Finite Element Methods for Nonlinar Problems", August 12-15, 1985, Trondheim, Norway.

9. Engseth, A.: "Finite Element Collapse Analysis of Tubular Steel Offshore Structures", Doctoral Thesis, Division of Marine Structures, The Norwegian Institute of Technology, Trondheim, December, 1984.

10. American Petroleum Institute (API): "Recommended Practice for Planning, Designing and Constructing Fixed Offshore Platforms", API RP 2A, Twelfth Edition, January 1981.

11. Lloyd, J.R.; Clawson, W.C.: "Reserve and Residual Strength of Pile Founded, Offshore Platforms, Design-Inspection-Redundancy", An International Symposium on the Role of Design, Inspection and Redundancy in Marine Structual Reliability, Williamsburg, Virginia, 1983. Working Papers.

12 Westin L.H.: "Random Analysis of a Riser for a Floating Production System (FPS) in a Nothern North Sea Environment", OTC 4560, 1983

Constraints of Stresses in Hybrid Plate and Shell Elements

T. H. H. PIAN
Massachusetts Institute of Technology, Cambridge, USA

Summary
Multivariable variational principles for incremental finite element analyses are derived using shallow shell problems as examples. In the new versions new variables for generalized strains and stresses are defined for the initial stress terms. The forms of the resulting variational principles are the same as the conventional variational principles of linear problems. It is more convenient in using these versions to assumed stresses to satisfy equilibrium conditions in variational sense in the element level. Options in the constraints for assumed stresses are discused.

1. Introduction

The paper is concerning the formulation of geometrically nonlinear problems by incremental finite element methods based on multivariable variational principles. Finite element solutions which contain only nodal disnlacements as unknowns, are discussed here.

The two basic mixed variational principles are the Hu-Washizu nrinciple which contains displacements, strains and stresses as field variables and the Hellinger-Reissner principle which contains displacements and stresses as field variables [1,2]. In the original forms for these variational principles, the variables in each of these principles are supnosed to be unconstrained. However, in the finite element formulation if stresses and strains are unconstrained the resulting element stiffness matrices will tend to that obtained by the assumed displacement method, according to the limitation principle of Fraeijs de Veubeke [3]. For the finite element formulation some constraints to the field variables are often imposed. For example, in the finite element method by the Hellinger-Reissner principle the exact equilibrium conditions can be imposed to the assumed stresses. Then the resulting element stiffness matrices will be identical to that of the early version of assumed stress hybrid elements which are obtained by the principle of minimum complementary energy [4,5].

Finite Element Methods for Nonlinear Problems
Europe-US Symposium, Trondheim, Norway 1985
Editors: Bergan, Bathe, Wunderlich

However, the satisfaction of stress equilibrium conditions is often a difficult task to meet, in particularly, for general shell elements and for nonlinear problems. The current thought, however, is that the stress equilibrium constraints are introduced through Lagrange multiplier method hence, in general, are satisfied in the element level only in the variational sense [6,7]. In so doing the assumed stresses can be expressed in terms of natural isoparametric coordinates and the resulting element behaviors are subjected to less deterioration when element geometries are distorted. This is a desirable feature for updated Lagrange formulation when element shapes are distorted under large strain conditions. It has also been found that when a desirable stress state in Cartesian coordinates for a regular shaped element has been obtained, the stress state in natural coordinates for a distorted shape can also be obtained by a simple tensor transformation between these two coordinates [7,8].

For geometrically nonlinear problems the element stiffness matrix also includes the contribution from initial stresses. This paper is to show that the initial stress matrix can be derived based on a new stress variable but not on displacement gradients. The latter is the basis for both the ordinary assumed displacements and hybrid finite element methods. With these new versions of incremental variational principles, the approaches that are currently in use in the assumed stress hybrid method can be extended to that for geometrically nonlinear problems.

To illustrate this new approach the problem of shallow shells is treated. The formulations are based on Marqurre's theory and total Lagrange coordinate system.

Discussions at the end are centered at the options in choice of constraints for the assumed stresses in finite element formulations. Here thin structure under large deflections has the same features as general shell, in that explicit representations of membrane and bending strains are desirable.

2. Incremental Variational Principles for Shallow Shells based upon Marqurre's Theory

Considering the x_1-x_2 coordinates on the global base plane as the reference coordinates, the geometry of a shallow shell element is defined by its vertical coordinates $z(x_1,x_2)$. Using the total (stationary) Lagrange formulation and assuming that at the beginning a particular incremental loading, the equilibrium and compatibility conditions are satisfied, the potential energy functional for his incremental loading problem is

$$\pi_p = \sum_n \left\{ \iint_{A_n} \left[U(\Delta\varepsilon_{\alpha\beta},\Delta\kappa_{\alpha\beta}) + \frac{1}{2} N_{\alpha\beta}\, \Delta w_{,\alpha}\, \Delta w_{,\beta} - \Delta\bar{F}_\alpha\, \Delta u_\alpha - \Delta\bar{F}_3\, \Delta w \right] dA - \int_{s_{\sigma_n}} (\Delta\bar{N}_{\alpha_\nu}\, \Delta\tilde{u}_\alpha + \Delta Q_\nu\, \Delta\tilde{w} - \Delta M_{\alpha_\nu}\, \Delta\tilde{w}_{,\alpha}) ds \right\} \tag{1}$$

where U = strain energy density

$\varepsilon_{\alpha\beta}$ = membrane strains

$\kappa_{\alpha\beta}$ = curvature strains

$N_{\alpha\beta}$ = membrane stress resultants

u_α = in-plane displacements

w = lateral displacement

ΔN_{α_ν} = boundary membrane traction

ΔQ_ν = boundary transverse shear

ΔM_{α_ν} = boundary couple

$\Delta\bar{F}_\alpha$, $\Delta\bar{F}_3$ = prescribed body force increments

Here A_n is the area of the nth element and s_{σ_n} is the boundary of the element over which tractions are prescribed. The strains $\varepsilon_{\alpha\beta}$ and $\kappa_{\alpha\beta}$ are expressed in terms of u_α and w according to linearized strain displacement relations

$$\Delta\varepsilon_{\alpha\beta} = \frac{1}{2} [\Delta u_{\alpha,\beta} + \Delta u_{\alpha,\beta} + (z_{,\alpha}+w_{,\alpha})\Delta w_{,\beta} + (z_{,\beta}+w_{,\beta})\Delta w_{,\alpha}] \tag{2}$$

$$\Delta\kappa_{\alpha\beta} = \Delta w_{,\alpha\beta} \tag{3}$$

and along the element boundary ∂A_n

$$u_\alpha = \tilde{u}_\alpha \tag{4}$$

$$w = \tilde{w} \tag{5}$$

where $(\tilde{\ })$ indicates values at the boundary. It is assumed that

$$\tilde{u}_\alpha = \bar{u}_\alpha \quad ; \quad \tilde{w} = \bar{w} \quad \text{on } S_{u_n} \tag{6}$$

The Hu-Washizu principle can be created when the equations of constraint for π_p are included and corresponding Lagrange multipliers are introduced. The quasi-conforming elements initiated by Tang [9] are based on assumed strains which are made to satisfy the strain displacement relation in an weighted average sense. It has been shown that such element is equivalent to that based on the Hu-Washizu principle [10]. In applying the quasi-conforming element to geometrically nonlinear problems Jiang [11] introduced additional strain variables Δe_α for the displacement gradients $\Delta w_{,\alpha}$ for the initial stress terms. There are two possible versions for the variational principle which is equivalent to the Hu-Washizu principle:

(1) $$\pi_G = \sum_n \left(\pi_{G_n} + \int_{A_n} \frac{1}{2} N_{\alpha\beta} \Delta w_{,\alpha} \Delta w_{,\beta}\, dA\right) \tag{7}$$

(2) $$\overline{\pi}_G = \sum_n \left(\pi_{G_n} + \frac{1}{2} \int_{A_n} (\frac{1}{2} N_{\alpha\beta}\, \Delta e_\alpha\, \Delta e_\beta - \Delta N^*_\alpha\, \Delta e_\alpha + \Delta N^*_\alpha\, \Delta w_{,\alpha}) dA\right) \tag{8}$$

where
$$\begin{aligned}\pi_{G_n} = & \int_{A_n} \{ U(\Delta\varepsilon_{\alpha\beta}, \Delta\kappa_{\alpha\beta}) - \Delta N_{\alpha\beta}\, \Delta\varepsilon_{\alpha\beta} + \Delta M_{\alpha\beta}\, \Delta\kappa_{\alpha\beta} \\ & + \frac{1}{2} \Delta N_{\alpha\beta}[\Delta u_{\alpha,\beta} + \Delta u_{\beta,\alpha} + (z_{,\alpha} + w_{,\alpha})\Delta w_{,\beta} + (z_{,\beta} + w_{,\beta})\Delta w_{,\alpha}] \\ & - \Delta M_{\alpha\beta}\, \Delta w_{,\alpha\beta} - \Delta\overline{F}_\alpha\, \Delta u_\alpha - \Delta\overline{F}_3\, \Delta w \}\, dA \\ & - \int_{\partial A_n} [\Delta N_{\alpha_\nu} (\Delta u_\alpha - \Delta\tilde{u}_\alpha) + \Delta Q_\nu (\Delta w - \Delta\tilde{w}) - \Delta M_{\alpha_\nu} (\Delta w_{,\alpha} - \Delta\tilde{w}_{,\alpha})] ds \\ & - \int_{s_{\sigma_n}} [\Delta\overline{N}_{\alpha_\nu}\, \Delta\tilde{u}_\alpha + \Delta\overline{Q}_\nu\, \Delta\tilde{w} - \Delta\overline{M}_{\alpha_\nu}\, \Delta\tilde{w}_{,\alpha}] ds\end{aligned} \tag{9}$$

Here $\Delta N_{\alpha\beta}$, $\Delta M_{\alpha\beta}$ and ΔN^*_α are introduced as Lagrange multipliers, ΔN^*_α being conjugate with Δe_α with

$$\Delta N_\alpha{}^* = N_{\alpha\beta}\, \Delta w_{,\beta} \tag{10}$$

The boundary tractions are related to these stresses by

$$\Delta N_{\alpha_\nu} = \Delta N_{\alpha\beta}\, \nu_\beta \tag{11}$$

$$\Delta Q_\nu = [\Delta N_{\alpha\beta}(z_{,\beta} + w_{,\beta}) + \Delta N^*_\alpha + \Delta M_{\alpha\beta,\beta}]\nu_\alpha \tag{12}$$

and $$\Delta M_{\alpha_\nu} = \Delta M_{\alpha\beta}\, \nu_\beta \tag{13}$$

ν_β being the direction cosines of the boundary normal.

By applying the Green's theorem two other alternative versions of this variational principles can be obtained. They are

(1) $$\pi_G{}^* = \sum_n \left\{ \pi_{G_n}^* + \int_{A_n} \frac{1}{2} N_{\alpha\beta} \Delta w_{,\alpha} \Delta w_{,\beta}\, dA \right\} \tag{14}$$

(2) $$\bar{\pi}_G{}^* = \sum_n \left\{ \pi_{G_n}^* + \frac{1}{2} \int_{A_n} \left(\frac{1}{2} N_{\alpha\beta}\, \Delta e_\alpha\, \Delta e_\beta - \Delta N_\alpha^*\, \Delta e_\alpha - \Delta N^*_{\alpha,\alpha}\, \Delta w \right) dA \right\} \tag{15}$$

where
$$\begin{aligned} \pi_{G_n}^* = \int_{A_n} & \Big\{ U(\Delta\varepsilon_{\alpha\beta}, \Delta\kappa_{\alpha\beta}) - \Delta N_{\alpha\beta} \Delta\varepsilon_{\alpha\beta} + \Delta M_{\alpha\beta}\, \Delta\kappa_{\alpha\beta} \\ & - (\Delta N_{\alpha\beta,\beta} + \Delta\bar{F}_\alpha) \Delta u_\alpha - \left\{ [\Delta N_{\alpha\beta}(z_{,\beta} + w_{,\beta})]_{,\alpha} + \Delta M_{\alpha\beta,\alpha\beta} + \Delta\bar{F}_3 \right\} \Delta w \Big\}\, dA \\ & + \int_{\partial A_n} (\Delta N_{\alpha_\nu}\, \Delta\tilde{u}_\alpha + \Delta Q_\nu\, \Delta\tilde{w} - \Delta M_{\alpha_\nu}\, \Delta\tilde{w}_\alpha) ds \\ & - \int_{S_{\alpha_n}} [\Delta\bar{N}_{\alpha_\nu}\, \Delta\tilde{u}_\alpha + \Delta\bar{Q}_\nu\, \Delta\tilde{w} - \Delta\bar{M}_{\alpha_\nu}\, \Delta\tilde{w}_{,\alpha}] ds \end{aligned} \tag{16}$$

From $\delta\bar{\pi}_G{}^* = 0$, one can obtain the equilibrium equations for the stress increments as follows:

$$\Delta N_{\alpha\beta,\beta} + \Delta\bar{F}_\alpha = 0 \tag{17}$$

$$[\Delta N_{\alpha\beta}\, (z_{,\beta} + w_{,\beta})]_{,\alpha} + \Delta N^*_{\alpha,\alpha} + \Delta M_{\alpha\beta,\alpha\beta} + \Delta\bar{F}_3 = 0 \tag{18}$$

It is seen that when ΔN_α^* are used as independent variables, the incremental displacement Δw no longer appears in the equations of equilibrium. A term which involves the displacement w at the beginning of the loading increment however, still exist. When updated coordinates are used, w will not appear but the coordinates z of the shell will need update for each incremental loading.

In deriving the corresponding Hellinger-Reissner functionals from π_G, $\bar{\pi}_G$ the procedure is to eliminate the strain increment terms by replacing

$U(\Delta\varepsilon_{\alpha\beta},\Delta\kappa_{\alpha\beta}) - \Delta N_{\alpha\beta}\,\Delta\varepsilon_{\alpha\beta} + \Delta M_{\alpha\beta}\,\Delta\kappa_{\alpha\beta}$ by $-U^*(\Delta N_{\alpha\beta},\Delta M_{\alpha\beta})$, which is the negative of the complementary strain energy. The terms in $\bar{\pi}_G$ and $\bar{\pi}_G{}^*$, $\frac{1}{2} N_{\alpha\beta}\,\Delta e_\alpha\,\Delta e_\beta - \Delta N_\alpha{}^*\,\Delta e_\alpha$ can be replaced by $-\frac{1}{2} Z^*_{\alpha\beta}\,\Delta N_\alpha{}^*\,\Delta N^*_{\alpha\beta}$, where the matrix of $Z^*_{\alpha\beta}$ is the inverse of the matrix of $N_{\alpha\beta}$. There are cases that such matrix, is singular or very close to singular, although the original problem is regular. In such case, a small perturbation to $N_{\alpha\beta}$ can be applied which will permit the construction of $Z^*_{\alpha\beta}$ but will maintain essentially the same numerical solution. The expressions for π_R and $\bar{\pi}_R$ are:

$$(1)\qquad \pi_R = \sum_n \left\{\pi_{R_n} + \frac{1}{2} N_{\alpha\beta}\,\Delta w_{,\alpha}\,\Delta w_{,\beta}\right\} \tag{19}$$

$$(2)\qquad \bar{\pi}_R = \sum_n \left\{\pi_{R_n} - \frac{1}{2} Z^*_{\alpha\beta}\,\Delta N_\alpha{}^*\,\Delta N_\beta{}^* + \Delta N_\alpha{}^*\,\Delta w_{,\alpha}\right\} \tag{20}$$

where
$$\begin{aligned}\pi_{R_n} = &\int_{A_n} \Big\{-U^*(\Delta N_{\alpha\beta},\Delta M_{\alpha\beta}) + \frac{1}{2}\Delta N_{\alpha\beta}[\Delta u_{\alpha,\beta} + \Delta u_{\beta,\alpha} \\ &+ (z_{,\alpha}+w_{,\alpha})\Delta w_{,\beta} + (z_{,\beta}+w_{,\beta})\Delta w_{,\alpha}] - \Delta M_{\alpha\beta}\,\Delta w_{,\alpha\beta} \\ &- \Delta\bar{F}_\alpha \Delta u_\alpha - \Delta\bar{F}_3 \Delta w\Big\}\, dA \\ &- \int_{\partial A_n} [\Delta N_{\alpha_\nu}(\Delta u_\alpha - \Delta\tilde{u}_\alpha) + \Delta Q_\nu(\Delta w - \Delta\tilde{w}) - \Delta M_{\alpha_\nu}(\Delta w_{,\alpha} - w_{,\alpha})]ds \\ &- \int_{S_{\sigma_n}} (\Delta\bar{N}_{\alpha_\nu}\,\Delta\tilde{u}_\alpha + \Delta\bar{Q}_\nu\,\Delta\tilde{w} - \Delta\bar{M}_{\alpha_\nu}\,\Delta\tilde{w}_{,\alpha})ds\end{aligned} \tag{21}$$

When the assumed stresses satisfy Eqs. (17) and (18) $-\pi_R$ and $-\bar{\pi}_R$ are reduced to the equivalent complementary energy functionals

$$(1)\qquad \pi_c = \sum_n [\pi_{c_n} - \frac{1}{2} N_{\alpha\beta}\,\Delta w_{,\alpha}\,\Delta w_{,\beta}] \tag{22}$$

$$(2)\qquad \bar{\pi}_c = \sum_n [\pi_{c_n} + \frac{1}{2} Z^*_{\alpha\beta}\,\Delta N_\alpha{}^*\,\Delta N_\beta{}^*] \tag{23}$$

where
$$\begin{aligned}\pi_{c_n} = &\int_{A_n} U^*(\Delta N_{\alpha\beta},\Delta M_{\alpha\beta})dA - \int_{\partial A_n} (\Delta N_{\alpha_\nu}\,\Delta\tilde{u}_\alpha + \Delta Q_\nu \Delta\tilde{w} - \Delta M_{\alpha_\nu}\,\Delta\tilde{w}_{,\alpha})ds \\ &+ \int_{S_{\sigma_n}} (\Delta\bar{N}_{\alpha_\nu}\,\Delta\tilde{u}_\alpha + \Delta\bar{Q}_\nu \Delta\tilde{w} - \Delta\bar{M}_{\alpha_\nu}\,\Delta\tilde{w}_{,\alpha})ds\end{aligned} \tag{24}$$

Here again the displacement increment Δw in the element does not appear in $\bar{\pi}_c$.

In earlier applications of π_R for geometrically nonlinear problems for plates and shells [12-14], it was suggested to use inconsistent model for which the terms that contain the incremental displacements are excluded in the constraint equations for stresses. Obviously using Eqs. (20) and (22) it is not necessary to take such action.

It should be pointed out that in the finite element formulations, the equilibrium conditions for initial stresses and the satisfaction of strain-displacement relations at the beginning of the load increment can be added in the variational functional with appropriate Lagrange multipliers. These terms are then lead to the corresponding correction terms for the iterative process in the finite element solution [12].

3. Finite Element Formulations

For convenience in the finite element formulation the following matrix and vector notations will be used in place of the tensor notation i.e.,

$$\begin{aligned}
\Delta \underset{\sim}{\varepsilon} &= \{\Delta\varepsilon_{\alpha\beta}, \Delta\kappa_{\alpha\beta}\} \\
\Delta \underset{\sim}{\sigma} &= \{\Delta N_{\alpha\beta}, \Delta M_{\alpha\beta}\} \\
\Delta \underset{\sim}{e} &= \{\Delta e_{\alpha}\} \\
\Delta \underset{\sim}{\sigma}^* &= \{\Delta N_{\alpha}^*\} \\
\Delta \underset{\sim}{u} &= \{\Delta u_{\alpha}, \Delta w\} \qquad \Delta \underset{\sim}{\tilde{u}} = \{\Delta \tilde{u}_{\alpha}, \Delta \tilde{w}\} \\
\Delta \underset{\sim}{T} &= \{\Delta N_{\alpha_\nu}, \Delta Q_{\nu}, \Delta M_{\alpha_\nu}\}
\end{aligned}$$

In the following only the formulations by $\bar{\pi}_G$, $\bar{\pi}_G^*$ and $\bar{\pi}_R$ are discussed. For $\bar{\pi}_G$ and $\bar{\pi}_G^*$ formulations, the key is to use the same functions for $\Delta\underset{\sim}{\varepsilon}$ and $\Delta\underset{\sim}{\sigma}$ and for $\Delta\underset{\sim}{e}$ and $\Delta\underset{\sim}{\sigma}^*$ i.e.

$$\Delta \underset{\sim}{\varepsilon} = \underset{\sim}{P}_1 \, \underset{\sim}{\alpha}_1 \tag{25}$$

$$\Delta \underset{\sim}{\sigma} = \underset{\sim}{P}_1 \, \underset{\sim}{\beta}_1 \tag{26}$$

$$\Delta \underset{\sim}{e} = \underset{\sim}{P}_2 \, \underset{\sim}{\alpha}_2 \tag{27}$$

$$\Delta \underset{\sim}{\sigma}^* = \underset{\sim}{P}_2 \, \underset{\sim}{\beta}_2 \tag{28}$$

By using Eqs. (11) - (13) the boundary tractions $\Delta\underset{\sim}{T}$ can also be expressed in terms of $\underset{\sim}{\beta}_1$ and $\underset{\sim}{\beta}_2$ i.e.

$$\Delta \underset{\sim}{T} = \underset{\sim}{R}_1 \underset{\sim}{\beta}_1 + \underset{\sim}{R}_2 \underset{\sim}{\beta}_2 \tag{29}$$

The element displacements $\Delta\underset{\sim}{u}$ and boundary displacements $\Delta\underset{\sim}{\tilde{u}}$ are all interpolated in terms of nodal displacements $\Delta\underset{\sim}{q}$, i.e.

$$\Delta\underset{\sim}{u} = \underset{\sim}{N}\,\Delta\underset{\sim}{q} \tag{30}$$

$$\Delta\underset{\sim}{\tilde{u}} = \underset{\sim}{L}\,\Delta\underset{\sim}{q} \tag{31}$$

and the nodal displacements at the beginning of the loading increment are $\underset{\sim}{q}$.

Then both $\bar{\pi}_G$ and $\bar{\pi}_G{}^*$ can be written in the form of

$$\bar{\pi}_G = \sum_n \left\{ \frac{1}{2}\,\underset{\sim}{\alpha}_1{}^T \underset{\sim}{J}_1 \underset{\sim}{\alpha} + \frac{1}{2}\,\underset{\sim}{\alpha}_2{}^T \underset{\sim}{J}_2 \underset{\sim}{\alpha}_2 - \underset{\sim}{\beta}_1{}^T \underset{\sim}{H}_1 \underset{\sim}{\alpha}_1 - \underset{\sim}{\beta}_2{}^T \underset{\sim}{H}_2 \underset{\sim}{\alpha}_2 \right.$$

$$\left. + \underset{\sim}{\beta}_1{}^T\, [\underset{\sim}{G}_1 + \underset{\sim}{G}_2(q)]\Delta\underset{\sim}{q} + \underset{\sim}{\beta}_2{}^T \underset{\sim}{G}_3 \Delta\underset{\sim}{q} - \Delta\underset{\sim}{Q}_n{}^T \Delta\underset{\sim}{q} \right\} \tag{32}$$

where

$$\underset{\sim}{J}_1 = \int_{A_n} \underset{\sim}{P}_1{}^T \underset{\sim}{C}\, \underset{\sim}{P}_1 \, dA \tag{33}$$

$$\underset{\sim}{J}_2 = \int_{A_n} \underset{\sim}{P}_2{}^T \underset{\sim}{N}^o \underset{\sim}{P}_2 \, dA \tag{34}$$

$$\underset{\sim}{H}_1 = \int_{A_n} \underset{\sim}{P}_1{}^T \underset{\sim}{P}_1 \, dA \tag{35}$$

$$\underset{\sim}{H}_2 = \int_{A_n} \underset{\sim}{P}_2{}^T \underset{\sim}{P}_2 \, dA \tag{36}$$

and $\underset{\sim}{C}$ is the matrix of stiffness coefficients and $\underset{\sim}{N}^o$ is the matrix of the initial stresses $N_{\alpha\beta}$.

The matrices $\underset{\sim}{G}_1$, $\underset{\sim}{G}_2$ and $\underset{\sim}{G}_3$ include expressions that are integrated both over the elements and around the boundary. The matrix $\underset{\sim}{G}_2(q)$ includes terms that are dependent on $w_{,\alpha}$ hence must be updated for each increment. The nodal loads $\Delta\underset{\sim}{Q}_n$ are from both prescribed body forces and boundary tractions.

From $\partial\bar{\pi}_G/\partial\underset{\sim}{\beta}_1 = 0$ and $\partial\bar{\pi}_G/\partial\underset{\sim}{\beta}_2 = 0$ one can obtain

$$\underset{\sim}{\alpha}_1 = \underset{\sim}{H}_1{}^{-1}(\underset{\sim}{G}_1 + \underset{\sim}{G}_2)\Delta\underset{\sim}{q} \tag{37}$$

$$\underset{\sim}{\alpha}_2 = \underset{\sim}{H}_2{}^{-1}\, \underset{\sim}{G}_3 \Delta\underset{\sim}{q} \tag{38}$$

Substituting into Eq. (32)

$$\bar{\pi}_G = \sum_n [\frac{1}{2} \Delta \underset{\sim}{q}^T \underset{\sim}{k} \Delta \underset{\sim}{q} - \Delta \underset{\sim}{Q}_n^T \Delta \underset{\sim}{q}] \quad (39)$$

where $\underset{\sim}{k}$ = element stiffness matrix

$$= \underset{\sim}{k}_1 + \underset{\sim}{k}_2 + \underset{\sim}{k}_3 \quad (40)$$

$$\underset{\sim}{k}_1 = \underset{\sim}{G}_1^T \underset{\sim}{H}_1^{-1} \underset{\sim}{J}_1 \underset{\sim}{H}_1^{-1} \underset{\sim}{G}_1^T \quad (41)$$

$$\underset{\sim}{k}_2 = \underset{\sim}{G}_1^T \underset{\sim}{H}^{-1} \underset{\sim}{J}_1 \underset{\sim}{H}^{-1} \underset{\sim}{G}_2 + \underset{\sim}{G}_2^T \underset{\sim}{H}^{-1} \underset{\sim}{J}_1 \underset{\sim}{H}^{-1} \underset{\sim}{G}_1 + \underset{\sim}{G}_2^T \underset{\sim}{H}^{-1} \underset{\sim}{J}_1 \underset{\sim}{H}^{-1} \underset{\sim}{G}_2 \quad (42)$$

and

$$\underset{\sim}{k}_3 = \underset{\sim}{G}_3^T \underset{\sim}{H}_2^{-1} \underset{\sim}{J}_2 \underset{\sim}{H}_2^{-1} \underset{\sim}{G}_3 \quad (43)$$

Here $\underset{\sim}{k}_1$ remains the same for all increments, $\underset{\sim}{k}_2$ is dependent on the initial displacement and $\underset{\sim}{k}_3$ is the initial stress matrix. This is basically the approach used for the quasi-conforming element [11].

In $\underset{\sim}{P}_1$ and $\underset{\sim}{P}_2$ the different strain and stress components are not coupled, hence $\underset{\sim}{H}_1$ and $\underset{\sim}{H}_2$ consist of submatrices located only along the diagonals. The inversion of $\underset{\sim}{H}_1$ and $\underset{\sim}{H}_2$ are very simple and, in fact, in many cases $\underset{\sim}{H}_1^{-1}$ and $\underset{\sim}{H}_2^{-1}$ can be expressed analytically.

In the π_R formulation again Eqs. (26), (28), (30) and (31) are used, the expressions are

$$\bar{\pi}_R = \sum_n \left\{-\frac{1}{2} \underset{\sim}{\beta}_1^T \underset{\sim}{\bar{H}}_1 \underset{\sim}{\beta}_1 - \frac{1}{2} \underset{\sim}{\beta}_2^T \underset{\sim}{\bar{H}}_2 \underset{\sim}{\beta}_2 + \underset{\sim}{\beta}_1^T [\underset{\sim}{G}_1 + \underset{\sim}{G}_2(\underset{\sim}{q})] \Delta \underset{\sim}{q} + \underset{\sim}{\beta}_2^T \underset{\sim}{G}_3 \Delta \underset{\sim}{q} - \Delta \underset{\sim}{Q}_n^T \Delta \underset{\sim}{q}\right\} \quad (44)$$

where

$$\bar{H}_1 = \int_{A_n} \underset{\sim}{P}_1^T \underset{\sim}{S} \underset{\sim}{P}_1 \, dA \quad (45)$$

$$\bar{H}_2 = \int_{A_n} \underset{\sim}{P}_2^T \underset{\sim}{Z}^* \underset{\sim}{P}_2 \, dA \quad (46)$$

and $\underset{\sim}{S}$ is the matrix of compliance coefficients and $\underset{\sim}{Z}^*$ is the matrix of $\underset{\sim}{Z}_{\alpha\beta}^*$.

From $\partial\bar{\pi}_R/\partial\beta_1 = 0$ and $\partial\bar{\pi}_R/\partial\beta_2 = 0$ one can write

$$\beta_1 = \underset{\sim}{\bar{H}}_1^{-1} (\underset{\sim}{G}_1 + \underset{\sim}{G}_2) \Delta q \quad (47)$$

$$\underset{\sim}{\beta}_2 = \underset{\sim}{\bar{H}}_2^{-1} \underset{\sim}{G}_3 \Delta \underset{\sim}{q} \quad (48)$$

Substituting into Eq. (44)

$$\bar{\pi}_R = \sum_n \left(\frac{1}{2} \Delta q^T \bar{k} q - Q_n^T \Delta q\right) \tag{49}$$

where $\bar{k} = \bar{k}_1 + \bar{k}_2 + \bar{k}_3$ (50)

$$\bar{k}_1 = G_1^T \bar{H}_1^{-1} G \tag{51}$$

$$\bar{k}_2 = G_1^T \bar{H}_1^{-1} G_2 + G_2^T \bar{H}_1^{-1} G_1 + G_2^T \bar{H}_1^{-1} G_2 \tag{52}$$

and $\bar{k}_3 = G_3^T \bar{H}_2^{-1} G_3$ (53)

Here $\bar{k}_2$ and $\bar{k}_3$ are again, respectively the initial displacement and initial stress stiffness matrices.

In the formulation by Hu-Washizu and Hellinger-Reissner principles it is possible to introduce constraints in stresses by introducing the equations of equilibrium Eqs. (17) and (18) and the corresponding Lagrange multipliers in the integral over A_n. The Lagrange multipliers are internal displacements which are independent from one element to the other. When such displacements are represented by finite number of displacement parameters which are statically condensed in the element level, a set of linear constraint equations for the stress parameters β_1 and β_2 can be obtained. Reference [6] has presented the corresponding expressions for element stiffness matrix when such constraint conditions are included. Alternatively when the constraint equations for β_1 and β_2 are available they can be utilized to modify the assumed stresses. The results will be of reduced number of independent stress parameters. However, the different stress components will be coupled and, in fact, in general, $\Delta\sigma$ and $\Delta\sigma^*$ will be coupled as well.

These formulations by multivariable variational principles will lead to the incremental finite element solution

$$K \Delta q = Q \tag{54}$$

When the initial inbalance in stresses and the initial incompatibility effects are taken into account, additional nodal loads should be added to the right hand side of Eq. (54). These loads are needed in the incremental-iterative procedure in nonlinear finite element solutions.

4. Discussions

In the formulation of hybrid elements for plane problems, 3-D solids and flat plates, a procedure has been established for the choice of assumed stress to provide desirable element properties [15]. The method is based on initially uncoupled stresses, but with stress equilibrium constraints included by the introduction of element internal displacements as Lagrange multipliers. When the number of parameters representing the internal displacements are larger than the number of equilibrium equations, additional conditions of constraint can be resulted [7,16]. A rational procedure is to choose the terms for the internal displacements such that the total displacements are now complete polynomials and that the corresponding strains are of the same order as that of the assumed stresses which are also initially chosen as complete polynomials. This procedure, in general, provides not only the stress equilibrium conditions but also additional constraints to the assumed stresses.

For example, for a 4-node rectangular plane stress element with the reference axes parallel to the sides, the constraint conditions obtained by this method limit the shear stress to only the constant term. The resulting number of independent stress parameters β is exactly the minimum required to suppress kinematic deformation modes. The resulting elements, indeed, will eliminate the locking difficulty under the condition of bending. For geometrically distorted elements, the assumed stresses can be expressed in natural isoparametric coordinates and the stress equilibrium condition is satisfied only in the variational sense within the element [7]. Another example is a quadrilateral 16-DOF semiLoof element [15] for which the boundary displacements w are quadratic and normal derivatives $w_{,n}$ are linear along the boundaries. The resulting elements derived by this procedure have been demonstrated to yield the best convergence properties in comparison to those by other assumed stress terms. The number of β-terms in this case is 23, while the minimum number required for the suppression of kinematic deformation mode is only 13.

For the formulation of general shell elements and for plates and shells in large deflections there is, however, another important consideration.

It is desirable to represent explicitly the membrane strains, the bending strains and the rigid body modes. In the conventional assumed displacement methods, if displacements are expressed as polynomial expansions in the shell coordinates, the in-plane and lateral displacements are coupled. In the formulation by multivariable variational principles, however, more options are available for the assumed stress and strain variables in order to obtain more desirable element behaviors.

The capability of explicit representation of rigid body modes, may also be classified as the satisfaction of zeroth order patch test. This can be accomplished automatically when an element is formulated based on equilibrating stresses. Based on principle of virtual work, if the element stresses are in equilibrium, there will be no work done due to rigid body motion. Thus, an element which is based on the satisfaction of homogeneous equilibrium conditions will always satisfy the rigid body motion requirement.

The importance of explicit representation of membrane and bending action for shell elements was pointed out by Morley [17]. A suggested "sensitive" and simple test problem is one with stress couple action under the inextensional condition. This is the problem of torsion of a split cylindrical shell under constant torsional moment. A similar and simple problem to test this decoupling behavior for geometrically nonlinear problem is a thin plate under cylindrical bending. This may be simply a thin rectangular plate clamped at one edge and loaded by uniformly distributed transverse shear along the opposite edge. The solution should be essentially the same as bending of slender beams, hence for moderate deflection the load deflection curve is linear. However, if in the element formulation the membrane and bending actions cannot be decoupled due to large deflection, the resulting curve will be nonlinear with increasing rigidity at larger deflections.

The problem has been solved using several rectangular shaped hybrid plate elements:

(A) Four-node element with 20 DOF formulated by Hu-Washizu principle π_G^*. The nodal displacements are w, $w_{,x}$, $w_{,y}$, u, and v at the four corners. The assumed stresses are initially uncoupled with 21 β-parameters but are constrained through the use of 3 internal displacement parameters as

Lagrange multipliers such that the linear equilibrium equations are satisfied exactly.

(B) SemiLoof plate element with 32-DOF formulated by Hu-Washizu principle π_G^*. The nodal displacements are w, u and v at the corners and the 4 mid-side points and $w_{,n}$ at the two 1/3 points of each side. The initially uncoupled stresses contain 38β-parameters and 7 internal displacement parameters are used to maintain the satisfaction of the linear equilibrium equations except in the cubic order moment terms.

(C) SemiLoof plate element with 32-DOF by Hellinger-Reissner principle π_R. The assumed stresses satisfy the linear equilibrium equations exactly and are subjected to additional constraints determined by consistently assumed internal displacements as Lagrange multipliers [18]. A total of 37 independent stress terms are used, 23 for $\Delta M_{\alpha\beta}$ and 14 for $\Delta N_{\alpha\beta}$.

A comparison of the solutions by these elements is given by the plots of the deflections of the mid-point of the outer edge vs the applied shear load as shown in Fig. 1. The reference solution is that of a corresponding cylindrical bending problem for which the load-deflection curve is linear. For element A the solutions by three mesh sizes are shown. Only a single element was used in the case of B and C.

It is seen that all solutions follow the initial solution initially but become nonlinear at large deflections. In the case of element A, the use of finer mesh reduces the non-linearity of the solution at large deflections. This is an indication that the exact solution of the problem should approach the referenced cylindrical bending solution, however the coupling of bending and membrane actions in large deflection range induces the nonlinear phenomena. The solutions by the two semiLoof elements apparently show less coupling effects. For the solutions by element C two different shape functions for the element displacements are used for the construction of the initial stress matrices. One of these is in terms of only the eight nodal displacements w along the boundary while the other one also depends on the 8 normal rotations $w_{,n}$. It is clear that the element derived by the latter provides the most accurate solution. A comparison of the spanwise bending moment distributions also indicate that element C yields the most accurate results.

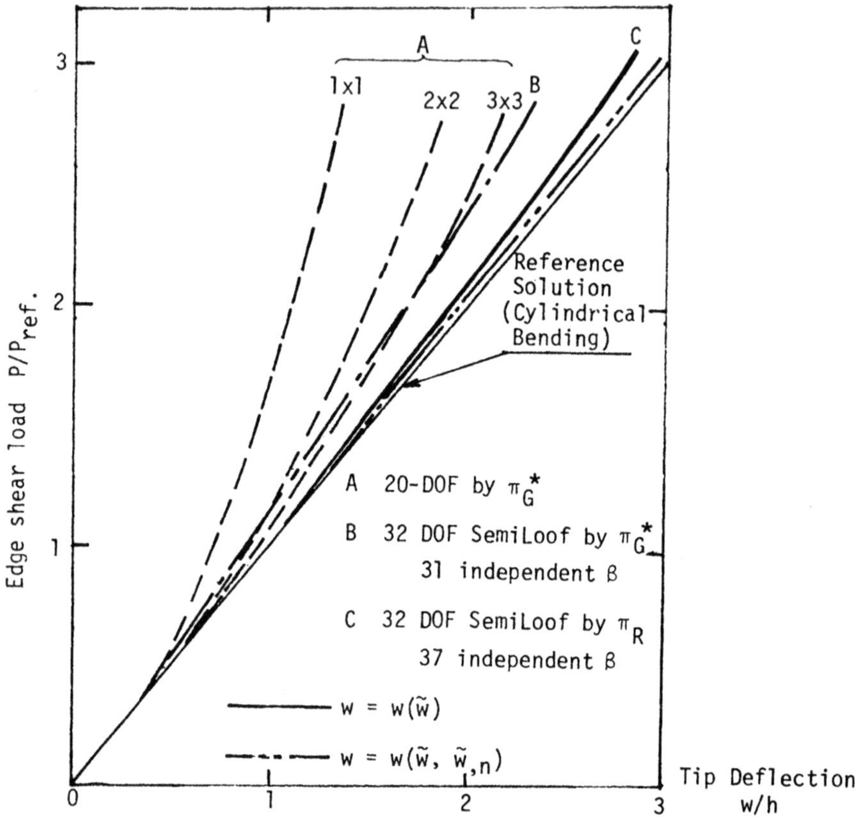

Figure 1. Edge Shear Load vs Tip Deflection - Cantilever Square Plate

5. Conclusions

Many alternative approaches can be used in finite element formulations of large deflection problems based on multi-variable variational principles. It is shown that the choice of constraint conditions for the assumed stresses is crucial for the success of the resulting solutions. Also the coupling of membrane and bending action can be a source of error for the solutions of geometrically nonlinear problems.

Acknowledgement: Numerical solutions were obtained by K. Sumihara, under the support of Boeing Military Aircraft Co. Wichita, Kansas under P.O. No. AA0045 and by Cheng Wang under the support of U.S. Air Force Materials Laboratory Contract No. F33615-83-K-5016.

References

1. K. Washizu, Variational Methods in Elasticity and Plasticity, 3rd. Ed. Pergamon Press 1982.

2. G. Horrigmoe and P.G. Bergan, "Incremental Variational Principles and Finite Element Models for Nonlinear Problems", Computer Meth. in App. Mech. and Engng., Vol. 7, pp. 201-217 (1976).

3. B. Fraeijs de Veubeke, "Displacement and Equilibrium Models in the Finite Element Method", Stress Analysis, Ed. by O.C. Zienkiewicz and G.S. Hollister, John Wiley and Sons Ltd. London, pp. 145-197.

4. T.H.H. Pian, "Derivation of Element Stiffness Matrices by Assumed Stress Distributions", AIAA J., Vol. 2, pp. 1333-1336 (1964).

5. T.H.H. Pian, "Finite Element Methods by Variational Principles with Relaxed Continuity Requirement", Variational Methods in Engineering Edited by C.A. Brebbia and H. Tottenham, Southampton University Press, 1973, pp. 3/1-3/24.

6. T.H.H. Pian and D.P. Chen, "Alternative Ways for Formulation of Hybrid Stress Elements", Int. J. Num. Meth. Engng, Vol. 18, 1982, pp. 1679-1684.

7. T.H.H. Pian and K. Sumihara, "Rational Approach for Assumed Stress Finite Elements", Int. J. Num. Meth. Engng, Vol. 20 (1984), pp. 1685-1695.

8. T.H.H. Pian and P. Tong, "Relations between Incompatible Displacement Model and Hybrid Stress Model", to be published in Int. J. Num. Meth. Engng.

9. Limin Tang, Wanji Chen and Yingxi Liu, "String Net Function Approximation and Quasi-Conforming Technique", Hybrid and Mixed Finite Element Methods, Edited by S. Atluri, et al, Wiley, pp. 173-188, 1983.

10. Limin Tang, Wanji Chen and Yingxi Liu, "Formulation of Quasi-conforming Element and Hu-Washizu Principle", Computers and Structures, Vol. 19, pp. 247-250, 1984.

11. Heyang Jiang, "Nonlinear Finite Elements by Quasi-conforming Model", (in Chinese), Doctoral Thesis Dalian Institute of Technology, June 1984.

12. P.L. Boland and T.H.H. Pian, "Large Deflection Analysis of Thin Elastic Structures by the Assumed Stress Hybrid Finite Element Method", Computers and Structures, Vol. 7, 1977, pp. 1-12.

13. T.H.H. Pian and P.L. Boland, "Formulations of Large Deflection Shell Analysis by Assumed Stress Finite Element Method", Formulations and Computational Algorithms in Finite Element Analysis, Edited by K.J. Bathe, J.T. Oden and W. Wunderlich, M.I.T. Press, 1977, pp. 241-263.

14. G. Horrigmoe, "Hybrid Stress Finite Element Model for Non-linear Shell Problems", Int. J. Num. Meth. Engng. Vol. 2, pp. 1819-1839 (1978).

15. T.H.H. Pian, "Finite Elements Based on Consistently Assumed Stresses and Displacements," to be published in Finite Elements in Analysis and Design (1985).

16. T.H.H. Pian, "On the Equivalence of Non-Conforming Elements and Hybrid Stress Elements", Applied Mathematics and Mechanics, EUST Press, Wuhan, China, Vol. 3, 1982, pp. 773-776.

17. L.S.D. Morley, "Analysis of Developable Shells with Special Reference to the Finite Element Method and Circular Cylinders", Philosophical Transactions of the Royal Society of London, Vol. 281, pp. 113-180, 1976.

18. K. Sumihara, "Thin Shell and New Invariant Elements by Hybrid Stress Method", Ph.D. Thesis, Department of Aeronautics and Astronautics, M.I.T., 1983.

A Family of C^0 Shell Elements Based on Generalized Hrennikoff's Method and Assumed Natural-Coordinate Strains

K. C. PARK, G. M. STANLEY and H. CABINESS
Lockheed Palo Alto Research Laboratory, USA

ABSTRACT

This paper presents a family of c^0 shell elements that are based on a generalized Hrennikoff's method and assumed natural-coordinate strains along the element lattice lines. In the present elements, the Cartesian strain-displacement matrices are obtained by a tensorial transformation of the natural-coordinate strains instead of the standard isoparametric transformation of natural-coordinate derivatives via the isoparametric Jacobian matrix. The elements thus derived preserve correct rank without transverse shear locking, consistent membrane strain interpolation that admits inextensional bending without reduced integration, and adequate representation of curvature effects to capture the important membrane-bending coupling for curved elements. The elements can be constructed either by full integration rules or by reduced integration rules if augmented with the proper rank compensating terms. Numerical experiments with the present family of elements on several benchmark problems indicate that the elements yield accurate and reliable solutions, especially for distorted grids, and are made work-horse elements for NICE-based nonlinear shell analysis modules.

1. Introduction

This paper presents an overview of basic formulation for the assumed natural-coordinate strain(**ans**) shell elements[1]: **ans-4** and **ans-9**. Our emphasis here is on physical insight that led to the formulation of the elements and on deriving strain-displacement relations. For those who are interested in theoretical aspects of the elements and their performance, we refer them to [1-3]. The genesis of the **ans**-elements essentially goes back to the Hrennikoff lattice[4] for discretizing plates by straight beams that are rigidly

Finite Element Methods for Nonlinear Problems
Europe-US Symposium, Trondheim, Norway 1985
Editors: Bergan, Bathe, Wunderlich

linked at the grid points. The Hrennikoff beams surrounding a plate element are to satisfy the conditions for pure bending states in two Cartesian directions and a pure twist state of the plate element. Satisfaction of these bending conditions then yields the appropriate moments of inertia for the surrounding beams. However, the original developers of finite elements, Turner, Clough, Martin and Topp[5], abandonded the Hrennikoff framework analogy from the following considerations: *"The determination of a lattice representation for a rectangular plate is straightforward; however, plate elements of nonrectangular form present basic difficulties. For example, if one attempts to apply the rectangular gridwork to a nonrectangular plate, difficulties arise in attempting to satisfy boundary conditions. On the other hand, if one goes to nonrectangular lattice forms, difficulties arise when attempting to satisfy the stress relations in the interior of the plate. Considerations such as these led to eventual abandonment of this approach."*

The present **ans**-elements are designed to overcome the above mentioned difficulties, 1) by replacing the representation of the rigidities of lattice beams with that of strains along the lattices, 2) by representing the strains in the interior of a shell element by interpolating the lattice beam strains, and 3) by constructing the desired Cartesian strains *via* the appropriate tensorial transformation of the natural-coordinate strain compoments. In other words, the present **ans**-elements adopt basically similar beam lattice arrangements as in the Hrennikoff lattice. A key difference in the present formulation is that, instead of assigning an equivalent bending rigidity for each lattice beam, strain distributions are specified along the beam-lattice lines which are designated here as ***natural-coordinate reference lattice lines*** or simply ***lattice lines***. Once the strains along the lattice lines are specified, the strains in the interior of the element in the natural-coordinate system are then obtained by interpolating the lattice-line strains. Another key feature of the **ans**-elements is in the transformation of the natural-coordinate strains to a Cartesian coordinate system. In the present shell elements, we abandon the standard isoparametric procedure for transforming the natural-coordinate derivatives to their corresponding Cartesian components. Instead, a tensorial strain transformation is adopted for this transformation.

The present **ans**-elements also make the so-called reduced integration and the associated rank-compensation stabilization unnecessary in that the elements neither locks nor suffer from rank deficiency. Specifically, we would like to reveal the consequences of a classical finite element dilemma, *viz.*, reduced integration in shell elements: a fix (reduced integration) creates another problem(spurious oscillation) while attempting to partially solve a problem(numerical locking). By relying on the results from symbolic element evaluations[6-9], we will attempt to make clear that reduced integration is not needed if locking is all that matters in element performance and that there is a consistent way of either recovering the attributes of fully integrated elements or of improving the element performance if reduced integration is desired for computational economy.

Of course, there are several steps starting from laying out Hrennikoff's lattices to strain interpolations and finally to the desired Cartesain strain-displacement matrices. The remainder of the present tour is to take you through these steps.

2. Basic Kinematical Definitions

For subsequent developments, let us introduce a position vector, a displacement vector and their derivatives, and a coordinate transformation matrix from a non-Cartesain to a Cartesian basis.

The position vector, $\mathbf{r}$, on the middle of the element surface may be expressed by

$$\mathbf{r} = x(\xi,\eta)\mathbf{e}_x + y(\xi,\eta)\mathbf{e}_y + z(\xi,\eta)\mathbf{e}_z \tag{1}$$

where

$$\mathbf{e} = (\mathbf{e}_x, \mathbf{e}_y, \mathbf{e}_z)^T \tag{2}$$

are the unit vectors in a local Cartesian coordinate system whose coordinates, (x, y, z), are interpolated by the two-dimensional Lagrange shape functions:

$$(x, y, z) = \sum_{i=1}^{n} N_i(\xi,\eta)(x_i, y_i, z_i) \tag{3}$$

in which $N_i(\xi,\eta)$ are the isoparametric interpolation shape functions, (x_i, y_i, z_i) are the nodal coordinates and n is the number of noded adopted within an element as shown in Fig. 1.

The unit vectors of the natural-coordinate system, $\mathbf{a}_\xi$ and $\mathbf{a}_\eta$, are given by

$$\mathbf{a}_\xi = \frac{1}{A_\xi}\frac{\partial \mathbf{r}}{\partial \xi}\,, \quad \mathbf{a}_\eta = \frac{1}{A_\eta}\frac{\partial \mathbf{r}}{\partial \eta} \tag{4}$$

in which A_ξ and A_η are the first fundamental coefficients given by

$$A_\xi = (\frac{\partial \mathbf{r}}{\partial \xi}\cdot\frac{\partial \mathbf{r}}{\partial \xi})^{\frac{1}{2}}\,, \quad A_\eta = (\frac{\partial \mathbf{r}}{\partial \eta}\cdot\frac{\partial \mathbf{r}}{\partial \eta})^{\frac{1}{2}} \tag{5}$$

The unit vector $\mathbf{a}_\varsigma$ that is normal to $(\mathbf{a}_\xi, \mathbf{a}_\eta)$ is then obtained by

$$\mathbf{a}_\varsigma = \frac{\mathbf{a}_\xi \times \mathbf{a}_\eta}{|\mathbf{a}_\xi \times \mathbf{a}_\eta|} \tag{6}$$

The displacement vector expressed in a local Cartesian coordinate system $\mathbf{e}$ may be given by

$$\mathbf{U} = (u_x + \bar{\varsigma}\hat{u}_x)\mathbf{e}_x + (u_y + \bar{\varsigma}\hat{v}_y)\mathbf{e}_y + u_z\mathbf{e}_z \tag{7}$$

where both the translations, u_x, u_y and u_z, and the rotations, $\hat{u}$ and $\hat{v}$, are assumed to depend only on the natural surface coordinates, (ξ, η). Therefore, the displacement components along the natural coordinate lines are then obtained by projecting $\mathbf{U}$ on the natural coordinate lines:

$$u_\xi = \mathbf{U}\cdot\mathbf{a}_\xi\,, \quad u_\eta = \mathbf{U}\cdot\mathbf{a}_\eta\,, \quad u_\varsigma = \mathbf{U}\cdot\mathbf{a}_\varsigma \tag{8}$$

We will also adopt the convention that the unit normal vector for the $\mathbf{e}$ basis and the one for the $\mathbf{a}$ basis coincides; in other words, the planes formed by $(\mathbf{e}_x, \mathbf{e}_y)$ and $(\mathbf{a}_\xi, \mathbf{a}_\eta)$ are coplanar. This convention leads to the follwoing simple transformation

$$\mathbf{u_a} = \mathbf{T}^{-1}\mathbf{u_e} \tag{9}$$

where

$$\mathbf{u_a} = \lfloor u_\xi, u_\eta, u_\varsigma \rfloor\,, \quad \mathbf{u_e} = \lfloor u_x, u_y, u_z \rfloor \tag{10}$$

and the transformation matrix $\mathbf{T}$ is given by

$$\mathbf{T} = \begin{bmatrix} t_{x\xi} & t_{x\eta} & 0 \\ t_{y\xi} & t_{y\eta} & 0 \\ 0 & 0 & 1 \end{bmatrix} = \frac{1}{|\mathbf{a}_\xi \times \mathbf{a}_\eta|}\begin{bmatrix} \mathbf{e}_y\cdot\mathbf{a}_\eta & -\mathbf{e}_y\cdot\mathbf{a}_\xi & 0 \\ -\mathbf{e}_x\cdot\mathbf{a}_\eta & \mathbf{e}_x\cdot\mathbf{a}_\xi & 0 \\ 0 & 0 & |\mathbf{a}_\xi \times \mathbf{a}_\eta| \end{bmatrix} \tag{11}$$

3. Strains Along A Lattice Line

As stated in **Introduction** the first step for formulating the strain-displacement relationships is to model strains along the Hrennikoff lattice lines. To this end, let us consider a lattice line ($\eta = 1$) from Fig. 1. There are three strain components: the membrane strain, ϵ_ξ, the bending strain, κ_ξ and the transverse shear strain, γ_ξ. Approximations of these three strains will be presented in this section.

3.1 Membrane Strain Along A Lattice Line

The membrane strain along a curved lattice line requires a careful observation of the interplay between the curvature of the geometry and the in-plane component of the beam displacements. By definition the membrane strain vector is tangent along the curved lattice line, which implies that the unit vector associated with the membrane strain also varies along the curved line. This is accounted for in continuum case by the relation in the shell coordinate system as

$$\epsilon_\xi = \frac{\partial u}{\partial s} + \frac{w}{R_\xi} \tag{12}$$

where s designates the shell coordinate tangent to the shell surface and R_ξ is the radius of curvature along the latice line ξ.

In the finite element approximation, the strains are expressed in a Cartesian system; hence the counterpart to the above expression becomes

$$\epsilon_\xi = \frac{1}{A_\xi}\frac{\partial \bar{u}}{\partial \xi} \tag{13}$$

where A_ξ is the fundamental surface coefficient given by

$$A_\xi = \left(\frac{\partial r}{\partial \xi} \cdot \frac{\partial r}{\partial \xi}\right)^{\frac{1}{2}} \tag{14}$$

and $\bar{u}$ is the tangent component of u at ξ, whose transformation brings the necessary curvature effect to the resulting strain-displacement relations. The question of how well the Cartesian expression, $\frac{\partial \bar{u}}{\partial \xi}$, that is used to model the membrane strain in the finite element approximation approximate the curvilinear membrane strain can be greatly influenced by the impelementation detail. We will now examine the membrane strain approximations of both the two and three-noded beam elements.

3.1.1 Modeling of Membrane Strain by Two-Noded Element

The membrane strain along a straight line(see Fig. 2a) is obtained as

$$\epsilon_\xi = (\bar{u}_2 - \bar{u}_1)/\ell_\xi \tag{15}$$

where ℓ_ξ is the element length. Let us consider now the membrane strain approximation of an arch by the two-noded beam element as shown in Fig.2b, which can be expressed in terms of the shell-coordinate displacement components, (u, w), as

$$\epsilon_\xi = \cos\theta(u_2 - u_1)/\ell_\xi + \sin\theta(w_2 + w_1)/\ell_\xi \tag{16}$$

where θ is the angle subtending half of the element cord length. Rearranging the above equation to compare it with the continuum case, Eq.(12), one obtains

$$\epsilon_\xi = \frac{\theta}{\tan\theta}\frac{(u_2 - u_1)}{\triangle s} + \frac{1}{R_\xi}\frac{(w_2 + w_1)}{2} \tag{17}$$

where $\triangle s$ is the arc length given by $2R_\xi\theta$.

Remark 1: Discretization of the constant membrane strain state in an arch subjected to uniform internal pressure $(u = 0)$ yields exact solutions whereas the case subjected to uniform stretching $(w = 0)$ gives rise to an error e by the amount

$$e = 1 - \frac{\theta}{\tan\theta} \tag{18}$$

This error can be corrected if the following modifications are introduced:

$$\ell_\xi \Leftarrow \triangle S, \qquad \cos\theta \Leftarrow 1, \qquad \sin\theta \Leftarrow \triangle S/2R_\xi \tag{19}$$

These modifications should improve the membrane accuracy when regular four-noded elements are adopted for discretizing cylindrical shells.

3.1.2 Modeling of Membrane Strain by Three-Noded Element

For three-noded beam element shown in Fig. 2c, , one may assume that the membrane strain and its direction vary linearly along the ξ-coordinate line. Hence, one may assume the following form:

$$\epsilon_{\xi} = \epsilon_0 + \xi \cdot \epsilon_1 \tag{20}$$

where $\epsilon_{(0,1)}$ are obtained by combinations of the Barlow point strain states as derived in [1,2], viz.,

$$\epsilon_0 = \frac{1}{2}\{\epsilon_{\xi_l} + \epsilon_{\xi_l}\}, \qquad \epsilon_1 = \frac{\sqrt{3}}{2}\{\epsilon_{\xi_r} - \epsilon_{\xi_l}\} \tag{21}$$

where $\xi_{(l,r)} = \pm 1/\sqrt{3}$ and ϵ_{ξ_l} is given by

$$\epsilon_{\xi_l} = \frac{1}{\triangle s}(\bar{u}_3^{\xi_l} - \bar{u}_1^{\xi_l}) + \frac{2\xi_l}{\triangle s}(\bar{u}_3^{\xi_l} - 2\bar{u}_2^{\xi_l} + \bar{u}_1^{\xi_l}) \tag{22}$$

where $\triangle s$ is the curved beam length and $(\bar{u})_1^{\xi_l}$ refers to the component of u_1 along the $\mathbf{a}_{\xi_l}$ basis vector. Similarly, ϵ_{ξ_r} can be obtained. This interpretaion plays a key role in deriving locking-free strain approximations. For example, $(\bar{u})_i^{\xi_l}$ stands for:

$$(\bar{u})_i^{\xi_l} = \cos(\theta_{\xi_l} - \theta_i)u_i + \sin(\theta_{\xi_l} - \theta_i)w_i \tag{23}$$

It can be shown that the above strain interpolation(20) models the membrane strain state at the three-noded element center (at node 2 in Fig. 2c)

$$\epsilon(0)_{\xi} = \frac{\cos(\theta_{\bar{\xi}})}{\triangle s}(u_3 - u_1) + \frac{\sin(\theta_{\bar{\xi}})}{\triangle s}\tilde{w}_2 \tag{24}$$

where

$$\tilde{w}_2 = \frac{w_3 + w_2 + w_1}{3} \tag{25}$$

and $\theta_{\bar{\xi}}$ is the tangent angle at the Barlow point relative to the local Cartesian coordinate system at the element center. Notice that the w/R term is modeled by a uniform averaging of the three normal displacements given by Eq.(25), hence the quadratic approximation accrues no additional advantage over the linear beam case. In other words, $\tilde{w}$ term in the natural coordinate-based strain ϵ_{ξ} is approximated by a constant $\tilde{w}$ state whereas the approximation of the first term in Eq.(12) is modeled to vary linearly. This inconsistency was examined in Park[1] and the desirable representation of $\tilde{w}/R$ was determined to be

$$\tilde{w}(\xi) = (\frac{1}{6} - \frac{\xi}{2})w_1 + \frac{2}{3}w_2 + (\frac{1}{6} + \frac{\xi}{2})w_3 = \sum_i^3 \bar{N}_i(\xi)w_i \tag{26}$$

which yields at the three nodes for $\tilde{w}_i$ as

$$\tilde{w}_1 = \frac{4w_1 + 4w_2 - 2w_3}{6}, \quad \tilde{w}_2 = \frac{w_3 + 4w_2 + w_1}{6}, \quad \tilde{w}_3 = \frac{-2w_1 + 4w_2 + 4w_3}{6} \tag{27}$$

3.2 Transverse Shear Strain Along A Lattice Line

The transverse shear strain, γ_ξ, along a lattice beam can be expressed as

$$\gamma_\xi = \hat{u} + \frac{1}{A_\xi}\frac{\partial \bar{w}}{\partial \xi} \tag{28}$$

where $\hat{u}$ is the rotational displacement as defined in Fig. 2a. It is noted that, for consistency reasons, $\bar{w}$ must be approximated one order higher than $\hat{u}$. A violation of this consistency requirement occurs when both $\bar{w}$ and $\hat{u}$ are interpolated with the same isoparametric shape function; a subject much studied in the literature starting with Zienkiewicz, Taylor and Too[10] and Pawsey and Clough[11] as as it can cause a serious pathology of c^0 elements. The conventional means to alleviate this inconsistency that is known to causes transverse shear locking has been to evaluate the transverse-shear strain energy term with reduced integration rules. However, this numerical trick often gives rise to spurious mechanisms. We shall now show that full integration does not cause locking provided the transverse strain along the lattice line is *consistently* approximated.

3.2.1 Modeling of Transverse Shear Strain by Two-Noded Element

The term $\partial \bar{w}/\partial \xi$ in the transverse shear strain displacement relation(28) is approximated as that of the membrane strain case:

$$\frac{1}{A_\xi}\frac{\partial \bar{w}}{\partial \xi} = (\bar{w}_2 - \bar{w}_1)/\ell_\xi \tag{29}$$

which reamins constant along the lattice line. Hence, $\hat{u}$ must be approximated by a constant term for consistency. This consistency requirement is met if one approximates $\hat{u}$ by

$$\hat{u} = \frac{1}{2}(\hat{u}_1 + \hat{u}_2) \tag{30}$$

The transverse shear strain along a constant η-line is therefore obtained by substituting Eqs.(29) and (30) into Eq.(28).

3.2.2 Modeling of Transverse Shear Strain by Three-Noded Element

The strain-displacement relations along a quadratic lattice beam can vary at most linearly. This means that, in order to avoid transverse-shear locking the interpolation of the rotation component $\hat{u}$ in Eq.(28) is modified as follows.

If the rotation, $\hat{u}$, were allowed to vary quadratically, the corresponding transverse displacement, $\bar{w}$, must vary cubically, viz.,

$$\bar{w} = \sum_{i=1}^{3} N(\xi)_i \bar{w}_i + \xi(1-\xi^2)\triangle\bar{w} \tag{31}$$

where $\triangle\bar{w}$ may be viewed as a hierachical displacement.

Substituting the above modified interpolation for $\bar{w}$ and the assumed quadratic interpolation of $\hat{u}$ into the transverse shear strain expression(28) and constraining the resulting transverse shear strain to be linear along ξ-line, one obtains for the hierarchical displacement $\triangle\bar{w}$ as

$$\triangle\bar{w} = \frac{1}{6}(\hat{u}_1 - 2\hat{u}_2 + \hat{u}_3) \tag{32}$$

Finally, collecting the constant and linear terms and rearranging the resulting expression for the transverse shear strain expression, one can obtain $\hat{u}$ as

$$\hat{u} = \sum_{i=1}^{3} \bar{N}(\xi)_i \hat{u}_i \tag{33}$$

A similar approach was successfully used in the development of plate bending elements by Crisfield[12].

To complete the derivation of the transverse-shear strains along the natural coordinate lines, care must be also exercised in the interpolation of u contained in $\frac{\partial \bar{w}}{\partial \xi}$ for the final transverse shear strain. Since for this case one approximates

$$\frac{\partial \bar{w}}{\partial \xi} \approx \frac{\partial w}{\partial s} - \frac{u}{R_\xi} \tag{34}$$

the axial displacement contribution due to the curvature, i.e., u/R_ξ, must follow a similar modification as that used to interpolate $\tilde{w}$ for the membrane strain case (see Eq.(26)), viz.,

$$(\bar{w})_i^{\xi_l} = \cos(\theta_{\xi_l} - \theta_i)w_i - \sin(\theta_{\xi_l} - \theta_i)\tilde{u}_i \tag{35}$$

where $\tilde{u}_i$ must be interpolated by

$$\tilde{u} = \sum_{i=1}^{3} \bar{N}(\xi)_i u_i \tag{36}$$

4. Natural-Coordinate Strains in Element Interior

The four and nine-noded shell elements can be constructed with four and six Hrennikof's lattice beams, respectively, as shown in Figs. 1a and 1b. The strains interpolated along the two and three-noded lattice beams presented in the preceding section can now be used to interpolate the natural-coordinate strains in the interior of the four and nine-noded elements. We first present the natural-coordinate strains for the four-noded shell element, which is followed by those of the nine-noded shell element.

4.1 Assumed Natural-Coordinate Strains for Four-Noded Element

The membrane strain, $\epsilon_{\xi\xi}$, in the direction of ξ everywhere in the element can be constructed by interpolating the membrane strains along the two lattice beams, namely, at $\eta = \pm 1$:

$$\epsilon_{\xi\xi} = N_1(\eta)\epsilon_\xi\Big|_{\eta=-1} + N_2(\eta)\epsilon_\xi\Big|_{\eta=+1} \tag{37}$$

where

$$\lfloor N_1(\eta), N_2(\eta) \rfloor = \lfloor \tfrac{1}{2}(1-\eta), \tfrac{1}{2}(1+\eta) \rfloor \tag{38}$$

and $\epsilon_\xi\Big|_{\eta=-1}$ and $\epsilon_\xi\Big|_{\eta=+1}$ are the membrane strains for lattice beams along $\eta = \pm 1$ as obtained by Eq.(15). Note that the direction of $\epsilon_{\xi\xi}$ at any $\eta = constant$ also coincides with the direction of ξ-line, which means that the membrane strain along a Hrennikoff lattice beam at any ξ-line is obtained by interpolating those of the two *reference* lattice beams. This interpretation enables us to bypass the usage of the isoparametric mapping from the natural to Cartesian coordinate systems and to adopt the present tensorial transformation, instead, for expressing the Cartesain strains in terms of the assumed natural-coordinate strains.

Similarly, $\epsilon_{\eta\eta}$ is obtained by

$$\epsilon_{\eta\eta} = N_1(\xi)\epsilon_\eta\Big|_{\xi=-1} + N_1(\xi)\epsilon_\eta\Big|_{\xi=+1} \tag{39}$$

It therefore is evident that interpolations of the bending strains, $\kappa_{\xi\xi}$ and $\kappa_{\eta\eta}$, and the transverse shear strains, $\gamma_{\varsigma\xi}$ and $\gamma_{\varsigma\eta}$, follows the membrane strains, Eqs.(37) and (39). However, the inplane bending strain, $\epsilon_{\xi\eta}$, and the curvature twist strain, $\kappa_{\xi\eta}$, must adopt a special procedure as discussed below. To accomplish this special discretization, the inplane bending strain in the natural-coordinate system is expressed as

$$\epsilon_{\xi\eta} = \frac{1}{A_\xi}\frac{\partial \bar{v}}{\partial \xi} + \frac{1}{A_\eta}\frac{\partial \bar{u}}{\partial \eta} \tag{40}$$

The above expression for $\epsilon_{\xi\eta}$ implies that, since the two natural-coordinate derivatives, $\frac{\partial \bar{v}}{\partial \xi}$ and $\frac{\partial \bar{u}}{\partial \eta}$, are constant along the ξ and η-lines, respectively, so must be their interpolations in the interior of the element. This homogeneity requirement is satisfied if $N_i(\xi)$ and $N_i(\eta)$ are replaced by $\bar{N}_i(\xi)$ and $\bar{N}_i(\eta)$:

$$\lfloor \bar{N}_1(\eta), \bar{N}_2(\eta) \rfloor = \lfloor \tfrac{1}{2}, \tfrac{1}{2} \rfloor \tag{41}$$

The inplane bending strain can now be expressed as

$$\epsilon_{\xi\eta} = \sum_{j=1}^{2} \bar{N}_j(\eta)\frac{1}{\ell_{\xi_j}}\bar{v}_{,\xi}\Big|_{\eta=(2j-3)} + \sum_{i=1}^{2} \bar{N}_i(\xi)\frac{1}{\ell_{\eta_i}}\bar{u}_{,\eta}\Big|_{\xi=(2i-3)} \tag{42}$$

where $\bar{v}_{,\xi}\Big|_{\eta=(2j-3)}$ and $\bar{u}_{,\eta}\Big|_{\xi=(2i-3)}$ are the derivatives of $(\bar{v}, \bar{u})$ with respect to (ξ, η) along the constant reference (ξ, η)-lines, respectively.

Similarly, $\kappa_{\xi\eta}$ can be constructed by substituting $\hat{u}$ and $\hat{v}$ for $\bar{u}$ and $\bar{v}$, respectively, into Eq.(42).

4.2 Assumed Natural-Coordinate Strains for Nine-Noded Element

The construction of the natural-coordinate strains for nine-noded shell element follows the same procedure adopted for the foru-noded element case with one exception that the strains are now interpolated quadratically using three appropriate lattice-line components. For example, the membrane strain, $\epsilon_{\xi\xi}$, is obtained as

$$\epsilon_{\xi\xi} = N_1(\eta) \cdot \epsilon_\xi \Big|_{\eta=-1} + N_2(\eta) \cdot \epsilon_\xi \Big|_{\eta=0} + N_3(\eta) \cdot \epsilon_\xi \Big|_{\eta=1} \tag{43}$$

where $N_i(\eta)$ are the one-dimensional quadratic shape functions along the natural coordinate η-line and $\epsilon_\xi \Big|_{\eta=-1,0,1}$ are the membrane strains along the $\eta = -1, 0, 1$ lines, respectively, given by Eq.(20).

Other strain components are similarly constructed except the inplane bending and the curvature twist strains as noted for the four-noded element case. Here, the inplane bending strain is obtained as

$$\epsilon_{\xi\eta} = \sum_{j=1}^{3} \bar{N}_j(\eta) \frac{1}{\Delta s_{\xi_j}} \bar{v}_{,\xi} \Big|_{\eta=(j-2)} + \sum_{i=1}^{3} \bar{N}_i(\xi) \frac{1}{\Delta s_{\eta_i}} \bar{u}_{,\eta} \Big|_{\xi=(i-2)} \tag{44}$$

where $\bar{N}_j(\xi)$ are given by the modified shape functions given by Eq.(26) and $\bar{v}_{,\xi}\Big|_{\eta=(j-2)}$ are the derivatives of $\bar{v}$ with respect to ξ along the constant reference η-lines.

Remark 2: The present formulation is easily extended to the sixteen-noded C^0 element if one adopts the following modifications. First, the consistent interpolations for $(\hat{u}, \hat{v})$, and the curvature terms, $(\tilde{w}/R, \tilde{u}/R, \tilde{v}/R)$ are modified according to [2], e.g., for $\tilde{u}$:

$$\tilde{u} = \frac{(-1 + 6\xi + 9\xi^2)}{16} u_1 + \frac{(9 - 42\xi - 9\xi^2)}{16} u_2 + \frac{(9 + 42\xi - 9\xi^2)}{16} u_3 + \frac{(-1 - 6\xi + 9\xi^2)}{16} u_4 \tag{45}$$

Second, the strains along the four element sides($\xi = \pm 1, \eta = \pm 1$) and the four interior natural coordinate lines($\xi = \pm 1/3, \eta = \pm 1/3$) are assumed to vary quadratically. Finally, the natural coordinate strains at an interior of the element are obtained by cubic interpolations of the strains along the eight reference lattice lines.

This completes the derivation of the natural coordinate strains at an interior point of the quadratic element. To evaluate the strain energy at the numerical integration points, these natural coordinate strains must be transformed to a local Cartesian coordinate system, which is addressed in the next section.

5. Cartesian Strain-Displacement Matrix

The Cartesian strain-displacement matrix needed to compute the element stiffness matrix, internal force and Cartesian strains can now be obtained by the tensor transformation of the natural-coordinate strains to a fixed local Cartesian system, which was derived in terms of the components of the transformation matrix given by Eq.(11) as (see [1] for details):

$$\begin{pmatrix} \varepsilon_{xx} \\ \varepsilon_{yy} \\ \varepsilon_{xy} \\ \varepsilon_{zx} \\ \varepsilon_{zy} \end{pmatrix} = \begin{bmatrix} t_{x\xi}^2 & t_{x\eta}^2 & t_{x\xi}t_{x\eta} & 0 & 0 \\ t_{y\xi}^2 & t_{y\eta}^2 & t_{y\xi}t_{y\eta} & 0 & 0 \\ 2t_{x\xi}t_{y\xi} & 2t_{x\eta}t_{y\eta} & t_{x\xi}t_{y\eta}+t_{x\eta}t_{y\xi} & 0 & 0 \\ 0 & 0 & 0 & t_{x\xi} & t_{x\eta} \\ 0 & 0 & 0 & t_{y\xi} & t_{y\eta} \end{bmatrix} \begin{pmatrix} \varepsilon_{\xi\xi} \\ \varepsilon_{\eta\eta} \\ \varepsilon_{\xi\eta} \\ \varepsilon_{\varsigma\xi} \\ \varepsilon_{\varsigma\eta} \end{pmatrix} \tag{46}$$

which, after appropriate substitutions, yields the desired strain-displacement matrix operator, **B**, as

$$\epsilon^c = \mathbf{B} \cdot \mathbf{u} \tag{47}$$

where

$$\epsilon^c = \lfloor \epsilon_{xx}, \epsilon_{yy}, \epsilon_{xy}, \kappa_{xx}, \kappa_{xy}, \kappa_{yy}, \gamma_{zx}, \gamma_{zy} \rfloor \tag{48}$$

$$\mathbf{u} = \lfloor u, v, w, \hat{u}, \hat{v} \rfloor \tag{49}$$

that are expressed in an orthogonal shell coordinate system. Details for arriving at the above **B** matrix may be found in [1]. It is now possible to generate the stiffness matrix, the internal force vector and the strains and stresses of the present elements.

6. Discussions

Formulation of the strain-displacement relationships for four and nine-noded c^0 shell elements have been presented. The present family of elements pass the patch test, if the element sides are straight, for arbitrary quadrilateral element shapes. The tensorial transformation of the natural-coordinate strains to obtain the Cartesian strains enables one to bypass the standard isoparametric transformation of the natural-coordinate derivatives to Cartesian components and has greatly improved element reliability for distorted mehses. The elements thus possess correct rank without transverse shear locking, consistent membrane strain interpolation that admits inextensional bending without reduced integration, adequate representation of curvature effects to capture the important membrane-bending coupling.

The elements have been implemented into NICE-based shell element libraries[3] and have been extensively tested for their performance. In this paper, the performance of the present **ans-9** element will be demonstrated as applied to solve the classical postbuckling analysis of long cylinder as shown in Fig. 3. When the meshes are uniform, most elements perform acceptably well. However, if the meshes are distorted as shown in the figure, the **ans-9** element still maintains its robustness while other elements yield deteriorated results. Additional extensive numerical experiments so far indicate that the elements maintain robustness for a variety of applications and for severe mesh distortions, hence is recommended for production applications.

ACKNOWLEDGEMENTS

This work was partially sponsored under NASA/Langley *Computational Structural Mechanics Initiave* Research Contract, NAS1-17660, with Lockheed and under the Discrete Element Technology Task of Lockheed's Independent Research Project. We thank Dr. Jeff Stroud for his encouragement and keen interest for the present work reported herein.

REFERENCES

1. Park, K. C. and Stanley, G. M., "A Curved C^0 Shell Element Based on Assumed Natural-Coordinate Strains," Lockheed Palo Alto Research laboratory, Rep. No. LMSC-F035235, Nov. 1984; Submitted for publication.

2. Park, K. C., 1985, "An Improved Strain Interpolation for Curved C^0 Elements," to appear in the Bruce Irons Memorial Issue of *Int. j. num. meth. engr.*.

3. Stanley, G. M., 1985, *Continuum-Based Shell Analysis*, PhD Thesis, Stanford University, Stanford, Calif.

4. Hrennikoff, A., 1941, "Solution of Problems of Elasticity by the Framework Method," *Journal of Applied Mechanics*, **8**, A169-175.

5. Turner, M. J., Clough, R. W., Martin, H. C. and Topp, L. J., "Stiffness and Deflection Analysis of Complex Structures," *J. Aero. Sci.*, **23**, Sept. 1956, pp.805-823, 854.

6. Park, K. C. and Flaggs, D. L., 1984a, "An Operational Procedure for the Symbolic Analysis of the Finite Element Methods," *Computer Methods in Applied Mechanics and Engineering* Vol. 42, 37-46.

7. Park, K. C. and Flaggs, D. L., 1984b, "A Fourier Analysis of Spurious Mechanisms and Locking in the Finite Element Method," *Comp. Meth. Appl. Mech. & Engr.*, **46**, 65-81.

8. Park, K. C. and Flaggs, D. L., 1985, "A Symbolic Fourier Synthesis of a One-Point Integrated Quadrilateral Plate Element," *Comp. Meth. Appl. Mech. Engr.*, **48**, (1985) 203-236.

9. Park, K. C., 1984, "Symbolic Fourier Analysis Procedures for C^0 Finite Elements," in: *Innovative Methods for Nonlinear Analysis*, Pineridge Press, Swansea, 1984, 269-293.

10. Zienkiewicz, O. C., Taylor, R. C. and Too, J. M., 1971, "Reduced Integration Technique in General Analysis of Plates and Shells," *Int. J. Num. Meth. Engr.*, **3**, 275-290.

11. Pawsey, S. F. and Clough, R. W., 1971, "Improved Numerical Integration of Thick Shell Finite Elements," *Int. J. Num. Meth. Engr.*, **3**, 575-586.

12. Crisfield, M. A., 1984, "A Quadratic Mindlin Element Using Shear Constraints," *Comp. Meth. Appl. Mech. & Engr.*, **18**, 833-852.

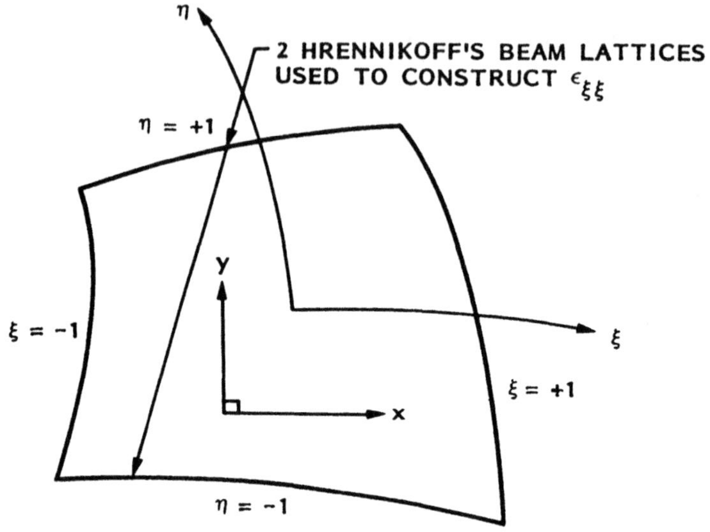

a. 4-NODE ELEMENT

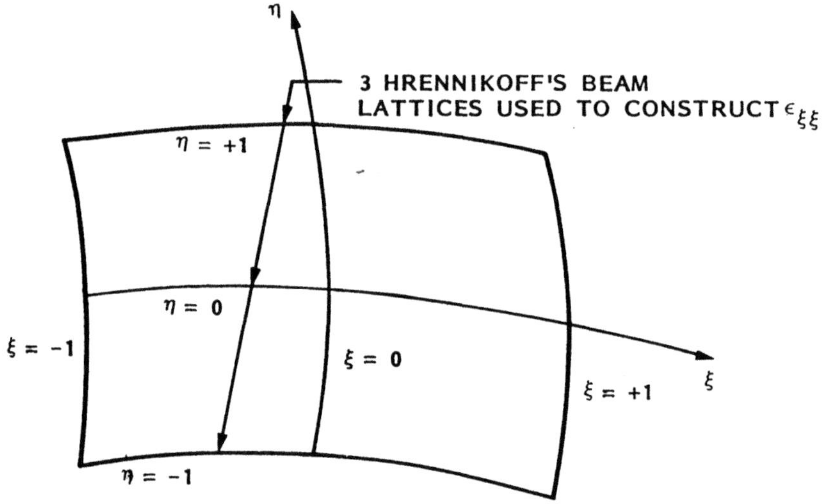

b. 9-NODE ELEMENT

Fig. 1 Element Definitions and Hrennikoff's Beam Lattices

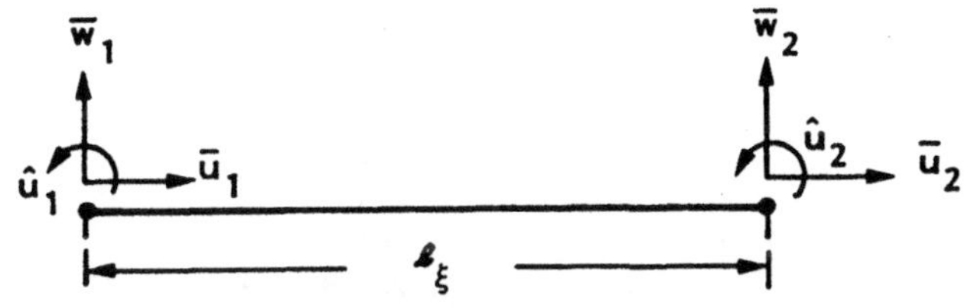

a. 2-NODE LATTICE BEAM

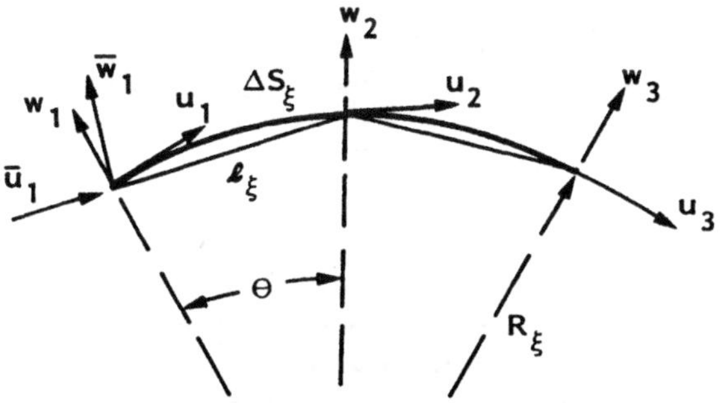

b. DISCRETIZATION OF ARCH BY STRAIGHT BEAM

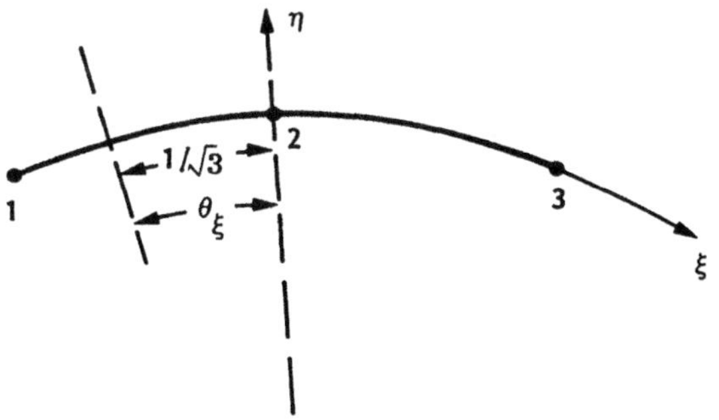

c. DISCRETIZATION OF ARCH BY QUADRATIC BEAM

Fig. 2 Straight and Curved Hrennikoff's Beam Lattices

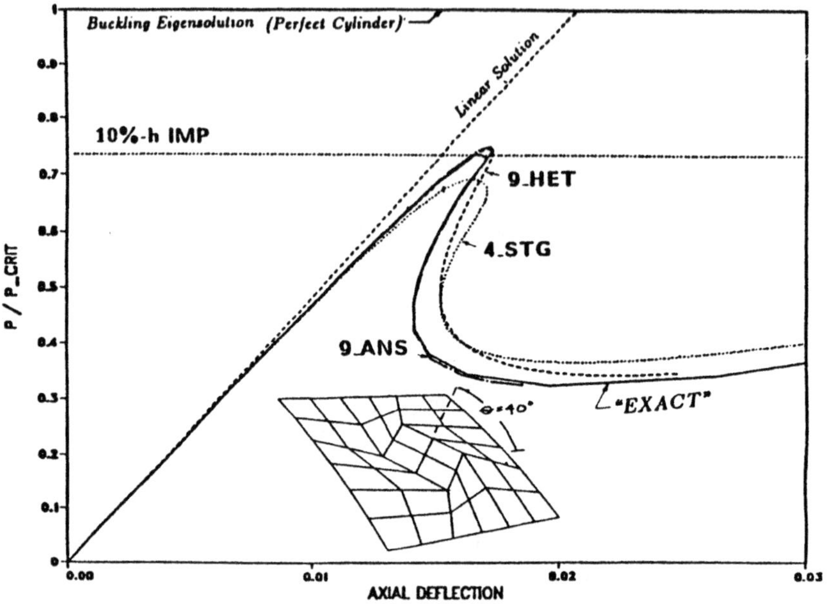

Figure 3 Post-Buckling of Axially Compressed Cylinder:
Mesh Sensitivity with Fixed Imperfection

Elastic-Plastic and Geometrically Nonlinear Analysis of Plates and Shells Using a New Nine Node Element

H. C. HUANG, E. HINTON*
Tianjin University, China
*University College of Swansea, U. K.

Summary

This paper is concerned with the elasto-plastic and geometrically nonlinear analysis of both thick and thin plates and shells using a new nine node element developed by Huang and Hinton. In the formulation of the new element an enhanced interpolation of the transverse shear strains in the natural coordinate system is used to overcome the shear locking problem and an enhanced interpolation of the membrane strains in the local coordinate system is applied to avoid the membrane locking behaviour. In the geometrically nonlinear analysis both Total and Updated coordinate systems are used. Some numerical examples are included to illustrate the good performance of the element for nonlinear analyses of plate and shell structures with both isotropic and anisotropic material behaviours.

1. Introduction

Of all the elements developed for plate and shell structures, the degenerated shell element family is widely considered to be the most efficient and economic [1]. However, for thin plate and shell problems reduced or selective integration schemes have to be used to eliminate the shear locking problem [2]. Unfortunately, this procedure induces unreliability of use due to the presence of extra zero energy modes. Furthermore, the occurrence of membrane locking for the deep thin shell structures gives rise to overstiff solutions.

In earlier studies of plate and shell elements [2],[3], the authors developed a new element with enhanced transverse shear and membrane strain interpolation, which has the requisite number of zero eigenvalues and associated rigid body modes. The element exhibits neither membrane nor shear locking for structures with large span/thickness ratios.

Finite Element Methods for Nonlinear Problems
Europe-US Symposium, Trondheim, Norway 1985
Editors: Bergan, Bathe, Wunderlich

In some previous linear static studies [2],[3], it was shown that the performance of the new element is almost identical to that of the 9-node Lagrange element with selective integration for plates or shallow shells. However, it should be recalled that the selectively integrated Lagrangian element suffers from the presence of mechanisms. The objective of this paper is to present the formulation, implementation and application of the new element for elasto-plastic and geometrically nonlinear analysis of plates and shells.

Both Updated and Total Lagrangian (U.L and T.L) geometrically nonlinear formulations are used in the present study. The U.L formulation is much more suitable for use with the present interpolation concept while the T.L formulation involves an enhanced interpolation for the nonlinear transverse shear strains.

2. New degenerated shell element formulation

A brief review of the new element is now given. The new element is based on the Ahmad degenerated shell element in which the displacements are expressed as

$$u_i = \sum_{k=1}^{n} N^k(\xi,\eta)\,u_i^k + \frac{\zeta}{2}\sum_{k=1}^{n} N^k(\xi,\eta)\,h^k(\underline{V}_{1i}^k\alpha_1^k - \underline{V}_{2i}^k\alpha_2^k) \qquad (1)$$

$$i=1,2,3$$

where $\underline{V}_{1i}^k$ and $\underline{V}_{2i}^k$ are mutually orthogonal unit vectors which are contained in the plane normal to the $\underline{V}_{3i}^k$ direction which is constructed from the normal coordinate of the top and bottom surfaces at node i as shown in Figure 1(a)

2.1 Elimination of shear locking

For thin shells, the transverse shear strains expressed in the natural coordinate system (ξ,η,ζ) $\gamma_{\xi\zeta} = \gamma_{\eta\zeta} \to 0$.

To avoid shear locking in the new 9-node degenerated shell element Huang and Hinton [2] replaced the shear strains by substitute shear strains of the form

$$\gamma_{\xi\zeta} = b_1+b_2\xi+b_3\eta+b_4\xi\eta+b_5\eta^2+b_6\xi\eta \qquad (2)$$

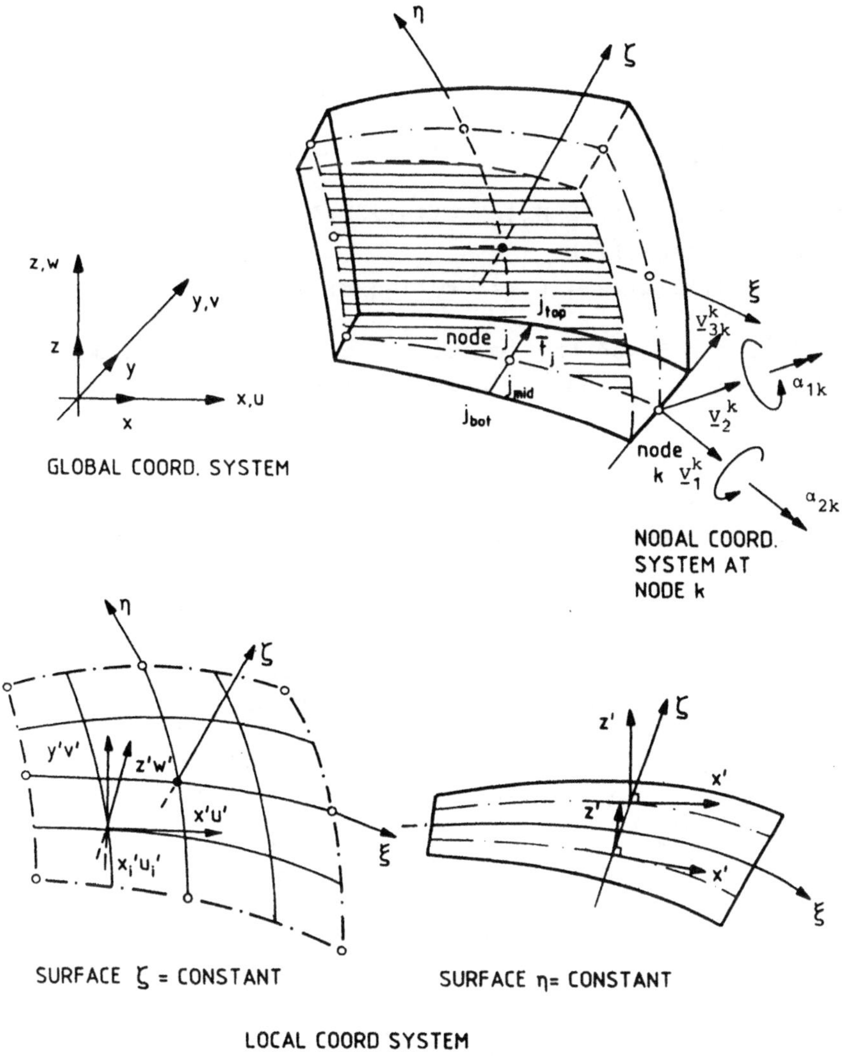

Fig.1(a). Coordinate system

$$\gamma_{\eta\zeta} = c_1+c_2\xi+c_3\eta+c_4\xi\eta+c_5\xi^2+c_6\xi^2\eta \tag{3}$$

The substitute strain fields are chosen as

$$\bar{\gamma}_{\xi\zeta} = \sum_{i=1}^{3} \sum_{j=1}^{2} L_j(\xi)\; H_i(\eta)\; \bar{\gamma}_{\xi\zeta}^{ij}$$

$$\bar{\gamma}_{\eta\zeta} = \sum_{i=1}^{3} \sum_{j=1}^{2} H_i(\xi)\; L_j(\eta)\; \bar{\gamma}_{\eta\zeta}^{ij} \tag{4}$$

where

$$H_1(z) = z(z/b+1)/2b \quad H_2(z) = 1-(z/b)^2$$
$$H_3(z) = z(z/b-1)/2b \tag{5}$$
$$L_1(z) = (1+z/a)/2 \quad L_2(z) = (1-z/a)/2$$

where $a = 3^{-\frac{1}{2}}$, $b = 1$ and the locations of i and j are shown in Figure 1(b). Thus $\gamma_{\xi\zeta}$ and $\gamma_{\eta\zeta}$ have the form suggested in References [2] and [3]. The six unknown parameters for $\bar{\gamma}_{\xi\zeta}$ are the values $\bar{\gamma}_{\xi\zeta}^{ij}$ (i=1,3, j=1,2) at the two Gauss point locations ($\xi = \pm a$) on the three lines $\eta=b$, $\eta=0$ and $\eta=-b$. Thus $\gamma_{\xi\zeta}$ is linear in ξ and quadratic in η. Similarly, the six un-

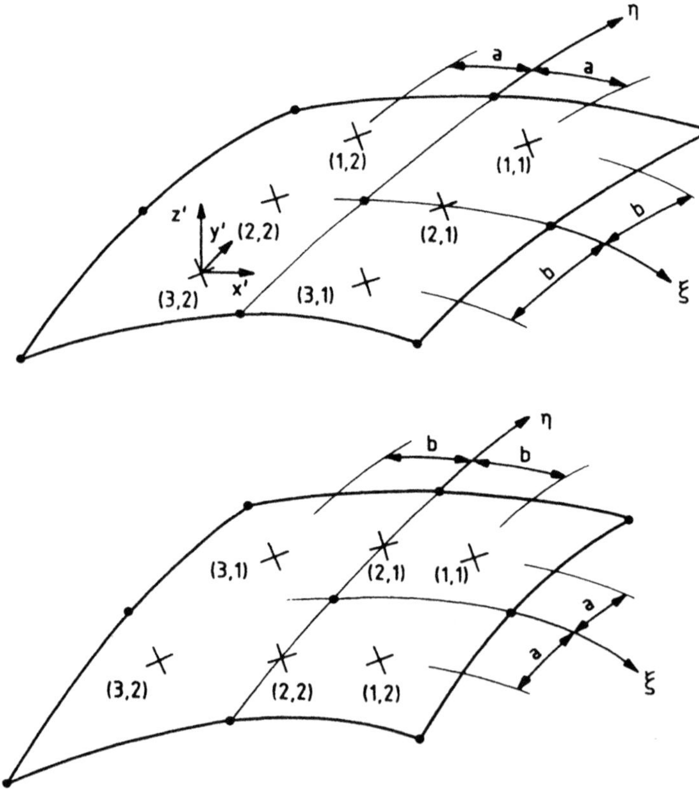

Fig.1(b). Interpolation points (i,j)

known parameters defining $\bar{\gamma}_{\eta\zeta}$ are the values $\bar{\gamma}_{\eta\zeta}^{ij}$ (i=1,3, j=1,2) on the two Gauss point locations ($\eta=\pm a$) on the three lines $\xi=b$, $\xi=0$ and $\xi=-b$. Thus $\bar{\gamma}_{\eta\zeta}$ is linear in η and quadratic in ξ. See Figure 1(c).

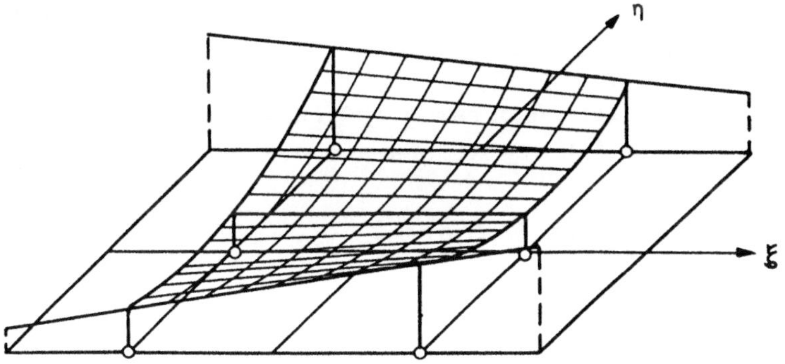

INTERPOLATION FOR $\gamma_{\xi z'}$, $\varepsilon_{mx'x'}$ AND $\frac{1}{2}\varepsilon_{mx'y'}$

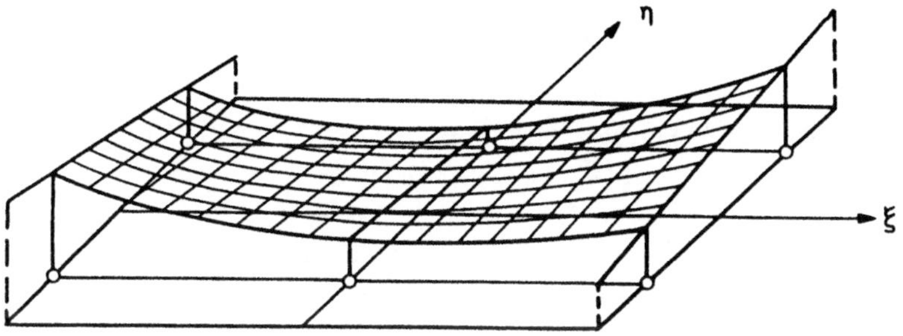

INTERPOLATION FOR $\gamma_{\eta z'}$, $\varepsilon_{my'y'}$ AND $\frac{1}{2}\varepsilon_{mx'y'}$

Fig.1(c). Interpolation functions ($a=3^{-\frac{1}{2}}$, $b=1$)

2.2 Enhanced interpolation of the membrane strains

Both the membrane stiffness and the shear stiffness are usually of the same order of magnitude. However, membrane locking differs somewhat from shear locking in which the transverse shear strains obtained using the finite element method cannot tend towards zero with decreasing thickness as they should according to thin shell theory. In shells the stiffness associated with the membrane strain energy, though important, especially when the shell thickness becomes very small, should not dominate the total stiffness. The membrane strains can only be separated from the bending strains in the local Cartesian coordinate system. Therefore it is expected that the use of an enhanced interpolation of the membrane strains in the local Cartesian coordinate system can help to eliminate membrane locking behaviour.

For the membrane strains expressed in the local coordinate system the following interpolations are used [3]

$$\bar{\varepsilon}_{mx'x'} = \sum_{i=1}^{3} \sum_{j=1}^{2} L_j(\xi)\, H_i(\eta)\, \bar{\varepsilon}^{ij}_{mx'x'} \tag{6}$$

$$\tfrac{1}{2}\bar{\varepsilon}_{mx'y'} = \sum_{i=1}^{3} \sum_{j=1}^{2} L_j(\xi)\, H_i(\eta)\, \bar{\varepsilon}^{ij}_{mx'y'}/2 \tag{7}$$

$$\bar{\varepsilon}_{my'y'} = \sum_{i=1}^{3} \sum_{j=1}^{2} H_i(\xi)\, L_j(\eta)\, \bar{\varepsilon}^{ij}_{my'y'} \tag{8}$$

$$\tfrac{1}{2}\bar{\bar{\varepsilon}}_{mx'y'} = \sum_{i=1}^{3} \sum_{j=1}^{2} H_i(\xi)\, L_j(\eta)\, \bar{\varepsilon}^{ij}_{mx'y'}/2 \tag{9}$$

where $H_i(z)$ and $L_j(z)$ have the same meaning as in (4) and $a = 3^{-\frac{1}{2}}$ and $b = 1$.

Thus $\bar{\varepsilon}_{mx'x'}$ and $\frac{1}{2}\bar{\varepsilon}_{mx'y'}$ are linear in x' and quadratic in y', $\bar{\varepsilon}_{my'y'}$ and $\frac{1}{2}\bar{\bar{\varepsilon}}_{mx'y'}$ are linear in y' and quadratic in x' (see Figure 1(b)). Finally, the in-plane shear strain is quadratic in both directions, that is

$$\begin{aligned}\bar{\varepsilon}_{mx'y'} = {} & \tfrac{1}{2} \sum_{i=1}^{3} \sum_{j=1}^{2} L_j(\xi)\, H_i(\eta)\, \bar{\varepsilon}^{ij}_{mx'y'} \\ & + \tfrac{1}{2} \sum_{\ell=1}^{3} \sum_{k=1}^{2} H_\ell(\xi)\, L_k(\eta)\, \bar{\bar{\varepsilon}}^{\ell k}_{mx'y'}\end{aligned} \tag{10}$$

Therefore in the formulation of the new 9-noded degenerated shell element the membrane and shear strains are interpolated from identical sampling points, but they are expressed in different coordinate systems - one is in the local Cartesian coordinate system, and the other is in the natural curvilinear coordinate system.

3. Geometrically nonlinear analysis

In the geometrically nonlinear analysis both Updated Lagrangian and Total Lagrangian formulations are considered. The local Cartesian coordinate system is used for both geometrically nonlinear formulations. The Green strain tensor can be expressed as follows

$$E_{ij} = 1/2(u_{i,j} + u_{j,i} + u_{k,i}u_{k,j}) \tag{11}$$

In the new element formulation the total shear strains have to be interpolated at every iteration in the Total Lagrangian formulation. It appears that the use of the Updated Lagrangian formulation is more appropriate when the interpolation concept is used for the new element. An incremental form of the principle of virtual work, based on the Updated Lagrangian formulation, is as follows

$$\int \Delta_t S_{ij}\delta\Delta_t E_{ij}dv_t + \int \tau_{ij}\delta\Delta\eta_{ij}dv_t = \delta W - \int \tau_{ij}\delta\Delta_t E_{ij}dv_t \tag{12}$$

where δW is the external incremental vitual work and the Green strains, $\Delta_t E_{ij}$ and 2nd Piola-Kirchhoff stresses, $\Delta_t S_{ij}$ are referred to the configuration at time t. The Cauchy stresses are represented as τ_{ij}. The incremental Green strains can be simply written as

$$\begin{aligned} \Delta_t E_{ij} &= \Delta_t \varepsilon_{ij} + \Delta_t \eta_{ij} \\ \Delta_t \varepsilon_{ij} &= 1/2(\Delta_t u_{i,j} + \Delta_t u_{j,i}) \\ \Delta_t \eta_{ij} &= 1/2(\Delta_t u_{k,i}\Delta_t u_{k,j}) \end{aligned} \tag{13}$$

where $\Delta_t \varepsilon_{ij}$ and $\Delta_t \eta_{ij}$ are the linear and quadratic parts of $\Delta_t E_{ij}$. As the quadratic strain expressions can be neglected in the first term of (12), the interpolations for shear strains and membrane strains are the same as in the linear case.

4. The constitutive relation

For specially anisotropic material, if the reference system of orthogonal axes (x', y', z') is parallel to the principal material axes (1, 2, 3), the incremental modulus matrix $\underline{D}$ and the yield criterion, etc. have been given in a transient study by the authors [4] in which Von-Mises yield function was used for both isotropic and anisotropic materials.

5. Numerical applications

5.1 Geometrical nonlinear analysis - long cantilever

Five QUAD9** elements are used to idealise a cantilever subjected to an end moment as shown in Figure 2. This problem is

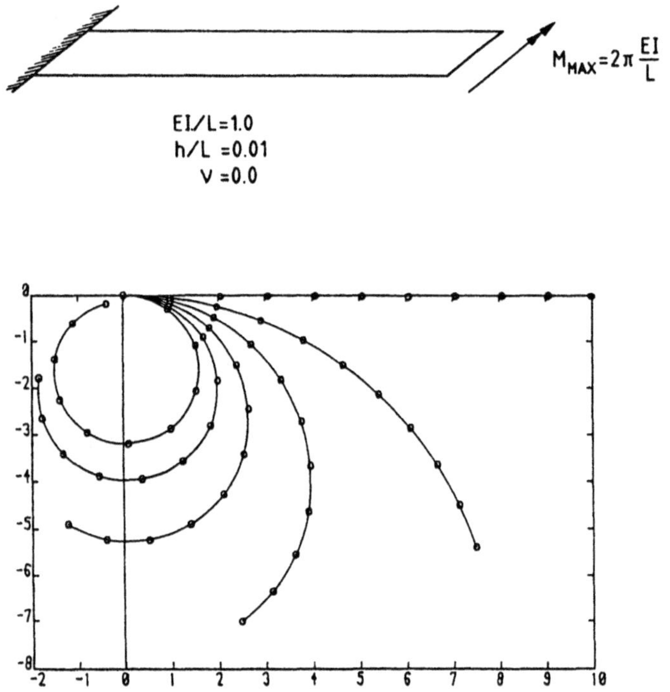

Fig.2. Elastic cantilever beam, deformed shape

chosen to examine the behaviour of the QUAD9** when used for a geometrically nonlinear problem with large displacements and rotations. Here, an Updated Lagrangian formulation is adopted. In the analytical solution to this problem the slope is prescribed at the free end by $\theta = ML/EI$ and the neutral axis is deformed into a circle with radius $R = EI/M$. L is the beam length, M is the end moment and EI is the flexural rigidity. For a cantilever of length $L = 10$, the analytical solution indicates that the radius of the final circle is 3.183, whereas the radius obtained using the new element, QUAD9**, is 3.185. In Figure 2 it is shown that the error is reasonably

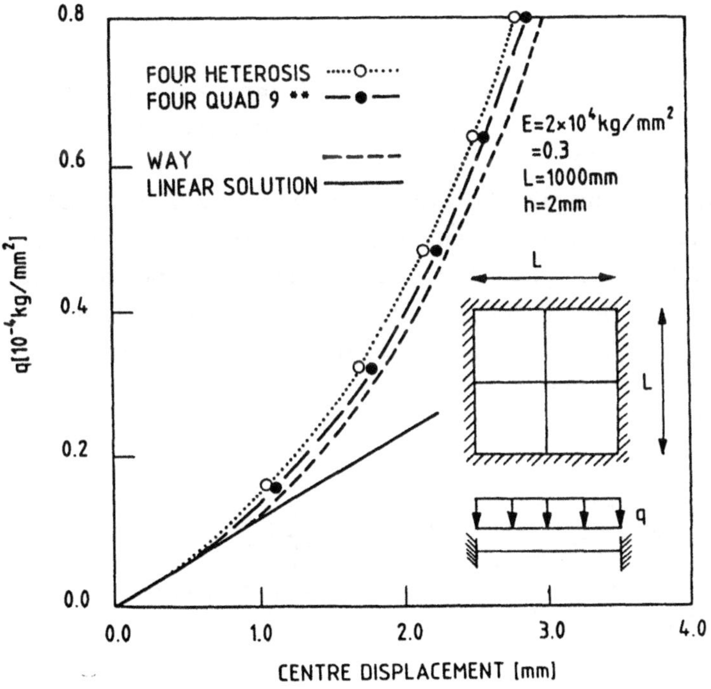

Fig.3. Clamped square plate under uniform load

small when the load approaches the value of 2π EI/L where only 5 new elements are used.

5.2 Geometrically nonlinear analysis - clamped square plate

Four QUAD9** are used to idealise a symmetric quadrant of a clamped square plate subjected to a uniformly distributed load as shown in Figure 3. Lateral displacements at the centre obtained using the new element are compared with the results obtained by the classical Rayleigh-Ritz solution presented by Way [5]. Figure 3 shows that the central displacements obtained in the present study lie between those obtained in the classical solution and those obtained using heterosis elements [6]. It is worth noting that the results obtained using the Largrangian nine node element with selective integration, QUAD9S, are almost the same as those obtained using QUAD9**.

5.3 Geometrically nonlinear analysis - hinged cylindrical shell

The hinged cylindrical shell, shown in Figure 4, has been studied by Horrigmoe [7]. The straight edges are hinged and the curved edges are free. The shell is subjected to a concentrated central load on the convex side. A 2x2 mesh of QUAD9** elements is used to represent a symmetric quadrant of the shell. In Figure 4 the load-displacement response obtained using QUAD9** elements is compared with that given in Horrigmoe's study in which a maximum load of 2.22 kN was obtained. In the present study the maximum load is 2.21 kN compared with a value of 2.10 kN obtained by Hughes [6] using 4 heterosis elements.

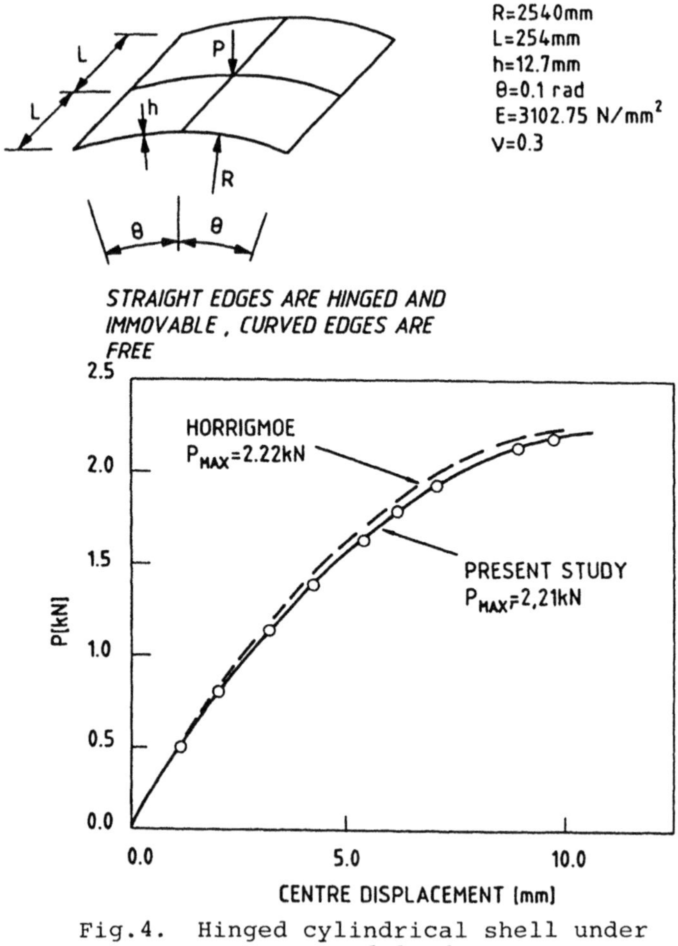

Fig.4. Hinged cylindrical shell under concentrated load

5.4 Elasto-plastic analysis - clamped square plate

A 3x3 mesh of QUAD9** elements is used to idealise a symmetric quadrant of a clamped uniformly loaded square plate. The side length of plate is taken as L = 6 and the thickness as h = 0.2. Elasto-plastic analyses are undertaken using both isotropic and anisotropic constitutive models. The following material properties are considered.

Isotropic material properties:

$E_x = E_y = 30000.0$ psi, $\nu = 0.3$,

$G_{xy} = G_{xz} = G_{yz} = 11540.0$ psi,

$\bar{\sigma}_o = \sigma_{x_o} = \sigma_{y_o} = \sigma_{45_o}$, $\tau_{12_o} = \tau_{13_o} = \tau_{23_o} = 17.32$ psi,

$H' = 300.0$ psi

Anisotropic material properties:

$\sigma_{y_o} = 40.0$ psi, $\sigma_{45_o} = 35.0$ psi,

$\tau_{12_o} = 20.2$ psi. The remaining values are the same as for the isotropic case.

Using 6 layers through the plate thickness the load-central deflection curve shown in Figure 5 is obtained. The spread of

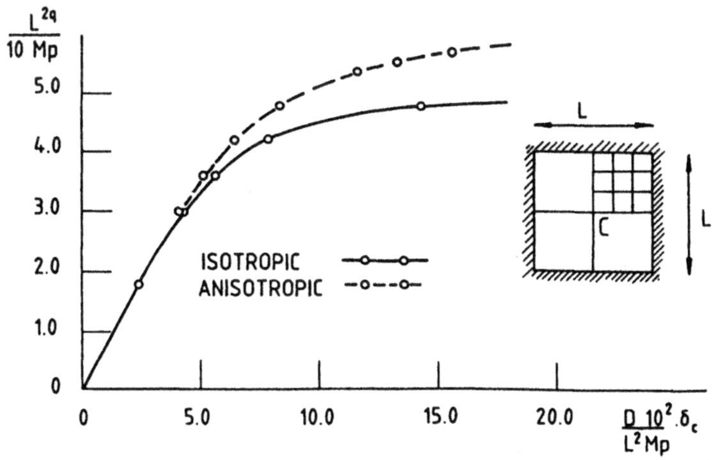

Fig.5. Load-deflections of uniformly loaded clamped square plate (plastic behaviour with 6 layers)

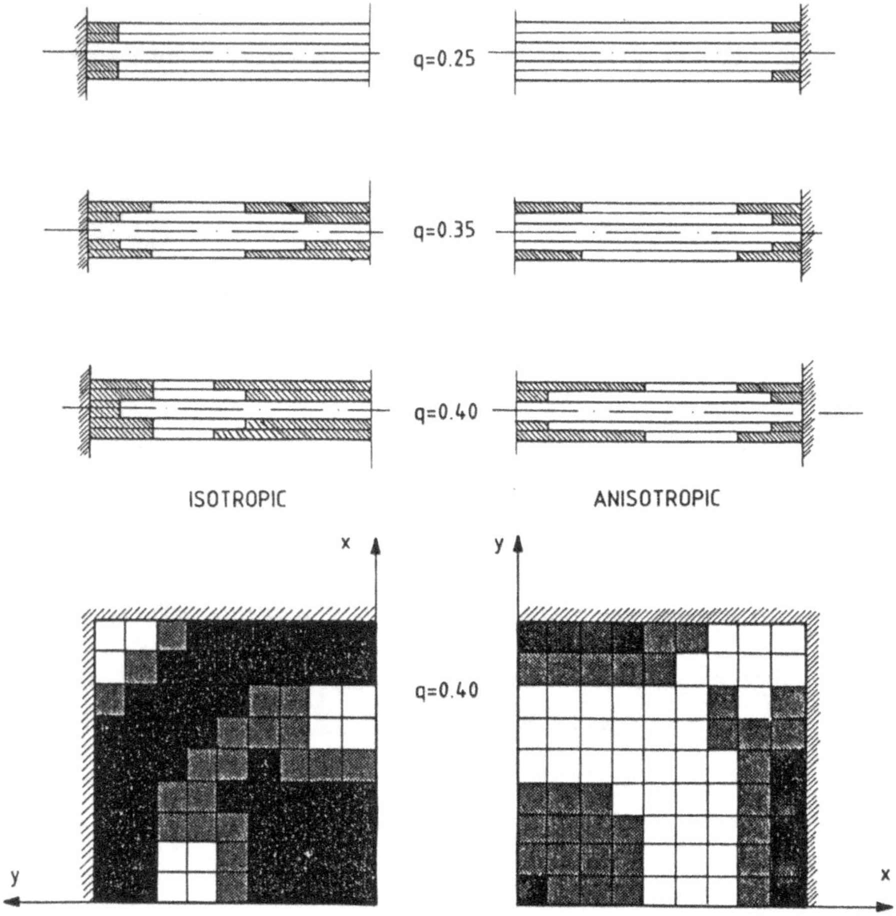

Fig.6. Uniformly loaded clamped square plate, spread of plastic zones

plastic zones is shown in Figure 6. Differences in the spread of plasticity for the isotropic and anisotropic materials are highlighted. The results obtained are almost identical to those obtained using the selectively integrated 9-node element QUAD9S. Furthermore, there is good agreement with results obtained using the selectively integrated 8-node element and the semiloof shell element [8].

5.5 Elasto-plastic analysis - clamped quadratic shell

Figure 7 shows the load deflection curves for the elasto-plastic analysis of a clamped quadratic shell with a centrally placed lateral point load. The material properties are the same as those used in the clamped plate example. The shell geometry is defined in Figure 7, where the thickness is taken h = 0.2 and the span as L = 6.0. Figure 7 shows the load-central deflection relationship. There is a very slight difference between the solutions obtained using QUAD9** elements and QUAD9S elements (which are prone to membrane locking behaviour).

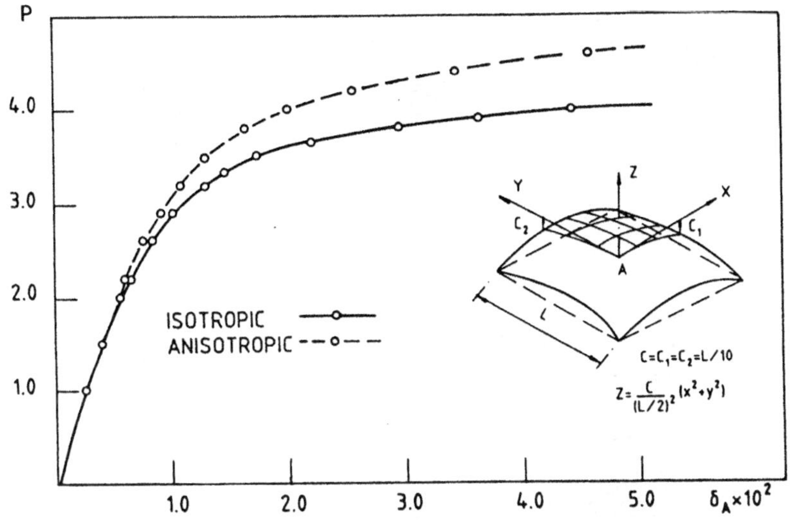

Fig.7. Load-central deflections of clamped quadratic shell (plastic behaviour with 6 layers)

In Figure 8 the plastic zone distributions are illustrated. Differences between the isotropic and anisotropic material behaviour are highlighted.

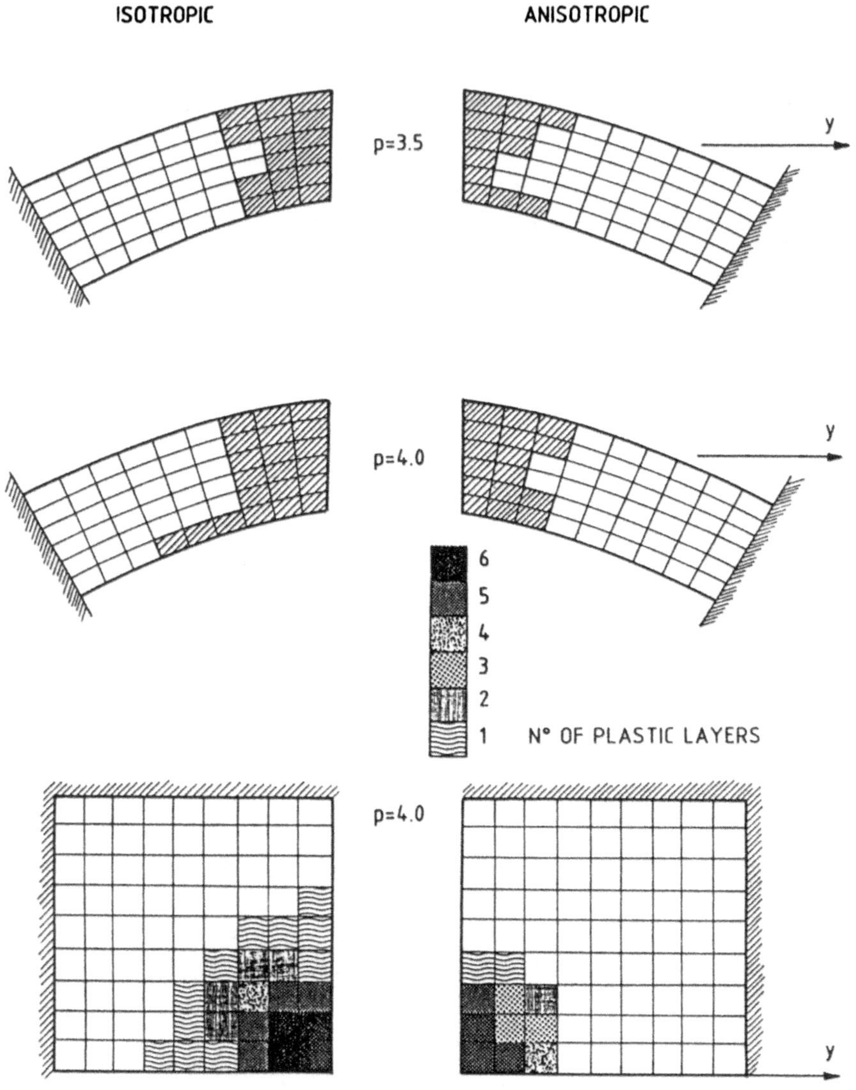

Fig.8. Centrally point load clamped quadratic shell, spread by plastic zones

6. Conclusions

This paper demonstrates the use of the new nine node Lagrangian element developed by Huang and Hinton in geometrically nonlinear and elasto-plastic analysis. For the geometrically nonlinear analysis an Updated Lagrangian formulation is adopted as it is more suitable for use with the new element. For elasto-plastic nonlinear analyis, the new element performs well and is convenient to use for both isotropic and anisotropic materials.

References

1. Ahmad, S.: Curved finite elements in the analysis of solid, shell and plate. Ph.D. thesis, University College of Swansea, C/Ph/7/69.

2. Huang, H.C.; Hinton, E.: A nine node Lagrangian Mindlin plate element with enhanced shear interpolation. Eng. Comput., 1, 369-379, 1984.

3. Huang, H.C.; Hinton, E.: A new nine node degenerated shell element with enhanced membrane and shear interpolation. Research Report University College of Swansea, C/R/500/84.

4. Huang, H.C.; Hinton, E.: Elasto-plastic dynamic analysis of plate and shell structures using a new nine node element. ASME Symposium in Material Nonlinearity in Vibration Problems, to be held in Florida, U.S.A.

5. Way, S.: Uniformly loaded, clamped, rectangular plates with large deformation. Proc. 5th International Congress of Applied Mechanics, Cambridge, MA, 1938.

6. Hughes, T.J.R.; Liu, W.K.: Nonlinear finite element analysis of shells: part I. three dimensional shells. Comp. Meth. Appl. Mech. Engng., 26, 331-362, 1981.

7. Horrigmoe, G.: Finite element instability analysis of free-form shells. Report 77-2 Division of Structural Mechanics, Norwegian Institute of Technology, University of Trondheim, Norway (May, 1977).

8. Figueiras, J.A.: Ultimate load analysis of anisotropic and reinforced concrete plates and shells. Ph.D. thesis University College of Swansea, C/Ph/72/83.

Improved Spurious Mode Control Through Mixed Variational Principles

T. BELYTSCHKO, W. K. LIU
Northwestern University, Evanston, USA

Abstract

Some features of consistent procedures for control of spurious singular modes in the 4 node and 9 node isoparametric elements are described. Properties which enable stabilization matrices to be developed in a variationally consistent fashion are given and their application to these elements are described. Finally, some examples of the performance of these stabilized elements are given.

1. Introduction

In order to accelerate the slow rates of convergence which are associated with fully integrated C^0 elements due to their susceptibility to shear and membrane locking, it is common practice to use reduced or selective reduced integration [1,2]. Unfortunately, these methodologies are not completely trouble free: for certain boundary conditions, reduced selective integration is associated with the presence of singular or near singular modes which cause large oscilations in the solution or even singularity of the assembled total stiffness matrix. In recent years, we have developed procedures for eliminating this deficiency by introducing additional generalized strains [3-10] which are constructed so that the consistency conditions and hence the patch test are not violated [11,12]. These methods have collectively been called γ-methods because the form of the projection operator which provides these generalized strains is identical in both two and three-dimensional elements and 4-node and 9-node elements.

One aspect of the γ-methods which was unappealing to some users is that it required non-physical constants in its application. The reason for this was that the γ-method was initially developed completely in terms of algebraic arguments without recourse to any variational principles. Although the importance of variational consistency is often unduly exaggerated, (see Ref. 13, 14 where it is shown that any B-bar method is

Finite Element Methods for Nonlinear Problems
Europe-US Symposium, Trondheim, Norway 1985
Editors: Bergan, Bathe, Wunderlich

variationally consistent and elements which will not converge and those with more subtle flaws are variationally consistent), nevertheless, variational consistency is desirable from two viewpoints: 1) It provides a framework for extending an element to material and geometric nonlinearities - without variational consistency it is difficult to define appropriate stress fields for the construction of the internal nodal force matrix or the geometric stiffness matrix. 2) It provides a partial basis for convergence and stability arguments.

An interesting aspect of the γ-methods is that in fact they can be developed from mixed variational principles of the Hellinger-Reissner or Hu-Washizu type (see Ref. 4) In Ref. 4, it was shown for example that the γ-operator provides the coefficient of the bilinear term in a quadrilateral in the dependent variable and that the mixed variational principle provides a means of interpreting the parameters of the γ-method entirely in terms of physical constants and geometrical constants of the element. It has subsequently been found that the method also provides an extremely efficient implementation for the flexure-superconvergent elements of Wilson and Taylor [15-16] which employ bubble modes [17].

However, the development of a variationally consistent framework for the 9-node element has proven to be far more elusive. In the 9-node element, the structure of the interpolants associated with 2 and 3 point quadrature are distorted by the reduced quadrature, and the orthogonality properties, which are so useful in the 4-node element, are minimal. Furthermore, whereas in the 4-node element the primitive annihilator operator which is the basis of the γ-projection is identical to the singular mode, in the 9-node element there are many candidates for this annihilator and some alternative annihilators seem to be numerically more accurate. These findings have recently been presented in Ref. [18].

In this paper, we will summarize basic relationship between γ-methods and mixed variational principles for the stabilization of elements which are rank defficient due to underintegration. The concepts will be presented in the framework of the 2-dimensional elastic continuum element for purposes of simplicity. In Section 2, the outline of these concepts for the 4-node element is given, whereas Section 3 presents the same framework for the 9-node element. In Section 4, some representative results, including both linear and nonlinear problems, are summarized.

2. Four Node Quadrilateral

The element stiffness matrix for a plane, two-dimensional element is given by

$$\underset{\sim}{K}_e = \underset{\sim}{K}_1 + \underset{\sim}{K}_{stab} \tag{2.1}$$

where $\underset{\sim}{K}_1$ is the stiffness obtained by one-point quadrature and

$$\underset{\sim}{K}_{stab} = \begin{bmatrix} c_1 \underset{\sim}{\gamma}\underset{\sim}{\gamma}^T & c_3 \underset{\sim}{\gamma}\underset{\sim}{\gamma}^T \\ c_3 \underset{\sim}{\gamma}\underset{\sim}{\gamma}^T & c_2 \underset{\sim}{\gamma}\underset{\sim}{\gamma}^T \end{bmatrix} \tag{2.2}$$

$$\qquad \underset{\sim}{u}_x \qquad\qquad \underset{\sim}{u}_y$$

where $\underset{\sim}{x}_i$ are the nodal coordinates and

$$\underset{\sim}{\gamma} = \frac{1}{4}\,[\underset{\sim}{h} - (\underset{\sim}{h}^T\,\underset{\sim}{x}_i)\underset{\sim}{b}_i] \tag{2.3}$$

$$\underset{\sim}{b}_i = \frac{\partial \underset{\sim}{N}^T(0)}{\partial x_i} \tag{2.4}$$

and $\underset{\sim}{h}$ is given in Table 1; the constants c_i can be determined from mixed variational principles, or can be set arbitrarily; for any values of c_i these elements will converge for they automatically satisfy the patch test [17].

Remark 1. The separability of the element stiffness into a 1-point quadrature stiffness and stabilization stiffness results from the orthogonality of the derivatives of the basic shape functions and the associated biorthogonal vector basis $\underset{\sim}{\psi}_i$, $\underset{\sim}{\phi}_i$ which has the property (see Table 1)

$$\underset{\sim}{\psi}_i^T\,\underset{\sim}{\phi}_j = \delta_{ij} \tag{2.5}$$

Remark 2. The element stiffness given by Eqs. (1) and (2) can be considered a three-parameter canonical form of the element stiffness. Any stiffness matrix which satisfies the patch test must be of the form given by Eqs. (1-5) with the free parameters being c_1, c_2 and c_3. For variationally consistent elements, the number of parameters is limited to 2 [17]. For a rectangular element, only a single free parameter occurs in a variationally consistent stiffness.

Remark 3. The vector $\underset{\sim}{h}$ can be replaced by any other primitive annihilator $\underset{\sim}{a}$ as long as $\underset{\sim}{a}$ satisfies the following condition

$$\underset{\sim}{a}^T \underset{\sim}{h} \neq 0 \tag{2.6}$$

Remark 4. The $\underset{\sim}{\gamma}$ matrix can be constructed numerically by a Gram-Schmidt orthogonalization by starting with any primitive annihilator $\underset{\sim}{a}$ and orthogonalizing this vector to $\underset{\sim}{x}$, $\underset{\sim}{y}$ and $\underset{\sim}{s}$. However, the explicit form of $\underset{\sim}{\gamma}$ as given in (2.3) is computationally more efficient and gives better insight into the element.

The constants c_1, c_2 and c_3 are identified by using the Hu-Washizu variational princple in conjunction with stress and strain fields which suppress undesirable (parasitic) shear and volumetric strains. Examples of such strain fields are given in Table 2 (from Ref. [17]). The arguments underlying these choices of strain fields are as follows:

1. the optimal bending (OB) field is constructed by suppressing the quasilinear shear strain terms;
2. the optimal incompressible (OI) field is constructed by suppressing any quasilinear terms in the dilatational (volumetric) strain.

3. Nine Node Element

The element stiffness matrix of a 9-node element when numerically integrated by 2x2 quadrature is given by

$$\underset{\sim}{K}_{(2x2)} = \sum_{\alpha=1}^{4} \rho_\alpha \underset{\sim}{B}^T (\underset{\sim}{x}_\alpha) \underset{\sim}{D} (\underset{\sim}{x}_\alpha) \underset{\sim}{B} (\underset{\sim}{x}_\alpha) \tag{3.1}$$

where $\underset{\sim}{x}_\alpha$ are the quadrature points and ρ_α are products of the quadrature weights and the Jacobian. The matrix $\underset{\sim}{B}$ is given by

$$\underset{\sim}{B}^T = \begin{bmatrix} \underset{\sim}{b}_x^\alpha & \underset{\sim}{0} & \underset{\sim}{b}_y^\alpha \\ \underset{\sim}{0} & \underset{\sim}{b}_y^\alpha & \underset{\sim}{b}_x^\alpha \end{bmatrix} \tag{3.2a}$$

$$\underset{\sim}{b}_x^\alpha = \frac{\partial \underset{\sim}{N}^T(\underset{\sim}{x}_\alpha)}{\partial x} \tag{3.2b}$$

$$\underset{\sim}{b}_y^\alpha = \frac{\partial \underset{\sim}{N}^T(\underset{\sim}{x}_\alpha)}{\partial y} \tag{3.2c}$$

$$\underset{\sim}{x}_\alpha = \underset{\sim}{x}(\xi = \pm \frac{1}{3}, \eta = \pm \frac{1}{3}) \tag{3.2d}$$

An alternative expression for (3.1) is

$$\underset{\sim}{K}^e{}_{(2x2)} = \overset{0}{\underset{\sim}{B}}{}^T \overset{0}{\underset{\sim}{D}} \overset{0}{\underset{\sim}{B}} \tag{3.3}$$

$$\overset{0}{\underset{\sim}{B}}{}^T = [\underset{\sim}{B}_o^T(\underset{\sim}{x}_1), \underset{\sim}{B}^T(\underset{\sim}{x}_2), \underset{\sim}{B}^T(\underset{\sim}{x}_3), B^T(\underset{\sim}{x}_4)] \tag{3.4}$$

$$\overset{0}{\underset{\sim}{D}} = \begin{bmatrix} \rho_1 \underset{\sim}{D}(\underset{\sim}{x}_1) & 0 & \vdots & 0 \\ & & \vdots & \\ \underset{\sim}{0} & & & \rho_4 \underset{\sim}{D}(\underset{\sim}{x}_4) \end{bmatrix} \tag{3.5}$$

Because $\overset{0}{\underset{\sim}{B}}$ is a 12x18 matrix and $\overset{0}{\underset{\sim}{D}}$ is a 12x12 matrix, $\underset{\sim}{K}_{(2x2)}$ is at most of rank 12. Since $\underset{\sim}{K}^e{}_{(2x2)}$ is 18x18 and there are 3 rigid body modes, its rank deficiency is at least 3. Two of the spurious singular modes can be shown to be

$$\underset{\sim}{u} = \begin{Bmatrix} \underset{\sim}{u}_x \\ \\ \underset{\sim}{u}_y \end{Bmatrix} = \begin{bmatrix} \underset{\sim}{h} & \underset{\sim}{0} \\ \\ \underset{\sim}{0} & \underset{\sim}{h} \end{bmatrix} \tag{3.6}$$

$$\underset{\sim}{h}^T = [+1,\ -1,\ +1,\quad -1,\ 0,\ -1,\quad +1,\ -1,\ +1]$$

$$\tag{3.7}$$

The projection χ is then given by

$$\chi = \underset{\sim}{a} - \underset{\sim}{a}^T \underset{\sim}{x}_i \sum_{\alpha=1}^{4} c_\alpha \underset{\sim}{b}_{i\alpha} \tag{3.8a}$$

where

$$\sum_{\alpha=1}^{4} c_\alpha = 1 \tag{3.8b}$$

Here $\underset{\sim}{a}$ is what we call the primitive annihilator: any matrix which is not orthogonal to $\underset{\sim}{h}$. Whereas nothing can be gained in the 4 node element by choosing $\underset{\sim}{a}$ to be any other vector than $\underset{\sim}{h}$, in the 9-node element it is advantageous to use other vectors. Equation (3.8b) is identical to that given in Ref. [7] except that it has been generalized through the addition of c_α; in Ref. [7] $c_\alpha = \frac{1}{4}$. Furthermore, in Ref. [7], it was assumed that $\underset{\sim}{a} = \underset{\sim}{h}$.

A set of interpolation functions for a rectangular element with orthogonal derivatives is given in Table 3. Note that the derivatives of ϕ_9 vanish at the quadrature points and are orthogonal to all of the other interpolants, so a separable stabilization matrix as in the 4-node element would appear to be obtainable.

However, when the stabilization matrix is developed through Eqs. (2.1-2) (with $\underset{\sim}{K}_1$ replaced by $K_{(2x2)}$ and $\underset{\sim}{a} = \underset{\sim}{h}$), the resulting total stiffness becomes much too stiff. The difficulty arises from the fact that whereas the derivatives of the functions are orthogonal to $\phi_{9,x}$ and $\phi_{9,y}$, the vectors associated with these functions are not all orthogonal to the dual to χ when $\underset{\sim}{a} = \underset{\sim}{h}$. Thus certain functions which comprise the stiffness that is obtained by the 2x2 stiffness matrix are repeated in the stabilization matrix.

To avoid this difficulty, it is necessary to choose another primitive annihilator. A procedure which we have found successful is to start with $\underset{\sim}{h}$ and then orthogonalize this vector sequentially with $\underset{\sim}{x}^2$, $\underset{\sim}{xy}$, $\underset{\sim}{y}^2$ by a Gram-Schmidt procedure ($\underset{\sim}{x}^2 = (x_1^2, x_2^2, x_3^2 \dots x_9^2)$ where x_I are the nodal x-coordinates of the element). This primitive annihilator yields better results for most problems. We will call the resulting vector $\underset{\sim}{a}$. It can be used in (3.8a) directly.

Remark 5. This procedure can easily be extended to obtain an annihilator for the noncommunicable 8-node spurious mode which apears in the 9-node element; Oden and Jacquotte [19] discuss the difficulty in suppressing this node. It is only necessary to start with a primitive annihilator orthogonal to the two $\underset{\sim}{h}$ modes which contains some of the 8-node mode. A suitable candidate is

$$\underset{\sim}{a}^T = [\underbrace{+1,\ 0,\ -1,\ -2,\ 0,\ +2,\ +1,\ 0,\ -1,}_{\underset{\sim}{u}_x}\quad \underbrace{-1,\ +2,\ -1,\ 0,\ 0,\ 0,\ +1,\ -2,\ +1}_{\underset{\sim}{u}_y}]$$

Remark 6. This implementation of spurious mode control requires no numerically adjusted factors; the exact same functions as for the 4-node element as given in Table 2 are used with $\underset{\sim}{h}$ simply replaced by ϕ_9.

Remark 7. Although the function given in Table 2 are orthogonal only for rectangular elements, their loss of orthogonality in a skewed element can be neglected without violating the patch test.

Remark 8. In Ref. [7] it was noted that efforts to impose consistency for the quadratic fields were unsuccessful. The procedure described here of orthogonalizing $\underset{\sim}{a}$ to the 3 quadratic fields does not provide strict quadratic consistency but it does improve performance since is increases the relationship between χ and $\underset{\sim}{\phi}_9$.

Results

Example 1. The first example is taken from [17] and concerns a cantilever beam which is solved by a 1x4 mesh shown in Fig. 1. The results are given for this problem in Table 4 for the 3 different strain fields which have been developed for the γ-method elements. The following features of the results can be noted:

1. The QBI provides exceptional accuracy with only one element through the thickness; compared with selective reduced integration (volumetric underintegration), it is substantially more accurate.
2. The OB element works well as long as the material behavior is compressible; as the material becomes incompressible, the element locks.

Example 2. The second example concerns a pinched cylinder supported by diaphragms [9]. It was shown in Ref. [9] that this is one of the most difficult and discriminating shell test problems. Figure 2 gives the displacement under the load for 4 different elements:

1. 9-node SRI (selective reduced integration with the transverse shear underintegrated).
2. 9-node 3x3 (full integration of all stresses).
3. 9-node γ-method (2x2 integration on all terms with γ-stabilization).
4. 4-node SRI.

Note that for a 17x17 node mesh for a 1/8 model (with all symmetries exploited), the 9-node SRI element is still more than 10% in error. The 4-node SRI element fares only slightly better; the 4-node γ-element incidentally, for which results are not presented, performs no better than the SRI element in this case. Figure 2 also brings out the comparative effects of shear and membrane locking in this element. The difference between the 3x3 and SRI elemnts indicates the effect of shear locking, since the improvement is achieved by shear underintegration. The difference between the SRI and γ-element indicates the effects of membrane locking. The 4-node element does not improve in going from SRI to γ because it is not subject to membrane locking

Example 3. Figure 3 gives the problem definition and gives the center displacement for a geometrically nonlinear analysis of a barrel roof subjected to a point load. In this case the results of the 3 different elements are almost indistinguishable.

Example 4. Figure 4 shows a buckling analysis of a cylindrical shell performed with the 4-node element. Five different bifurcation configurations are shown. Two of these exhibit diamond shaped buckles.

Acknolwedgement

The support of the Air Force under Grant F-49620-82-K-0013 is gratefully acknowledged.

References

1. Zienkiewicz, O.C.; Taylor, R.L.; Too, J.M.: Reduced integration technique in general analysis of plates and shells. Int. J. Numer. Meths. Engrg. 3 (1971) 275-290.

2. Hughes T.J.R.; Liu, W.K.: Nonlinear finite element analysis of shells: Part 1. Three-dimensional shells. Comp. Meths. Appl. Mech. and Engrg. 26 (1981) 331-362.

3. Flanagan D.P.; Belytschko, T.: A uniform strain hexahedron and quadrilateral with orthogonal hourglass control. Int. J. Numer. Meths. Engrg. 17 (1981) 679-706.

4. Belytschko, T.; Ong, J.S-J.; Liu, W.K.; Kennedy, J.M.: Hourglass control in linear and nonlinear problems. Comp. Meths. Appl. Mechs. and Engrg. 43 (1984) 251-276.

5. Belytschko, T.; Tsay, C.S.: A stabilization procedure for the quadrilateral plate element with one-point quadrature. Int. J. Numer. Meths. Engrg. 19 (1983) 405-420.

6. Liu, W.K.; Belytschko, T.: Efficient linear and nonlinear heat conduction with a quadrilateral element. Int. J. Numer. Meths. Engrg. 20(5) (1984) 931-948.

7. Belytschko, T.; Ong, J.S-J.; Liu, W.K.: A consistent control of spurious singular modes in the 9-node Lagrange element for the Laplace and Mindlin plate equations. Comp. Meths. Appl. Mechs. and Engrg. 44 (1984) 269-295.

8. Belytschko, T.; Liu, W.K.; Ong, J.S-J.; Lam, D.: Implementation and application of a 9-node Lagrange shell element with spurious mode control. Computers and Structures. 20(1-3) (1985) 121-128.

9. Belytschko, T.; Stolarski, H.; Liu, W.K.; Carpenter N.; Ong, J.S-J.: Stress projection for membrane and shear locking in shell finite elements. Presented at the FENOMECH conference, Stuttgart, Germany, Sept. 1984; to appear in Comp. Meths. Appl. Mech. and Engrg.

10. Liu, W.K.; Belytschko, T.; Ong, J.S-J.; Law, S.E.: The use of stabilization matrices in nonlinear analysis. in Innovative Methods for Nonlinear Problems, Liu, W.K. et al. (eds.), Pineridge, Swansea, U.K. 233-258 1984.

11. Irons B.M.; Razzaque, A.: Experience with the patch test for convergence of finite elements. The Mathematical Foundations of the Finite Element Method with Applications to Partial Differential Equations. Aziz, A.K. (ed.) Academic Press, New York: 557-588 1972.

12. Strang, G.: Variational crimes in the finite element method. The mathematical foundations of the finite element method with application to partial differential equations. Aziz, A.K. (ed.) Academic Press, New York: 689-710 1972.

13. Simo J.C.; Hughes, T.J.R.: On the variational foundations of assumed strain methods," to appear in Comp. Meths. Appl. Mech. and Engrg.

14. Hughes, T.J.R.; Simo, J.C.; Belytschko, T.; Stolarski, H.: Foundations of assumed strain methods in finite element analysis. Presented at the Joint ASCE/ASME Mechanics Conference, June, 1985; to appear in Comp. Meths. Appl. Mech. Engrg.

15. Doherty, W.P.; Wilson, E.L.; Taylor, R.L.: Stress analysis of axisymmetric solids utilizing higher order quadrilateral finite elements. SESM Report No. 69-3, Dept. of Civil Engineering, Univ. California, Berkeley, 1969.

16. Taylor, R.L.; Beresford, P.J.; Wilson, E.L.: A non-conforming element for stress analysis. Int. J. Numer. Meths. Engrg. 10 (1976) 1211-1219.

17. Belytschko, T.; Bachrach, W.: Simple quadrilaterals with high coarse-mesh accuracy. To appear in Comp. Meth. Appl. Mech. Eng.

18. Belytschko, T.; Liu W.K.; Ong, J.S-J.: On mixed methods and spurious mode control. Presented at the Joint ASCE/ASME Mechanics Conference, Albuquerque, June, 1985; to appear in Comp. Meths. Appl. Mechs. and Engrg.

19. Oden, J.T.; Jacquotte, B.P.: Convergence and stability of underintegrated finite element methods. Ibid.

Table 1

Biorthogonal Basis and Associated Shape Functions for 4-Node Quadrilateral

α	ϕ_α	$\phi_{\alpha,x}$	$\phi_{\alpha,y}$	$\underset{\sim}{\phi}_\alpha$	$\underset{\sim}{\psi}_\alpha$
1	1	0	0	$\underset{\sim}{s}$	$\underset{\sim}{\Delta}$
2	x	1	0	$\underset{\sim}{x}$	$\underset{\sim}{b}_x$
3	y	0	1	$\underset{\sim}{y}$	$\underset{\sim}{b}_y$
4	$h \equiv \xi\eta$	$h_{,x}$	$h_{,y}$	$\underset{\sim}{h}$	$\underset{\sim}{\gamma}$

$$\underset{\sim}{s} = [1, 1, 1, 1]$$

$$\underset{\sim}{\Delta} = \underset{\sim}{s} - (\underset{\sim}{s}^T \underset{\sim}{x})\underset{\sim}{b}_i$$

$$u_j = (\underset{\sim}{\Delta}^T \underset{\sim}{u}) + (\underset{\sim}{b}_i^T \underset{\sim}{u}_j)\underset{\sim}{x}_i + (\underset{\sim}{\gamma}^T \underset{\sim}{u}_j)\xi\eta$$

Table 2

Strain and Stress Field[3] for the γ-Element

	Optimal Bending (OB-γ)	Quintessential Bending/ (QIB-γ)	Frame-Invariant Bending (FIB-γ)	Optimal Incompressible (OI-γ)
Strain Field				
	$\bar{\varepsilon}_x+q_x\boldsymbol{h},_x$	$\bar{\varepsilon}_x+q_x\boldsymbol{h},_x-\bar{\nu}q_y\boldsymbol{h},_y$	$\bar{\varepsilon}_x+q_x\boldsymbol{h},_x+q_y\boldsymbol{h},_y$	$\bar{\varepsilon}_x+q_x\boldsymbol{h},_x-q_y\boldsymbol{h},_y$
	$\bar{\varepsilon}_y+q_y\boldsymbol{h},_y$	$\bar{\varepsilon}_y-\bar{\nu}q_x\boldsymbol{h},_x+q_y\boldsymbol{h},_y$	$\bar{\varepsilon}_y+q_x\boldsymbol{h},_x+q_y\boldsymbol{h},_y$	$\bar{\varepsilon}_y-q_x\boldsymbol{h},_x+q_y\boldsymbol{h},_y$
	$\bar{\gamma}_{xy}$	$\bar{\gamma}_{xy}$	$\bar{\gamma}_{xy}$	$\bar{\gamma}_{xy}$
Stress Field				
	$\bar{\sigma}_x+Q_x\boldsymbol{h},_x$	$\bar{\sigma}_x+Q_x\boldsymbol{h},_x$	$\bar{\sigma}_x+Q_x\boldsymbol{h},_x+Q_y\boldsymbol{h},_y$	$\bar{\sigma}_x+Q_x\boldsymbol{h},_x$
	$\bar{\sigma}_y+Q_y\boldsymbol{h},_y$	$\bar{\sigma}_y+Q_y\boldsymbol{h},_y$	$\bar{\sigma}_y+Q_x\boldsymbol{h},_x+Q_y\boldsymbol{h},_y$	$\bar{\sigma}_y+Q_y\boldsymbol{h},_y$
	$\bar{\sigma}_{xy}$	$\bar{\sigma}_{xy}$	$\bar{\sigma}_{xy}$	$\bar{\sigma}_{xy}$
c_1	$\bar{\lambda}+2\mu$	$\dfrac{\bar{E}H_xH_y}{H_xH_x-\bar{\nu}^2H_{xy}}$	$\frac{1}{2}(\bar{\lambda}+\mu)$	$\dfrac{4\mu H_xH_y}{H_xH_y-H_{x^2y}}$
c_3	$\bar{\lambda}$	$\bar{\nu}c_1$	c_1	c_1

$\bar{\varepsilon}_x$, $\bar{\varepsilon}_y$, $\bar{\varepsilon}_{xy}$ are the constant portions of the strain fields

$\bar{\sigma}_x$, $\bar{\sigma}_y$, $\bar{\sigma}_{xy}$ are the constant portions of the stress fields

$H_{ij} = \int \boldsymbol{h},_i\boldsymbol{h},_j d\Omega$, $\bar{E} = E/(1-\nu^2)$ and $\bar{\nu} = \nu/(1-\nu)$ for plane strain

$\bar{E} = E$ and $\bar{\nu} = \nu$ for plane stress

Table 3

Basis with Orthogonal Derivatives for 9-Node Lagrange Element

	α	ϕ_α	$\phi_{\alpha,\xi}$	$\phi_{\alpha,\eta}\ \underset{\sim}{\ell}_\alpha$
1	1	0	0	$\underset{\sim}{s}$
2	ξ	1	0	[-1,0,+1,-1,0,+1,-1,0,+1]
3	η	0	1	[-1,-1,-1,0,0,0,+1,+1,+1]
4	$\xi^2-1/3$	2ξ	0	
5	$\xi\eta$	η	ξ	
6	$\eta^2-1/3$	0	2η	
7	$\eta(\xi^2-1/3)$	$2\xi\eta$	$\xi^2-1/3$	
8	$\xi(\eta^2-1/3)$	$\eta^2-1/3$	$2\xi\eta$	
9	$(\xi^2-1/3)(\eta^2-1/3)$	$2\xi(\eta^2-1/3)$	$2\eta(\xi^2-1/3)$	$\underset{\sim}{h}$=[+1,-1.+1,-1,0,-1,+1,-1,+1]

Table 4

Results for Example 1 Cantilever Beam Problem

	w_{FEM}/w_{EXACT}			
	Plane Strain		Plane Stress	
	$\nu = 0.25$	$\nu = 0.4999$	$\nu = 0.25$	$\nu = 0.4999$
Optimal Bending (OB - γ)	0.884	0.0596	0.929	0.749
Frame Invariant Bending (FIB - γ)	1.30	0.0596	1.46	0.984
Optimal Incom- Pressible (OI - γ)	0.671	0.98	0.63	0.75
QBI - γ	0.989	0.980	0.988	0.984
SRI	1.12	1.64	1.12	1.25
2x2 Quadrature	0.676	0.0596	0.688	0.609

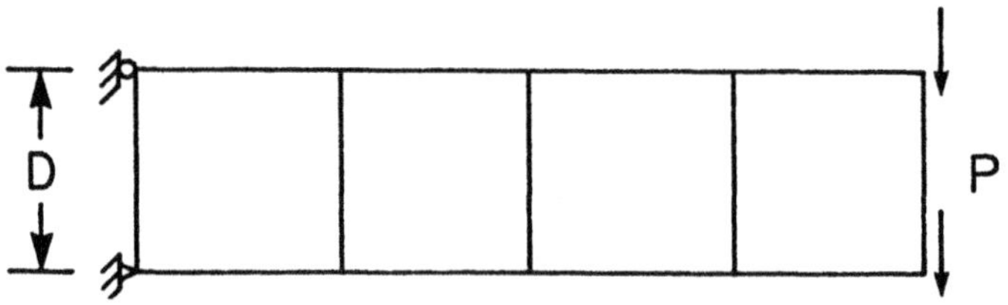

Fig. 1. Beam finite element model

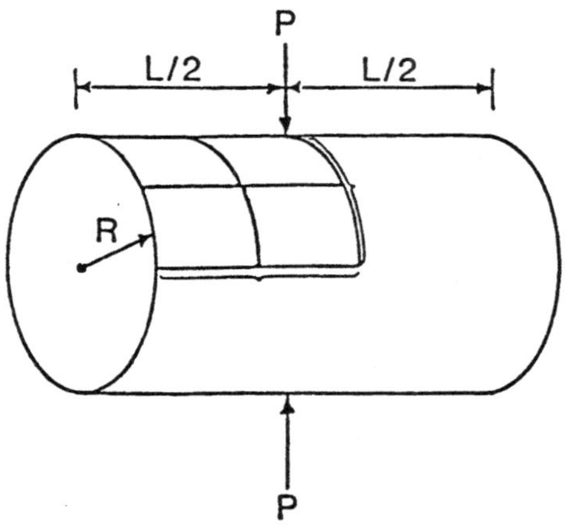

Pinched Cylinder with Diaphragm

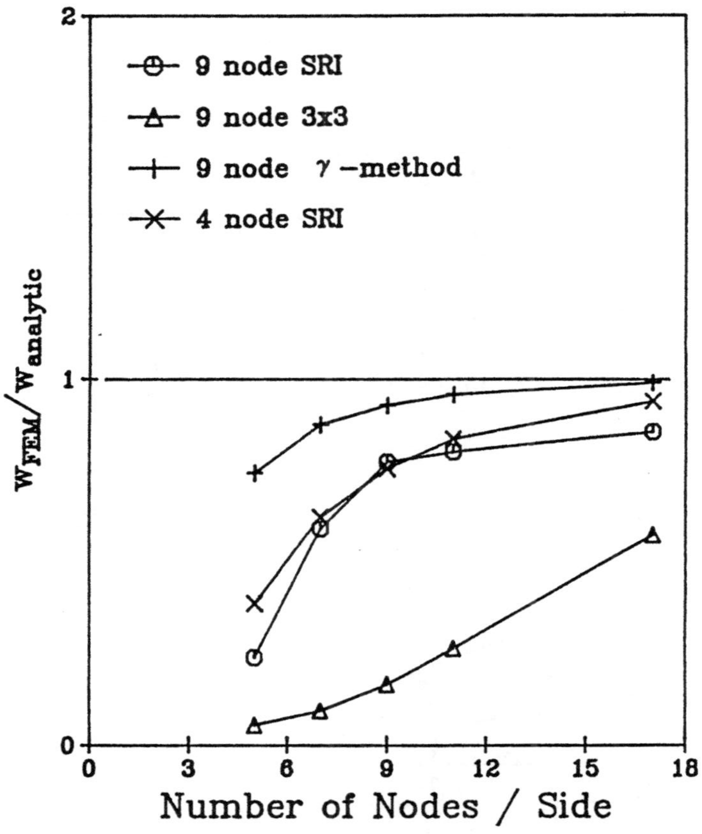

Fig. 2. Pinched cylindrical shell problem - definition and result

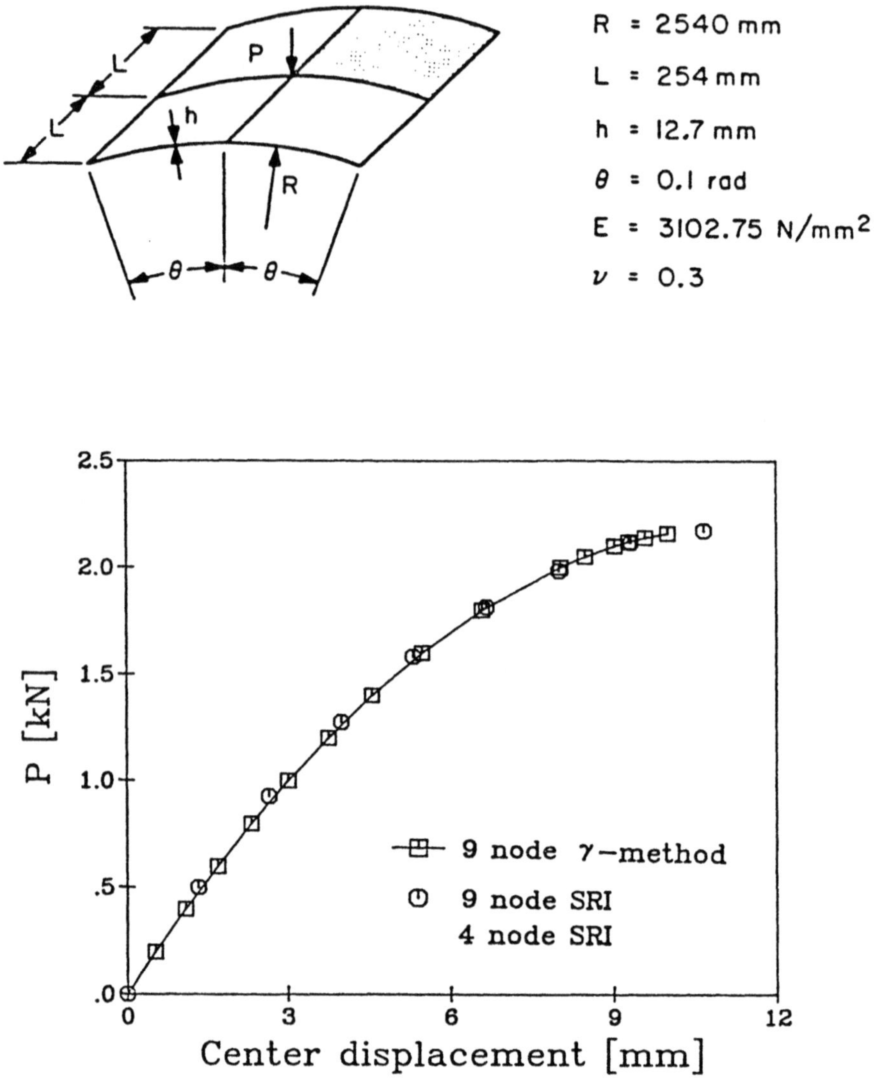

Fig. 3. Barrel vault problem - definition and result

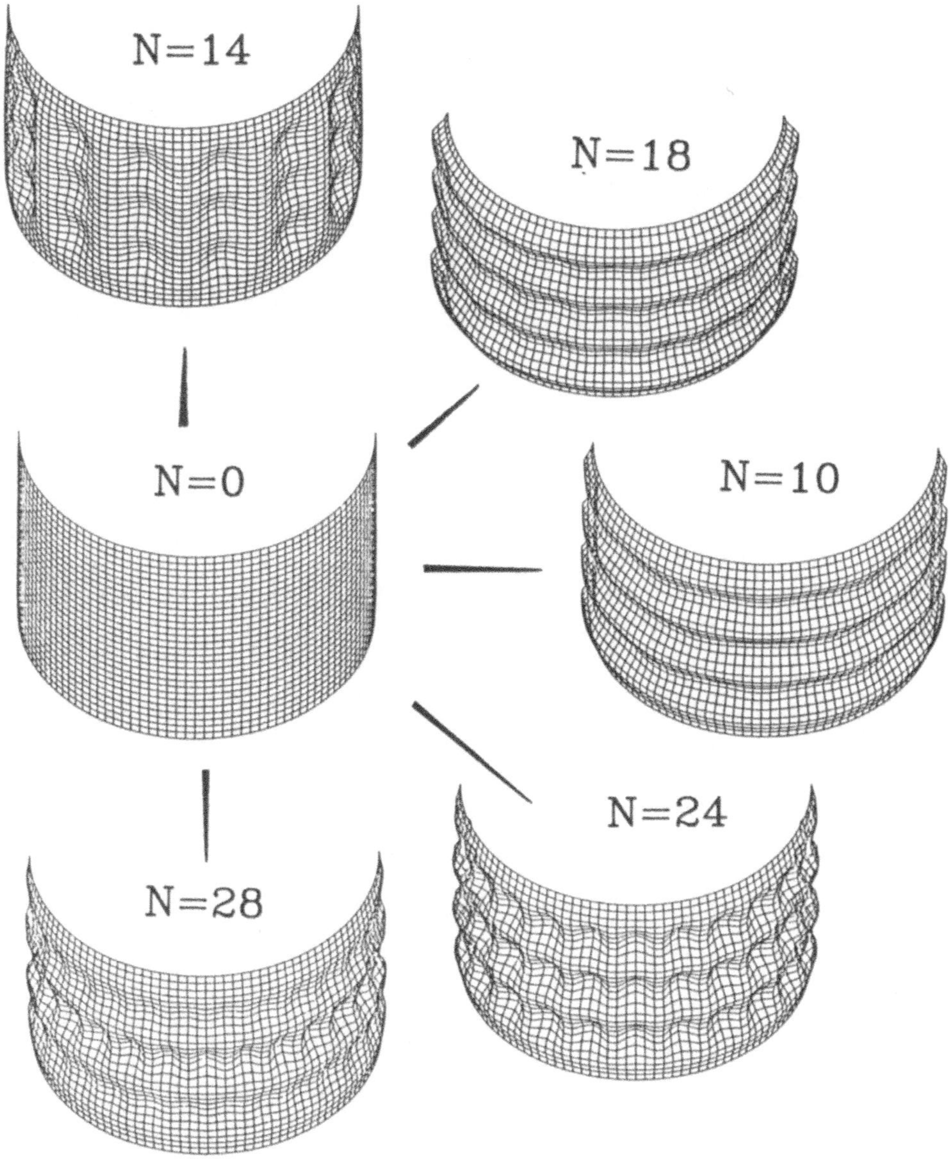

Fig. 4. Buckling of cylindrical shell under axial load

Nonlinear Shell Analysis Using Free Formulation Finite Elements

P. G. BERGAN, M. K. NYGÅRD*
The Norwegian Institute of Technology, Trondheim, Norway
*A. S, Veritec, Høvik, Norway

SUMMARY

The main principles of "free formulation" theory for finite elements with large displacements and material nonlinearities are outlined. It is shown how thin shell elements may be derived this way via use of a co-rotated reference description. An efficient method for handling large rotations in space is also described. Some new triangular and quadrilateral thin shell elements are suggested. These elements are initially planar and have three translational and three rotational freedoms per corner node. The "drilling" rotations perpendicular to the element plane yield highly improved in-plane response. The elements are nonconforming, but they satisfy the patch test/individual element test. Numerical examples involving large displacements, instability and inelastic behaviour of plates and shells are shown. It is concluded that the present elements have excellent accuracy.

1. INTRODUCTION

Finite elements based on the "Free formulation" by Bergan and Nygård [1] are, in most respects, similar to standard displacement elements. However, they have the added advantage that the displacement patterns need not satisfy interelement compatibility (conformity). The convergence of the free formulation elements is assured through satisfaction of the so-called "individual element test". This test, which was proposed by Bergan and Hanssen [2], is expressed as a set of linear constraints that the element stiffness matrix must satisfy, and it implies intrinsically that the "patch test" by Irons [3] is satisfied.

Up till now, the free formulation theory has been applied for deriving various types of linear membrane and plate elements, [1,4,5,6]. The excellent performance of these elements has motivated the present development of the free formulation theory for application to nonlinear problems with large displacements and inelastic materials. This paper also demonstrates the use of the free formulation theory for thin shell elements. An important aspect of these elements is the introduction of rotational freedoms normal to the element plane ("drilling freedoms"). Bergan and Felippa [4] have shown how these freedoms improve the in-plane bending response substantially over that of similar elements without such freedoms. The drilling freedoms are also convenient in that they eliminate the problem of stiffness singularity for co-planar elements.

Finite Element Methods for Nonlinear Problems
Europe-US Symposium, Trondheim, Norway 1985
Editors: Bergan, Bathe, Wunderlich

2. FREE FORMULATION THEORY

2.1 Linear Formulation

The displacement field $\mathbf{u}$ within an element may be approximated by a set of linearly independent displacement modes

$$\mathbf{u} = \mathbf{N}_{rc}\mathbf{q}_{rc} + \mathbf{N}_h\mathbf{q}_h \tag{1}$$

$\mathbf{N}_{rc}$ is a complete set of rigid/body and constant strain modes (rc-modes) with associated coefficients $\mathbf{q}_{rc}$, and $\mathbf{N}_h$ is a set of "higher order" displacement modes (h-modes) with coefficients $\mathbf{q}_h$. The number of rc-modes plus h-modes must be equal to the number of element freedoms.

The relationship between the generalized coefficients $\mathbf{q}$ and the element freedoms $\mathbf{v}$ is easily obtained by substitution of nodal coordinates into (1) and, for rotational freedoms, into derivatives of $\mathbf{N}$

$$\mathbf{v} = \mathbf{v}_{rc} + \mathbf{v}_h = \mathbf{G}_{rc}\mathbf{q}_{rc} + \mathbf{G}_h\mathbf{q}_h = \mathbf{G}\mathbf{q} \tag{2}$$

The generalized modes must be separable in terms of nodal freedoms, i.e., $\mathbf{G}$ must be regular and invertible

$$\mathbf{q} = \begin{vmatrix} \mathbf{q}_{rc} \\ \mathbf{q}_h \end{vmatrix} = \mathbf{G}^{-1}\mathbf{v} = \begin{vmatrix} \mathbf{H}_{rc} \\ \mathbf{H}_h \end{vmatrix}\mathbf{v} \tag{3}$$

A standard potential energy or virtual work stiffness formulation based on (1) would give an element that satisfies completeness but not, in most cases, interelement compatibility. Such an element would then not pass the patch test and it would fail to converge. As an alternative, Bergan and Nygård [1] suggested that the element stiffness should be formed from

$$\mathbf{k} = \frac{1}{V}\mathbf{L}\mathbf{C}\mathbf{L}^T + \mathbf{H}_h^T\mathbf{k}_{qh}\mathbf{H}_h \tag{4}$$

The first contribution to $\mathbf{k}$ is denoted "basic stiffness" and it is only associated with the constant strain/stress behaviour of the element. $\mathbf{L}$ is the so-called "lumping matrix" which transfers surface traction from constant stress states to consistent nodal forces. $\mathbf{C}$ is the constitutive matrix and V is the element volume. The second contribution to $\mathbf{k}$ is the "higher order stiffness" which is based on the stiffness matrix for the generalized higher order modes

$$\mathbf{k}_{qh} = \int_V \mathbf{B}_{qh}^T\mathbf{C}\mathbf{B}_{qh}\,dV \tag{5}$$

$\mathbf{B}_{qh}$ is the strain-producing matrix associated with the h-modes. It was demonstrated by the authors [1] that the stiffness in (4) passes the individual element test/the patch test even for incompatible displacement functions. This relaxation of constraints when choosing displacement functions opens the door to a new class of finite elements. The free formulation (4) is particularly useful for C^1 type of elements, such as plate and shell elements.

Equation (4) may also be written on traditional form

$$\mathbf{k} = \int_V \mathbf{B}^T \mathbf{C} \mathbf{B} dV \tag{6}$$

in which the strain-producing matrix is

$$\mathbf{B} = \frac{1}{V}\mathbf{L}^T + \widetilde{\mathbf{B}}_{qh}\mathbf{H}_h \tag{7}$$

It should be noted that $\mathbf{L}$ is constant and that $\widetilde{\mathbf{B}}_{qh}$ is "energy orthogonal" [7] with respect to the rc-modes

$$\int_V \widetilde{\mathbf{B}}_{qh}\, dV = 0 \tag{8}$$

This orthogonality property is obtained by an adjustment of the strain-producing matrix $\mathbf{B}_{qh}$ of the original h-modes

$$\widetilde{\mathbf{B}}_{qh} = \mathbf{B}_{qh} - \frac{1}{V}\int_V \mathbf{B}_{qh}\, dV \tag{9}$$

This adjustment is unnecessary if the h-modes already satisfy energy orthogonality.

After having solved the finite element problem with respect to the nodal displacements, the strains and stresses are easily obtained via (7) and the constitutive matrix $\mathbf{C}$.

It is worth noting that the stiffness matrix defined by (6) and (7) is exactly the same as one gets from virtual work or potential energy provided the displacement modes satisfy energy and force orthogonality [1].

2.2. Nonlinear Formulation

The free formulation theory may be extended to nonlinear situations with displacements and strains of arbitrary magnitude, see Nygård [8]. For instance, the lumping matrix in (4) may be found by use of virtual work for "updated", "co-rotated" and "total" Lagrangian descriptions. Unfortunately the scope of the present article does not allow for presentation of this.

The derivation of free formulation elements for nonlinear situation with small deformational strains is rather straight-forward. In cases of fixed or co-rotated coordinate frames the internal, element nodal reaction forces $\mathbf{S}_{int}$ is found from the usual expression

$$\mathbf{S}_{int} = \int_V \mathbf{B}^T \sigma dV \tag{10}$$

$\mathbf{B}$ is defined in (7). The Cauchy stress σ may be found from the current strain, or, for inelastic materials, the strain history.

The incremental equilibrium is expressed through the relation

$$\Delta \mathbf{S}_{int} = \mathbf{k}_T \Delta \mathbf{v} = (\mathbf{k}_{mat} + \mathbf{k}_{geom}) \Delta \mathbf{v} \tag{11}$$

The tangential or incremental stiffness $\mathbf{k}_T$ is comprised of material and geometric stiffness contributions. The material stiffness is obtained from the conventional expression

$$\mathbf{k}_{mat} = \int_V \mathbf{B}^T \mathbf{C} \mathbf{B} dV \tag{12}$$

where $\mathbf{C}_T$ is the tangential or incremental material law. $\mathbf{k}_{geom}$ is the usual geometric stiffness matrix; its form depends somewhat on the type of application. It is found from the element displacement field defined by (1) and (3).

3. SHELL ELEMENTS

3.1 Formulation of Plate and Shell Elements

The plate and shell elements are initially flat, and the usual Kirchhoff thin plate assumption are adopted. Introducing an orthogonal x,y,z coordinate system with corresponding displacements u,v,w when x and y follow the plane of the plate, the strain producing displacements are

$$\begin{aligned} u &= u_o - z w_{,x} \\ v &= v_o - z w_{,y} \end{aligned} \tag{13}$$

Index o indicates reference plane value. The in-plane (membrane) displacement w are interpolated between the nodal freedoms in the normal fashion

$$\mathbf{u}_m = \begin{bmatrix} u_o \\ v_o \end{bmatrix} = \begin{bmatrix} \mathbf{N}_u \\ \mathbf{N}_v \end{bmatrix} \mathbf{v}_m = \mathbf{N}_m \mathbf{v}_m \,, \qquad w = \mathbf{N}_b \mathbf{v}_b \tag{14}$$

The interpolation functions $\mathbf{N}_m$ and $\mathbf{N}_b$ will be dealt with later under specific elements. The membrane and bending freedoms for an element with n nodes are

$$\mathbf{v}_m = \begin{bmatrix} \mathbf{v}_{m1} \\ \mathbf{v}_{m2} \\ \vdots \\ \mathbf{v}_{mn} \end{bmatrix} , \qquad \mathbf{v}_b = \begin{bmatrix} \mathbf{v}_{b1} \\ \mathbf{v}_{b2} \\ \vdots \\ \mathbf{v}_{bn} \end{bmatrix} \tag{15}$$

The freedoms at node i are defined by

$$\mathbf{v}_{mi} = \begin{bmatrix} u_o \\ v_o \\ \frac{1}{2}(v_{o,x} - u_{o,y}) \end{bmatrix}_i = \begin{bmatrix} u_o \\ v_o \\ \Theta_z \end{bmatrix}_i , \qquad \mathbf{v}_{bi} = \begin{bmatrix} w \\ w_{,y} \\ -w_{,x} \end{bmatrix}_i = \begin{bmatrix} w \\ \Theta_x \\ \Theta_y \end{bmatrix}_i \tag{16}$$

The rotational freedom Θ_z about the z-axis ("drilling freedom") is defined as the mechanical rotation derived from the in-plane displacement gradients. This quantity is invariant with respect to the reference coordinates. However, there is no unique relationship between the freedom Θ_z and the rotations of the adjacent element sides, see discussion by Bergan and Felippa [4]. This kind of freedom can therefore not be used in connection with the usual displacement formulation which requires C^o interelement continuity for the membrane action. This is, however, no object for the free formulation.

It is now assumed that the deformational strains are small and follow linear theory

$$\varepsilon = \begin{bmatrix} \varepsilon_x \\ \varepsilon_y \\ \gamma_{xy} \end{bmatrix} = \begin{bmatrix} u_{,x} \\ v_{,y} \\ u_{,y} + v_{,x} \end{bmatrix} = \begin{bmatrix} u_{o,x} \\ v_{o,y} \\ u_{o,y} + v_{o,x} \end{bmatrix} - z \begin{bmatrix} w_{,xx} \\ w_{,yy} \\ 2w_{,xy} \end{bmatrix}$$

$$= \mathbf{B}_m \mathbf{v}_m - z\mathbf{B}_b \mathbf{v}_b = \begin{bmatrix} \mathbf{B}_m & -z\mathbf{B}_b \end{bmatrix} \begin{bmatrix} \mathbf{v}_m \\ \mathbf{v}_b \end{bmatrix} = \mathbf{B}\mathbf{v} \qquad (17)$$

The strain producing matrices are found in accordance with (7)

$$\mathbf{B}_m = \frac{1}{A}\mathbf{L}_m^T + \widetilde{\mathbf{B}}_{qhm}\mathbf{H}_{hm} \qquad (18a)$$

$$\mathbf{B}_b = \frac{1}{A}\mathbf{L}_b^T + \widetilde{\mathbf{B}}_{qhb}\mathbf{H}_{hb} \qquad (18b)$$

A is the area of the element. The lumping matrices $\mathbf{L}_m$ and $\mathbf{L}_b$ for membrane and bending action of polygon-shaped elements are given in Appendix.

Bergan and Felippa [4] used a special scaling factor, α, for the influence of the "drilling freedoms". The effect of this factor is such that increasing α-values give a more flexible element. If nothing else is stated, $\alpha = 1.5$ is used for the present study.

The internal reaction forces are derived following (10)

$$\mathbf{S}_{int} = \begin{bmatrix} \mathbf{S}_m \\ \mathbf{S}_b \end{bmatrix} = \int_V \mathbf{B}^T \sigma dV = \int_A \begin{bmatrix} \mathbf{B}_m^T \mathbf{N} \\ \mathbf{B}_b^T \mathbf{M} \end{bmatrix} dA \qquad (19)$$

Where the cross-sectional reaction forces are found by integration through the thickness h

$$\mathbf{N} = \int_h \sigma dz \quad , \qquad \mathbf{M} = \int_h -z\sigma dz \qquad (20)$$

The material stiffness, which is required for an incremental formulation, is defined in (12)

$$\mathbf{k}_{mat} = \int_V \mathbf{B}^T \mathbf{C}_T \mathbf{B} dV = \int_A \begin{bmatrix} \mathbf{B}_m^T \mathbf{N}_T \mathbf{B}_m & \mathbf{B}_m^T \mathbf{M}_T \mathbf{B}_b \\ \mathbf{B}_b^T \mathbf{M}_T \mathbf{B}_m & \mathbf{B}_b^T \mathbf{D}_T \mathbf{B}_b \end{bmatrix} dA \qquad (21)$$

where the tangential material matrices for the cross-sectional forces are

$$\mathbf{N}_T = \int_h \mathbf{C}_T dz \quad , \qquad \mathbf{M}_T = \int_h -z\mathbf{C}_T dz \quad , \qquad \mathbf{D}_T = \int_h z^2\mathbf{C}_T dz \tag{22}$$

The geometric stiffness matrix $\mathbf{k}_{geom}$ (11) for a flat shell element may be derived from the incremental form of the virtual work principle. This matrix may be partitioned as follows

$$\mathbf{k}_{geom} = \int_A \begin{vmatrix} \mathbf{k}_{gm} & \mathbf{0} \\ \mathbf{0} & \mathbf{k}_{gb} \end{vmatrix} dA \tag{23}$$

The geometric stiffness for in-plane (membrane) action is

$$\mathbf{k}_{gm} = \int_A (N_{xx}(\mathbf{N}_{u,x}^T\mathbf{N}_{u,x} + \mathbf{N}_{v,x}^T\mathbf{N}_{v,x}) + N_{yy}(\mathbf{N}_{u,y}^T\mathbf{N}_{u,y} + \mathbf{N}_{v,y}^T\mathbf{N}_{v,y})$$

$$+ N_{xy}(\mathbf{N}_{u,x}^T\mathbf{N}_{u,y} + \mathbf{N}_{u,y}^T\mathbf{N}_{u,x} + \mathbf{N}_{v,x}^T\mathbf{N}_{v,y} + \mathbf{N}_{v,y}^T\mathbf{N}_{v,x}))dA \tag{24}$$

$\mathbf{N}_u$ and $\mathbf{N}_v$ are in-plane interpolation functions defined in (14). The geometric stiffness for the out-of-plane (bending) action is

$$\mathbf{k}_{gb} = \int_A (N_{xx}\mathbf{N}_{w,x}^T\mathbf{N}_{w,x} + N_{yy}\mathbf{N}_{w,y}^T\mathbf{N}_{w,y} + N_{xy}(\mathbf{N}_{w,x}^T\mathbf{N}_{w,y} + \mathbf{N}_{w,y}^T\mathbf{N}_{w,x}))dA \tag{25}$$

3.2 Triangular Element

Membrane Action

The membrane element suggested by Bergan and Felippa [4] is adopted here. As shown in Fig. 1a, nondimensional coordinates with origin at the centroid (x_c,y_c) is introduced

$$\xi = \lambda(x - x_c) \quad , \qquad \eta = \lambda(y - y_c) \tag{26}$$

where the length scaling factor is

$$\lambda = \frac{1}{\sqrt{A}} \tag{27}$$

The complete set of six rigid-body and linear strain modes may be written

$$\mathbf{N}_{rcm} = \begin{vmatrix} 1 & 0 & -\eta & \xi & 0 & \eta \\ 0 & 1 & \xi & 0 & \eta & \xi \end{vmatrix} \tag{28}$$

According to equations (14), (15) and (16) there are nine freedoms controlling the membrane displacements; three additional higher order modes are thus required. Following [4], the best choice is the in-plane "bending" modes oriented along the median lines of the triangle (see Fig. 1a)

$$\mathbf{N}_{hm,i} = \begin{bmatrix} \bar{\xi}_i \bar{\eta}_i \\ -\frac{1}{2}\bar{\xi}_i^2 \end{bmatrix} \quad , \quad i = 1,2,3 \tag{29}$$

These natural modes are objective with respect to the reference system. They also give a **G** matrix which is regular.

Bending action

The bending modes are expressed most conveniently in terms of area coordinates ζ_i . The complete set of rigid-body and constant curvature modes may be written

$$\mathbf{N}_{rcb} = \begin{bmatrix} \zeta_1 & \zeta_2 & \zeta_3 & \zeta_1\zeta_2 & \zeta_2\zeta_3 & \zeta_3\zeta_1 \end{bmatrix} \tag{30}$$

The three higher order modes suggested in [9] are used. These modes are force orthogonal and have proved to give very good results

$$\mathbf{N}_{hb} = \begin{bmatrix} \zeta_1\zeta_2(\zeta_1-\zeta_2) & \zeta_2\zeta_3(\zeta_2-\zeta_3) & \zeta_3\zeta_1(\zeta_3-\zeta_1) \end{bmatrix} \tag{31}$$

The **H** -matrix for these bending modes has also been found explicitly [8].

Integration

The membrane and bending stiffnesses in (21) are completely uncoupled in case of a linear, homogeneous material. However, these stiffnesses become coupled for inelastic materials and can thus only be computed by use of numerical integration. It has been found that a three-station numerical integration based on the mid-side points gives sufficient accuracy. In addition, the cross-sectional quantities (20) and (22) are computed using either the two point Gaussian or the three to ten point Lobatto integration schemes. For nonlinear material the Lobatto integration has an advantage over Gaussian integration since it is a closed integration rule that includes the important surface points.

3.3 Quadrilateral Element

Membrane Action

A set of nondimensional coordinates ξ and η with origin at the centroid is introduced, see Fig. 1b. As for the triangular element, ξ and η are defined by (26), however, the scaling factor is here chosen as

$$\lambda = \frac{2}{\sqrt{A}} \tag{32}$$

The basic rc-modes are the same as in (28). The six, additional higher order modes are chosen as

$$\mathbf{N}_{hm} = \begin{bmatrix} \xi\eta & \eta^2 & 0 & 0 & \eta(\xi-\eta)^2 & -\eta(\xi+\eta)^2 \\ 0 & 0 & \xi\eta & \xi^2 & -\xi(\eta-\xi)^2 & \xi(\eta+\xi)^2 \end{bmatrix} \tag{33}$$

A main consideration in the selection of these modes is that they do not give a singular **G** regardless of the geometry. It is seen that these h-modes do not correspond to complete polynomials of a certain degree, hence, these modes have a directional dependency in accordance with the orientation of the ξ,η -system. The ξ,η -system is therefore chosen such that it follows "natural" directions of the quadrilateral. The origin of the ξ,η -system is placed at the centroid of the quadrilateral, and the coordinate directions are taken as the orthogonal directions of a "best fit" imaginary rectangle, see Fig. 1b. This fitting is done on the basis of the deviatoric element side angles α_{ij} , see Fig. 1b. This leads to the angle of rotation in relation to side 1-2 for the ξ,η -system.

Bending action

The rc-modes for the out-of-plane displacement are easily expressed as the six terms of a complete second degree polynomial. The following functions are chosen for higher order modes

$$\mathbf{N}_{bh} = \left| \xi^3 \quad \xi^2\eta \quad \xi\eta^2 \quad \eta^3 \quad F_1^3 F_2 \quad F_1 F_2^3 \right| \tag{34}$$

The four first terms form a complete and objective third degree polynomial. The last two functions make use of

$$F_1 = f_{41} - f_{23} \quad , \qquad F_2 = f_{12} - f_{34} \tag{35}$$

where f_{ij} are objective "line functions" going through corner nodes [6]

$$f_{ij} = \frac{1}{A} \left| x_i y_j - x_j y_i + (y_i - y_j)x + (x_j - x_i)y \right| \tag{36}$$

Integration

The forces and stiffness matrices may be found using a two by two Gaussian integration over the area of the quadrilateral. This order of integration is sufficient for the bending stiffness, however, it renders the membrane stiffness with a defect of one (one spurious zero energy mode). This mode can result in singularity only for a single element with minimum boundary conditions. However, it cannot propagate through neighbouring elements. Two by two integration is thus safe and sufficient for all practical applications. The defect disappears for three by three integration. As for the triangle, either two Gaussian or three to ten Lobatto points are used for the integration through the thickness.

4. ANALYSIS OF SHELLS

4.1 Ghost reference Description

The large displacement motion of the shell is defined in a Lagrangian description. The deformation of each element is referred to the initial configuration of each element,

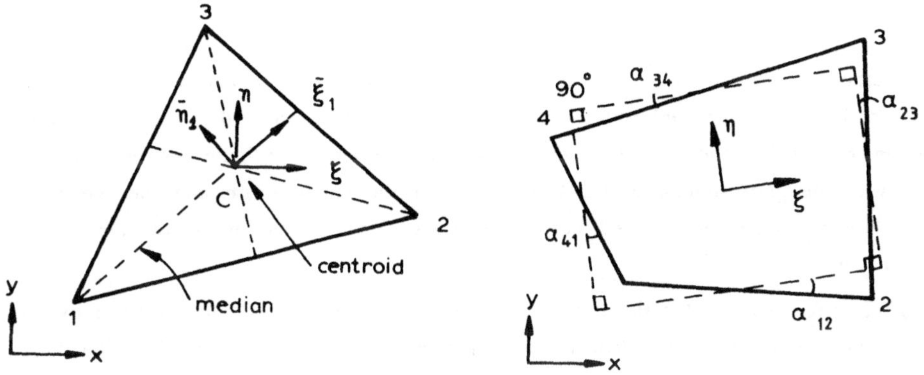

Figure 1. Local coordinate systems for triangular and quadrilateral elements.

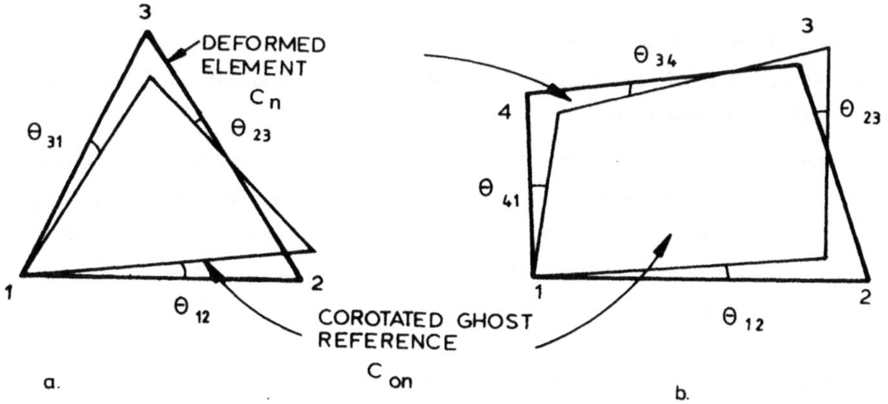

Figure 2. Establishment of a ghost reference for triangular and quadrilateral elements.

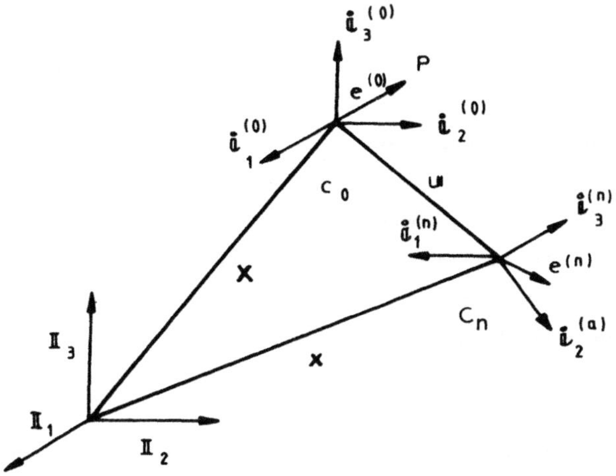

Figure 3. Motion of a nodal point with associated eccentricity vector.

however, this reference configuration is closely displaced and rotated along with the deformed configuration in order to minimize the rigid-body part of the rotational motion. This method has been used by several investigators, and it is here denoted "ghost reference description". It is applicable to large strains, as well as small deformational strains. This description has the advantage that, when the element coordinate system is moved along with the ghost reference, the components of the Green strain and the symmetric Piola-Kirchhoff stress tensor need not be transformed each time the ghost reference is updated [10].

The ghost reference C_{on} for the triangular element is placed in the plane formed by the corner nodes of the deformed element C_n with a common node 1, see Fig. 2a. The direction of the local x-axis, which goes through nodes 1 and 2 of the ghost reference, is chosen such that the sum of the squared deviatoric angles Θ_{12} , Θ_{23} and Θ_{31} is minimized. This yields a simple linear equation for determining Θ_{12} . It is crucial that such an objective method is used for rotating the ghost reference. Introduction of directional bias, such as use of a common side 1-2, has proved to give asymmetric components in displacement patterns that ought to be symmetric.

A similar method is used for establishing the ghost reference for the quadrilateral element. The normal to the plane of the ghost reference is taken as the vector formed by the cross-product of the vectors formed by the two diagonals of the deformed element. Node 1 is taken to be the same for the deformed element and its reference. The other corners of the warped quadrilateral are projected onto the reference plane. The ghost reference C_{on} is then rotated in its plane such that the sum of the squared deviatoric angles Θ_{ij} from the projected element C_n is minimized, see Fig. 2b. Again, this is a method for establishing a ghost reference which is objective with respect to rotations.

4.2 Treatment of Large Rotations

Each nodal point is thought of as carrying along with it a rigid mass or "potato". The adjacent finite elements are then attached directly to the nodal point or to a surface point of the potato. The off-set of the attachment point from the node is defined by an eccentricity vector $\mathbf{e}$, see Fig. 3. The translational motion of the node is defined by the displacement vector $\mathbf{v}$. The large rotations, which are not true vectorial quantities, are defined in terms of an orthogonal base vector triad $\mathbf{i}_i$ which is initially parallel to the global base vectors $\mathbf{I}_i$. The base vectors $\mathbf{i}_i$ translate and rotate along with the rigid nodal potato. The rotation of a node at configuration C_n is defined by

$$\mathbf{i}_i^{(n)} = T_{ij}^{(n)} \mathbf{I}_j \tag{37}$$

Note that the rotation is represented by an orthogonal rotation matrix $T_{ij}^{(n)}$ rather than angles. Going incrementally from configuration C_n to a neighbouring configuration C_{n+1}, the rotational relationship is updated

$$\mathbf{i}_i^{(n+1)} = T_{ij}^{(n+1)}\mathbf{I}_j = T_{ik}^{(n)}\Delta T_{kj}^{(n+1)}\mathbf{I}_j \tag{38}$$

This updating follows a factorial and not an additive law. $\Delta T_{ij}^{(n+1)}$ depends on the incremental rotations $\Delta\Theta_i$ about the global axes and are given by

$$\Delta T_{ij}^{(n+1)} = \delta_{ij}\cos(\Delta\Theta) + \frac{\Delta\Theta_i}{\Delta\Theta}\frac{\Delta\Theta_j}{\Delta\Theta}(1 - \cos(\Delta\Theta)) + e_{ijk}\frac{\sin(\Delta\Theta)}{\Delta\Theta}\Delta\Theta_k \tag{39}$$

where

$$\Delta\Theta = \sqrt{\sum_{i=1}^{i=3}\Delta\Theta_i^2} \tag{40}$$

δ_{ij} and e_{ijk} are the Kronecker delta and permutation symbols, respectively, see e.g. [11]. The present method corresponds to the methods described by Argyris [12], and Rankin and Brogan [13], however, it is believed that the present form (39) is new and simpler than the previous expressions. Further details about the derivation of (39) may be found in [8].

Rotational freedoms are only used incrementally in the global finite element equations, the total rotations are defined by rotational matrices (37). On local level, however, total rotations are used to define the deformational rotations between the deformed configuration C_n and the ghost reference configuration C_{on} for an element. The use of rotational freedoms is in fact based on the existence of such "vectorial" rotations. It is imagined that each corner node, say corner "a", of an element carries a separate orthogonal base vector system $\mathbf{i}_i^a$ which is initially parallel to the planar ξ,η -system in C_o . The deformational rotation at node a appears through the rotation matrix $T_{ij}^{a(n)}$ between the local corner system $\mathbf{i}_i^{a(n)}$ in configuration C_n and base system $\mathbf{i}_i^{(on)}$ that follows the ξ,η -axes of the ghost reference C_{on}

$$\mathbf{i}^{a(n)} = \widetilde{T}_{ij}^{a(n)}\mathbf{i}_j^{(on)} \tag{41}$$

The relatively small deformational rotations $\theta_i^{(n)}$ at corner a about the axes $\mathbf{i}_i^{(on)}$ are found to be

$$\begin{aligned}
\Theta_1 &= \frac{1}{2}\frac{\Theta}{\sin(\Theta)}(\widetilde{T}_{23}^a - \widetilde{T}_{32}^a) \\
\Theta_2 &= \frac{1}{2}\frac{\Theta}{\sin(\Theta)}(-\widetilde{T}_{13}^a + \widetilde{T}_{31}^a) \\
\Theta_3 &= \frac{1}{2}\frac{\Theta}{\sin(\Theta)}(\widetilde{T}_{12}^a - \widetilde{T}_{21}^a)
\end{aligned} \tag{42}$$

where

$$\Theta = arccos\left|\frac{1}{2}(trace(\widetilde{T}_{ij}^a) - 1)\right| \tag{43}$$

Superscript (n) is omitted in these equations. Note also that these equations are objective in the sense that a certain order of taking the components $\Theta_i^{(n)}$ has not been assumed.

4.3 System Equations and Solution Methods

The equilibrium equations for the total shell system is established from external loads and element reaction forces (19) using standard assemblage procedures. Similarly, the incremental equilibrium equations for the system is assembled from the material stiffness (21) and the geometric stiffness (23) for all elements. Three incremental rotations are introduced at all nodes, however, the total rotations only appear through the rotational matrices (37).

The nonlinear equations are solved in an incremental-iterative fashion using a simple load incrementation (Euler Method) combined with "true" or "modified" Newton-Raphson iteration, see e.g. [14], [15]. Adaptive solution schemes which can automatically set the load steps and perform unloading at extremum points are also used. In particular, the so-called hyperplane displacement control method [16], [17] has been adopted in the present studies.

5. NUMERICAL EXAMPLES

5.1 Computer Programs

The element and methods described herein are implemented in the general purpose nonlinear finite element program FENRIS. This program performs all the standard tasks of a nonlinear program, such as assemblage of loads, forces and stiffnesses as well as solution of nonlinear equations.

The FENRIS program has a highly modular structure which allows for combination of substructures of different types of structures. For instance, the present shell elements may be combined with truss, frame and solid type substructures.

The free formulation triangular and quadrilateral shell elements described herein are denoted FFT and FFQ.

5.2 Shear-Loaded Cantilever Beam

The shear-loaded cantilever beam in Fig. 4a is commonly used for evaluating membrane elements, and give a good test on the elements ability to represent in-plane bending modes. The cantilever was analysed using regular nxm meshes as typified for the triangular element in Fig. 5.1b. The results for the tip deflection, v_c , are tabulated in Tab. 1. This table reveals excellent performance by the free formulation elements. Bergan and Felippa [4] have shown that the accuracy of the FFT element may be increased further by utilizing a scaling factor for the higher order stiffness, however, this possibility has not

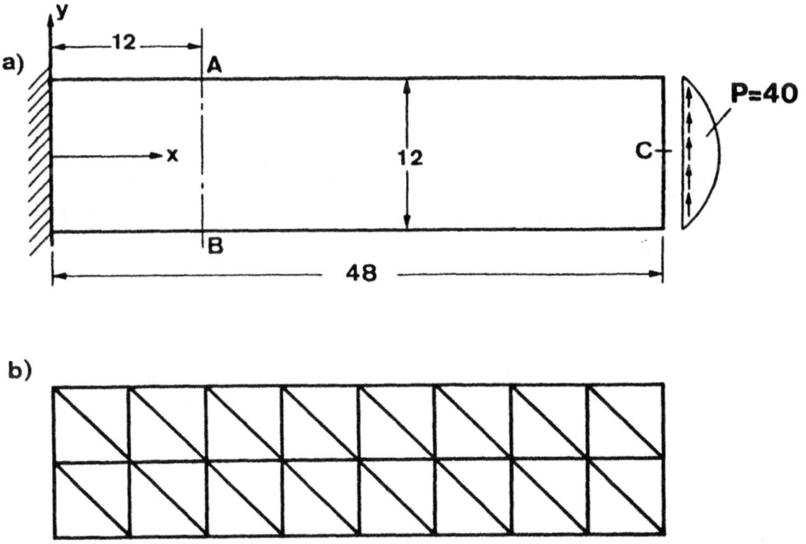

Figure 4. (a) Cantilever beam with load P parabolically distributed along right edge ($E=30000$, $\nu=\frac{1}{4}$); (b) regular 8x2 mesh.

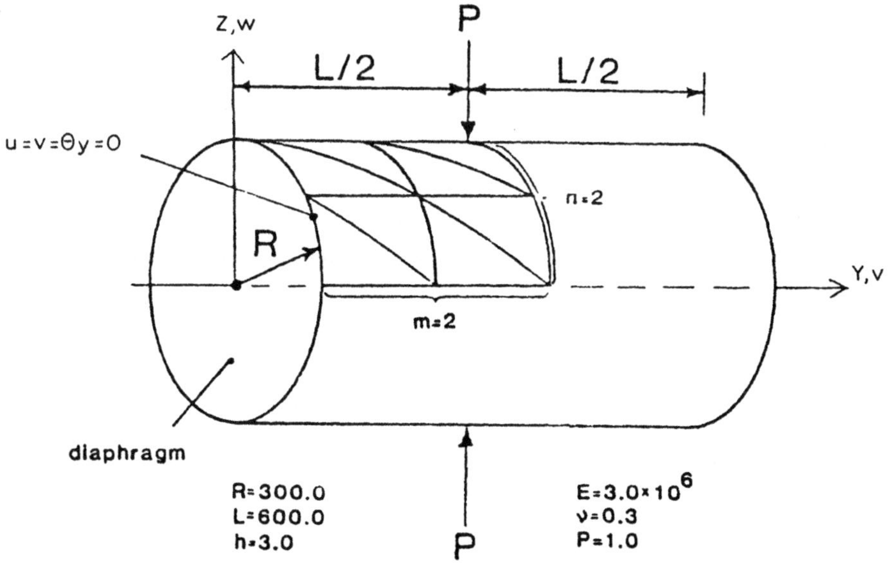

Figure 5. Pinched cylinder with end diaphragms.

been investigated yet for the present nonlinear elements.

The quadrilateral element is used in three versions, that is 2x2 integration with $\alpha = 1.5$ and 1.0 and 3x3 integration with $\alpha = 1.5$. The table shows that all cases give good results, and that the difference between them is rather small. In fact, the difference between the 2x2 and 3x3 integration is negligible, and the former thus appears as a winner because of computational efficiency. As for the FFT-element [4], an α-value of 1.5 instead of 1.0 slightly increases the elements flexibility.

Mesh	CST	FFT	FFQ	FFQ(α=1.)	FFQ(3x3)	QSEL
4x1	35.48	75.23	92.19	92.19	92.18	68.13
8x2	55.24	92.12	100.36	98.89	100.07	88.91
16x4	82.66	97.69	100.01	99.58	99.92	96.91
32x8	94.96	99.36	99.97	99.88	99.95	99.30
64x16	98.71	99.86	100.01	100.00		

Tab. 1 Tip deflection (v_C) for shear-loaded cantilever beam (exact solution equal to 100.).

The other elements in Tab. 1 are the constant strain triangle (denoted CST, see e.g. [21]), and a linear four-noded quadrilateral with selective reduced shear integration (denoted QSEL, [22]) which corresponds to the incompatible rectangle with Wilson modes [23]. Both these are completely inferior to the free formulation elements.

5.3 Linear Analysis of Pinched Cylinder with End Diaphragms

The pinched cylinder problem shown in Fig. 5 is a classical test case for shell elements, and different variations of this problem have been used by other investigators. The present problem has been studied in references [24], [25] and it is a good test of both membrane and bending action. A convergence study is performed using regular, nxn meshes for 1/8 of the total cylinder, (three symmetry planes). Tab. 2 gives the normal deflection under the point load. The results are presented as the percentages of the exact result which is $1.8248 \cdot 10^{-4}$ [24]. The FFT- and FFQ-element give both good results. Comparisons are made with a four-node continuum mechanics based element by Dvorkin and Bathe [25], and a three-node discrete Kirchhoff theory element by Stolarski et. al [24], who developed both flat and curved versions of their element. It is seen that the FFQ element is the best performer among the flat elements, while the DKT-element [24] increases it's accuracy considerably by utilizing a curved element concept. The free formulation shell elements can also be extended to account for curved shapes. It is believed that this would be extra important for element that have "drilling freedoms" since one would thereby avoid the direct coupling between in-plane and out-of-plane rotational stiffnesses at nodes that the faceted surface approximation leads to.

Mesh on octant	FFT	FFQ	Stolarski et al.		Dvorkin and Bathe
			Flat	Curved	
4x4	53.5	63.2	46.2	77.7	
5x5					51.
8x8	85.7	95.2	86.0	94.8	
10x10			93.0	98.8	83.
16x16	98.0	99.2			
20x20					96.

Tab. 2 Central deflection for a pinched cylinder, P=1. (exact solution equal to 100.).

5.4 Cylindrical Shell Roof Subjected to Gravity Loading

The cylindrical shell roof supported on end diaphragms suggested by Scordelis and Lo [26], has become a classical problem for testing shell elements. The geometry of the shell is given in Fig. 6. The uniform gravity force is $4.375 \cdot 10^{-3}$ N/mm^2 (4.375 kN/m^2), and the vertical deflection at the midpoint of the free edge is given for several linear analyses with uniform nxn meshes for one quarter of the roof. The exact results is 91.93 mm, while the tabulated results are normalized by the number 100.

Mesh on quadrant	FFT	FFQ	Stolarski et al.		Dvorkin and Bathe	MacNeal and Harder QUAD4
			Flat	Curved		
2x2	114.11	150.46	97.4	86.3		137.6
4x4	99.00	108.83	70.7	64.4		105.0
5x5					96.0	
8x8	99.63	102.34	86.6	84.5	98.1	100.8
12x12					99.2	
16x16	99.89	100.52				

Tab. 3 Linear analysis of the Scordelis-Lo shell. Vertical deflection at midpoint of free edge (exact deflection equal to 100.).

The results are compared to the same elements as for the pinched cylinder problem, in addition to the 4-noded QUAD4-element [27]. Again the FF-elements perform very well, especially the FFT element which is within 1 percent error already for a 4x4 subdivision.

Nonlinear analyses with elastic-perfectly-plastic material with yield limit $\sigma_y = 4.2 N/mm^2$ where also performed. Von Mises yield criterion was assumed, and ten integration points

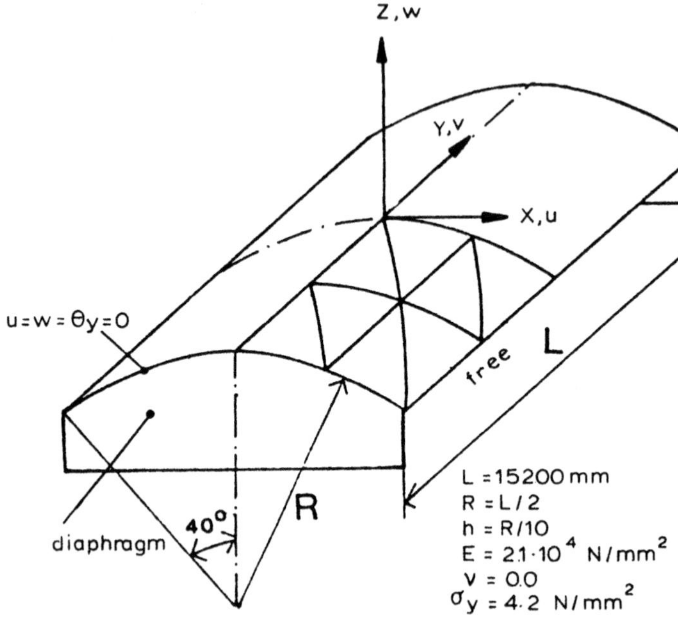

Figure 6. Cylindrical shell roof.

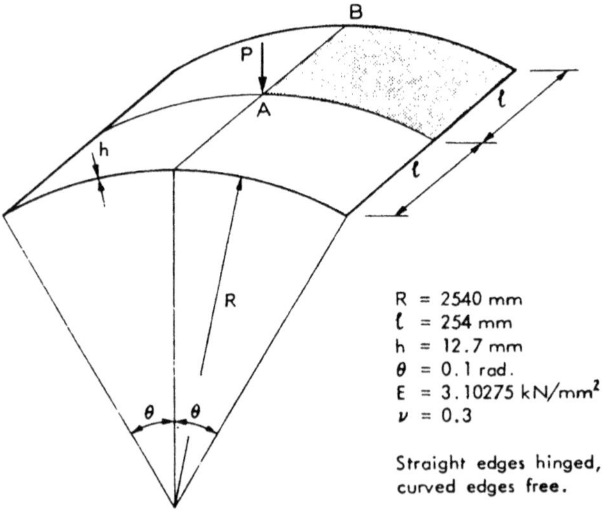

Figure 8. Hinged cylindrical shell.

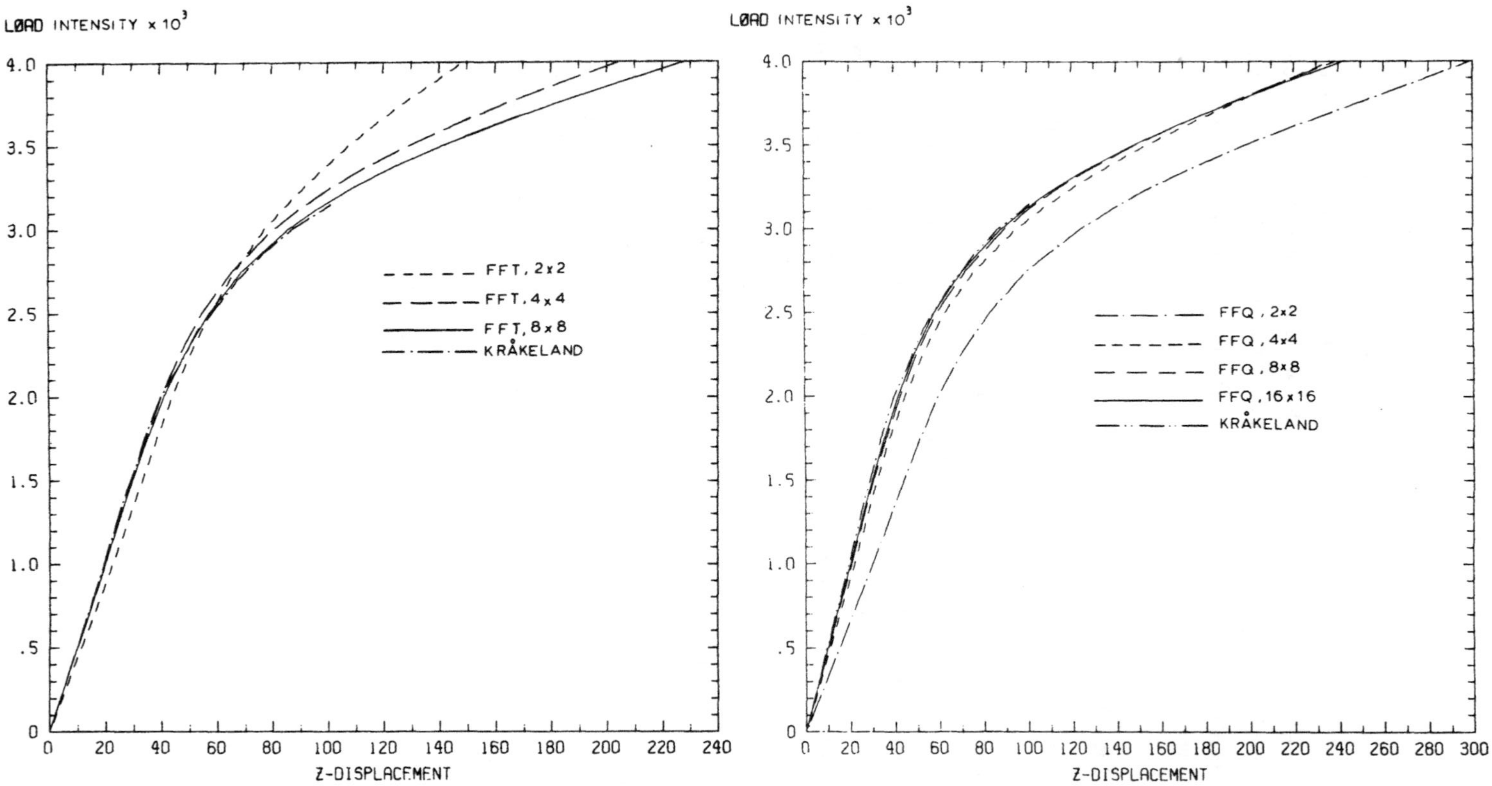

Figure 7. Nonlinear analysis of Scordelis-Lo shell. Deflection at midpoint of free edge obtained with (a) triangular elements; (b) quadrilateral elements.

were used for integration through the thickness of the shell. Fig. 7a and 7b show the vertical deflection at the midpoint of the free edge against the intensity of the vertical load. The results are compared to those achieved by Kråkeland [28], who employed a curved 8-node isoparametric element in a 4x4 mesh. In this nonlinear case, the FFQ-element is more accurate than the FFT-element. It is quite remarkable how well the present "low order" elements perform. Similar results are presented in [25] for the previous mentioned four-node element by Dvorkin and Bathe [25]. However, even if it is stated that these results are identical to those by Kråkeland [28], it is evident that this element is much more flexible for high load levels.

5.5 Hinged Cylindrical Shell Under Concentrated Load

The hinged cylindrical shell with a concentrated load shown in Fig. 8 is a good test case for geometric nonlinearities. Assuming symmetrical response only one quarter of the shell is modeled. Thicknesses $h = 12.7$ mm and $h = 6.35$ mm was considered.

The deflections, w_A and w_B at points A and B (Fig. 8), are plotted versus the applied load in Fig. 9a and 9b for the thick and thin cases, respectively. A 4x4 mesh was used for the triangle whereas 4x4 and 8x8 meshes were used for the rectangle. The thick case was analysed using prescribed displacement increments at point A and a modified Newton-Raphson procedure with equilibrium iteration. The results compared are in good agreement with other results reported in the literature (e.g. [29]).

A much more severe test case is the snap-through behaviour of the thin shell. It is seen in Fig. 9b that point A moves upwards again after the first instability point of the path has been passed. Such snap-back problems requires more advanced solution algorithms. The present example was solved with automatic load application/reversal in a single solution sequence using the so-called hyperplane displacement control method in [17]. There is some discrepancy between the three cases, particularly near the extremum points. The results obtained with the 4x4 mesh for the FFQ-element are very close to the flat element solution by Horrigmoe and Bergan [29], while the 8x8 FFQ results compare very well with the higher order curved element solution by Oliver and Onate [30].

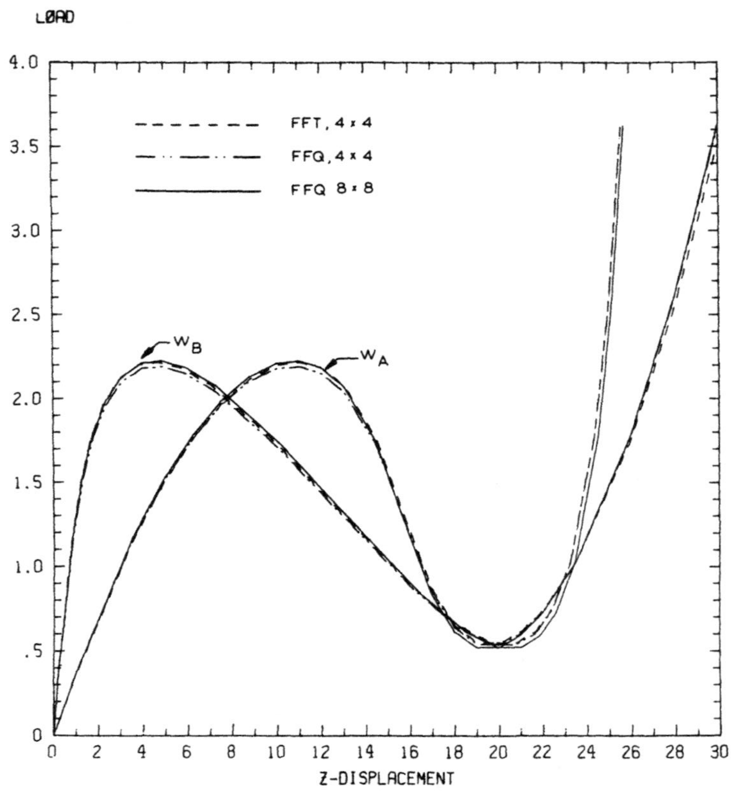

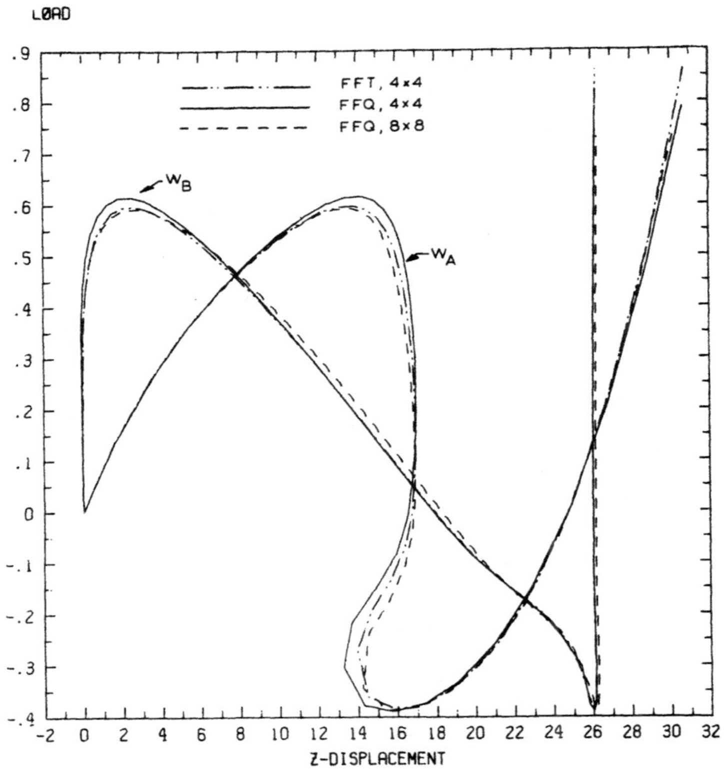

Figure 9. Load-deflection curve for hinged cylindrical shell. (a) thick case (h=12.7mm); (b) thin case (h=6.35mm).

CONCLUSION

The theory of free formulation finite elements has been extended to general nonlinear analysis, including geometric and material nonlinearities. This theory has been applied for development of some new, planar thin shell elements with membrane type rotational ("drilling") freedoms. It is demonstrated that these freedoms have a major influence on the accuracy. The new elements are without doubt significantly better than comparable, traditional displacements based elements. Several numerical examples are presented, including some linear problems, buckling of a cylindrical panel and elasto-plastic collapse of a shell. The results obtained are extremely encouraging.

ACKNOWLEDGEMENT

The authors would like to express their appreciation to their colleagues in the FENRIS project for their contribution in developing the computer code that was used. A research grant to the second author from the Royal Norwegian Council for Technical and Scientific Research (NTNF) is also acknowledged.

REFERENCES

1 Bergan, P.G. and Nygård, M.K.: Finite elements with increased freedom in choosing shape functions. Int. J. of Num. Meth. in Engng., 20 (1984) 643-664.

2 Bergan, P.G. and Hanssen, L.: A new approach for deriving "good" finite elements. In J.R. Whiteman (ed.), The Mathematics of Finite Elements and Applications. Vol. II. 483-498, London: Academic Press 1976.

3 Irons, B.M. and Razzaque, A.: Experience with the patch test for convergence of finite elements. In K. Aziz (Ed.), Mathematical Foundations of the Finite Element Method with Application to Partial Differential Equations. 557-587, New York, London: Academic Press 1972.

4 Bergan, P.G. and Felippa, C.A.: A triangular membrane element with rotational degrees of freedom., Comput. Meth. in Appl. Mech. and Engng., 50 (1985) 25-69.

5 Bergan, P.G., Nygård, M.K. and Wang, X.: A class of quadrilateral plate bending elements. In H. Guangquian and Y.K. Cheung (Eds.), Proc. Int. Conference on Finite Element Methods. 112-117, Beijing: Science Press 1982.

6 Bergan, P.G. and Wang, X.: Quadrilateral plate bending elements with shear deformations, Comput. and Struct., 19 (1984) 25-34.

7 Bergan, P.G.: Finite elements based on energy orthogonal functions. Int. J. of Num. Meth. in Engng., 15 (1980) 1541-1555.

8 Nygård. M.K.: The Free Formulation for Nonlinear Finite Elements with Applications to Shells. Dr. Ing. Thesis, To be published, Div. of Struct. Mech., Norwegian Inst. of Technology, Trondheim.

9 Hanssen, L., Syvertsen, T.G. and Bergan, P.G.: Stiffness derivation based on element convergence requirements. In J.R. Whiteman (ed.), The Mathematics of Finite Elements and Applications. Vol. III. 83-96, London: Academic Press 1979.

10 Bergan, P.G.: Lecture Notes in Nonlinear Finite Element Analysis. Trondheim 1981.

11 Malvern, L.E.: Introduction to the Mechanics of a Continuous Medium. Prentice-Hall, Inc. 1969.

12 Argyris, J.H.: An excursion into large rotations. Comput. Meth. in Appl. Mech. and Engng., 32 (1982) 85-155.

13 Rankin, C.C. and Brogan, F.A.: An element-independent corotational procedure for the treatment of large rotations. In L.H. Sobel and K. Thomas (Eds.) Collapse Analysis of Structures. 85-100, New York, ASME 1984.

14 Bergan, P.G., Horrigmoe, G., Kråkeland, B. and Søreide, T.: Solution techniques for nonlinear finite element problems. Int. J. of Num. Meth. in Engng., 12 (1978) 1677-1696.

15 Bergan, P.G.: Automated incremental-iterative solutions schemes. In C. Taylor et al. (Eds.), Proc. of First Int. Conf. on Num. Meth. for Nonlinear Problems. Vol. 1 291-306, Swansea, Pineridge Press 1980.

16 Bergan, P.G. and Mollestad, E.: Static and dynamic solution strategies in nonlinear analysis. In C. Taylor, E. Hinton and D.R.J. Owen (Eds.), Numerical Methods for Nonlinear Problems. Vol. 2, 3-17. Swansea, Pineridge Press 1984.

17 Simons, J.F., Bergan, P.G. and Nygård, M.K.: Hyperplane displacement control methods in nonlinear analysis. In W.K. Liu, T. Belytschko and K.C. Park (Eds.), Proc. Int. Conf. on Innovative Methods for Nonlinear Problems. 345-364, Pineridge Press, 1984.

18 Bergan, P.G. and Arnesen, A.: FENRIS - A general purpose nonlinear finite element program. In Proc. 4th Int. Conf. Finite Element Systems. Southampton 1983.

19 Bergan, P.G.: Nonlinear finite element analysis by use of FENRIS. In J.F. Gouldman et al. (Ed.), Proc. 7th Int. Seminar on Computational Aspects of the F. E. Method. Chicago, CAFEM-7 1983.

20 Arnesen, A., Engseth, A. and Eskesen, S.D.: Nonlinear Finite Element Analysis of offshore Structures. In P.G. Bergan, K.J. Bathe ang W. Wunderlich (Eds.), Finite Element Methods for Nonlinear Problems. Springer Verlag 1985.

21 Zienkiewicz, O.C.: The Finite Element Method. 3rd edition, London, McGraw-Hill 1977.

22 Bergan, P.G. et al.: FENRIS Satellite 1 Manual, Theory - Program Outline - Data Input, A.S. Veritec, Høvik, Norway 1984.

23 Wilson, E.L., Taylor, R.L. Doherty, W.P. and Ghaboussi, J.: Incompatible displacement modes. In S.J. Fenves et al. (Ed.) Numerical and Computer Methods in Structural Mechanics. New York: Academic Press 1973.

24 Stolarski, H., Belytschko, T., Carpenter, N., and Kennedy, J.K.: A simple triangular curved shell element. Eng. Comput., 1 (1984) 210-218.

25 Dvorkin, E.N. and Bathe, K.J.: A continuum mechanics based four-node shell element for general nonlinear analysis. Eng. Comput., 1 (1984) 77-88.

26 Scordelis, A.C. and Lo, K.S.: Computer analysis of cylindrical shells. J. Amer. Concr. Inst. 61 (1969) 539-561.

27 MacNeal, R.H. and Harder, L.: A proposed standard set of problems to test finite element accuracy. Finite Elements in Analysis and Design 1 (1985) 3-20.

28 Kråkeland, B. : Large Displacement Analysis of Shells Considering Elasto-Plastic and Elasto-Viscoplastic Materials. Report No. 77-6, Div. of Struct. Mech., Norwegian Inst. of Technology, Trondheim (1977).

29 Horrigmoe, G. and Bergan, P.G.: Nonlinear analysis of free-form shells by flat shell elements. Comput. Meth. in Appl. Mech. and Engng., 16 (19) 85-155.

30 Oliver, J and Onate, E.: A total Lagrangian formulation for the geometrically nonlinear analysis of structures using finite elements. Part I: Two-dimensional problems: Shell and plate structures. Int. J. of Num. Meth. in Engng., 20 (1984) 2253-2281.

The membrane lumping matrix associated with one node is

$$\mathbf{L}_{mj} = \frac{1}{2}\begin{vmatrix} y_{ki} & 0 & x_{ik} \\ 0 & x_{ik} & y_{ki} \\ \frac{\alpha}{6}(y_{ji}^2 - y_{kj}^2) & \frac{\alpha}{6}(x_{ij}^2 - x_{jk}^2) & \frac{\alpha}{3}(x_{ij}y_{ji} - x_{jk}y_{kj}) \end{vmatrix}$$

The value of α is set equal to 1.5 in the present analysis.
The corresponding lumping matrix for the bending behaviour is

$$\mathbf{L}_{bj} = \frac{1}{2}\begin{vmatrix} 0 & 0 & 0 \\ 0 & x_{ik} & y_{ki} \\ y_{ik} & 0 & x_{ki} \end{vmatrix}$$

Ultimate Load Analysis of Thin Shells Under Pressure Loads

E. RAMM, K. SCHWEIZERHOF, H. STEGMÜLLER
University of Stuttgart, West Germany

Abstract

At first, a proper load definition is given for deformation dependent pressure load allowing to identify when a problem is conservative and when it is not. The second part is concerned with the performance of the entire family of degenerated isoparametric shell elements with linear, quadratic and cubic interpolation schemes. Next the authors' experience with a combination of the arc-length-method with Quasi-Newton and line search schemes is discussed. Finally, several numerical examples are given, among these are stability and ultimate load analyses of liquid filled conical and toroidal shells used for example in sewer plants.

1. Introduction and Scope of the Study

In finite element literature deformation dependent loads such as follower loads are often falsely related to non-conservative problems with nonsymmetric load stiffness matrices. However, for the majority of corresponding problems a potential exists so that they are conservative, for example pressurized shells with proper boundary conditions. Despite this fact some authors use an inconsistent formulation, supposedly simplifying the analysis which may result in a wrong interpretation of the problem /2/. In the present study the conditions are discussed when pressure loading leads to conservative problems and when it does not /7/.

Degenerated isoparametric elements are the most often used elements in linear and nonlinear shell analysis /9/. It is well known that these models may show a rather slow convergence rate due to either shear or membrane locking caused by parasitic stresses in pure bending modes /8/. Many schemes have been proposed to improve the performance /10/, /11/, /12/, among these

Finite Element Methods for Nonlinear Problems
Europe-US Symposium, Trondheim, Norway 1985
Editors: Bergan, Bathe, Wunderlich

the reduced integration is the most popular technique. This, in turn, may lead to internal zero energy modes polluting the overall solution. In this part the authors' experience with linear, quadratic and cubic Lagrangian and serendipity elements using either full or uniform reduced integration is described.

The third part focusses on the combination of the arc-length-method /19/ with Quasi-Newton and/or line search schemes. Because the variables defining the arc-length parameter have different dimensions a scaling procedure is introduced /21/.

Catastrophic failures of liquid filled conical shells indicated that these may buckle in the lower part despite the stabilizing effect of the internal pressure. Unfortunately, very little is known about the interaction of hoop tensions and meridional compression combined with material failure in this region. From an extended parametric investigation /13/ two numerical examples are selected for this study.

2. Deformation Dependent Pressure Loads

The action of a medium like gas or fluid on a structure can be represented by deformation dependent loads. This is certainly a simplification neglecting the real interaction, but it seems to be an appropriate model when the deformation of the structure shows only a minor influence on the overall loading behaviour.

In statics deformation dependence of loading may result from a change in direction, magnitude and/or loaded area (Fig. 1).

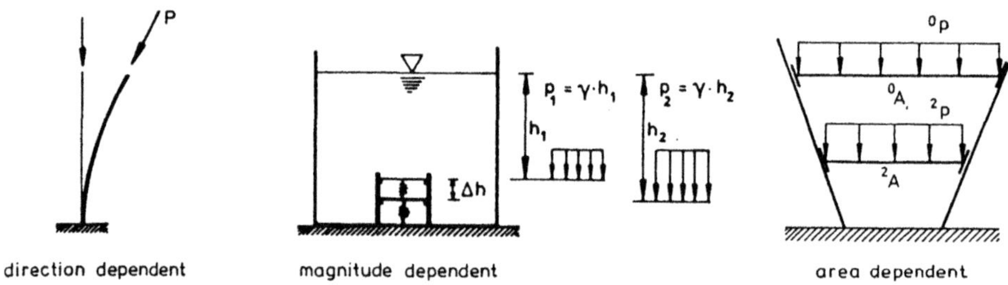

Fig. 1: Displacement dependent loads

In the classification given in Fig. 2 it is assumed that the load depends on the change of area in all three models.

	constant directional load (k = 0, l = 0)	follower normal load (k = m)	
		body attached (l = 0)	space attached (l = m)
	①	②	③
load $^{m}p_i$	$^{m}\lambda \cdot {}^{0}f \cdot {}^{0}n_i$	$^{m}\lambda \cdot {}^{0}f \cdot {}^{m}n_i$	$^{m}\lambda \cdot {}^{m}f \cdot {}^{m}n_i$
load in configuration m	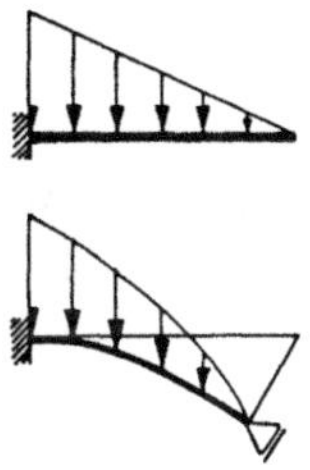	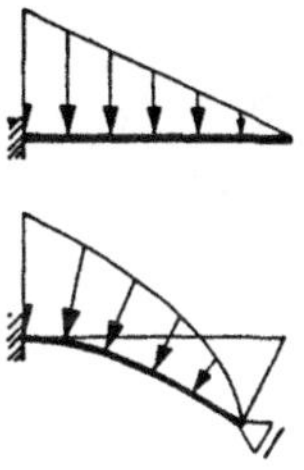	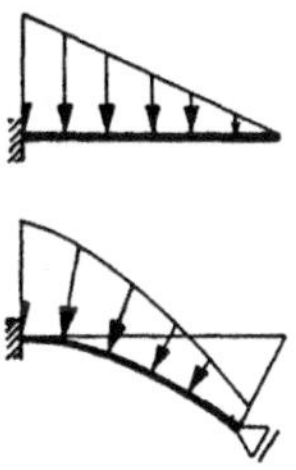

Fig. 2: Load definition

Then the pressure component in the direction i of a deformed state m is defined as:

$$^{m}p_i = {}^{m}\lambda \;\; {}^{l}f \;\; {}^{k}n_i$$

with the load multiplier $^{m}\lambda$, the function $^{l}f = {}^{l}f\ ({}^{l}x_j)$ of the load distribution which may depend on the coordinates of either the initial configuration (l = 0) or the deformed current configurations (l = m). The product $^{m}\lambda \cdot {}^{l}f$ determines the load magnitude at a point of the structure and $^{k}n_i$ is the component i of the surface normal in configuration k = 0 or k = m.

Whereas the classical non-deformation dependent load is shown as case 1 for reference, the discussion of follower force effects has to distinguish between body attached loads (case 2) and space attached loads (case 3). Both are identical for uniform load, but for non-uniform loading case 2 leads to a non-conservative problem, whereas case 3 gives a conservative problem, resulting in an unsymmetric or a symmetric structural "load stiffness" matrix, respectively.

The external virtual work defined for a deformed position 2 is given by:

$$^{2}\delta\, W_{ext} = \int_{^{2}A} {}^{2}p_i\ \delta u_i\ d^{2}a = \int_{^{2}A} {}^{2}\lambda\ {}^{1}f\ {}^{2}n_i\ \delta u_i\ d^{2}a \qquad i = 1,2,3 \tag{2.1}$$

Using curvilinear coordinates r, s the linearized incremental expression from state 1 to state 2 is:

$$ {}^{2}\delta W_{ext} = {}^{2}\delta W^{L}_{ext} + e_{ijk}\, {}^{2}\lambda \iint_{{}^{1}A} {}^{1}f \frac{\partial\, {}^{1}x_{j}}{\partial r} \frac{\partial\, {}^{1}x_{k}}{\partial s} \delta u_{i}\, dr ds \qquad (2.2) $$

see Table 1

with the permutation symbol e_{ijk}.

The integral defines the load vector, whereas the terms in Table 1 represent the load stiffness matrix which is discussed in the following.

	space attached $l = 1$	body attached $l = 0$	matrix notation
domain terms (I)	$-\frac{1}{2} e_{ijk}\, {}^{2}\lambda \iint_{rs} {}^{l}f \left[\frac{\partial\, {}^{1}x_j}{\partial r} \left(u_i \frac{\partial \delta u_k}{\partial s} + \delta u_i \frac{\partial u_k}{\partial s} \right) + \frac{\partial\, {}^{1}x_k}{\partial s} \left(u_i \frac{\partial \delta u_j}{\partial r} + \delta u_i \frac{\partial \delta u_j}{\partial r} \right) \right] dr\, ds$		$\iint_{rs} \delta u^T \overset{D}{K}{}^{I}_{L}\, u\, dr\, ds$
(II)	$-\frac{1}{2} e_{ijk}\, {}^{2}\lambda \iint_{rs} {}^{1}f_{,n} \left(\frac{\partial\, {}^{1}x_n}{\partial r} \cdot \frac{\partial\, {}^{1}x_j}{\partial s} - \frac{\partial\, {}^{1}x_n}{\partial s} \cdot \frac{\partial\, {}^{1}x_j}{\partial r} \right) \cdot (u_k \cdot \delta u_n + u_n \cdot \delta u_k)\, dr\, ds$ $- e_{njk} \cdot {}^{2}\lambda \iint_{rs} {}^{1}f_{,n} \frac{\partial\, {}^{1}x_j}{\partial r} \cdot \frac{\partial\, {}^{1}x_k}{\partial s} \cdot u_n\, \delta u_n\, dr\, ds$	$-\frac{1}{2} e_{ijk} \cdot {}^{2}\lambda \iint_{rs} {}^{0}f_{,n} \left(\frac{\partial\, {}^{0}x_n}{\partial r} \cdot \frac{\partial\, {}^{1}x_j}{\partial s} - \frac{\partial\, {}^{0}x_n}{\partial s} \cdot \frac{\partial\, {}^{1}x_j}{\partial r} \right) u_k\, \delta u_j\, dr\, ds$	$\iint_{rs} \delta u^T \overset{D}{K}{}^{II}_{L}\, u\, dr\, ds$
boundary terms	$-\frac{1}{2} e_{ijk}\, {}^{2}\lambda \left[\int_b {}^{l}f\, u_j \frac{\partial\, {}^{1}x_k}{\partial s} \delta u_i\, ds - \int_b {}^{l}f\, u_j \frac{\partial\, {}^{1}x_k}{\partial r} \delta u_i\, dr \right]$		$\int_b \delta u^T \overset{D}{K}{}^{III}_{L}\, u\, db$

Table 1: Load stiffness for deformation dependent pressure loads

By interchanging the indices it can be easily verified that the matrices $\overset{D}{\mathbf{K}}_{L}$ are symmetric except the second term for non-uniform body attached load and the boundary terms for either kind of loading. These are skew-symmetric /7/.

If the <u>boundary term</u> is written in a vector notation

$$ {}^{2}\delta W^{L,b}_{ext} = \frac{1}{2}\, {}^{2}\lambda \int_{b} {}^{1}f\, (\vec{u} \times \delta \vec{u})\, {}^{1}\vec{t}\, db \qquad (2.3) $$

the conditions for symmetry - vanishing boundary terms - are easily obtained.

1) The boundary is not loaded in all deformed configurations ($^{1}f = 0$); or
2) at least two displacement components are prescribed on the boundary; or
3) the u, δu and t vectors are located in the same plane; or
4) the total sum of the boundary terms is zero.

The physical meaning of condition 1 - 3 is rather obvious. It requires a "closed" system in which energy is preserved. Whereas condition 2 or 3 is satisfied for most boundary conditions except the free boundary; in this case condition 1 might become necessary. Condition 4 is only randomly satisfied. For body attached loads symmetry can only be achieved for uniform load distribution which, in turn, requires that boundary condition 2 or 3 or 4 is satisfied. Condition 1 cannot be satisfied in this case.

Discussion

The symmetry of the stiffness matrix is a formal proof for the existence of a potential and is in clear coincidence with physical reasonings. This is also confirmed by the fundamental papers of Pearson /4/ and Romano /5/ where the conditions for the existence of a potential are given for the case of uniform pressure. Following the ideas in /6/, Bufler /1/ extended the classical proofs to non-uniform pressure.

Very often deformation dependent loads are automatically classified as non-conservative. Obviously this is misleading because the definition of conservativeness is connected to the entire problem, looking at energy conservation.

In many papers a supposedly simplified load definition is assumed neglecting the dependence of the load on the displacements and only taking into account the deformation gradients or using a wrong linearization. A non-conservativeness is then concluded for problems which are clearly conservative in reality /2/.

Finite Element Formulation

The finite element formulation for conservative problems does require only the domain terms of equation (2.2) for an element. The boundary terms of each element cancel each other during the assembly process of the global stiffness matrix. For further details is referred to /7/. The storage schemes and equations solvers of the conventional F.E. programs can be used. This is not possible for non-conservative problems where the unsymmetry remains after the assemblage and unsymmetric matrices have to be stored and factorized. In the general case twice the storage and a general equation solver are necessary.

For a nonlinear analysis it is often not necessary to use unsymmetric stiffness matrices during the iterations either because of symmetrization /3/ or by simply neglecting the load stiffness terms and the deformation dependent load is taken into account only on the right hand side via the load vector.

3. Performance of Degenerated Shell Elements

Isoparametric degenerated C_o - elements have been successfully applied in linear and nonlinear plate and shell analyses /8/ - /12/. However, using full integration schemes, the displacement models may perform rather poorly. Mainly there are two reasons:

* Shear locking: due to excessive shear stresses in bending, in particular for thin structures with a large element length to thickness ratio (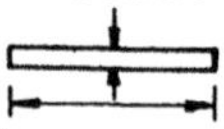).
* Membrane locking in curved structures: due to membrane stresses in pure inextensional modes depending on the curvature ratio of the element (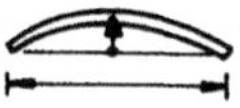), in particular for low order elements.

The locking attributes are not restricted to degenerated elements but are likewise typical for similar "classical" shell elements. Both phenomena lead to parasitic stresses because of non-vanishing transverse shear and membrane energy, respectively, in pure bending modes. Despite a rather slow convergence rate the solution ultimately converges to the exact solution when the mesh size is refined. Locking diminishes when the interpolation order is increased. The problems with locking are sometimes falsely attributed to inadequate representation of rigid body modes. These are properly included in the formulation.

The shear locking has been extensively investigated in the literature mainly for plate bending. This explains why the membrane locking of low order elements known from "classical" models is not discussed so often for degenerated elements. For curved shells exhibiting inextensional deformations it is an important characteristic.

Many techniques have been reported in order to avoid either kind of locking resulting in a better convergence rate. All methods have in common to suppress the parasitic shear and membrane energy terms. Among these are the use of additional constraint equations /11/, /12/ or stress projection methods /10/ and - most popular - uniform and selective reduced integration schemes. It need to be stressed that in shell analysis the quadrature of transverse shear <u>and</u> membrane energy has to be reduced. It is well known that reduced integrated elements may lead to internal kinematics or zero-energy modes (ZEM) on the element but sometimes also on the structural level after assembly. For some reduced integrated elements a relationship to mixed models can be pointed out /14/.

The following statements are based on the authors' experiences with linear, quadratic and cubic element versions, either with full or with uniform reduced integration /13/. The selective integrated elements are deliberately omitted, since in view of computational efficiency the energy split does not seem to pay off.

integration	S 4 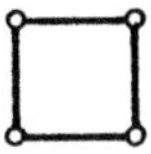	S8 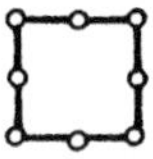	S9 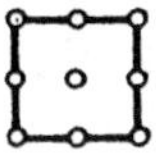	S 12 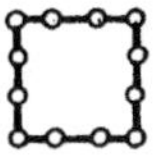	S 16 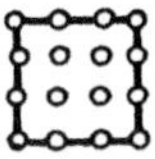
exact	2 × 2	3 × 3	3 × 3	4 × 4	4 × 4
reduced(R)	1	2 × 2	2 × 2	3 × 3	3 × 3

Table 2: Quadrature for degenerated plate/shell elements

Corresponding triangular element versions also available are not discussed herein.

As a benchmark the linear problem of a pinched cylinder with diaphragms on both sides is studied. It includes inextensional bending modes and complex membrane states (Fig. 3). All fully integrated elements including the cubic version show more or less locking but finally converge. All reduced integrated models, including too the linear S 4 R element, also converge but with a much better performance for the bending deformation. Except the 8 node element S 8 R all reduced versions show a kind of zero-energy mode for the membrane deformation. However, the average axial deformation is also represented by these models.

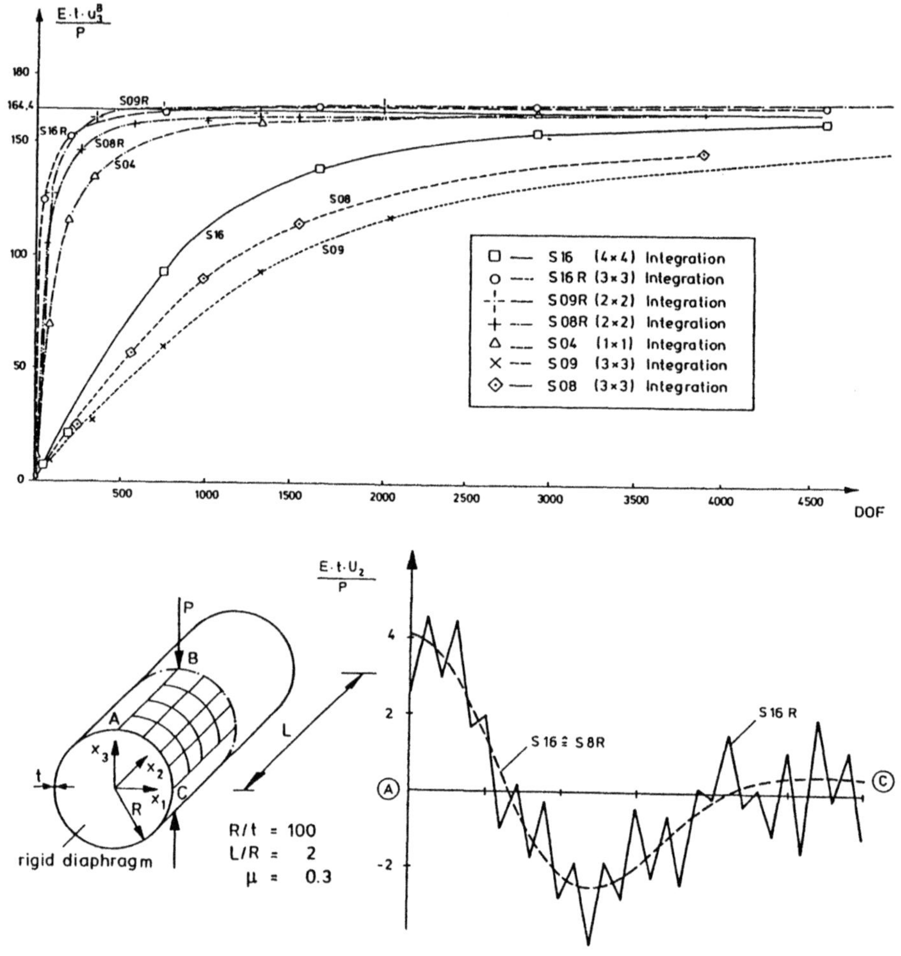

Fig. 3: Pinched cylinder - convergence study and axial deformation plot

The overall experiences with the elements in Table 2 are summarized as follows:

S 16: shows little locking, a reliable, robust element which is preferred if very little is known about the structural response and if the solution of other elements has to be checked.

S 16 R: shows unpredictable ZEM (6 on element level), poor stresses if distorted.

S 12/S 12 R: not recommended because of considerable shear locking.

S 8/S 9: because of considerable shear and membrane locking need reduced integration.

S 9 R: shows unpredictable ZEM (7 on element level).

S 8 R: ZEM on element level (2) disappear after assembly in nearly all cases. If structure has many displacement constraints some shear locking present; in essence good performance.

S 4 R: extremely efficient if ZEM (6 on element level) do not show up.

If a program has several element models it is recommended to compare results from different element versions, for example to check for ZEM. In particular, this is useful if the topology of the net has not to be changed, i.e. for the 8 and 9 node element or 16 and 4 node element. It is found that the solution with 4 node elements can easily be checked by the 16 node element in nonlinear analysis at certain load levels.

4. Numerical Solution Procedures

Reestablishment and factorization of the stiffness matrix during iteration can be avoided by resorting to Quasi-Newton schemes (QN), e.g. /15/, /17/ or acceleration procedures as described in /16/. The load deformation behaviour can be traced into the limit and postbuckling range using additional contraints like the displacement control method or the constant-arc-length method (CALM) with an iteration on a sphere /19/ or on tangent planes /18/. The stiffness equation for iteration (j) in step (i) is:

$$ {}^{i}\mathbf{K}^{(j)}\ {}^{i}\Delta\mathbf{u}^{(j+1)} = {}^{i}\mathbf{P}^{(j)} - {}^{i}\mathbf{F}^{(j)} = {}^{i}\mathbf{R}^{(j)} \qquad (4.1) $$

with the current stiffness matrix ${}^{i}\mathbf{K}^{(j)}$, the displacement increments ${}^{i}\Delta\mathbf{u}^{(j+1)}$, the out-of-balance forces ${}^{i}\mathbf{R}^{(j)}$, external forces ${}^{i}\mathbf{P}^{(j)} = {}^{i}\lambda^{(j)}\ \mathbf{P}$ and internal forces ${}^{i}\mathbf{F}^{(j)}$. The two-step procedure of the CALM /18/:

1) first iterate in step i

$$ \Delta{}^{i}\mathbf{u}^{(1)T}\ \Delta{}^{i}\mathbf{u}^{(1)} + {}^{i}\Delta\lambda^{(1)2} = ds_i^2 \qquad (4.2) $$

scaling the arc-length ds_i to a given length ds_o

2) iteration on tangent plane

$$ {}^{i}\Delta\mathbf{u}^{(1)T}\ {}^{i}\Delta\mathbf{u}^{(j+1)} + {}^{i}\Delta\lambda^{(1)}\ {}^{i}\Delta\lambda^{(j+1)} = 0 \tag{4.3} $$

with $$ {}^{i}\Delta\mathbf{u}^{(j+1)} = \Delta{}^{i}\lambda^{(j+1)}\ {}^{i}\Delta\mathbf{u}^{(j+1)I} + {}^{1}\Delta\mathbf{u}^{(j+1)II} \tag{4.4} $$

$$ {}^{i}\Delta\mathbf{u}^{(j+1)I} = ({}^{i}\mathbf{K}^{(j)})^{-1}\ \mathbf{P} \tag{4.4 a} $$

$$ {}^{i}\Delta\mathbf{u}^{(j+1)II} = ({}^{i}\mathbf{K}^{(j)})^{-1}\ {}^{i}\mathbf{R}^{(j)} \tag{4.4 b} $$

The incremental load factor is:

$$ \Delta{}^{i}\lambda^{(j+1)} = - \frac{\Delta\mathbf{u}^{(1)T}\ {}^{i}\Delta\mathbf{u}^{(j+1)II}}{\Delta\mathbf{u}^{(1)T}\ {}^{i}\Delta\mathbf{u}^{(j+1)I} + {}^{i}\Delta\lambda^{(1)}} \tag{4.5} $$

If the tangent plane is updated from iterate to iterate, ${}^{i}\Delta\mathbf{u}^{(1)}$ has to be replaced by the total incremental displacements ${}^{i}\Delta\mathbf{u}^{(j)}$. The combination of QN and CALM using an updated inverse $({}^{i}\mathbf{K}^{(j)})^{-1}$ in equations (4.4 a/b) may improve the efficiency of the procedure /19/, /20/.

A line search (LS) can be performed in each iteration with a line search direction lying in the actual iteration plane.

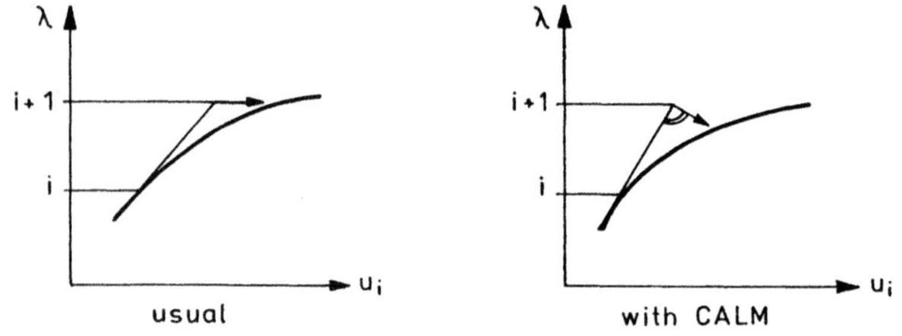

Fig. 4: Line search direction

To make the method independent from the measure quantities used in the analysis scaling factors for the displacement, rotation and load terms can be introduced /21/. If independent quantities are taken for every degree of freedom and the load term, also local changes and possible failure can be taken into account. Written in matrix notation the scaling factors are forming a diagonal matrix $\mathbf{V}_s$ and the proportional load factor is determined via:

$$\mathbf{V}_s = \begin{bmatrix} v_{s1} & & \\ & v_{s2} & \\ & & \ddots v_{sn} \end{bmatrix} \; ; \quad \Delta^i\lambda^{(j+1)} = - \frac{{}^i\Delta\mathbf{u}^{(1)T}\,\mathbf{V}_s\,\mathbf{V}_s\,\Delta^i\mathbf{u}^{(j+1)II}}{{}^i\Delta\mathbf{u}^{(1)T}\,\mathbf{V}_s\,\mathbf{V}_s\,\Delta^i\mathbf{u}^{(j+1)I} + \Delta^i\lambda^{(1)}/S_L} \tag{4.6}$$

The latter scheme improves the ability of the CALM to adapt the step size according to the nonlinearity of the system.

The advantage of the Quasi-Newton methods stems from the fact that within the solution process the inverse is never updated. Instead the update vectors of the low rank updates are stored for the iterations and then directly used in the solution process. All popular methods like BFGS, DFP, Davidon's and Broyden's method can be cast into this so-called "vectorized" form. The nonsymmetry of Broyden's update is not entering the algorithm. The combination with the CALM does not require additional forward/backward substitutions to solve equation (4.4 a); but note that the form of the BFGS scheme, given in /15/, has to be modified herefore. The numerical advantage of the Quasi-Newton over the modified Newton scheme, both combined with the CALM, was less pronounced than in the pure versions, at least for the geometrically nonlinear examples tested in this study (see Table 3 for the conical shell defined in section 5.2).

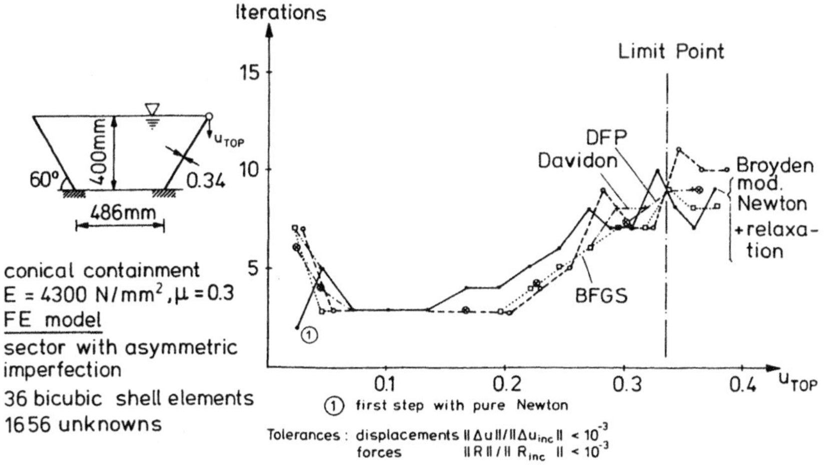

Table 3: Combination of Quasi-Newton with constant-arc-length methods

The higher rank updates BFGS and DFP were not superior to the low rank updates - Broyden and Davidon - which require less storage and less numerical operations. This confirms the statements made in /17/.

Combinations with line searches did not improve the convergence rate of the schemes tested on geometrically nonlinear problems in the prebuckling range. This is certainly different in the case of material nonlinearities like reinforced concrete structures.

5. Numerical Examples and Application

The above mentioned elements are incorporated in the computer code NISA 80 /22/ which was recently extended by the described algorithms. The elements are formulated for arbitrarily large displacements and rotations. Through a layered approach elasto - plasticity and a RC - material model are included. Except in example 1 no follower force effects are considered.

5.1 Cylindrical Shell under Wind Load

The clamped/free shell shown in Fig. 5 together with the wind load definition was experimentally investigated in /27/. The

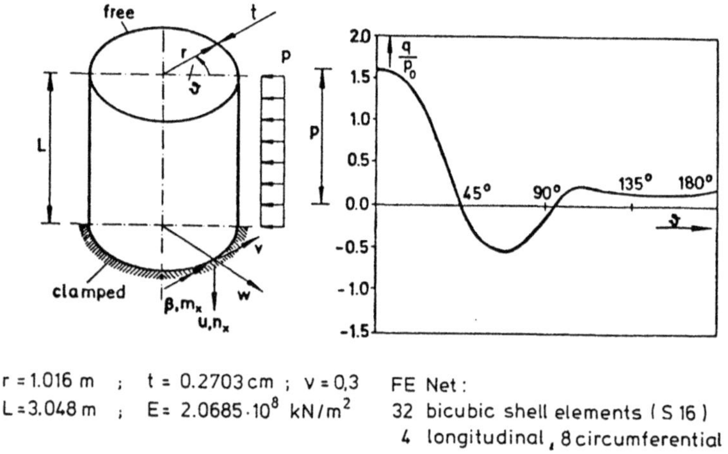

Fig. 5: Cantilever shell - geometry and load

open cantilever shell is analysed by a nonlinear elastic analysis assuming either constant directional or follower load. The buckling load found from the F.E. analysis with follower forces correlates excellently to the experimental results in /25/ (Fig.6).

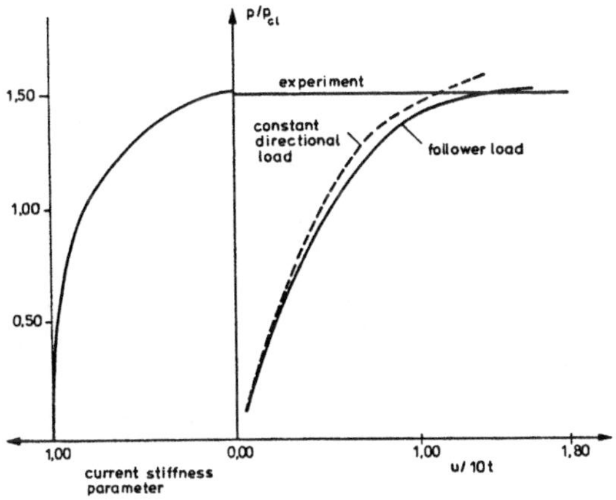

Fig. 6: Load deflection diagram

The final difference of the failure load between either kind of loading amounts to 4.4 %. Despite the low number of buckling waves in the failure mode (Fig. 7) the difference is negligible for practical purposes. With the warping restraint of the lower boundary condition relaxed this difference grows and follower forces are more important /6/.

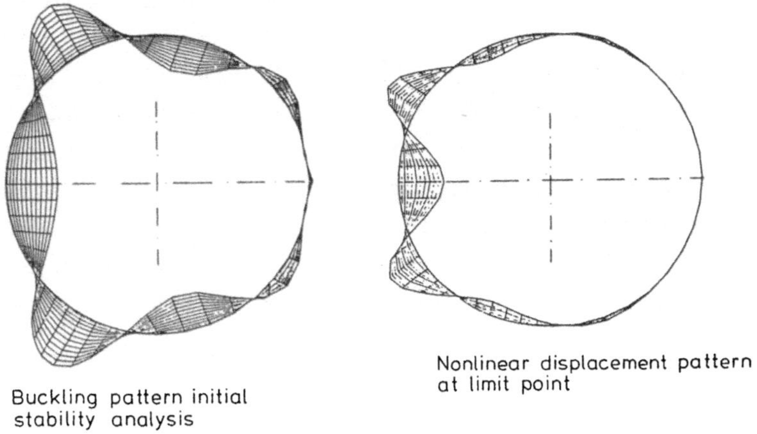

Fig. 7: Displacement and buckling patterns - top view

5.2 Liquid Filled Elastic Conical Shell

Conical or cone like structures are used as water tanks or bio-reactors. Due to high compression the thin shells tend to buckle near the lower boundary. The influence of the liquid pressure, leading to tension in hoop direction, is a matter of current experimental and analytical research. The elastic model shell shown in Fig. 8 was tested in /23/ and buckled at a water level of

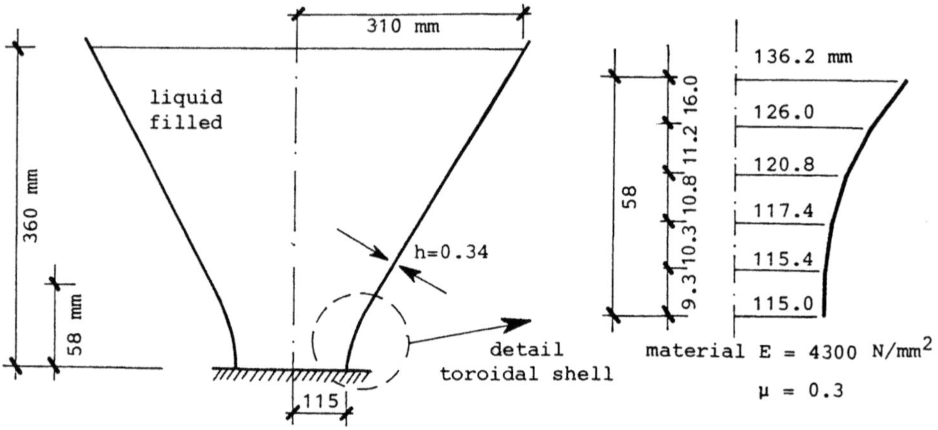

Fig. 8: Liquid filled conical shell - geometry

360 mm. In the analysis the specific weight g is taken as load factor for the above mentioned water level. A linear initial buckling analysis, based on a sector corresponding to 13 waves in circumference, leads to a bifurcation load g = 1,03. In order to study the influence of the internal pressure two load cases are investigated in nonlinear analysis: the real load by the liquid and a fictitious tangential load only, the latter leading to the same meridional forces as the real load but with the normal pressure removed. Note that in the second case the hoop tension stresses in the cone vanish but the compression in the torus part is increased (Fig. 9). In both studies initial geometrical

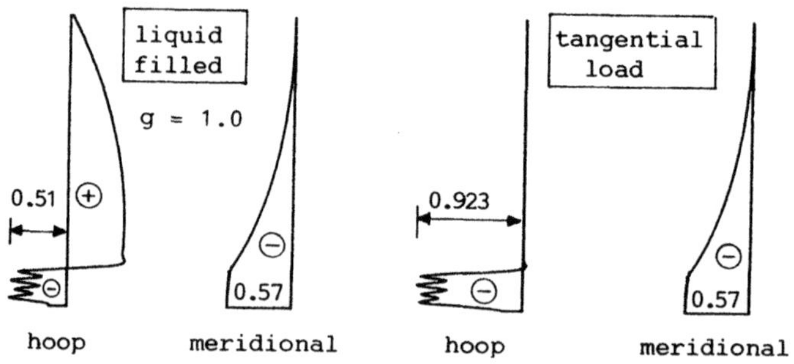

Fig. 9: Membrane forces in conical shell (N/mm)

imperfections corresponding to the buckling mode (Fig. 10, right) with a maximum amplitude equal to the thickness is chosen. In Fig. 10 the load (max. n_x) displacement diagram of the nonlinear response is given together with the initial buckling values. The considerable stabilizing influence of the internal pressure can be seen. Furthermore, the experimental failure load is amazingly good represented by the analysis. The structure buckles in the toroidal part which is known to be not imperfection sensitive.

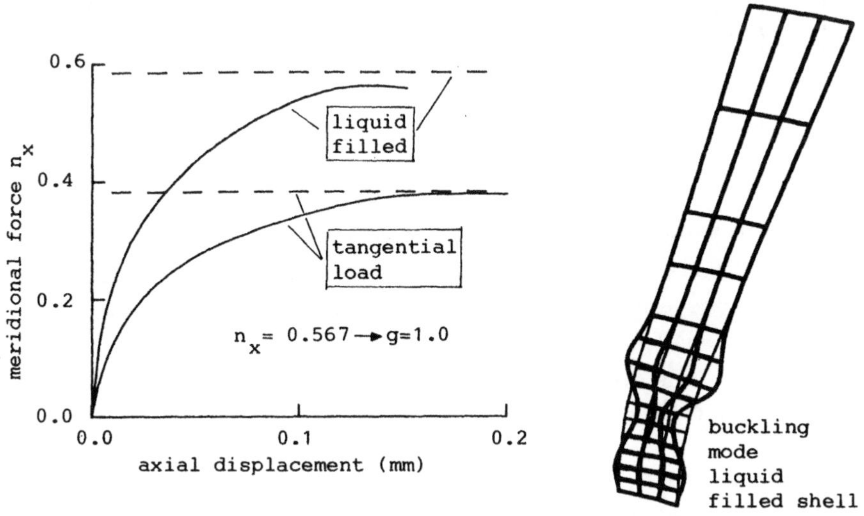

Fig. 10: Force - deflection curve and buckling mode - conical shell

5.3 Pressurized Containment

This structure is a model shell /24/ out of mild steel representing a typical bioreactor. The system consists of three parts: a cone, a torus made out of three conical frustra and a cylinder which is clamped at the base. As in 5.2, the influence of internal pressure on the failure load is investigated. In the experiment the shell is loaded by uniform internal pressure first and then additional axial forces are applied at the upper boundary. In the analysis a sector of 15° with symmetry boundary conditions is modelled (Fig. 11) which corresponds to one half of a buckling wave.

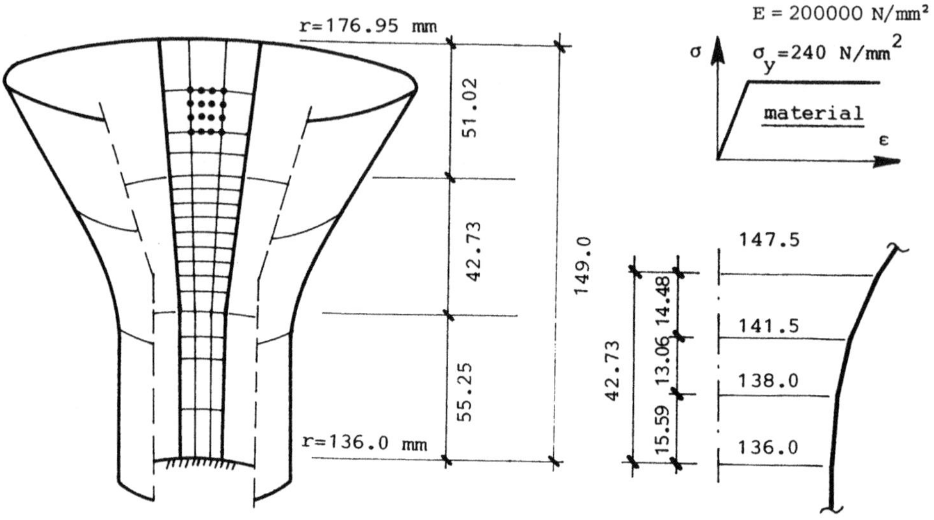

Fig. 11: Pressurized containment - geometry and F.E. model

In the first elastic analysis only axial forces are applied. The linear initial buckling load is P = 32.9 kN. In Figure 12 the elastic nonlinear response for the perfect and imperfect shell is given together with the symmetric displacement and unsymmetric

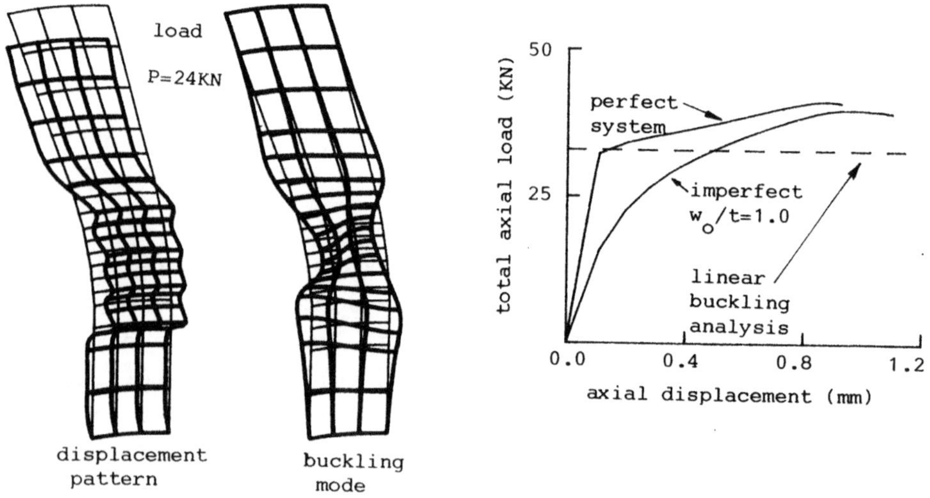

Fig. 12: Elastic nonlinear response under axial pressure

buckling pattern. The failure is located in the toroidal part. The perfect shell shows a stable postbuckling response and the imperfection sensitivity is low. In the second study an elasto-plastic material model is introduced (Fig. 13). A considerable

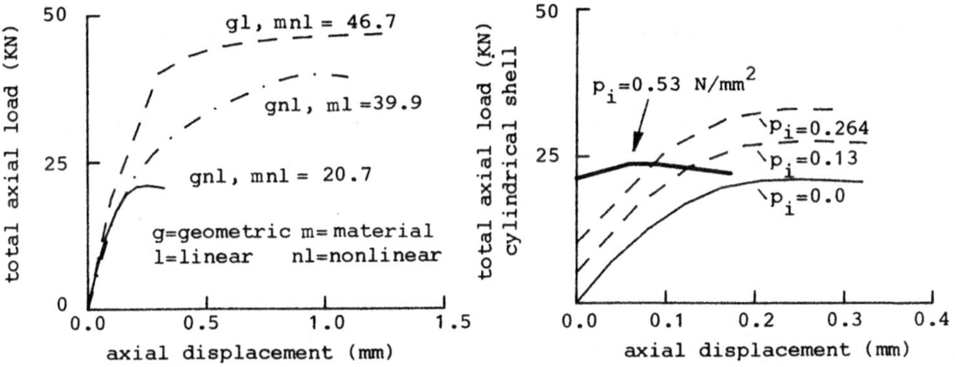

Fig. 13: Elasto - plastic analysis -
axial pressure - axial pressure and internal pressure

reduction of the failure load occurs when both nonlinearities interact. This is typical for structures with intermediate slenderness. In the third part the influence of additional internal pressure is investigated (Fig. 13, right). The pressure also reduces the axial force acting on the structure. In the diagram the total force in the cylindrical shell is plotted versus the axial displacement. The figure shows first an increase in the ultimate load, i.e. a stabilizing effect of the internal pressure. If the pressure is further increased to p = 0,53 N/mm² the structure is already near collapse under internal pressure alone. Therefore, only a small additional axial force can be carried by the shell.

6. Conclusions

The authors' experience in the nonlinear analysis of pressurized shells using degenerated elements can be summarized as follows:

* It is necessary to distinguish between body and space attached loads for deformation dependent loading; this clearly allows the definition of conservative and non-conservative problems.

* In most buckling problems of shells the deformation dependence of the pressure load can be neglected.

* Fully and reduced integrated shell elements have been classified. The most robust element is the fully integrated 16 node version. It is strongly recommended to compare results obtained from different elements based on the same mesh topology.

* The combination of Quasi-Newton with arc-length procedures does improve the convergence rate only slightly. Rank one updates should be preferred.

References

/1/ Bufler, H.: Pressure Loaded Structures under Large Deformations. ZAMM 64 (1984) 287-295.

/2/ Frey, F.; Cescotto, S.: Some New Aspects of the Incremental Total Lagrangian Description in Nonlinear Analysis. Int. Conf. 'Finite Elements in Nonlinear Mechanics', Vol. 1, CO. 5.1/5.20, Geilo, Norway, 1977.

/3/ Mang, H.A.: Symmetricability of Pressure Stiffness Matrices for Shells with Loaded Free Edges. Int. J. Num. Meth. Eng. 15 (1980) 981-990.

/4/ Pearson, G.E.: General Theory of Elastic Stability. Quart. Appl. Math. 14 (1956) 133-144.

/5/ Romano, G.: Potential Operators and Conservative Systems. Meccanica 7 (1972) 141-146.

/6/ Schweizerhof, K.: Nichtlineare Berechnung von Tragwerken unter verformungsabhängiger Belastung mit finiten Elementen. Dissertation, Universität Stuttgart, 1982.

/7/ Schweizerhof, K.; Ramm, E.: Displacement Dependent Pressure Loads in Nonlinear Finite Element Analyses. Computers & Structures, Vol. 18 (1984) 1099-1114.

/8/ Ramm, E.; Stegmüller, H.: The Displacement Finite Element Method in Nonlinear Buckling Analysis of Shells. In: 'Buckling of Shells' (ed. E. Ramm), Springer Verlag, 1982.

/9/ Ramm, E.: A Plate/Shell Element for Large Deflections and Rotations. Proc. U.S. - Germany Symp., MIT Press, 1977.

/10/ Belytschko, T.; Stolarski, H.; Liu, W.K.; Carpenter, N.; Ong, J. S.-J.: Stress Projection for Membrane and Shear Locking in Shell Finite Elements.
To appear in: Comp. Meth. Appl. Mech. Eng. (1985).

/11/ Li Xi-Kui; Guo-Qiang Liu; Owen, D.R.J.: Geometrically Nonlinear Analysis of Thin Plates and Shells Using a Generalized Displacement Method. Eng. Comput. 1 (1984) 318-323.

/12/ Dvorkin, E.N.; Bathe, K.J.: A Continuum Mechanics Based on Four-node Shell Element for General Non-linear Analysis. Eng. Comput. 1 (1984) 77-78.

/13/ Stegmüller, H.: Grenzlastberechnungen flüssigkeitsgefüllter Schalen mit degenerierten Schalenelementen. Dissertation, Universität Stuttgart, 1985.

/14/ Shimodaira, H.: Equivalence between Mixed Models and Displacement Models Using Reduced Integration. Int. J. Num. Meth. Eng. 21 (1985) 89-104.

/15/ Matthies, H.; Strang, G.: The Solution of Nonlinear Finite Element Equations. Int. J. Num. Meth. Eng. 14 (1979) 1613-1626.

/16/ Crisfield, M.A.: Accelerating and Damping the Modified Newton-Raphson Method. Comp. Struct. 18 (1984) 395-407.

/17/ Geradin, M.; Idelsohn, S.; Hogge, M.: Computational Strategies for the Solution of Large Nonlinear Problems via Quasi-Newton Methods. Comp. Structures 13 (1981) 73-81.

/18/ Ramm, E.: Strategies for Tracing Nonlinear Response Near Limit Points. Europe - U.S. Workshop 'Nonlinear Finite Element Analysis in Structural Mechanics', Bochum, 1980.

/19/ Crisfield, M.A.: A Fast Incremental/Iterative Solution Procedure that Handles Snap-through. Comp. Structures 13 (1981) 55-62.

/20/ Zhu Runxiang; Parisch, H.: A Study of the Different Solution Methods for Nonlinear Problems. Report No. 304, Institut für Statik und Dynamik der Luft- und Raumfahrtkonstruktionen, Universität Stuttgart, 1982.

/21/ Simo, J.; Wriggers, P.; Schweizerhof, K.; Taylor, R.L.: Finite Deformation Postbuckling Analysis Involving Inelasticity and Contact Constraints. In: 'Innovative Methods for Nonlinear Problems', eds. W.K. Liu, T. Belytschko, K.C. Pork, Pineridge Press, Swansea, 1984.

/22/ Stegmüller, H.; Ramm, E.; Sättele, J.M.; Häfner, L.: Theoretische Grundlagen des Programmsystems NISA 80. Mitteilung Nr. 1, Inst. für Baustatik, Universität Stuttgart, 1983.

/23/ Hey, H.; Grein, W.: Stabilitätsversuche an Kegel- und Torusmodellen. Stahlbau 11 (1984) 338-344.

/24/ Schulz, U.: Das Stabilitätsverhalten von torusförmigen Schalen. Stahlbau 11 (1983) 321-328.

/25/ Prabhu, S.K.; Gopalacharyulu, S.; Johns, D.J.: Stability of Cantilever Shells under Wind Loads. J. Eng. Mech. Div. Proc. ASCE 101 (1975) 517/530

Computational Procedures for Postbuckling of Composite Shells

G. M. STANLEY, C. A. FELIPPA
Lockheed Palo Alto Research Laboratory, USA

SUMMARY

A recently developed finite-element capability for general nonlinear shell analysis, featuring the use of three-dimensional constitutive equations within an efficient resultant-oriented framework, is employed to simulate the postbuckling response of an axially-compressed composite cylindrical panel with a circular cutout. The problem is a generic example of modern composite aircraft components for which postbuckling strength (i.e., fail-safety) is desired in the presence of local discontinuities such as holes and cracked stiffeners. While the computational software does a reasonable job at predicting both the buckling load and the qualitative aspects of postbuckling (compared both with experiment and another code) there are some discrepancies due to: (1) uncertainies in the nominal layer material properties, (2) structural sensitivity to initial imperfections, and (3) the neglect of dynamic and local material delamination effects in the numerical model. Corresponding refinements are suggested for the realistic continuation of this type of analysis.

§1. INTRODUCTION

Advanced composite materials, due to their superior strength-to-weight ratios and stiffness tailorability, have become key ingredients in the design of modern aerospace vehicles. However, the complex structural response associated with such materials coupled with the intricacy of their fabrication create harsh requirements for numerical simulation.

Specifically, a problem that is of current interest to NASA/LaRC is the determination of the post-buckling strength of thin laminated composite shells comprising the "skin" of stiffened air-transport fuselages [14]. These shells are required to maintain safe load-carrying capability substantially beyond the point at which skin buckling (*i.e.*, *wrinkling*) occurs. To complicate matters, aircraft fuselages typically feature local discontinuites, such as fasteners, stiffeners and *cutouts*, which can induce high local stress gradients that tend to *delaminate* the composite material. In the presence of buckling, these local delaminations can propagate throughout — and hence fail — a composite structure.

It is NASA's ultimate goal to be able to predict such phenomena analytically. To this end, they have asked us to match some experimental data obtained for a

Finite Element Methods for Nonlinear Problems
Europe-US Symposium, Trondheim, Norway 1985
Editors: Bergan, Bathe, Wunderlich

representative test specimen [14].* We have completed the first phase of the *global* analysis, which is described in the following sections. It employs a recently developed finite-element shell analysis capability, featuring the use of three-dimensional constitutive equations within an efficient resultant-oriented framework, to determine the buckling load and explore the postbuckling regime of an elastically in-tact composite model. The more ambitious *global/local* analysis, which involves the prediction of local material delamination and its interaction with global structural response to determine *postbuckling strength*, is still in the planning phase.

§2. PROBLEM DESCRIPTION

§2.1 Setup

The focal problem mentioned above is depicted in Figure 1. Shown is a moderately deep (55.6 deg.) cylindrical panel with a circular cutout. The panel has a 14 in. square planform, a 15 in. radius of curvature and a radius-to-thickness ratio of 150. The hole is centrally located and measures 2 in. in diameter. The composite shell wall consists of 16 layers of unidirectional graphite fibers in epoxy-resin. Each layer is .0056 in. thick (for a total of .086 in.) and the layers are arranged in the symmetric, *quasi-isotropic*** stacking sequence: $\{\pm 45/90/0/0/90/\mp 45\}$ degrees — repeated twice. The orthotropic-elastic material properties for each layer are listed in the figure. Note that these properties represent *nominal* values and will require some adjustment in the sequel (see §5).

Surrounded by a metallic test frame, the appropriate boundary conditions for the cylindrical panel are (i) fully clamped on the bottom edge, (ii) clamped except for axial motion on the top edge and (iii) simply-supported on the vertical edges.

§2.2 Experimental Results

The test conducted by NASA consisted of statically imposing a uniform end-shortening, δ, to determine the load-carrying capability of the panel beyond the initial buckling load. Experimental results are shown in Figure 2.

Figure 2a represents a normalized "load versus end-shortening" curve. Note that buckling occurs abruptly, followed by a rapid drop in the axial load. Somewhere

---

* This problem was first suggested by Dr. Norman F. Knight, Jr. at NASA Langley Research Center. We amusingly refer to it as "Knight's problem" both as an acknowledgement to the originator and as a reminder of the many pitfalls obstructing its numerical solution.

** The term *quasi-istropic* refers to the fact that the resultant constitutive matrix is essentially isotropic, due to a balanced sequence of fiber angles through the thickness. However, in contrast to truly isotropic materials, there is some additional coupling between bending and twisting deformations.

between the top and bottom of this vertical branch, which spans a period of *milli-seconds* in the experiment, delamination occurs near the hole (Fig. 2b). Although the delamination gets progressively worse and eventually distorts the hole, there is a secondary (postbuckling) stiffening branch in the load/displacement curve. The test was stopped at the point on this secondary branch labeled "collapse", at which point extensive delamination is evident (see [14] for details).

§3. COMPUTATIONAL APPROACH

§3.1 Formulation

The formulation of the governing equations and associated solution algorithms used to analyze the above problem is outlined in Figure 3. A detailed description of this approach may be found in [20]. Briefly, it employs *continuum-based* (CB) shell elements, similar to the Ahmad element [1], but extended to the nonlinear regime and reduced to an economical *resultant* form in which element stiffness and force operators are *pre-integrated* through the thickness.

Shell Equations

Thickness pre-integration of the CB shell equations is achieved by augmenting the standard, Mindlin-type [15] hypotheses of *straight normals* and *zero normal stress* with two additional hypotheses, namely: *small transverse-shear strains* and *mild taper*. As argued in [20], these additional hypotheses do not significantly alter the range of applicability of the original formulation. The resulting theory, which is expressed in terms of stress-resultants rather than pointwise (continuum) stresses, is referred to as a *continuum-based resultant* (CBR) shell formulation.

Note that unlike earlier efforts to pre-integrate the CB shell equations, which typically assume a constant through-thickness variation of the surface metric (*e.g.*, [17]), the present CBR formulation bypasses this assumption and hence is not restricted to very thin shells.

Shell Elements

To spatially discretize the CBR shell equations, a variety of shell finite-elements have been implemented within the above framework. However, on the basis of the numerical evaluation conducted in [20], only the following two shell elements were considered for the present analysis: (i) the nine-node Heterosis (HET) element [10], and (ii) a new nine-node assumed natural strain (ANS) shell-element [18]. While both elements are parabolically curved (Fig. 3) and use standard isoparametric interpolation as a starting point; each departs from the basic recipe in order to properly represent inextensional bending deformation for thin shells.

In particular, the 9_HET element *selectively under-integrates* all stiffness and force terms involving membrane strains to avoid membrane "locking", and uses a mixture of Lagrange shape functions (for rotations) and Serendipity shape functions (for translations) to avoid "spurious modes" otherwise evoked by reduced integration.

In contrast, the 9_ANS element assumes an appropriate (inextensionally accurate) strain field from the outset, using a modified set of Lagrange shape functions and employing full numerical integration throughout. Due to the fact that the strains are assumed in the generally *non-orthogonal* isoparametric coordinate basis, an apparent advantage of the ANS approach is its decreased sensitivity to element mesh distortion.

Nonlinear Solution Procedures

After performing an *updated-Lagrange* linearization of the equilibrium equations, a modified Newton-Raphson (NR) version of the Riks-Crisfield (RC) arc-length control algorithm [4] is used to trace incrementally the load-displacement curve. The RC procedure was adopted as a convenient means of statically traversing the bifurcation points that arise in shell postbuckling analysis. However, such methods are not foolproof, and special attention by the analyst (*e.g.*, in the selection of imperfections, step-sizes and error tolerances) is often required in the vicinity of closely-spaced or *multiple* bifurcation points.

Two kinds of update procedure are required to advance the solution of the discrete nonlinear shell equations from one load-step to the next: (i) a *kinematic* update that "accumulates" incremental *nodal displacement* — both translational and rotational components; and (ii) a *constitutive* update that generates new *element stresses* from the corresponding displacement field.

Only the *rotational* part of the kinematic update is non-trivial.* In the present approach, the rotation increments are used to update *orthogonal triads* defined at each shell node. The triad-update algorithm involves no trigonometric functions, maintains orthogonality at each step and provides a shell-oriented coordinate system in which the *normal rotational* degree-of-freedom may be eliminated at all shell nodes (except at junctures). Furthermore, once the nodal triads and reference-surface coordinates have been so updated, the *current* element configuration is completely and uniquely defined (see [20] and [19] for details).

The *stress update* is handled via an incrementally objective algorithm [11] that

* Since the translational components of displacement are *vectorial*, translation increments are simply *added* to obtain total displacements and hence update the nodal coordinates.

features a midpoint-rule numerical integration of rate-type constitutive equations. For finite-strain analysis, the constitutive algorithm additionally involves shell ***thickness updates*** that account for large Poisson effects. These are computed (as in [13]) by recovering the normal strain increments from the consitutive equations via the zero normal stress (ZNS) hypothesis.

Summary of Computational Features

The computational features of the above approach may be summarized as follows:

- Applicable to both Thin and Moderately-Thick Shells
- Rotations may be Arbitrarily Large
- Strains may be Large (except transverse shears)
- Resultant Format Yields | Cost Savings $\propto$ Number of Layers |
- Shell Elements are Fairly Robust
  - — No Locking
  - — No Spurious Mechanisms
  - — Low Sensitivity to Mesh Distortion (esp. ANS elements)

Finally, note that while the formulation allows for finite strains and inelastic material behavior, the present analysis simply employs *orthotropic linear-elasticity* within each layer and *does not* account for the material delamination (*i.e.*, damage) observed in the NASA experiment.

§3.2 Implementation (NICE Software Architecture)

The shell-element capabilities mentioned above have been implemented in a modular fashion to facilitate research and transferral to other finite element codes. Some of the specific shell-element functions that are available via independent FORTRAN77 subroutine calls are shown in Figure 4a. A complete description of the shell-element software (including listings) is given in Appendix S of reference [20].

The ***host finite-element program*** used by the authors is actually a ***network*** of independently executable programs (or ***processors***) which are coordinated via high-level ***procedures*** written in a mathematically oriented command-language (Fig. 4b). In such an environment, the shell-element software is embedded in a single ***processor*** and the global solution algorithms are implemented as ***procedures***; examples are given in [7].

The software architecture (*i.e.*, utilities) used to construct this particular analysis system is known as NICE (for *Network of Interactive Computational Elements*, [6,7]). Due to the flexibility provided by the NICE architecture and its suitability for nonlinear and coupled-field problems, it is currently being explored by NASA as the basis for a standard generic *testbed* system for Computational Structural Mechanics (see [9] and other presentations therein). One of the motivating factors for developing such a system is the implementational complexity associated with a comprehensive global/local analysis of the present composite-shell problem.

§4. FINITE-ELEMENT MODELS

Due to the physics of the problem, a *full* numerical model of the test specimen is required (Fig. 5). The slight anisotropy emanating from the composite material stacking sequence is only partly responsible for the lack of available symmetry.* As will become apparent, the nonlinear postbuckling response is *inherently* non-symmetric due to the presence of the hole and the participation of many diverse mode shapes.

Several combinations of shell-element type and mesh density were employed during the course of the linear (pre-buckling), stability (buckling eigenvalue) and nonlinear (postbuckling) analyses. Figure 5 shows three representative grids, involving 300, 1500 and 5000 degrees-of-freedom, respectively. These grids correspond to 16, 80 and 256 nine-node elements (or alternatively to 64, 320 and 1024 four-node elements), respectively. Note that there is intrinsic element mesh-distortion in these models — both in-plane and out-of-plane — due to the focus on the hole and the curvature of the shell. However, the elements nearest the hole (where it counts most) have the most regular shapes.

The coarsest grid (1) was used to *verify* the modeling procedure, the finest grid (3) was used exclusively to check convergence of the *linear* and *eigen* solutions, and the intermediate grid (2) became the workhorse for *nonlinear* analysis. Furthermore, as little difference was observed between the 9_HET and 9_ANS elements (§3.1) during the early stages of analysis, the 9_HET element (which is slightly less expensive) is featured in the analytical results that follow.

Boundary conditions were imposed as described in Section 2 and illustrated in Figure 5. To simulate end-shortening, an axial *force* was applied in conjunction with a *degree-of-freedom equivalence* among all axial displacements on the loaded edge. This was done to avoid the use of *specified displacements*, which tend to complicate the adaptive (RC) nonlinear solution algorithm.

---

* This has been confirmed via numerical experiments with an *isotropic* model.

§5. LINEAR (PRE-BUCKLING) ANALYSIS

Results for the linear pre-buckling analysis are shown in Figure 6. The deformations due to an applied axial compressive load of 22480 lbs. (1 kN) are shown magnified by a factor of 10 in the top half of the figure — for grid 2. The corresponding distribution of the axial stress resultant, N_x, along the panel circumference at mid-span is shown in the bottom half of the figure — for both grids 2 and 3.

Regarding the *displacement* solution, convergence of the axial end-shortening, δ, was achieved with grid 2. (Grid 3 yielded less than a 1% increase in end-shortening.) However, the converged end-shortening solution, $\delta = .0316$, is approximately 15% lower than the experimental value; as deduced from the linear portion of the experimental load-displacement curve (Fig. 2). It is presumed that this over-estimation of the axial stiffness is due to the uncertainty in the *nominal* material properties, a conclusion that is reinforced in §8. Thus, to compensate for the mismatch, the principle layer elastic modulus, E_1, is reduced by a corresponding factor in the subsequent nonlinear analysis (§7).

From a qualitative perspective, the solution shows substantial bending deformation in the vicinity of the hole (Fig. 6a). This suggests that geometric nonlinearity may be important even at relatively low load levels, and diminishes the credibility of linear response and buckling-eigenvalue analyses.

Regarding the linear *stress* solution, note that the compressive axial resultant, N_x, is distributed evenly along most of the panel circumference except for a very localized region near the hole. While grid 2 was adequate for the *displacement* solution, grid 3 provides much more accurate resolution of this *stress concentration*. In particular, grid 2 yielded a peak stress concentration factor (SCF) of 2.8; about 14% less than the SCF obtained with grid 3.

As experimental data was not available for verification of the computed stresses, the convergence of the grid 3 stress solution was inferred via analytical agreement with a closed-form (asymptotic) solution due to Steele [21]. While the closed-form solution pertained to a purely *isotropic* panel, the linear finite-element stress solutions for isotropic and quasi-isotropic panels were found to be quite similar. It is also interesting that the SCFs for both the *isotropic* model (3.1) and the *quasi-isotropic* model (3.25) are not very different from the classical SCF for a *flat plate* with a circular hole (3.0).

Finally, note that by strongly *biasing* the mesh towards the hole, it was possible to obtain grid-3 accuracy with grid-2 for the local stress gradients. Such biasing, however, was found to be unnecessary in the subsequent, globally-oriented buckling and postbuckling analyses.

§6. STABILITY (BUCKLING) ANALYSIS

Figure 7 shows the first 5 buckling eigenmodes for grid 2. These results represent perturbations about the linear pre-buckling solution described above. The eigenvalues, $\lambda = 1.084, 1.106, 1.181, 1.432, 1.582$, are the ratios of the corresponding buckling loads to the axial load applied in the linear pre-buckling analysis. As before, grid 2 seems to provide adequate resolution, with grid 3 giving only a 2% reduction in the first two eigenvalues, and a 4% reduction in the remaining three.

The following observations are important for subsequent computational purposes: (i) the eigenvalues are closely spaced; (ii) the eigenmodes are vastly different in character; (iii) there is no *single* form of symmetry to be exploited computationally; (iv) the first buckling mode is symmetric and bears the most resemblance to the linear pre-buckling solution; (iv) the second and third modes possess *skew* symmetries; (v) the fourth and fifth modes are symmetric; the latter mode featuring practically no distortion of the circular hole; and (vi) higher modes (not shown) look much like those for a cylindrical panel *without* a hole, though the values remain closely spaced.

Finally, it was found that the first (*i.e.*, critical) buckling load is approximately 25% lower than that of an identical cylindrical panel without a hole. Hence, while the influence of the hole on the buckling *load* is only moderate (*i.e.*, relative to the stress concentration factor), its influence on the buckling *modes* is profound. As we shall see, the hole has an even stronger influence on the postbuckling response.

§7. NONLINEAR (POSTBUCKLING) ANALYSIS

In practice, more than just a linearly converged model and an adaptive solution strategy were necessary to obtain a reasonable *nonlinear* solution. One additional pre-requisite was a 15% reduction in the nominal elastic modulus, E_1 (from 19.6×10^6 to 17.1×10^6), to match the linear branch of the load vs. end-shortening curve (as explained in §5). Another important ingredient was the specification of *initial imperfections*. In this regard, three versions of the analysis were run: (1) one involving no imperfections; (2) one with an imperfection amplitude of 1% of the thickness applied to each of the first 4 buckling modes (see Fig. 7); and (3) one with an imperfection amplitude of 10% of the thickness applied to each of the first 4 buckling modes. A discussion of the results for these three cases follows.

§7.1 No Imperfections

One would think that the out-of-plane "imperfections" introduced by the linear pre-buckling solution (Fig. 6) would be sufficient to trigger a realistic buckling response. However, this was not the case. With no imperfections, the computed solution path resembled the experiment only up to the descending part of the load-displacement curve (Fig. 8a). The computed curve then rolled back onto itself with the stiffening branch of the postbuckling curve practically aligned with the *pre-buckling* curve.

To gain further insight, it is useful to look at the deformation and stress histories portrayed in Figure 8b. Shown is a sequence of computational "snapshots" taken at various points (*i.e.*, load steps) on the nonlinear load-displacement curve.* The variation of the axial stress resultant along the mid-span circumference is plotted below each frame.

Note that the deformation starts out (at load step 10) much like the first linear buckling mode (Fig. 7), with inward dimples both fore and aft of the hole, then *articulates* through the second and third modes during the initial postbuckling phase (steps 15–20). This *rotation* of the two dimples about the hole is probably triggered by the bending/twisting coupling inherent in the composite stacking sequence. The dimples continue to rotate and broaden until, at step 40, the pattern begins to resemble the *fifth* linear buckling mode. Evidently, it is this "locking" into mode 5 that is responsible for the excessive secondary stiffening in the load-displacement curve (Fig. 8a). Clearly, mode 5 is an unrealistically *stiff* one, resembling what might occur if a ring *stiffener* had been placed around the hole. This is also evident in the axial stress distribution, where the stress concentration has practically been *shed* by step 40.

§7.2 Large Imperfections

To avoid the unrealistic mode-5 locking observed in the preceding analysis, a fairly large initial imperfection was introduced in the numerical model. This was accomplished by adding an equal measure of each of the first 4 buckling modes to the initial geometry, such that the maximum radial (*i.e.*, shell-normal) displacement in each mode was equal to 10% of the panel thickness. Thus, the magnitude of the combined radial imperfection approached 40% of the thickness at some points. It is emphasized that this rather arbitrary choice of imperfections was designed primarily to minimize the influence of mode 5.

The computed nonlinear response for the imperfect panel is shown in Figure 9. Note that the secondary stiffening branch bears more resemblance to the experiment than did the imperfection-less analysis. Unfortunately, it is also true that the buckling (*i.e.*, peak) load has dropped by about 15%, and now underestimates the *experimental* buckling load by more than 20%. A heuristic explanation is

* Note that the pre-buckling phase of the analysis appears *linear* with respect to the *axial* displacement, δ. However, due to the rapid growth in *radial* displacements (not shown), the analysis is actually quite nonlinear from the outset; which explains the relatively large number of load-steps required on the "linear" branch of Fig. 8.

provided by considering the deformation and stress histories.

In Figure 9b, we see that with the 10%-h imperfections, the buckling pattern leaves mode 1 almost immediately and develops an intensified inward dimple on one side of the panel. Consequently, the full axial load is re-distributed to the *other* side of the panel (see N_x plots in Fig. 9b), accounting for the reduction in both the buckling load and the postbuckling stiffness (fig. 9a).

§7.3 Small Imperfections (the "Bottom Line")

Finally, the "best of both worlds" was obtained with a 1%-h imperfection in each of the first 4 modes. Compared with the previous analysis, the computed load-displacement curve (Fig. 10) shows both an increase in the maximum load and a decrease in the minimum load, thus bringing the solution more in line with experiment.

The improved performance, obtained by reducing the imperfections, may again be related to the deformation history (Fig. 10b). Here, as in the case without imperfections, two inward dimples develop and proceed to rotate about the hole. Just after buckling, however, one dimple tends to deepen while the other diminishes, and eventually there is a *double* snap-through. This accounts for the double dip in the load-displacement curve (Fig. 10a) and seems to explain why a lower minimum load is obtained with the smaller imperfection.

Still, there are some serious descrepancies between analysis and experiment, namely: (i) a 7% under-estimation of the buckling load, (ii) a 25% over-estimation of the minimum load, (iii) a 30% under-estimation of the postbuckling end-shortening, and (iv) a 10% over-estimation in the postbuckling stiffness. These will be addressed in §9.

§8. CORROBORATION WITH ANOTHER CODE

To support the above results, obtained via the computational procedures described in Section 3, parallel analyses were performed with another finite-element computer code. For this purpose, we employed the STAGS code [2], which has been used for more than a decade by varous government agencies and industrial firms (*e.g.*, NASA and Lockheed) to analyze difficult nonlinear shell problems. Another reason for using STAGS is that it features finite-element computational procedures that are substantially different from those used in the present approach, thus adding strength to an analytical comparison.

For the *linear* (pre-buckling and buckling) analysis, excellent agreement was obtained between the STAGS and NICE-based solutions. For completeness, the STAGS runs were performed with two radically different shell-element types, which both converged to the same solution as the NICE/9_HET element, albeit at slower rates and from *below* in stiffness. (Note that while the axial pre-buckling stiffness of the STAGS elements converged from below, the *buckling loads* converged from *above*).

In particular, the comparison included: (i) the commonly used STAGS/410 element — a flat quadrilateral plate element based on Kirchhoff-Love theory and cubic membrane/bending displacement interpolation; and (ii) the less frequently used STAGS/422 element — a quadrilateral composed of two Kirchhoff-based triangular plate sub-elements with cubic bending interpolation and quadratic membrane interpolation. It is believed that the relatively slow convergence from-below of the STAGS/410 element is due to *warping sensitivity*, while that of the STAGS/422 element is probably due to the *incompatibility* between membrane and bending displacement fields for non-flat quadrilateral element shapes.*

For the *nonlinear* comparisons with NICE/9_HET, the STAGS/410 element was used exclusively. The resulting load-displacement curves (for 1%-h and 10%-h imperfections) are summarized in Figure 11. The dashed curves represent the NICE/9_HET solutions and the dotted curves represent the STAGS/410 solutions; both were obtained with Grid 2 (Fig. 5).

Note that although the STAGS and NICE-based solutions use different finite-element types, large-rotation update procedures and nonlinear solution strategies, the correlation is remarkable — especially during the postbuckling phase. Even the STAGS zero-imperfection analysis (not shown) resulted in the same excessive postbuckling stiffness as displayed in Figure 8a. One other point: While the STAGS/410 element consistently shows about a 5% higher buckling load than the NICE/9_HET element, thus coming closer to the experimental peak; STAGS/410 is actually *less* accurate — with respect to *discretization* errors — than NICE/9_HET. This follows from the fact that both STAGS/410 and NICE/9_HET converge from *above* in the buckling load, and was confirmed by running the NICE/9_HET element with a *coarser grid*, for which it too showed a 5% higher peak. That non-converged solutions compare better with experiment than converged ones suggests that spatial discretization is not the only source of error here (e.g., see §9).

---

* It is interesting to note that the STAGS/422 element was used in the related study conducted in [14], where it yielded a 17% more flexible linear solution, and thus agreed better with the linear portion of the experiment. Nevertheless, it has since been found that the boundary conditions were not consistently applied to the element's *mid-side* freedoms in that analysis. By correcting this implementation error, the 17% discrepancy with the other elements has been completely eliminated. Thus, it appears that the "accuracy" obtained in [14] with the STAGS/422 element is due to compensating errors in the nominal material properties.

§9. CONCLUSIONS

§9.1 Summary

The present study may be summarized as follows:

- PURPOSE:
  - — Validate continuum-based resultant (CBR) shell formulation
  - — Evaluate new shell elements
  - — Gain experience with composite-shell postbuckling analysis
- RESULTS:
  - — Good agreement in pre-buckling/buckling range
  - — Good *qualitative* agreement in postbuckling range
  - — Discrepancies due to:
    - —— *Material properties*
    - —— *Imperfection sensitivity*
    - —— *Dynamic effects*
    - —— *Delamination*

The "good" agreement obtained between the present shell-element formulation and experiment in the pre-buckling and buckling range was possible only after adjusting the nominal material properties so that the linear axial stiffnesses coalesced. The material-property modification was further justified via corroboration with the STAGS finite-element code, which features a substantially different computational approach.

The "best" solution for the nonlinear response was obtained by introducing small (1%-thickness) imperfections corresponding to each of the first four buckling modes. The computed load-displacement curve (Fig. 10a), which again compared well with STAGS (Fig. 11), still showed major discrepancies with the experiment: The discrepancy in the *buckling load* (which is relatively small) may be due to the inadequacy of adjusting only the principal layer elastic modulus, E_1, rather than the complete set of orthotropic material constants. The discrepancy in the *unloading* phase of postbuckling (*i.e.*, the computed end-shortening *reversal*) is attributed to the quasi-static approximation of what is, in reality, a *dynamic* phenomenon. Finally, the discrepancy in the *stiffening* phase of postbuckling is clearly dominated by damage, *i.e.*, delamination observed in the experiment but not represented in the model.

§9.2 Recommendations

The goal is to eliminate the discrepancies listed above with a minimum of computational cost and complexity. To this end, the following steps are recommended:

1) PERFORM ADDITIONAL EXPERIMENTS. First, more experimental data is required to verify existing computational capabilities for composite shell postbuckling. For example, panel imperfections should be carefully measured and selected strains and overall deformation patterns should be monitored at frequent intervals. Additionally, an isotropic panel should be tested in order to eliminate the material identification problem, and also to provide a standard *benchmark for shell-element evaluation*. The isotropic problem would be valuable for screening out geometrically sensitive elements.

2) REFINE NONLINEAR GRID. The nonlinear analysis should be repeated with a finer grid (*e.g.*, Grid 3), as convergence in the linear regime is no guarantee of convergence in the nonlinear regime. Moreover, a study of modal participation both in the imperfections and in the nonlinear response may help establish modeling guidelines for future analysis.

3) INCLUDE DYNAMIC EFFECTS. Dynamic effects, which are relatively straightforward to incorporate, should be assessed at the first opportunity. The analysis could be started in a quasi-static mode, switched to an *explicit* transient response algorithm during the unstable phase, and switched again to an *implicit* algorithm during the stable postbuckling phase.

4) REPRESENT LOCAL FAILURE (DELAMINATION). To account for the composite delamination mechanism, appropriate *failure criteria* and *progressive-failure modeling procedures* need to be developed, implemented and evaluated. For shell-based failure criteria, *accurate stress recovery* is essential. Improved stress resolution is required both *along the surface* (*e.g.*, via stress/displacement iterations and adaptive refinement) and *through-the-thickness*, as both normal and transverse-shear stresses can play a dominant role in delamination. Progressive failure may then be simulated by methods ranging from a simplified shell model that *selectively degrades layer properties*, to a full *3D analysis near the hole* with an evolving 2D/3D transition. An intermediate approach is to "split" shell elements along delaminating boundaries. (See Fig. 12.) The simplest approach, however, has obvious implementational advantages.

5) DEVELOP EFFICIENT GLOBAL/LOCAL ALGORITHMS. Finally, there is a need to reduce the cost of nonlinear analysis for such problems. The cost of the *global* analysis is dominated by the large number of iterations/steps in the linear-to-postbuckling transition regime. Possible approaches include the reduced-basis technique [16,22], Thurston's method [24], and improved extrapolators. For the combined *global/local* problem, where material "properties" are changing rapidly during postbuckling, additional features such as line-searches [3], quasi-Newton stiffness updates [5] and nonlinear substructuring [23] may greatly improve efficiency.

We are presently acting on recommendations (2)–(3) and will report the outcome in a forthcoming paper.

ACKNOWLEDGEMENTS

The preparation of this paper was supported by the Independent Research Program of Lockheed Missiles and Space Company, Inc. Ongoing research is being supported by NASA/Langley Research Center as part of their newly launched program in Computational Structural Mechanics.

REFERENCES

[1] Ahmad, S., Irons, B.M. and Zienkiewiecz, O.C. (1970), "Analysis of Thick and Thin Shell Structures by Curved Finite Elements", *International Journal for Numerical Methods in Engineering*, **2**, 419–451.

[2] Almroth, B.O., Brogan, F.A. and Stanley, G.M. (1979), "Structural Analysis of General Shells, Vol. II: User Instructions for the STAGS(C-1) Computer Code", Lockheed Report LMSC-D33873.

[3] Crisfield, M.A. (1982), "An Arc-Length Method Including Line Searches and Acceleration", *International Journal for Numerical Methods in Engineering*, **19**, 1269–1289.

[4] Crisfield, M.A. (1983), "A Fast Incremental/Iterative Solution Procedure that Handles Snap-Through", *Computers and Structures*, **13**, 55–62.

[5] Dennis, J.E. and Moré, J.J. (1977), "Quasi-Newton Methods, Motivation and Theory", *SIAM Review*, **19**, 46–89.

[6] Felippa, C.A. (1981), "Architecture of a Distributed Analysis Network for Computational Mechanics", *Computers and Structures*, **13**, 405–413.

[7] Felippa, C.A. and Stanley, G.M. (1985), "NICE: A Utility Architecture for Computational Mechanics", proceedings of US-Europe Symposium on Finite-Element Methods for Nonlinear Problems, Univ. of Trondheim, Norway (ed. by P. Bergan).

[8] Flaggs, D.L. (1985), "Prediction of Tensile Matrix Failure in Composite Laminates", *Journal of Composite Materials*, **19**, 29–50.

[9] Geers, T.L. (1984), "Comments on Proposed Research Program in Computational Structural Mechanics", proceedings of *Workshop on Computational Structural Mechanics*, held 5–7 Sep 1984 at NASA/LaRC, edited by W.J. Stroud.

[10] Hughes, T.J.R., and Cohen, M. (1979), "The 'Heterosis' Finite Element for Plate Bending", *Computers and Structures*, **9**, 445–450.

[11] Hughes, T.J.R. and Winget, J. (1980), "Finite Rotation Effects in Numerical Integration of Rate Constitutive Equations Arising in Large-Deformation Analysis", *International Journal for Numerical Methods in Engineering*, **15**, 1862–1867.

[12] Hughes, T.J.R. and Liu, W.K. (1981), "Nonlinear Finite-Element Analysis of Shells; Part I: Three-Dimensional Shells", *International Journal for Numerical Methods in Engineering*, **26**, 331–362.

[13] Hughes, T.J.R. and Carnoy, E. (1981), "Nonlinear Finite Element Shell Formulation Accounting for Large Membrane Strains", pp. 193–208 in *Nonlinear Finite Element Analysis of Plates and Shells* (eds. T.J.R. Hughes et al.), ASME, New York.

[14] Knight, N.F, Jr. and Starnes, J.H., Jr. (1984), "Postbuckling Behavior of Axially Compressed Graphite-Epoxy Cylindrical Panels with Circular Holes", presented at the *1984 ASME Joint Pressure Vessels and Piping / Applied Mechanics Converence*, San Antonio, Texas.

[15] Mindlin, R.D. (1951), "Influence of Rotatory Inertia and Shear on Flexural Motions of Isotropic, Elastic Plates", *Journal of Applied Mechanics*, **18**, 31–38.

[16] Noor, A.K. and Peters, J.M. (1983), "Recent Advances in Reduction Methods for Instability Analysis of Structures", *Computers and Structures*, **16**, 67–80.

[17] Parisch, H. (1981), "Nonlinear Analysis of Shells Using Isoparametric Elements", pp. 47–63 in *Nonlinear Finite Element Analysis of Plates and Shells*, (eds. T.J.R. Hughes et al.), ASME, New York.

[18] Park, K.C. and Stanley, G.M. (1985d), "A Curved C^0 Shell Element based on Assumed Natural-Coordinate Strains", submitted for publication.

[19] Rankin, C.C. and Brogan, F.A. (1984), "An Element-Independent Corotational Procedure for the Treatment of Large Rotations", pp. 85–100 in *Collapse Analysis of Structures* (eds. L.H. Sobel and K. Thomas), ASME, New York.

[20] Stanley, G.M. (1985), *Continuum-Based Shell Elements*, Stanford University Ph.D. thesis, also Lockheed report LMSC-F035839.

[21] Steele, C.R. (1985), *private communication.*

[22] Stehlin, P. and Brogan, F.A. (1984), "Analysis of Structural Collapse by the Reduced Basis Technique", pp. 69–84 in *Collapse Analysis of Structures* (eds. L.H. Sobel and K. Thomas), ASME, New York.

[23] Stehlin, P. (1985), "Nonlinear Substructuring Using a Mixed Global-Local Formulation", submitted for publication.

[24] Thurston, G.A., Brogan, F.A. and Stehlin, P. (1985), "Postbuckling Analysis Using a General Purpose Code", presented at AIAA 26th Structures, Structural Dynamics and Materials Conference, Orlando, Florida, AIAA Paper No. 85-0719-CP.

[25] Wang, A.S.D. and Crossman, F.W. (1980), "Initiation and Growth of Transverse Cracks and Edge Delamination in Composite Laminates, Part I: An Energy Method", *Journal of Composite Materials*, **14**, 71...

COMPOSITE CYLINDRICAL PANEL WITH CIRCULAR HOLE

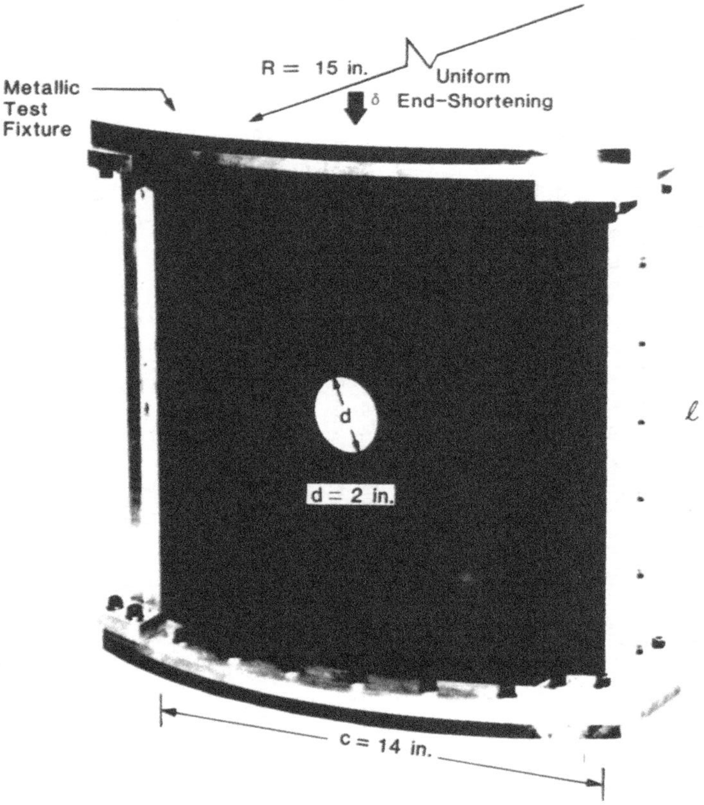

NOMINAL COMPOSITE MATERIAL PROPERTIES:

16 Layers

Graphite-Epoxy

$h_{layer} = 0.0056$ in.

$h_{total} = 0.09$ in.

$E_1 = 19.6 \times 10^6$ psi

$E_2 = 1.89 \times 10^6$ psi

$\nu_{21} = 0.38, \ \nu_{12} = 0.037$

$G_{12} = 0.93 \times 10^6$ psi

$\{\pm 45/90/0/0/90 \mp 45\}^\circ$ sym

Figure 1 Knight's Problem, NASA Test Specimen

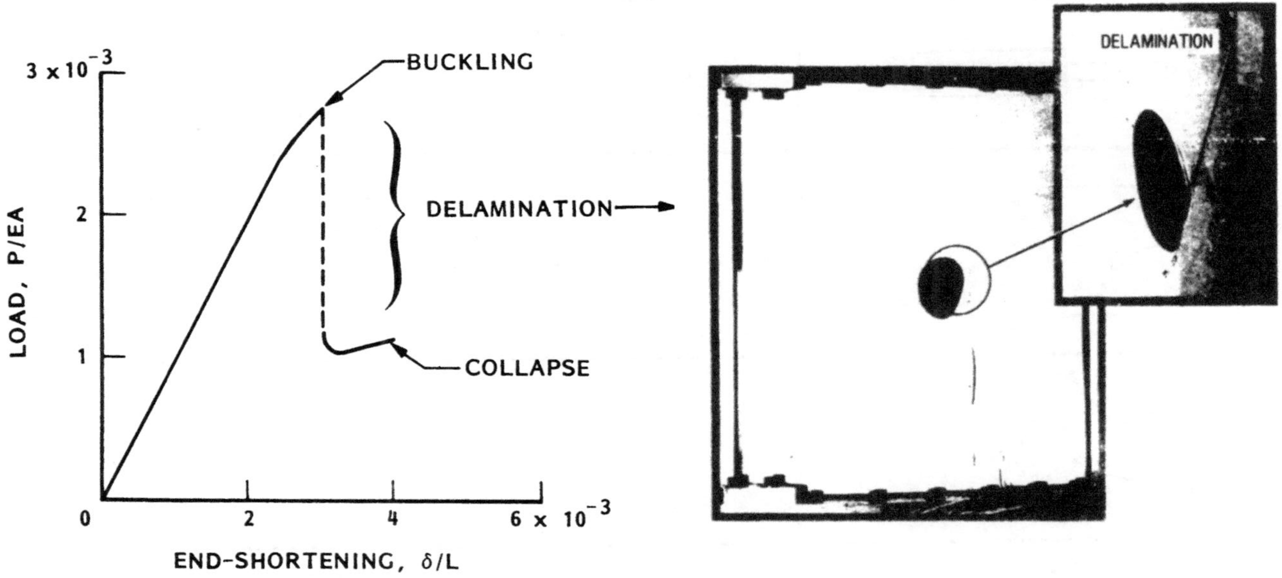

Figure 2 Experimental Results

• Continuum-based (CB) shell equations – 3-D Continuum equations (equilibrium/constitutive) – Embed shell hypotheses (straight normals, zero normal stress)	
• Reduce to "resultant" (CBR) form – Assume: small transverse-shear strains, mild taper – Preintegrate through-the-thickness	
• Discretize via curved "isoparametric" elements – Selective/reduced integration (SRI) – Assumed natural strain (ANS)	
• Solve nonlinear matrix equations via: – Linearization w.r.t. currrent configuration (UL) – Modified NR algorithm with adaptive (RC) strategy – Nodal triad-updates for large rotations	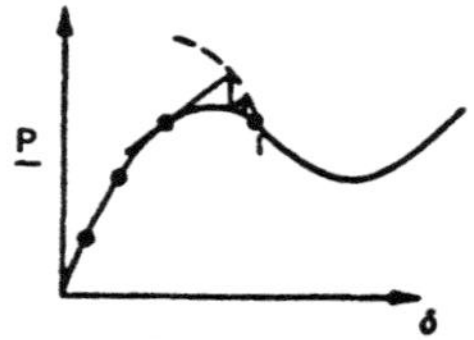
• Solve rate-constitutive equations via: – "Midpoint rule" incremental algorithm – ZNS recovery of normal strains (thickness updates)	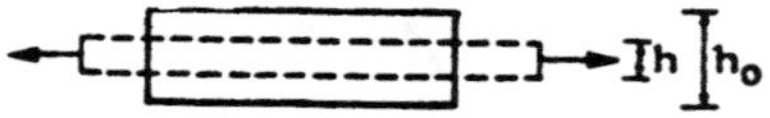

Figure 3 Computational Approach

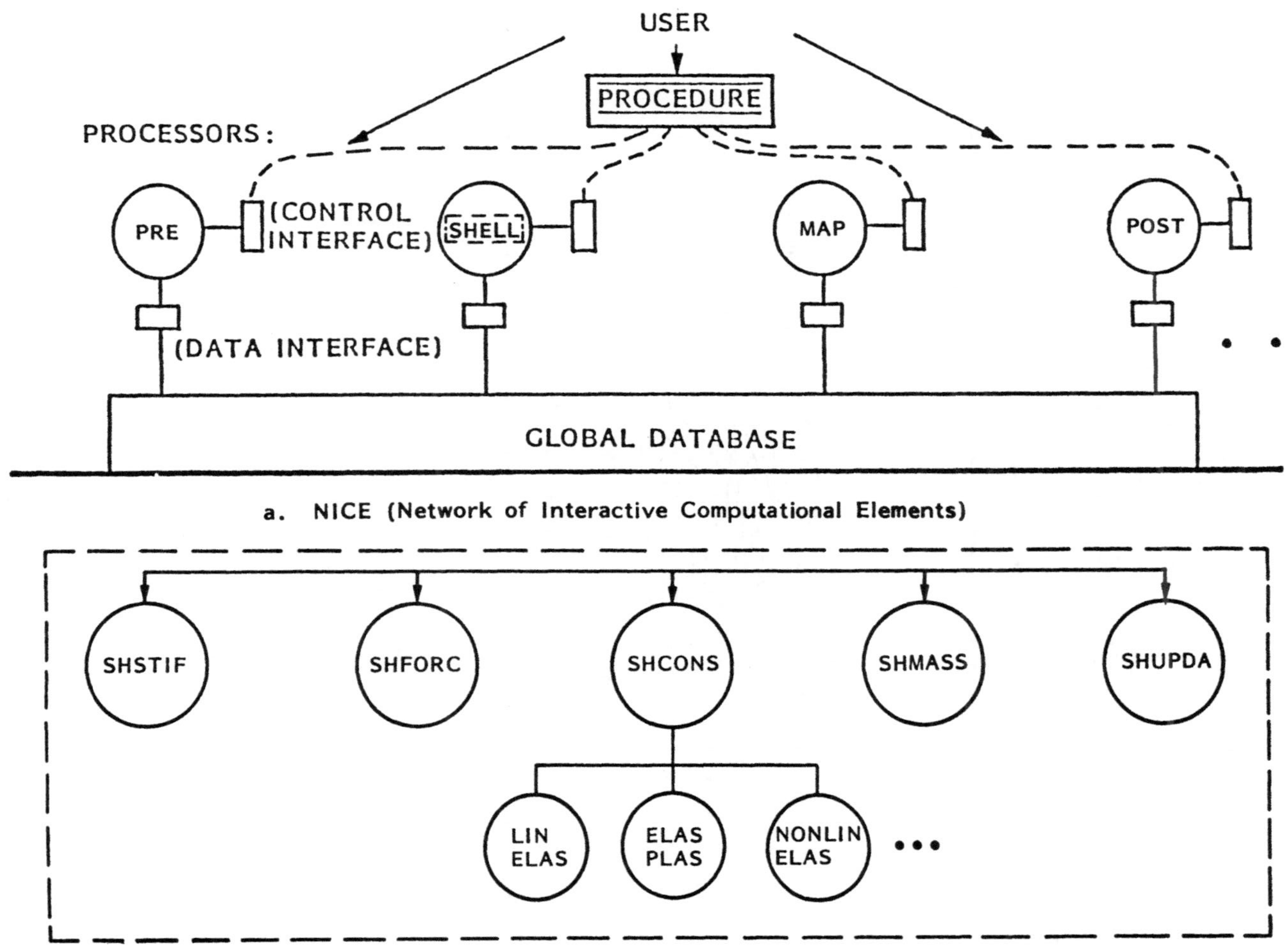

a. NICE (Network of Interactive Computational Elements)

b. Shell Element Processor

Figure 4 Implementation

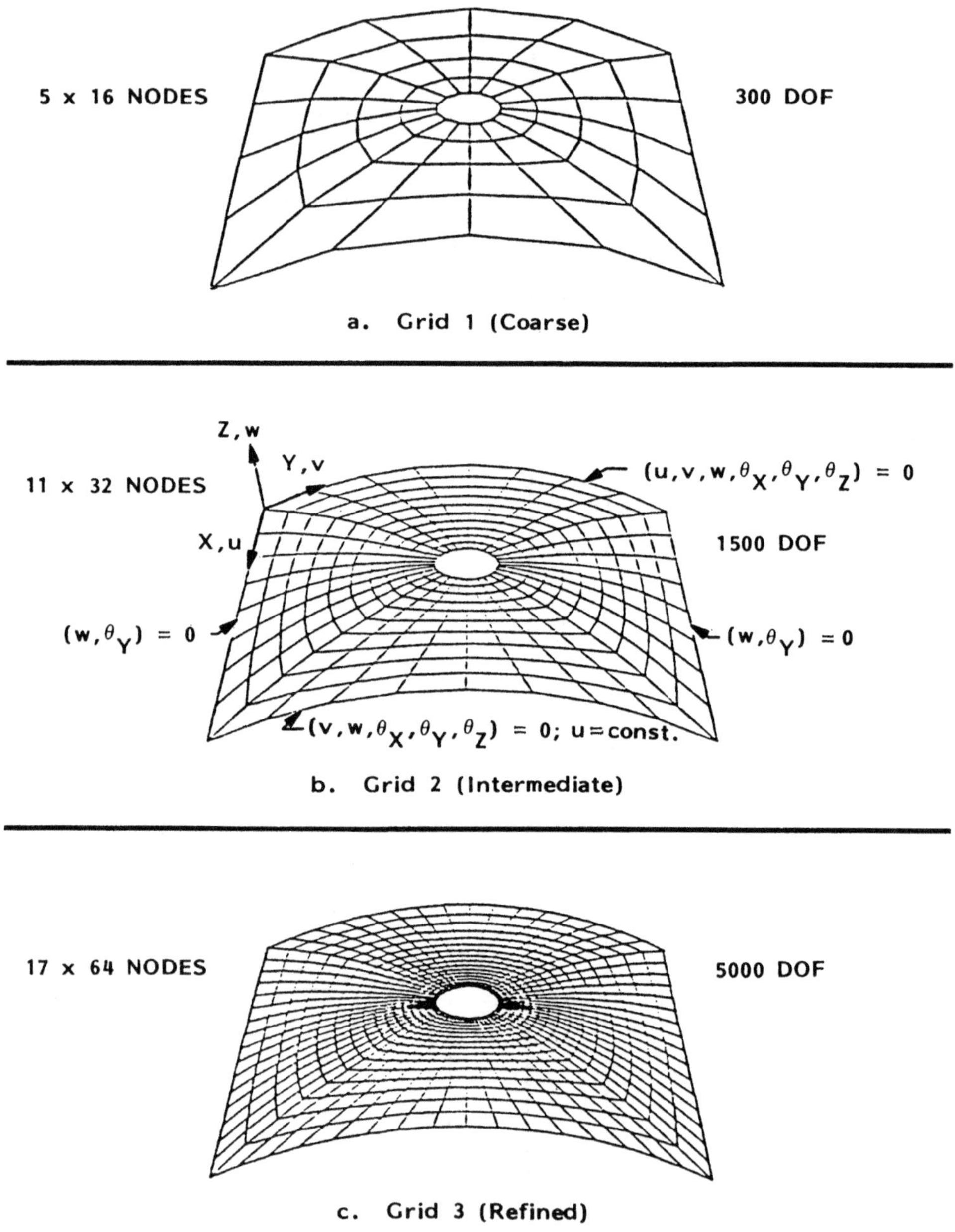

Figure 5 Finite-Element Models

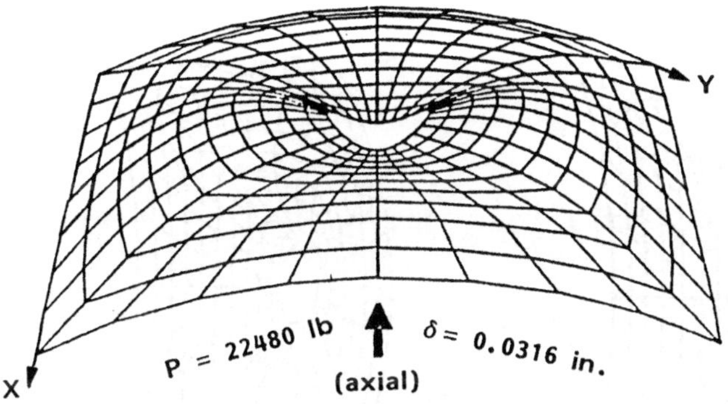

a. Linear Displacement Solution
(Magnified by 10)

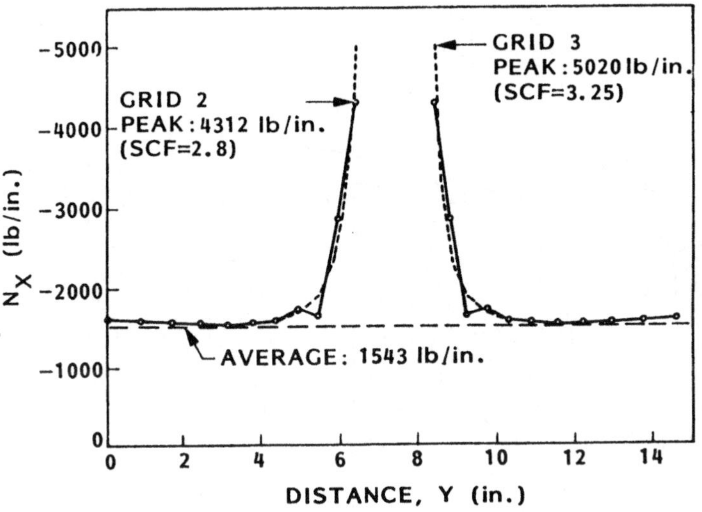

b. Linear Stress-Resultant Solution: N_X versus Y (X=L/2)

Figure 6 Linear (Prebuckling) Analysis

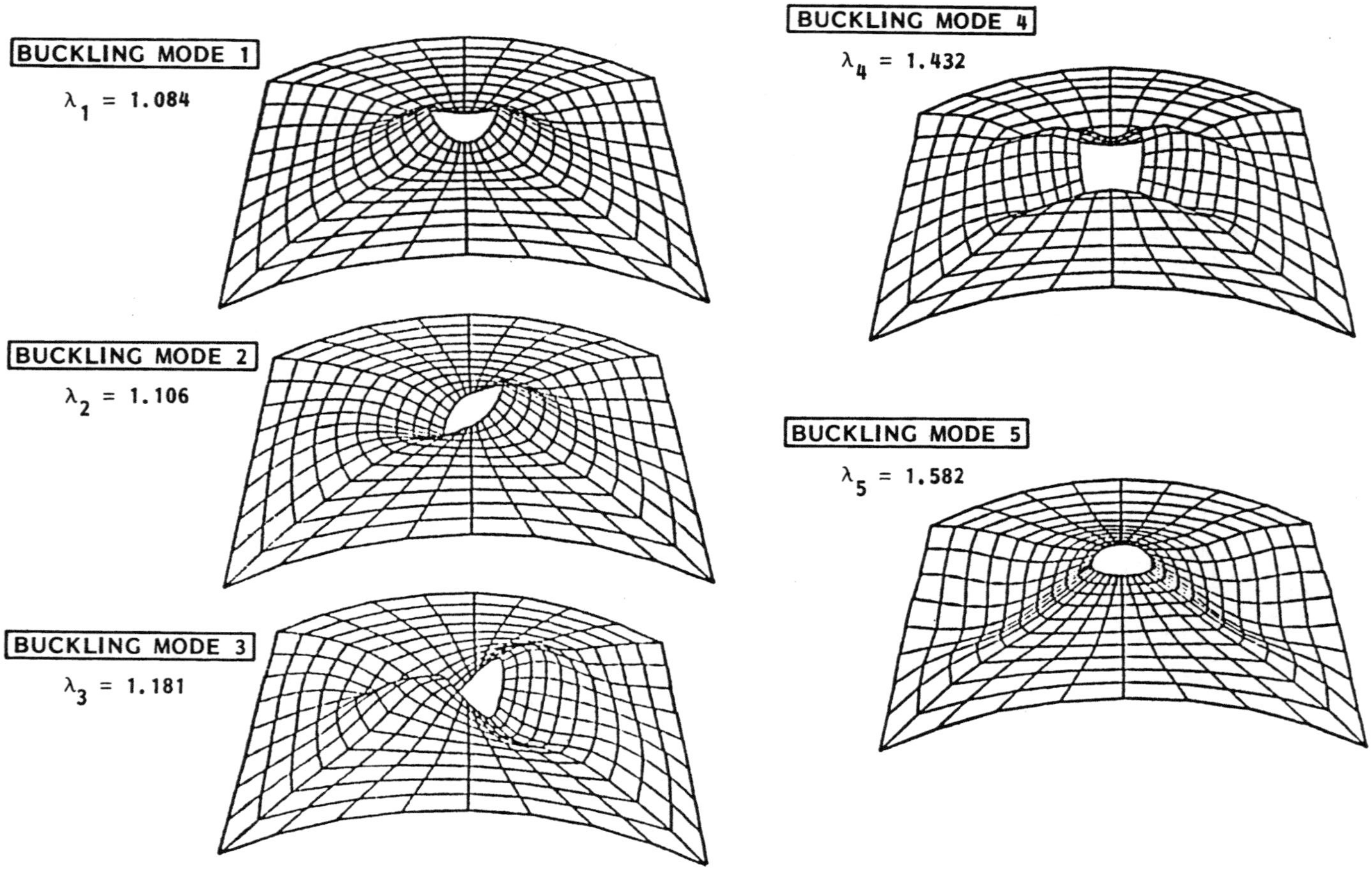

Figure 7 Stability (Buckling Eigenvalue) Analysis

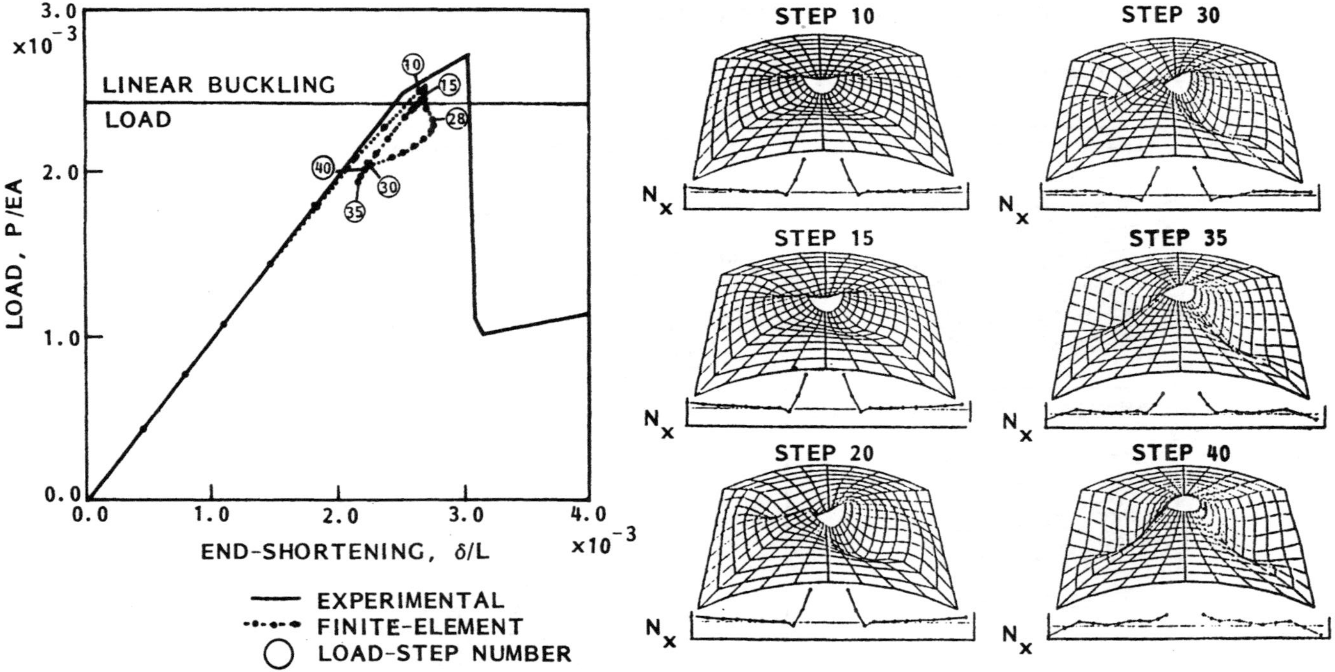

a. Load versus End-Shortening

b. Deformations and Stresses

Figure 8 Nonlinear (Postbuckling) Analysis; No Imperfections

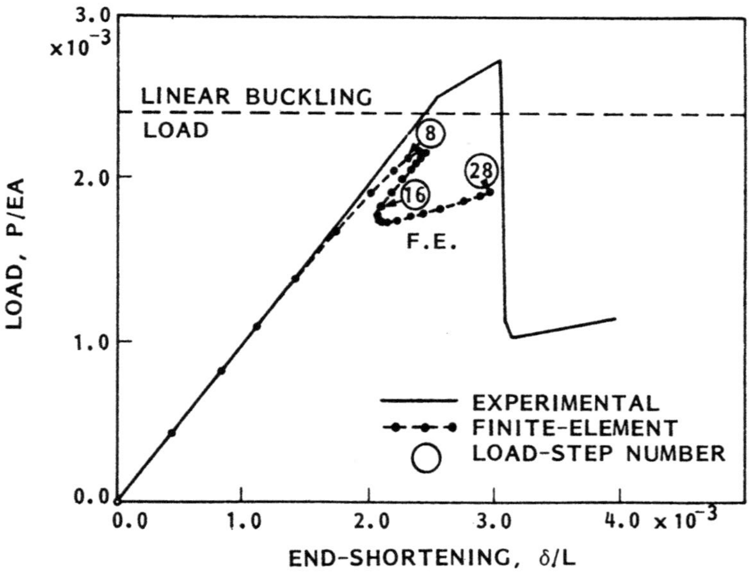

a. Load Versus End Shortening

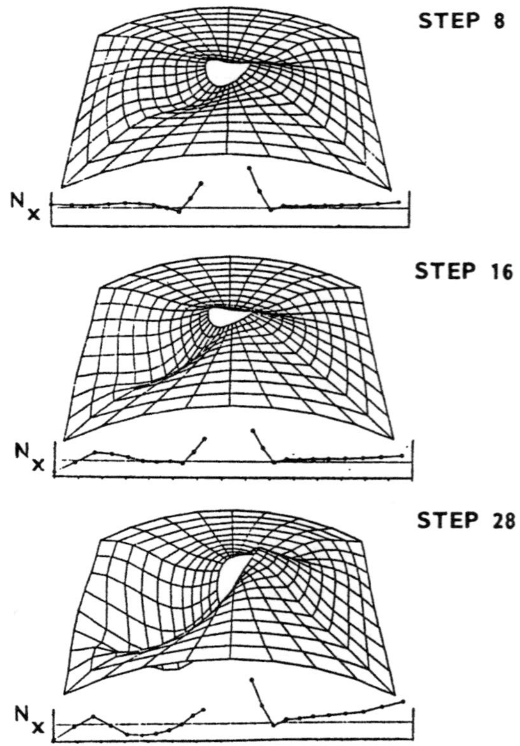

b. Deformations and Stresses

Figure 9 Nonlinear (Postbuckling) Analysis; 10%-h Imperfections

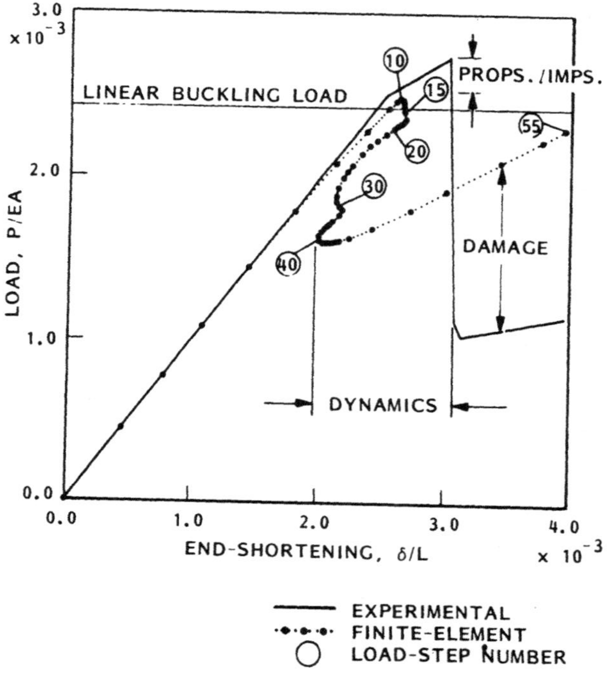

a. Load versus End-Shortening

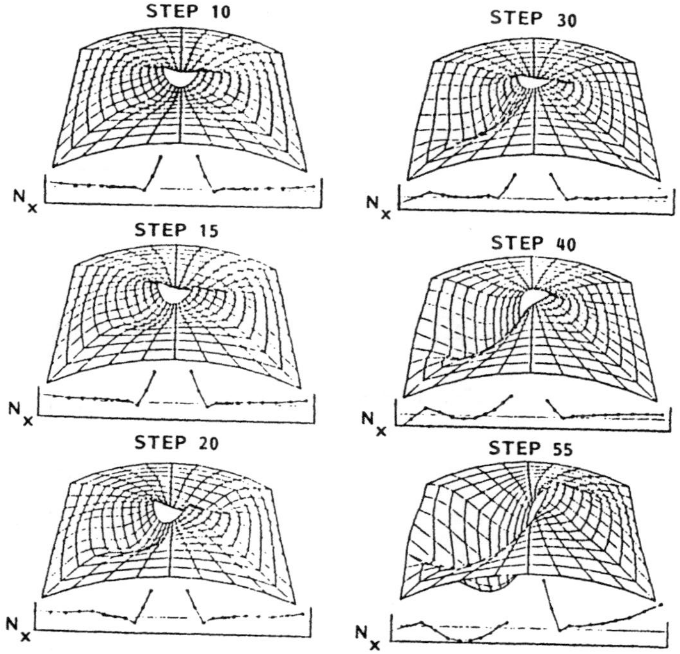

b. Deformations and Stresses

Figure 10 Nonlinear (Postbuckling) Analysis; 1%-h Imperfections

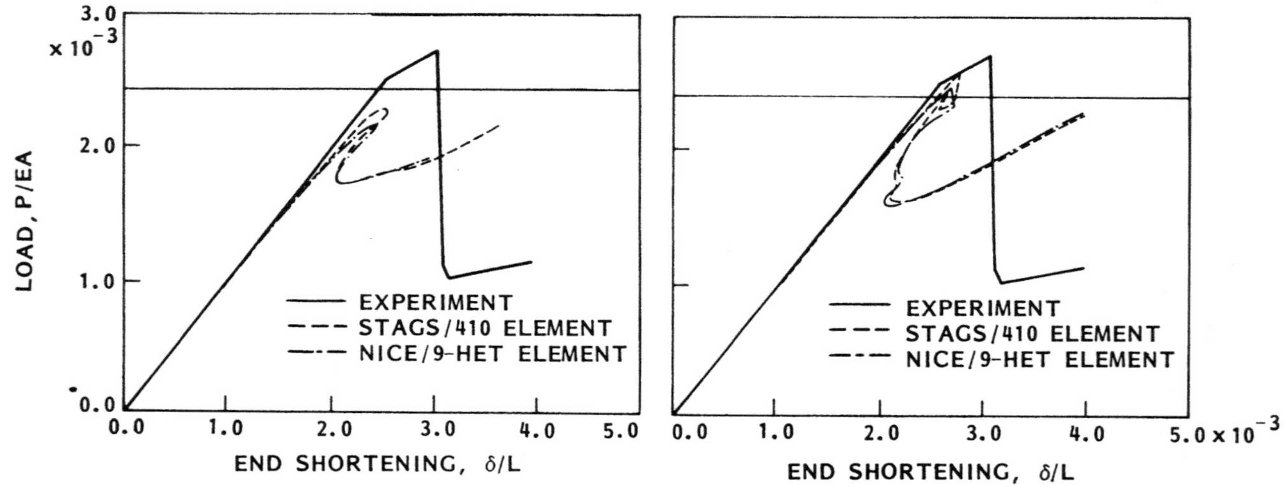

a. 10%-h Imperfections

b. 1%-h Imperfections

Figure 11 Corroboration with the STAGS Code, Postbuckling Analyses

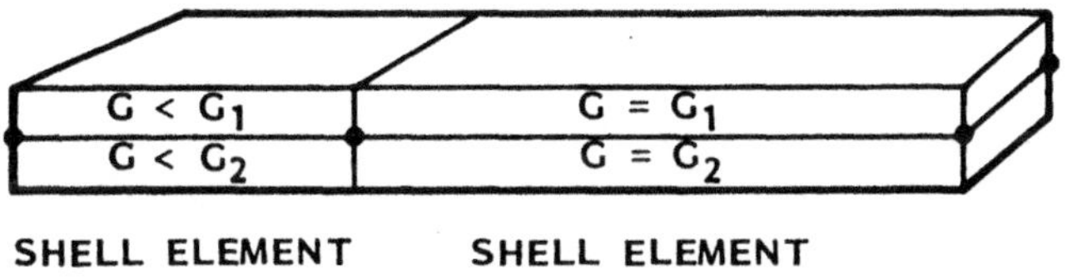

a. Simple: Shell-Element Material-Property Degradation

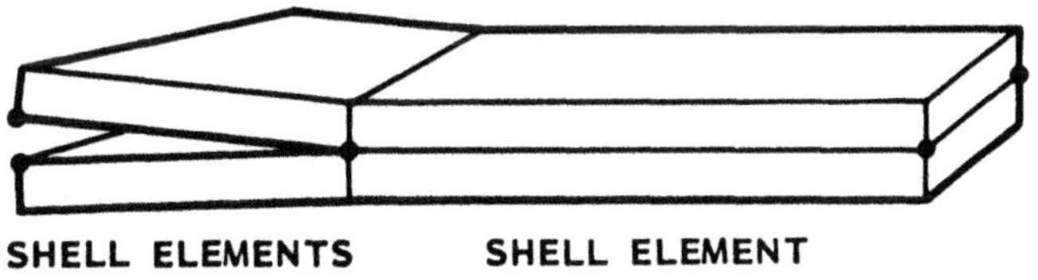

b. Intermediate: Shell-Element "Splitting"

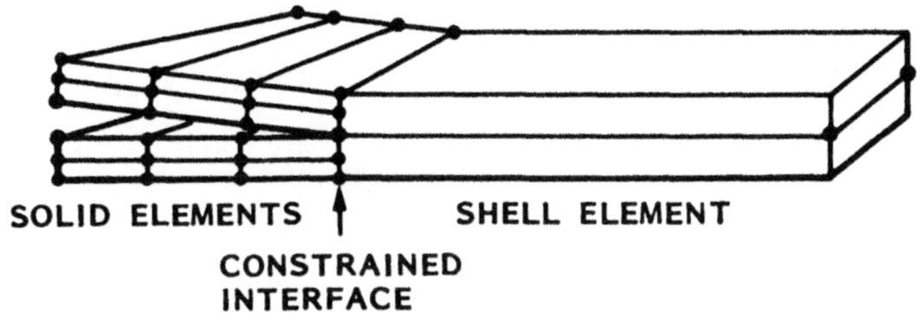

c. Complex: 3D/2D Transition

Figure 12 Composite Failure-Progression Models

Part IV: Solution Techniques and Computer Methods

Reduction Method for the Nonlinear Analysis of Symmetric Anisotropic Panels

A. K. NOOR
George Washington University Center at NASA, Hampton, USA

Summary

Reduction method and computational procedures are presented for reducing the size of the analysis model and the number of degrees of freedom used in predicting the nonlinear response of symmetric anisotropic panels. The key elements of the method are: a) decomposition of the characteristic arrays of the finite element model into the sum of orthotropic and nonorthotropic contributions; b) combined use of the finite element method and the classical Rayleigh-Ritz technique. The finite element method is first used to generate few global approximation vectors (or modes). Then the amplitudes of these modes are computed by using the classical Rayleigh-Ritz technique.

The global approximation vectors are selected to be those commonly used in single (or multiple) parameter perturbation technique; namely, a nonlinear solution corresponding to zero nonorthotropic arrays and a number of its derivatives with respect to an anisotropic tracing parameter (and possibly, to a load or arc-length parameter in the solution space). The size of the analysis model used in generating the global approximation vectors is identical to that of the corresponding orthotropic structure.

The effectiveness of the proposed reduction method is demonstrated by means of a numerical example, and its potential for solving quasi-symmetric nonlinear problems of anisotropic panels is discussed.

Nomenclature

E_L, E_T	Elastic moduli of the individual layers in the direction of fibers and normal to it, respectively
F_{IJK}, G_{IJKL}	Nonlinear stiffness coefficients of the discretized panel
$\mathcal{F}_{IJK}$, F_{IJK}, $\mathcal{G}_{IJK\ell}$, G_{IJKL}	Orthotropic and nonorthotropic parts of the arrays F_{IJK} and G_{IJKL}
$\tilde{\mathcal{F}}_{ijk}$, $\tilde{F}_{ijk}$, $\tilde{\mathcal{G}}_{ijk\ell}$, $\tilde{G}_{ijk\ell}$	Orthotropic and nonorthotropic parts of the nonlinear stiffness coefficients of the reduced system
G_{LT}, G_{TT}	Shear moduli in the plane of the fibers and normal to it, respectively
h	Total thickness of panel

Finite Element Methods for Nonlinear Problems
Europe-US Symposium, Trondheim, Norway 1985
Editors: Bergan, Bathe, Wunderlich

K_{IJ}	Linear stiffness coefficients of the discretized panel
K_{IJ}, $\mathcal{K}_{IJ}$	Orthotropic and nonorthotropic parts of the K_{IJ} array
$\tilde{\mathsf{K}}_{ij}$, $\tilde{\mathcal{K}}_{ij}$	Orthotropic and nonorthotropic parts of the linear stiffness coefficients of the reduced system
$\overset{*}{\mathsf{K}}_{IJ}$, $\overset{*}{\mathcal{K}}_{IJ}$, $\bar{\mathsf{K}}_{IJ}$	Stiffness coefficients defined in Eqs. 10, 11 and 30, respectively
L_1, L_2	Side lengths of the panel
N	Total number of displacement degrees of freedom
p_o	Intensity of uniform normal loading
Q_I, $\bar{Q}_I$	External load components and normalized components, respectively
$\tilde{Q}_i$	Load components of the reduced system
q	Load parameter
R	Radius of curvature of the panel
r	Number of global approximation vectors
U	Total strain energy of the panel
U_o	Strain energy corresponding to λ=0
u_1, u_2, w	Displacement components in the coordinate directions
X_I	Nodal displacement parameters of the anisotropic panel
$\mathbf{X}_I$	Displacement parameters for the case λ=0 (nonorthotropic terms neglected)
x_1, x_2, x_3	Orthogonal curvilinear coordinate system (see Fig. 1)
Γ_{Ji}	$= \dfrac{\partial^{i-1} \mathbf{X}_J}{\partial \lambda^{i-1}}$
λ	Parameter identifying all the nonorthotropic (anisotropic) contributions in the stiffness arrays
ρ_ℓ	(ℓ=1 to r-1) Parameters measuring the sensitivity of the global response of the structure to nonorthotropic material coefficients
ν_{LT}	Major Poisson's ratio of the individual layers
ϕ_1, ϕ_2	Rotation components of the middle surface of the panel (see Fig. 1)
ψ_i	Unknowns of the reduced equations (amplitudes of global approximation vectors)

Range of Indices

Uppercase Latin indices: 1 to N

Lowercase Latin indices: 1 to r

Introduction

In comparison with the literature devoted to isotropic and orthotropic panels, little is available on anisotropic panels. Interest in the use of filament-winding techniques for manufacturing composite panels has recently been expanded in aircraft, shipbuilding and other industries. Therefore, an understanding of the nonlinear and postbuckling responses of anisotropic panels is desirable.

The limited studies of anisotropic panels may be attributed to the high computational expense involved in the nonlinear analysis of these panels. As has been reported in the literature, anisotropy has an adverse effect on the accuracy and convergence of both numerical and analytic solutions (see, for example, Refs. [1] to [3]). In an attempt to reduce the cost of analyzing anisotropic panels, the different symmetries that exist in composite plates and shells having symmetric geometry, loading and boundary conditions were identified in Refs. [4] and [5]. Also, techniques have been developed for exploiting these symmetries in the finite element analysis (see Refs. [4] and [5]). However, in spite of these developments, the computational expense of analyzing highly anisotropic panels, particularly in the presence of nonlinearities, remains high. The present study focuses on this problem. Specifically, the objectives of this paper are:

1) to present a simple and efficient reduction method and computational procedures for reducing the size of the analysis model and the number of degrees of freedom used in predicting the nonlinear response for symmetric anisotropic panels;

2) to demonstrate the effectiveness of the proposed reduction technique by means of numerical examples; and,

3) to discuss the potential of the proposed reduction method and its application to other quasi-symmetric problems.

The analytical formulation is based on a form of the geometrically nonlinear shallow shell theory with the effects of transverse shear deformation, anisotropic material behavior, and bending-extensional coupling included.

Basic Idea of Model-Size Reduction Method

Governing Finite Element Equations:

A total Lagrangian formulation is used and the panel is discretized by using displacement finite element models. The governing finite element equations can be cast in the following form:

$$f_I = K_{IJ}\, X_J + \frac{1}{2}\, F_{IJK}\, X_J X_K + \frac{1}{3}\, G_{IJKL}\, X_J X_K X_L - Q_I = 0 \qquad (1)$$

where K_{IJ} are the linear stiffness coefficients; F_{IJK} and G_{IJKL} are nonlinear stiffness coefficients; X_J are unknown nodal displacement parameters; and Q_I are external load components. The range of the indices I, J, K, L is 1 to N where N is the total number of displacement degrees of freedom, and a repeated index denotes summation over its full range. The coefficients K_{IJ}, F_{IJK} and G_{IJKL} are completely symmetric with respect to permutation of indices. The loading is assumed to be conservative and independent of the displacement parameters X_I.

The proposed model-size reduction method can be conveniently divided into the following two distinct steps: 1) generation of global approximation vectors (or modes) using the finite element method; and 2) computation of the amplitudes of the approximation vectors via the classical Rayleigh-Ritz technique. Two possible computational procedures for use in conjunction with this method are described subsequently.

Generation of Global Approximation Vectors:

For the purpose of generating global approximation vectors (or modes), the arrays K_{IJ}, F_{IJK} and G_{IJKL} are decomposed into the sum of orthotropic and nonorthotropic parts as follows:

$$K_{IJ} = \mathbf{K}_{IJ} + \lambda\, \mathcal{K}_{IJ} \qquad (2)$$

$$F_{IJK} = \mathfrak{F}_{IJK} + \lambda\, \mathcal{F}_{IJK} \qquad (3)$$

$$G_{IJKL} = \mathbf{G}_{IJKL} + \lambda\, \mathcal{G}_{IJKL} \qquad (4)$$

where λ is a tracing parameter which identifies *all* the nonorthotropic (anisotropic) contributions contained in the arrays K_{IJ}, F_{IJK} and G_{IJKL}.

The governing finite element equations of the panel, Eqs. 1, are now embedded in a single-parameter family of equations of the form:

$$f_I(\lambda) = \mathbf{K}_{IJ}\, X_J + \frac{1}{2}\, \mathcal{F}_{IJK}\, X_J X_K + \frac{1}{3}\, \mathbf{G}_{IJKL}\, X_J X_K X_L - Q_I$$

$$+ \lambda (K_{IJ}\, X_J + \frac{1}{2}\, F_{IJK}\, X_J X_K + \frac{1}{3}\, G_{IJKL}\, X_J X_K X_L) = 0 \tag{5}$$

Note that Eqs. 5 reduce to Eqs. 1 when the parameter $\lambda=1$.

The global approximation vectors are selected to be the vector $\mathbf{X}_I$, obtained by solving Eqs. 5 for the case $\lambda=0$, and its various-order derivatives with respect to λ (evaluated at $\lambda=0$). Henceforth, the derivatives of $\mathbf{X}_I$ with respect to λ will be referred to as *path derivatives*. The path derivatives are obtained by successive differentiation of the governing finite element equations, Eqs. 5. The recursive relations for the first four approximation vectors have the following form:

$$\mathbf{K}_{IJ}\, \mathbf{X}_J + \frac{1}{2}\, \mathcal{F}_{IJK}\, \mathbf{X}_J \mathbf{X}_K + \frac{1}{3}\, \mathbf{G}_{IJKL}\, \mathbf{X}_J \mathbf{X}_K \mathbf{X}_L - Q_I = 0 \tag{6}$$

$$\overset{*}{\mathbf{K}}_{IJ} \frac{\partial \mathbf{X}_J}{\partial \lambda} = -(K_{IJ}\, \mathbf{X}_J + \frac{1}{2}\, F_{IJK}\, \mathbf{X}_J \mathbf{X}_K + \frac{1}{3}\, G_{IJKL}\, \mathbf{X}_J \mathbf{X}_K \mathbf{X}_L) \tag{7}$$

$$\overset{*}{\mathbf{K}}_{IJ} \frac{\partial^2 \mathbf{X}_J}{\partial \lambda^2} = -2\, \overset{*}{K}_{IJ} \frac{\partial \mathbf{X}_J}{\partial \lambda} - (\mathcal{F}_{IJK} \frac{\partial \mathbf{X}_J}{\partial \lambda} \frac{\partial \mathbf{X}_K}{\partial \lambda} + 2\, \mathbf{G}_{IJKL} \frac{\partial \mathbf{X}_J}{\partial \lambda} \frac{\partial \mathbf{X}_K}{\partial \lambda} \mathbf{X}_L) \tag{8}$$

$$\overset{*}{\mathbf{K}}_{IJ} \frac{\partial^3 \mathbf{X}_J}{\partial \lambda^3} = -3\, \overset{*}{K}_{IJ} \frac{\partial^2 \mathbf{X}_J}{\partial \lambda^2} - 3\, \mathcal{F}_{IJK} \frac{\partial^2 \mathbf{X}_J}{\partial \lambda^2} \frac{\partial \mathbf{X}_K}{\partial \lambda} - \mathbf{G}_{IJKL}\, (6 \frac{\partial^2 \mathbf{X}_J}{\partial \lambda^2} \frac{\partial \mathbf{X}_K}{\partial \lambda} \mathbf{X}_L$$

$$+ 2 \frac{\partial \mathbf{X}_J}{\partial \lambda} \frac{\partial \mathbf{X}_K}{\partial \lambda} \frac{\partial \mathbf{X}_L}{\partial \lambda}) - 3(F_{IJK} \frac{\partial \mathbf{X}_J}{\partial \lambda} \frac{\partial \mathbf{X}_K}{\partial \lambda} + 2\, G_{IJKL} \frac{\partial \mathbf{X}_J}{\partial \lambda} \frac{\partial \mathbf{X}_K}{\partial \lambda} \mathbf{X}_L) \tag{9}$$

where

$$\overset{*}{\mathbf{K}}_{IJ} = \mathbf{K}_{IJ} + \mathcal{F}_{IJK}\, \mathbf{X}_K + \mathbf{G}_{IJKL}\, \mathbf{X}_K \mathbf{X}_L \tag{10}$$

and

$$\overset{*}{K}_{IJ} = K_{IJ} + F_{IJK}\, \mathbf{X}_K + G_{IJKL}\, \mathbf{X}_K \mathbf{X}_L \tag{11}$$

Note that the generation of the vector $\mathbf{X}_I$ requires the solution of a non-linear system of equations, Eqs. 6. However, *the generation of the path derivatives involves the solution of a system of linear equations*, with the same left hand side array $\overset{*}{\mathbf{K}}_{IJ}$. Moreover, if Newton-Raphson iterative technique is used in generating $\overset{*}{\mathbf{X}}_I$, the Jacobian matrix used in the last iteration is the same array $\overset{*}{K}_{IJ}$.

For symmetric loading, Q_I, the right hand sides in the recursion formulas for the even-order derivatives of $\mathbf{X}_I$ with respect to λ are symmetric; while those for the odd-order derivatives are antisymmetric. Hence, *the size of the analysis region used in evaluating the approximation vector* $\mathbf{X}_I$ *and its path derivatives is the same as that used for the orthotropic case* (λ=0).

The efficient generation of the approximation vectors can be accomplished by a) partial decomposition of the portion of the matrix $\overset{*}{K}_{IJ}$ which does not include the symmetry and antisymmetry conditions once; then b) incorporating the symmetry (and antisymmetry) conditions; completing the decomposition for each of the two cases; evaluating the right hand sides; and successively generating the vectors.

Computation of Amplitudes of Global Approximation Vectors:
The vector $\mathbf{X}_I$ and its path derivatives are now chosen as approximation vectors and the vector of nodal displacements, X_I, for the anisotropic panel is expressed as a linear combination of these vectors as follows:

$$X_J = \Gamma_{Ji}\,\psi_i \quad ; \qquad J=1 \text{ to } N \text{ and } i=1 \text{ to } r \tag{12}$$

where

$$\Gamma_{Ji} = [\mathbf{X}_J \; \frac{\partial \mathbf{X}_J}{\partial \lambda} \; \frac{\partial^2 \mathbf{X}_J}{\partial \lambda^2} \; \cdots \; \frac{\partial^{r-1} \mathbf{X}_J}{\partial \lambda^{r-1}}] = \frac{\partial^{i-1} \mathbf{X}_J}{\partial \lambda^{i-1}} \tag{13}$$

and ψ_i (i=1 to r; r << N) is a vector of unknown parameters which represent the amplitudes of the approximation vectors (or modes).

A Rayleigh-Ritz technique is used to replace Eqs. 5 by the following reduced system of nonlinear equations in the unknown parameters ψ_i:

$$\tilde{f}_i(\lambda) = \tilde{K}_{ij}\,\psi_j + \frac{1}{2}\tilde{\mathcal{F}}_{ijk}\,\psi_j\psi_k + \frac{1}{3}\tilde{\mathbf{G}}_{ijk\ell}\,\psi_j\psi_k\psi_\ell - \tilde{Q}_i$$

$$+ \lambda(\tilde{K}_{ij}\,\psi_j + \frac{1}{2}\tilde{F}_{ijk}\,\psi_j\psi_k + \frac{1}{3}\tilde{G}_{ijk\ell}\,\psi_j\psi_k\psi_\ell) = 0 \tag{14}$$

where

$$\tilde{K}_{ij} = K_{IJ}\,\Gamma_{Ii}\,\Gamma_{Jj} \tag{15}$$

$$\tilde{\mathcal{F}}_{ijk} = \mathcal{F}_{IJK}\, \Gamma_{Ii}\, \Gamma_{Jj}\, \Gamma_{Kk} \tag{16}$$

$$\tilde{\mathcal{G}}_{ijk\ell} = \mathcal{G}_{IJKL}\, \Gamma_{Ii}\, \Gamma_{Jj}\, \Gamma_{Kk}\, \Gamma_{L\ell} \tag{17}$$

and

$$\tilde{Q}_i = Q_I\, \Gamma_{Ii} \tag{18}$$

with similar expressions for $\tilde{K}_{ij}$, $\tilde{F}_{ijk}$ and $\tilde{G}_{ijk\ell}$.

Comments on the Generation of the Reduced Arrays:

For symmetric loading the generation of the reduced arrays $\tilde{\mathbf{K}}_{ij}$, $\tilde{\mathcal{F}}_{ijk}$, $\tilde{\mathcal{G}}_{ijk\ell}$, $\tilde{K}_{ij}$, $\tilde{F}_{ijk}$, $\tilde{G}_{ijk\ell}$ and $\tilde{Q}_i$ can be greatly simplified by noting the following expressions for their typical elements.

$$\tilde{\mathbf{K}}_{ij} = \mathbf{K}_{IJ} \frac{\partial^{i-1} \mathcal{X}_I}{\partial \lambda^{i-1}} \frac{\partial^{j-1} \mathcal{X}_J}{\partial \lambda^{j-1}} \tag{19}$$

$$= 0 \quad \text{if } i+j \text{ is odd.}$$

$$\tilde{\mathcal{F}}_{ijk} = \mathcal{F}_{IJK} \frac{\partial^{i-1} \mathcal{X}_I}{\partial \lambda^{i-1}} \frac{\partial^{j-1} \mathcal{X}_J}{\partial \lambda^{j-1}} \frac{\partial^{k-1} \mathcal{X}_K}{\partial \lambda^{k-1}} \tag{20}$$

$$= 0 \quad \text{if } i+j+k \text{ is even.}$$

$$\tilde{\mathcal{G}}_{ijk\ell} = \mathcal{G}_{IJKL} \frac{\partial^{i-1} \mathcal{X}_I}{\partial \lambda^{i-1}} \frac{\partial^{j-1} \mathcal{X}_J}{\partial \lambda^{j-1}} \frac{\partial^{k-1} \mathcal{X}_K}{\partial \lambda^{k-1}} \frac{\partial^{\ell-1} \mathcal{X}_L}{\partial \lambda^{\ell-1}} \tag{21}$$

$$= 0 \quad \text{if } i+j+k+\ell \text{ is odd.}$$

$$\tilde{Q}_i = Q_I \frac{\partial^{i-1} \mathcal{X}_I}{\partial \lambda^{i-1}} \tag{22}$$

$$= 0 \quad \text{if } i \text{ is even.}$$

where a repeated uppercase Latin index denotes summation over its full range. The expressions of the typical elements of the arrays $\tilde{K}_{ij}$, $\tilde{F}_{ijk}$ and $\tilde{G}_{ijk\ell}$ are given by Eqs. 19-21 after replacing $\tilde{\mathbf{K}}$, $\tilde{\mathcal{F}}$ and $\tilde{\mathcal{G}}$ by $\tilde{K}$, $\tilde{F}$ and $\tilde{G}$, respectively. Moreover, the following relations exist:

$$\tilde{K}_{ij} = 0 \qquad \text{if } i+j \text{ is even.} \tag{23}$$

$$\tilde{F}_{ijk} = 0 \qquad \text{if } i+j+k \text{ is odd.} \tag{24}$$

$$\tilde{G}_{ijk\ell} = 0 \qquad \text{if } i+j+k+\ell \text{ is even.} \tag{25}$$

Therefore, the nonzero $\tilde{\mathbf{K}}_{ij}$, $\tilde{\mathcal{F}}_{ijk}$ and $\tilde{\mathbf{G}}_{ijk\ell}$ correspond to the zero $\tilde{K}_{ij}$, $\tilde{F}_{ijk}$, and $\tilde{G}_{ijk\ell}$ and vice versa. Note that all the reduced arrays are obtained by using the same size model as for the orthotropic case.

Computational Procedure

For a given symmetric loading, Q_I, two possible approaches can be used for the efficient generation of the nonlinear response of the anisotropic panel. In both approaches the arrays $\mathbf{K}_{IJ}$, $\mathcal{F}_{IJK}$, $\mathbf{G}_{IJKL}$, Q_I, K_{IJ}, F_{IJK} and G_{IJKL} are evaluated using the same size model as for the orthotropic panel. Also, advantage is taken of the symmetry of the different arrays. The two approaches are described subsequently for the case of proportional loading $Q_I = q\,\bar{Q}_I$ where q is a load parameter and $\bar{Q}_I$ are normalized external load components.

The first approach consists of the successive application of two single-parameter reduction methods. The first reduction method is used to generate the vector $\mathbf{X}_I$ (nonlinear solution corresponding to $\lambda=0$) at the given value of q. The method is described in Refs. [6] and [7] and the global approximation vectors used in this method are $\frac{\partial \mathbf{X}_I}{\partial q}$, $\frac{\partial^2 \mathbf{X}_I}{\partial q^2}$, ... Then the path derivatives $\frac{\partial \mathbf{X}_I}{\partial \lambda}$, $\frac{\partial^2 \mathbf{X}_I}{\partial \lambda^2}$, ... are generated, using Eqs. 7 to 9, and the foregoing reduction method is subsequently used to generate the nonlinear solution of the anisotropic panel, X_I, corresponding to $\lambda=1$ and to the given value of q (using Eqs. 14 and 12).

The second approach consists of a single application of a two-parameter reduction method with the two parameters selected as q and λ. The governing finite element equations of the panel are embedded in a two-parameter family of equations of the form:

$$f_I(\lambda,q) = \mathbf{K}_{IJ}\, X_J + \frac{1}{2}\mathcal{F}_{IJK}\, X_J X_K + \frac{1}{3}\mathbf{G}_{IJKL}\, X_J X_K X_L - q\,\bar{Q}_I$$

$$+ \lambda (K_{IJ}\, X_J + \frac{1}{2} F_{IJK}\, X_J X_K + \frac{1}{3} G_{IJKL}\, X_J X_K X_L) = 0 \tag{26}$$

The global approximation vectors are chosen to be $\frac{\partial \mathbf{X}_I}{\partial \lambda}$, $\frac{\partial \mathbf{X}_I}{\partial q}$, $\frac{\partial^2 \mathbf{X}_I}{\partial \lambda^2}$, $\frac{\partial^2 \mathbf{X}_I}{\partial \lambda \partial q}$,

$\frac{\partial^2 \mathbf{X}_I}{\partial q^2}$, ... which are obtained by successive differentiation of Eqs. 26. For simplicity, these vectors are evaluated at $q=\lambda=0$. The reduced nonlinear equations corresponding to $\lambda=1$ and the chosen value of q, are solved for ψ_i and the anisotropic response is obtained using Eqs. 12.

Note that the global approximation vectors used in the second approach differ from those of the first approach in the following:

1. The presence of various-order mixed partial derivatives with respect to λ and q.

2. The point of generation of all the global approximation vectors is $\lambda=q=0$. By contrast, the global approximation vectors in the two reduction methods of the first approach are generated at two different values of q.

As a consequence of the aforementioned differences, a reduced system of equations, similar to Eqs. 14, with new reduced arrays has to be generated for each new value of q in the first approach. By contrast, only a single reduced system of equations is generated in the second approach.

Sensitivity of the Nonlinear Response of the Panel to Nonorthotropic Material Coefficients

The chosen set of global approximation vectors provide a direct measure of the sensitivity of the different response quantities to nonorthotropic (anisotropic) material coefficients of the structure. Moreover, the global approximation vectors can be used to compute the derivatives of the total strain energy U with respect to λ. The expressions of the first three derivatives of U are as follows:

$$\left.\frac{\partial U}{\partial \lambda}\right|_o = \bar{K}_{IJ} \mathbf{X}_I \frac{\partial \mathbf{X}_J}{\partial \lambda} + \frac{1}{2} (K_{IJ} \mathbf{X}_I \mathbf{X}_J + \frac{1}{3} F_{IJK} \mathbf{X}_I \mathbf{X}_J \mathbf{X}_K + \frac{1}{6} G_{IJKL} \mathbf{X}_I \mathbf{X}_J \mathbf{X}_K \mathbf{X}_L) \tag{27}$$

$$\left.\frac{\partial^2 U}{\partial \lambda^2}\right|_o = \bar{K}_{IJ} \mathbf{X}_I \frac{\partial^2 \mathbf{X}_J}{\partial \lambda^2} + \overset{*}{K}_{IJ} \frac{\partial \mathbf{X}_I}{\partial \lambda} \frac{\partial \mathbf{X}_J}{\partial \lambda} + 2 \bar{K}_{IJ} \mathbf{X}_I \frac{\partial \mathbf{X}_J}{\partial \lambda} \tag{28}$$

$$\left.\frac{\partial^3 U}{\partial \lambda^3}\right|_o = \bar{K}_{IJ} \mathbf{X}_I \frac{\partial^3 \mathbf{X}_J}{\partial \lambda^3} + 3 \overset{*}{K}_{IJ} \frac{\partial \mathbf{X}_I}{\partial \lambda} \frac{\partial^2 \mathbf{X}_J}{\partial \lambda^2} + 3 \bar{K}_{IJ} \mathbf{X}_I \frac{\partial^2 \mathbf{X}_J}{\partial \lambda^2}$$
$$+ 3 \overset{*}{K}_{IJ} \frac{\partial \mathbf{X}_I}{\partial \lambda} \frac{\partial \mathbf{X}_J}{\partial \lambda} + (\mathcal{F}_{IJK} + 2 \mathbf{G}_{IJKL} \mathbf{X}_L) \frac{\partial \mathbf{X}_I}{\partial \lambda} \frac{\partial \mathbf{X}_J}{\partial \lambda} \frac{\partial \mathbf{X}_K}{\partial \lambda} \tag{29}$$

where $\overset{*}{K}_{IJ}$ and $\overset{*}{K}_{IJ}$ are defined in Eqs. 10 and 11, and

$$\bar{K}_{IJ} = K_{IJ} + \frac{1}{2} \mathcal{F}_{IJK} \mathbf{X}_K + \frac{1}{3} \mathbf{G}_{IJKL} \mathbf{X}_K \mathbf{X}_L \tag{30}$$

The expression of $\bar{K}_{IJ}$ is given by Eqs. 30 after replacing $\mathbf{K}$, $\mathbf{F}$ and $\mathbf{G}$ by K, F and G, respectively.

The derivatives of U in Eqs. 27 to 29 can be used as quantitative measures for the sensitivity of the global response of the structure to the nonorthotropic material coefficients. Specifically, dimensionless parameters ρ_ℓ (ℓ=1 to r-1) are introduced such that:

$$\rho_\ell = \left.\frac{\partial^\ell U}{\partial\lambda^\ell}\right|_o / U_o \tag{31}$$

where

$$U_o = \frac{1}{2} \mathbf{K}_{IJ} \mathbf{X}_I \mathbf{X}_J + \frac{1}{6} \mathbf{F}_{IJK} \mathbf{X}_I \mathbf{X}_J \mathbf{X}_K + \frac{1}{12} \mathbf{G}_{IJKL} \mathbf{X}_I \mathbf{X}_J \mathbf{X}_K \mathbf{X}_L \tag{32}$$

The smaller the values of ρ_ℓ the less sensitive the response is to the non-orthotropic material coefficients of the structure and vice versa. Also, small values of ρ_ℓ (less than 0.05) indicate that only few global approximation vectors (e.g., three) are needed for approximating X_I; and that Taylor series expansion can provide an acceptable approximation for X_I. On the other hand, large values of ρ_ℓ (greater than 0.25) indicate that more global approximation vectors (e.g., five or six) are needed for approximating X_I. Note that all the terms on the right hand sides of Eqs. 27 to 29, apart from a numerical coefficient, are identical to the coefficients of the reduced arrays (see Eqs. 19 to 21).

Numerical Studies

To evaluate the effectiveness of the proposed technique for model-size reduction, a number of nonlinear problems of symmetric anisotropic panels have been solved by using this technique. For each problem the solutions obtained by the proposed technique were compared with Taylor series expansion based on the same set of global approximation vectors, and with the direct solution of the anisotropic panel. Herein the solutions for a typical problem of a fifteen-ply cylindrical panel subjected to uniform normal loading are discussed. The panel is made of graphite-epoxy material and to amplify the effect of anisotropy, the fiber orientation is assumed to be $[+50_5/35_5/-50_5]$ (see Fig. 1). The numerical results obtained are summarized in Figures 2 through 5 and in Tables 1 and 2. As expected, the normal displacement w exhibits inversion symmetry. The in-plane displacements u_1 and u_2, and the rotation components ϕ_1 and ϕ_2 exhibit inversion antisymmetry. These symmetry

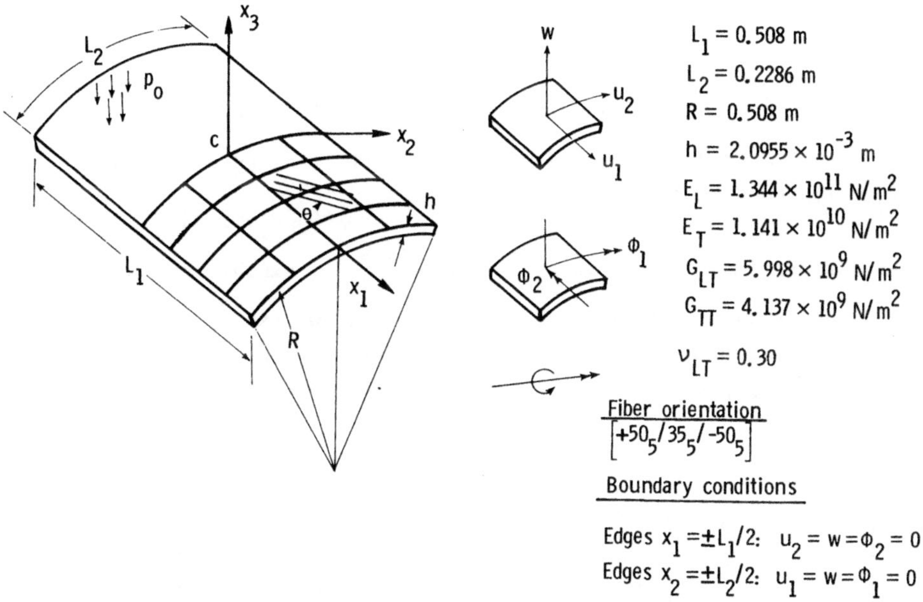

Figure 1 - Cylindrical panel used in the present study.

properties were used in conjunction with the procedure outlined in Ref. [4] to *reduce the analysis model to half the shell.* A 4 × 4 uniform grid of nine-noded Lagrangian elements was used with biquadratic interpolation functions for each of the displacement and rotation components (a total of 309 nonzero degrees of freedom). Reduced integration (with four quadrature points in the element domain) was used in generating the elemental arrays.

The highly anisotropic response of the panel is depicted in Figures 2 and 3. Figure 2 shows the variations of the normal displacement w_c and the total strain energy U with the loading parameter $q = \frac{p_o R}{E_T h}$; for the two cases λ=1 (anisotropic panel) and λ=0 (nonorthotropic arrays set equal to zero). Figure 3 shows contour plots for the generalized displacements, obtained by direct solution of the anisotropic shell at $q = 2 \times 10^{-4}$.

The analysis of the panel by the proposed technique was accomplished by successive application of two single-parameter reduction methods. In the first reduction method, the global approximation vectors $\frac{\partial \mathbf{X}_I}{\partial q}$, $\frac{\partial^2 \mathbf{X}_I}{\partial q^2}$, ...

were generated at q=0 (and λ=0) and were used to generate the nonlinear solution $\mathbf{X}_I$ at $q = 2 \times 10^{-4}$. The vector $\mathbf{X}_I$ and its path derivatives with respect to λ; namely, $\frac{\partial \mathbf{X}_I}{\partial \lambda}, \frac{\partial^2 \mathbf{X}_I}{\partial \lambda^2}, \ldots$ were then used as global approximation vectors and the foregoing reduction method was used to generate the anisotropic solution X_I corresponding to $q = 2 \times 10^{-4}$ and λ=1.

The global approximation vectors $\mathbf{X}_I, \frac{\partial \mathbf{X}_I}{\partial \lambda}, \frac{\partial^2 \mathbf{X}_I}{\partial \lambda^2}, \ldots$ exhibit reflection symmetry (and/or antisymmetry) with respect to the two axes of symmetry of the shell. These symmetry relations are depicted in Figure 4 and are represented by the following relations:

$$\frac{\partial^m}{\partial \lambda^m}\begin{Bmatrix} u_1 \\ u_2 \\ w \\ \phi_1 \\ \phi_2 \end{Bmatrix}_{(x_1,x_2)} = \frac{\partial^m}{\partial \lambda^m}\begin{Bmatrix} -(-1)^m u_1 \\ (-1)^m u_2 \\ (-1)^m w \\ -(-1)^m \phi_1 \\ (-1)^m \phi_2 \end{Bmatrix}_{(-x_1,x_2)} = \frac{\partial^m}{\partial \lambda^m}\begin{Bmatrix} (-1)^m u_1 \\ -(-1)^m u_2 \\ (-1)^m w \\ (-1)^m \phi_1 \\ -(-1)^m \phi_2 \end{Bmatrix}_{(x_1,-x_2)} \qquad (33)$$

where $m = 0,1,2, \ldots$. The maximum values of the generalized displacements and their derivatives with respect to λ are given in Table 1.

Table 1 - Maximum values of generalized displacements and their derivatives with respect to λ. Anisotropic cylindrical shell subjected to uniform normal loading (see Fig. 1); $q = 2 \times 10^{-4}$.

	X	$\mathbf{X}$	$\frac{\partial \mathbf{X}}{\partial \lambda}$	$\frac{\partial^2 \mathbf{X}}{\partial \lambda^2}$	$\frac{\partial^3 \mathbf{X}}{\partial \lambda^3}$
$10^{-2} \times u_1 E_T/(p_o R)$	3.0422	2.1598	-0.1845	0.6982	-0.5719
$10^{-2} \times u_2 E_T/(p_o R)$	-8.1788	5.5496	0.4436	2.2501	1.2146
$10^{-3} \times w E_T/(p_o R)$	-5.834	-3.650	0.3322	-1.797	-1.2203
$10^{-4} \times \phi_1 E_T/p_o$	-2.0351	1.2618	0.4860	-0.6055	-1.0658
$10^{-5} \times \phi_2 E_T/p_o$	4.4773	-2.7511	0.7453	-1.1172	2.4061

Note: Global sensitivity measures are ρ_1=0, ρ_2=0.388, ρ_3=0 and ρ_4=2.558.

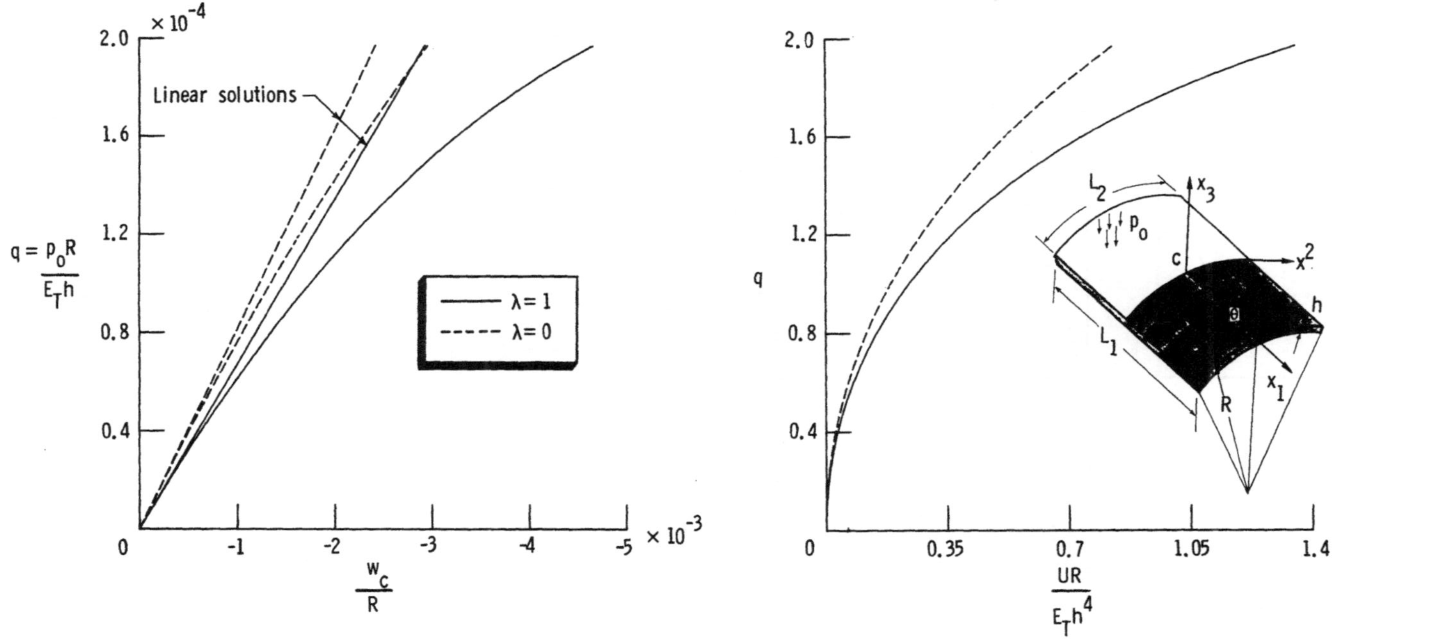

Figure 2 - Variation of normal displacement w_c and total strain energy U with loading. Cylindrical panel subjected to uniform normal loading (see Fig. 1).

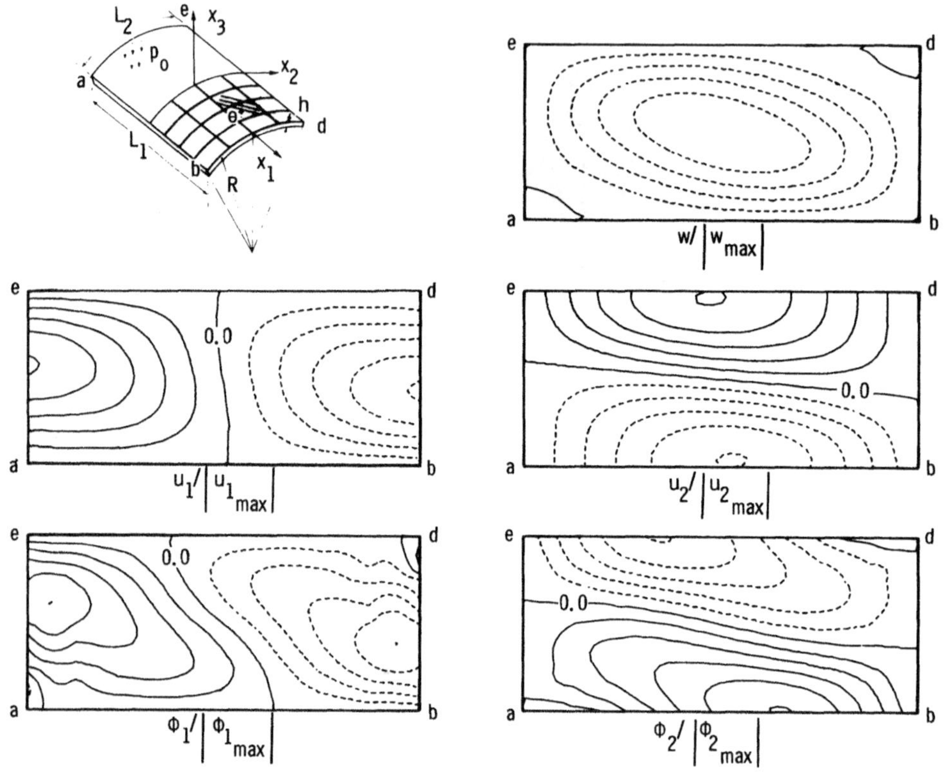

Figure 3 - Normalized contour plots for generalized displacements. Anisotropic cylindrical panel subjected to uniform normal loading ($q = 2 \times 10^{-4}$) (see Fig. 1). Dashed lines represent negative contours.

The symmetry relations, Eqs. 33, clearly demonstrate the fact that the global approximation vectors can each be *obtained by analyzing only one quadrant of the shell.*

The numerical values of the amplitudes of the global approximation vectors in the reduction method are given in Table 2. The corresponding amplitudes of the same vectors in the Taylor series expansion are also given in Table 2.

An indication of the accuracy and convergence of the solutions obtained by the proposed technique and the Taylor series expansion is given in Figure 5. The standard of comparison is taken to be the direct nonlinear finite element solution of the anisotropic shell. As can be seen from Figure 5, the solutions obtained by the proposed reduction method are considerably more accurate than those obtained by the corresponding Taylor series expansions (based on the same global approximation vectors). As an example to this the

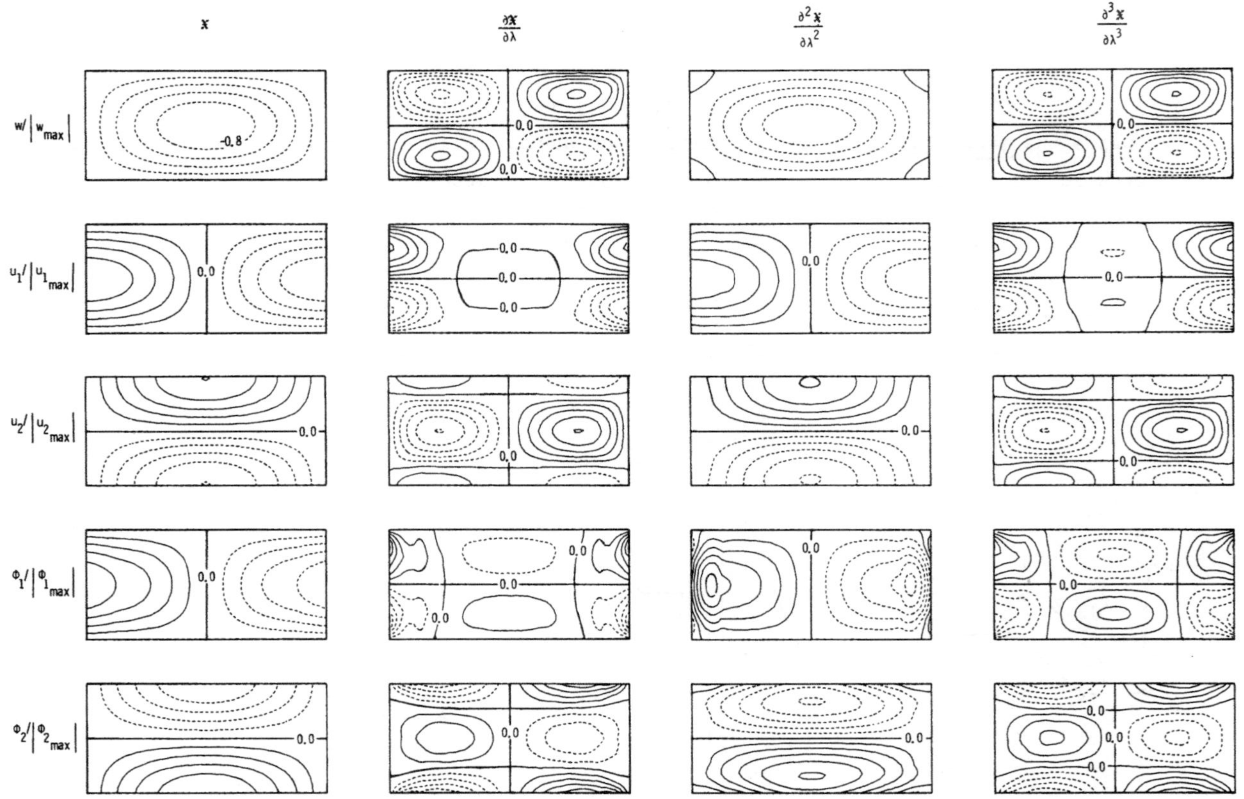

Figure 4 - Normalized contour plots for the global approximation vectors. Anisotropic cylindrical shell subjected to uniform normal loading ($q = 2 \times 10^{-4}$) (see Fig. 1). Dashed lines represent negative contours.

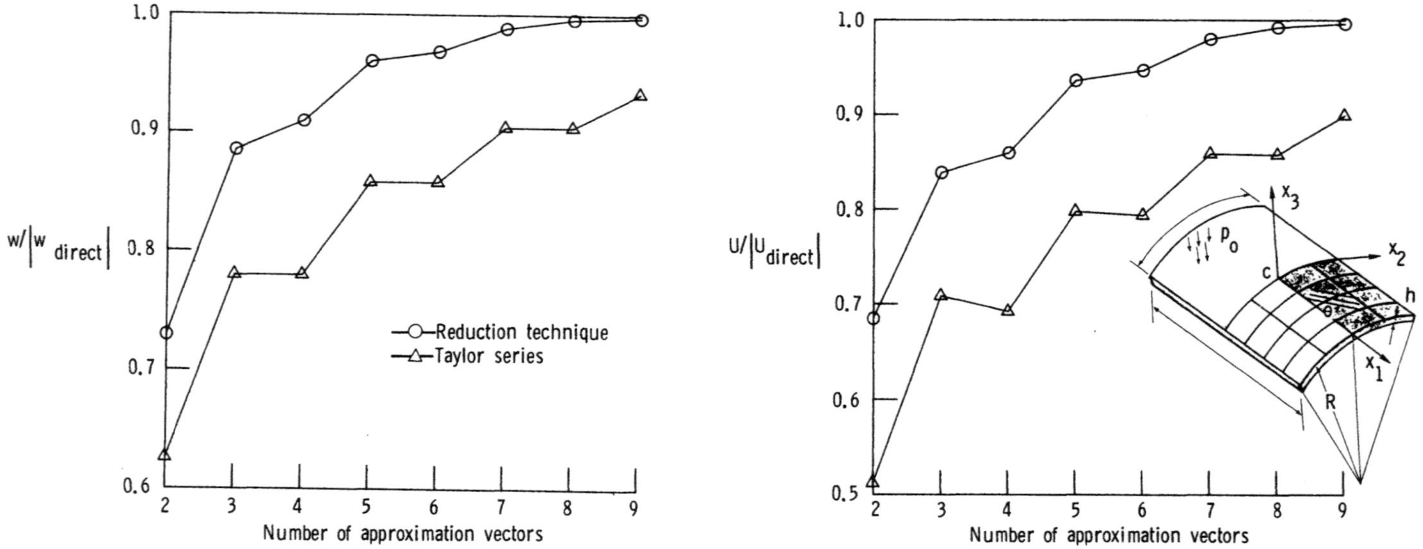

Figure 5 - Convergence of normal displacement w_c and total strain energy U obtained by the proposed reduction method and Taylor series expansion. Anisotropic cylindrical panel subjected to uniform normal loading ($q = 2 \times 10^{-4}$) (see Fig. 1).

errors in the total strain energy obtained using the Taylor series with five and nine global approximation vectors were 20.25% and 10.1%. The corresponding errors in the strain energy obtained by the proposed technique were only 6.4% and 0.41%.

Table 2 - Amplitudes of global approximation vectors in proposed technique. Anisotropic cylindrical shell subjected to uniform normal loading (see Fig. 1).

r	ψ_1	ψ_2	ψ_3	ψ_4	ψ_5	ψ_6	ψ_7	ψ_8
2	1.1658	1.1359						
3	0.9676	1.6578	0.9084					
4	0.9248	0.9075	1.0747	0.3363				
5	1.0022	0.5924	0.1310	0.5041	0.1557			
6	1.0101	0.9761	-0.0561	-0.0066	0.1879	0.0363		
7	0.9892	1.1011	0.7140	-0.1472	-0.1058	0.0465	0.0105	
8	0.9846	0.9440	0.9319	0.3777	-0.1868	-0.0475	0.0134	0.0021
Taylor Series	1.0	1.0	0.5	0.1667	0.0417	8.333×10^{-3}	1.389×10^{-3}	1.984×10^{-4}

Potential of the Proposed Reduction Method

The proposed reduction method appears to have high potential for the nonlinear analysis of symmetric anisotropic panels. The numerical studies conducted clearly demonstrate the effectiveness of the technique. In particular, the following points are worth mentioning:

1. The proposed reduction method can be thought of as either of the following:

a) Generalized single- and two-parameter Taylor series expansions of the response in terms of λ (and q) in which i) the Taylor series expansions contain free parameters rather than fixed coefficients, and ii) the parameter(s) λ (and q) need not be small (corresponding to slightly anisotropic structure, and small loading parameter q).

b) Extended Rayleigh-Ritz technique in which i) the approximation vectors are generated by using the finite element method rather than chosen *a priori*, and ii) only a small number of approximation vectors is used. The size of the model required for the generation of the approximation vectors is the same as that used for the corresponding orthotropic structure.

2. The reduction method provides a general computational strategy in which the nonlinear response of a complex structural system (e.g., anisotropic

panel) is generated using (small or large) perturbations from the response of a simpler system (e.g., orthotropic panel). It is also possible to use a hierarchy of simpler systems in generating the response of the original complex structural system by choosing a number of perturbation parameters, and successively applying a single-parameter reduction method with each of the parameters.

3. The proposed reduction method can be used to reduce the size of the analysis model and the number of degrees of freedom required for predicting the nonlinear response of a quasi-symmetric anisotropic panel such as symmetric panels with unsymmetric imperfections. This can be accomplished by decomposing the finite element arrays, corresponding to the panel with unsymmetric imperfections, into symmetric and antisymmetric parts, and using the procedure outlined in the second and third sections.

Concluding Remarks

Reduction method and computational procedures are presented for reducing the size of the analysis model and the number of degrees of freedom used in predicting the nonlinear response of symmetric anisotropic panels. The key elements of the procedure are: a) decomposition of the characteristic arrays of the finite element model into the sum of orthotropic and nonorthotropic contributions; b) combined use of the finite element method and the classical Rayleigh-Ritz technique. The finite element method is first used to generate few global approximation vectors (or modes). Then the amplitudes of these modes are computed by using the classical Rayleigh-Ritz technique.

The global approximation vectors are selected to be those commonly used in single (or multiple) parameter perturbation technique; namely, a nonlinear solution corresponding to zero nonorthotropic arrays and a number of its derivatives with respect to an anisotropic tracing parameter (and possibly, to a load or arc-length parameter in the solution space). The size of the analysis model used in generating the global approximation vectors is identical to that of the corresponding orthotropic structure.

Numerical example of the nonlinear response of an anisotropic cylindrical panel subjected to uniform normal loading is used to demonstrate the effectiveness of the proposed reduction method.

The results of the study suggest the following conclusions relative to the selection of global approximation vectors and to the potential of the proposed reduction method. These conclusions are as follows:

1. The use of path derivatives as global approximation vectors leads to accurate solutions with a small number of vectors. Therefore, the time required to solve the reduced equations is relatively small, and the total analysis time to a first approximation equals the time required to evaluate the global approximation vectors and generate the reduced equations.

2. The global approximation vectors provide a direct measure of the sensitivity of the different response quantities to the nonorthotropic (anisotropic) material coefficients of the structure. The sensitivity of the global response can also be assessed by using these vectors.

3. The proposed reduction method exploits the best elements of the finite element method and the Rayleigh-Ritz technique as follows:

a) The finite element method is used as a general approach for generating global approximation vectors using the same size model as that of the corresponding orthotropic panel.

b) The Rayleigh-Ritz technique is used as an efficient procedure for minimizing and distributing the error throughout the structure.

4. The proposed method extends the range of applicability of the Taylor series expansion by relaxing the requirement of using small parameter(s) λ (and q) in the expansion.

References

1. Whitney, J.M.: Analysis of anisotropic rectangular plates. AIAA Journal, 10 (1972) 1344-1345.

2. Sun, C.T.: Double Fourier series solution to general anisotropic Plates. Journal of Math. and Phys. Sci. 6 (1972) 205-223.

3. Chia, C.Y.: Nonlinear analysis of plates. McGraw-Hill Book Co. 1980.

4. Noor, A.K.; Camin, R.A.: Symmetry considerations for anisotropic shells. Computer Methods in Applied Mechanics and Engineering 9 (1976) 317-335.

5. Noor, A.K.; Mathers, M.D.; Anderson, M.S.: Exploiting symmetries for efficient postbuckling analysis of composite plates. AIAA Journal 15 (1977) 24-32.

6. Noor, A.K.; Peters, J.M.: Reduced basis technique for nonlinear analysis of structures. AIAA Journal 18 (1980) 455-462.

7. Noor, A.K.; Peters, J.M.: Recent advances in reduction methods for instability analysis of structures. Computers and Structures 16 (1983) 67-80.

Global Constraints in Nonlinear Solution Strategies

B. KRÖPLIN, D. DINKLER*, H. HILLMANN*
University of Dortmund, West Germany
*University of Braunschweig, West Germany

Summary

The global constraint technique is presented as a general mean for nonlinear structural analysis. A unified formulation is lined out, which allows to construct a number of iteration schemes including Quasi Newton methods, path tracing methods and perturbation methods in order to investigate the stability of a fundamental solution path. Examples are given for thin walled structures.

1. Introduction

Constraints are a often used remedy in structural analysis procedures. They have been used in element context either as local constraints (plasticity) /1/, as element constraints /2/, or as global constraint, which is the topic of this paper.

A well known group of algorithms of the global kind are the so called constant arc methods, which use a length constraint in the parameter space in order to trace the structural response in static unstable situations. Some authors use work or creep type algorithms for the same purpose /3/.

In general nonlinear structural analysis the task is to solve the equation (1)

$$\delta \underline{z}^T \quad : \quad \underline{A}_{(\underline{z})} \cdot \underline{z} \quad = \quad \underline{p} \qquad , \tag{1}$$

which may be found from the variation of a potential energy. The features of the structure are represented by the structural matrix $\mathbf{A}_{(\underline{z})}$. The solution of (1) is obtained by an iteration procedure, where a sequence of linearized equations is solved. Iteraton methods are the

Finite Element Methods for Nonlinear Problems
Europe-US Symposium, Trondheim, Norway 1985
Editors: Bergan, Bathe, Wunderlich

standard Newton-Raphson-iteration or Quasi Newton-iteration, which may be stabilized, around singular points by artificial damping or other strategies.

By splitting $\mathbf{A}_{(\underline{z})}$ into the linear and the nonlinear part the nonlinearity can be formulated as additional constraint to the linear problem. This leads to iteration algorithms, which iterate only the constraint equation. Depending on the content of the constraint the Quasi-Newton-matrix updates or other updates e.g. a tangent Newton method or a secant method may be formulated in this form.

Another sort of application of global constraints are creep type methods. Additional constraints may be added in order to let the iteration proceed in the vicinity of the solution path without incrementation and iteration within the increments. The solution creeps along the solution path, while the load is applied all at once.

Further, constraints can be used in order to detect and calculate solution pathes of the nonlinear solution manifold in the surrounding of a fundamental path. This is important, if unexpected perturbations or initial imperfections let the structure snap from the fundamental path to another solution path which does not intersect with the fundamental path, but may be close to it. A perturbation energy is defined as a measure for the distance to the secondary solution path. The secondary path is reached by applying an energy constraint to the solution in the actual point.

2. A formulation of constraints

First we assume the existence of a potential for the considered problem. This assumption holds for elastic problems except nonconservative forces. If path dependencies shall be regarded, we have to consider a differential potential. Equation (1) and following equations are in this case incremental relationships.

The matrix notation used herein is based on a mixed formulation with displacements, axial forces and bending moments as independed variables. The mixed formulation reduces the order of nonlinearity in comparison to the displacement formulation. The potential π , of which an extremum is sought is given in equation (2).

$$\pi = \frac{1}{2}\,\underline{z}^T \cdot \underline{A}_{(\underline{z})} \cdot \underline{z} - \underline{z}^T \cdot \underline{p} \longrightarrow \textbf{extremum} \qquad (2)$$

$\underline{A}_{(\underline{z})}$: systemmatrix
$\underline{z}$: nodal variables
$\underline{p}$: nodal forces, strains and curvatures

The potential is extended by a constraint condition using the Lagrange-multiplier method. A constraint couples the nodal variables $\underline{z}$ with the set of Lagrange multipliers $\underline{\lambda}$, (3).

$$\pi^* = \pi + \pi^z \longrightarrow \textbf{extremum} \qquad (3)$$

$$\pi^z = -\,\underline{\lambda}^T \left\{ \underline{B}^T_{(\underline{z})} \cdot \underline{z} - \frac{1}{2}\,\underline{H}_{(\underline{z})} \cdot \underline{\lambda} - \underline{r} \right\} \qquad (4)$$

π^* : extended potential
π^z : constraint
$\underline{B}_{(\underline{z})}$: operator on $\underline{z}$ and $\underline{\lambda}$
$\underline{H}_{(\underline{z})}$: operator on $\underline{\lambda}$
$\underline{\lambda}$: Lagrange multipliers
$\underline{r}$: external constraint

Equation (4) represents a global constraint, which may affect all variables in the extended potential. The extremum condition leads after variation of $\underline{z}$ and $\underline{\lambda}$ to the set of equations (5).

$$\begin{matrix} \delta\underline{z}^T : \\ \delta\underline{\lambda}^T : \end{matrix} \quad \begin{bmatrix} \underline{A}_{(\underline{z})} & -\underline{B}_{(\underline{z})} \\ -\underline{B}^T_{(\underline{z})} & \underline{H}_{(\underline{z})} \end{bmatrix} \cdot \begin{bmatrix} \underline{z} \\ \underline{\lambda} \end{bmatrix} = \begin{bmatrix} \underline{p} \\ \underline{r} \end{bmatrix} \qquad (5)$$

In order to keep the linearity of the formulation $\underline{A}_{(\underline{z})}$, $\underline{B}_{(\underline{z})}$ and $\underline{H}_{(\underline{z})}$ are kept constant in the variation. The extremum is then sought in iterative manner. The matrices which depend on $\underline{z}$ have to be updated in every iteration step until $\underline{z}$ and $\underline{\lambda}$ do not change any more. Normally $\underline{A}_{(\underline{z})}$ is banded while $\underline{B}_{(\underline{z})}$ and $\underline{H}_{(\underline{z})}$ are fully populated. The solution of (5) can be obtained with an algorithm after Zielke /4/, which is advantageous for small numbers of Lagrange parameters $\underline{\lambda}$. The solution scheme consists of two steps (6).

$$\underline{\lambda} = \left[\underline{H}_{(\underline{z})} - \underline{B}^T_{(\underline{z})} \cdot \underline{A}^{-1}_{(\underline{z})} \cdot \underline{B}_{(\underline{z})}\right]^{-1} \cdot \left[\underline{r} + \underline{B}^T_{(\underline{z})} \cdot \underline{A}^{-1}_{(\underline{z})} \cdot \underline{p}\right]$$

$$\underline{z} = \underline{A}^{-1}_{(\underline{z})} \cdot \left[\underline{p} + \underline{B}_{(\underline{z})} \cdot \underline{\lambda}\right] \tag{6}$$

$\underline{B}_{(\underline{z})} \cdot \underline{\lambda}$ is considered as a second load case. It should be noted that for a constant $\underline{A}_{(\underline{z})}$ the iteration can be performed by inverting $\underline{A}_{(\underline{z})}$ only once. The scheme holds also for nonsymmetric sets of equations, which may appear for constraints of other than energy form. The elimination of $\underline{\lambda}$ is possible if $\underline{H}_{(\underline{z})}$ is a definite matrix (7).

$$\underline{z} = \left[\underline{A}_{(\underline{z})} - \underline{B}_{(\underline{z})} \cdot \underline{H}^{-1}_{(\underline{z})} \cdot \underline{B}^T_{(\underline{z})}\right]^{-1} \cdot \left[\underline{p} + \underline{B}_{(\underline{z})} \cdot \underline{H}^{-1}_{(\underline{z})} \cdot \underline{r}\right] \tag{7}$$

The solution scheme (6) is direct comparable with the inverse update schemes (8) given for example in /5/.

$$\left[\underline{A} - \underline{B} \cdot \underline{H}^{-1} \cdot \underline{B}^T\right]^{-1} = \underline{A}^{-1} + \underline{A}^{-1} \cdot \underline{B}\left[\underline{H} - \underline{B}^T \cdot \underline{A}^{-1} \cdot \underline{B}\right]^{-1} \underline{B}^T \cdot \underline{A}^{-1} \tag{8}$$

3. Quasi-Newton-Method as constraint

The iterative solution of equation (1) leads to a sequence of solutions of linearized equations. Dependent on the chosen algorithm either the left hand side $\underline{A}_{(\underline{z})}$ or the right hand side $\underline{p}$ are successively adapted to the actual values of $\underline{z}$. Classical methods are

- Newton-Raphson-Method where $\underline{A}_{(\underline{z})}$ and $\underline{p}$ are adapted,
- the modified Newton-Raphson-Method where only $\underline{p}$ is adapted and
- the secant Newton method where $\underline{A}_{(\underline{z})}$ is adapted.

Especially the Secant Newton method takes advantage of the earlier mentioned inverse update (8), which allows to invert the structural matrix only once. Since the suggested solution scheme for constraint problems (5) and the inverse update are identical, all rank 1 (also the unsymmetric ones) and the rank 2 (BFGS and DFP) updates can be formulated by constraint conditions. However rank 1 updates leads to one constraint per iteration step and rank 2 updates to two constraints.

Since the Zielke-Algorithm (6) is advantageous only for small numbers of extra equations, two formulations of Quasi-Newton-Methods are given herein, which use only one global constraint.

3.1 Secant iteration

In order to obtain an iteration with constant $\underline{A}_{(\underline{z})}$ the nonlinear potential energy π (2) is splitted into the linear part (9)

$$\pi = \frac{1}{2}\,\underline{z}_i^T \cdot \underline{A}^L \cdot \underline{z}_i - \underline{z}_i^T \cdot \underline{p} \qquad (9)$$

and the nonlinear part

$$\pi^z = \lambda_i \left[\underline{z}_{i-1}^T \cdot \underline{A}^N_{(\frac{1}{2}\underline{z}_{i-1})} \cdot \underline{z}_i - \frac{1}{2}\,\underline{z}_{i-1} \cdot \underline{A}^N_{(\frac{1}{2}\underline{z}_{i-1})} \cdot \underline{z}_{i-1} \cdot \lambda_i \right] \qquad (10)$$

$\underline{A}^L$: linear part of the structural matrix

$\underline{A}^N_{(\frac{1}{2}\underline{z}_{i-1})}$: nonlinear (secant) part of the structural matrix

$\underline{z}$: vector of nodal unknowns

λ : Lagrange parameter

The extended potential π^* (3) contains the geometric nonlinearities and the contraint $\underline{z}_i \sim \underline{z}_{i-1} \cdot \lambda_i$. The geometric nonlinear matrix $\underline{A}^N_{(\frac{1}{2}\underline{z}_{i-1})}$ can be obtained in the mixed principle as half of the tangent matrix /6/. For plasticity the matrix is given for point wise constraint in /1/. The variation of (9) and (10) with respect to $\underline{z}$ and $\underline{\lambda}$ gives (11).

$$\begin{matrix} \delta \underline{z}_i : \\ \delta \lambda_i : \end{matrix} \quad \begin{bmatrix} \underline{A}^L & \underline{A}^N_{(\frac{1}{2}\underline{z}_{i-1})} \cdot \underline{z}_{i-1} \\ \underline{z}_{i-1}^T \cdot \underline{A}^N_{(\frac{1}{2}\underline{z}_{i-1})} & -\underline{z}_{i-1}^T \cdot \underline{A}^N_{(\frac{1}{2}\underline{z}_{i-1})} \cdot \underline{z}_{i-1} \end{bmatrix} \cdot \begin{bmatrix} \underline{z}_i \\ \lambda_i \end{bmatrix} = \begin{bmatrix} \underline{p} \\ 0 \end{bmatrix} \qquad (11)$$

i means the iteration index.

The iteration converges for $\underline{z}_{i-1} \rightarrow \underline{z}_i$ and $\lambda_i \rightarrow 1.0$ towards the solution of the nonlinear problem. The iteration can be interpreted as successive change of the structural matrix. In the converged case the constraint contains the difference between $\underline{A}^L$ and $\underline{A}^S_{(\underline{z})}$, see fig. 1. The iteration is faster than the modified Newton-Raphson-Iteration but the convergence slows down in the vicinity of the fixpoint.

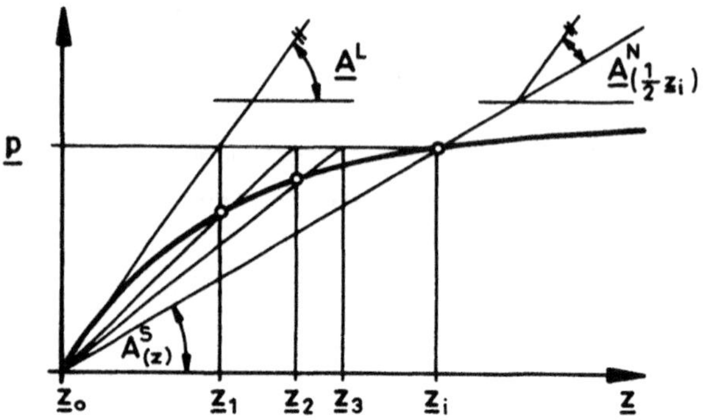

Fig. 1. Secant method

3.2 Tangent iteration

The performance of the secant iteration can be improved, if instead of the secant matrix the tangent matrix is approximated by the constraint (12).

$$\pi = \frac{1}{2}\Delta\underline{z}_i^T \cdot \underline{A}^L \cdot \Delta\underline{z}_i - \Delta\underline{z}_i^T \cdot \Delta\underline{p}_{i-1}^u$$

$$\pi^z = \Delta\lambda_i \left[\Delta\underline{z}_{i-1}^T \cdot \underline{A}^N_{(\underline{z}_{i-1})} \cdot \Delta\underline{z}_i - \frac{1}{2}\Delta\underline{z}_{i-1}^T \cdot \underline{A}^N_{(\underline{z}_{i-1})} \cdot \Delta\underline{z}_{i-1} \cdot \Delta\lambda_i\right] \qquad (12)$$

$\Delta\underline{z}_i$:increment of $\underline{z}$ (like N-R-Method)

$\Delta\underline{p}_{i-1}^u$:unbalanced force, in step 1 : $\Delta\underline{p}_1^u = \underline{p}$

The variation of $\Delta\underline{z}_i$ and $\Delta\lambda_i$ leads to (13)

$$\begin{array}{ll} \delta\Delta\underline{z}_i^T : \\ \delta\Delta\lambda_i : \end{array} \begin{bmatrix} \underline{A}^L & \underline{A}^N_{(\underline{z}_{i-1})}\ \underline{z}_{i-1} \\ \underline{z}_{i-1}\cdot\underline{A}^N_{(\underline{z}_{i-1})} & -\underline{z}_{i-1}\cdot\underline{A}^N_{(\underline{z}_{i-1})}\cdot\underline{z}_{i-1} \end{bmatrix} \cdot \begin{bmatrix} \Delta\underline{z}_i \\ \Delta\lambda_i \end{bmatrix} = \begin{bmatrix} \Delta\underline{p}_{i-1}^u \\ 0 \end{bmatrix} \qquad (13)$$

Since the tangent matrix is approximated by only one constraint the convergence is quadratic if $\underline{z}_i$ and $\underline{z}_{i-1}$ are of the same shape. In the fixpoint $\Delta\underline{p}_{i-1}^u$, $\Delta\underline{z}_i$ and $\Delta\lambda_i$ tend to zero. In the fixpoint the constraint contains the tangent matrix minus the linear matrix, see fig. 2.

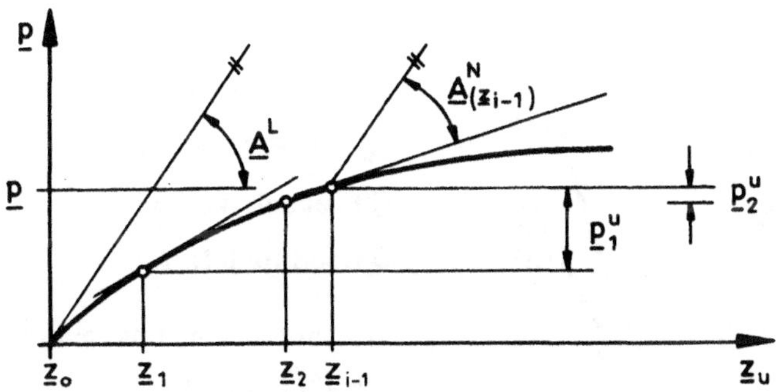

fig. 2. Tangent method

The linear matrix $\underline{A}^L$ is decomposed only once. In every iteration two right hand sides have to be solved.

3.3 Examples

The performance of both methods is compared in relation to the Newton-Raphson-Iteration for a cylindrical shell roof, given in fig. 3. Fig. 4 and fig. 5 give the comparison of the CPU time for different discretisations and the CPU time dependent on the relative error.

The effectivity of the constraint methods increases as the system increases (fig. 4), the effectivity decreases as smaller as the tolerated error is (fig. 5).

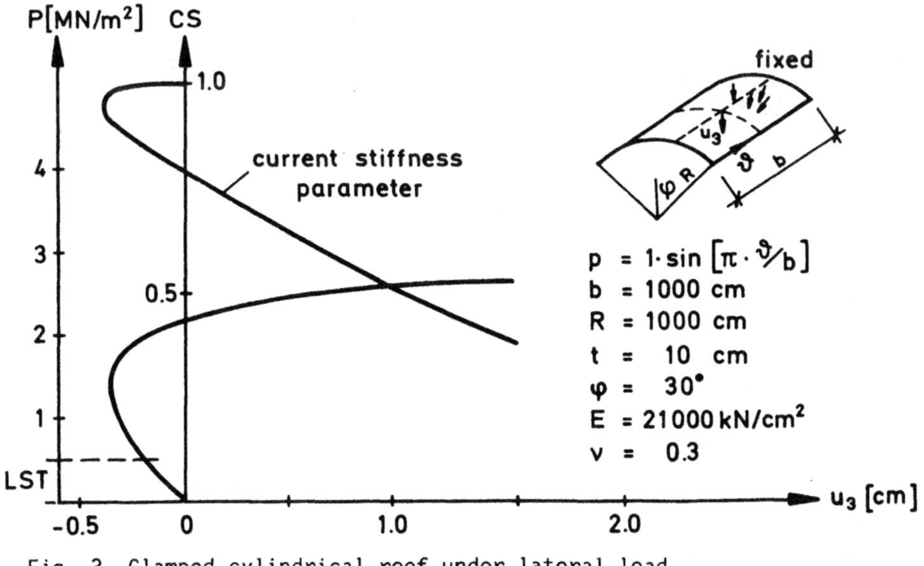

Fig. 3. Clamped cylindrical roof under lateral load

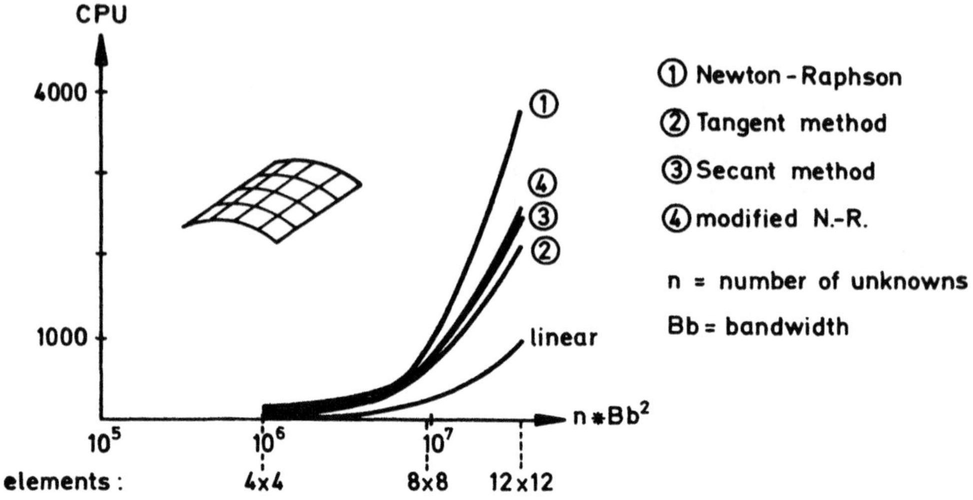

Fig. 4. CPU versus effort at load level LST (fig. 3)

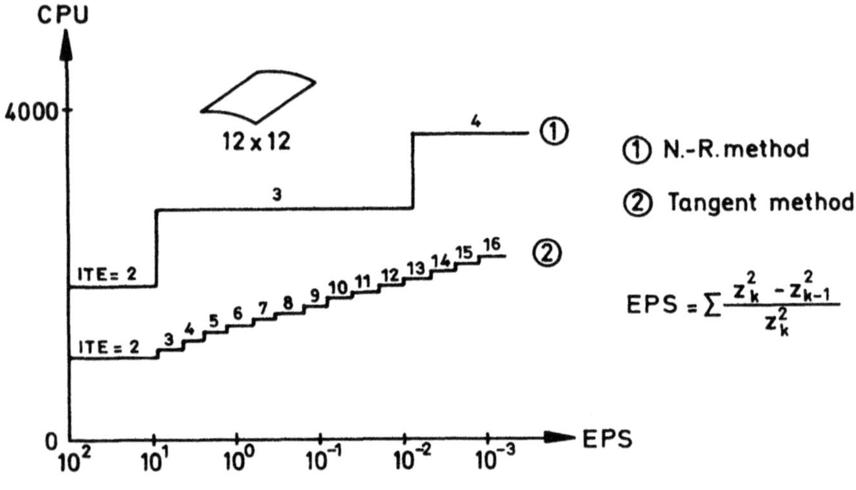

Fig. 5. CPU versus relative error

4. Load path approximation by constraints

In many cases it is efficient to follow the solution path in the vicinity of the exact solution. This can be achieved by using a set of constraints instead of an iterative correction in every increment. The ingredients of such a method are a step length parameter, and a selfcorrecting feature, which corrects the error of the previous step automaticly. The nonlinear potential is the one of equation (1). The step length is given by an energy step $\Delta\pi$. The constraint then says, that the unbalanced force $\underline{p}^u$ times $\Delta\underline{z}$ should be of size $\Delta\pi$, (14).

$$\pi_\lambda^z = -\lambda \left(\underline{p}^{u^T} \cdot \Delta\underline{z}_i - \Delta\pi \right) \tag{14}$$

λ :load factor

$\underline{p}^u$:unbalanced force, for $i = 1$: $\underline{p}^u = \underline{p}$

$\Delta\pi$:energy step.

For rapidly changing nonlinearities $\Delta\pi$ can be taken as variable e.g. dependent on the unbalanced forces. In order to make the solution selfcorrecting a decomposition of the unbalanced forces in the load space is considered. The unbalanced force is decomposed into an orthogonal part $\underline{p}^N$ and a tangential part $\underline{p}^T$. Two constraints regulate the direction of $\underline{p}^T$ and $\underline{p}^N$ such that a applied load increment lies between $\underline{p}^u$ and $\underline{p}^N$, see fig. 6. In the solution space then a spiral around the solution path is obtained, see fig. 6.

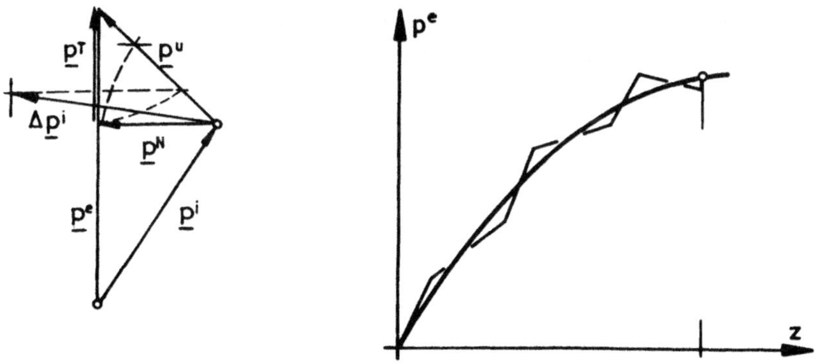

Fig. 6. Decomposition in the load space

The further constraints read (15).

$$\pi_\alpha^z = \alpha \cdot \left[\underline{p}^{T^T} \cdot \Delta\underline{z}_i - \frac{1}{2} \underline{p}^{u^T} \cdot \Delta\underline{z}_{i-1} \cdot \alpha \right]$$

$$\pi_\beta^z = \beta \cdot \left[\underline{p}^{N^T} \cdot \Delta\underline{z}_i + \frac{1}{2} \underline{p}^{u^T} \cdot \Delta\underline{z}_{i-1} \cdot \beta + \underline{p}^{u^T} \cdot \Delta\underline{z}_{i-1} \right] \tag{15}$$

$\underline{p}^u$: unbalanced forces

$\underline{p}^T$: component of $\underline{p}^u$ in direction of the external force

$\underline{p}^N$: component of $\underline{p}^u$ orthogonal to the external force

$\Delta\underline{p}^i$: increment of the internal forces

The variation of the extended potential (16)

$$\pi^* = \pi + \pi^z_\lambda + \pi^z_\alpha + \pi^z_\beta \longrightarrow \textbf{extremum} \qquad (16)$$

leads to the following set of equation (17)

$$\begin{array}{ll} \delta\Delta\underline{z}_i^T : \\ \delta\lambda_i : \\ \delta\alpha_i : \\ \delta\beta_i : \end{array} \begin{bmatrix} \underline{A}_{(z)} & -\underline{p}^u & \underline{p}^T & \underline{p}^N \\ -\underline{p}^{u^T} & & & \\ \underline{p}^{T^T} & & -\underline{p}^{u^T}\cdot\Delta\underline{z}_{i-1} & \\ \underline{p}^{N^T} & & & \underline{p}^{u^T}\cdot\Delta\underline{z}_{i-1} \end{bmatrix} \cdot \begin{bmatrix} \Delta z_i \\ \lambda_i \\ \alpha_i \\ \beta_i \end{bmatrix} = \begin{bmatrix} 0 \\ -\Delta\pi \\ 0 \\ -\underline{p}^{u^T}\cdot\Delta\underline{z}_{i-1} \end{bmatrix} \qquad (17)$$

An application of the method is given in fig. 7. The shallow arch is compared to the solution obtained by the constant arc length method. In the whole 20 steps were given. Maximal 6 iterations were necessary to iterate back to the solution path in an arbitrary point, .
With a suitable energy increment adjustment a considerably of computing time can be expected.

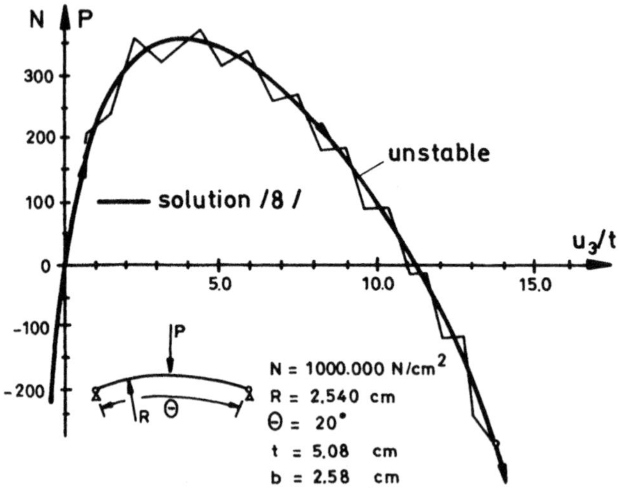

Fig. 7. Load path approximation by constraints

5. Energy constraint

5.1 Constant arc length as energy constraint

The common constant arc length methods are not based on an energy formulation. Hence unsymmetric extra equations occur e.g. /7/. However this has no disadvantage if the numerical solution scheme (6) is used. The arc length methods can be put into the context of energy formulations, if instead of one two constraint conditions are used. The energy π and the two constraints are given in equation (18).

$$\begin{aligned}
\pi &= \frac{1}{2}\,\underline{z}_i^T \cdot \underline{A}_{(\underline{z}_o)} \cdot \underline{z}_i \; - \; \underline{z}^T \cdot \underline{p}_{i-1}^u \\
\pi_1^z &= -\lambda_{1i}\left[\underline{N}_i^T \cdot \underline{z}_i + H_i \cdot \lambda_{2i} - R\right] \qquad (18) \\
\pi_2^z &= -\lambda_{2i}\left[\underline{B}_i^T \cdot \underline{z}_i - \frac{1}{2}\,\underline{B}_i^T \cdot \underline{A}_{(\underline{z}_o)}^{-1} \cdot \underline{B}_i \cdot \lambda_{2i} + H_i \cdot \lambda_{1i} - \underline{B}^T \cdot \underline{A}_{(\underline{z}_o)}^{-1} \cdot \underline{p}_{i-1}^u\right]
\end{aligned}$$

$\underline{B}_i^T$: reference load vector

$\underline{N}_i^T$: displacement vector, to which the displacement $\underline{z}_i$ has to be orthogonal

R : quadrat of arc length

$\underline{p}_{i-1}^u$: unbalanced force

H_i : zero in standard constant arc length method

The variation of the extended energy leads to (19).

$$\begin{array}{ll}
\delta \underline{z}_i^T : \\
\delta \lambda_{1i} : \\
\delta \lambda_{2i} :
\end{array}
\left[\begin{array}{c|c|c}
\underline{A}_{(\underline{z}_o)} & -\underline{N}_i & -\underline{B}_i \\
\hline
-\underline{N}_i^T & 0 & -H_i \\
\hline
-\underline{B}_i^T & -H_i & +\underline{B}_i^T \cdot \underline{A}_{(\underline{z}_o)}^{-1} \cdot \underline{B}_i
\end{array}\right]
\cdot
\left[\begin{array}{c}
\underline{z}_i \\ \hline \lambda_{1i} \\ \hline \lambda_{2i}
\end{array}\right]
=
\left[\begin{array}{c}
\underline{p}_{i-1}^u \\ \hline -R \\ \hline -\underline{B}_i^T \cdot \underline{A}_{(\underline{z}_o)}^{-1} \cdot \underline{p}_{i-1}
\end{array}\right]
\qquad (19)$$

From the given algorithm (2) can be seen, that the parameter λ_{1i} is always zero. Hence the related column in equation (19) can be dropped. The third equation in (19) is in every step identically satisfied. Therefore it is also dropped. The remaining relevant subset of the energy formulation are the two marked equations of (19). For $H = 0$ they are identical with the wellknown constant arc length method in the displacement space. For $H_i \neq 0$ is the formulation of an arc length method in the load displacement space possible. An application is given in fig. 7.

5.2 Constant volume constraint

For some purposes it may be neccessary to increment the global volume parameter of the structure. This can be done by using a constraint (20) which relates the surface of the structure and the displacements to a given volume increment.

$$\begin{aligned} \pi_2^z &= 0 \\ \pi_1^z &= -\lambda_1 \left[\underline{N}_i^T \cdot \Delta \underline{z}_i - \Delta V \right] \end{aligned} \tag{20}$$

$\underline{N}_i^T$: surface of the structure
$\Delta \underline{z}_i$: deflections of the structure
ΔV : volume
λ_1 : pressure

This constraint has been used in order to model experimental conditions.

5.3 Simultaneous approximation of given displacements fields and resultant forces

This combination is also often used in experiments, when e.g. a cylindrical shell is uniformly compressed in axial direction but the resultant force is imposed. The constraint condition for this case may be written as (21).

$$\pi_1^z = -\lambda_1 \left[\underline{N}_i^T \cdot \Delta \underline{z}_i - C \right] \tag{21}$$

$\underline{N}_i^T$: strain distribution forces
λ_1 : multiplier for given strains
C : forces times displacements (external energy)

An example of this application is also given in fig. 8.

6. Perturbation as constraint

In general nonlinear problems a number of possible solution pathes may exist. If the solution pathes intersect, the branches may be followed by path tracing methods for example arc length method. In order to detect branching points by accompanying measures for example eigenvalue calculations have to be applied. When branches do not intersect but come close to each other, such methods are not useful. For practical purposes this cases are also of interest since an initial imperfection, which is not known before or an unexpected perturbation of the load may indicate a snap of the real structure to a solution path in the neighbourhood of the fundamental path. In order to handle such cases a perturbation method is lined out /8/, which allows to calculate the size related to the direction of a perturbation and to apply this perturbation energy as a constraint to the nonlinear equation (1).

6.1 Size and direction of the perturbation

The accompanying eigenvalue calculation provides with its eigenmodes a set of potential perturbation patterns in every point of the solution path. The eigenvalue λ_z gives the distance from the actual state to the next singularity with a buckling pattern of the same shape as the one in the actual point (22).

$$\left(\underline{A}_{(z)} - \lambda_z \cdot \overset{N}{\underline{A}}_{(z)} \right) \cdot \underline{\Phi}_1 = 0 \tag{22}$$

Is the distance to a singularity sought in a direction orthogonal to the buckling mode of the actual state the related eigenvalue problem takes the form (23).

$$\left(\underline{A}_{(z)} - \lambda_\Phi \cdot \overset{N}{\underline{A}}_{(\underline{\Phi}_1)} \right) \cdot \underline{\Phi}_2 = 0 \tag{23}$$

The eigenvector $\underline{\Phi}_2$ is not important. The eigenvalue λ_Φ gives a measure for the distance to the solution in direction of $\underline{\Phi}_1$. With a normalisation of the quadratic form of the eigenvectors (24)

$$\underline{\Phi}_1^T \cdot \overset{N}{\underline{A}}_{(z)} \cdot \underline{\Phi}_1 = \underline{\Phi}_2^T \cdot \overset{N}{\underline{A}}_{(\underline{\Phi}_1)} \cdot \underline{\Phi}_2 = 1.0 \tag{24}$$

λ_Φ becomes an energetic measure for the perturbation which is neccessary to initiate the switch to the secondary path in direction of

$\underline{\Phi}_1$ (25).

$$\underline{\Phi}_2^T \cdot \underline{A}_{(\underline{z})} \cdot \underline{\Phi}_2^T = \lambda_\Phi \cdot \underline{\Phi}_2^T \cdot \underline{A}_{(\underline{\Phi}_1)} \cdot \underline{\Phi}_2 = \lambda_\Phi \tag{25}$$

Now the potential energy (26)

$$\pi = \frac{1}{2} \Delta \underline{z} \cdot \underline{A}_{(\underline{z} + \Delta \underline{z}_{i-1})} \cdot \Delta \underline{z} \tag{26}$$

can be extended by a constraint, which imposes the perturbation $\pi^P = \lambda_\Phi$ to the structure (27).

$$\pi^z = \Delta\lambda \left[\Phi_1^T \cdot A_{(\underline{z})} \cdot \Delta \underline{z}_i - \pi^P \right] \tag{27}$$

A few iterations with the constraint shift the solution to the neighbourhood of the secondary path. Then the constraint is dropped and the iterations settle at the new solution path. A stable iteration algorithm is necessary for this. In the example the quasi viscous damping was used /9/.

An example is given for the calculation of a perfect cylinder subjected to perturbations under constant axial forces n and under constant axial displacements $\bar{u}$. In the first case the displacements are kept constant along the boundary by means of equation (21), while the external force is imposed. The perturbation of the stable states of equilibrium $P_{1,2,3}$ was performed with constant axial forces. The applied perturbations are given in fig. 8. In fig. 8 also the perturbation energy $\bar{\pi}^P$ calculated from the eigenvalue calculation (23) is given.

Point P_4 shows the application of the perturbation with given axial displacements. Following the solution branch after the snap through a change of the buckling mode appears in point 6 and a snap back from point 7 to point 8 occurs to the prebuckling mode of the perfect cylinder. If the branch is followed with adjustable load level (21), the snap back from point 7 to point 8 does not appear. The structure turns back to the bifurcation point B.

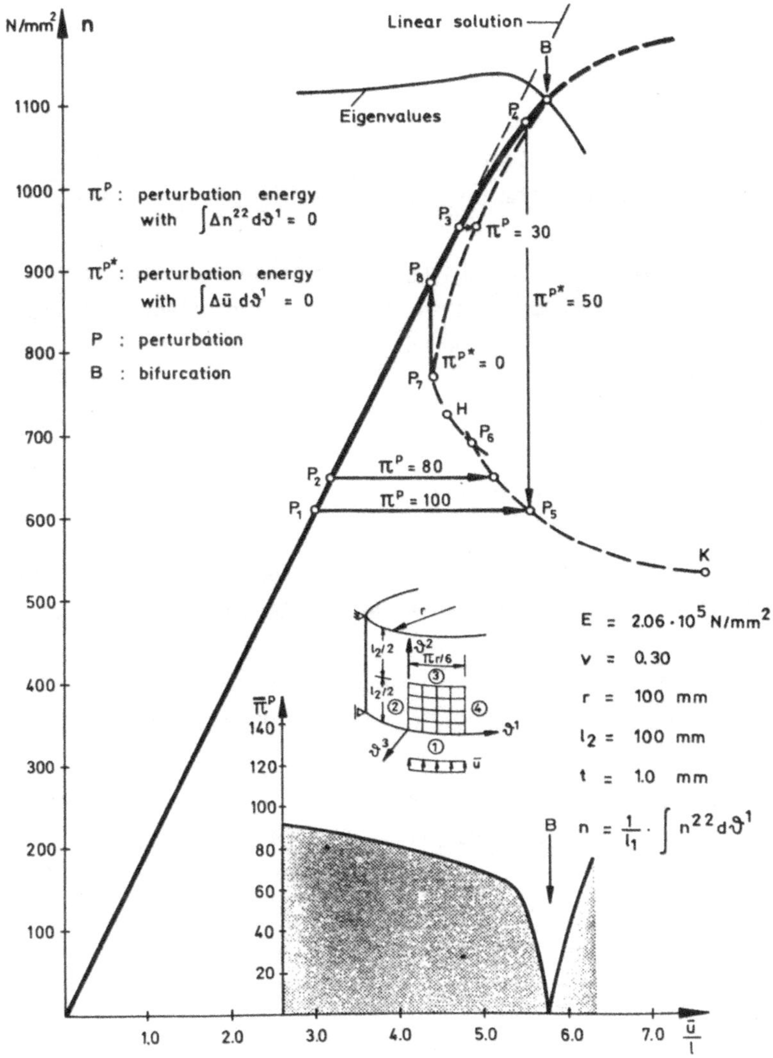

Fig. 8. Perturbation technique in application on the cylinder buckling

References

/1/ Eggers, H.; Kröplin, B.: Yielding of plates with hardening and large deformations. Int. Journal of Numerical Methods in Engineering 12, (1978) 738-750

/2/ Harbord, R.: Berechnung dünner Schalentragwerke mit finiten Elementen Vergleichende Untersuchung unterschiedlicher Diskretisierungsvarianten, Bericht Nr. 77-21 aus dem Institut für Statik der TU Braunschweig, (1977)

/3/ Kröplin, B.-H.; Dinkler, D.: A Creep Type Strategy Used for Tracing the Load Path in Elastoplastic Post Buckling Analysis. Computer Methods in Applied Mechanics and Engineering 32 (1982) 365-376

/4/ Zielke, G: Numerische Berechnung von benachbarten inversen Matrizen und linearen Gleichungssystemen. Schriften zur Datenverarbeitung Band 2, Friedr. Vieweg u. Sohn, Braunschweig (1970)

/5/ Geradin, M.; Idelsohn, S.; Hogge, M.: Computional Strategies for the Solution of Large Nonlinear Problems Via Quasi-Newton Methods. Computers & Structures 13 (1981) 73 -81

/6/ Kröplin, B.-H.: Postbuckling Instability Analysis of Shells Using the Mixed Method. Buckling of Shells, ed. E. Ramm, (1982) 175-199

/7/ Riks, E.: An Incremental Approach to the Solution of Snapping and Buckling Problems. Int. J. Solids Struct. 15 (1979) 529 -551

/8/ Kröplin, B.; Dinkler, D.; Hillmann, J: An Energy Perturbation Applied to Nonlinear Structural Analysis. To appear in: Computer Methods in Applied Mechanics and Engineering (Proceedings of FENOMECH 84)

/9/ Kröplin, B.: Implicit Relaxation Applied to Postbuckling Analysis of Cylindrical Shells. Flexible Shells, eds. E. L. Axelrad and F. A. Emmerling, (1984) 163-174

On Some Efficient Solution Algorithms for Nonlinear Finite Element Analysis

R. KOLAR,[†] H. A. KAMEL[‡]
University of Arizona, Tucson, USA

Solution algorithms are presented for tracing the load-configuration path for structures exhibiting snap-through behavior. The algorithms are based on arc-length methods, where the load parameter is treated as a variable, in addition to the unknown displacements. A constraint equation is constructed comprising the displacements and loads. Unlike the arc-length methods proposed by several other researchers, the load term contribution to the constraint equation is not neglected and the disparity arising due to different magnitudes of loads and displacements is recognized. This issue is addressed and implications of introducing a scaling parameter are presented. The algorithms include automatic means of computing the scaling parameter. Further, an efficient approach for load incrementation is proposed. The efficacies of the proposed algorithms are demonstrated by application to problems exhibiting large displacement characteristics.

† Post-Doctoral Research Associate

‡ Professor, Aerospace and Mechanical Engineering, University of Arizona, Tucson, AZ 85721

Finite Element Methods for Nonlinear Problems
Europe-US Symposium, Trondheim, Norway 1985
Editors: Bergan, Bathe, Wunderlich

Introduction

An increasing number of researchers are devoting their efforts in finding solution algorithms for studying the nonlinear behavior of structures. In particular, complete structural response beyond the limit points is of interest in analysis and design procedures. Newton-Raphson and Modified Newton-Raphson methods have been applied widely for studying the nonlinear structural behavior with loads approaching the buckling loads. In the present paper, we restrict our discussion to modified Newton method and its improvements.

The Newton type methods fail to provide a solution in the neighborhood of limit points. The solution to problems with such complex behavior necessitates the discovery of new algorithms. An artificial spring was suggested by Sharifi and Popov [1] while several investigators [2-6] have recommended a displacement incrementation in the neighborhood of limit points. A novel approach, referred to as arc-length methods, was proposed independently by Riks [7] and Wempner [8]. In this method the load parameter is treated as a variable in addition to the unknown displacements. The method has attracted attention recently after Crisfield [9] and Ramm [10] cast it in a form that is easily adopted into existing finite element codes using the Modified Newton procedure.

An important aspect in arc-length methods is the construction of a constraint equation in the load-configuration space. Several investigators [10,11,12,13] point out that displacements and loads have different magnitudes, but in the applications, neglect the load term contribution. This issue has been addressed in this research and a scaling parameter that nondimentionalizes the constraint equation is proposed. The parameter is shown to have a significant impact on the overall solution efficiency.

Another important issue in a nonlinear structural analysis performed by incremental methods centers on the selection of load increment size. Bergan [14] had proposed a method based on the current stiffness parameter (CSP), while Crisfield [10] and Ramm [9] adopt an approach based on the ratio of the number of iterations needed for convergence at a given step and the number of iterations prescribed by the user. The former approach results in the load step being too small near the limit points, thereby requiring a large number of load steps for tracing the complete response. Furthermore, it fails near vertical tangents to the load-configuration curve. The later approach encounters difficulties whenever the number of iterations for a step becomes large, which results in very small load steps; thus requiring a large number of load steps to trace the complete response.

The cost of nonlinear analysis increases, in general, as the number of load steps required to trace the load-displacement path increases. Herein a procedure is proposed that calculates the load step automatically and adaptively, resulting in substantial overall efficiency for nonlinear structural analysis. Two typical problems exhibiting highly nonlinear characteristics are included to demonstrate the proposed algorithms.

Problem Statement

Adopting the total Lagrangian formulation, the linearized incremental finite element relations may be expressed as [15,16]

$$\mathbf{K}_{T_N}\,\Delta\mathbf{p} = {}^{N+1}\mathbf{R} - {}^{N}\mathbf{F} \tag{1}$$

where $\mathbf{K}_{T_N}$ is the tangent stiffness matrix or the Jacobian matrix at the Nth step, $\Delta\mathbf{p}$ is the incremental displacement vector, ${}^{N+1}\mathbf{R}$ is the total applied load vector at the (N+1)th step, and ${}^{N}\mathbf{F}$ is the vector of nodal point forces equivalent to internal stresses. It is to be noted that the tangent stiffness matrix is given by

$$\mathbf{K}_{T_N} = \mathbf{K}_0 + \mathbf{K}_1 + \mathbf{K}_2 \tag{2}$$

where $\mathbf{K}_0$ is the small displacement stiffness matrix, $\mathbf{K}_1$ is the initial displacement matrix, and $\mathbf{K}_2$ is the initial stress matrix. The finite element equations may be derived by invoking the principle of virtual displacements and utilizing the appropriate strain displacement equations. For simplicity, elastic material stiffness relations and deformation independent loading is assumed in the above derivations.

Methods of Solution

Very often complete response well beyond the critical load is desired for structural problems. The Newton-Raphson(NR) and the Modified Newton-Raphson(MNR) methods are widely used to obtain solution [17,18]. However, the NR approach requires evaluation of the Jacobian at each iteration, with a high penalty on the computational effort, while in the MNR method the Jacobian is evaluated at each load step, thereby lessening the computational effort as compared to NR method. However, both methods encounter convergence difficulties or even diverge near the limit points, as the Jacobian tends to become singular. The disadvantages of discontinuing load incrementation and switching to displacement incrementation are well known. Other methods, such as quasi-Newton methods, also exist, but suffer similar drawbacks as the NR and the MNR methods. In the subsequent sections, we restrict our discussion to arc-length methods in conjunction with the MNR procedure.

Arc-length Methods

In arc-length methods, the load parameter is treated as a variable in addition to the unknown displacements at each iteration of a load step, as opposed to the NR and MNR methods where the load level is kept constant for all iterations of a given load step. The means to calculate the additional variable is achieved by appending a constraint equation comprising the displacements and loads.

The arc-length methods may be classified according to the type of constraint equation used. A spherical constraint equation is used in the spherical constraint arc-length method (SPHM), a normal plane constraint equation is adopted in the initial normal plane arc-length method (INPM), while an updated normal plane constraint equation is constructed in the updated normal plane arc-length method (UNPM).

Spherical Arc-length Method

The arc-length is constrained to lie on an (N+1)-dimension hyper-sphere of displacements and load parameter, given by

$$\mathbf{\Delta p}_i^T \mathbf{\Delta p}_i + \gamma \Delta\lambda_i^2 \mathbf{R}^T \mathbf{R} = dS^2 \tag{3}$$

where $\mathbf{\Delta p}_i$ is the vector of incremental displacements at the ith iteration, $\Delta\lambda_i$ is the incremental load parameter at the ith iteration, $\mathbf{R}$ is the fixed applied load vector, dS is the arc-length at the current step, and γ is a scaling parameter to be defined later. It may be noted that by assigning a value of zero to γ in equation (3), one obtains the constraint equation proposed by Crisfield [10]. On examination of equation (3), we may infer that the arc-length in the (N+1)-dimension is constructed by N-dimensional displacements and a load parameter. As the displacements and loads have different magnitudes, it seems reasonable to introduce the scaling parameter γ in the constraint equation.

In order to facilitate the book-keeping of the incremental quantities, the following notation is introduced:

$$\mathbf{\Delta p}_{i+1} = \mathbf{\Delta p}_i + \mathbf{\Delta \hat{p}}_i \tag{4}$$

and

$$\Delta\lambda_{i+1} = \Delta\lambda_i + \Delta\hat{\lambda}_i \tag{5}$$

where $\mathbf{\Delta p}_{i+1}$ is the vector of incremental displacements from the the last converged solution, $\Delta\lambda_i$ is the incremental load parameter reckoned from the last converged solution; and $\mathbf{\Delta \hat{p}}_i$ and $\Delta\hat{\lambda}_i$ are the corrective displacements and change in the incremental load parameter at the *i*th iteration respectively.

The set of equations (1) and (3) constitute (N+1) equations in (N+1) unknowns comprising displacements and load parameter. It may be noted that by appending the later equation the symmetric properties of the stiffness matrix are destroyed. Keller [20] has described a method using the 'bordering algorithm' to solve this set of equations. We follow Crisfield [10] and Ramm [9], by adapting the displacement incrementation as proposd by Batoz and Dhatt [6]. By defining the residual forces as

$$\mathbf{h}_{i+1} = \lambda_{i+1}\mathbf{R} - \mathbf{F}_{i+1} \tag{5a}$$

we may expand $\mathbf{h}$ as a Taylor's Series about the configuration at the *i th* iteration. And by requiring the equilibrium at the *(i+1) th* iteration there results

$$\frac{\partial \mathbf{h}}{\partial \mathbf{p}}\mathbf{\Delta p} = \lambda_i\mathbf{R} - \mathbf{F}_i \tag{6}$$

On recognizing that the right side of the equation is linear in λ, we may compute the displacements [11] as follows:

$$\mathbf{\Delta\hat{p}}_i = \Delta\hat{\lambda}_i\mathbf{\Delta p}_T \tag{7}$$

where,

$$\mathbf{\Delta p}_{i_g} = \mathbf{K}^{-1}\mathbf{h} \tag{8}$$

and

$$\mathbf{\Delta p}_T = \mathbf{K}^{-1}\mathbf{R} \tag{9}$$

We observe that the corrective displacements $\mathbf{\Delta\hat{p}}$ are obtained by calculating the displacements, $\mathbf{\Delta p}_{i_g}$, due to the residual forces and the dispalcements $\mathbf{\Delta p}_T$, due to the fixed applied load vector. Further, displacements due to $\mathbf{R}$ need to be calculated at the first iteration of the current step and need not be computed for subsequent iterations of that step.

On substituting equation (7) into the constraint equation (3) and rearranging terms, one obtains a quadratic equation in $\Delta\hat{\lambda}$, given by

$$(\Delta \mathbf{p}_T^T \Delta \mathbf{p}_T + \gamma \mathbf{R}^T \mathbf{R})\ \Delta\hat{\lambda}_i^2 + 2(\Delta \mathbf{p}_{i_g}^T \Delta \mathbf{p}_T - \gamma\, \Delta\lambda_i \mathbf{R}^T \mathbf{R})\ \Delta\hat{\lambda}_i + (\Delta \mathbf{p}_{i_g}^T\ \Delta \mathbf{p}_{i_g} + \gamma\, \Delta\lambda_i^2 \mathbf{R}^T \mathbf{R} - dS^2) = 0 \quad (10)$$

Of the two roots that result from the equation (10), that root is selected which does not double back on the load - configuration path [10]. This criterion is determined by computing the angle between the displacement vectors at the *i th* and *i+1 th* iterations for each $\Delta\hat{\lambda}$ and selecting that $\Delta\hat{\lambda}$ which results in a positive angle. However, in the case when both roots yield positive values, the root closer to the linear solution to equation (10) is chosen. In the event of the roots becoming a complex pair, the iteration is discontinued and a new stiffness matrix is formed at the last converged solution, with the step size halved for the next load step.

The displacements and the load parameters are updated using

$$\Delta \mathbf{p}_{i+1} = \mathbf{p}_i + \Delta \hat{\mathbf{p}}_i \quad (11)$$

and

$$\lambda_{i+1} = \lambda_i + \Delta\lambda_i \quad (12)$$

while the incremental quantities are updated as given by equations (5) and (6). For $i = 1$, the updates are formed as follows:

$$\Delta \hat{\mathbf{p}}_1 \equiv \Delta \mathbf{p}_1 \quad (13)$$

$$\mathbf{p}_1 = \mathbf{p}_0 + \Delta \mathbf{p}_1 \quad (14)$$

$$\lambda_1 = \lambda_0 + \Delta\lambda_1 \quad (15)$$

$$\Delta\hat{\lambda} \equiv 0 \quad (16)$$

where $\mathbf{p}_0$ is the converged solution at the previous load step.

Scaling Parameter

As discussed earlier, the scaling parameter is introduced in the constraint equation to take into account the fact that the displacements and loads have different dimensions. In order to compute the parameter and render it nondimensional, equal contribution of loads and displacements are assumed in constructing the arc-length which is given by

$$\gamma_c = \frac{\Delta \mathbf{p}_1^T\ \Delta \mathbf{p}_1}{\Delta\lambda_0^2 \mathbf{R}^T \mathbf{R}} \quad (17)$$

where $\mathbf{\Delta p_1}$ is the vector of incremental displacements at the first load step, while $\Delta\lambda_0\mathbf{R}$ is the incremental load at the first step. γ_c is computed at the first step and is kept constant for all subsequent computations using equation (10).

However, it is perfectly feasible to consider the scaling parameter as varying from one step to another. To this end, the current stiffness parameter [14] as defined by

$$\beta_M = \left(\frac{\Delta\lambda_M}{\Delta\lambda_1}\right)\frac{\mathbf{\Delta p}_M^T\mathbf{R}}{\mathbf{\Delta p}_1^T\mathbf{R}} \tag{18}$$

is utilized in the definition of variable scaling parameter, γ_{M_v}, at step M by,

$$\gamma_{M_v} = \beta_M\gamma_c \tag{19}$$

The current stiffness parameter (CSP), β_M, varies from unity at the undeformed configuration to zero at the limit points and changes signs as the limit points are passed. It may be noted, in passing, that the CSP decreases as the structure softens and increases as the structure stiffens. Also, the configurations corresponding to vertical tangents to the load-displacement path result in division by zero in the computation of CSP. Such an eventuality is handled, in the present analysis, by incorporating a suitable logic that discontinues the usage of the variable scaling parameter and continues the computation by switching to constant scaling.

Automatic Load Incrementation

The selection of load step size is an important decision in an incremental analysis. Too small a step size results in a large number of steps, thereby making the cost of analysis prohibitive, while too large a step size may produce convergence difficulties or even diverge in some cases. Bergan [14] had proposed a load incrementation based on the CSP concept, while Crisfield [10] and Ramm [9] advocate an arc-length calculation based on the ratio of the number of iterations desired and the number of iterations required in a step for convergence, given by

$$dS_{M+1} = dS_M\sqrt{\frac{I_D}{I_M}} \tag{20}$$

where, I_D is the number of iterations desired, prescribed usually to be 4-5, and I_M is number of iterations required for convergence at step M, while dS_{M+1} and dS_M are the arc-lengths at step M and $M+1$ respectively.

In the present paper, it is proposed to keep the arc-length unchanged. That is,

$$dS_{M+1} = dS_M \tag{21}$$

The incremental load parameter for the first iteration is calculated using

$$\Delta\lambda^2_{M,_1} = \frac{dS^2_M}{(\Delta\mathbf{p}^T_{M,_1}\ \Delta\mathbf{p}_{M,_1} + \gamma_M \mathbf{R}^T\mathbf{R})} \tag{22}$$

where the suffix $_{M,_1}$ denotes the first iteration at the M th step. It may be noted from the equation (22) that the incremental load is computed adaptively. The sign is determined according to the sign of the CSP.

Numerical Examples

In this section the algorithms proposed above are evaluated by application to selected examples that display pronounced geometrically nonlinear characteristics. An interactive computer program is developed incorporating the solution algorithms. Various options and computational features are outlined in detail in Reference [16]. Numerical solutions have been obtained by employing nine noded biquadratic Lagrangian elements.

Rigid Jointed Plane Frame

Analytical and experimental solutions for the plane toggle frame shown in Figure 1 have been reported by Williams [20], while a finite element solution based on displacement incrementation was presented, among others, by Wood and Zienkiewicz [21]. Five biquadratic Lagrangian elements are employed to model half the structure, owing to the symmetry of the geometry and loading.

Figures 2-4 portray the variation of the apex displacement with load for different values of the initial incremental load parameter, $\Delta\lambda_0$, of 0.1, 0.3, and 0.5 respectively with no scaling. The load step incrementation is calculated using equations (20 and 22). Figures 5-7 portray the variation employing a constant scaling (equation 19) and equations (21 and 22) for computing the increment size. The plots reveal that constant scaling together with a constant arc-length for all steps traces the complete load-displacement path in a lesser number of load steps than with no scaling. Thus it results in a significantly smaller number of Jacobian evaluations and subsequent decompositions, which constitutes a major computational component in nonlinear analyses of structures. This is further evidenced (Table I) by the computer times for both approaches for different initial step sizes. An overall computational savings of 35-50 percent may be observed from this table.

Shallow Circular Arch

The procedure is illustrated next in the analysis of a shallow circular arch under a concentrated apex load. Figure 8 shows the geometry, loads and idealization using seven elements. Analytical results have been reported by Sharifi and Popov [1], while finite element solutions are given by Bergan and Soreide [14], among others. Figures 9-11 depict the variation of the apex displacement with the load, for initial step sizes of 0.1, 0.3, and 0.5 respectively with no scaling. Figures 12-14 show the variation using constant scaling and keeping the arc-length constant through all the steps. The robustness of the proposed algorithms is again displayed by these plots. The relative performance is given in the Table II, which show the computational savings obtained.

Some observations on the performance of the proposed algorithms on the cited examples are in order. The SPRM with the scaling parameter in conjunction with MNR results in a robust algorithm capable of tracing the complete post-buckling response of geometrically nonlinear structures. While the arc-length is kept constant for all steps, the incremental load parameter is computed adaptively, depending on the state of the deformed structure, independent of the number of iterations taken for convergence. Since the major share of the cost of analysis is devoted to the evaluation of the Jacobian and its decomposition, particularly in incremental nonlinear analyses, where repeated evaluations are evidently indispensible, the proposed algorithms result in an efficient and reliable procedure. The variable scaling parameter is seen to differ little from the constant scaling parameter in terms of computational efficiency.

As the procedure results in fewer load steps, only few equilibrium points on the load-displacement path are obtained. The analysis yields only an overall trend of the response. It may be desirable to obtain a detailed response close to some selected crucial equilibrium states. Currently efforts are under way to modify the program to restart and obtain refined response about any desired load step(s), without the need to recompute from the beginning.

Conclusions

Solution algorithms are proposed for tracing load-displacement path for structures exhibiting pronounced geometrically nonlinear behavior. The arc -length method in conjunction with the Modified Newton-Raphson method is adopted and the constraint equation is constructed by the inclusion of a scaling parameter. A scheme to compute adaptive load incrementation is delineated. The efficacies of the proposed algorithms are demonstrated by examples.

Acknowledgements

The present work has been supported in part by the Office of Naval Research under the contract N00014-75-C-0837, and is greatfully acknowledged. Special acknowledgements are due to GIFTS Users' Group and CASA/GIFTS Inc. for the financial support provided. The immaculate efforts and help of Ms. Ruth Makepeace in the preparation of this paper is greatly appreciated.

References

1). Sharifi, P., and Popov, E.P., "Nonlinear Buckling Analysis of Sandwich Arches," *Journal for Engineering Mechanics Division*, ASCE, Vol.97, 1971, pp.1397-1412.

2). Lock, A.C., and Sabir, A.B., "Algorithms for Large Deflection Geometrically Nonlinear Plane and Curved Structures," in *Mathematics for Finite Elements and Applications*, (ed. Whiteman, J.R.), Academic Press, N.Y., 1973, pp.483-494.

3.) Argyris, J.H., "Continua and Discontinua," *Proceedings of Conference on Matrix Methods for Structural Mechanics*, AFFDL-TR-66-80, Wright Patterson AFB, Ohio,1965.

4). Zienkiewicz, O.C., "Incremental Displacement in Nonlinear Analysis," *International Journal for Numerical Methods in Engineering*, Vol.3, 1971, pp.587-588.

5). Haisler, W., Stricklin, J., and Key, J., "Displacement Incrementation in Nonlinear Structural Analysis," *International Journal for Numerical Methods in Engineering*, Vol.11, 1977, pp.3-10.

6). Batoz, J.L., and Dhatt, G., "Incremental Displacement Algorithms for Nonlinear Problems," *International Journal for Numerical Methods in Engineering*, Vol.14, 1979, pp.1262-1266.

7). Riks, E., "The Application of Newton's Method to the Problems of Elastic Stability," *Journal of Applied Mechanics*, Vol.39, 1972, pp.1060-1066.

8). Wempner, G.A., "Discrete Approximations Related to Nonlinear Theories," *International Journal of Solids & Structures*, Vol.7,1971, pp.1581-1599.

9). Ramm, E., "Strategies for Tracing the Nonlinear Response Near Limit Points," *Proceedings Europe-U.S. Workshop*, editors Wunderlich et al, 1980.

10). Crisfield, M.A., "A Fast Incremental/Iterative Solution Procedure that Handles 'Snap-Through'," *Computers & Structures*, Vol.13, 1981, pp.55-62.

11). Crisfield, M.A., "An Arc-Length Method Including Line-Searches and Accelerations," *International Journal for Numerical Methods in Engineering*, Vol.19, 1983, pp.1269-1289.

12). Park, K.C., "A Family of Solution Algorithms for Nonlinear Structural Analysis Based on Relaxation Equations," *International Journal for Numerical Methods in Engineering*, Vol.18, No.5, 1982, pp.1337-1347.

13). Myoung Jin, G., "Post-Buckling Analysis of Elastic Space Frames," Dissertation submitted in partial fulfillment for the degree of Doctor of Philosophy, University of Arizona, 1983.

14). Bergan, P.G., and Soreide, T.H., "Solution of Large displacement and Instability Problems using the Current Stiffness Parameter," in *Finite Elements in Nonlinear Mechanics*, Tapir Publishers, Trondheim, Norway, 1977, pp.647-669.

15). Bathe, K.J., Ramm, E., and Wilson, E.L., "Finite Element Formulations for Large Deformation Dynamic Analysis," *International Journal for Numerical Methods in Engineering*, Vol.9, 1975, pp.353-386.

16). Kolar, R., "Efficient Solution Methods for Nonlinear Finite Element Analysis," Dissertation submitted in partial fulfillment for the degree of Doctor of Philosophy, University of Arizona, 1984.

17). Strickland, J.A., Haisler, W.E., and Von Reismann, W.A., "Evaluation of Procedures for Material and/or Geometric Nonlinear Structural Analysis," *American Institute of Aeronautics and Astronautics Journal*, Vol.11, 1977, pp.292-299.

18). Endo, T., Oden, J.T., Becker, E.B., and Miller, T., "A Numerical Analysis of Contact and Limit-point Behavior of a Class of Problems of Finite Elastic Deformation," *Computers & Structures*, Vol.18, No.5, 1984, pp.899-910.

19). Padovan, J, and Touichakchaikul, S., "Self-Adaptive Predictor-Corrector Algorithms for Static Nonlinear Structural Analysis," *Computers & Structures*, Vol.15, No.4, 1982,pp.365-377.

20). Keller, H.B. "Practical Procedures in Path Following Near Limit Points," in *Computing Methods in Applied Sciences and Engineering* , edited by Glowinski and Lions, North- Holland, Amsterdam, 1982.

21). Williams, J.W., "An Approach to the Nonlinear Behavior of the Members of a Rigid Jointed Plane Framework with Finite Deflections," *Quarterly Journal of Mechanics and Applied Mathematics* , Vol.17, No.4, 1964, pp.451-469.

22). Wood, R.D., and Zienkiewicz, O.C., "Geometrically Nonlinear Analysis of Beams, Frames, Arches, and Axisymmetric Shells," *Computers & Structures* , Vol.17, 1977,pp.725-735.

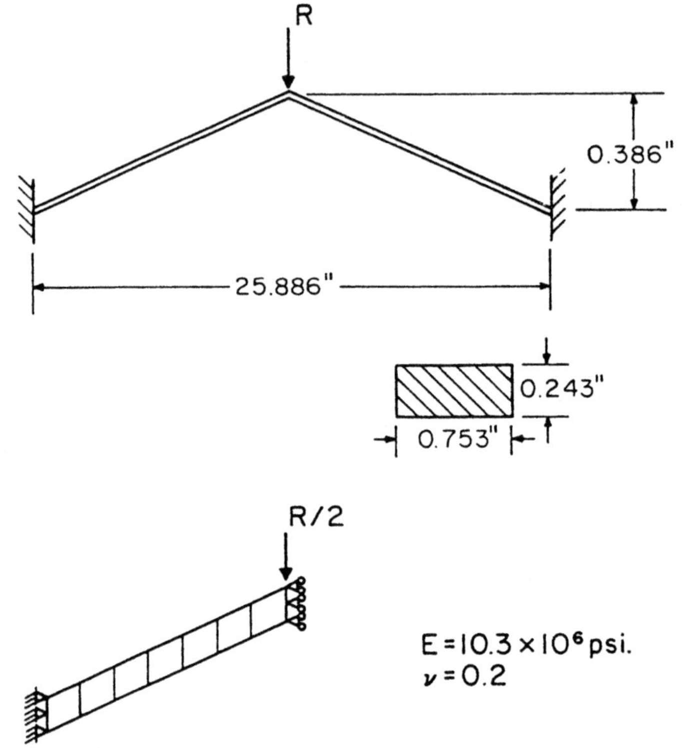

Figure 1. Geometry and material properties for Toggle frame

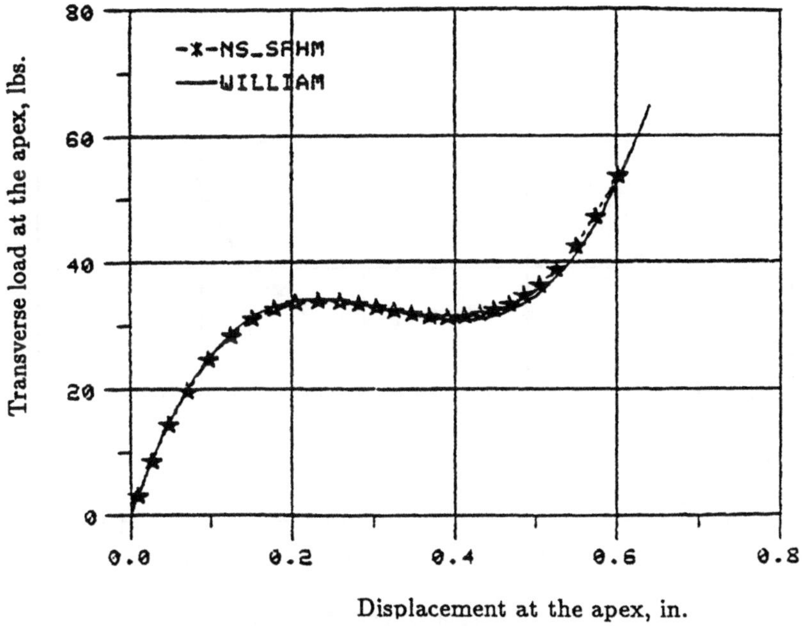

Figure 2. Load-maximum displacement curve for Toggle using the NS_SPHM with $\Delta\lambda_0=0.1$

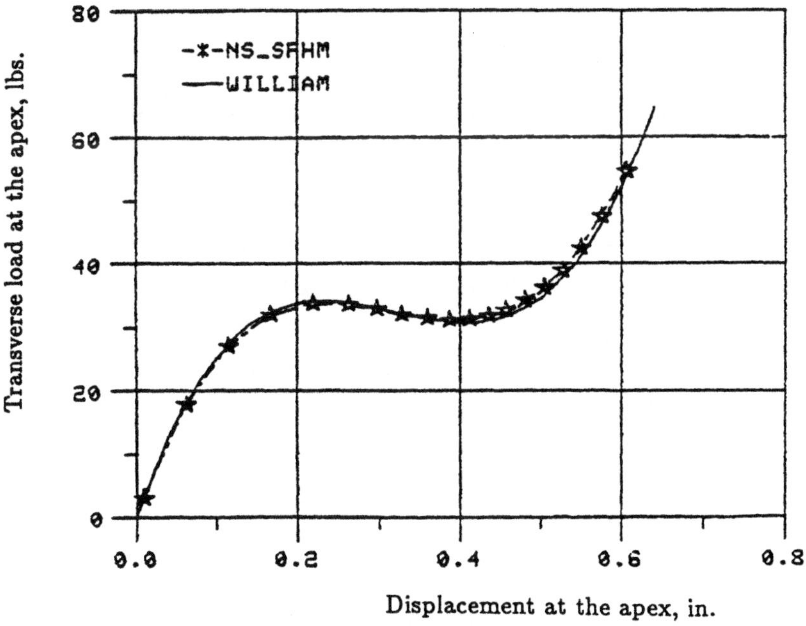

Figure 3. Load-maximum displacement curve for Toggle using the NS_SPHM with $\Delta\lambda_0=0.3$

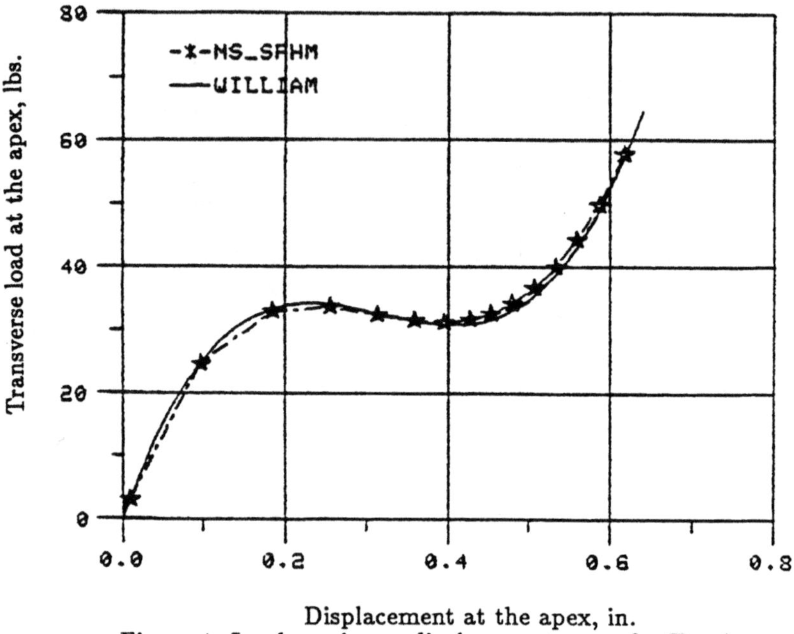

Figure 4. Load-maximum displacement curve for Toggle using the NS_SPHM with $\Delta\lambda_0$=0.5

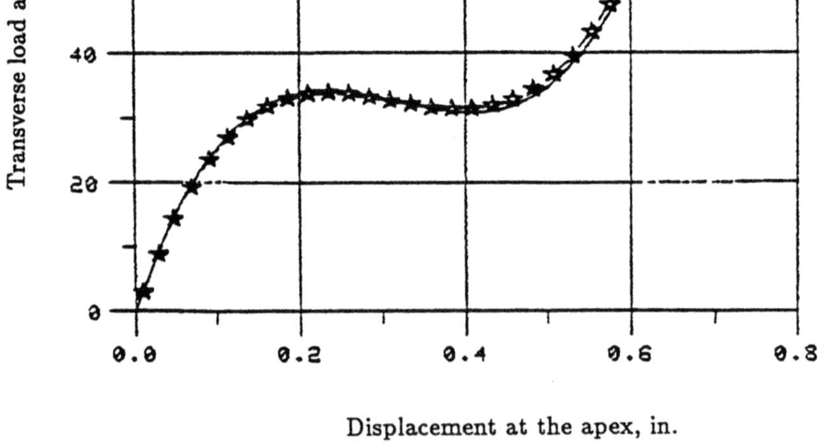

Figure 5. Load-maximum displacement curve for Toggle using the CS_SPHM with $\Delta\lambda_0$=0.1

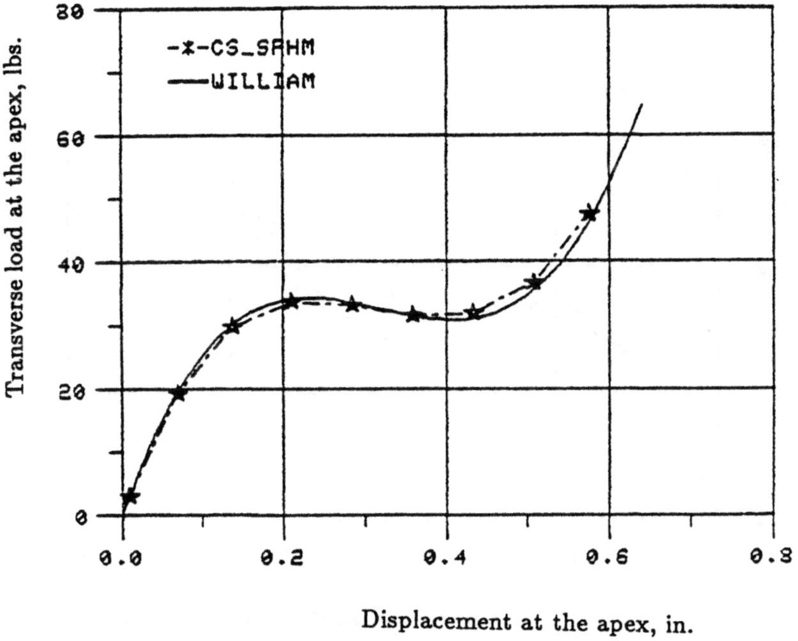

Figure 6. Load-maximum displacement curve for Toggle using the CS_SPHM with $\Delta\lambda_0$=0.3

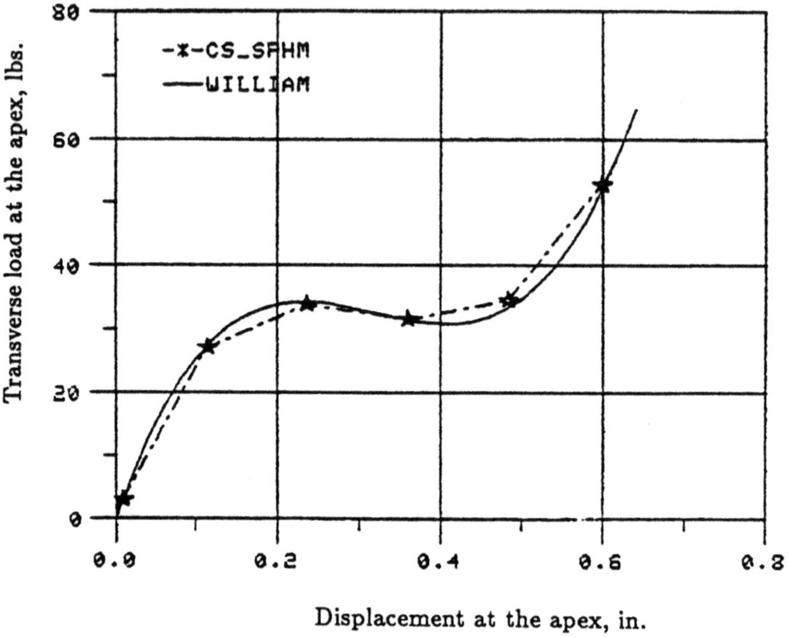

Figure 7. Load-maximum displacement curve for Toggle using the CS_SPHM with $\Delta\lambda_0=0.5$

$\Delta\lambda_0$	SPHM		
	C.S.	V.S.	N.S.
0.1	78.21	78.15	76.81
0.3	31.72	31.34	56.04
0.5	26.78	26.55	45.77

Table I. Comparison of computer time, in seconds, using Spherical Method for Toggle Frame.

$\Delta\lambda_0$–Initial Incremental Load Parameter
SPHM–Spherical Method
C.S.–Constant Scaling ; V.S.–Variable Scaling
N.S.–No Scaling

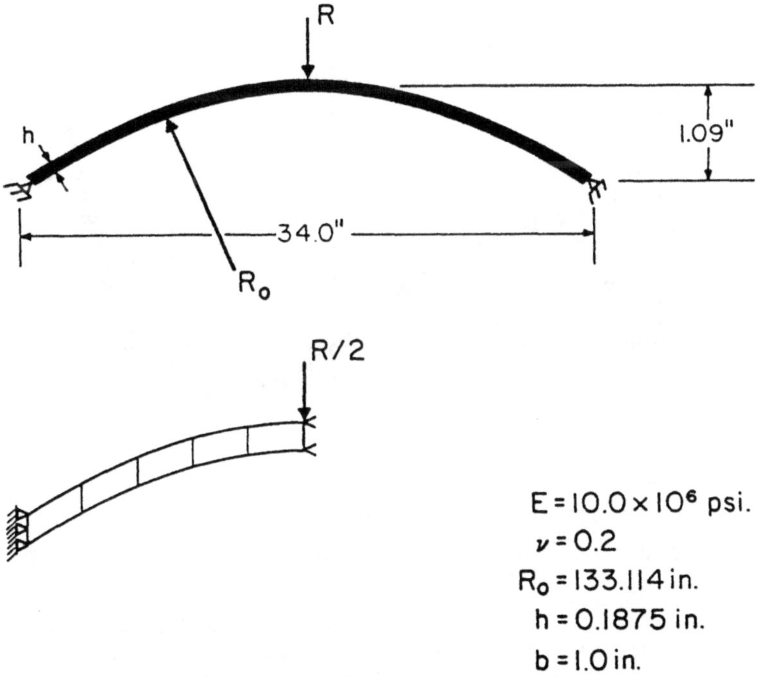

Figure 8. Geometry and material properties for a clamped shallow circular arch

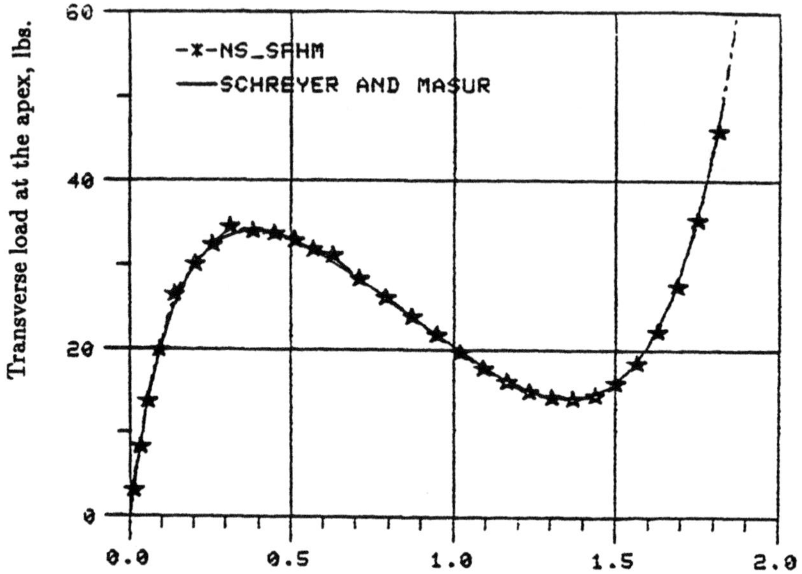

Displacement at the apex, in.
Figure 9. Load-maximum displacement curve for arch using the NS_SPHM with $\Delta\lambda_0=0.1$

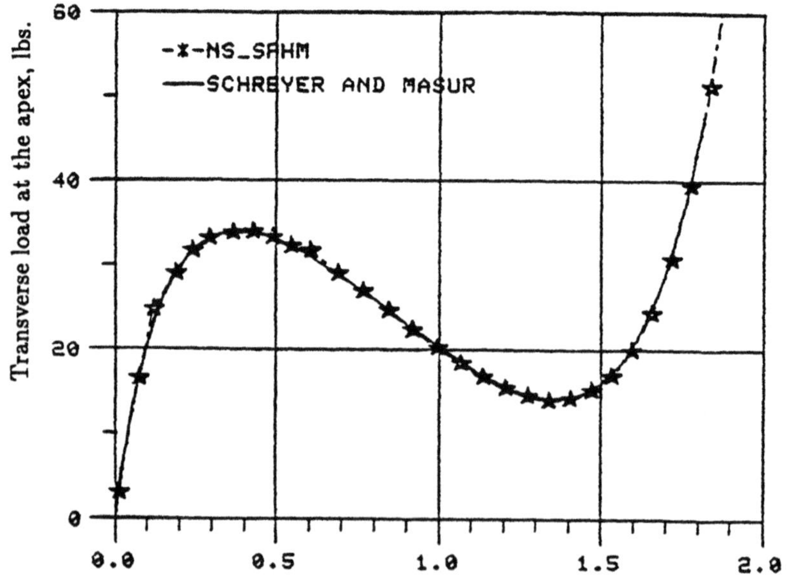

Displacement at the apex, in.
Figure 10. Load-maximum displacement curve for arch using the NS_SPHM with $\Delta\lambda_0=0.3$

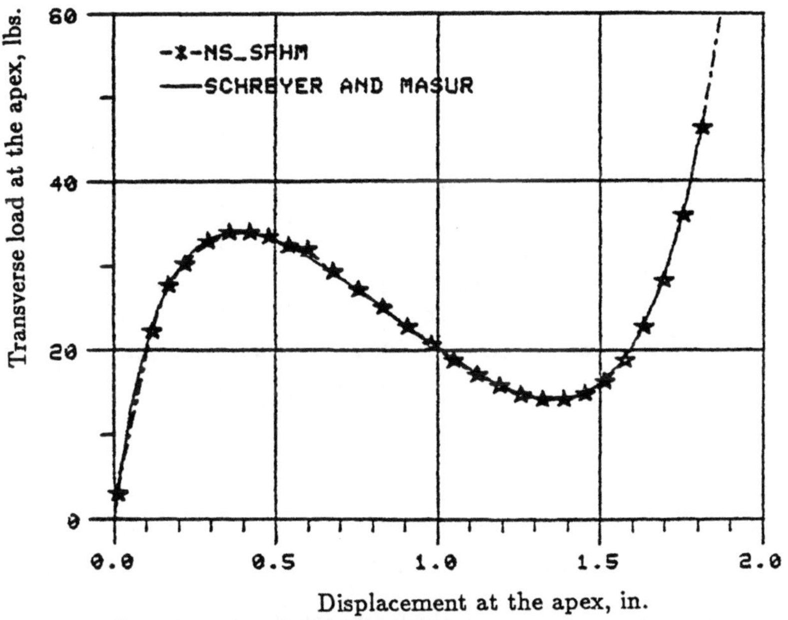

Figure 11. Load-maximum displacement curve for arch using the NS_SPHM with $\Delta\lambda_0$=0.5

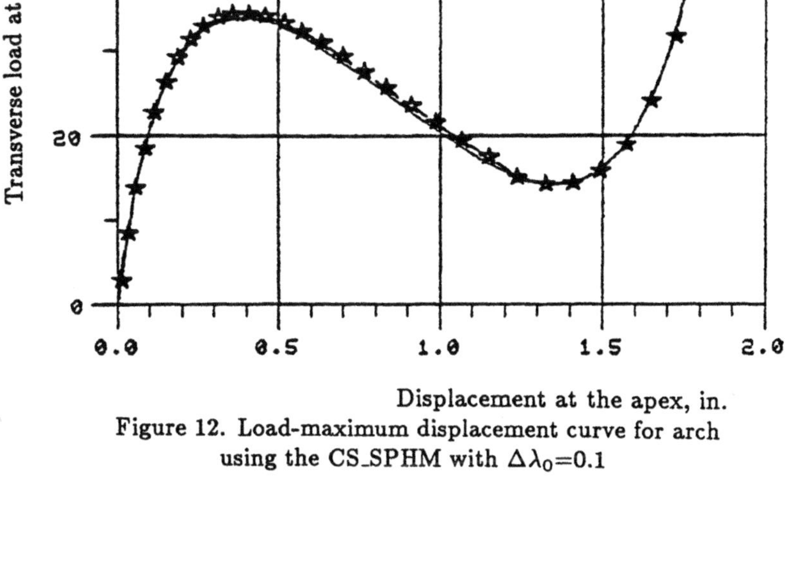

Figure 12. Load-maximum displacement curve for arch using the CS_SPHM with $\Delta\lambda_0$=0.1

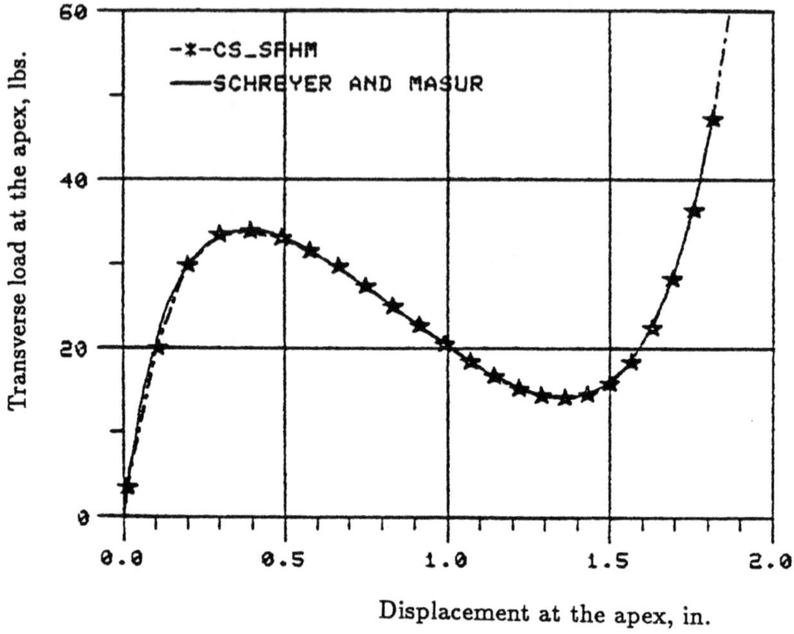

Figure 13. Load-maximum displacement curve for arch using the CS_SPHM with $\Delta\lambda_0=0.3$

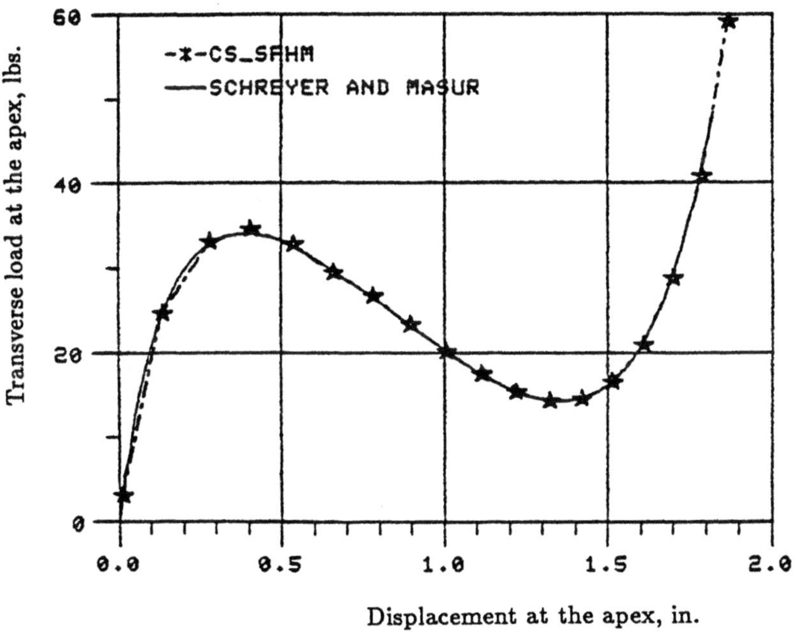

Figure 14. Load-maximum displacement curve for arch using the CS_SPHM with $\Delta\lambda_0$=0.5

$\Delta\lambda_0$	SPHM		
	C.S.	V.S.	N.S.
0.1	61.32	57.90	62.90
0.3	54.93	51.80	60.23
0.5	44.72	46.64	63.84

Table II Comparison of computer time, in seconds, using Spherical Method for a Shallow Circular Arch.

$\Delta\lambda_0$–Initial Incremental Load Parameter
SPHM–Spherical Method
C.S.–Constant Scaling ; V.S.–Variable Scaling
N.S.–No Scaling

NICE: A Utility Architecture for Computational Mechanics

C. A. FELIPPA, G. M. STANLEY
Lockheed Palo Alto Research Laboratory, USA

SUMMARY

An integrated software system called NICE (Network of Interactive Computational Elements) has been developed during the past five years at the Applied Mechanics Laboratory. The system is characterized by emphasis on interactive operation, reliance on advanced software techniques, a distinct separation of processors and architecture, and the selection of a utility, as opposed to executive, architecture. The system is configured as a network of independently-executable processors that share a common database and are driven by an extendible command language. Operational compatibility is achieved by replicating two architectural components: database manager and command language interpreter, in each processor. The paper gives an overview of the NICE architecture and illustrates, using "walk through" scenarios, the flavor of software based on this architecture. The last scenario deals with nonlinear static analysis, which has been one of the major applications built on the NICE architecture.

§1. INTRODUCTION

NICE is an acronym for Network of Interactive Computational Elements. The name identifies a long-term software-research project started in 1980 at the Applied Mechanics Laboratory. The purpose of the project has been to provide a state-of-the-art framework for the development of integrated computational-mechanics software in an interactive environment of linked computers [1]. The system was therefore called upon to accommodate the conflicting requirements of methods research at one end and production engineering analysis at the other.

Design guidelines for the project are covered in more detail in §2. In accordance with these guidelines, NICE consists of a *generic architecture* that implements what computer scientists call a "virtual machine," and of independently-executable, application-specific *processors*. The architecture consists of three software components that serve all processors:

1. The command language interpreter *CLIP*.
2. The global database manager *GAL*.
3. Miscellaneous tools to preprocess, scan and maintain source code.

The processors perform the useful work, and communicate through a global database that survives processor execution. The implementation language for both architecture and processors is Fortran 77.

Finite Element Methods for Nonlinear Problems
Europe-US Symposium, Trondheim, Norway 1985
Editors: Bergan, Bathe, Wunderlich

A key point elaborated upon below is that NICE *has no central executive.* Each processor, being independently executable, has its own main program and its own command-interpretation shell. Operational compatibility is achieved by *replicating* two architecture components: the command interpreter *CLIP* and the data manager *GAL*, in each processor.

Individual processors are designed to perform limited functions well. System power is achieved by combining processors in either of two configurations, which are described in more detail in [1]. A group of related processors tightly bundled under an ephemeral main program constitute a *macroprocessor.* A group of related processors loosely connected via a command procedure but maintaining independence of execution is called a *network.* The latter is the most general configuration allowed under the NICE architecture and has obvious appeal for the forthcoming generation of parallel-computing machines.

System evolution

Following preliminary system design in 1979, the implementation of the NICE architecture began in February 1980. The first version was up and running on a VAX 11/780 six months later. To exercise this version a three-processor prototype "mininetwork" was built and exercised. Lessons learned were incorporated in a new version, thus commencing a regular feedback cycle.

Versions of the architecture appeared regularly during the period 1981 to date, and were used to build production-level processors for engineering analysis. It is estimated that less than a third of the architecture software survived the criticism of processor developers and users, which is not unusual for a long-term software project. What is somewhat surprising is that the original system design discussed in [1] survived so many implementation changes unscathed.

Utility vs. executive architecture

A key assertion of this paper is that survival of the original design was made possible by the fact that NICE makes use of a *utility* architecture, a term contrasted here to *executive* architecture. A utility architecture has been likened to a restaurant buffet as opposed to a fixed menu: a customer (processor developer or user) takes whatever is necessary, and no more.

In more technical terms, a utility architecture is accessed by a processor via selective linking to an object library so as to satisfy the processor external references, and the processor remains the master. In an executive architecture, processors are linked to satisfy the external references of an executive program, which becomes the master.

The technical differences between these two concepts are not crucially important. And yet there is a deep psychological angle involved, one that has made the difference between acceptance and rejection of NICE in an open-shop environment. Processor developers like the idea of remaining masters of their own houses, and are comforted by the realization that if a processor works with an old but stable release of the architecture, there is no need to relink with newer versions of the architecture unless additional capabilities are required. Or more succintly: if it works, don't touch it.

On the other hand, working under an executive system is not unlike sitting under the sword of Damocles. *All* processors are immediately affected by major changes in the executive, which can be a traumatic experience for everyone involved. Although executive systems permit stricter control of the software and are irreplaceable in time- or life-critical applications (think, for example, of an air-traffic control system), they don't fit well the "laid-back" environment for which NICE was intended.

Giving users control

The utility-architecture philosophy of "take only what you need" also applies to the way processor users grow into the NICE system at their own pace. Novice users are comfortable with the step-by-step, *conversational* mode of operation in which everything that happens is visible. As experience and confidence grows, users learn to deal with the database and to hide processing details into scripts and command procedures, run loosely-coupled processor networks, and ultimately construct tightly-coupled processor assemblies called macroprocessors. But the path may be reversed at any moment; going from "black box" to "white box" selectively has been crucial in solving difficult problems in nonlinear mechanics.

§2. THE GENESIS OF NICE*

After two decades of uncoordinated development of scientific and engineering software at the Applied Mechanics Laboratory (AML) a decision was made in 1979 to construct an integrated software system for computational mechanics starting in 1980. The decision was prompted by factors such as software development becoming a major component of in-house activities, technology contracts calling for machine-transportable production-level codes, and perception of a rapidly changing computational environment. (As to the last factor, the projected 1979 advent of distributed and personal computing paled in comparison to the reality.)

A key challenge in building the integrated software system has been to accommodate the following conflicting usage requirements.

Research environment. Flexible and visible "white box" analysis flow, conversational interaction with programs, easy access to intermediate results, simple ways for branching "out of the system" into one's software and back again, human flow-time critical; computational efficiency, extensive error crosschecking and problem adaptivity unimportant.

Production environment. Ability to guide analysis details in "black box" fashion; aids for minimizing input-preparation efforts; extensive data validity and consistency checking; elaborate display facilities; ability to dispatch jobs to networked computers; reasonably efficient utilization of computer resources; problem adaptivity desirable.

Design guidelines

A real danger in trying to meet both extremes (with growth potential thrown in) is that an overcomplex monster may result. This would have crippled the new system from birth. To forestall this danger, the following design guidelines were agreed upon.

1. Clear-cut separation between computational and architectural components, with the latter being developed in advance.

---

* The material for this Section is taken largely from [1] with annotations where appropriate.

2. A network organization for independently executable computational elements called *processors.*
3. Processors data-coupled through a common global database.
4. Multilevel control system based on a command language.
5. Modularity enforced by replicating architectural components in each processor and by establishing processor configuration standards.

Rationale for design guidelines

Simple economics suggests that many processors will be prepared by the time honored "divide and conquer" technique, i.e., cannibalization of useful portions of obsolescent monolithic codes. The heterogenous origin of processors suggests an executive-less, database-coupled organization.

High operational flexibility calls for a system organization in which processor execution order is unconstrained, that is, a network. This "macro flow" is generally unknown in advance and may change in response to the application. A similar requirement dictates the use of a problem-oriented language for "micro flow,", *i.e.*, detailed guidance of a processor. The two basic control levels are subsequently referred to as *interprocessor* and *intraprocessor* control, respectively.

Finally, adaptation to the production end as well as growth provisions suggests an extendible multilevel control structure. The underlying idea is that a user should not have to know more about software operation than that necessary for accomplishing his tasks. Since one starts with a basic command language at the research end, a logical way of moving toward the "smart black box" end relies on providing user-selectable procedural levels of increasing detail-concealing power.

The remainder of the paper emphasizes processor control as presently provided by the NICE command language.

§3. THE COMMAND LANGUAGE

All NICE processors, without exception, are driven by a *command language.* The following example, taken from [2], should convey the flavor of what a command language is. Suppose you are running interactively from a video terminal a NICE processor that shows the results of a finite element analysis. On response to a prompt, you type

```
PRINT ELEMENT STRESSES 12 TO 35
```

and obediently the processor responds by tabulating the stress data it has on those elements. Once the display is over, another prompt appears on the screen and the processor waits for your next input.

What you have typed at the terminal is called a *command.* A command received by the processor, such as the one above, is called an *ordinary command* to distinguish it from another type discussed below. The actions specified by an ordinary command are performed *before* the user is asked to enter another command. This do-it-now mode is well suited to conversational interactive work because the user can enter commands in response to the observed effect of previous commands.

All commands entered by the user go through the *CLIP* (Command Language Interface Program) component of the NICE architecture. As noted in §1, *CLIP* is replicated in each individual processor. A drastic 20:1 condensation of *CLIP*, called *TinyCLIP*, is presented with its source code in [2].

Not all commands are equal

What makes the NICE command language different from other problem-oriented languages? Two distinguishing features are *freedom* and *extendibility*. The developer of a processor is given complete freedom to design the processor language. At the same time the processor user is given the ability to extend the language through generic features called directives and messages, as described below.

Processor Language. All ordinary commands are parsed by *CLIP* into items and then are passed along to the processor. A segment of the processor called the command-interpretation shell has the job of translating commands into specific actions. The set of valid commands constitutes the processor language. This language is designed and implemented by the processor developer, who is given a free hand in this regard. The rationale behind this freedom is that the developer is supposed to be an expert in what the processor does. Consequently, he or she is presumed to be best qualified to translate the processor actions into a domain-specific language for the users of the processor.

The architecture supplies only generic utilities to help the language implementation task. For example, there is nothing in the NICE architecture which says that the `PRINT` example has to have the format shown. This point is taken up further in §7.

Directives. Certain actions of general usefulness known as "services," can be executed directly by the command interpreter *CLIP*. These services are requested by special commands called *directives*, which are distinguished by a leading asterisk-prefixed keyword. Two examples:

```
*TYPE INPUT.DAT
*ADD  INPUT.DAT
```

The `TYPE` directive lists the contents of image file `INPUT.DAT` on the screen whereas the `ADD` directive "redirects" the input so that *CLIP* begins reading commands from that file. (In the NICE nomenclature, a file of commands accessed through an `ADD` directive is called a *script*.) Directives are available to *all* processors linked to the NICE architecture.

Messages. Commands or directives may be entered by either the user or the processor. User commands are entered at the terminal, read from script files (such as `INPUT.DAT` in the example above), or extracted from the database. On the other hand, commands entered by the processor are called *messages*. These are submitted by calling a "mailbox" extry point. For example, the Fortran statement

```
CALL CLPUT ('*ADD INPUT.DAT')
```

sends the message shown in quotes via `CLPUT`, which is one of three mailbox entry points provided to processor writers. The message has exactly the same effect as if the user had typed the message text, and as a result *CLIP* will begin reading commands from script `INPUT.DAT`.

Serendipity helps

Developers engaged in similar long-term software efforts may find it instructive to know how the three features discussed above came to be. None is actually the carefully planned product of top-down design. Prior experience, roll-of-the-dice, and users' feedback had a bigger part.

The freedom given processor developers in language matters is technically questionable, as such freedom can easily degenerate into anarchy. But the psychological benefits were found to clearly outweigh the risks. At various times during the NICE project rigid constraints were imposed in command syntax and data structures, but such standardization attempts were fruitless: developers merely ignored them.

Directives initially evolved by accident. When command procedures were implemented in *LODREC* (*CLIP*'s ancestor) in 1976, it was found necessary to use a header and terminator line to delimit the body of the procedure. These lines were treated as "invisible" or "transparent" commands, and to facilitate recognition the leading keyword was prefixed with an asterisk. For example, here is the definition of a one-line procedure:

```
*PROCEDURE SOLVE (A; B; X)
 SOLVE [A] [X] = [B]
*END
```

The procedure is invoked by another "transparent" command such as*

```
*CALL SOLVE (A=STIFFNESS.MATRIX; --
    X=INCREMENTAL.DISPLACEMENTS; --
    B=RESIDUAL.FORCES)
```

The CALL invocation is supposed to be magically replaced by the procedure body with formal arguments A, B and C replaced by the text supplied after the equals signs. It should be stressed that the processor is not supposed to "see" the *PROCEDURE, *END and *CALL lines. These are handled by the command interpreter.

The idea of transparent commands proved so sucessful that when the first version of *CLIP* was written in 1980 it was generalized to commands with arbitrary keywords, and the concept of directive was born. The present *CLIP* implements 72 directives, which together account for over 80% of its 18,000 lines of source code. The availability of so many services has given the command interpreter the flavor of a ***customized operating system*** that supports interactive programming.

Messages are also a serendipitous feature. As discussed in [1], the first implementation of messages had an exotic purpose: synchronization of data-coupled independently executing processors. As often happens, the original intention was not successful: better and simpler ways were found to accomplish that task. But processor developers

---

* Syntactical notes: (1) the CALL directive shown above is a single command continued over 3 lines, the double minus being the standard hyphenation mark; (2) arguments are keyworded and so they may appear in any order or be omitted; (3) arguments are separated by semicolons to simplify the specification of comma-linked item lists, such as A=1,2,3,5,8,13, as replacement text; (4) formal arguments are identified in the procedure body by being enclosed in balancing square brackets.

began using messages routinely to request services via directives that they wanted to keep hidden from novice users (§4 below gives a specific example). And so the power of the message facility came to be realized on a different track.

System perception by scenarios

The feel of a complex software system such as NICE can be best given by going through a set of concrete scenarios rather than through a formal description. In the following narrative, 'you' may refer to a novice user, an advanced user, or a processor developer, depending on the scenario. The distinction is roughly as follows.

A novice user is knowledgeable about the physics of the problem to be solved, knows something about the processors that are to be used, and some rudiments of the command language. An advanced user is in addition familiar with aspects of the command language such as directives, macrosymbols and command procedures, through which the processor language may be extended and customized. A processor developer is of course intimately familiar with the internal workings of the processor and is expected to know the processor-architecture interface.

Note that the user-processor dichotomy of ordinary application programs does not apply. For such programs the user's role is largely confined to preparing input according to detailed instructions given in the appropriate manual. In a system such as NICE, the user can exercise a much higher degree of control by developing expertise in the command language without having to descend to the inner sanctum of the Fortran code.

§4. Scenario 1: HELPING THE USER

You are a novice user who has to solve a fairly simple problem, and are told that processor ***MUP*** can do it. You skim through the user's manual but after half an hour you are saturated and discard it. That night you log into the computer from home, start up ***MUP***, and are greeted by a '`MUP command:`' prompt. Having forgotten all about the manual, you try something:

```
MUP command:  hi there!
*** Error *** Illegal command HI. Try HELP
MUP command:
```

So you try `HELP` and suddenly the "root section" of a help file is shown on the screen giving a list of valid commands. Looking at that list you remember from the Manual that the first command was `BEGIN`. Since you don't have a Manual at home you try `HELP BEGIN` and you get the root section of command `BEGIN`, and there is enough there to get you started.

Now let us look at the help problem from the point of view of the *developer* of ***MUP*** trying to implement a `HELP` command. It turns out that since the help facility is obviously a service that may be used by any processor, the NICE architecture furnishes a generic directive-driven procedure for getting online help. The help text is not part of the processor itself but resides on text files that the developer can prepare with an ordinary text editor. The first column of each text record is used to specify the *tree structure* of the help information, a technique stolen from the VAX/VMS help file facility. The most interesting aspect, however, is how the help file is accessed by *directives* and *messages.*

Let us assume that the help file for processor MUP on a particular VAX computer is on file NICE$HLP:MUP.HLP, and that you want to get information on topic BEGIN. To do that you may run any processor (not necessarily MUP) and enter the directives

```
*SET HFILE = NICE$HLP:MUP.HLP
*HELP BEGIN
```

The SET HFILE directive declares the help filename. Subsequent HELP directives will reference that file. The HELP BEGIN directive tells *CLIP* to open the help file, search for the first-level topic called BEGIN, and if found to list that topic on the terminal. Since *CLIP* is replicated in all processors, this means that from processor ZZZ, say, you can get help on processor MUP by typing the previous sequence. But should the user be expected to do that? An advanced user would know how to enter directives, but such users do not normally need much help — it's the novice that does. A novice user should not be bothered about such things as help filenames and directives; instead what he or she would like to type is

```
HELP BEGIN
```

which is an ordinary command. Here is where messages come in handy. On receiving the above command the processor sends a two-directive message:*

```
CALL CLPUT ('*SET HFILE=NICE$HLP:MUP.HLP ; *HELP BEGIN')
```

The message line is easily constructed using the character-string facilities of Fortran 77. Note that the default help file declared by the SET directive is for the processor that is running.†

Messages are normally invisible to the user (directive expertise is required to make them visible), and so the help display looks like a direct response to what was typed. This example illustrates the general principle of *information hiding.*

§5. Scenario 2: GRID GENERATION

Grid generation for discrete element models (finite element, finite difference or boundary element) provides a good illustration of "procedure power". The concept emphasized here is that an advanced user can augment the power of a NICE processor by extending the *command language* rather than the underlying Fortran code. This strategy has obvious appeal when the Fortran source code is difficult to modify, or is inaccesible.

---

* Several short commands may be placed on the same line if separated by individual semicolons

† One could get help on topic SIESTA for processor ZZZ, say, while running MUP by implementing a command format such as HELP SIESTA /PROCESSOR=ZZZ ; here keyword PROCESSOR is called a *command qualifier* and /PROCESSOR=ZZZ a qualifier phrase. A command qualifier component may always be omitted, in which case a default interpretation (in the present case, the running processor name) must exist.

Suppose that you have to use processor GEOM to define the node points of a finite element model of a slack cable structure. The model has 4 elements and 5 nodes initially located on the parabola

$$z = 4\delta(1 - \frac{x}{s})(\frac{x}{s}) \tag{1}$$

where $s = 400$ is the cable span and $\delta = 80$ is the initial sag. The nodes are equally spaced along the x axis. The input data expected by GEOM is

```
BEGIN NODES
 1 0., 0., 0.
 2 100., 0., -60.
 3 200., 0., -80.
 4 300., 0., -60.
 5 400., 0., 0.
END NODES
```

Aside from the BEGIN and END commands and the free-field format, the input does not look terribly different from that of an old-fashioned card-reading, batch-oriented finite element program. The nth node is defined by

$$n \quad x_n, y_n, z_n$$

The fact that this input looks old-fashioned is comforting to a novice user. This is tied up to the common observation that such users shy away from constructions that involve work-saving statements (such as looping) and prefer spelling out everything in detail.

But even a novice is soon confronted with the task of changing the model by adding more elements, changing the cable length, etc. And realistic finite element models may contain hundreds of such parameters. Having to recalculate everything by hand time after time rapidly becomes annoying. Let us examine how advanced command-language features can be utilized to make the task more pleasant.

First, *CLIP* accepts inline *arithmetic expressions.* For example, the z coordinate of node 2 could be specified by a literal expression such as

```
2 100., 0., (4* 80.*(1.-100./400.)*(100./400))
```

in which the value $x_2 = 100$ has been replaced in the parabola equation (1). In general, arithmetic expressions in the NICE command language are similar to Fortran's with two differences: the whole expression must always be surrounded by parentheses, and the exponentiation operator is ^ instead of **.

But, as in programming languages, literal arithmetic expressions are of marginal utility. Substantial power is gained only when those expressions can be used in conjunction with variable-like items. The latter capability is provided by the *macrosymbol* facility of CLIP, which allows the definition of symbols that stand for a string of characters known as *definition text.* For example, the number of elements may be parametrized by the DEFINE directive:

```
*DEFINE NELEMS = 4
```

NELEMS is a macrosymbol (a term often abbreviated to macro) whose definition text is the character 4. Once defined, a macrosymbol may be later *referenced* by surrounding it with balancing angle brackets, as in

```
NUMBER OF ELEMENTS = <NELEMS>
```

Here <NELEMS> is replaced by what NELEMS stands for, namely, 4, so the command to be interpreted is 'NUMBER OF ELEMENTS = 4'. This replacement mechanism has some points in common with the treatment of variables in conventional programming languages, but is more general in the sense that *evaluation follows replacement.* To clarify this point, imagine the hypothetical command

```
OPEN FILE = SOLUTION<NELEMS>.DAT
```

The last item expands to SOLUTION4.DAT, which is a filename and not an integer. So macrosymbols are more general than programming language variables, although for many applications (as in the example that follows) they function similarly.

The number of nodes is one more than the number of elements, a fact that can be expressed by defining next a NNODES macrosymbol:

```
*DEFINE NNODES = (<NELEMS>+1)
```

which illustrates the fact that the macro definition text may contain macrosymbol references to itself or to other macros. This macro "nesting" can proceed to very deep levels.

The last facility we need here is a looping construction. CLIP provides DO-ENDDO and WHILE-ENDWHILE blocks, both of which are implemented with the aid of directives. Both forms can only be used inside command procedures. For the present example, the DO form, which is very similar to the Fortran DO loop, is simpler:

```
*DO $N = 1,<NNODES>
  node generation statements
*ENDDO
```

The loop control variable $N is an implicitly-defined integer macrosymbol* that takes the values 1,2,...<NNODES> as the loop is executed, and may be materialized inside the loop body by writing <$N>.

With the help of arithmetic expressions, macrosymbols, and the DO-ENDDO construct, we are now in a position to write a command procedure called GENERATE that generates the nodal coordinates automatically:

---

* The leading dollarsign conventionally identifies a *local* macrosymbol analogous to the concept of local variable in Fortran. Such variables are invisible outside the procedure. This avoids potential conflicts should the loop body contain CALLs to other procedures which in turn may have loops controlled by a $N variable.

```
*PROCEDURE GENERATE
  BEGIN NODES
  *DEF NELEMS = 4
  *DEF NODES = (<NELEMS>+1)
  *DO $N = 1,<NNODES>
    *DEF X = (100.*(<$N>-1))
    *DEF Z = (4*80.*(1.-<X>/100.)*(<X>/100.))
    <$N> <X>, 0., <Z>
  *ENDDO
  END NODES
*END
```

Note that most commands (8 out of 11) in this example are directives, which work silently in the background but are not seen by the processor. Accordingly if you say

```
*CALL GENERATE
```

the processor GEOM will receive exactly the same seven commands (BEGIN NODES and END NODES, plus the five generated by the DO) of the input data given earlier. Of course the advantage of the more elaborate procedural form clearly emerges should the number of elements be changed, for in that case you need only to redefine the NELEMS macro. This not only cuts preparation time but, more importantly, lessens dramatically the chances for error.

As a final touch, let us make the number of elements, cable-span and initial-sag formal arguments of the procedure, with the preceding values (4 elements, $s = 400$ and $\delta = 80$) specified as default values:

```
*PROCEDURE GENERATE (ELEMENTS=4; SPAN=400.; SAG=80.)
  BEGIN NODES
  *DEF NELEMS = [ELEMENTS] ; *DEF S = [SPAN] ; *DEF DELTA = [SAG]
  *DEF NODES = (<NELEMS>+1)
  *DO $N = 1,<NNODES>
    *DEF X = (<S>*(<$N>-1)/<NELEMS>)
    *DEF Z = (4*<DELTA>*(1.-<X>/<S>)*(<X>/<S>))
    <$N> <X>,0.,<Z>
  *ENDDO
  END NODES
*END
```

Now the directive

```
*CALL GENERATE (ELEMENTS=100; SAG=124.8)
```

will generate the nodes for a 100-element model over a parabola with $\delta = 124.8$. Since the SPAN argument does not explicitly appear, the default value of 400 specified in the PROCEDURE header applies.

This has been a fairly trivial example, but brings forward some points that persist through more complex cases: *you can add capabilities without touching the Fortran source code.* Think about that. How many times have you wanted some foreign program to do something which is not in the User's Manual, and either you don't have access to the source code, or the code is so complex that you don't dare touch it? But preparing procedures is within the grasp of any user that master directives, and directives are the same for all NICE processors.

What are the disadvantages? Primarily a loss of efficiency. A hardwired Fortran program can generate nodes in microseconds, but processing procedure lines takes milliseconds. This would be hardly noticeable for most model-preparation tasks, which as a rule are heavily dominated by labor. But there are exceptions; for example if you habitually do things such as modelling pipe intersections (which involves lots of iterations), or extensive regriding. Should efficiency be of prime concern, Fortran coding is inevitable. If the processor source code is accessible, the solution is to go and implement new commands, or to get the developer to do it. If this is not possible, you may want to write a processor to do the job yourself, after learning how to extract inputs and place results in the database.

§6. Scenario 3: NONLINEAR STRUCTURAL ANALYSIS

Nonlinear static analysis, unlike its linear counterpart, cannot be considered a black box when phenomena such as bifurcation and collapse are considered. Of course, the same qualification applies even more forcefully if dynamic effects are included, or if the analysis is part of an automated-optimization or man-in-the-loop design process.

Because of these complications, the ability to "drive" NICE processors in a higher-level language has proven extremely helpful in solving difficult nonlinear problems. The advantages noted in the grid generation example become more powerful here, in that *solution strategies can be implemented at the command language level.* There are additional concerns, however, that did not appear in that example:

1. Multiple processors running cyclically.
2. Interaction with the global database.
3. Concerns about computational and data-transfer (I/O) efficiency.

Because of space limitations, only some key aspects of the first and last concerns are discussed below.

General description

Figure 1 shows the mathematical statement of the nonlinear structural dynamics problems and a numerical algorithm for solving it. Figure 2 shows a schematic of the NICE implementation of the numerical algorithm. The circles in the figure represent NICE processors (*e.g.* `ELT` = finite element generator, `ASM` = matrix assembler, etc.), which are logically connected only through the global database managed by *GAL*. The processors are driven by a command-language procedure, an outline of which is given to the left of the processor flowchart. Note that the commands are simple statements involving a verb (*e.g.* `FORM`) and an object (*e.g.*, `K`, the stiffness matrix). The "adjectives and adverbs" required to manipulate the data are contained in the database and are dealt with at the Fortran level, *i.e.* within processors.

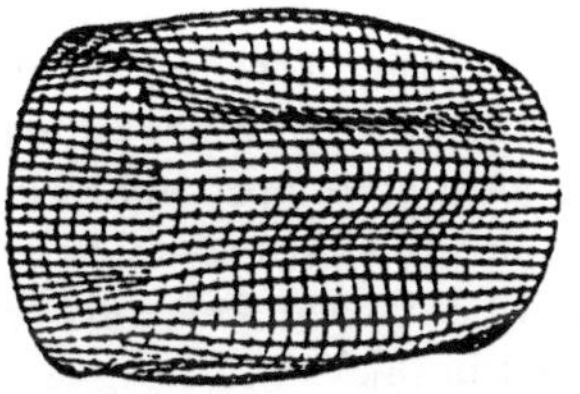

Problem Statement:

(Finite-Element Equations of Motion)

$$M\ddot{d} + F^{int}(d, \sigma) = F^{ext}(t)$$

Algorithm: (Iterative/Incremental Implicit Integration)

$$(\alpha M + \beta K)\,\Delta d^{(i+1)}_{(n+1)} = F^{ext}(t_n + 1) - F^{int}\left(d^{(i)}_{n+1}\right) - M\ddot{d}^{(i)}_{n+1}$$

with

$$d^{(i+1)}_{n+1} = d^{(i)}_{n} + \Delta d^{(i+1)}_{n+1}$$

where

n = time-step index,

i = iteration index,

and

$$\left.\begin{aligned} K &= A^{N_{elt}}_{e=1}\,[k_e]; \\ M &= A^{N_{elt}}_{e=1}\,[m_e]; \\ F^{int} &= A^{N_{elt}}_{e=1}\,[f^{int}_e] \end{aligned}\right\}\ \text{Element Assembly}$$

Fig. 1 Typical Nonlinear Structural Dynamics Problem and Numerical Algorithm

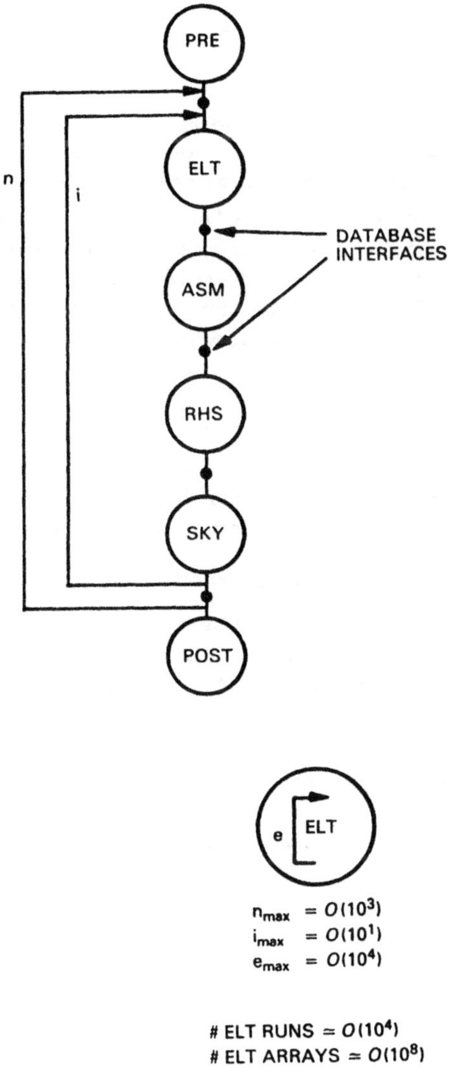

Fig. 2 NICE Implementation of a Nonlinear Structural Analysis Algorithm

A static analysis procedure

A closer look at the advantages and nuances of command-language procedures is provided in Tables 1 and 2, which show excerpts from two alternate versions of a procedure called NLSTATIC2. This particular procedure performs nonlinear static collapse analysis via an adaptive Newton-Raphson strategy based on the Riks-Crisfield arclength control algorithm. This algorithm allows complex shell response problems to be followed well into the post-buckling range, and was used in the research work presented in [3].

Table 1 represents a "hierarchical" version of NLSTATIC2, in which numerous calls to sub-procedures are made by the master procedure. Notice that arguments are nominal (named and order-independent) in contrast to the positional (order dependent) calling sequences of Fortran. Furthermore, because the NICE processors are invoked by subprocedures (not shown), this procedure form is well suited for adding new processor capabilities.

In contrast to the hierarchical version of NLSTATIC2, Table 2 shows an algorithmic, or math-language, version of the same procedure, in which the algebraic command language of a VECTOR processor is exploited, and processor/subprocedure invocations are minimized. This procedure form has several advantages: it is easy to follow; it is easy to construct, say from a published algorithm; and finally it is more efficient than the hierarchical procedure because of the reduction in interprocessor overhead.

Computational efficiency

Nonlinear and dynamic structural analysis pertain to the class of *cyclic* problems, where massive amounts of data must be transferred back and forth between processors and the global database (see the lower portion of Figure 2).

There are two basic sources of computational overhead associated with naive use of NICE: *intraprocessor overhead* due to direct transactions with the disk-resident global database, and *interprocessor overhead*, which includes processor start-up and shut-down time, repeated opening and closing of global database files, and redundant transfer of intermediate data to and from the global database to allow next-processor continuation.

The cure for high intraprocessor overhead is efficient local data management, wherein a *local data manager* is used to minimize actual disk transfers. This is particularly important in processors that work on "small granularity" data structures, such as those that carry out element stiffness, residual, and stress calculations.

The most effective solution to the interprocessor overhead problem is the macroprocessor mentioned in §1. A macroprocessor is a short-lived entity that is constructed by bundling together a collection of individual processors into a single executable module. A utility program called *MCPASS* (**M**a**C**ro**P**rocessor **ASS**embler) is used to build a macroprocessor automatically by performing the following steps:

1. The master command procedure for the analysis is scanned for processor names.
2. A main Fortran program is generated that calls the processors as subroutines.
3. Operating system commands are generated to link the individual processors into an executable macroprocessor.

There are two secondary preprocessing steps: one in which name conflicts between various processors are resolved, and another in which a subroutine counterpart of each processor's main program is extracted.

Table 1. **Hierarchical Version of NLSTATIC2**

```
*procedure NLSTATIC2 ( Database; max_step; beg_load; path_length ... )

    *DO :ITER_LOOP $iter = 1, [max_iters]
      . -------------------------------------------------------------
      . COMPUTE DISPLACEMENT INCREMENT BASED ON PREVIOUS LOAD FACTOR
      . -------------------------------------------------------------
        *call SOLVE ( db = [Database]   ; fac_matrix  = <K_fac>;
                      rhs_vector = <R_i>; soln_vector = <delta_S> )
      . ---------------------------------------
      . SUBTRACT CONTRIBUTION OF EXTERNAL FORCES
      . ---------------------------------------
        *call VECADD ( db = [Database];
                       vec_1    = <delta_S> ; vec_2    = <delta_T>;
                       scale_2  = -<load_i> ; vec_out  = <delta_B>)
      . -------------------------------------------------------------
      . SOLVE PATH_LENGTH CONSTRAINT EQUATION TO GET NEW LOAD FACTOR
      . -------------------------------------------------------------
        *call VECDOT ( db=[Database]; vec_1=<delta_T>; vec_2=<d_inc]  ;
                       DOT_PROD=d1 )
        *call VECDOT ( db=[Database]; vec_1=[delta_T]; vec_2=<delta_B>;
                       DOT_PROD=d2 )
        *call VECDOT ( db=[Database]; vec_1=[delta_B]; vec_2=[delta_B];
                       DOT_PROD = d3 )
        *call VECDOT ( db=[Database]; vec_1=[delta_B]; vec_2=[d_inc]  ;
                       DOT_PROD = d4 )
        *def a  =        <c1>                 . ( coefficients of
        *def b  = ( (2.*<d1>) + (2.*<d2>) )   .   quadratic equation
        *def c  = ( <d3> + (2.*<d4>) )        .   defining path_length )

        *call LOADFACTOR ( a = <a>; b = <b>; c = <c>; p = [path_length];
                           load_factor = load_ip1 )
      . --------------------------------------
      . UPDATE SOLUTION (TOTAL AND INCREMENTAL)
      . --------------------------------------
        *call VECADD ( db        = [Database];
                       vec_1     = <delta_B> ; vec_2    = <delta_T> ;
                       scale_2   = <load_ip1>; vec_out  = <delta> )
        *call VECADD ( db        = [Database];
                       vec_1     = <d_i>     ; vec_2    = <delta> ;
                       scale_2   =  1.0      ; vec_out  = <d_ip1> )
        *call VECADD ( db        = [Database];
                       vec_1     = <d_inc_i> ; vec_2    = <delta> ;
                       scale_2   = 1.0       ; vec_out  = <d_inc_ip1> )
      . ---------------------------------------------------------
      . UPDATE RIGHT-HAND-SIDE AND CHECK CONVERGENCE CRITERIA
      . ---------------------------------------------------------
        *call RESIDUAL ( db    = [Database]             ;
                         d_old = <d_old>; dnew = <d_ip1>;
                         F_ext = <F_ext>; R    = <R_ip1> )
        *call CHKCONV  ( db=[Database]; c_FLG=CONVERGE; d_flg=DIVERGE )
    :ITER_LOOP
*end
```

Table 2. **Algorithmic Version of** NLSTATIC2

```
*procedure NLSTATIC2 ( Database; max_step; beg_load; path_length ... )

    *run VECTOR
    *DO :ITER_LOOP $iter = 1, [max_iters]

        *run SKYPUL
            SOLVE <K_fac> <R_i>   <delta_S>
        *stop

         <delta_B> =  <delta_S>  PLUS  -<load_i> TIMES <delta_T>

         d1  =  <delta_T> DOT <d_inc>
         d2  =  <delta_T> DOT <delta_B>
         d3  =  <delta_B> DOT <delta_b>
         d4  =  <delta_B> DOT <d_inc>

        *def a  =        <c1>                     . ( coefficients of
        *def b  = ( (2.*<d1>) + (2.*<d2>) )       .   quadratic equation
        *def c  = ( <d3> + (2.*<d4>) )            .   defining path_length )

        *call LOADFACTOR ( a = <a>; b = <b>; c = <c>; p = <path_length>;
                           load_factor = load_ip1 )

         <DELTA>        =   <delta_B>  PLUS  <load_ip1> TIMES <delta_T>
         <d_ip1>        =   <d_i>      PLUS  <DELTA>
         <d_inc_ip1>    =   <d_inc_i>  PLUS  <DELTA>

        *run SHELL
          *def d_new = <d_ip1>
          *def F_ext = <F_ext>
          *def R     = <R_ip1>
           FORM FORCE/RESIDUAL
        *stop

        *call CHKCONV  ( db=[Database]; c_FLG=CONVERGE; d_flg=DIVERGE )

        *if   <CONVERGE>  go to :CONVERGED
        *if   <DIVERGE>   go to :DIVERGED

    :ITER_LOOP
    *stop VECTOR
*end
```

A preliminary evaluation of the concept of macroprocessors for static structural analysis was performed in 1984. This evaluation revealed two things: (a) the run-time difference between network and macroprocessor analysis mode for linear analysis is insignificant, and (b) for nonlinear analysis run-time can be reduced by as much as a factor of 5 for small (100-element) problems, but no more than a factor of 2 for moderate-size (1000-element) problems.

Nevertheless, upon closer examination of the macroprocessor assembly process, it becomes clear that substantially more speed is yet to be gained. While *MCPASS* does eliminates processor startup delays, it does nothing to alleviate the global database bottleneck. For this, the solution is once again local data management. By introducing a *common* local data manager it is possible to share data within a macroprocessor and obtain maximal efficiency. The potential of this latter feature, however, has not yet been exploited.

§7. CONCLUDING REMARKS

It is clear now that NICE is a successful system with its own momentum. Although the system has incorporated advanced ideas from computer science, it appears that psychological issues have been as important as technical ones in determining its acceptance. A key point in this regard is that the user always feels in control [4].

Following five years of internal development, the system was recently given a major boost with its adoption by NASA Langley Research Center as candidate for a "generic software bus" that can be used to integrate existing application programs of interest to NASA, and to serve as a basis for developing more advanced future software systems. Such future systems will have to account for evolving hardware configurations (for example, parallel processing), and evolving computer-to-computer communications (for example, moving databases between workstations and supercomputers). As part of the recently launched CSM program [5], the NICE architecture together with a public-domain "testbed" structural analyzer will be distributed next year to US aerospace companies for evaluation.

ACKNOWLEDGEMENT

The preparation of this paper was supported by the Independent Research Program of Lockheed Missiles and Space Company, Inc.

REFERENCES

1. C. A. Felippa: Architecture of a distributed processing network for computational mechanics, *Computer and Structures*, **13** (1981) 405–413

2. C. A. Felippa: A command reader for interactive programming. To appear in *Engineering Computations*

3. G. M. Stanley: *Continuum Based Shell Elements*, Ph. D. Thesis, Dept. of Applied Mechanics, Stanford University, 1985

4. P. Heckel: *The Elements of Friendly Software Design*, Warner Books: New York 1982

5. W. J. Stroud (ed.): *Proc. Workshop on Computational Methods for Structural Mechanics and Dynamics*, NASA Langley Research Center, Hampton, Va. 1985

Implementation of a Modified Hughes-Liu Shell into a Fully Vectorized Explicit Finite Element Code

J. O. HALLQUIST, D. J. BENSON, G. L. GOUDREAU
Lawrence Livermore National Laboratory, USA

Summary

Operation counts are given for our fully vectorized explicit implementation of the shell element developed by Hughes and co-workers. We have dropped reduced/selective integration in favor of reduced integration augmented by an hourglass viscosity. As a consequence, for four integration points through the thickness, execution speed is 100μs per surface element cycle on the Cray-1 compared to 140μs for four brick elements. In addition, the explicit time step will be from four to many times larger than for brick modeling.

INTRODUCTION

During the past year we have implemented shell and beam elements into the explicit finite element code DYNA3D [1,2]. Since most applications of DYNA3D involve large deformations either caused by impact or high explosives, we sought elements that would admit finite strains and retain compatibility with our 8 node solid hexahedron element which is fully vectorized and integrated with one point quadrature [3]. We had succeeded in obtaining speeds of 35μs per element cycle on the Cray-1s with the brick element (140μs for four bricks through the thickness) and were very hopeful in obtaining speeds in the same ballpark with shell or beam elements. The greatest advantage of a structural element is the insensitivity of the time step size on the thickness admitting large time step size relative to multiple solid elements through the thickness. Courant stability limits the step size of solids to a value that is less than the transient time of a stress wave across its thinnest dimension. With efficient structural elements, applications of DYNA3D to crashworthiness problems involving shell structures become feasible.

We chose to implement the "degenerate" shell and beam elements proposed by Hughes and Liu [4,5] with modifications for explicit dynamics [6,7] and

Finite Element Methods for Nonlinear Problems
Europe-US Symposium, Trondheim, Norway 1985
Editors: Bergan, Bathe, Wunderlich

thickness change [8]. When our initial implementation of these elements into NIKE3D [9] was transferred into DYNA3D we found the execution speed on the Cray-1 to be disappointing. For example, the 4 node shell with selective reduced integration, 4 integration points through the thickness, and an iterative plane stress plasticity model executed at an unbelievable 16000 μs per element cycle. The dismal speed of our initial implementation was a result of our mindless approach of getting the shell to work regardless of cost. In the process we used matrix algebra in coding the equations verbatim, ignored multiplies by zero, and gave no thought to vectorization.

Since completely recoding these elements we have reduced their cost by more than two orders of magnitude. We now believe our implementation approaches the optimum for structural elements of this type. Because all our effort has concentrated on vectorization and reducing operation counts, the elements should be competitive in speed with less sophisticated formulations with unvectorized implementations on vector hardware. In the paper we briefly describe our implementation concentrating on the details that provided us with the greatest cost reduction. First we shall justify our choice of one point integration. Next we will describe our implementation of the Jaumann stress rate. Though equivalent to the Hughes-Liu formulation, it is somewhat more cost effective. Finally, we shall discuss the plane stress radial return plasticity algorithm, the shell brick interface, and the hourglass viscosity. A simple example problem is shown.

IMPLEMENTATION

Element Integration

In our first implementation of the Hughes-Liu shell, we used reduced/selective integration. This approach requires an assembly of the strain displacement matrix, $\underset{\sim}{B}$, by adding a partial strain displacement matrix for the transverse shear strains, computed at the centroid, to its complement at the Gauss integration points. Therefore, with reduced/selective integration we face the added cost of the centroidal calculations for $\underset{\sim}{B}$. The cost of $\underset{\sim}{B}$ plus the cost of the constitutive model dominate the operation counts as can be seen in Table 1. Our choice of one point integration has reduced cost by nearly a factor of four over the reduced/selective option. Our operation counts take advantage of the simplifications that are possible at the element centroid to reduce the operation count, i.e. an integration point computed away from the centroid requires additional operations. The same simplification for one point integration of the brick has already been discussed [2,3].

Rotate fiber vectors and compute miscellaneous quantities including Hughes-Winget [10] rotation matrix		238 [127]	
Hourglass control		356 [77]	
Transform force components and update global force vector		60 [24]	
For each integration point through thickness	$\underset{\sim}{B}$ matrix leading to strains in lamina system		351 [207]
	Apply Jaumann rate and rotate stresses into lamina system		108 [60]
	Elastic stress evaluation		22 [9]
	Global stresses		79 [42]
	Nodal forces		122 [61]
TOTAL		654 [228] +	682 [379] *(Integ. pnts.)

Table 1. Operation count for the elastic Hughes-Liu shell in DYNA3D. This operation count includes adds, subtracts, multiplies, and divides and is independent of vectorization. Quantities in square brackets are a subtotal that includes only multiplication operations. The cost is dominated by the cost of the element integration which makes one point integration with multiple thickness points very attractive relative to 2 × 2 reduced selective integration.

Jaumann Rate

In [4], stresses are computed and stored in the lamina system, but we believe that operation counts can be minimized by storing stresses in the global system and updating in the lamina system. We will briefly outline our approach.

The material time derivative of Cauchy stress is given as

$$\dot{\sigma}_{ij} = \sigma_{ij}^{\nabla} + \sigma_{ik}\,\Omega_{kj} + \sigma_{jk}\,\Omega_{ki} \tag{1}$$

where

$$\sigma_{ij}^{\nabla} = C_{ijk\ell}\,\dot{\varepsilon}_{k\ell} \tag{2}$$

is the Jaumann stress rate. The spin and strain rate tensors are given as:

$$\begin{aligned} \Omega_{ij} &= \frac{1}{2}\,(\partial v_j/\partial x_i - \partial v_i/\partial x_j) \\ \dot{\varepsilon}_{ij} &= \frac{1}{2}\,(\partial v_i/\partial x_j + \partial v_j/\partial x_i) \end{aligned} \tag{3}$$

where v_i is the velocity.

Before updating, we rotate the stress, σ_{ij}^{n}, at time t^n into the configuration at t^{n+1}

$$\hat{\sigma}_{ij}^{n+1} = \sigma_{ij}^{n} + (\sigma_{ip}^{n}\Omega_{pj}^{n+1/2} + \sigma_{jp}^{n}\Omega_{pi}^{n+1/2})\,\Delta t^{n+1/2} \tag{4}$$

and transform this result and the strain rates into the n+1 lamina system in Eq. (5) where R_{ij} is an orthogonal transformation matrix between the global and lamina coordinate systems.

$$\begin{aligned} \hat{\tau}_{ij}^{n+1} &= R_{ik}\,R_{j\ell}\,\hat{\sigma}_{k\ell}^{n+1} \\ d_{ij}^{n+1/2} &= R_{ik}\,R_{j\ell}\,\dot{\varepsilon}_{k\ell} \end{aligned} \tag{5}$$

The stress tensor can now be updated in the lamina system where $C_{ijk\ell}$ is the plane stress constitutive matrix in the lamina system:

$$\tau_{ij}^{n+1} = \hat{\tau}_{ij}^{n+1} + C_{ijk\ell}\,d_{k\ell}^{n+1/2}\,\Delta t^{n+1/2} \tag{6}$$

A rotation is performed to return the stress tensor to the global system:

$$\sigma_{ij}^{n+1} = R_{ki}\,R_{\ell j}\,\tau_{k\ell}^{n+1} \tag{7}$$

We have found from our experience with explicit codes that the foregoing procedure preserves objectivity in nearly all computations. However, in implicit implementations where Δt is typically 2 orders of magnitude greater, we believe it is preferable to split the Jaumann rate rotation into two parts. In the first, we rotate the stress into the n+ 1/2 configuration, then update the stress, and, in the second, we rotate the stress from the n+ 1/2 configuration to the n+1 configuration. The Jaumann stress rate in 3D implicit calculations is implemented with the rotation formulas in [10].

Rather than updating the stress in the lamina system, we could also transform the plane stress-strain constitutive law to the global system and update there; a possibility alluded to by Bathe [11]. Based on operation counts, we have decided our implementation is more cost effective. Transforming $C_{ijk\ell}$ into the global system cost nearly 750 operations for the most general case which includes computing the transformation matrix. The significance of this cost is apparent in light of Table 1 where the cost of each integration point is just 682 operations. The transformation of $C_{ijk\ell}$ in implicit codes may be advantageous since the stiffness matrix must also be computed.

Plane Stress Plasticity Model

We have modified the isotropic-kinematic radial return bilinear plasticity model of Krieg and Key [12] to accommodate plane stress. We will discuss the implementation for shells since the same procedures are applied to the beam elements with only slight modifications. We perform the stress update in the lamina system and seek a final stress state such that the normal stress

$$\tau_{33} = 0 \tag{8}$$

In applying this model we first solve for the normal strain increment assuming an elastic solution where i is the iterate, and λ and μ are Lame's constants.

$$\Delta d_{33} = \frac{-\hat{\tau}_{33} - \lambda(\Delta d_{11} + \Delta d_{22})}{\lambda + 2\mu} \tag{9}$$

where

$$\Delta d_{ij} = d_{ij}\,\Delta t$$

Equation (9) accounts for the possibility that after the Jaumann rate rotation, Eq. (4), the normal stress does not necessarily equal zero [4]. We update the stress state elastically to obtain the trial stress ${}^{*}\tau_{ij}^{n+1}$ where ${}^{*}\tau_{33}^{n+1} = 0.$

$$ {}^{*}\tau_{ij}^{n+1} = \hat{\tau}_{ij}^{n+1} + C_{ijk\ell}\, \Delta d_{k\ell} \qquad (10) $$

Let α_{ij} denote the back stress, σ_y the current yield stress and

$$ \xi_{ij} = {}^{*}\tau_{ij}^{n+1} - \frac{1}{3}\, {}^{*}\tau_{kk}^{n+1}\, \delta_{ij} - \alpha_{ij} \qquad (11) $$

If $\quad \Lambda^2 - \sigma_y^2 \leq 0 \qquad (12)$

where

$$ \Lambda^2 = \frac{3}{2}\, \xi_{ij}\, \xi_{ij} $$

the state is elastic, and we rotate the stress back to the global system and proceed to the next element. If not, we update the plastic strain ε_{eff}

$$ \varepsilon_{eff}^{n+1} = \varepsilon_{eff}^{n} + \frac{\Lambda - \sigma_y}{3\mu + E_p} = \varepsilon_{eff}^{n} + \Delta\varepsilon_{eff}^{i} \qquad (13) $$

where E_p is the plastic hardening modulus and solve for the normal stress

$$ {}^{i}\tau_{33}^{n+1} = {}^{*}\tau_{33}^{n+1} - \frac{3\mu\Delta\varepsilon_{eff}^{i}}{\Lambda}\, \xi_{33} \qquad (14) $$

If i=1, the first iterate, set

$$ \Delta d_{33}^{i} = - \Delta d_{11} - \Delta d_{22} \qquad i=2 \qquad (15) $$

and repeat Eqs. (10), (11), (13) and (14) for i=2. We now iterate for Δd_{33} such that ${}^{i}\tau_{33}^{n+1}$ approaches zero.

$$ \Delta d_{33}^{i+1} = \Delta d_{33}^{i-1} - {}^{i-1}\tau_{33}^{n+1}\, \frac{\Delta d_{33}^{i} - \Delta d_{33}^{i-1}}{{}^{i}\tau_{33}^{n+1} - {}^{i-1}\tau_{33}^{n+1}} \qquad (16) $$

Convergence is assumed if

$$ \frac{|\Delta d_{33}^{i} - \Delta d_{33}^{i-1}|}{|\Delta d_{33}^{i+1}|} < 10^{-5} \qquad (17) $$

If Eq. (17) is unsatisfied, i=i+1 and Eqs. (10), (11), (12), (13) and (14) are repeated. After convergence is achieved, we update ε_{eff}^{n+1} τ_{ij}^{n+1}, and α_{ij}^{n+1} :

$$\tau_{ij}^{n+1} = {}^*\tau_{ij}^{n+1} - \frac{3\mu \; \Delta\varepsilon_{eff}}{\Lambda} \xi_{ij} \tag{18}$$

$$\alpha_{ij}^{n+1} = \alpha_{ij}^{n} + \frac{(1-\beta)E_p \; \Delta\varepsilon_{eff}}{\Lambda} \xi_{ij} \tag{19}$$

$$\varepsilon_{eff}^{n+1} = \varepsilon_{eff}^{n} + \Delta\varepsilon_{eff}, \tag{20}$$

where β is a hardening parameter, equal to 1 for isotropic hardening and 0 for kinematic hardening, i.e., $0 \leq \beta \leq 1$.

Usually fewer than 5 iterations are required for convergence. With the exception of the operations in the iteration loop, the model fully vectorizes. In Table 2 we tabulate the operations count for the plasticity model. For localized plasticity and isotropic hardening this model adds very little to the shell cost given in Table 1.

Operations outside iteration loop	78	[33]
Operations inside iteration loop per iteration	53	[16]
Additional operations to rotate α_{ij}. Done only if $\beta < 1$	173	[99]

Table 2. Operation count for the isotropic/kinematic radial return plasticity model. This operation count includes adds, subtracts, multiplies, and divides. Quantities in square brackets are a subtotal that includes only multiplication operations. The number of additional operations that are required for kinematic hardening, $\beta < 1$, are significant.

Shell Brick Interface

Tied slidelines and surfaces are used in the DYNA and NIKE programs to join regions that are coarsely zoned with more finely zoned regions. The tied slideline capability has been extended in DYNA3D to tie regions of hexahedron elements to regions of shell elements. A shell node may be tied to an arbitrary number of nodes lying along the tangent vector to the nodal fiber. During the calculation, nodes thus constrained must lie along the fiber but can move relative to each other in the fiber direction.

The equations of motion for the rotational degrees of freedom of a shell node are in terms of the angular acceleration about the global axes of a local reference frame attached to the normal vector. Integrating the acceleration with time gives the angular velocity of the local reference frame, but the integral of the angular velocities does not result in a meaningful angular displacement measure. The derivation of the shell-brick slidelines will therefore be based on virtual power.

The principle of virtual power is simply the weak form of the equations of motion with virtual velocities instead of virtual displacements.

$$VP = \int_{\Omega} F_i \, \delta v_i \, d\Omega + \int_{\Gamma} f_i \, \delta v_i \, d\Gamma = 0 \tag{21}$$

where

Ω and Γ are the interior and boundary of the domain.

F_i and f_i are the forces, including the stress divergence, D'Alembert forces, body forces, and surface tractions.

δv_i are the components of the virtual velocities.

The discrete form of Eq. (21) is derived by introducing the usual finite element approximations.

$$VP = F_{\alpha i} \, \delta V_{\alpha i} = 0 \tag{22}$$

where

$F_{\alpha i}$ are the nodal force components.

$\delta V_{\alpha i}$ are the nodal velocities.

For solid hexahedron elements, i ranges from 1 to 3 for the three translational velocity components, while for shell elements, i ranges from 1 to 6 with 4, 5, and 6 corresponding to the additional angular velocities.

The problem of tying shell elements to solid elements reduces to finding a consistent velocity transformation from solid element nodal velocities to shell nodal velocities. Once such a transformation is available, it can be substituted in Eq. (22) to derive a consistent transformation between the force components at the nodes along the fiber and the shell nodal force.

$$V_{\alpha i} = \Phi_{\alpha i \beta j} v_{\beta j} \tag{23}$$

$$F_{\alpha i} \delta V_{\alpha i} = f_{\beta j} \delta v_{\beta j} = F_{\alpha i} \Phi_{\alpha i \beta j} \delta v_{\beta j} \tag{24}$$

Therefore:

$$f_{\beta j} = F_{\alpha i} \Phi_{\alpha i \beta j} \tag{25}$$

where

$F_{\alpha i}$ and $f_{\beta j}$ are solid and shell nodal forces respectively.

$V_{\alpha i}$ and $v_{\beta j}$ are solid and shell nodal velocities respectively.

$\Phi_{\alpha i \beta j}$ is the transformation from shell velocities to solid velocities.

The summation over α is performed over all nodes that are tied to the shell node β. The solid element velocities are linear functions of the shell velocities which gives us a particularly simple form for Φ. The generalization of Φ to constrain beam to solid elements is obvious.

$$V_{\alpha i} = v_{\beta i} + \ell e_{ijk} v_{\beta j+3} n_k \tag{26}$$

where

ℓ is the distance along the normal from α and β.

n_k are the components of the shell normal at β.

Substituting Eq. (26) into Eq. (25) gives the force contributions from the solid nodes to the shell nodes.

$$f_{\beta j} = f_{\beta j} + \textstyle\sum_\alpha F_{\alpha j} \qquad \text{for } j = 1,2,3 \tag{27}$$

$$f_{\beta i+3} = f_{\beta i+3} + \textstyle\sum_\alpha \ell e_{ijk} n_j F_{\alpha k} \qquad \text{for } i = 1,2,3 \tag{28}$$

An approach similar to the above would give a consistent contribution to the shell mass matrix, but we use a simple lumping procedure in practice in order to keep the mass matrix diagonal.

$$M_{\beta j} = M_{\beta j} + \sum_\alpha M_{\alpha j} \qquad \text{for } j = 1,2,3 \tag{29}$$

$$M_{\beta j+3} = M_{\beta j+3} + \sum_\alpha M_{\alpha j} \ell^2 \qquad \text{for } j = 1,2,3 \tag{30}$$

where

$M_{\beta j}$ is the lumped mass matrix.

After the shell displacements are updated, the brick nodes along the slideline are updated.

$$x_{\alpha i} = x_{\beta i} + \ell n_{\beta i} \tag{31}$$

where

α is a solid element node and β is a shell node.

The velocities of the solid nodes along the slideline are determined by differencing the displacements.

Hourglass Control

The major drawback of using reduced integration is the presence of hourglass modes in the solution. These modes were first noticed and controlled by Maenchen and Sack while developing TENSOR [13] and their hourglass viscosity is one of the methods that is optional in DYNA2D. Other methods we currently use include those by Flanagan and Belytschko [14], and Hancock [15]. In our experience, hourglass modes tend to form over a time duration that is much shorter than the duration of the structural response. Since we wish to preserve hourglass modes that are a stable kinematic component of the global deformations, we attempt to stop the hourglassing with a small amount of damping that has little affect on the stable global deformation. In keeping with the spirit of degenerating a continuum element into a shell element, the hourglass control for the shell element is degenerated from the hourglass control for the solid element by applying the virtual power equations presented in the previous section.

There are four hourglass modes which are repeated in each of the three dimensions for a solid element. The magnitude of the viscous damping force is proportional to the magnitude of the modes.

$$F_{\beta i} = - a_h h_{ij} \Gamma_{\beta j} \tag{32}$$

$$h_{ij} = V_{\alpha i}\Gamma_{\alpha j} \tag{33}$$

$$a_h = \frac{1}{4} Q_{hg} \rho hc \tag{34}$$

where

$\Gamma_{\beta j}$ are the hourglass base vectors.

h_{ij} is the magnitude of hourglass mode j in direction i.

Q_{hg} is a user specified constant, usually between .05 and .15.

ρ is the density.

h is an element dimension currently set to cubic root of the volume.

c is the material sound speed.

The hourglass control is implemented by the following procedure:

- The velocities of the four corners of the upper and lower surfaces of the shell element are calculated using Eq. (26).
- Viscous damping forces are calculated using Eq. (32) and (33).
- Equivalent shell damping forces are calculated using Eq. (27) and (28).

The hourglass control for the shell elements is more expensive than for the solid elements, but it is still less expensive than competing methods. While we do not believe that the above procedure is an optimal one for shell elements, it does allow us to try various hourglass control methods on both our solid and shell elements in a consistent manner.

EXAMPLE

Igarashi [16] has provided us with geometry definitions of a mild steel frame member that Suzuki has dynamically crash tested. Frame members are attached to a sled that accelerates to 30km/hr before impact into a fixed barrier, see Fig. 1. Experimental output ceases at 35ms when the sled impacts a bumper that stops its motion. Force time histories are obtained by load cells at the frame-barrier interface, and movies record the deformation time history.

We have performed preliminary calculations on coarsely meshed members to study the feasibility of this type of explicit simulation. We generated the mesh, Fig. 2 shows a close-up view of one with 1600 shell elements, using our mesh generator INGRID [17]. Figure 3 shows the cross section of the 1000mm beam which is fabricated by spot welding along the flange.

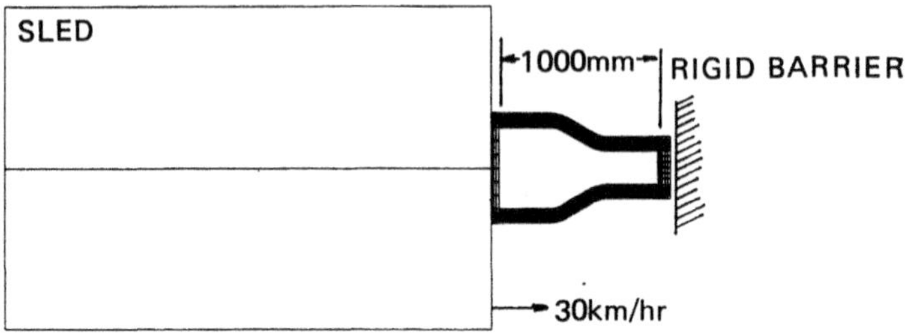

Figure 1. Finite element model of sled test.

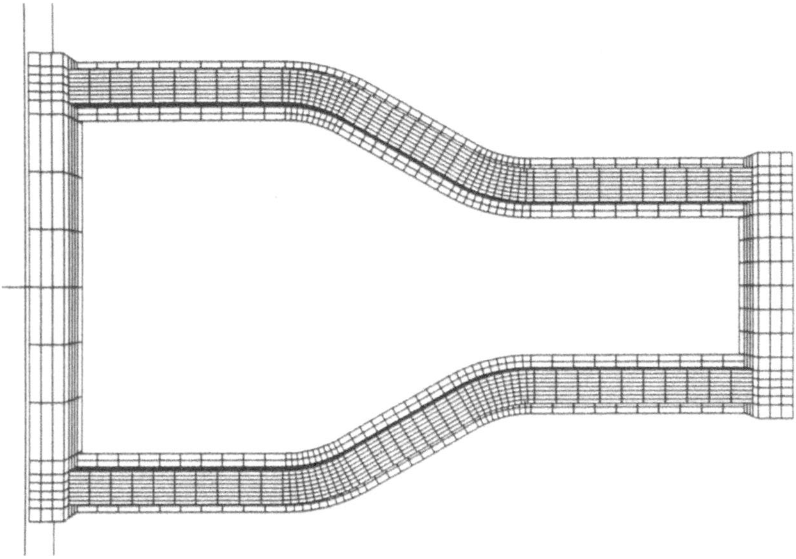

Figure 2. Close-up view of the finite element mesh. The mesh contains 1600 shell elements.

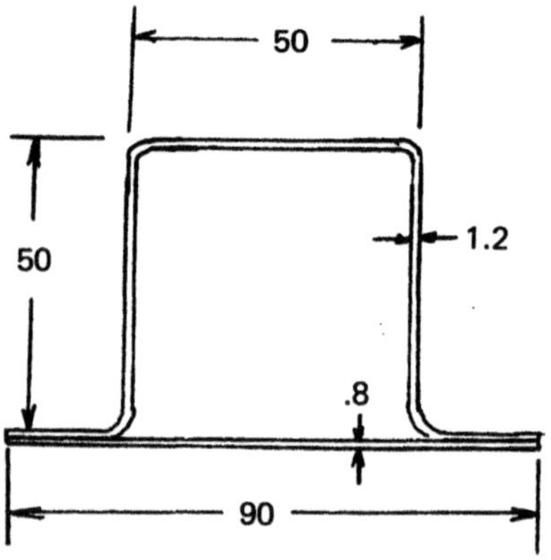

Figure 3. Cross section of frame member. All dimensions are in millimeters.

Although we did not model the spot welds, we believe that such modeling detail is important in capturing the details of the force versus time curves. Deformed shapes are plotted at 10μs intervals in Fig. 4. A more refined mesh must be used if the highly localized buckling is to be resolved. Although we accurately predicted the peak force, we under predicted the pulse width. In the near future, we intend to provide comparisons of DYNA3D results with experimental data as part of a publication by Igarashi [16].

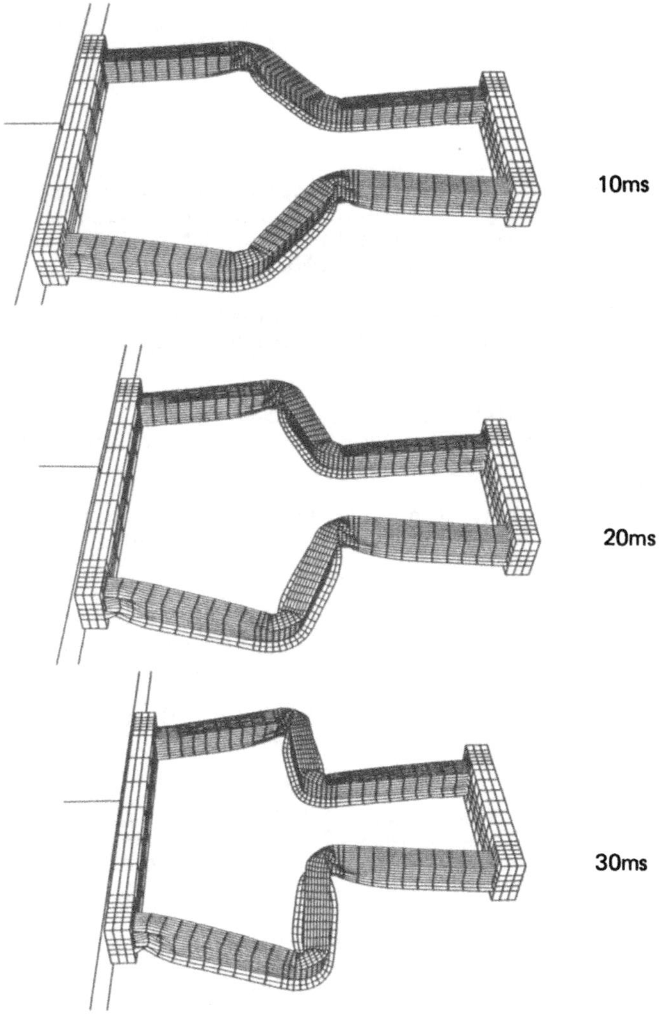

Figure 4. Deformed shapes at 10ms output intervals.

Computation time for a 640 element mesh run out to 35ms of simulation time required 80 minutes of CPU on the Cray-1 and for the refined mesh of Fig. 2, 252 minutes. Since the peak force occurs almost immediately and decays rapidly, the interesting part of the force versus time curve occurs well before 5ms. Simulations of 5ms duration can be done very cheaply making parameter studies feasible.

Acknowledgement

This work was performed under the auspices of the U.S. Department of Energy by the Lawrence Livermore National Laboratory under contract number W-7405-Eng-48.

References

1. Hallquist, J.O.: Preliminary user's manuals for DYNA3D and DYNAP (nonlinear dynamic analysis of solids in three dimensions), University of California, Lawrence Livermore National Laboratory, Rept. UCID-17268 (1976) and Rev. 1 (1979).

2. Hallquist, J.O.: Theoretical manual for DYNA3D, University of California, Lawrence Livermore National Laboratory, Rept. UCID-19401 (1982).

3. Goudreau, G.L.; Hallquist, J.O.: Recent developments in large scale finite element Lagrangian hydrocode technology. J. Comp. Meths. Appl. Mechs. Eng. 30 (1982).

4. Hughes, T.J.R.; Liu, W.K.: Nonlinear finite element analysis of shells: part I. Three-dimensional shells. Comp. Meths. Appl. Mechs. 27 (1981) 331-362.

5. Hughes, T.J.R.; Liu, W.K.: Nonlinear finite element analysis of shells: part II. Two-dimensional shells. Comp. Meths. Appl. Mechs. 27 (1981) 167-181.

6. Hughes, T.J.R.; Cohen, M.; Haroun: Reduced and selective integration techniques in the finite element analysis of plates. Nucl. Eng. Design 46 (1978) 203-222.

7. Hughes, T.J.R.; Liu, W.K.; Levit, I.: Nonlinear dynamics finite element analysis of shells. Nonlinear Finite Element Analysis in Struct. Mechs. eds. W. Wunderlich, E. Stein, and K.J. Bathe, Springer-Verlag, Berlin, (1981) 151-168.

8. Hughes, T.J.R.; Carnoy, E.: Nonlinear finite element shell formulation accounting for large membrane strains. Nonlinear Finite Element Analysis of Plates and Shells 48 (1981).

9. Hallquist, J.O.: NIKE3D: An implicit, finite-deformation, finite element code for analyzing the static and dynamic response of three-dimensional solids. University of California, Lawrence Livermore National Laboratory, Rept. UCID-18822, Rev. 1 (1984).

10. Hughes, T.J.R.; Winget, J.: Finite rotation effects in numerical integration of rate constitutive equations arising in large-deformation analysis. Int. J. Numer. Meths. Eng. 15 (1980) 1862-1867.

11. Bathe, K.J.: Finite element procedures in engineering analysis. Englewood Cliffs, N.J.: Prentice-Hall, Inc. (1982) 258-260.

12. Krieg, R.D.; Key, S.W.: Implementation of a time dependent plasticity theory into structural computer programs. Vol. 20 of Constitutive Equations in Viscoplasticity: Computational and Engineering Aspects (ASME, NY) (1976) 125-157.

13. Maenchen, G.; Sack, S.: The Tensor code. Meths. Comp. Phys. 3, New York: Academic Press (1964).

14. Flanagan, D.P.; Belytschko, T.: A uniform strain hexahedron and quadrilateral and orthogonal hourglass control. Int. J. Numer. Meths. Eng. 17 (1981) 679-706.

15. Hancock, S.L.: An hourglass subtraction procedure. TCAM Tech. Memo 73-6, Physics International Co., San Leandro, California (1973).

16. Igarashi, M.: private communication, Suzuki Motor Co., Ltd., Hamamatsu, Japan, May 1985.

17. Stillman, D.W.; Hallquist, J.O.: INGRID - A three dimensional mesh generator for modeling nonlinear systems. To be published, University of California, Lawrence Livermore National Laboratory.

Adaptive Substructuring Techniques for Finite Element Analysis with Localized Nonlinearities

T. Y. HAN, J. F. ABEL
Cornell University, Ithaca, USA

Summary
The application of substructuring techniques to the analysis of localized nonlinear problems has been known for years. To substructure before starting nonlinear analysis, the analyst must know the locations and sizes of the possible nonlinear regions in advance. Should nonlinearity develop in the linear substructures, the analysis must stop; otherwise, inaccurate results are obtained. These difficulties can be eliminated by adaptive substructuring techniques. In adaptive substructure analysis, the nonlinear substructure grows when the nonlinear zone expands or moves, or when new nonlinearity arises in different places. This paper describes both interactive- and self-adaptive substructuring.

Introduction

In many problems involving such nonlinearities as plasticity, fracture propagation, and progressive contact, the nonlinearity is localized; therefore, distinguishing the linear and nonlinear portions of a structure in the analysis may be significant. Substructuring techniques are one efficient way to do so. The part of structure which is known to remain linear during the loading process is defined as either one or several linear substructures, while the part of the structure which experiences nonlinear behavior is defined as one nonlinear substructure. The advantage of this substructuring is obvious. The linear elastic substructures can be treated as superelements and represented by condensed substructures. During the incremental/iterative analysis of the complete structure, only the stiffness of the nonlinear substructure need be modified. Owen and Goncalves [1] give a clear description of this process for the case of material nonlinearity. References [2-4] are additional examples of the application of substructuring for localized nonlinearity.

In substructure analysis of locally nonlinear problems, if the analyst knows the locations of probable nonlinear zones in advance and if the nonlinearities remain within relatively small portions of the structure, he may define

Finite Element Methods for Nonlinear Problems
Europe-US Symposium, Trondheim, Norway 1985
Editors: Bergan, Bathe, Wunderlich

the substructures before analysis. These substructures must remain valid throughout the loading process. In some problems, however, the analyst does not know the location of the nonlinear zones in advance, and there is no way for him to substructure before starting analysis. In some other cases, the nonlinearities may start in a relatively small portion of the structure and progress during the loading process; new nonlinearity may arise at different places. To define a nonlinear substructure including all the possible nonlinear regions of the structure through the complete loading process is not efficient. Adaptive substructuring techniques [5] are a better way to handle these problems. When new nonlinearity occurs in the linear substructures the analysis program modifies the substructures and continues the analysis.

Adaptive Substructuring

In adaptive substructure analysis, the substructures may change their size, and even the number of substructures may change. The nonlinear substructure adapts by growing as nonlinear zones expand or new nonlinearity occurs. Once an element has been incorporated into the nonlinear substructure this element belongs to the nonlinear substructure throughout the remainder of loading processes, i.e., the nonlinear substructure does not contract. There is only one level of substructuring in the present work. The entire structure is subdivided into one nonlinear substructure and one or several linear substructures. Noncontiguous linear portions of the structure are divided into separate linear substructures according to their connectivity to preserve computational efficiency.

Two adaptive substructuring algorithms are presented. The self-adaptive algorithm is employed throughout the loading process, while the interactive-adaptive alogrithm is an option for the analyst upon his intervention in the nonlinear analysis process. Both methods attempt to avoid nonlinearity in linear substructures; however, if it does occur, the analysis system can make the necessary adjustment and continue the analysis.

Although the ultimate intent is to extend adaptive substructuring to a variety of types of localized nonlinearity, including fracture nucleation and propagation, the first implementation has been to elasto-plastic analysis [5]. The remainder of this paper focuses on this type of problem, but the techniques and criteria presented can readily be modified for other problem types.

As will be demonstrated below, adaptive substructuring analysis has proven to be efficient. The relative efficiency of different problems compared to direct analysis is illustrated schematically in Fig. 1. The CPU time saving is problem dependent and depends on the relative size of the nonlinear substructure, the distribution of plastification, the number of load steps during which the substructures do not change, the numerical method used to perform iteration, the load step size, and the number of iterations to achieve convergence. In general, substructure analysis is more efficient when the plastic zone is small (in terms of fraction of the total number of degrees of freedom) and restricted to one place. If the plastic zone becomes large or if separate, multiple zones occur, the efficiency is reduced. At a certain stage, substructure analysis probably becomes less efficient than direct analysis; this is shown schematically by the dashed lines for schemes S1 and S2 in Fig. 1. Scheme S2 becomes less efficient in the local sense when the slope of its curve becomes greater than that of the direct analysis. Not only does scheme S1 exhibit the same local inefficiency, but

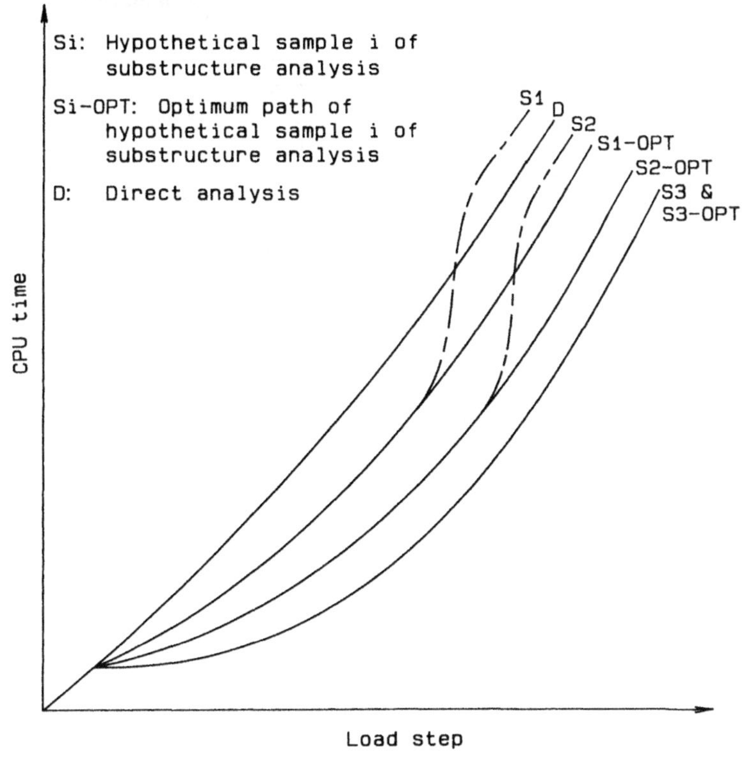

Fig. 1. Schematic relative efficiency of direct and substructure analysis.

also, because its curve crosses that of the direct analysis, it becomes globally less efficient. To avoid the onset of local inefficiency, when most of the structure yields the whole structure is incorporated into the nonlinear substructure, and the substructure analysis has the same local efficiency as direct analysis, i.e., a slope parallel to D in Fig. 1.

The adaptive substructuring algorithm tries to follow a nearly optimum path. A criterion is necessary to ascertain when substructuring has reached diminishing returns and the entire structure is to be incorporated into the nonlinear substructure. In the present program the criterion consists of a CPU time estimate based on a series of procedures in which empirical coefficients are employed [5]. Figure 2 shows typical curves of CPU time versus load step for a direct and substructure analysis. The load steps at the beginning of which the substructures change such as n+1 consume more CPU time than the subsequent load steps with the same

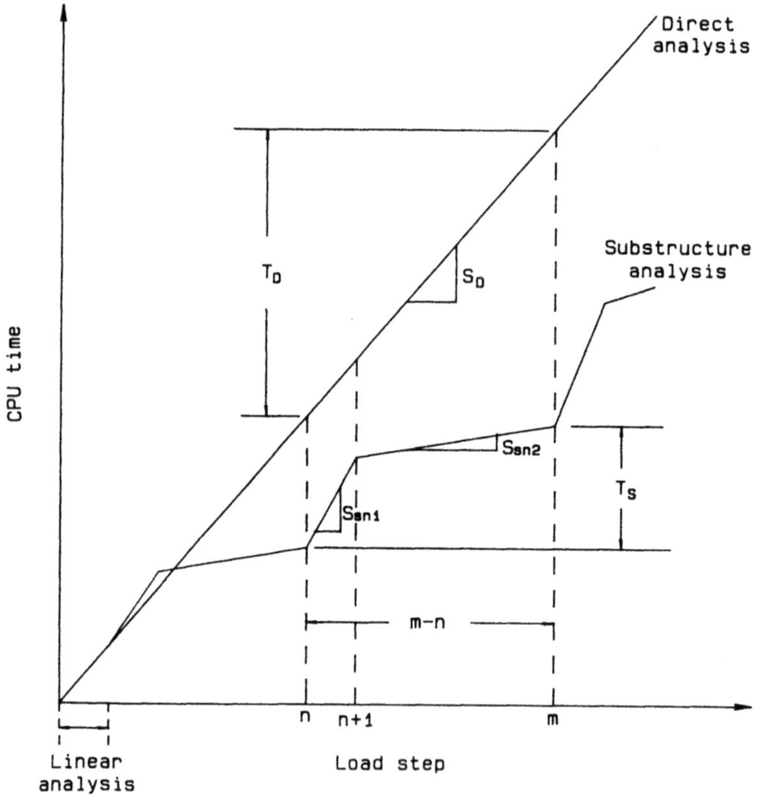

Figure 2. Typical curves of CPU time versus load step for a direct and substructure analysis.

substructures, e.g., n+2 to m. Whenever the substructures change, the program compares the estimates of CPU time between direct and substructure analysis for the load interval from n+1 to m, i.e., compares the estimated values of T_S and T_D. If T_S is greater than T_D, the program incorporates the entire structure into the nonlinear substructure.

Re-substructuring occurs whenever the effective stress at any point in the linear substructures reaches the second highest stress level as described below. To estimate the number of load steps (NSTEP = m-n, as shown in Fig. 2) during which the substructures do not change, it is assumed that in linear substructures the effective stress increases almost linearly. Because of stress re-distribution, however, the effective stress increment may be greater than that obtained by linear incrementation. Therefore, the linear effective stress increment is multiplied by an empirical factor greater than unity, 1.05.

It is assumed that the major difference between CPU time of direct and substructure analysis is due to the different equation solving times; other operations consume about the same CPU time in the two analyses. It is also assumed that the number of iterations (NITER) of each load steps n+1 to m is the same as that of the load step n. The CPU time spent for solving a system of N equations with an average semi-bandwidth of B is then to be proportional to NB^2. If the BFGS method is employed the relative computational times are estimated as

$$T_D = N_e B_e^2 * NSTEP * (1.0 + 0.08 * (NITER - 1)) \tag{1}$$

and

$$T_S = \sum_{i=2}^{NS} (N_l B_l^2)_i + N_n B_n^2 * NSTEP * (1.0 + 0.08 * (NITER - 1)) \tag{2}$$

in which the subscripts e, l, and n denote the entire structure, the linear substructure, and the nonlinear substructure, respectively; and NS is the number of substructures. The empirical factor 0.08 is based on the observation that to modify the stiffness matrix by the BFGS method requires only about 8% of the CPU time required for re-computing and decomposing the stiffness matrix. If the Newton-Raphson method is employed the estimated CPU times can be obtained by

$$T_D = N_e B_e^2 * NSTEP * NITER \tag{3}$$

and

$$T_S = \sum_{i=2}^{NS} (N_1 B_1^2)_i + N_n B_n^2 * NSTEP * NITER \tag{4}$$

Self-Adaptive Substructuring

A self-adaptive substructuring algorithm restricts plastification to the nonlinear substructure. The basic concern of this algorithm is the analysis efficiency, under which two conditions are considered, the state of stress and the geometric connectivity. Nodal and element connectivity tables must be prepared for this algorithm. The former gives the node numbers that connect to individual nodes, and the latter gives the element numbers that connect to individual elements. Nodes belonging to the same element are considered connected, while elements having nodes in common are considered connected.

Criteria for substructuring. Self-resubstructuring is invoked only when necessary, not in every load step. The structure is re-subdivided when plastification occurs or almost occurs in a previously all-elastic zone or when the plastic zone progresses too close to the boundary of the nonlinear substructure.

For elasto-plastic self-substructuring the state of stress is divided into four levels according to the magnitude of the effective stress. The first level is yielding, the second is F1 times the yielding stress, the third is F1*F2 times the yielding stress, and the fourth level is below the third. Here F1 and F2 are factors less than unity and generally greater than about 0.7. F1 is determined by an empirical formula which takes into account the load-step sizes [6], while F2 is a user-specified parameter. F1 should be small enough so that the largest load step will not increase effective stress from level three to yielding.

The highest stress level among the integration points of one element represents the stress level of that element. Self-substructuring takes place whenever the stress level of any element in a linear substructure reaches level one or two. If the stress level of any element in the linear substructures reaches level one, iteration of the same load step continues after resubstructuring until equilibrium is attained.

The stiffness matrix of individual linear substructures is condensed at the beginning of analysis. The reduced stiffness matrix and load vector remain valid as long as no plastification occurs in this substructure. The criteria for self-substructuring are selected so that the nonlinear substructure is sufficiently large that it may remain valid for several load steps, but is not so large that efficiency is reduced.

Upon substructuring, elements with the three highest stress levels are immediately considered to belong to the nonlinear substructure. In addition, to avoid narrow "peninsulae" in the linear mesh, elements with all of their corner nodes at the interface between the nonlinear and linear substructures are absorbed into the nonlinear substructure.

Subdivision algorithm. The subdivision procedure assigns elements previously in linear substructures to the nonlinear substructure according to the two criteria described above. The overall procedures of the algorithms have been summarized in Ref. [6], while the details of each step are expanded in comprehensive flow charts presented in Ref. [5].

Interactive-Adaptive Substructuring.

The interactive-adaptive substructuring technique relies heavily on interactive computer graphics. Man-machine communication is accomplished through a digitizing tablet and a pen. A high-speed refress vector terminal with dynamic display capability and a raster color display terminal provide necessary information about the stress distribution and substructuring. The stress levels and substructuring information are displayed either on a vector graphic terminal or on a color raster graphic terminal at the end of every load step in an interactive analysis.

Interactive-adaptive substructuring is useful for problems in which the analyst knows the locations of probable plastic zones in advance. In addition, it may be used to override or modify partitioning suggested by the self-adaptive substructuring algorithm. The interactive procedure entails designation of elements which are to be incorporated into the nonlinear substructure. This is accomplished simply by pointing to individual elements or groups of elements. Once the analyst finishes this designation, the same scheme as used in the self-adaptive algorithm is applied to distinguish between boundary and interior nodes and elements and to subdivide the linear substructures.

Numerical Example

A number of numerical examples have been performed to test the self-adaptive substructing algorithm and to compare relative efficiencies of different analysis approaches [5]. Two examples have already been published [6], and one additional example is presented here. In all examples the von Mises yield criterion is employed, and a strain hardening material model with a linear isotropic hardening rule is used. Only material nonlinearity is considered. The examples were performed on a VAX 11/780 with an in-house general-purpose finite element analysis program incorporating both the Newton-Raphson and BFGS methods.

The finite element model of a quarter of the plate is shown in Figure 3.

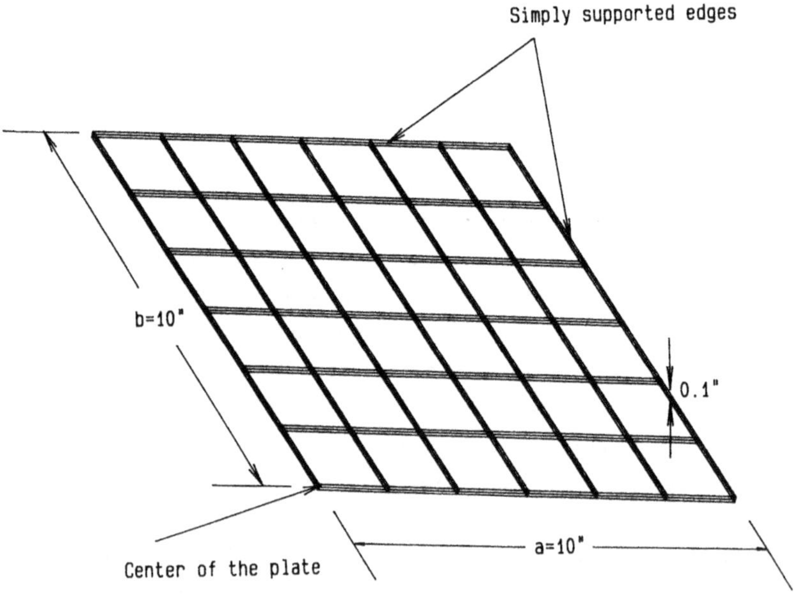

Fig. 3. Finite element model of one quarter of a simply supported square plate. ($E = 10^7$ psi, $\nu = 0.3$, $\sigma_y = 30{,}000$ psi, $H' = 0$.)

Two layers of twenty-node solid elements are used to provide four layers of integration points to detect the growth of plastification from the top and bottom of the plate. This mesh, consisting of 72 20-node elements, 497 nodes, and 1360 free d.o.f., is probably the coarsest useful subdivision through the thickness. A fine mesh model (8x8x3) has been employed to obtain a better solution. Although geometric nonlinearities are not considered in the analysis, the central deflection versus the applied load compares reasonably well with the finite element results from Armen et al. [7] and with the finite difference results from Bhaumik and Hanley [8]. With either the coarse mesh shown in Figure 3 or the fine mesh, the growth of the

Corner of the plate

				12	12	9	10	5	7	2	5
		14	13	10	10	7	8	3	4	0	2
		14	14	11	11	9	9	6	8	4	7
	14	12	11	9	9	7	8	5	6	3	5
13	13	11	11	9	9	8	9	8	9	8	10
11	10	9	9	8	9	8	8	7	9	7	9
9	9	9	9	8	9	9	9	9	11	10	12
8	8	8	8	8	8	8	9	9	11	10	12
6	7	7	7	8	9	9	11	11	14	13	
6	6	6	7	8	9	9	11	12	14	14	
5	5	6	7	8	9	10	13	14			
5	5	6	6	8	9	11	13				

Center of the plate

The label i at a point indicates that the point yields at the end of nonlinear load step i.

i=0 corresponds to initial yielding at the end of the linear analysis.

Fig. 4. Yielding sequence of the simply supported plate. Load step size is 5% of the initial yield load.

plastic zone is similar. The analysis results presented here are obtained with the coarse mesh.

Initial yielding occurs near the corner of the plate at a load $f = pa^2/6M_O$ = 0.498, in which p is the uniform load intensity, a = 10 in. is the half length of the plate, and M_O = 90 lb-in/in. represents the fully plastic moment. The small area closest to the corner yields later. This differs from the results obtained by using plate bending elements in the analysis, in which the initial yielding occurs right at the corner. This difference arises due

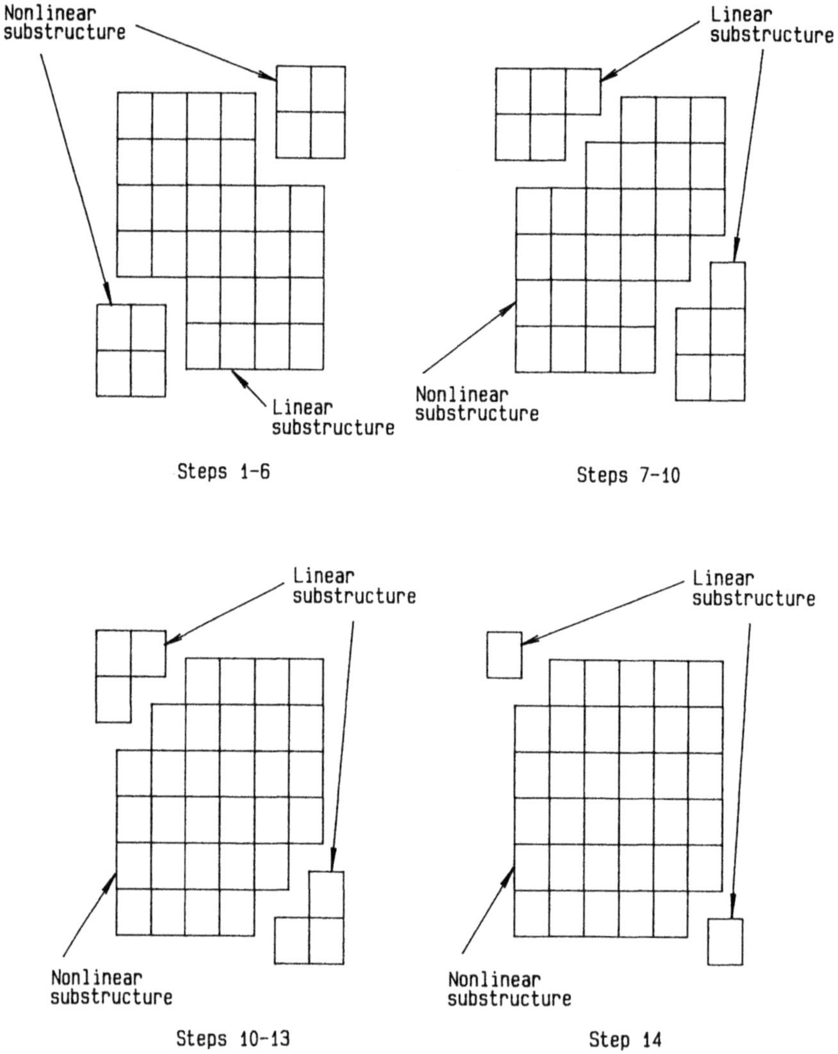

Fig. 5. Self-adaptive substructures of the simply supported plate (F2=0.8).

Anal. Tech.	BFGS Method			Newton-Raphson Method		
	CPU time (min.)	Total no. of iter.	Sav-ings (%)	CPU time (min.)	Total no. of iter.	Sav-ings (%)
Direct	215	120	---	426	44	---
Substruc. F2=0.8	156	95	27.6	305	41	28

Table 1. Statistics of analysis CPU time for the simply supported plate.

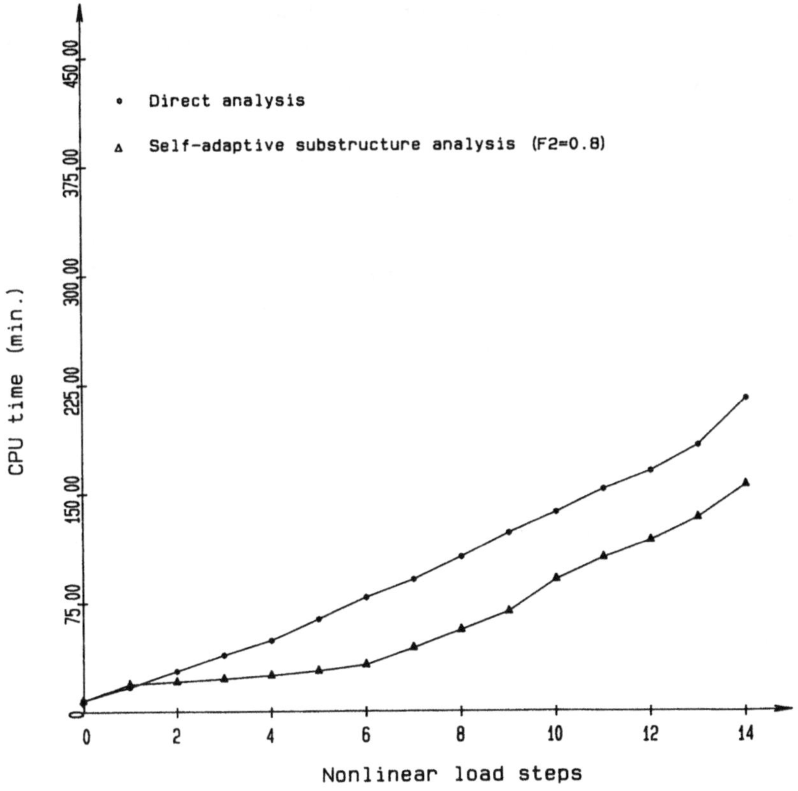

Fig. 6. **Analysis CPU time for the simply supported plate. Comparison between direct and substructure analysis by the BFGS method.**

to the three-dimensional stress state of the present analysis as opposed to the two-dimensional state implicit in the plate bending elements. An additional plastic region starts to develop at the center of the plate at f = 0.623. As the applied load increases, the two plastic zones propagate from the vicinity of the corner and the center of the plate until collapse occurs. The sequence of yielding for fourteen load steps with increments equal to 0.05 times the yield load is shown in Figure 4.

The self-adaptive substructures with F2 equal to 0.8 are illustrated in Figure 5. In load step 10 plastification occurs in the linear substructures. This is detected and the substructures are modified by the self-adaptive substructuring algorithm. Numerical iterations continue after the modification until equilibrium is reached.

The analysis CPU time and the total number of iterations are listed in Table 1. Comparisons of efficiency between direct and substructure analysis by the BFGS and the Newton-Raphson methods are given in Figures 6 and 7, respectively. Substructure analysis with F2 equal to 0.8 is about 28% faster than direct analysis in both cases.

Conclusion

Adaptive substructuring techniques provide an efficient way to perform finite element analysis of problems with localized nonlinearities. The analysis results obtained from this approach are effectively identical to the results obtained from direct analysis. The self-adaptive substructuring algorithm performs well in all the examples [5,6]. By use of this technique the analysis computation can be decreased by anywhere from 5-50%. Moreover, by the ability to switch to direct analysis when the local efficiency of substructuring deteriorates, the net efficiency of the adaptive substructuring approach remains greater than direct analysis. Whenever the nonlinearities of the problems are localized or the nonlinearities initiate in a small subgroup of elements and expand or move gradually during the loading process, adaptive substructure analysis is generally more efficient than direct analysis. However, the CPU time saved with substructure analysis is problem dependent.

The current substructuring algorithm allows only single-level subdivision. The algorithm can be more flexibile if multi-level substructures are available. In some cases the optimum distribution of non-zero terms in the stiffness matrix of a linear substructure is destroyed due to the rearrangement of boundary degrees of freedom. Consequently, the static condensation of the linear stiffness matrix may consume much more computational time than desired. Multi-level substructures can reduce the size of the linear stiffness matrices which need to be reformulated and condensed. Similarly, the storage and retrieval of the unchanged submatrices of the stiffness matrix of the linear substructure obtained from single-level subdivision can also improve the efficiency.

The rapid improvement of computational machine capabilities makes further increases in efficiency feasible. Adaptive substructuring techniques become more attractive when array processing is available. The disadvantage in

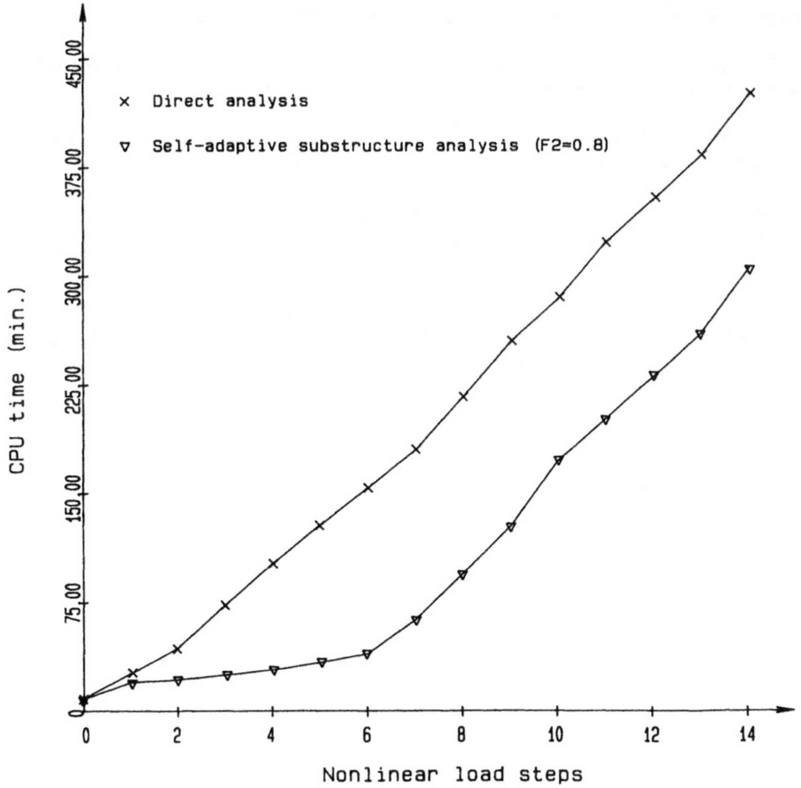

Fig. 7. Analysis CPU time for the simply supported plate. Comparison between direct and substructure analysis by the N-R method.

substructure analysis of destroying the optimum distribution of nonzero terms in the linear stiffness matrices becomes less significant, because the computational time for decomposition is reduced due to the use of vector processing. In addition, with parallel processing, computation and decomposition of individual substructure stiffness matrices can be performed simultaneously; thus, the analysis efficiency can be further increased.

Acknowledgments

Portions of this work were sponsored by the U.S. National Science Foundation under Grant No. CME-79-16818 to Cornell University, and by the Program of Computer Graphics. This support is gratefully acknowledged. The opinions expressed are those of the writers and not of any sponsoring organization.

References

1. Owen, D.R.J.; Goncalves, O.J.A.: Substructuring techniques in material nonlinear analysis. Comput. Structures 15 (1982) 205-213.

2. Dodds, Jr., R.H.; Lopez, L.A.: Substructuring in linear and nonlinear analysis. Int. J. Num. Meth. Engng. 15 (1980) 583-597.

3. Row, D.G.; Powell, G.H.: A substructure technique for nonlinear static and dynamic analysis. Report UCB/EER/C-78/15, Univ. of Calif. Berkeley (1978).

4. Clough, R.W.; Wilson, E.L.: Dynamic analysis of large structural systems with local nonlinearities. Comp. Meth. Appl. Mech. Engnr. 17 (1979) 107-129.

5. Han, T.Y.: Adaptive substructuring and interactive graphics for three-dimensional elasto-plastic finite element analysis. Ph.D. Thesis, Cornell Univ., Ithaca, NY (1984).

6. Han, T.Y.; Abel, J.F.: Adaptive substructuring techniques in elasto plastic finite element analysis. Comput. Structures 20 (1985) 181-192.

7. Armen, H.; Pifko, A.; Leone, H.S.: Finite element analysis of structures in the plastic range. NASA Contractor Report 1649 (1971).

8. Bhanmik A.K.; Hanley, J.T.: Elasto-plastic plate analysis by finite differences. J. Struc. Div., ASCE 93 (1967) 279-294.

Part V: Dynamics

Improvements in Transient Dynamic Time Integration with Application to Spent Nuclear Fuel Shipping Cask Impact Analyses

S. W. KEY
RE/SPEC Inc., Albuquerque, USA

Summary

The central difference method is commonly used in finite element method computer programs designed to calculate the large deformation, elastic and inelastic, transient dynamic response of two-dimensional and three-dimensional solids. An explicit time integration scheme, such as the central difference method, is required to operate with a time step below a critical value related to the transit time for information traveling between adjacent nodes. When applications are encountered that involve the combination of thin sections made of materials with high sound speeds such as steel and thick sections made of materials with low sound speeds such as lead, local variations in transit time between adjacent nodes over the mesh can exceed 100 to 1. Spent nuclear fuel shipping cask impact analyses frequently present this circumstance. Here an explicit-explicit partitioning of the central difference time integrator is introduced as a means of reducing the total effort required. A partitioning of the nodal equations of motion into groups with like critical time steps is used. The nodes controlled by the smallest critical time step are integrated every time step while nodes controlled by larger time steps are periodically processed. Example applications of the algorithm are presented.

1 Introduction

The central difference method is commonly used in finite element method computer programs designed to calculate the large deformation, elastic and inelastic, transient dynamic response of two-dimensional and three-dimensional solids. An explicit time integration scheme, such as the central difference method, is required to operate with a time step below a critical value related to the transit time for information traveling between adjacent nodes. When applications are encountered that involve the combination of thin sections made of materials with high sound speeds such as steel and thick sections made of materials with low sound speeds such as lead, local variations in transit time between adjacent nodes over the mesh can exceed 100 to 1. Spent nuclear fuel shipping cask impact analyses frequently present this circumstance.

Typically, a computer program based on an explicit time integration scheme selects the time step on the basis of the smallest critical time step occurring anywhere

Finite Element Methods for Nonlinear Problems
Europe-US Symposium, Trondheim, Norway 1985
Editors: Bergan, Bathe, Wunderlich

in the mesh. As a result, the response of significant portions of the mesh are computed using time steps well below what local stability would require. There are a number of approaches which may be used to relax the constraint imposed by small values of the critical time step found in what is frequently only a small portion of the mesh.

The use of shell or membrane elements in place of solid elements to model slender sections eliminates the thickness dimension of the section in the calculation for critical time step. As a result, the distance between nodes in the slender element is increased with a corresponding increase in the critical time step. Alternatively, unconditionally stable time integration schemes with iterative corrections for the nonlinearities can be used, the technique employed by ADINA, MARC, and ABAQUS, for example. Though typical of the requirements for accurate results when using implicit integrators for impact calculations, the users of these codes select time steps comparable to the critical time step limit found in an explicit time integration scheme, see for example, Yagawa, *et al* [1].

A third technique is to use a partitioning of the nodal points into groups with like critical time steps and to proceed with an explicit time integrator which subcycles on those nodal points with small critical time steps, periodically processing those nodes with larger critical time steps. This is a technique introduced by Belytschko [2] and later implemented by Belytschko and Robinson [3]. This paper is concerned with the discussion of a fully automated implementation of the explicit-explicit partitioning procedure. As an initial demonstration of the use of explicit-explicit partitioning of the central difference time integrator, three problems with attributes found in spent nuclear fuel shipping cask impact analyses will be discussed.

2 Underlying Mechanics

The treatment in this section is general for the sake of brevity and completeness. The treatment of continuum mechanics found in Truesdell and Toupin [4] is used. The nonpolar case is considered.

2.1 Equations of Motion

A body V is given which occupies a finite region of Euclidean space. Subjected to prescribed body forces and surface tractions, the body V undergoes the motion $x^i \equiv \chi^i(X^\alpha, t)$. The particles of the body are identified by the coordinates X^α. They are referred to as material coordinates, and the relations of the particles to the coordinates X^α does not change in time. The places in space which the particles occupy during the motion are identified by the coordinates x^i. The function χ^i describes the motion of the particles X^α through space as a function of time t. It is the motion χ^i which is sought.

The place occupied by the body at $t = 0$ is taken as the reference configuration. In this configuration, the body is assumed to be strain free, though not necessarily stress free. Only material coordinates X^α which coincide with the spatial coordinates x^i in the reference configuration are considered. Thus, in the reference configuration, $\chi^i(X^\alpha, 0) \equiv X^\alpha$.

The problem is stated in terms of the principle of virtual work. The differential form

$$\delta\pi = \int_V \rho\ddot{x}^k \delta x_k dv + \int_V t^{km} \delta x_{k,m} dv - \int_V \rho f^k \delta x_k dv - \oint_{S^1} s^k \delta x_k da \tag{1}$$

is to vanish at all points along the path of motion for all variations of δx_k satisfying the displacement boundary conditions on S^2. The integration is performed over the current configuration of the body V, where ρ is the mass density in the configuration, $\ddot{x}^k$ is the acceleration, t^{km} is the Cauchy stress — the stress in the current configuration, f^k is the body force density in the current configuration, and s^k is the surface traction which is acting on S^1. The comma in $\delta x_{k,m}$ denotes covariant differentiation.

The divergence theorem is employed to reveal the differential equations of motion. In anticipation of using the finite element method to generate approximate solutions, the case where $\delta x_{k,m}$ is only piecewise continuous, is considered, Jones [5], Prager [6], and Key [7]. Interior surfaces where the discontinuities of $\delta x_{k,m}$ occur are denoted by S^0. Only surfaces S^0 which are stationary with respect to the material are considered. The situation is pictured in Figure 1 where n_k is the normal to S^0 and the symbols + and − denote the respective sides of the surface. The result is

$$\int_V (\rho\ddot{x}^k - t^{km}_{,m} - \rho f^k)\delta x_k dv + \oint_{S^0} (t^{km}_{+} - t^{km}_{-}) n_m \delta x_k da + \oint_{S^1} (t^{km} n_m - s^k)\delta x_k da = 0. \tag{2}$$

The differential form will vanish if and only if the respective integrands vanish. The resulting expressions are the equations of motion,

$$t^{km}_{,m} + \rho f^k = \rho\ddot{x}^k \text{ in } V, \tag{3}$$

the jump condition at a contact discontinuity,

$$(t^{km}_{+} - t^{km}_{-}) n_m = 0 \text{ on } S^0, \tag{4}$$

and the traction boundary conditions,

$$t^{km} n_m = s^k(t) \text{ on } S^1. \tag{5}$$

The displacement boundary conditions are

$$\chi^i(X^\alpha,t) = \kappa^i(t) \text{ on } S^2. \tag{6}$$

Only initial conditions which are homogeneous in place at time zero are considered. Thus, the initial conditions are given by

$$\chi^i(X^\alpha,0) = X^\alpha \text{ in } V,\ \ \dot{\chi}^i(X^\alpha) = \nu^i(X^\alpha) \text{ in } V. \tag{7}$$

It is important to realize that these remarks are completely general with regard to the scale of deformations being considered.

While at this point it appears as if the spatial or Eulerian description of the problem is being employed, actual usage is in terms of the material description. That is, all of the quantities are treated as functions of X^α and t and the finite element discretization is fixed with respect to the material points. This approach has already been used with good results.

2.2 Kinematics

In finite deformations there are many strain measures which are useful. The majority of them can be computed from the deformation gradient F^k_α defined by

$$F^k_\alpha \equiv \frac{\partial \chi^k}{\partial X^\alpha}(X^\beta,t). \tag{8}$$

The velocity v^k is defined as

$$v^k(X^\alpha,t) \equiv \frac{\partial \chi^k}{\partial t}(X^\alpha,t). \tag{9}$$

The stretching is given by

$$d_{km} = \frac{1}{2}(v_{k,m} + v_{m,k}), \tag{10}$$

with the spin given by

$$w_{km} = \frac{1}{2}(v_{k,m} - v_{m,k}). \tag{11}$$

For use in the invariant time derivatives, the polar decomposition of the deformation gradient,

$$F^k_\alpha = R^k{}_m V^m_\alpha, \tag{12}$$

gives the rotation R_{kn} from which the rotation rate ω_{km} is defined by

$$\omega_{km} = \dot{R}_{kn} R^{n}{}_{m}. \tag{13}$$

3 Finite Strain Plasticity

Plasticity is the behavior characteristic of ductile metals. Figure 2 shows results which are typical of the behavior of a metal bar loaded first in uniaxial tension followed by uniaxial compression. The straight line representation in Figure 2 is an idealization of this behavior. This is the approximation which results from the plasticity relations employed here [8]. Assuming the material to be initially isotropic, they translate uniaxial data into general triaxial behavior by using the notion of a universal hardening curve. A hypoelastic description is used where an invariant stress rate is related to the stretching. Thus,

$$\check{t}^{rs} = C^{rsmn} d_{mn}. \tag{14}$$

When no yielding is occurring, C^{rsmn} is the isotropic tensor $\lambda g^{rs} g^{mn} + 2\mu g^{rm} g^{sn}$, where λ and μ are the Lamé parameters. When plastic yielding is occuring, that is, when $\frac{1}{2}(t'_{rs} - \alpha'_{rs})(t'^{rs} - \alpha'^{rs}) - k^2 = 0$ and $(t'_{rs} - \alpha'_{rs})(\check{t}'^{rs} - \check{\alpha}'^{rs}) > 0$, the moduli are replaced by

$$C^{rsmn} = \lambda g^{rs} g^{mn} + 2\mu (g^{rm} g^{sn} - n^{rs} n^{mn}), \tag{15}$$

where

$$n^{rs} = (t'^{rs} - \alpha'^{rs}) / \left[2k^2(1 + H/3\mu)\right]^{\frac{1}{2}},$$

$$\sqrt{2}k = \sqrt{2/3}\, t_0 + \beta(2/3)H \int_0^t |d^p_{rs}| dt,$$

$$\check{\alpha}_{rs} = (1 - \beta)(2/3) H d^p_{rs}.$$

The prime denotes deviatoric components and the superscript p denotes the plastic strain. The isotropic hardening description ($\beta = 1$) preserves the approximate symmetry in tension and compression of the true stress versus the logarithmic strain curves. Kinematic hardening is included in a rather obvious way by letting the center of the yield surface α_{ij} move according to $\check{\alpha}_{ij} = (1 - \beta)\frac{2}{3} H d^p_{rs}$. When kinematic hardening is introduced ($\beta \neq 1$) the theory departs from a hypoelastic description.

Referring to Equation (14), two separate invariant time derivatives of the Cauchy stress will be considered. The invariant time derivative of the Cauchy stress

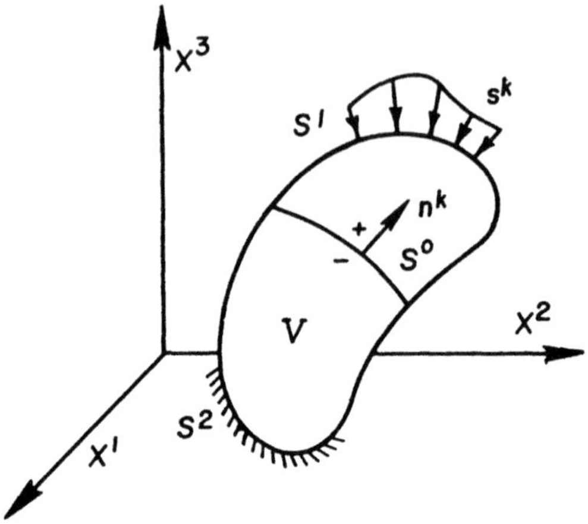

Figure 1. The body V with surface tractions s^k on the boundary S^1 and a prescribed motion on the boundary S^2. An interior boundary S^0 with a unit normal vector n^k is pictured.

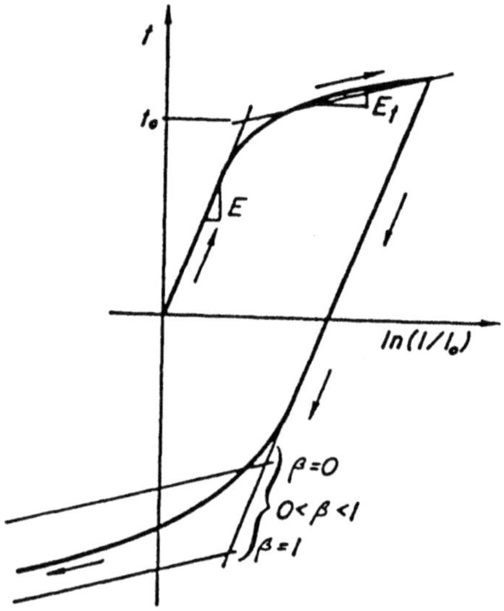

Figure 2. The typical behavior of a ductile metal bar loaded first in uniaxial tension followed by uniaxial compression. The straight line approximation is characterized as an elastic modulus E, a yield stress t_0, a strain hardening modulus E_t, and a hardening parameter β, where kinematic hardening is obtained with $\beta = 0$, isotropic hardening is obtained with $\beta = 1$, and a linear combination of the two is obtained for β between zero and one.

is composed of the material derivative of the Cauchy stress plus two additional terms involving rotation rates. The first form of the invariant time derivative to be considered is the Jaumann rate. It is based on the spin w_{km}. Until very recently, the Jaumann derivative was considered the appropriate invariant time derivative to use in finite strain plasticity. However, in recent papers by Dienes [9], and Nagtegaal and de Jong [10], it was observed that the Jaumann derivative, as the basis of a plasticity model, possessed an undesirable trait in simple shear. Namely, when kinematic hardening is present, the back stress α_{rs} will oscillate when the material is subjected to large magnitude shears. The Jaumann derivative is given by

$$\check{t}^{rs} = \dot{t}^{rs} - w^{r}{}_{m} t^{ms} + t^{rm} w_{m}{}^{s}. \tag{16}$$

Dienes identified the Green-McInnis time dervative of the Cauchy stress as a derivative leading to a far more suitable prediction of material behavior under finite strains. In simple shear, the Green-McInnis rate results in a model of material behavior with monotone increasing shear stress. The Green-McInnis rate is given by

$$\check{t}^{rs} = \dot{t}^{rs} - \omega^{r}{}_{m} t^{ms} + t^{rm} \omega_{m}{}^{s}. \tag{17}$$

For examples of the behavior of these two implementations of plasticity, Key [11] may be consulted. In the calculations which follow, the two-dimensional results are based on the Green-McInnis invariant derivative, Equation (17), while the three-dimensional results are based on the Jaumann derivative, Equation (16).

4 Finite Element Spatial Discretization

In the two-dimensional examples to follow, the finite element is a 4-node quadralateral for which a 2×2 Gaussian quadrature is used on the element level to evaluate the stress divergence integral. The strain rate calculation within the element incorporates a spatially constant bulk strain rate approximation to eliminate "locking," which would otherwise occur in incompressible or almost incompressible large strain, large deformation distortions of elastic-plastic materials [12], [13].

In the three-dimensional examples to follow, the finite element is an 8-node hexahedron in which a one point quadrature is used on the element level to evaluate the stress divergence integral. The element is based directly on the work of Belytschko, *et al* [14]. Incorporated in the element are elastic forces resisting the twelve hourglass modes accompanying the one point quadrature. The forces resisting the hourglass modes are orthogonal to the constant strain states evaluated by the one point quadrature and, therefore, result in minimal interference with the dominant response.

Within each element the motion is assumed to vary bilinearly on the basis of the values at the nodes. Thus, in terms of the shape functions N_k^α, the motion is approximated within an element by

$$x^k = \sum_\alpha \chi^{k\alpha}(t) N_k^\alpha(\eta^i(X^\beta)), \quad \text{(no sum on } k). \tag{18}$$

5 Central Difference Time Integration

Using the spatial discretization discussed above, the Principle of Virtual Work, Equation (1), yields the following set of coupled ordinary differential equations in time,

$$\mathbf{M}\ddot{\mathbf{q}} + \mathbf{F}_{int}(t) = \mathbf{F}_{ext}(t), \tag{19}$$

where,

$$\ddot{\mathbf{q}} = \left[\ddot{\chi}^{k\beta}(t)\right],$$

$$\mathbf{M} = \left[\sum_e \int_{V_e} \rho N_k^\beta dv\right],$$

$$\mathbf{F}_{int} = \left[\sum_e \int_{V_e} t^{km} N_{k,m}^\beta dv\right], \quad \text{(no sum on } k),$$

$$\mathbf{F}_{ext} = \left[\sum_e \int_{V_e} f^k N_k^\beta dv\right] + \left[\sum_e \oint_{S_e^1} s^k N_k^\beta da\right], \quad (\text{ no sum on } k).$$

At time t_n central difference expressions are introduced for the acceleration in terms of the velocity $\mathbf{p}$ and for the velocity in terms of the configuration $\mathbf{q}$ to obtain

$$\mathbf{p}_{n+\frac{1}{2}} = \mathbf{p}_{n-\frac{1}{2}} + \frac{1}{2}(\Delta t_{n+\frac{1}{2}} + \Delta t_{n-\frac{1}{2}})\mathbf{M}^{-1}(\mathbf{F}_{ext}^n - \mathbf{F}_{int}^n) \tag{20}$$

$$\mathbf{q}_{n+1} = \mathbf{q}_n + \Delta t_{n+\frac{1}{2}} \mathbf{p}_{n+\frac{1}{2}} \tag{21}$$

Thus, on the basis of the applied loads at time t_n, $\mathbf{F}_{ext}^n$, and the divergence of the stress state at time t_n, $\mathbf{F}_{int}^n$, nodal forces are calculated. Dividing the nodal forces by nodal masses produces accelerations at time t_n which are then used to compute the velocity at time $t_{n+\frac{1}{2}}$, $\mathbf{p}_{n+\frac{1}{2}}$, from the velocity at time $t_{n-\frac{1}{2}}$, $\mathbf{p}_{n-\frac{1}{2}}$. The new configuration at time $t_{n+\frac{1}{2}}$, $\mathbf{q}_{n+\frac{1}{2}}$, is obtained by updating the configuration at time t_n, $\mathbf{q}_n$, with the velocity at time $t_{n+\frac{1}{2}}$, $\mathbf{p}_{n+\frac{1}{2}}$.

An explicit-explicit partitioning of these equations is introduced by using the following rules:

1. The critical time step at each nodal point n is the minimum critical time step found in those elements "attached" to node n. The critical time step in an element is the time it takes for information to propagate across the element. The calculation utilizes the sound speed in the element and a generalized distance between the nodes of the element developed by Flanagan and Belytschko [15].

2. The equations of motion at each nodal point are integrated with an integer multiple of the minimum critical time step found in the mesh. The partitioning of the respective time steps within the mesh uses bins separated by a multiple of two. Thus, all those nodes with critical time steps between $2^n(\Delta t_{crit})_{min}$ and $2^{n+1}(\Delta t_{crit})_{min}$ are integrated with a time step equal to $2^n(\Delta t_{crit})_{min}$ for $n = 0, 1, 2, 3....$

3. When the nodal point with the maximum critical time step is reached, the entire system of equations is repartitioned on the basis of updated critical time steps throughout the mesh.

4. Within an element, when any nodal point clock corresponds to the time t_n of the master clock, it is necessary to advance the stresses from the time t_e, the last time that the stresses were advanced to time t_n. Essential to this updating of the stresses is the proper calculation of the strain increment within the element. One can expect the configuration of one or more of the remaining nodes in the element to have been advanced in a previous time step beyond time t_n. Thus, prior to calculating the velocity gradient within the element, the configuration of any nodal point beyond time t_n must be interpolated back to time t_n. This step is essential to the satisfaction of the stability requirements of the central difference time integrator.

This technique was introduced by Belytschko [2] and later implemented by Belytschko and Robinson [3].

6 Calculations Using Explicit-Explicit Partitioning

The sudden pressurization and rotation of a cylinder is an elementary problem in which there is no circumferential spatial variation in the response. However, with sufficient pressurization and initial angular velocity, the response involves large strains, large rotations and large displacements. Since the calculation is conducted in a rectangular coordinate system which is "unnatural" to the response, the calculation exercises the full capability of the kinematics to handle multiaxial strain rates and large rotations, the full capacity of the constitutive model to handle multiaxial stress incremention, and the full capacity of the divergence operator to handle large displacements and multicomponent stresses.

The finite element model of the cylindrical cross-section at the start of the calculation is shown in Figures 3 and 4 along with the deformed cross-section at a latter point in time. The calculation without subcycling, Figure 3, where the same time step was used everywhere in the mesh, shows a circumferentially uniform

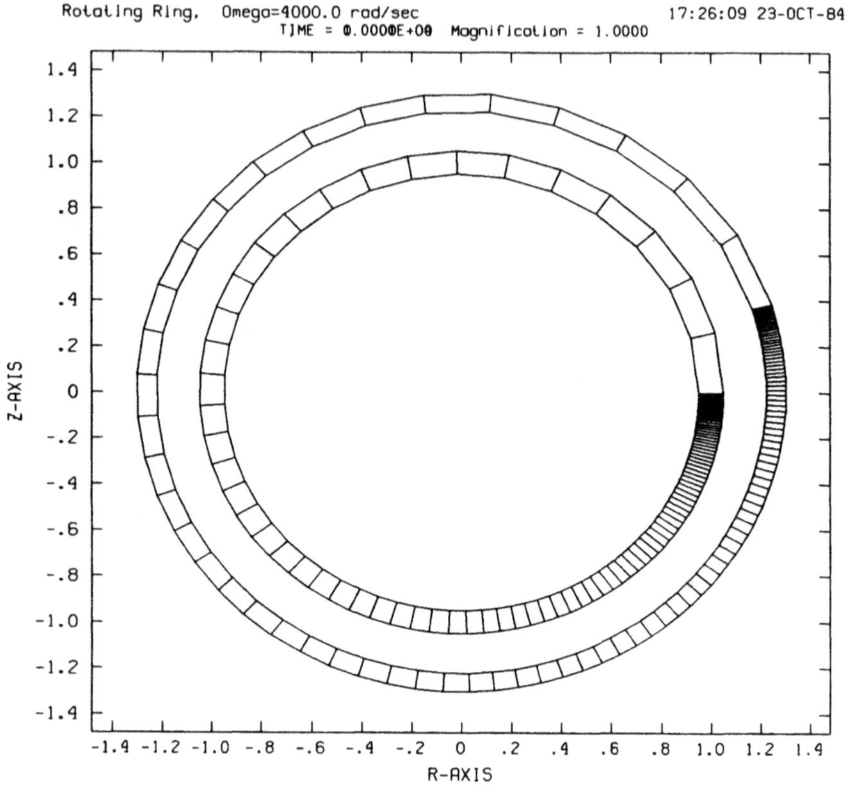

Figure 3. The initial and a deformed plane strain cross section at 0.1 msec's of a cylinder spinning about its axis and loaded by internal pressure, 4,000 rad/sec. Subcycling was suppressed in the computation. The pressure, 67.3 MPa (9,9760 psi), is applied as a step function in time and the spin, 4,000 rad/sec, as an initial condition. The material is aluminum with a modulus E of 71,000 MPa (1.03×10^7 psi), of a Poisson's ration ν of $\frac{1}{3}$, a strain hardening modulus E_t of 3,560 MPa, a yield stress T_o of 286 MPa, a density ρ of 2,860 kg/m^3 (2.508×10^{-4}lbf $-$ sec^2/in^4), and kinematic hardening. The inside radius is 24.1 mm (0.95 in) and the outside radius is 26.7 mm (1.05 in).

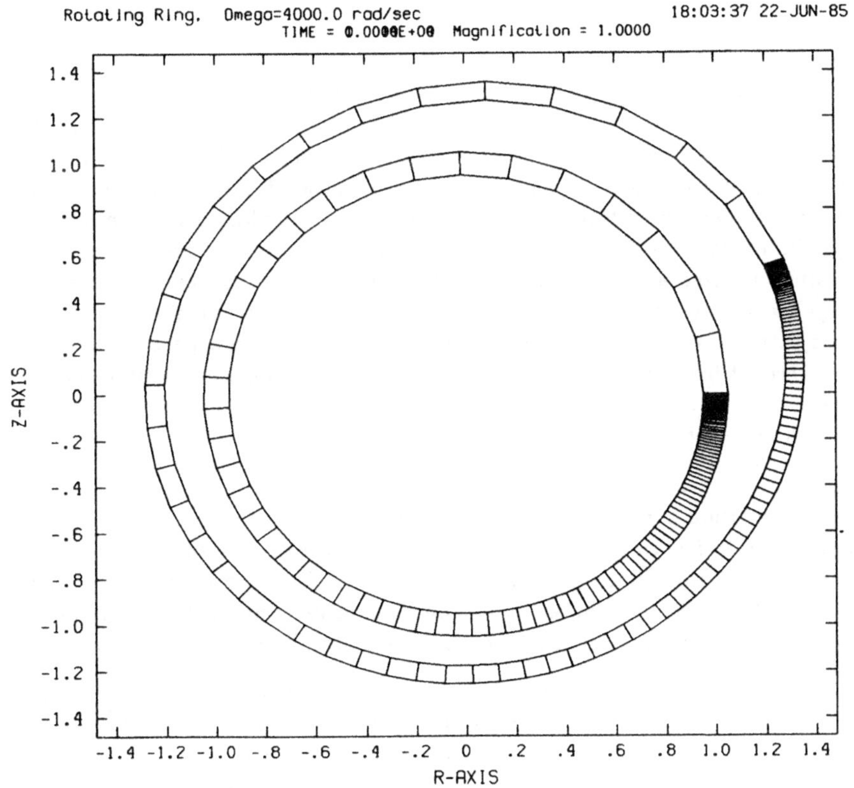

Figure 4. The initial and deformed plane strain cross section at 0.1 msec's of a cylinder spinning about its axis and loaded by internal pressure. Subcycling was used with bins separated by factors of 2 resulting in an initial partition of |32|34|36|44|54|. The pressure, 67.3 MPa (9,760 psi), is applied as a step function in time and the spin, 4,000 rad/sec, as an initial condition. The material is aluminum with a modulus E of 71,000 MPa (1.03×10^7 psi), a Poisson's ration of ν of $\frac{1}{3}$, a strain hardening modulus E_t of 3,560 MPa, a yield stress t_o of 268 MPa, a density ρ of 2,860 kg/m^3 (2.508×10^{-4}lbf $-$sec^2/in^4), and kinematic hardening. The inside radius is 24.1 mm (0.95 in) and the outside radius is 26.7 mm (1.05 in).

response. When subcycling is introduced, Figure 4, the development of localized thinning, suggestive of a structural instability appears to be developing.

In a recent article by Yagawa, *et al* [1], results of two benchmark calculations related to the impact response of spent nuclear fuel shipping casks sponsored by the Japan Society of Mechanical Engineers Subcommittee on Structural Analysis of Nuclear Shipping Casks were reported. The purpose of the benchmark study was to begin to identify suitable techniques and software which could be used in future studies on full-scale cask impacts. Two idealized impacts were considered. The first study utilized a right circular cylinder of lead subjected to a 9.0 m free fall onto a rigid target, while the second study utilized a stainless steel clad cylinder of lead subjected to the same impact conditions. A total of ten separate computer programs were utilized in the first study on the lead cylinder without cladding, and four separate computer programs were utilized in the second study on the lead cylinder with cladding.

The software used in these studies was a variety of finite difference method and finite element method computer programs utilizing both explicit central difference time integration and various implicit time integration schemes ranging from the Houbolt method to the Hilber-Hughes-Taylor method. In a subsequent article [16] additional results were contributed based on calculations emphasizing the above explicit-explicit partitioning procedure for the central difference time integrator.

The second study, concerned with the stainless steel clad cylinder, while seemingly simple in its statement, required significant computational capability because of the markedly different material properties of lead and stainless steel, the presence of very thin sections next to thick sections, and interface conditions between the two materials which can be either bonded or unbonded. The use of two separate interface conditions between the lead cylinder and the stainless steel cladding resulted in two separate physical behaviors. Figure 5 pictures the deformed meshes at the time of maximum axial shortening for each of the respective problems along with the undeformed mesh.

The cladding was modeled as an additional layer of four-node quadrilateral elements whose thickness dimension corresponded to that of the cladding. In this study, the differences between Young's modulus, density, and mesh dimensions for the cylinder and the cladding dictated by the physical dimensions of the problem result in a ratio of critical time step in the lead cylinder to the critical time step in the stainless steel cladding of 100 to 1. Thus, the cylinder with cladding potentially represents a calculation requiring 100 times the effort needed for the bare cylinder. The partitioning used bins separated by a multiple of two. The nodes defining the stainless steel clad occur in the first bin, while the nodes interior to the lead cylinder occur in the seventh bin which is integrated with a time step that is 64 times larger than the time step used in the first bin. The use of the explicit-explicit partitioning procedure results in a reduction of computational effort by a factor in excess of five.

Figure 6 depicts both an undeformed and a deformed three-dimensional model

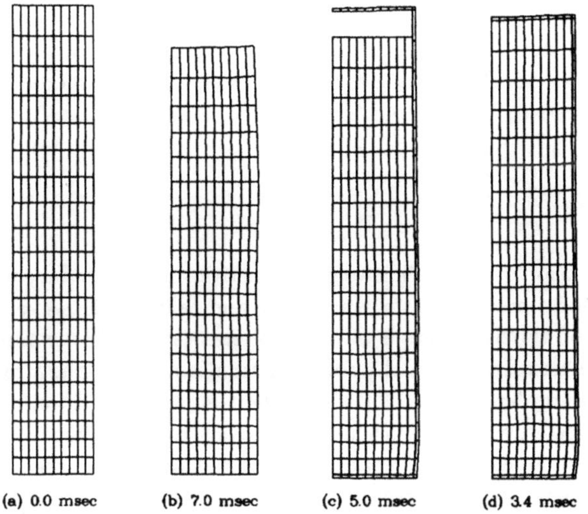

Figure 5. The undeformed shape for the (a) bare lead cylinder, 0 msec, and the deformed shape at maximum head settlement for the (b) bare lead cylinder, 7 msec, (c) unbonded stainless steel clad covered lead cylinder, 5 msec, and (d) bonded stainless steel clad lead cylinder, 3.4 msec, after Key [16].

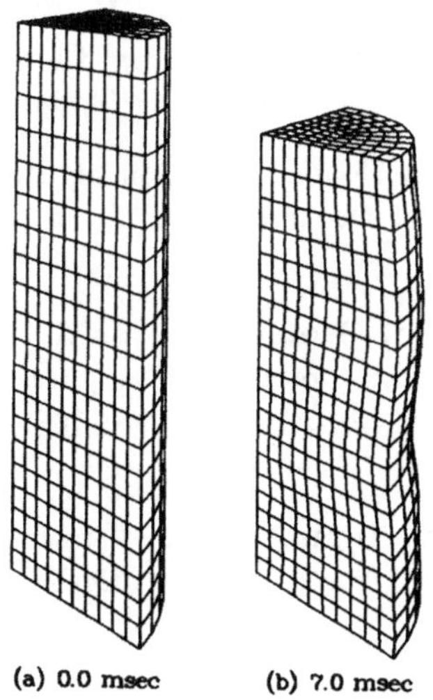

Figure 6. The undeformed and deformed shape of a bare lead cylinder at maximum head settlement, 7 msec. To simulate the axisymmetric impact, a 90° sector of the circular cylinder is modeled with 8-node hexahedrons and two symmetry planes. The magnitude of the displacements has been doubled to animate the distorted shape.

of a quadrant of the axisymmetric lead cylinder without the stainless cladding undergoing an axial impact. The displacements are magnified by two to animate the deformation. Since the mesh is very close to uniform, all of the nodal equations of motion were integrated with the same time step.

7 Conclusions

The implementation of an explicit-explicit partitioning procedure with the central difference method time integrator is effective and reliable. However, some work remains to be done on the finer details of the implementation. For example, it is not clear if the asymmetric response in the rotating ring is a true structural instability or an artifact of the implementation of the explicit-explicit partitioning. Based on the comparison in Reference [15] with other analysis tools in Yagawa, *et al* [1] it is clear the procedure is accurate, capable of automation, and successful in reducing the effort needed to compute results. To date, a stability proof has not been developed, but the procedure as implemented here preserves the physical interpretations advanced with rigorous stability proofs for the case without partitioning. The principal remaining contribution needed for the technique is a stability proof.

References

1. **Yagawa, G., H. Ohtsubo, H. Takeda, Y. Toi, T. Aizawa and T. Ikushima.** "A Round Robin on Finite Element Calculation for Impact Problems," *Nuclear Engineering and Design*, Vol. 78, pp. 377-388, 1984.

2. **Belytschko, T.** "Explicit Time Integration of Structure-Mechanical Systems," *Advanced Structural Dynamics*, Ed. by J. Donéa, Applied Science Publishers, London, 1980.

3. **Belytschko, T. and R. R. Robinson.** "SAMSON2: A Nonlinear Two-Dimensional Structure/Media Interaction Computer Code," U. S. Air Force Weapons Laboratory Report No. AFWL-TR-81-109, Kirtland Air Force Base, New Mexico, 1982.

4. **Truesdell, C. and R. A. Toupin.** "The Classical Field Theories," *Encyclopedia of Physics*, Vol. III, No. 1, Springer, Berlin, pp. 226-793, 1960.

5. **Jones, R. E.** "A Generalization of the Direct-stiffness Method of Structural Analysis," *AIAA Journal*, Vol. 2, pp. 821-826, 1964.

6. **Prager, W.** "Variational Principles for Linear Elastostatics for Discontinuous Displacements, Strains, and Stresses," *Recent Progress in Applied Mechancs*, F. Odqvist Volume, Wiley, New York, pp. 463-474, 1967.

7. **Key, S. W.** "A Specialization of Jones' Generalization of the Direct-stiffness Method of Structural Analysis," *AIAA Journal*, Vol. 9, pp. 984-985.

8. **Goel, R. P., and L. E. Malvern.** "Biaxial Plastic Simple Waves with Combined Kinematic and Isotropic Hardening," *Journal of Applied Mechanics*, Vol. 37, pp. 1100-1106, 1970.

9. **Dienes, J. K.** "On the Analysis of Rotation and Stress Rate in Deforming Bodies," *Acta Mechanica*, Vol. 32, pp. 217-232, 1979.

10. **Nagtegaal, J. C. and J. E. de Jong.** "Some Aspects of Nonisotropic Work-Hardening in Finite Deformation Plasticity," *Proceedings of the Workshop on Plasticity of Metals at Finite Strain: Theory, Computation and Experiment*, Ed. by E. H. Lee and R. L. Mallett, Division of Applied Mechanics, Stanford University, Stanford, California, 1981.

11. **Key, S. W.** "On an Implementation of Finite Strain Plasticity in Transient Dynamic Large-Deformation Calculations," *Theoretical Foundation for Large-Scale Computations of Nonlinear Material Behavior*, Ed. by S. Nemat-Nasser, R. J. Asaro, and G. A. Hegemier, Kluwer Academic Publishers, Hingham, Massachusetts, 1984.

12. **Nagtegaal, J. C., D. M. Parks, and J. R. Rice.** "On Numerically Accurate Finite Element Solutions in the Fully Plastic Range," *Computer Methods in Applied Mechanics and Engineering*, Vol. 4, pp. 153-177, 1974.

13. **Hughes, T. J. R.** "Equivalence of Finite Elements for Nearly-Incompressible Elasticity," LBL-5237, Lawrence Berkeley Laboratory, University of California, Berkeley, 1976.

14. **Belytschko, T., J. S. Ong, W. K. Liu, and J. M. Kennedy.** "Hourglass Control in Linear and Nonlinear Problems," *Computer Methods in Applied Mechanics and Engineering*, Vol. 43, pp. 251-276, 1984.

15. **Flanagan, D. P., and T. Belytschko.** "Eigenvalues and Stable Time Steps for the Uniform Hexahedron and Quadrilateral," *Journal of Applied Mechanics*, Vol. 51, pp. 35-40, 1984.

16. **Key, S. W.** "A Comparison of Recent Results from HONDO III with the JSME Nuclear Shipping Cask Benchmark Calculations," *Nuclear Engineering and Design*, Vol. 85, pp. 15-23, 1985.

Nonlinear Dynamic Analysis of Shell Structures Using General Single Step Algorithms

D. R. J. OWEN, G. Q. LIU*, X.-L. LI**
University College of Swansea, U. K.
*Tong Ji University, Shanghai, China
**Dalian Institute of Technology, China

Abstract

In this paper general single step explicit, implicit and explicit-implicit algorithms with predictor-corrector forms are considered for the nonlinear dynamic analysis of shell structures of arbitrary shape. The stability condition related to the general single step schemes (SS22) are deduced by invoking an established technique; namely the energy method. Geometrically nonlinear behaviour is taken into account using a total Lagrangian formulation. Numerical examples show the accuracy and efficiency of the general single step algorithm in comparison with other time integration methods.

Introduction

The purpose of this paper is to analyse arbitrary shell structures under such loadings as impact, explosive,or seismic conditions by using a general single step solution scheme. In contrast to the Newmark, Houbolt and Wilson θ-methods, the single step algorithm [1-3] is based on the assumption that the equations of motion are satisfied in a weighted average sense for each time interval. In this paper the general single step implicit, explicit and implicit-explicit algorithms with predictor-corrector forms are presented and implemented in a 'degenerated' thick shell element program. The stability condition relating to these general single step schemes is also derived by invoking an established energy principle [4].

The approach and the associated computer code developed are illustrated by the solution of two numerical examples which are compared with Newmark, Wilson θ-method and other available solution methods.

Finite Element Methods for Nonlinear Problems
Europe-US Symposium, Trondheim, Norway 1985
Editors: Bergan, Bathe, Wunderlich

Spatial discretisation

The discretised dynamic equilibrium equations for linear problems take on the following standard form

$$\underset{\sim}{M}\,\ddot{\underset{\sim}{d}} + \underset{\sim}{C}\,\dot{\underset{\sim}{d}} + \underset{\sim}{K}\,\underset{\sim}{d} = \underset{\sim}{f} \tag{1}$$

in which $\underset{\sim}{d}$ represents the vector of nodal displacements and a dot indicates differentiation with respect to time. The mass matrix $\underset{\sim}{M}$, damping matrix $\underset{\sim}{C}$, stiffness matrix $\underset{\sim}{K}$ and applied load vector $\underset{\sim}{f}$ have the usual finite element forms [12]. For nonlinear problems, the damping and stiffness matrices are no longer constant, so that (1) becomes

$$\underset{\sim}{M}\,\ddot{\underset{\sim}{d}} + \underset{\sim}{P}(\underset{\sim}{d}, \dot{\underset{\sim}{d}}) = \underset{\sim}{f} \tag{2}$$

in which $\underset{\sim}{P}$ denotes the internal restoring forces.

The "degenerated" isoparametric shell element is employed for analysis in which the displacements and rotations are independent nodal variables. The shape function and strain matrices $\bar{\underset{\sim}{N}}$, $\underset{\sim}{B}$ are given explicitly in Ref. [5,6].

Time discretisation

(i) The general single step algorithm (SSp2)

The general single step algorithm (which is termed the SSp2 algorithm) where p is the order of the polynomial expansion used for displacement and 2 stands for the order of equation (1), is based on the concept of satisfying equation (1) in a weighted average sense within a time interval Δt from time t_n to t_{n+1} [2,3] so that

$$\underset{\sim}{M}\,\ddot{\bar{\underset{\sim}{d}}}_{n+1} + \underset{\sim}{C}\,\dot{\bar{\underset{\sim}{d}}} + \underset{\sim}{K}\,\bar{\underset{\sim}{d}}_{n+1} = \bar{\underset{\sim}{f}}_{n+1} \tag{3}$$

where

$$\bar{\underset{\sim}{d}}_{n+1} = \tilde{\underset{\sim}{d}}_{n+1} + C_1\,\underset{\sim}{\alpha}^p_n \qquad \tilde{\underset{\sim}{d}}_{n+1} = \sum_{q=0}^{p-1} \overset{q}{\underset{\sim}{d}}_n \frac{\Delta t^q}{q!}\,\theta_q$$

$$\dot{\bar{\underset{\sim}{d}}}_{n+1} = \dot{\tilde{\underset{\sim}{d}}}_{n+1} + C_2\,\underset{\sim}{\alpha}^p_n \qquad \dot{\tilde{\underset{\sim}{d}}}_{n+1} = \sum_{q=1}^{p-1} \overset{q}{\underset{\sim}{d}}_n \frac{\Delta t^{q-1}}{(q-1)!}\,\theta_{q-1}$$

$$\ddot{\bar{d}}_{n+1} = \ddot{\tilde{d}}_{n+1} + C_3\, \alpha_n^p \quad (3.a) \qquad \ddot{\tilde{d}}_{n+1} = \sum_{q=2}^{p-1} d_n^q \frac{\Delta t^{q-2}}{(q-2)!}\theta_{q-2} \quad (3.b)$$

$$C_1 = \frac{\Delta t^p}{p!}\theta_p, \quad C_2 = \frac{\Delta t^{p-1}}{(p-1)!}\theta_{p-1}, \quad C_3 = \frac{\Delta t^{p-2}}{(p-2)!}\theta_{p-2} \qquad (3.c)$$

$$\bar{f}_{n+1} = \theta_1 f_{n+1} + (1-\theta_1) f_n \qquad (3.d)$$

with $q = 1, 2, \dots p$ and $\theta_o = 1$

Parameters θ_1 to θ_p must be specified in the above expressions. Starting from given initial conditions d_o, $\dot{d}_o \dots d_o^{p-1}$ *, then (3) can be solved by a time marching process. If the values of d_n to d_n^{p-1} at time t_n are known, the unknown vector α_n in the time interval between t_n and t_{n+1} is evaluated as

$$\alpha_n^p = (C_3 M + C_2 C + C_1 K)^{-1}\,(\bar{f}_{n+1} - M\,\ddot{\tilde{d}}_{n+1} - C\,\dot{\tilde{d}}_{n+1} - K\,\tilde{d}_{n+1}) \qquad (4)$$

and then the values of d_{n+1} to d_{n+1}^{p-1} at time t_{n+1} are obtained according to

$$d_{n+1} = \sum_{q=0}^{p-1} d_n^q \frac{\Delta t^q}{q!} + \alpha_n^p \frac{\Delta t^p}{p!}$$

$$\dot{d}_{n+1} = \sum_{q=1}^{p-1} d_n^q \frac{\Delta t^{q-1}}{(q-1)!} + \alpha_n^p \frac{\Delta t^{p-1}}{(p-1)!}$$

$$\ddot{d}_{n+1} = \sum_{q=2}^{p-1} d_n^q \frac{\Delta t^{q-2}}{(q-2)!} + \alpha_n^p \frac{\Delta t^{p-2}}{(p-2)!} \qquad (5)$$

With a polynomial of order p having been chosen to represent the unknown function d_{n+1}, in equation (5), the algorithm is p-order accurate and errors are generally $O(\Delta t^{p+1})$, [1]. In the case of $p = 2,3$ the general single step algorithm includes most of the currently used integration schemes, such as the Newmark family, Wilson θ-method etc. by appropriate choice of

---

* Superscript p-1 here denotes the $(p-1)^{th}$ time derivative.

the parameters θ_i (i = 1,2,...). The method also provides possibilities for the improved solution of dynamic differential equations.

(ii) General single step implicit algorithms

Nonlinear dynamic equations (2) can be written in a weighted average form in a time interval Δt from t_n to t_{n+1} as [3]

$$\underset{\sim}{M}\,\ddot{\bar{\underset{\sim}{d}}}_{n+1} + \underset{\sim}{P}_{n+1}(\dot{\bar{\underset{\sim}{d}}}_{n+1}, \bar{\underset{\sim}{d}}_{n+1}) = \bar{\underset{\sim}{f}}_{n+1} \tag{6}$$

Let

$$[\underset{\sim}{K}_T]_{n+1} = \partial \underset{\sim}{P}_{n+1}/\partial \bar{\underset{\sim}{d}}_{n+1} \qquad [\underset{\sim}{C}_T]_{n+1} = \partial \underset{\sim}{P}_{n+1}/\partial \dot{\bar{\underset{\sim}{d}}}_{n+1} \tag{7}$$

denote the tangent stiffness and damping matrices respectively. If the terms of $\tilde{\underset{\sim}{d}}_{n+1}$, $\dot{\tilde{\underset{\sim}{d}}}_{n+1}$ and $\ddot{\tilde{\underset{\sim}{d}}}_{n+1}$ in (3.b) are regarded as weighted average predicted values, then $\bar{\underset{\sim}{d}}_{n+1}$, $\dot{\bar{\underset{\sim}{d}}}_{n+1}$ and $\ddot{\bar{\underset{\sim}{d}}}_{n+1}$ in equation (3a) are the weighted average corrector values. Nonlinear equations (6) can then be transformed into an effective static problem in terms of the unknown vector $\underset{\sim}{\alpha}_n^p$ as

$$\underset{\sim}{\psi}(\underset{\sim}{\alpha}_n^p) \equiv \bar{\underset{\sim}{f}}_{n+1} - \underset{\sim}{M}\,\ddot{\bar{\underset{\sim}{d}}}_{n+1} - \underset{\sim}{P}_{n+1}(\dot{\bar{\underset{\sim}{d}}}_{n+1}, \bar{\underset{\sim}{d}}_{n+1}) = 0 \tag{8}$$

The incremental value $\underset{\sim}{\alpha}_n^p$ is computed and $\bar{\underset{\sim}{d}}_{n+1}$, $\dot{\bar{\underset{\sim}{d}}}_{n+1}$ and $\ddot{\bar{\underset{\sim}{d}}}_{n+1}$ are corrected within a time step until $\underset{\sim}{\psi}(\underset{\sim}{\alpha}_n^p) = 0$. The algorithm starts from the initial values $\underset{\sim}{d}_o$, $\dot{\underset{\sim}{d}}_o$ ... $\overset{p-1}{\underset{\sim}{d}}$ and the procedure for time interval t_n to t_{n+1} is described in Table I.

(iii) General single step explicit algorithms

In addition to the implicit scheme, we may construct an explicit algorithm for the general single step procedure. When dealing with nonlinear equations such as

$$\underset{\sim}{M}\,\ddot{\bar{\underset{\sim}{d}}}_{n+1} + \underset{\sim}{P}(\dot{\tilde{\underset{\sim}{d}}}_{n+1}, \tilde{\underset{\sim}{d}}_{n+1}) = \bar{\underset{\sim}{f}}_{n+1} \tag{9}$$

the weighted average predictor and corrector values of displacement, velocity and acceleration are defined by (3.a) and (3.b). Implementation of the explicit scheme is presented in Table II.

Stability analysis

General stability conditions for the SSp2 algorithm, for the case p=2, are considered, since the algorithm SS22 represents the simplest possible recurrence relation for dynamic equations. A technique termed the energy method, which was first presented in Ref. [4], is adopted to establish the stability criterion. Here we only consider linear differential equation systems.

Denoting the undivided forward difference and mean value operators of an arbitrary vector $\underset{\sim}{x}_n$ by

$$[\underset{\sim}{x}_n] = \underset{\sim}{x}_{n+1} - \underset{\sim}{x}_n \quad ; \qquad \langle\underset{\sim}{x}_n\rangle = (\underset{\sim}{x}_{n+1} + \underset{\sim}{x}_n)/2 \tag{10}$$

then if, and only if, the norm of vector $\underset{\sim}{x}_n$ satisfies the condition $\|\underset{\sim}{x}_n\| \leq C$, ($n = 0,1,2,...$ and C is a non-negative constant value) it can be said that the vector $\underset{\sim}{x}_n$ is uniformly bounded. This implies that if the vector $\underset{\sim}{x}_n$ is a solution to a set of equations, the solution will be stable.

(i) Implicit SS22 algorithms

The undivided forward difference of (3) and the weighted average displacement, velocity and acceleration vectors are expressed, for zero loading, as,

$$\underset{\sim}{M}\,[\ddot{\bar{\underset{\sim}{d}}}_n] + \underset{\sim}{C}\,[\dot{\bar{\underset{\sim}{d}}}_n] + \underset{\sim}{K}\,[\bar{\underset{\sim}{d}}_n] = 0 \tag{11}$$

with

$$[\ddot{\bar{\underset{\sim}{d}}}_n] = \ddot{\bar{\underset{\sim}{d}}}_{n+1} - \ddot{\bar{\underset{\sim}{d}}}_n$$

$$[\dot{\bar{\underset{\sim}{d}}}_n] = \dot{\bar{\underset{\sim}{d}}}_{n+1} - \dot{\bar{\underset{\sim}{d}}}_n = \Delta t\,\langle\ddot{\bar{\underset{\sim}{d}}}_n\rangle + \Delta t\left(\theta_1 - \tfrac{1}{2}\right)[\ddot{\bar{\underset{\sim}{d}}}_n]$$

$$\begin{aligned}[\bar{\underset{\sim}{d}}_n] = \bar{\underset{\sim}{d}}_{n+1} - \bar{\underset{\sim}{d}}_n &= \Delta t\left(\langle\dot{\bar{\underset{\sim}{d}}}_n\rangle - \frac{[\dot{\bar{\underset{\sim}{d}}}_n]}{2}\right) \\ &+ \frac{\Delta t^2}{2}\left(\langle\ddot{\bar{\underset{\sim}{d}}}_n\rangle - \frac{[\ddot{\bar{\underset{\sim}{d}}}_n]}{2}\right) \frac{\Delta t^2}{2}\,\theta_2[\ddot{\bar{\underset{\sim}{d}}}_n]\end{aligned} \tag{12}$$

Substituting (12) into (11) and pre-multiplying both sides of (11) by $(2[\dot{\bar{d}}_n]^T)/\Delta t$ results in

$$(\ddot{\bar{d}}_{n+1})^T B^I \ddot{\bar{d}}_{n+1} + (\dot{\bar{d}}_{n+1})^T K \dot{\bar{d}}_{n+1} = (\ddot{\bar{d}}_n)^T B^I \ddot{\bar{d}}_n + (\dot{\bar{d}}_n)^T K \dot{\bar{d}}_n$$

$$-2[\ddot{\bar{d}}_n] A^I [\ddot{\bar{d}}_n] - 2 \langle\ddot{\bar{d}}_n\rangle^T C \Delta t \langle\ddot{\bar{d}}_n\rangle \qquad (13)$$

where

$$B^I = G^I + \Delta t(\theta_1 - \tfrac{1}{2}) C$$

$$A^I = (\theta_1 - \tfrac{1}{2}) G^I$$

$$G^I = M + C \Delta t(\theta_1 - \tfrac{1}{2}) + \frac{\Delta t^2}{2} K (\theta_2 - \theta_1) \qquad (14)$$

For $\theta_1 \geq \frac{1}{2}$, $\theta_2 \geq \theta_1$, the matrices B^I, A^I and G^I will be positive definite, provided that the stiffness and damping matrices K, C are positive definite. In this case the vectors $\ddot{\bar{d}}_n$ and $\dot{\bar{d}}_n$ are uniformly bounded, i.e.

$$(\ddot{\bar{d}}_{n+1})^T B^I \ddot{\bar{d}}_{n+1} + (\dot{\bar{d}}_{n+1})^T K \dot{\bar{d}}_{n+1} \leq (\ddot{\bar{d}}_n)^T B^I \ddot{\bar{d}}_n + (\dot{\bar{d}}_n)^T K \dot{\bar{d}}_n$$

$$\cdots\cdots$$

$$\leq (\ddot{\bar{d}}_o)^T B^I \ddot{\bar{d}}_o + (\dot{\bar{d}}_o) K \dot{\bar{d}}_o$$

$$n = 0, 1, 2, 3, \ldots\ldots \qquad (15)$$

From (3)

$$\|\bar{d}_n\| \leq \|K^{-1}\| \left(\|M\| \, \|\ddot{\bar{d}}_n\| + \|C\| \, \|\dot{\bar{d}}_n\|\right) \qquad (16)$$

so that the weighted average displacement vector $\bar{d}_n$ is also uniformly bounded since $\ddot{\bar{d}}_n$ and $\dot{\bar{d}}_n$ are uniformly bounded. Thus the unconditional stability characteristic of SS22 algorithms is

$$\theta_2 \geq \theta_1 \geq \frac{1}{2} \tag{17}$$

when $\theta_2 = \theta_1 = 0.5$, the SS22 algorithm has identical stability properties to the Newmark average acceleration algorithm. For conditional stability, the critical time step length for viscous damping can be derived from the fact that matrix $\underset{\sim}{G}^I$ should be positive definite

$$\theta_1 \geq \frac{1}{2}, \quad \theta_2 < \theta_1$$

$$\Delta t_{cr} \leq \{-2\xi(\theta_1 - \tfrac{1}{2}) - [4\xi^2(\theta_1 - \tfrac{1}{2})^2 - 2(\theta_2 - \theta_1)]^{1/2}\} \Big/ (\theta_2 - \theta_1)\, w_{max} \tag{18}$$

and

$$w_{max} \leq \max_e \, (w_{max}^{(e)}) \tag{19}$$

where ξ is the damping ratio, $w_{max}^{(e)}$ is the maximum frequency of the eth element, which is usually the smallest element in the mesh.

If $\xi = 0$, $\theta_2 = 0$

$$\Delta t_{cr} \leq \frac{1}{w_{max}} \sqrt{\frac{2}{\theta_1}} \tag{20}$$

The maximum permissible time step length then occurs for $\theta_1 = \frac{1}{2}$, which coincides with the established result for the central difference method.

(ii) <u>Explicit SS22 algorithms</u>

The undivided forward difference of the linear form of (9) can be expressed as

$$\underset{\sim}{M}[\ddot{\bar{\underset{\sim}{d}}}_n] + \underset{\sim}{C}[\dot{\bar{\underset{\sim}{d}}}_n] + \underset{\sim}{K}[\tilde{\underset{\sim}{d}}_n] = 0 \tag{21}$$

where

$$[\ddot{\bar{\underset{\sim}{d}}}_n] = \ddot{\bar{\underset{\sim}{d}}}_{n+1} - \ddot{\bar{\underset{\sim}{d}}}_n$$

$$[\dot{\bar{\underset{\sim}{d}}}_n] = \dot{\bar{\underset{\sim}{d}}}_{n+1} - \dot{\bar{\underset{\sim}{d}}}_n = \Delta t(\langle \ddot{\bar{\underset{\sim}{d}}}_n \rangle - \frac{[\ddot{\bar{\underset{\sim}{d}}}_n]}{2})$$

$$[\tilde{d}_n] = \tilde{d}_{n+1} - \tilde{d}_n = \Delta t(<\dot{\bar{d}}_n> - \frac{[\dot{\bar{d}}_n]}{2}) + \frac{\Delta t^2}{2} (<\ddot{\bar{d}}_n> - \frac{[\ddot{\bar{d}}_n]}{2}) \tag{22}$$

Premultiplying both sides of (21) by $\frac{[\dot{\bar{d}}_n]^T}{\Delta t}$ and noting $\frac{[\dot{\bar{d}}_n]^T}{\Delta t} = <\ddot{\bar{d}}_n>^T + (\theta_1 - \frac{1}{2}) [\ddot{\bar{d}}_n]^T$ leads to the identity

$$(\ddot{\bar{d}}_{n+1})^T B^E \ddot{\bar{d}}_{n+1} + (\dot{\bar{d}}_{n+1})^T K \dot{\bar{d}}_{n+1} = (\ddot{\bar{d}}_n)^T B^E \ddot{\bar{d}}_n + (\dot{\bar{d}}_n)^T K \dot{\bar{d}}_n$$

$$- 2 [\ddot{\bar{d}}_n]^T A^E [\ddot{\bar{d}}_n] - 2 <\ddot{\bar{d}}_n>^T [\Delta t (C + \Delta t K)] <\ddot{\bar{d}}_n> \tag{23}$$

in which

$$A^E = G^E + \frac{\Delta t^2}{2} (\theta_1 - \frac{1}{2}) K$$

$$B^E = G^E + \Delta t(\theta_1 - \frac{1}{2}) C + \frac{3}{2} \Delta t^2 (\theta_1 - \frac{1}{2}) K$$

$$G^E = M - \frac{\Delta t}{2} C - \frac{\Delta t^2}{4} K \tag{24}$$

If $\theta_1 \geq \frac{1}{2}$ and G^E is positive definite, then $\ddot{\bar{d}}_n$ and $\dot{\bar{d}}_n$ are uniformly bounded. The weighted average displacement vector $\bar{d}_n$ is also uniformly bounded, which can be shown from (25)

$$\|\bar{d}_n\| \leq \frac{\Delta t^2}{2} |\theta_2| \; \|\ddot{\bar{d}}_n\| + \|K^{-1}\| \Big(\|M\| \; \|\ddot{\bar{d}}_n\| + \|C\| \; (\|\dot{\bar{d}}_n\| + \Delta t \theta_1 \|\ddot{\bar{d}}_n\|) \Big) \tag{25}$$

By a deductive process, it is seen that there are no unconditionally stable predictor-corrector schemes for explicit SS22 algorithms. The stability characteristics of explicit SS22 algorithms are dependent on the conditions that require

$$\theta_1 \geq \frac{1}{2}$$

$$\Delta t_{cr} \leq \left(2(\xi^2 + 1)^{1/2} - 2\xi\right)/w_{max} \tag{26}$$

When viscous damping is absent, i.e. $\xi = 0$

$$\Delta t_{cr} \leq \frac{2}{w_{max}} \tag{27}$$

which is consistent with the results for implicit SS22 algorithms, when conditional stability is considered. It should be noted from Table II that the stability conditions of explicit SS22 algorithms are independent of θ_2.

General single step implicit-explicit algorithms

An implicit-explicit method is also considered which is based on an element partition scheme [7,8] rather than a nodal partition approach [9,10]. Elements are divided into explicit and implicit groups corresonding to different subregions in the problem and the numerical algorithm employed is similar to the Newmark implicit-explicit scheme.

Stability analysis of the implicit-explicit SS22 algorithm indicates that stability is achieved when $\theta_1 \geq \frac{1}{2}$ and the critical time step length satisfies expressions (19) or (26) for $\xi \neq 0$ and expressions (20) or (27) when damping is absent.

Numerical Examples

The general single step algorithms discussed in the previous sections are applied to the solution of two numerical examples and the results are compared with those obtained by use of other schemes.

(i) Simply supported square plate

Fig. 1 shows a simply supported square plate subjected to a uniformly distributed Heaviside loading of intensity $q = 0.005$, which has been previously analysed by Bauer [11]. Four 8-node Serendipity elements are used to model one quarter of the plate. The plate material is assumed to be isotropic and linear elastic with $E = 10^5$, $\nu = 0.3$ and $\rho = 1.0$. For comparison purposes, Newmark implicit and general single step implicit algorithms (SS22, SS32) are employed with a consistent mass matrix.

The time step length is selected as 0.068 sec., which is 1/100 of the fundamental period of the plate. It can be seen that the two schemes give practically the same solution for a linear elastic analysis and are in good agreement with the analytical solution. For parameter values $\theta_1 = 1.4$, $\theta_2 = 1.96$, $\theta_3 = 2.744$ the SS32 algorithm is equivalent to the Wilson θ-method.

When $\theta_2 < \theta_1$, i.e. condition (17) is not satisfied, unstable solution is observed in Fig. 1.

(ii) Spherical shell cap with clamped edges

A spherical shell cap subjected to a distributed step pressure of 600 lb/in^2 is shown in Fig. 2, where a quarter of the shell is divided into twelve 8-node 'degenerated' isoparametric thick shell elements. The dimensions and material properties of the shell are the same as those of Ref. [13], with

internal radius	$R = 22.27$ in.
thickness of shell	$t = 0.41$ in.
semi angle	$\alpha = 26.67°$
elastic modulus	$E = 10.5 \times 10^6$ lb/in^2
Poisson's ratio	$\nu = 0.3$
yield stress	$\sigma_y = 0.024 \times 10^6$ lb/in^2
mass density	$\rho = 2.45 \times 10^{-4}$ lb-sec^2/in^4

The fundamental period of the shell is 0.55×10^{-3} sec. [13]. For implicit general single step algorithms (SS22 algorithm with $\theta_1 = \theta_2 = 0.5$, SS32 algorithm $\theta_1 = 1.4$, $\theta_2 = 1.96$, $\theta_3 = 2.744$) the time step length was chosen as $\Delta t = 0.6 \times 10^{-5}$ sec. Figs. 2 and 3 respectively show the vertical displacement of the crown for small elasto-plastic deformation analysis, in which the material is assumed to obey the von Mises yield criterion and large elastic deformation behaviour. Again, as observed in the previous example, the SS22 and SS32 implicit algorithms give practically identical results to the Newmark solution, and are also close to the dynamic curves given in References [12-14].

Finally, to demonstrate the capabilities of implicit-explicit general single step algorithms, a modified version of the spherical shell cap, which has one stiff element at the crown, is examined. The material properties for the stiff element are $E = 10.5 \times 10^7$ lb/in^2, $\nu = 0.3$, $\rho = 0.78 \times 10^{-2}$ lb.sec^2/in^4 and $\sigma_y = 0.5 \times 10^5$ lb/in^2. Inclusion of this stiff element does not influence the fundamental period of the shell significantly, but it does change the smallest element period quite dramatically. The smallest period is 1.98×10^{-6} sec. and hence the limiting time step length is taken, from (20), as 0.6×10^{-6} sec., which is smaller than the original value. Therefore the implicit-explicit integration scheme is employed, in which the stiff element is integrated implicitly and the remainder are treated explicitly, so that the time step length for the explicit element group should only be less than or equal to 0.1×10^{-5} sec.

The elasto-plastic dynamic results for the modified shell cap are shown in Fig. 4, in which different schemes (SS22 explicit-implicit, SS22 explicit, SS22 implicit and the Newmark explicit-implicit) are employed. The general single step explicit-implicit algorithms are in good agreement with the corresponding Newmark schemes.

References

[1] Zienkiewicz, O.C.; Wood, W.L.; Taylor, R.L.: An alternative single step algorithm for parabolic boundary value problems. Earthquake Eng. Struct. Dyn., vol. 8 (1980) 31-40.

[2] Zienkiewicz, O.C.; Wood, W.L.; Taylor, R.L.; Hine, N.W.: A unified set of single step algorithms. Int. J. Num. Meth. Eng., vol. 20 (1984) 1524-1552.

[3] Zienkiewicz, O.C.; Kui, Li Xi: A unified set of single step algorithms in dynamic analysis of saturated media (linear and nonlinear). C/R494/80 internal report of University College of Swansea, 1980.

[4] Hughes, T.J.R.; Liu, W.K.: Implicit-explicit finite elements in transient analysis: stability theory. J.Appl. Mech. 45 (1978) 371-374.

[5] Ahmad, S.; Irons, B.M.; Zienkiewicz, O.C.: Analysis of thick and thin shell structures by curved finite elements. Int. J. Num. Meth. Eng., vol. 2 (1970) 419-451.

[6] Ahmad, S.: Curved finite elements in the analysis of solid, shell and plate structures. Ph.D. Thesis, University of Wales, Swansea, 1969.

[7] Hughes, T.J.R.; Liu, W.K.: Implicit-explicit finite elements in transient analysis: Implementation and Numerical examples. J. App. Mech., 45 (1978) 375-378.

[8] Hughes, T.J.R.; Pister, K.S.; Taylor, R.L.: Implicit-explicit finite elements in nonlinear transient analysis. Comp. Meth. Appl. Mech. Eng. 17/18 (1979) 159-182.

[9] Belytschko, T.; Mullen, R.: Mesh partitions of explicit-implicit time integration: in Formulation and Comput. Algor. in F.E. Analysis, pp. 673-690. K.J. Bathe et al. (eds.), MIT, Cambridge, Ma., 1976.

[10] Belytschko, T.; Mullen, R.: Stability of explicit-implicit mesh partitions in time integration. Int.J. Num. Meth. Eng., 12 (1978) 1575-1586.

[11] Bauer, H.F.: Nonlinear response of elastic plates to pulse excitations. J. Appl. Mech. 35 (1968) 47-52.

[12] Owen, D.R.J.; Hinton, E.: Finite elements in plasticity, theory and practice. Pineridge Press, Swansea, U.K., 1980.

[13] Nagarajan, S.; Popov, E.P.: Elastic-plastic dynamic analysis of axisymmetric solids. Computer and Structures, 4 (1974) 1117-1134.

[14] Owen, D.R.J.; Dinis, L.M.S.: Transient dynamic analysis of thin shells, including visco-plastic and large displacement effects in Mathematics of F.E. and Appl. - III, ed. J.R. Whiteman, pp. 423-435, Academic Press, 1978.

Table I : General single step implicit algorithms

(1) Set the iteration counter i = 0

$$\underset{\sim}{\alpha}_n^{p(i)} = 0$$

(2) Enter the averaged predictor phase

$$\bar{\underset{\sim}{d}}_{n+1}^{(i)} = \tilde{\underset{\sim}{d}}_{n+1}$$

$$\dot{\bar{\underset{\sim}{d}}}_{n+1}^{(i)} = \dot{\tilde{\underset{\sim}{d}}}_{n+1}$$

$$\ddot{\bar{\underset{\sim}{d}}}_{n+1}^{(i)} = \ddot{\tilde{\underset{\sim}{d}}}_{n+1}$$

(3) Evaluate the weighted averaged residual forces

$$\underset{\sim}{\psi}^{(i)}\,(\underset{\sim}{\alpha}_n^{p(i)}) = \bar{\underset{\sim}{f}}_{n+1} - \underset{\sim}{M}\,\ddot{\bar{\underset{\sim}{d}}}_{n+1}^{(i)} - \underset{\sim}{P}_{n+1}(\dot{\bar{\underset{\sim}{d}}}_{n+1}^{(i)}, \bar{\underset{\sim}{d}}_{n+1}^{(i)})$$

If $\underset{\sim}{\psi}^{(i)}$ satisfy the convergence condition go to Step (6)

(4) Solve the incremental effective static problem

$$\underset{\sim}{\alpha}_n^{p(i+1)} = (\underset{\sim}{K}^*)^{-1}\,\underset{\sim}{\psi}^{(i)}$$

where

$$\underset{\sim}{K}^* = [C_3\,\underset{\sim}{M} + C_2\,\underset{\sim}{C}_T + C_1\,\underset{\sim}{K}_T\,(\bar{\underset{\sim}{d}}_{n+1}^{(i)})\,]$$

(5) Enter the averaged corrector phase

$$\bar{\underset{\sim}{d}}_{n+1}^{(i+1)} = \bar{\underset{\sim}{d}}_{n+1}^{(i)} + C_1\,\underset{\sim}{\alpha}_n^{p(i+1)}$$

$$\dot{\bar{\underset{\sim}{d}}}_{n+1}^{(i+1)} = \dot{\bar{\underset{\sim}{d}}}_{n+1}^{(i)} + C_2\,\underset{\sim}{\alpha}_n^{p(i+1)}$$

$$\ddot{\bar{\underset{\sim}{d}}}_{n+1}^{(i+1)} = \ddot{\bar{\underset{\sim}{d}}}_{n+1}^{(i)} + C_3\,\underset{\sim}{\alpha}_n^{p(i+1)}$$

and set i = i+1. Go to Step (3).

(6) Calculate values of $\underset{\sim}{d}_{n+1}$ to $\underset{\sim}{d}_{n+1}^{p-1}$ at time t_{n+1} according to equation (8).

(7) Replace n by n+1 and return to (1) for the next time step.

<u>Table II : General single step explicit algorithms</u>

(1) Begin the predictor phase by setting

$$\underset{\sim}{\alpha}_n^{p(o)} = \underset{\sim}{0}$$

$$\bar{\underset{\sim}{d}}_{n+1}^{(o)} = \tilde{\underset{\sim}{d}}_{n+1}$$

$$\dot{\bar{\underset{\sim}{d}}}_{n+1}^{(o)} = \dot{\tilde{\underset{\sim}{d}}}_{n+1}$$

$$\ddot{\bar{\underset{\sim}{d}}}_{n+1}^{(o)} = \ddot{\tilde{\underset{\sim}{d}}}_{n+1}$$

(2) Evaluate the weighted averaged residual forces as

$$\underset{\sim}{\psi}^{(o)} = \bar{\underset{\sim}{f}}_{n+1} - \underset{\sim}{M}\,\ddot{\bar{\underset{\sim}{d}}}_{n+1}^{(o)} - \underset{\sim}{P}_{n+1}(\dot{\bar{\underset{\sim}{d}}}_{n+1}^{(o)}, \bar{\underset{\sim}{d}}_{n+1}^{(o)})$$

(3) Solve the 'effective static equation'

$$\underset{\sim}{\alpha}_n^{p(1)} = (\underset{\sim}{K}^*)^{-1}\,\underset{\sim}{\psi}^{(o)} \quad , \qquad \underset{\sim}{K}^* = C_3\,\underset{\sim}{M}$$

Since the mass matrix $\underset{\sim}{M}$ does not change, $\underset{\sim}{K}^*$ will be formed once only.

(4) Calculate values $\underset{\sim}{d}_{n+1}$ to $\underset{\sim}{d}_{n+1}^{p-1}$ at time t_{n+1} according to equation (8)

(5) Set n = n+1 for the next time step.

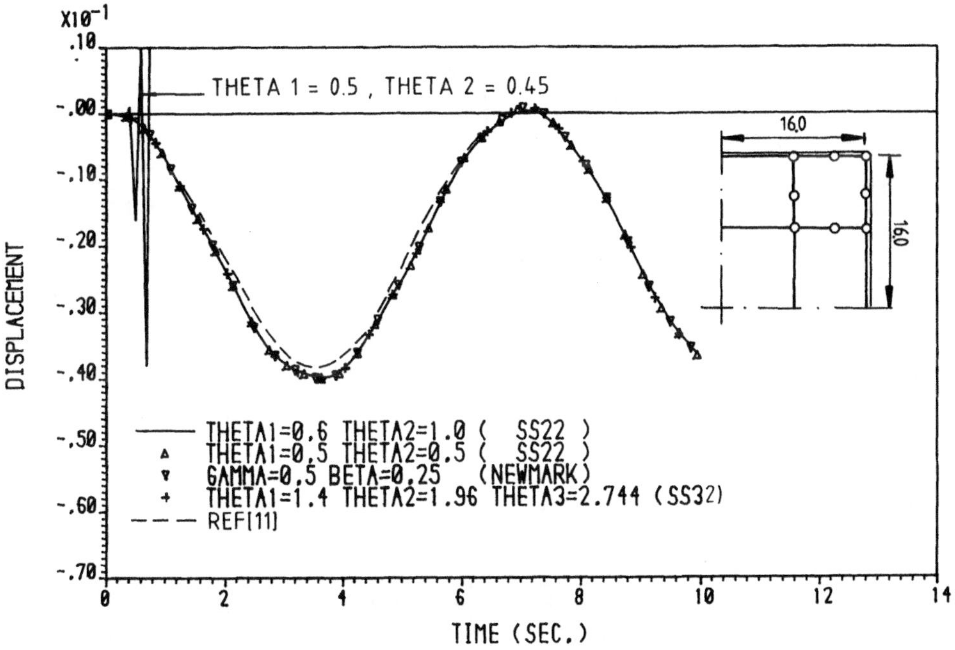

Fig. 1 DYNAMIC ELASTIC RESPONSE OF A SIMPLY SUPPORTED SQUARE PLATE

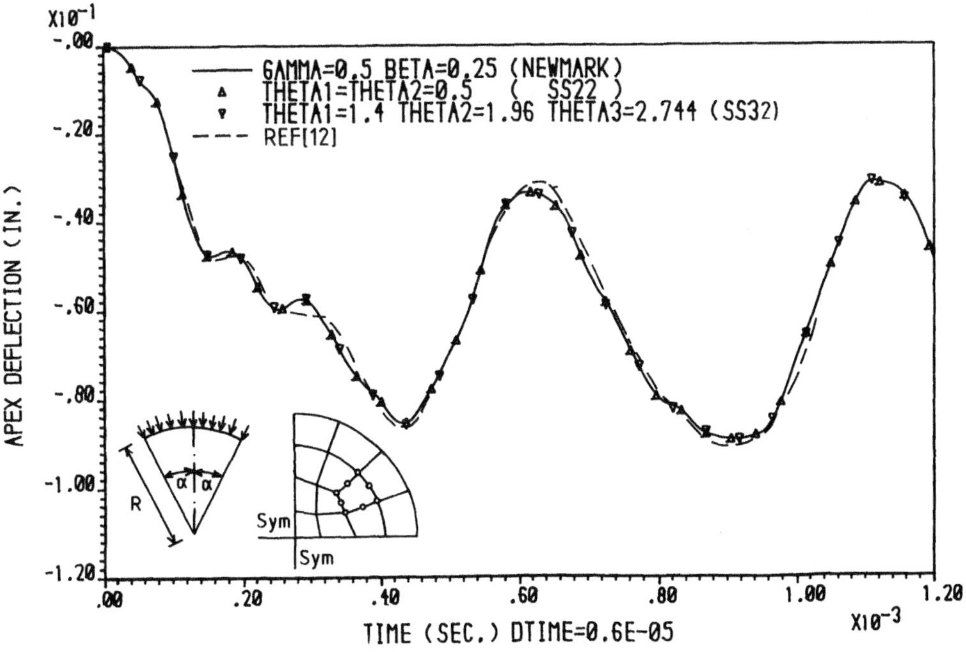

Fig. 2 ELASTO-PLASTIC RESPONSE OF A SPHERICAL SHELL CAP

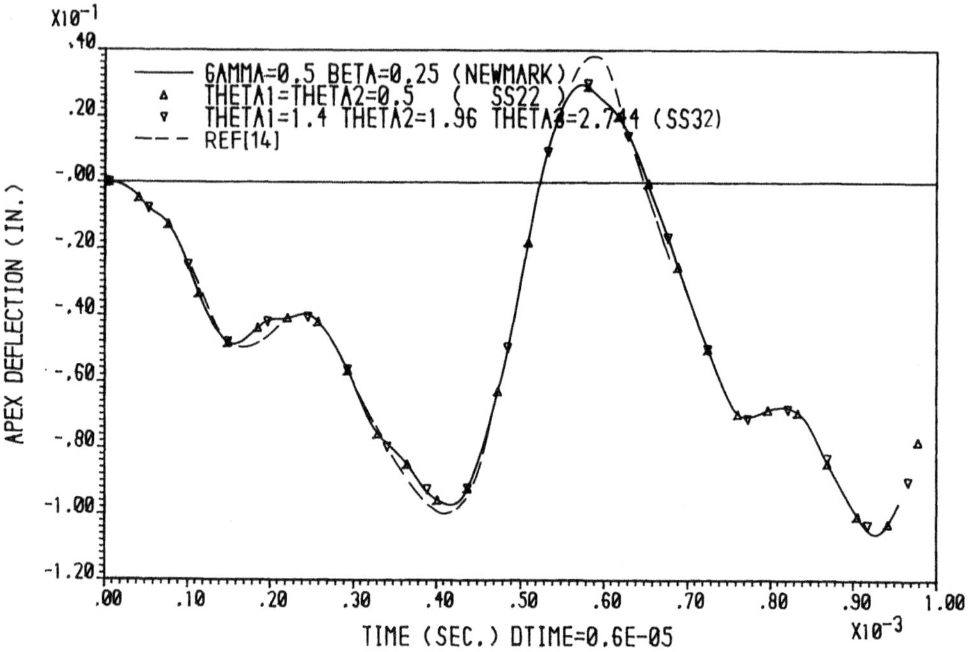

Fig. 3 LARGE DEFLECTION RESPONSE OF A SPHERICAL SHELL CAP

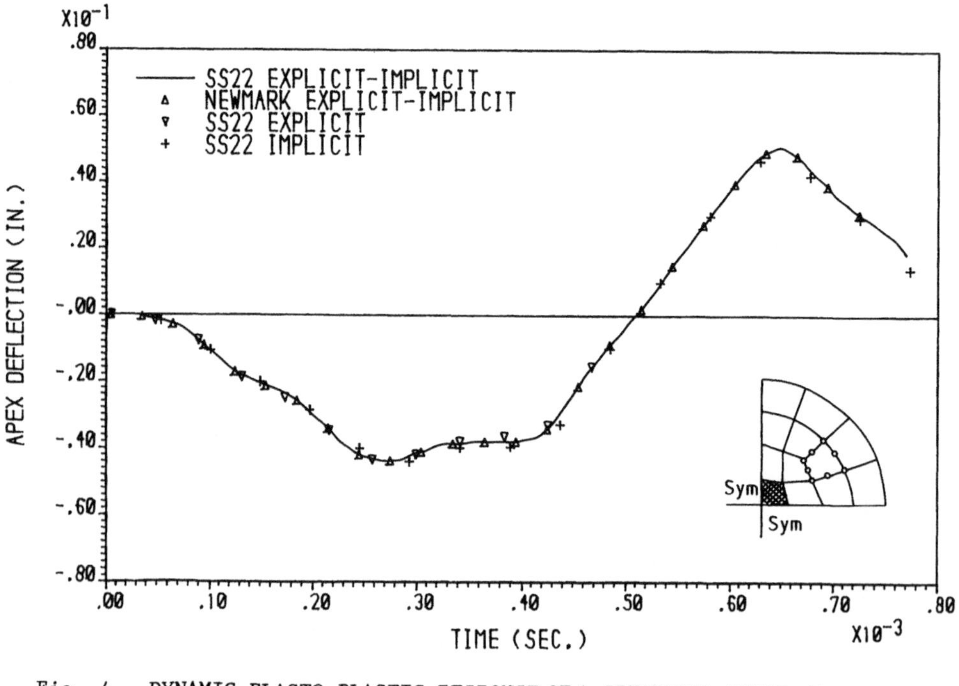

Fig. 4 DYNAMIC ELASTO-PLASTIC RESPONSE OF A SPHERICAL SHELL CAP WITH A CENTRAL STIFF REGION

Dynamic Quasi-Bifurcations in Structures Subjected to Step Loadings

M. KLEIBER, W. KOTULA*, M. SARAN*
Academy of Sciences, Warsaw, Poland
*Technical University of Lublin, Poland
**Technical University of Warsaw, Poland

Summary

A simple concept of the instability analysis in dynamically loaded structures is presented. An interesting phenomenon called a dynamic quasi-bifurcation is discussed in detail. The suggested analysis may yield important informations about the dynamic structural behaviour in the nonlinear range and can easily be incorporated into the existing finite element codes. Two numerical examples illustrate the theory.

Introduction

Practically oriented treatment of dynamic structural buckling has undoubtedly been hampered in the past by the lack of a simple and appealing algebraic representation in terms of the tangent stiffness matrix singularity so well known in static analysis. What has often been only pursued is for instance the repetitive solution of the structural response for increasing load parameters with the attempt to identify the so--called dynamic snap-through behaviour. The amount of computer time necessary, the lack of precisely defined stability criterion and the negligence of any solution path branching and related problems of imperfection sensitivity are clearly the major deficiencies of such an approach. The classical papers on dynamic buckling such as [1] are of limited value since they are based on a number of simplifying assumptions / inertia forces in the pre-buckling state are neglected, some nonlinear terms are omitted in the equations of motion, dynamic and static buckling shapes are assumed identical/ which are not applicable in the analysis to follow. Even if we have today a rich theoretical literature on the Liapunov-type ana-

Finite Element Methods for Nonlinear Problems
Europe-US Symposium, Trondheim, Norway 1985
Editors: Bergan, Bathe, Wunderlich

lysis of instability problems, the finite element applications in this field are still discouragingly limited.
It was observed in [2-4] that there might exist some "bifurcated" solutions in structural dynamics. Theoretical and numerical substantiation of this concept were given in[2]and [3, 4], respectively. In this paper a finite element approach is presented for estimating the buckling loads in structures subjected to dynamic step loadings. The quasi-bifurcation for a given load history is defined as the point along the fundamental unperturbed trajectory at which a perturbed motion initially oscillating in the small neighborhood of the fundamental trajectory departs monotonically from the neighbourhood. The present development aims at effective large scale computations in structural dynamics. Two truss examples are selected to illustrate the theory.

Fundamental concept

Let us consider the finite element representation of the nonlinear dynamic behaviour of structures without damping in the form

$$\mathbf{M}\ddot{\mathbf{r}}(\tau) + \mathbf{F}(\mathbf{r}(\tau)) = \mathbf{R}(\tau) \qquad /1/$$

where $\mathbf{M}$ is the mass matrix, $\mathbf{r}(\tau)$ the generalized displacement vector at time τ, $\mathbf{R}(\tau)$ the external nodal load vector at this time instant and $\mathbf{F}(\mathbf{r}(\tau))$ the internal nodal load vector calculated from stresses existing at time τ. The notation $\mathbf{F}(\mathbf{r}(\tau))$ has a symbolic rather than the precise significance- -the dependence of $\mathbf{F}$ on the history of the process is obviously not excluded. The initial conditions are

$$\mathbf{r}(0) = \mathbf{r}_0 \quad , \quad \dot{\mathbf{r}}(0) = \dot{\mathbf{r}}_0 \qquad /2/$$

$\mathbf{r}_0$, $\dot{\mathbf{r}}_0$ being given vectors. Let us assume for the future reference that the mass matrix is positive-definite, which is true for both the consistent and diagonal mass matrices provided the massless degrees of freedom have been statically condensed in the latter approach. We rewrite eq. /1/ as

$$\ddot{\mathbf{r}}(\tau) = \mathbf{R}(\tau) - \mathbf{F}(\mathbf{r}(\tau)) \qquad /3/$$

or, simpler still, as

$$\ddot{\mathbf{r}}(\tau) = \mathbf{P}(\tau; \mathbf{r}(\tau)) \qquad /4/$$

$\mathbf{P} = \mathbf{M}^{-1}(\mathbf{R} - \mathbf{F})$ being the generalized force vector. The solution $\mathbf{r}(\tau)$ of eq. /4/ may generally be written as

$$\mathbf{r}(\tau) = \mathbf{p}(\tau; \mathbf{r}_0, \dot{\mathbf{r}}_0) \qquad /5/$$

Changing now the initial values $\mathbf{r}_0$, $\dot{\mathbf{r}}_0$ to $\mathbf{r}_0 + \delta\mathbf{r}_0$, $\dot{\mathbf{r}}_0 + \delta\dot{\mathbf{r}}_0$ respectively, we have the resulting motion given by

$$\mathbf{r}(\tau) + \delta\mathbf{r}(\tau) = \mathbf{p}(\tau; \mathbf{r}_0 + \delta\mathbf{r}_0, \dot{\mathbf{r}}_0 + \delta\dot{\mathbf{r}}_0) \qquad /6/$$

with $\delta\mathbf{r}(\tau)$ being the perturbation of motion.
According to the Liapunov stability concept we say that the motion $\mathbf{p}(\tau; \mathbf{r}_0, \dot{\mathbf{r}}_0)$ is stable if a small distance between the points $(\mathbf{r}_0, \dot{\mathbf{r}}_0)$ and $(\mathbf{r}_0 + \delta\mathbf{r}_0, \dot{\mathbf{r}}_0 + \delta\dot{\mathbf{r}}_0)$ in the phase space implies the distance between the points $(\mathbf{r}(\tau), \dot{\mathbf{r}}(\tau))$, $(\mathbf{r}(\tau) + \delta\mathbf{r}(\tau), \dot{\mathbf{r}}(\tau) + \delta\dot{\mathbf{r}}(\tau))$ be small for all $\tau > 0$.
Knowing the explicit dependence of $\mathbf{r}(\tau)$ on $\mathbf{r}_0$, $\dot{\mathbf{r}}_0$ cf.eq. /5/, it is an easy matter to check the above criterion directly. Our goal is, however, to formulate a criterion which will make it possible to analyze the stability on the basis of the fundamental solution only, that is the solution obtained for one particular value of the initial values $\mathbf{r}_0$, $\dot{\mathbf{r}}_0$. Using eqs./4/ and /6/ we obtain the relationship

$$\delta\ddot{\mathbf{r}}(\tau) = \mathbf{P}(\tau; \mathbf{r} + \delta\mathbf{r}) - \mathbf{P}(\tau; \mathbf{r}) = \delta\mathbf{P}(\tau; \mathbf{r}, \delta\mathbf{r}) \qquad /7/$$

or, by eq. /3/ under the assumption that the load history is the same for both the fundamental and perturbed processes, the relationship

$$\delta\ddot{\mathbf{r}}(\tau) = \mathbf{M}^{-1}[-\mathbf{F}(\mathbf{r} + \delta\mathbf{r}) + \mathbf{F}(\mathbf{r})] \qquad /8/$$

Linearising the right hand-side of eq /8/ with respect to $\delta \mathbf{r}$ we arrive at

$$\delta \ddot{\mathbf{r}}(\tau) = \mathbf{M}^{-1}[-\frac{\partial \mathbf{F}}{\partial \mathbf{r}} \delta \mathbf{r}(\tau)] \tag{/9/}$$

or, introducing the concept of the tangent stiffness matrix $\mathbf{K}_T = \frac{\partial \mathbf{F}}{\partial \mathbf{r}}$, at

$$\delta \ddot{\mathbf{r}}(\tau) = -\mathbf{M}^{-1}\mathbf{K}_T(\mathbf{r}(\tau))\,\delta \mathbf{r}(\tau) \tag{/10/}$$

It can now be easily shown, similarly as in [2], that for any deviated motion, stable or not, there is the following identity

$$\begin{aligned} \frac{1}{2}\delta \mathbf{r}^T(\tau)\delta \mathbf{r}(\tau) = \frac{1}{2}\delta \mathbf{r}_0^T \delta \mathbf{r}_0 + \int_0^{\tau} [\delta \dot{\mathbf{r}}_0^T \delta \mathbf{r}_0 + \\ + \int_0^{\tau'} \delta \ddot{\mathbf{r}}^T(\tau'')\delta \mathbf{r}(\tau'')\,d\tau'' + \int_0^{\tau'} \delta \dot{\mathbf{r}}^T(\tau'')\,\delta \dot{\mathbf{r}}(\tau'')]\,d\tau' \end{aligned} \tag{/11/}$$

which results directly from the trivial relationship

$$\int_0^{\tau} \delta \ddot{\mathbf{r}}^T(\tau')\delta \mathbf{r}(\tau')d\tau' = \int_0^{\tau} \left\{ \frac{d}{d\tau'}\left[\delta \dot{\mathbf{r}}^T(\tau')\delta \mathbf{r}(\tau')\right] - \delta \dot{\mathbf{r}}^T(\tau')\delta \dot{\mathbf{r}}(\tau') \right\} d\tau' \tag{/12/}$$

Since the term $\delta \dot{\mathbf{r}}^T \delta \dot{\mathbf{r}}$ is positive for any τ, eq. /11/ implies that the boundedness of the deviated motion $\delta \mathbf{r}(\tau)$ /or the stability of the motion $\mathbf{p}$, eq. /5/ is governed by the history of the term $\delta \ddot{\mathbf{r}}^T \delta \mathbf{r}$. This term can be presented as, cf. eq. /10/

$$\delta \ddot{\mathbf{r}}^T(\tau)\delta \mathbf{r}(\tau) = -\delta \mathbf{r}^T(\tau)\mathbf{M}^{-1}\mathbf{K}_T(\mathbf{r}(\tau))\,\delta \mathbf{r}(\tau) \tag{/13/}$$

Now, if there exists a $\delta \mathbf{r}(\tau)$ such that for a certain time τ_{cr}

$$\delta \ddot{\mathbf{r}}^T(\tau)\,\delta \mathbf{r}(\tau) > 0 \quad \text{for } \tau > \tau_{cr} \tag{/14/}$$

then $\delta \mathbf{r}$ grows without bound and the fundamental motion is unstable. In other words, by using eq. /13/, if for any $\delta\mathbf{r}(\tau)$

$$\delta\mathbf{r}^T(\tau)\mathbf{M}^{-1}\mathbf{K}_T(\mathbf{r}(\tau))\delta\mathbf{r}(\tau) < 0 \qquad /15/$$

then one /or more/ of the vibration modes begins to grow in an unlimited manner. Clearly, the tangent stiffness matrix $\mathbf{K}_T$ for any realistic structural configuration is positive--definite at $\tau = 0$ and at a number of subsequent time instants $\tau < \tau_{cr}$. We have then at each instant the eigenproblem

$$[\mathbf{M}^{-1}\mathbf{K}_T - \lambda \mathbf{I}]\boldsymbol{\vartheta} = 0 \qquad /16/$$

or

$$[\mathbf{K}_T - \lambda \mathbf{M}]\boldsymbol{\vartheta} = 0 \qquad /17/$$

which yields the positive eigenvalues

$$\lambda_1(\tau) \leq \lambda_2(\tau) \leq \lambda_3(\tau) \leq \ldots \quad , \quad \tau < \tau_{cr} \qquad /18/$$

with the corresponding normalized eigenvectors

$$\boldsymbol{\vartheta}_1(\tau),\ \boldsymbol{\vartheta}_2(\tau),\ \boldsymbol{\vartheta}_3(\tau), \ldots \qquad /19/$$

At $\tau = \tau_{cr}$ the smallest eigenvalue $\lambda_1(\tau_{cr})$ changes its sign /becomes negative/ and the deviated motion mode $\boldsymbol{\vartheta}_1(\tau)$ begins its unlimited growth. The foregoing discussion indicates that the condition $\lambda_1 = 0$, corresponding to the singularity of the tangent stiffness matrix, plays a crucial role in the dynamic stability analysis of structures. Since the criterion bears striking resemblance to the static branching condition two important remarks have to be additionally made :

1. To discover singularity of $\mathbf{K}_T(\mathbf{r}(\tau))$ the dynamic solution path $\mathbf{r} = \mathbf{r}(\tau)$ must be followed so that generally the condition det $\mathbf{K}_T = 0$ corresponds to different displacements and stresses than those rendering the singularity in the static process. The dynamic analysis is clearly more general and may be considered to contain the static uniqueness

criterion due to Hill [5] as a special case.

2. Contrary to statics, the dynamic branching behaviour should be understood as the existance of at least one perturbed motion, which is initially in the small neighborhood of the fundamental undisturbed motion and which starts at $\tau = \tau_{cr}$ to depart monotonically from the neighborhood. This phenomenon is proposed to be referred to as a quasi--bifurcation phenomenon.

The computational procedure based on the above development can be applied to any nonlinear structural dynamics problem in the following way.

1. The fundamental solution /5/ is obtained by using any implicit method of direct integration and the smallest eigenvalue of /17/ is monitored at each step to discover at time $\tau = \tau_{cr}$ the critical point at which K_T ceases to be strictly positive-definite. Clearly, the effective stiffness matrix of the integration scheme is still non--singular.
2. The step-by-step analysis can easily be continued beyond this point but it must be remembered that only a slight modification in the initial conditions in accord with the discovered mode of unbounded growth will amplify the solution leading to instability. This can be checked numerically by repeating the step-by-step computations with the only difference being the appropriately perturbed initial conditions.

Examples

Let us consider the simple 3D structure shown in Fig.1. The truss was used to ilustrate the nonlinear static behaviour of structures in [6] . The geometric and material parameters as well as the load distribution are shown in the figure. The critical time vs load magnitude curve is plotted in Fig.2. For the loads leading to quasi-bifurcations the eigenvector corresponding to the zero eigenvalue is $\{0,a\}$, a - any real number. By calculating the nonlinear response for any no-critical-time case, say for P = 4,3 kN, Fig.3 shows that both the curves corresponding to the ideal and perturbed by the above eigenvector /with a = 0.0001 m/ geometries remain close to each other in the whole time interval considered.

$E = 2.1 \cdot 10^7$ (kN/m^2) , k = 40 (kN/m) , A = $6 \cdot 10^{-4}$ (m^2) , h = 0.22 (m) , m = 1 (kg)
l = 2.1888618 (m)

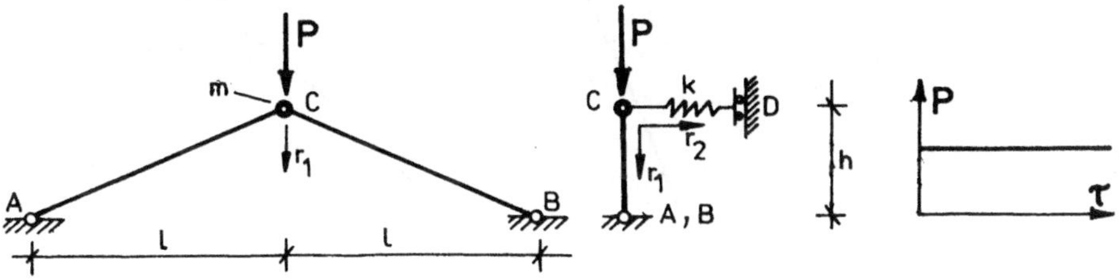

Fig.1. Simple 3D truss structure under step loading

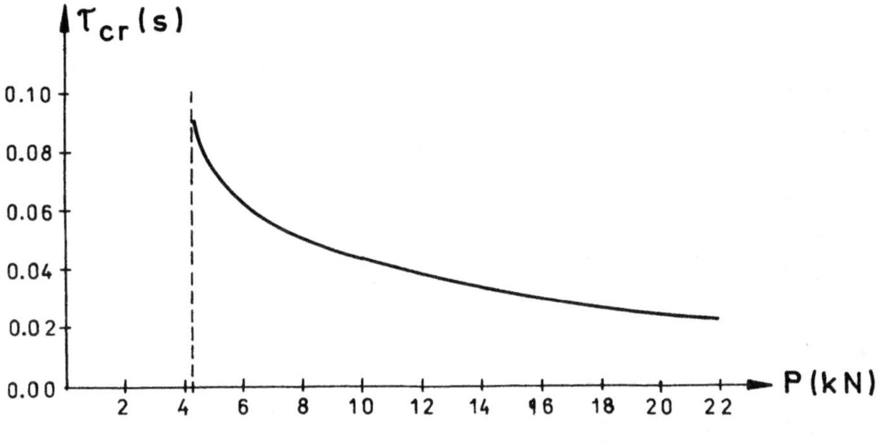

Fig.2. Critical time vs load magnitude curve

On the contrary and in full accord with our theoretical result, the perturbation in the initial geometry /by the eigenvector {0, 0.0001 m}/ for the loads P = 10 kN and P = 15 kN which

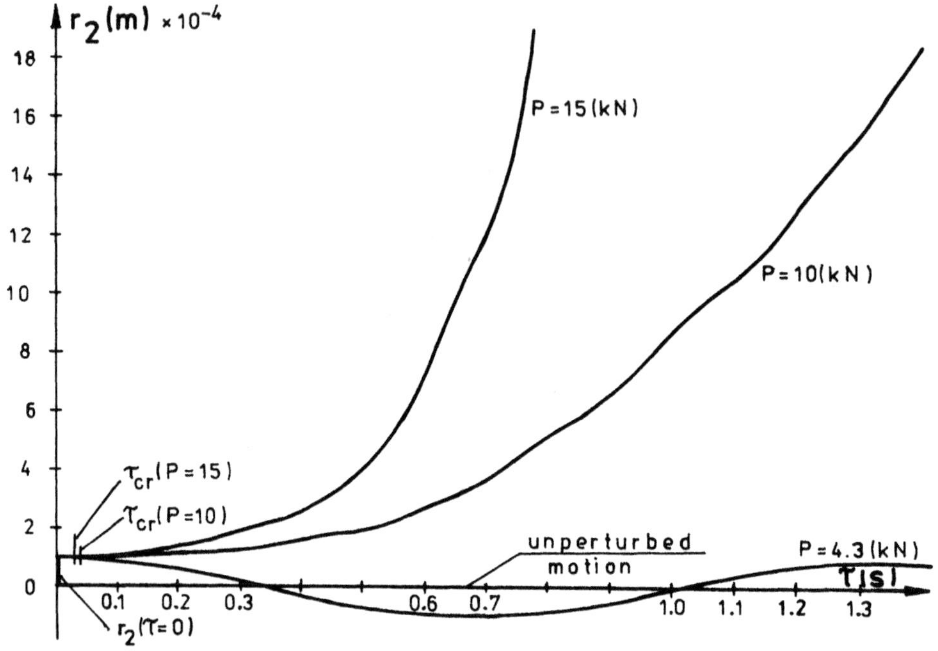

Fig.3. Perturbed and unperturbed response curves for different load magnitudes

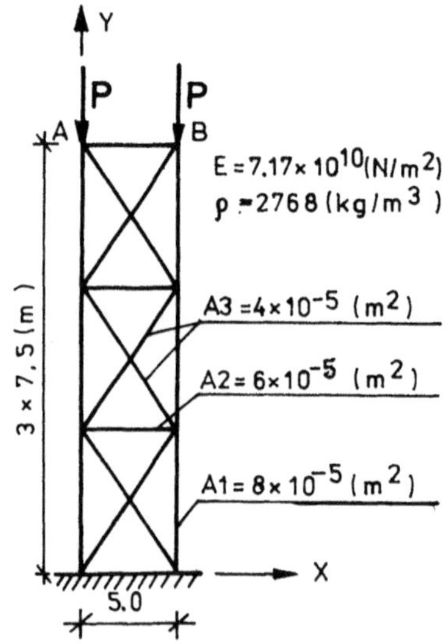

Fig.4. 2D truss structure under step loading

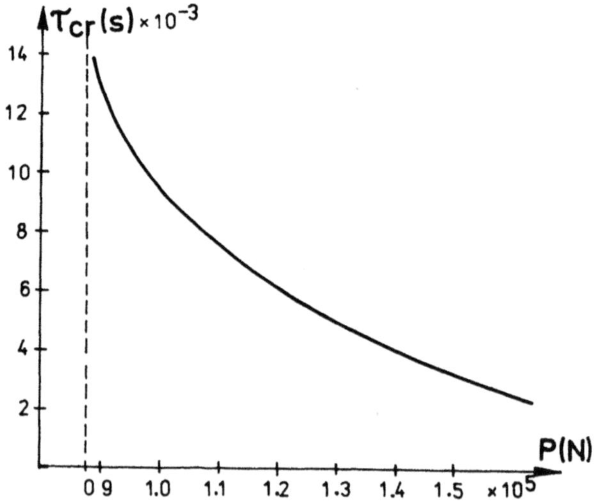

Fig.5. Critical time vs load magnitude curve

imply the existence of critical time, lead to the response curves departing in a distinct way from the fundamental solutions for the unperturbed configurations, Fig.3. Also, the points of departure along the time axis may by identified as the critical times for these particular load magnitudes.
The qualitatively similar results have been obtained for the 2D truss shown in Fig.4. The truss is constrained to deform in-plane only. The plot of the critical time vs load magnitude is shown in Fig.5.The perturbation to the initial geometry is introduced again by means of an eigenvector corresponding for the given load to the zero eigenvalue. The solid and broken lines in Fig.6 represent the fundamental and perturbed motions of the points A and B, respectively.

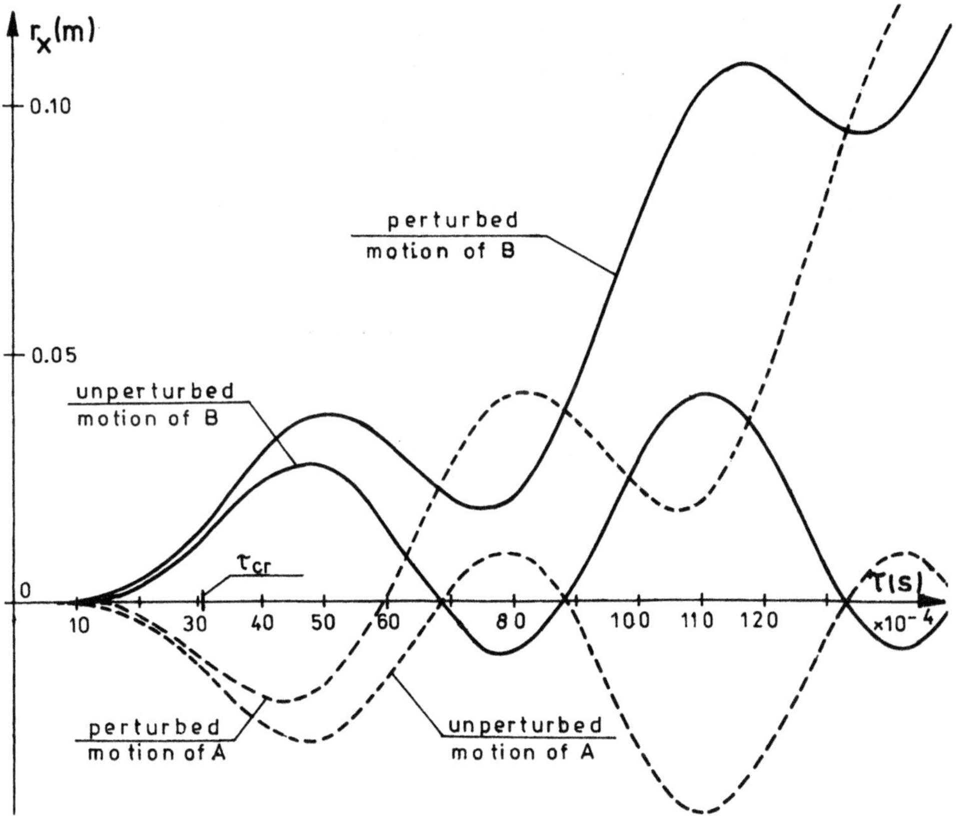

Fig.6. Perturbed and unperturbed response curves for points A and B for P = 150 kN

Concluding remarks

The results of this paper indicate the possible direction of further studies in the field of effective evaluation of dynamic buckling loads. The concept of quasi-bifurcations seems to offer an attractive framework for working out numerical algorithms to deal with this class of problems. However, to understand fully the phenomena described in this paper extensive further studies are required.

References

1. Budiansky, B. : Dynamic buckling of elastic structures : Criteria and Estimates in : Dynamic Stability of Structures, Herrmann, G. /eds./. Pergamon Press 1966.

2. Lee, L.H.N. : On dynamic stability and quasi-bifurcation. Int.J. Non-linear Mech. 16 /1981/ 79 - 87.

3. Lee, L.H.N. : Dynamic buckling of an inelastic column. Int.J.Sol.Struct. 17 /1981/ 271 - 279.

4. Kleiber, M. : Some results in the numerical analysis of structural instabilities, Part II : Dynamics. Eng. Trans. 30 /1982/ 355 - 367.

5. Hill, R. : A general theory of uniqueness and stability in elastic - plastic solids. J.Mech. Phys. Sol. 6 /1957/ 236 - 249.

6. Bergan, P.G. : Solution algorithms for nonlinear structural problems. Comp.Struct. 12 /1980/ 497 - 509.

Finite Strip Analysis of Blast Loaded Plates

D. S. MOFFLIN*, M. D. OLSON, D. L. ANDERSON
University of British Columbia, Vancouver, Canada
*University of Washington, Seattle, USA

Abstract

The large deflection, elastic-plastic response of plate type structures subjected to air-blast loading is analyzed via the finite strip method. The theoretical basis and computer code development to date is presented and preliminary results for a few examples are shown along with other known computed and experimental results for simple beams and plates. These preliminary finite strip results compare very favourably. The computational effort and cost is considerably less than for those obtained by the conventional general purpose programs.

1. INTRODUCTION

Recent years have seen the development of finite element [1,2] and finite difference [3,4] programs for predicting the static, elasto-plastic, large deflection behaviour of beams, plates, and stiffened panels [5,6]. Some finite element [7] and finite difference [8] programs have also been developed to study the dynamic behaviour of members under transient loads. ADINA [7] is one example. It can undertake a dynamic elasto-plastic, large deflection analysis of both plates and stiffened panels using finite elements. However, the program is so time consuming that it is expensive to undertake parametric studies, a necessary input to any design formulation.

For the full analysis of both plates and plate assemblages under in-plane and lateral loads, the finite strip method [9] has been shown to produce accurate results at the fraction of the cost of

* Now at Dept. of Civil Engineering, University of Washington, Seattle, Washington, U.S.A.

Finite Element Methods for Nonlinear Problems
Europe-US Symposium, Trondheim, Norway 1985
Editors: Bergan, Bathe, Wunderlich

finite element or finite difference methods. A finite strip program, LOBUC [10,11] has previously been developed by the first author for the static local and interactive buckling behaviour of plate assemblages [12] (eg. I sections and channels), and the versatility and speed of the method has been proven. The method is now extended to predict the behaviour of panels under dynamic lateral loads. Allowance has also been made to allow for more boundary conditions than previously possible.

2. THE FINITE STRIP APPROACH

The finite strip approach [9] is an adaptation of the finite element method for dealing with members that have constant cross-sectional properties down one axis. The member, instead of being divided into elements, is divided into a discrete number of strips. Each strip has constant cross-section down its length, and the ends of each strip are boundaries of the member. A typical strip division for a simple plate is shown in Fig. 1(a).

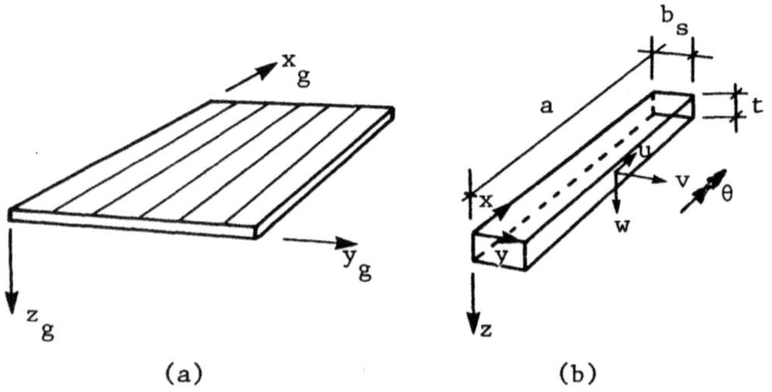

Figure 1. (a) Typical strip division for a plate.
(b) Definition of strip parameters.

Both the in and out of plane deflected shape within each strip is assumed to vary analytically down the length, and in the form of a polynomial across the width. This enables the overall deflected shape to be defined by a discrete number of displacement parameters, where the displacement parameters correspond to the amplitudes of the various displacements u, v, w, θ along the nodal lines (Fig. 1(b)). Having defined the deflected shape, stresses and strains can be specified in terms of the displacement parameters, which can consequently be found by satisfying dynamic equilbrium.

An important requirement of the method is that the displacement functions, defining the assumed deflected shape within each strip, be able to accurately describe the actual deflected shape. The displacement functions should also comply with the requirements of:

1) Satisfying the end boundary conditions of the strip;
2) Allowing a state of constant, non-zero strain to exist; and
3) Satisfying compatibility between neighbouring strips.

2.1 Displacement Functions

The displacement functions presented in [10,11] were developed for local buckling analysis, and were therefore formulated for simply supported transverse edges. If a structure is subjected to lateral rather than axial loading, a more realistic boundary condition is that the transverse edges are clamped. A clamped boundary condition can be accommodated by setting w and θ to vary in the form of a cosine function down the length of each strip as follows,

$$u = f_m^u(\eta)\sin(m\pi\xi)+ae(1/2-\xi) \qquad m = 2,4,6...$$

$$v = f_n^v(\eta)\cos(n\pi\xi)+b_s\nu e(\eta-1/2)+c_v(\eta-1/2) \qquad n = 2,4,6... \qquad (1a)$$

$$w = f_p^w(\eta)\,[1-\cos(p\pi\xi)] \qquad p = 2,4,6...$$

$$\theta = f_p^\theta(\eta)\,[1-\cos(p\pi\xi)] \qquad p = 2,4,6...$$

where $\eta = y/b_s$, $\xi = x/a$, e = end shortening/a, ν = Poisson's ratio, c_v = transverse pull-in variable, and

$$f_m^u(\eta) = (1-\eta)u_{1m} + \eta u_{2m}$$

$$f_n^v(\eta) = (1-\eta)v_{1n} + \eta v_{2n}$$

$$f_p^w(\eta) = (1-3\eta^2+2\eta^3)w_{1p} + (\eta-2\eta^2+\eta^3)b_s\theta_{1p} \tag{1b}$$

$$+ (3\eta^2-2\eta^3)w_{2p} + (\eta^3-\eta^2)b_s\theta_{2p}$$

$$f_p^\theta(\eta) = \partial f_p^w(\eta)/\partial y = (-6\eta+6\eta^2)(w_{1p}/b_s)$$

$$+ (1-4\eta+3\eta^2)\theta_{1p} + (6\eta-6\eta^2)(w_{2p}/b_s)$$

$$+ (3\eta^2-2\eta)\theta_{2p}$$

In the previous static buckling analyses [10,11]. an incremental approach similar to Little [14] was followed in which the equilibrium deflected shape for each increment of loading was found by minimizing an energy function F, where both the function and its first derivatives with respect to the displacement parameters were known. Although a knowledge of the first derivatives is not essential in a minimization procedure, the rate of convergence is vastly increased if they can be calculated. The function F is referred to as the live energy function, and is given by

$$F = \int_V \Psi \, dV \tag{2}$$

where

$\psi = [\{\sigma\} + 1/2\{\Delta\sigma\}]^T.\{\Delta\varepsilon\}$

$\{\sigma\}$ = stress at the start of the increment

$\{\Delta\sigma\},\{\Delta\varepsilon\}$ = incremental changes in the stress and strain

The first derivative of F with respect to the amplitude of the dislacement parameters is given by

$$F' = \int_V \psi' \, dV \tag{3}$$

where $\psi' = [\{\sigma\} + 1/2\{\Delta\sigma\}]^T \{(\Delta\varepsilon)'\} + 1/2 \{(\Delta\sigma)'\}^T \{\Delta\varepsilon\}$

For small increments, Eq. (3) can be simplified to

$$\psi' = \{\sigma\}^T \{(\Delta\varepsilon)'\} \tag{4}$$

which can readily be calculated.

2.2 Work-Energy Formulation

To satisfy dynamic equilibrium at any point of time a work-energy approach is adopted. For the present application to plates, it is necessary only to include pressure and inertia loads in the z_g direction. Consider displacement increments Δu, Δv and Δw. The contribution to the net work during an increment is given by

$$W = \int_A \left[\frac{\partial \phi}{\partial u} \Delta u + \frac{\partial \phi}{\partial v} \Delta v + \frac{\partial \phi}{\partial w} \Delta w - (P - ma)\Delta w\right] dA \tag{5}$$

where
- a = acceleration in z_g direction
- m = mass per unit area
- ϕ = increment in strain energy = $\int_t (\sigma + \frac{1}{2} \Delta\sigma)\Delta\varepsilon \, dz$
- P = pressure loading in z_g direction

2.3 Solution Procedure

The solution procedure involves integrating Eq. (5) with respect to the displacement increments, giving an energy function analogous to F. The function is then minimized in exactly the same manner as it was for the static in-plane loading case, but using Eq. (5) to generate the energy gradient.

When calculating the work due to inertia forces in Eq. (5), the acceleration is assumed to vary linearly across each strip. This reduces the number of terms required to describe the acceleration and simplifies the calculations, but is cruder than the assumed distribution of out-of-plane displacements. However the displacements have to be differentiated to calculate strains and hence to determine their influence on the internal energy. Accelerations do not need to be differentiated to calculate their contribution to the inertia of the system, enabling a cruder distribution to be assumed without loss of accuracy.

It is further assumed that the work due to inertia forces is only a function of Δw, and independent of the rotations $\Delta\theta$. This reduces the need to iterate within each time step, and greatly increases the computational efficiency of the method.

During each increment of time, the acceleration is assumed to be constant and equal to the average of the initial and final values. This is equivalent to Newmarks method with $\beta = 1/4$ [6]. This enables the acceleration at the end of a time increment to be expressed as a function of the velocity and acceleration at the start of the increment, and the displacement during the increment, namely

$$(\ddot{w})_{t+\Delta t} = \frac{4\Delta w}{(\Delta t)^2} - \frac{4(\dot{w})_t}{\Delta t} + (\ddot{w})_t \qquad (6)$$

where $(\ddot{w})_{t+\Delta t}$ = acceleration at the end of the time step
Δw = change in displacement during the time step
$(\ddot{w})_t, (\dot{w})_t$ = acceleration and velocity at the start of the time step
Δt = time step increment

This approach is unconditionally stable for linear systems [16], and allows time steps many times larger than the lowest natural period of vibration to be adopted. This is essential for plate analysis, because each step requires considerable computational effort, and it is not feasible to take a large number of very small time steps. However, the approach is not unconditionally stable for nonlinear systems, so some care is still required in choosing the time step.

3. OUTLINE OF COMPUTER PROGRAM

The input data required, and therefore parameters that can be varied, are as follows: (i) Young's modulus, Poisson's ratio, (ii) Proof stress for each strip, and knee factor if the material is strain hardening (see below), (iii) Geometry of problem, including strip division, number of monitoring points within each strip, and position and type of boundaries, (iv) The number of terms and associated wavelengths for each Fourier series term in the displacement functions, (v) Residual stresses and initial imperfections, (vi) The initial pressure, its duration, and decay constant, (vii) The density of the material, and (viii) The time step.

The overall strain energy of the system is found by monitoring stresses and strains at points down the length, across the width, and through the thickness of each strip, calculating the energy per unit volume at each point, and then integrating using Simpson's Rule. Symmetry is used whenever possible, both down the length and across the width. Strips can differ in both width and thickness across the plate, but have constant cross-sections down the length.

The work due to pressure and inertia forces are found by integrating analytically over each strip. The pressure is assumed to be uniform over the member, and to vary with time according to (Fig. 2)

$$P(t) = P_m \left(1 - \frac{t}{T}\right)^{(1/\alpha - 1)} \qquad 0 < t < T \qquad (7)$$
$$= 0 \qquad t > T$$

where P_m = initial pressure
t = time
T = load duration
α = decay constant

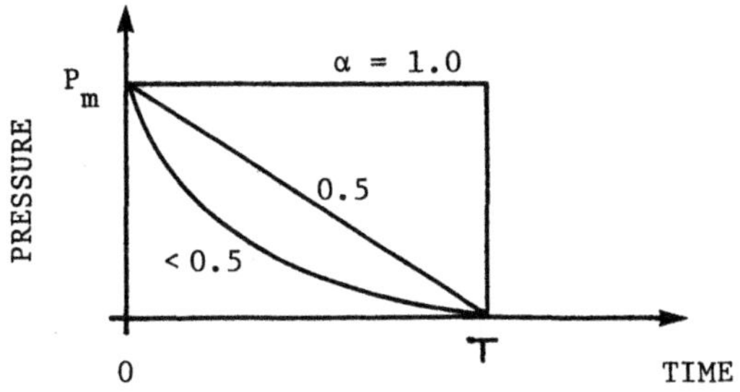

Figure 2. Pressure-time history.

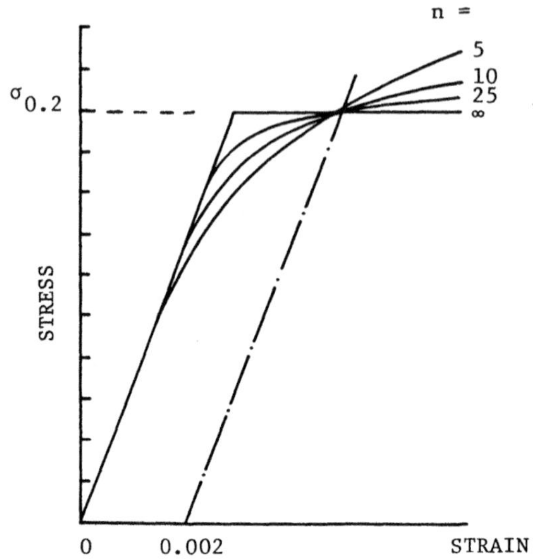

Figure 3. Ramborg-Osgood stress-strain curves.

Both elastic-perfectly plastic and strain hardening material can be modelled. For strain hardening materials, the stress-strain curve is described by the Ramberg-Osgood [17] equation (Fig. 3),

$$\varepsilon = \sigma/E + 0.002(\sigma/\sigma_{0.2})^n \tag{8}$$

where $\sigma_{0.2}$ = the 0.2% proof stress, E = Young's modulus, and n = knee factor, describing the roundness of the curve.

Behaviour beyond yield is assumed to follow the Prandtl-Reuss flow rule. During each increment of applied load or time, it is assumed that points which were initially elastic remain elastic. As a result, some points may wander past the yield surface during an increment. This is checked and corrected for at the end of each step. The incremental changes in stresses at each point where there has been wandering past yield are scaled down until the effective stress is equal to the yield stress, using the same scaling factor for each of the stress components. As a result of this adjustment, the dynamic equilibrium state obtained at the end of the step is disturbed. This is reflected in a change in the live energy function, F. If the change in F is small (i.e. less than 2%) it is assumed that dynamic equilibrium has been obtained to sufficient accuracy, and the next step is applied. Otherwise, iteration is performed.

4. RESULTS - COMPARISON WITH KNOWN BEHAVIOUR

To establish the validity of the method for describing the behaviour of plates under lateral and dynamic loading, the predictions have been compared with known theoretical and experimental results. In all the following examples, p = 1,3... implies that the original sine functions have been used for w and θ to satisfy simply supported edges.

The ability of LOBUC to predict the behaviour of beams under static lateral loads is shown in Fig. 4. Here both the elastic large deflection and the elasto-plastic large deflection behaviours of an axially constrained simply supported beam are shown. The predictions are compared with the results from Timoshenko (elastic), and Soreide et al [6] (elasto-plastic). It can be seen that using only one term in the sine series for describing the out of plane deflection gives good agreement.

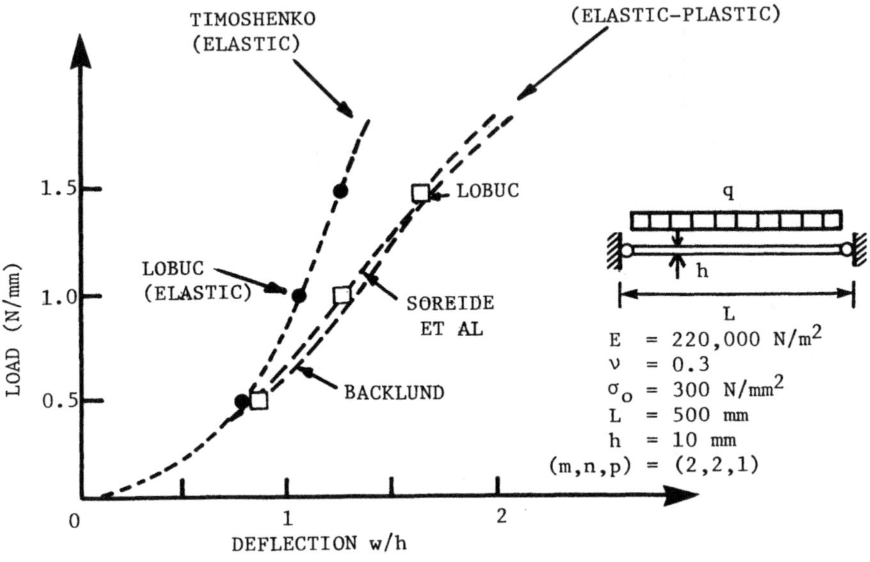

Figure 4. Comparison of LOBUC and known static beam behaviour.

The dynamic, elasto-plastic prediction of LOBUC for the behaviour of a simply supported beam under a step load is shown in Fig. 5. The predicted behaviour is in excellent agreement with that predicted by other theoretical models [19], even though a large time step and only one term in the Fourier series describing the deflected shapes have been used.

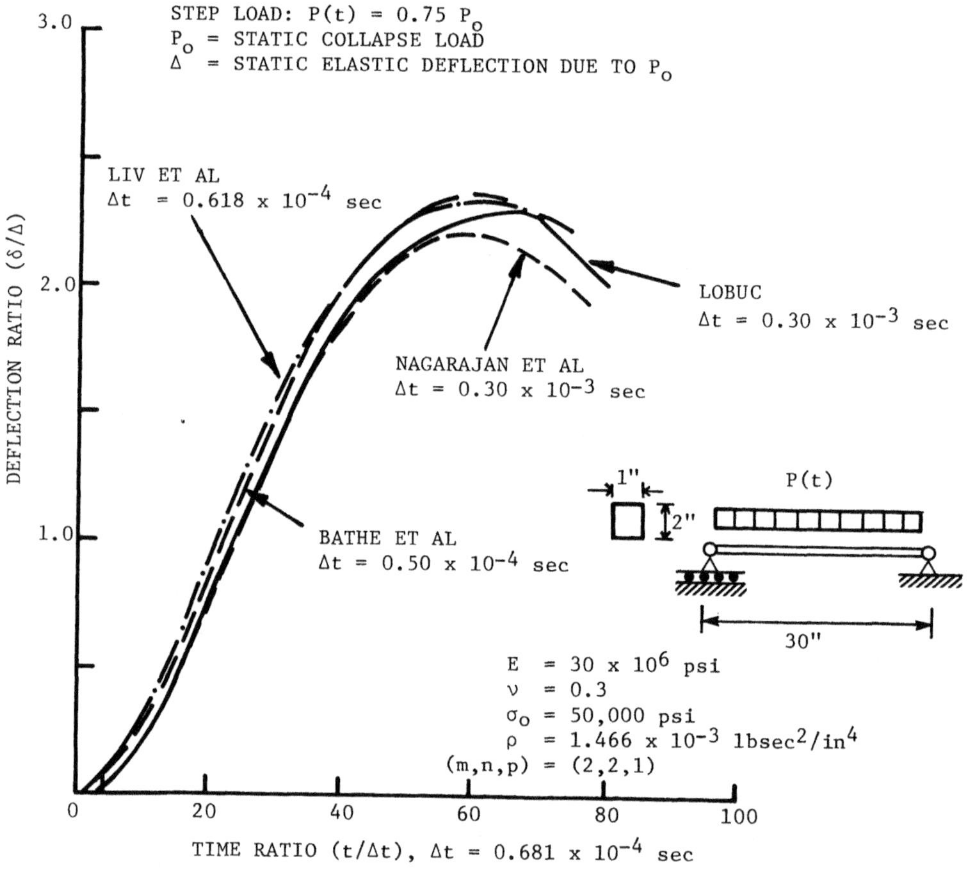

Figure 5. Comparison of LOBUC and known elasto-plastic dynamic beam response.

The elastic, small deflection behaviour of a simply supported plate under a step load is shown in Fig. 6. The predicted period and peak deflection are in close agreement with standard analytical results.

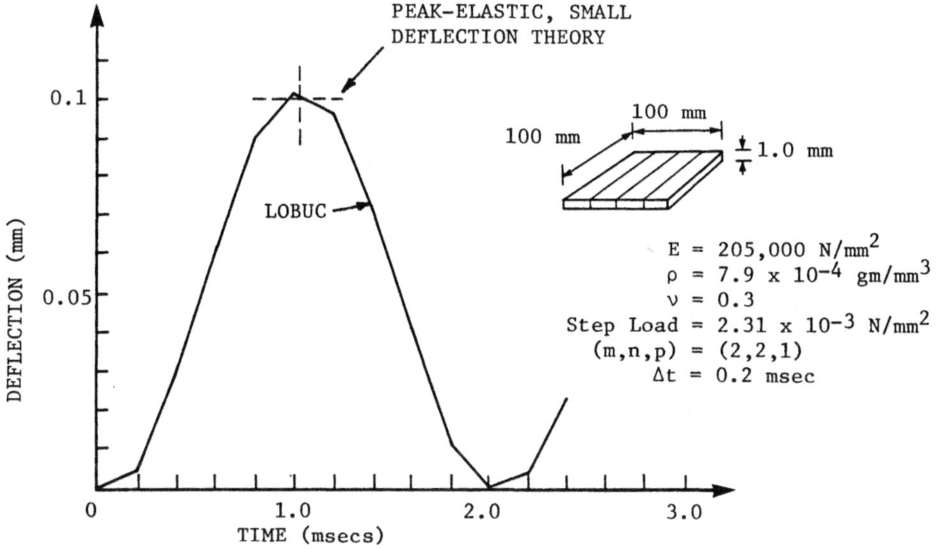

Figure 6. Comparison of LOBUC and linear, elastic plate theory.

Fig. 7 shows a comparison between the present predictions and those of PETROS, DEPROP, and ADINA [7] for a simply supported square plate with constrained edges subjected to a triangular pulse. The present program has been used with a time step 10 times larger than the other methods, and is between 6 and 30 times faster. There is close agreement between LOBUC and the other programs, except at very large deflections where LOBUC overpredicts the peak deflection by 10%. This is because LOBUC is based on moderately large deflection theory (i.e. larger than the plate thickness, but considerably smaller than the width),

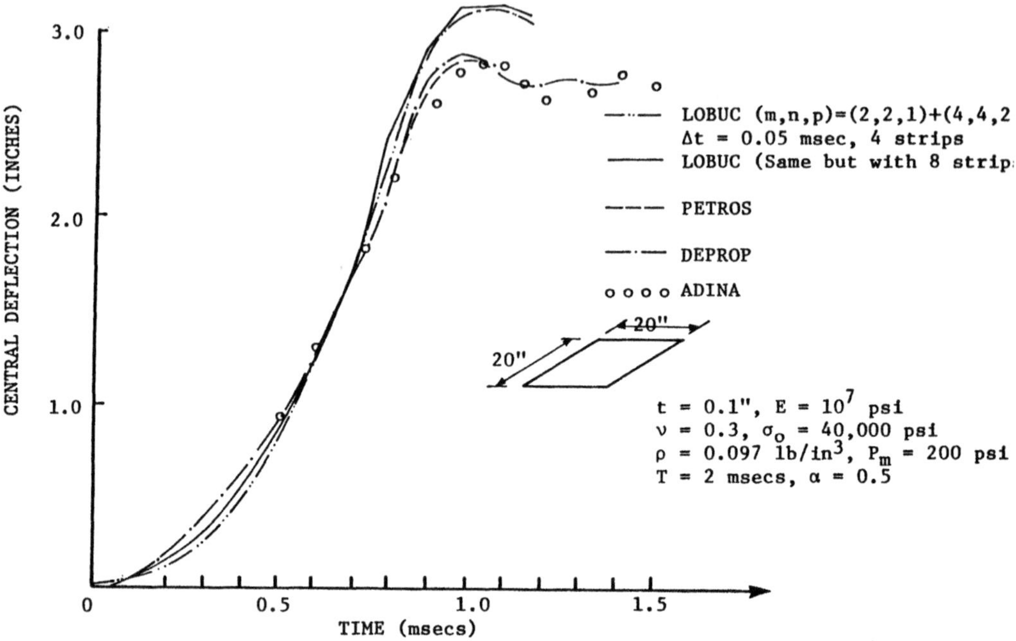

Figure 7. Comparison of LOBUC and known elasto-plastic, large deflection dynamic plate response.

whereas the other programs are based on very large deflection theory (i.e. they take into account geometry change during the deformation). It is intended that LOBUC will be modified to include this effect in the near future.

The elasto-plastic, large deflection predictions of a plate loaded by a small duration pressure pulse are shown in Fig. 8. The edges of the plate are totally constrained against in-plane motion, and simply supported. The plate resistance in this case comes almost entirely from membrane forces, explaining why the period of vibration (≈ 1 msec.) is so much shorter than the elastic, small deflection value (= 15.7 msecs.). Also shown for comparison is the final deflection measured experimentally for a similar plate [18]. The test plate material was mild steel,

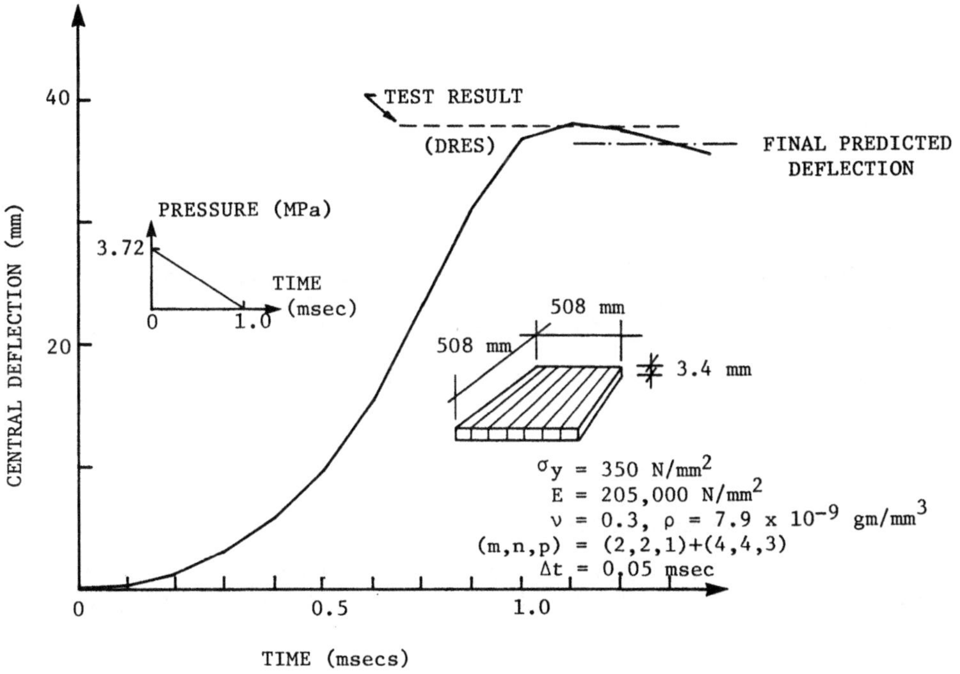

Figure 8. Dynamic response of fully constrained square plate.

with a nominal yield strength of 350 N/mm^2. It can be seen that the predicted peak and final deflections are very close but it should be noted that strain rate effects have not been included in the analysis. Further the test plate was observed to pull in significantly from its edge supports. These two effects have perhaps compensated each other.

5. CONCLUSIONS

There is a need for a full elasto-plastic, large deflection dynamic plate program that can produce results efficiently. The finite strip method is known to be efficient for static analysis, and has the versatility to analyze both stiffened and unstiffened plates. The method has now been extended to allow for dynamic loads. The predictions from the method agree closely with known theoretical and experimental results for simple beams and plates, except at very large deflections. Work is in progress to extend the method to stiffened plates as well.

6. ACKNOWLEDGEMENT

This work was supported by the Canadian Department of National Defence through a contract from the Defence Research Establishment Suffield (DRES).

7. REFERENCES

1. CHRISFIELD, M.A., "The automatic nonlinear analysis of stiffened plates and shallow shells using finite elements", Proc. Inst. of Civil Engng., Part 2, Vol. 69, Dec. 1980.

2. EKKEHARD, RAMM, "Geometrisch nichtlineare elastostatik und finite elemente", Univeristy of Stuttgart, Report 76-2, 1976.

3. FRIEZE, P.A., DOWLING, P.J. and HOBBS, R.E., "Ultimate load behaviour of plates in compression", Steel Plated Structures, London, 1976.

4. HARDING, J.E., HOBBS, P.E. and NEAL, B.G., "The elasto-plastic analysis of imperfect square plates under in-plane loading", Proc. Inst. of Civil Engng., Part 2, Vol. 63, 1977.

5. WEBB, S.E. and DOWLING, P.J., "Large deflection elasto-plastic behaviour of discretely stiffened plates", Proc. Inst. of Civil Engng., Part 2, Vol. 69, 1980.

6. SOREIDE, T.H., MOAN, T. and NORDSVE, N.T., "In the behaviour and design of stiffened plates in ultimate limit state", Journal of Ship Research, Vol. 22, No. 4, Dec. 1978.

7. STAGLIANO, T.R. and MENTE, L.J., "Large deflection, elastic-plastic dynamic structural response of beams and stiffened or unstiffened panels - A comparison of finite element, finite difference and modal solutions", Nonlinear Finite Element Analysis and Adina, Proc. ADINA Conference, 1979.

8. WU, R.W.H. and WITMER, E.A., "Nonlinear transient responses of structures by the spatial finite element method", AIAA Journal, Vol. 11, No. 8, August 1973.

9. CHEUNG, Y.K., "Finite strip method in structural analysis", Pergamon Press, 1976.

10. MOFFLIN, D.S., "A finite strip method for the collapse analysis of compressed plates and plate assemblages," University of Cambridge, Dept. of Engineering, Report CUED/D-Struct/TR101, 1983.

11. MOFFLIN D.S., "Manual for use of computer program LOBUC," University of Cambridge, Dept. of Engineering, Report CUED/D-Struct/TR105, 1984.

12. MOFFLIN, D.S., "Results obtained from a finite strip method for the ultimate compressed strength of plates and plate assemblages," University of Cambridge, Department of Civil Engineering, Report CUED/D-Struct/TR102, 1983.

13. BRADFIELD, C.D. and CHLADNY, E., "A review of the elastic-plastic analysis of steel plates loaded in in-plane compression", University of Cambridge, Department of Engineering, Report CUED/D-Struct/TR77, 1979.

14. LITTLE, G.H., "Rapid analysis of plate collapse by live energy minimization", Int. J. Mech. Sci., Vol. 19, 1974.

15. GILL, P.E. and MURRAY, W., "Quasi-Newton methods for unconstrained minimization", J. Inst. Math. and its Appl., Vol. 9, 1972.

16. NEWMARK, N.M., "A method of computation for structural dynamics", J. of Eng. Mech. Div., Proc. A.S.C.E., July, 1959.

17. RAMBERG, N. and OSGOOD, R., "Description of stress-strain curves by three parameters", NACA, Tech. Note 902, 1943.

18. RITZEL, D. HOULSTON, R. and RUDE, G., "Structural response of steel panels to air blast waves; Progress report 1", DRES Research Note 324, 1984.

19. LIU, S.C. and LIN, T.H., "Elastic-plastic dynamic analysis of structures using known elastic solutions", Earthquake Engineering and Structural Dynamics, Vol. 7, 1979.

Simplified Dynamic Analysis of Civil Engineering Structures

P. N. KASMAI, J. H. PREVOST*, J. ZARKA
Ecole Polytechnique, Palaiseau, France

Summary

We have recently introduced a method for the evaluation of the asymptotical limiting state of an inelastic structure under dynamical loadings in which therefore inertia terms are to be taken into account(J.ZARKA and as.(1985)).
The originality of this new method remains similar to that given for quasi-statical loadings (J.ZARKA and as. (1979)) i.e. only simple elastic calculus are made with "cleverly" chosen loadings and elastic constants.
Our aim ,in the present paper, will be to provide all the elements necessary for its practical application to civil engineering structures.
At first we shall give on one example the basic idea of the method.Then we shall summaryze its fundamental principles.At last we shall show, on the case of an earth dam,how to treat the seismic loading.

1.INTRODUCTION

We are concerned with the Fundamental Problem of the INELASTIC ANALYSIS OF STRUCTURES during CYCLIC STATICAL OR DYNAMICAL LOADINGS.

Attention has been given to Experimental Approaches which imply special experimental tecniques and apparatus
or Theoretical Approaches which are based on particular Constitutive Laws and drive to Analysis of Structures.
Both are very expensive and limited.

Our Main Purpose will be in the Description of a **NEW FRAMEWORK** where every point is SIMPLE , PRACTICAL and which may be easily used with the classical tools of engineers.
Indeed we shall show how to obtain the limiting States or Bounds of the Real State with only classical ELASTIC ANALYSIS

*University of Princeton

Finite Element Methods for Nonlinear Problems
Europe-US Symposium, Trondheim, Norway 1985
Editors: Bergan, Bathe, Wunderlich

2. EXAMPLE OF A SYSTEM

The system consists only of a particular assembly of on friction block (S) one mass (M) and two springs (E,h) which is classically given to illustrate Kinematical Hardening material (fig 1)

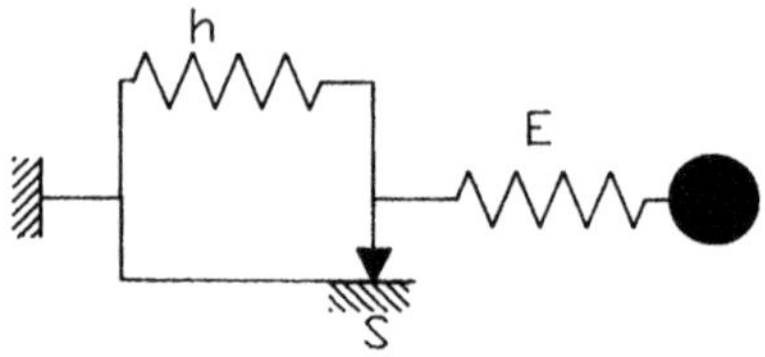

2.1. First loading

A **Constant** F suddenly applied to the extremity.
We must solve:

motion equation $-M\ddot{x} + E(x-\alpha) = F$ (2.1)

evolution eqution $-\dot{\alpha} \in \partial\psi_{C_0}(\sigma^1)$ (2.2)

i.e.

$|\sigma^1| = |E(x-\alpha) - h\alpha| \leq S$

with

if $|\sigma^1| < S$ then $\dot{\alpha} \equiv 0$

if $\sigma^1 = S$ then $\dot{\alpha} \geq 0$ (2.2)'

if $\sigma^1 = -S$ then $\dot{\alpha} \leq 0$

where σ^1 is the local stress on the friction block.

and the initial conditions

$x = x_0 \quad \alpha = \alpha_0 \quad \dot{x} = 0.$

It may be easily shown that the limiting state is an elastic shakedown state with if $F > S$

$$\boxed{\alpha_l = F/h} \tag{2.3}$$

2.2. Second loading

A **Dead** load F is at first applied quasi-statically on the extremity then an **Alternating** given dynamic motion $x_0(t)$ progressively increasing and then progressively decreasing towards 0 is given to the other extremity.

We must solve:

$$M\ddot{x} + E(x-\alpha) = F - M\ddot{x}_0''(t) \tag{2.4}$$

once again we found for the limiting state

$$\boxed{\alpha_l = F/h} \tag{2.5}$$

2.3. <u>Third loading</u>

A **Large** cyclic quasi-statical loading varying between F_{max} and F_{min} such that $F=(F_{max} + F_{min})/2$ is applied.

We get that the mean state of the plastic-shakedown state is such that

$$\boxed{\alpha_{mean} = F/h} \tag{2.6}$$

BASIC IDEA :

Given a structure subjected to a dead load,if a large dynamical symmetrical loading around 0 is superimposed on that dead load;then , the limiting dynamical state can be evaluated by subjecting the structure to a large radial quasi-statical cyclic loading alternating around the dead load.

This has only a meaning if a simplified method is available to obtain this mean limiting state.Such a method was proposed several years ago .We shall just give its principle elements.

3. REVIEW of the FRAMEWORK

3.1. General behaviours

Only particular materials obeying to the following description can be treated in this framework

3.1.1 Elementary examples

$$-S_0 \leq \sigma^1 \leq S_0 : \sigma^1 \in C_0$$

$$\dot{\alpha} \in \partial\Psi_{C_0}(\sigma^1)$$

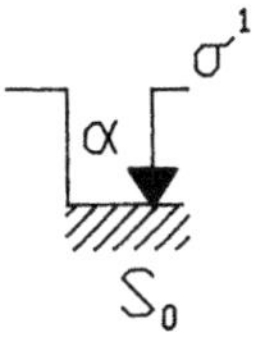

The Friction block is the symbol for the **<u>Perfect Plasticity</u>**

$$\dot{\beta} = \frac{\sigma^2}{\eta}$$

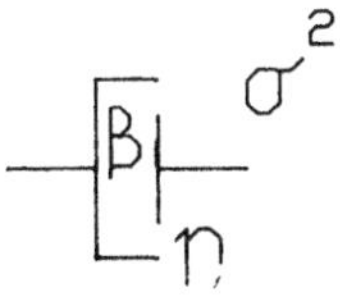

The Dashpot is the symbol for the linear **Viscoelasticity** (no Threshold)

$\dot{\gamma} \equiv 0$ iff $\sigma^3 \in D_0$

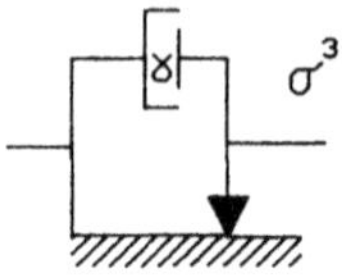

$$\dot{\gamma} = \frac{1}{\eta}\left(\langle \sigma^3 - S_0 \rangle + \rangle \sigma^3 + S_0 \langle \right)$$

This assembly is the symbol for the perfect **Viscoplasticity**

3.1.2 General constitutive laws

We assume that we may define for the representative volume element:

i)

$$\underline{X}^t = \begin{bmatrix} \underline{\alpha} \\ \underline{\beta} \\ \underline{\gamma} \end{bmatrix} \tag{3.1}$$

called the **INTERNAL PARAMETERS** and linked to the local inelastic strains of the inelastic mechanisms;

ii)it may be proved that ,when these mechanisms are set into a linear elastic matrix,the local stresses at the level of each mechanism are expressed by:

$$\begin{bmatrix} \underline{\sigma}^{(1)} \\ \underline{\sigma}^{(2)} \\ \underline{\sigma}^{(3)} \end{bmatrix} \equiv \begin{bmatrix} \underline{A}^{(1)} \\ \underline{A}^{(2)} \\ \underline{A}^{(3)} \end{bmatrix} \sigma - \begin{bmatrix} \hat{\underline{\alpha}} \\ \hat{\underline{\beta}} \\ \hat{\underline{\gamma}} \end{bmatrix} \tag{3.2}$$

where σ is the applied global stress tensor;

iii)and that the **tranformed internal parameters** are given by:

$$\hat{\underline{X}} \equiv \begin{bmatrix} \hat{\underline{\alpha}} \\ \hat{\underline{\beta}} \\ \hat{\underline{\gamma}} \end{bmatrix} \equiv \mathbf{B} \begin{bmatrix} \underline{\alpha} \\ \underline{\beta} \\ \underline{\gamma} \end{bmatrix} \equiv \mathbf{B}\,\underline{X} \tag{3.3}$$

where **B** is a non-negative symmetrical matrix.

$$\mathbf{B} \equiv \begin{bmatrix} \mathbf{B}^{(1)} & \mathbf{C}^t & \mathbf{D}^t \\ \mathbf{C} & \mathbf{B}^{(2)} & \mathbf{E}^t \\ \mathbf{D} & \mathbf{E} & \mathbf{B}^{(3)} \end{bmatrix}$$

two cases are possible:

---$\mathbf{B}$ is strictly positive ,there is one to one relation

$\underline{\hat{\mathbf{X}}} = \mathbf{B}\,\underline{\mathbf{X}} \iff \underline{\mathbf{X}} = \mathbf{B}^{(-1)}\underline{\hat{\mathbf{X}}}$

---$\mathbf{B}$ is singular ,$\underline{\hat{\mathbf{X}}}$ must belong to $\hat{\mathbf{L}}$,subspace of compatibility and for such a compatible $\underline{\hat{\mathbf{X}}}$ there may be an infinite number of $\underline{\mathbf{X}}$ satisfying $\underline{\hat{\mathbf{X}}} = \mathbf{B}\,\underline{\mathbf{X}}$;

iv)as a generalisation of the uniaxial behaviours in 3.1.1 the Evolution Laws for the inelastic mechanisms will be:

1) Instantaneous Plasticity

since $\sigma^1 \in C_0$ we have $\underline{\alpha} \in \hat{C}(\sigma)$ where

$$\boxed{\hat{C}(\sigma) = \hat{C}(0) + A^1\, \sigma} \tag{3.4}$$

which is a Locally built Convex ,we may write

$$\underline{\dot{\alpha}} \in \partial\Psi_{C_0}(\sigma^1) \implies \tag{3.5}$$

$$\underline{\dot{\alpha}} \in -\partial\Psi_{\hat{C}(\sigma)}(\underline{\hat{\alpha}}) \tag{3.6}$$

2) Viscosity without threshold

$$\dot{\beta} = \frac{\partial w}{\partial \sigma^{(2)}} \tag{3.7}$$

with $\dot{\beta} \equiv 0$ iff $\sigma^2 \equiv 0$

$\implies$

3) Viscoplasticity

$$\dot{Y} = \frac{\partial \Omega}{\partial \sigma^3} \qquad (3.8)$$

with $\dot{Y} \equiv 0$ iff $\sigma^3 \in D_0$

===>

$$\dot{\underline{Y}} = -\frac{\partial \underline{\underline{\Omega}}}{\partial \hat{\underline{Y}}} (\underline{A}^3 \ \sigma - \hat{\underline{Y}}) \qquad (3.8')$$

and $\dot{Y} \equiv 0$ Iff $\hat{\underline{Y}} \in \hat{D}(\sigma) = \hat{D}(0) + \underline{A}^3 \ \sigma$

v)the Global inelastic strain for the volume element is equal to:

$$\dot{\epsilon}^P \equiv \underline{A}^{1t} \ \dot{\underline{\alpha}} + \underline{A}^{2t} \ \dot{\underline{\beta}} + \underline{A}^{3t} \ \dot{\underline{Y}} \equiv \underline{A}^t \ \dot{\underline{X}} \qquad (3.9)$$

vi)the Global elastic strain is defined by

$$\dot{\epsilon}^e \equiv M \ \dot{\sigma} \qquad (3.10)$$

where M is the global elastic matrix.

Indeed the WORKHARDENING shown by the volume element ,similarly as in the overlay model,results of a Redistribution of the local Residual stresses.
It was shown that very efficient bounds could be obtained by defining the **statical** and the **dynamical** behaviours where respectively $\dot{\epsilon} \simeq 0$ and $\dot{\epsilon} \simeq \infty$ and where the physical time occurs no more.

3.2. GLOBAL EVOLUTION FOR THE STRUCTURE

3.2.1.Loading

Surface forces
---$\underline{F}_i^d$ (t) on ∂_{F_i} V
Body forces
---$\vec{\underline{X}}^d$ (t) in V
Initial strain
---$\underline{\epsilon}^I$ (t) in V
Displacement
---$\underline{u}_j^d$ (t) on ∂_{u_j} V

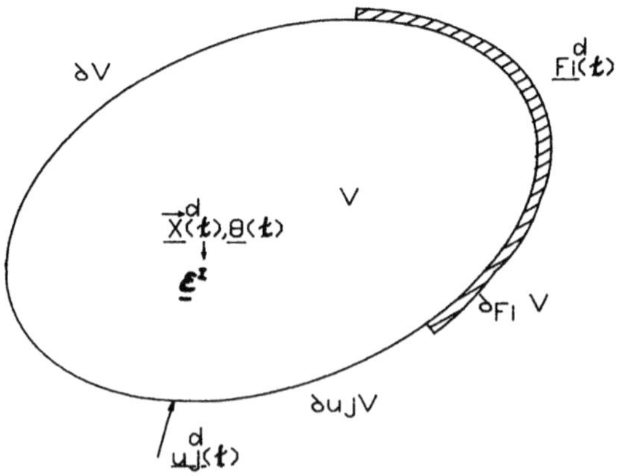

3.2.2. Elastic Response of the Structure

The Routine **ELAS** which may be based upon any analytical or numerical tool with the arguments:

$$\mathbf{ELAS}(V, \partial_{F_i} V, \partial_{u_j} V, \vec{\underline{X}}^d, \underline{F}_i^d, \underline{u}_j^d, \underline{\epsilon}^I, \underline{\mathbf{M}}) \tag{3.11$_a$}$$

will give the various classical fields

$\vec{\underline{u}}^{el}(t), \underline{\epsilon}^{el}(t)$ are Kinematically Admissible K.A. with $\underline{u}_j^d(t)$ on $\partial_{u_j} V$ (3.11)$_b$

and

$$\underline{\sigma}^{el}(t) = \underline{L}(\underline{\epsilon}^{el}(t) - \underline{\epsilon}^I(t)) \tag{3.11$_c$}$$

or

$$\underline{\epsilon}^{el}(t) = \underline{\mathbf{M}}\underline{\sigma}^{el}(t) + \underline{\epsilon}^I(t) \tag{3.11$_d$}$$

is Statically Admissible S.A. with $\vec{\underline{X}}^d(t)$ in V and $\underline{F}_i^d(t)$ on $\partial_{F_i} V$

3.2.3. Real Response of the Structure

$\vec{\underline{u}}(t), \underline{\epsilon}(t)$ real displacement and strain fields are K.A. with $\underline{u}_j^d(t)$ on $\partial_{u_j} V$

$\underline{\sigma}(t)$ real stress field $\equiv \underline{L}(\underline{\epsilon}(t) - \underline{\epsilon}^I(t) - \underline{\epsilon}^P(t))$ is S.A. with $\vec{\underline{X}}^d(t)$ in V and $\underline{F}_i^d(t)$ on $\partial_{F_i} V$

this implies that we may split the fieds into:

$$\vec{\underline{u}}(t) \equiv \vec{\underline{u}}^{el}(t) + \vec{\underline{u}}^{ine}(t)$$

$$\underline{\epsilon}(t) \equiv \vec{\underline{\epsilon}}^{el}(t) + \vec{\underline{\epsilon}}^{ine}(t) \tag{3.12$_a$}$$

$$\underline{\sigma}(t) \equiv \underline{\sigma}^{el}(t) + \underline{\rho}^{ine}(t)$$

with

$\vec{\underline{u}}^{ine}(t), \underline{\epsilon}^{ine}(t) = \underline{\mathbf{M}}\underline{\rho}(t) + \underline{\epsilon}^P(t)$ K.A. with 0 on $\partial_{u_j} V$ (3.12)$_b$

$\underline{\rho}$ S.A. with 0 in V and 0 on $\partial_{F_i} V$ (3.12)$_c$

or symbolically

$$\underline{\rho} \equiv \underline{\mathbf{B}}_0 \, \underline{\epsilon}^P$$

3.3. Kinematical Hardening Material

This material is very often introduced and moreover is the simplest one to illustrate the particular trick of our framework.

3.3.1

Classically it is assumed for

i)the Yield criterion:

$f(\boldsymbol{\sigma} - \hat{\boldsymbol{\alpha}}) \leq 0 \Longrightarrow$
$C(\hat{\boldsymbol{\alpha}})$ in $\boldsymbol{\sigma}$ space
or
$C(\boldsymbol{\sigma})$ in $\hat{\boldsymbol{\alpha}}$ space

and allmost always the MISES criterionis taken $f(\boldsymbol{\sigma} - \hat{\boldsymbol{\alpha}}) \equiv f(s - \hat{\boldsymbol{\alpha}}) \equiv$

$$\frac{1}{2}(s - \hat{\boldsymbol{\alpha}})^t (s - \hat{\boldsymbol{\alpha}}) - \sigma_0^2 \leq 0 \qquad (3.13)$$

with $s \equiv \mathrm{dev}\boldsymbol{\sigma}$
and

$$\hat{\boldsymbol{\alpha}} \equiv C\, \boldsymbol{\epsilon}^{\mathbf{P}} \iff \boldsymbol{\epsilon}^{\mathbf{P}} \equiv C^{-1}\, \hat{\boldsymbol{\alpha}} \qquad (3.14)$$

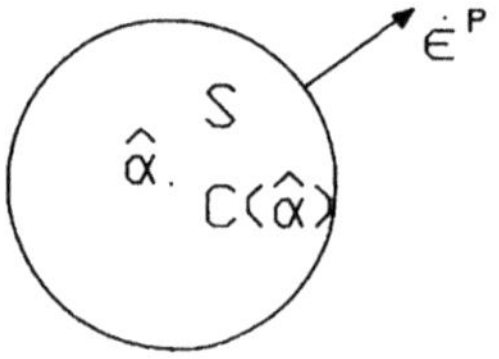

$C \equiv$ HARDENING Modulus
$\sigma_0 \equiv$ shear elastic limit

ii)and the Flow Rules:(normality rule)

$$\dot{\boldsymbol{\epsilon}}^{\mathbf{P}} \in \partial\Psi_{C(\hat{\boldsymbol{\alpha}})}(\boldsymbol{\sigma}) \qquad (3.15)$$

external normal to $C(\hat{\boldsymbol{\alpha}})$ in $\boldsymbol{\sigma}$

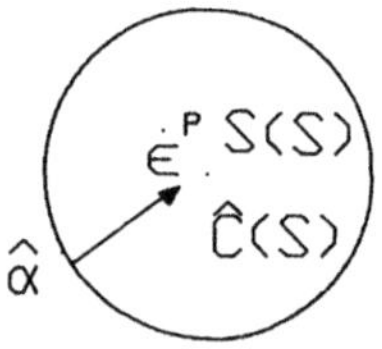

or

$$\dot{\boldsymbol{\epsilon}}^{\mathbf{P}} \in -\partial\Psi_{C(\boldsymbol{\sigma})}(\hat{\boldsymbol{\alpha}}) \text{ internal normal to } C(\boldsymbol{\sigma}) \text{ in } \hat{\boldsymbol{\alpha}} \qquad (3.15)'$$

3.3.2 New writing

$\hat{\boldsymbol{\alpha}}$ are the transformed Internal Parameters

we shall write

$$f(s - \hat{\boldsymbol{\alpha}}) \equiv f(s^{el} + \mathrm{dev}p - \hat{\boldsymbol{\alpha}}) \equiv f(s^{el} - \hat{\boldsymbol{\alpha}}_{\mathbf{y}}) \qquad (3.16)$$

with

$$\boxed{\hat{\boldsymbol{\alpha}}_{\mathbf{y}} \equiv \hat{\boldsymbol{\alpha}} - \mathrm{dev}p} \qquad (3.17)$$

$\hat{\boldsymbol{\alpha}}_{\mathbf{y}}$ are the Transformed Parameters **for the structure.**

$\Longrightarrow$

$\dot{\epsilon}^p \in -\partial\Psi_{\hat{C}(\sigma^{el})}(\hat{\alpha}_y)$ internal normal to $\hat{C}(\sigma^{el})$ in $\hat{\alpha}_y$ (3.18)$_a$

$\hat{C}(\sigma^{el}) \equiv \hat{C}(0) + s^{el}$ (3.18)$_b$

The Routine **ELAS**($V, \partial_{F_i} V, \partial_{u_j} V, \vec{\underline{0}}, \underline{0}, \underline{0}, \underline{\epsilon}^p = \underline{C}^{-1}\underline{\alpha}, \underline{M}$) will give:

$\vec{\underline{u}}^{ine} \Longrightarrow \underline{\epsilon}^{ine}$ and thus

$\underline{p} = \underline{L}(\underline{\epsilon}^{ine} - \underline{\epsilon}^p)$

and

$\hat{\underline{\alpha}}_y \equiv \hat{\underline{\alpha}} - \mathrm{dev}\underline{p}$

there is an univocal relation from $\hat{\underline{\alpha}}$ to $\hat{\underline{\alpha}}_y$

Conversely to any $\hat{\underline{\alpha}}_y$ since $\underline{\epsilon}^{ine} = \underline{M}\underline{p} + \underline{\epsilon}^p =$

$\underline{\epsilon}^{ine} = \underline{M}\underline{p} + \underline{C}^{-1}(\hat{\underline{\alpha}}_y + \mathrm{dev}\underline{p}) = (\underline{M} + \underline{C}^{-1})\underline{p} + \underline{C}^{-1}\hat{\underline{\alpha}}_y$

$\underline{\epsilon}^{ine} = \hat{\underline{M}}\underline{p} + \underline{C}^{-1}\hat{\underline{\alpha}}$ with $\hat{\underline{M}} \equiv \underline{M} + C^{-1}\mathrm{dev}$ (3.19)

This equation is reduced ,for isotropic elasticity ,to a simple Change (E, ν) into ($\hat{E}, \hat{\nu}$) with

$$\hat{E} = \frac{Eh}{E+h}, \quad \frac{\hat{\nu}}{\hat{E}} = \left(\frac{\nu}{E} + \frac{.5}{h}\right) \text{ and } h = \frac{3}{2}C \tag{3.20}$$

We deduce that the Routine **ELAS**($V, \partial_{F_i} V, \partial_{u_j} V, \vec{\underline{0}}, \underline{0}, \underline{0}, \underline{C}^{-1}\hat{\underline{\alpha}}_y, \hat{\underline{M}}$) will give

$\vec{\underline{u}}^{ine} \Longrightarrow \underline{\epsilon}^{ine}$ and thus

$\underline{p} = \hat{\underline{L}}(\underline{\epsilon}^{ine} - \underline{C}^{-1}\hat{\underline{\alpha}}_y)$

and then

$\hat{\underline{\alpha}} \equiv \hat{\underline{\alpha}}_y + \mathrm{dev}\underline{p}$

that means an univocal relation from $\hat{\underline{\alpha}}_y$ to $\hat{\underline{\alpha}}$

3.3.3.Global Evolution for the structure

the $\hat{\underline{\alpha}}_y$ are the Main Unknowns

i) $\hat{\underline{\alpha}}_y \in \hat{\underline{C}}(\underline{\sigma}^{el}(t))$ which is LOCALLY BUILT once known the elastic response of the structure.

ii) $\underline{\dot{\epsilon}}^p \in -\partial\Psi_{\hat{\underline{C}}(\underline{\sigma}^{el})}(\hat{\underline{\alpha}}_y)$ internal normal to $\hat{\underline{C}}(\underline{\sigma}^{el})$ in $\hat{\underline{\alpha}}_y$

iii)

$$\dot{\hat{\underline{\alpha}}}_y \in -\left[\underline{C} - \mathrm{dev}\,\underline{B}_0\right]\partial\Psi_{\hat{\underline{C}}(\underline{\sigma}^{el})}(\hat{\underline{\alpha}}_y(t))$$

3.3.4 PRACTICAL RULES during Radial Cyclic Loading

3.3.4.1 Basic Principles

the first stage is to to obtain

1) $\underline{\sigma}^{el}(t)$: Elastic Response of the structure which will be expressed for a radial cyclic loading by:

$$\underline{\sigma}^{el}(t) = \lambda(t)\underline{\sigma}^{el}_{max} + (1-\lambda(t))\,\underline{\sigma}^{el}_{min} \qquad (3.22)$$

with $\lambda(t)$ varying between 0 and 1 then 1 and 0

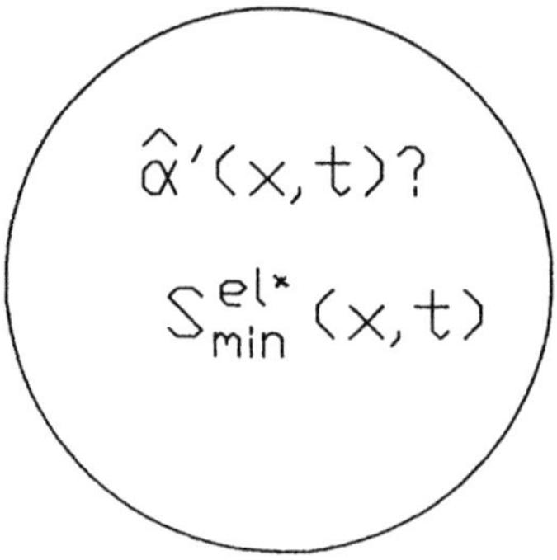

2) then locally in each point x we build:

$\hat{C}(S^{el}(x,t))$ necessarily $\hat{\underline{\alpha}}_y \in \hat{C}(\underline{S}^{el}(t))$

3) We look at the intersection of two extrem convex sets and we conclude that:

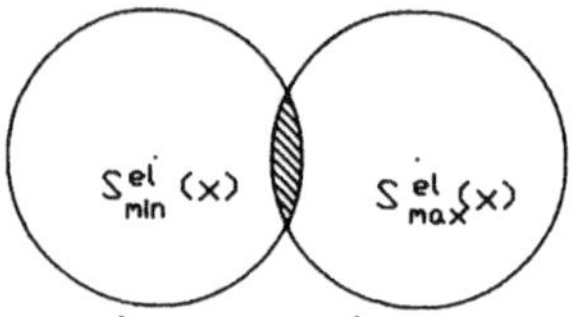

ii) if $\forall\, x$, $\hat{C}_1(x) = \hat{C}(S^{el}_{min}(x)) \cap \hat{C}(S^{el}_{max}(x)) \neq 0$ there will be **ELASTIC SHAKEDOWN**

ii) If in at least one x the intersection $\hat{C}(S^{el}_{min}(x)) \cap \hat{C}(S^{el}_{max}(x))$ is void ,there will be **PLASTIC SHAKEDOWN**(we recall that there is never Ratchetting)

4) The problem is to find for each region a "good" evaluation of $\hat{\underline{\alpha}}_{y_1}(t)$ for t going towards ∞ and to come back to the classical fields $\underline{p}_1$, $\underline{\epsilon}^p_1$.

Indeed in our method only an elastic tool based for example on the finite element method is needed . It is used with a special processor which takescare of the good choice of $\hat{\underline{\alpha}}_y$.

3.3.4.2 ELASTIC SHAKEDOWN

Various method are possible

a) 1st Method:

i) We perform the inelastic (incremental) analysis of the 1st half cycle

$\Rightarrow \hat{\underline{\alpha}}_y(1) \qquad \lambda_0 \nearrow 1$

with the initial conditions $\hat{\underline{\alpha}}_{y_0}$

ii) We take the Normal projection on $\hat{C}_1$ (this is a LOCAL Projection)

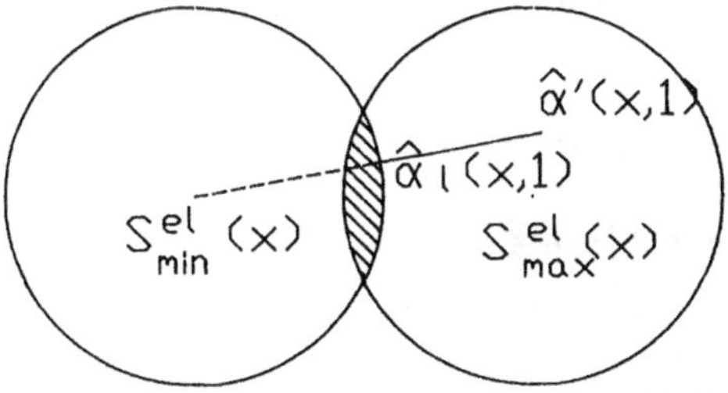

iii) $\Longrightarrow \underline{\alpha}_1 \ , \underline{\epsilon}_1^p \ , \underline{p}_1$

b) 2nd Method

i) From $\hat{\underline{\alpha}}_{y_0}$ (or $\underline{\epsilon}^p_0$) we define the volumes

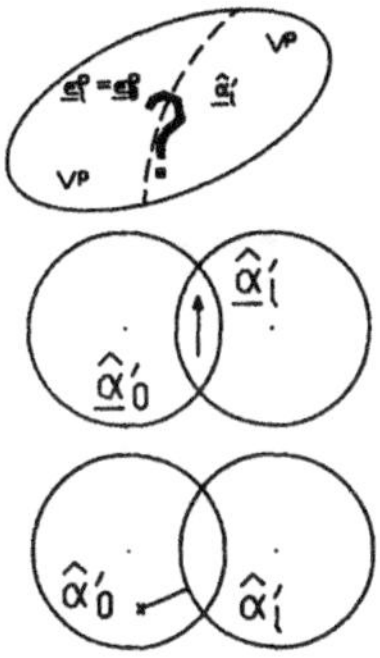

V^e where $\underline{\Delta\epsilon}^{ine} = \underline{M}\Delta\underline{p} + \underline{\theta}$

V^p where $\underline{\Delta\epsilon}^{ine} = (\underline{M} + \underline{C}^{-1}dev)\underline{\Delta p} + \underline{C}^{-1}(\hat{\underline{\alpha}}_{y_1} - \hat{\underline{\alpha}}_{y_0})$

ii) We Iterate untill a reasonnable convergency is reached for these volumes.

$$\text{ELAS}\left(V, \partial_{F_i} V, \partial_{u_j} V, \vec{\underline{0}}, \underline{0}, \underline{0}, \frac{\underline{M}}{V^e}, \frac{\underline{M}+\underline{C}^{-1}dev}{V^p}, \frac{\theta}{V^e}, \frac{\underline{C}^{-1}(\hat{\underline{\alpha}}'_1 - \hat{\underline{\alpha}}'_0)}{V^p}\right)$$

this implies at each iteration a variable rigidity matrix

c) 3rd Method

i) from any given $\hat{\underline{\alpha}}_{y0}$ we reach a state in $\hat{C}\,(S^{el}_{max})$ with the iterations

$$- \epsilon^{p}_{n+1} = \epsilon^{p}_{n} + \langle D_n - \sigma_0 \rangle \frac{S^{el}_{max} - \hat{\alpha}_{yn}}{(2\mu + C) D_n}$$

$$\text{with } D_n = \sqrt{\frac{1}{2} (S^{el}_{max} - \hat{\alpha}_n)^T (S^{el}_{max} - \hat{\alpha}_n)}$$

$\text{ELAS}(V, \partial_{F_i} V, \partial_{u_j} V, \underline{\vec{0}}, \underline{0}, \underline{0}, \underline{M}\epsilon^{p}_{n+1})$

after convergency

ii) we take the normal projection $\hat{\underline{\alpha}}_{y1}$ on $\hat{C}_1$.

iii) and we come back to $\underline{\epsilon}^{p}_{1}$ with $\hat{K}$

3.3.4.3 PLASTIC SHAKEDOWN

a) Preliminary remark

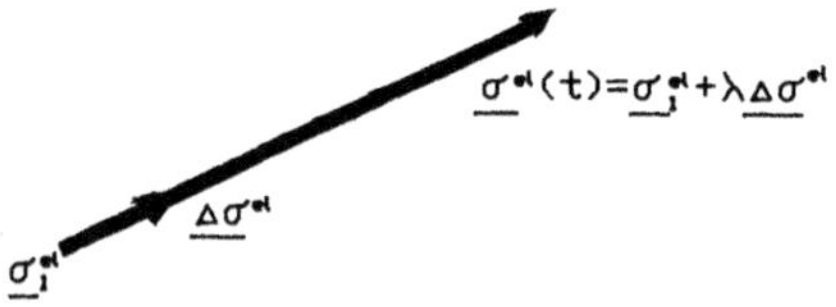

A Stationnary solution is reached during this large radial loading along the direction such that $\hat{\underline{\alpha}}_y = \hat{\underline{\Delta\alpha}}_y = \underline{\Delta S}^{el}$

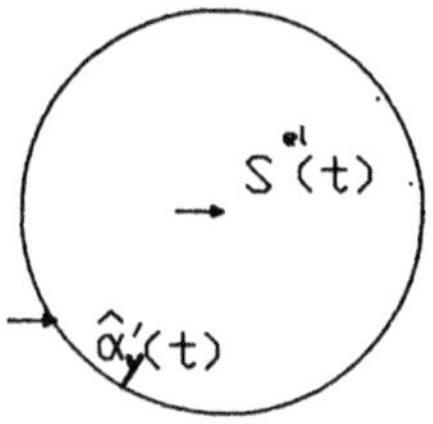

$\dot{\hat{\underline{\alpha}}} = B_0^{-1} \dot{\hat{\underline{\alpha}}}_y$ may thus be obtained with the help of the routine

$\text{ELAS}(V, \partial_{F_i} V, \partial_{u_j} V, \underline{\vec{0}}, \underline{0}, \underline{0}, \underline{\hat{M}}\, C^{-1} \dot{\hat{\underline{\alpha}}}_y)$

$\Rightarrow \vec{\underline{u}}, \dot{\underline{p}} \Rightarrow \dot{\underline{\alpha}} \Leftrightarrow \dot{\underline{\epsilon}}^p$

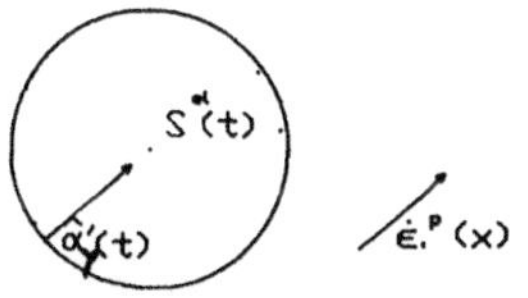

once $\dot{\underline{\alpha}}$ **or** $\dot{\underline{\epsilon}}^p$ known we deduce the relative position of $\hat{\underline{\alpha}}_y$.
The Practical rules are evident and for example the mean limiting state will correspond to :

$$\hat{\underline{\alpha}}_{y_{mean}} = \frac{\underline{S}^{el}_{min} + \underline{S}^{el}_{max}}{2}$$

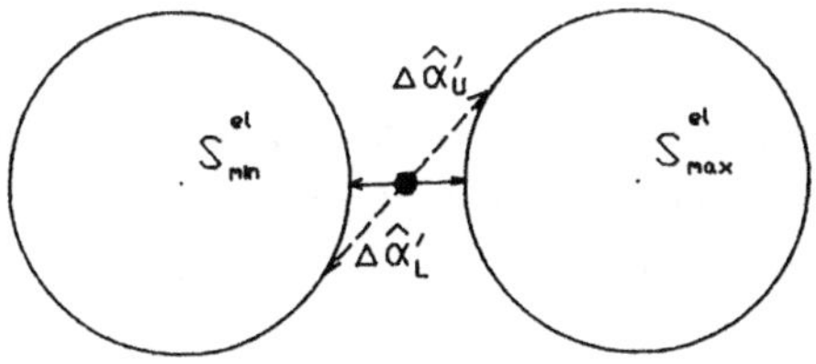

4. **PARTICULAR APPLICATIONS TO A SEISMIC LOADING**

4.1.Practical rules for the limiting state

Referring to the example given in the paragraphe 2. and our basic idea,we shall say that the loading on our structure is splitted into two parts:

i) a dead loading resulting from the weights,the constant loads and pressures of various sources(wind,water....) and so on..

ii) a dynamical seismic loading which is very important and that we don't want to know precisely and which will be simulated by a large cyclic quasi-statical loading.

Everything will be as if the total loading is cyclic around the previous dead loading .
From the rules given in the paragraphe 3. and more particularly in 3.3.4.3.,we propose to perform:

a) a first elastic analysis of the structure associated to the dead loading;

we apply

ELAS$(V, \partial_{F_i} V, \partial_{u_j} V, \vec{X}_{dead}, F_{idead}, u_{jdead}, \underline{\epsilon}^I{}_{dead}, \underline{M})$

and we deduce the fields $\vec{\underline{u}}^{el}{}_{dead}, \underline{\epsilon}^{el}{}_{dead}$ and $\underline{\sigma}^{el}{}_{dead}$

b) a second elastic analysis with the transformed parameters field
$\underline{\hat{a}}_y = \underline{\underline{S}}^{el}{}_{dead}$
and the transformed matrix $\hat{K}$.
That means we apply
ELAS$(V, \partial_{F_i} V, \partial_{u_j} V, \vec{\underline{0}}, \underline{0}, \underline{0}, \underline{C}^{-1}\underline{\hat{a}}_y, \underline{\hat{M}})$

to obtain the evaluation of the ultimate limiting state.

In a following paper(CISM workshop on Simplified analysis of structures, Ecole Polytechnique,Palaiseau October 1985)other rules to obtain extrema values of the various fields,associated to one given particular seismic excitation,will be given.

4.2EXAMPLE OF AN EARTH DAM

Here we shall only consider the case of a 2D earth dam.(other more complex examples have been solved too).
We have used the Finite Element code DYNAFLOW to perform the incremental dynamical inelastic analysis and we have added elementary subroutines to allow the step 4.1.b).
The geometry of this 2D dam is shown in the figure 1.

The material properties are defined by:
The elastic constants:Young modulus E=200000 MPA
Poisson ratio ν=0.3
The plastic constants:hardening modulus C=50000 MPA
elastic limit σ_0=1800 MPA
The loading is defined by:
i)the dead loading linked to the weight:
density 5.55 T/M3
ground acceleration 9.81 M.SEC-2

ii)the dynamical excitation on the basis of the dam given in the figure 2.

Using the spatial discretisation as shown in the figure 1,we have obtained its complete incremental response.
Some particular components of the plastic strains field are shown in the figures 3 and 4.

With our simplified method we have obtained an evaluation of the limiting state straightforwardly.
In the tables 1 and 2 ,it is possible to see a very nice comparaison of the two analysis.

5.REFERENCES

A.M.ELGAMAL,A.M.ABDEL-GHAFFAR,J.H.PREVOT (1984)"Nonlinear earthquake response analysis of earth dams".**Report n° 84-SM-14** Dept of Civil Engineering Princeton.NJ 08546 (USA0)

J.ZARKA,J.CASIER (1979)"Elastic plastic response of a structure to cyclic loading;Practical rules".**Mechanics today** Vol 6,S.NEMAT-NASSER editor, Pergamon Press.

J.ZARKA,G.INGLEBERT (1980)"Sur une nouvelle analyse simple des structures inelastiques".**Cours a l'Universite de Calabre**,Cosenza Italie.

J.ZARKA,K.NAVIDI (1985)"Simplified dynamical analysis of inelastic structures ".**8th SMIRT** Conference B 5/1* Bruxelles.

TABLE 1

Node	Dof	Static Displacement Dead Load	Step by Step Displacement Limiting State	Simplified Inelastic Displacement
11	2	- 1.245	- 2.312	- 1.925
18	2	- 2.327	- 3.190	- 3.940
25	2	- 3.206	- 7.300	- 5.950
29	2	- 4.254	- 10.00	- 8.550
31	2	- 5.354	- 11.500	- 11.310

1E-2 times displacements

TABLE 2

Element	Component	Static Strain Dead Load	Step by Step Strain Limiting State	Simplified Inelastic Strain
4	yy (vert.)	- 3.530	- 4.500	- 5.370
4	xx (horiz.)	0.705	0.950	0.805
16	yy	- 2.560	- 5.300	- 5.600
16	xx	0.630	3.100	3.170

1E-3 times strains

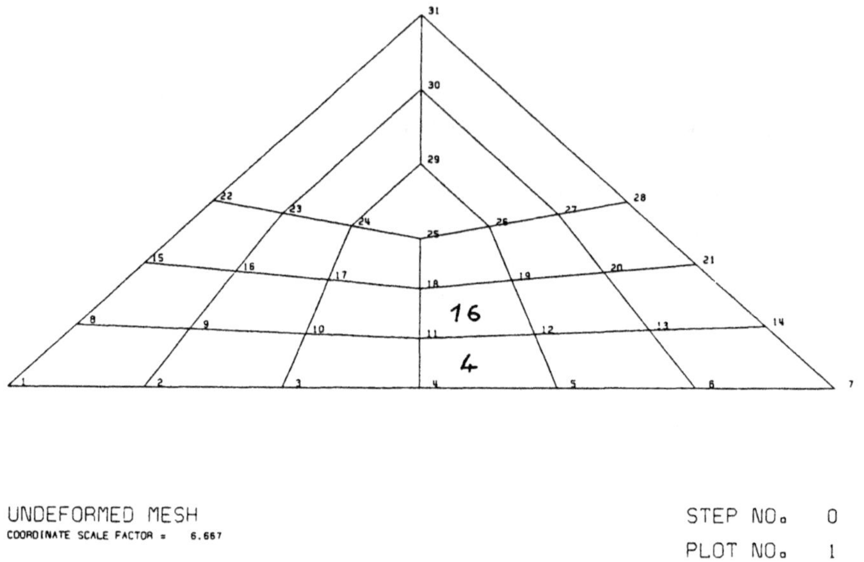
1
2
3
4
5
6
7
8
9
10
11
12
13
14
15
16
17
18
19
20
21
22
23
24
25
26
27
28
29
30
31
16
4
UNDEFORMED MESH
COORDINATE SCALE FACTOR = 6.667
STEP NO. 0
PLOT NO. 1

Fig.1

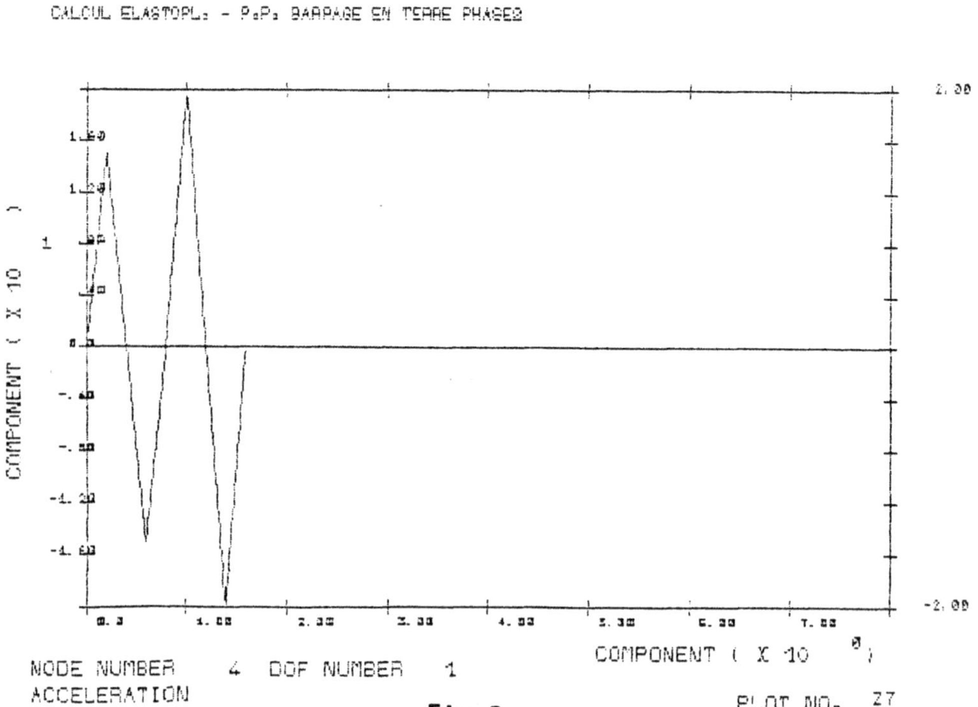
CALCUL ELASTOPL. - P.P. BARRAGE EN TERRE PHASE2
COMPONENT (X 10 1)
2.00
-2.00
COMPONENT (X 10 0)
NODE NUMBER 4 DOF NUMBER 1
ACCELERATION
PLOT NO. 27

Fig.2

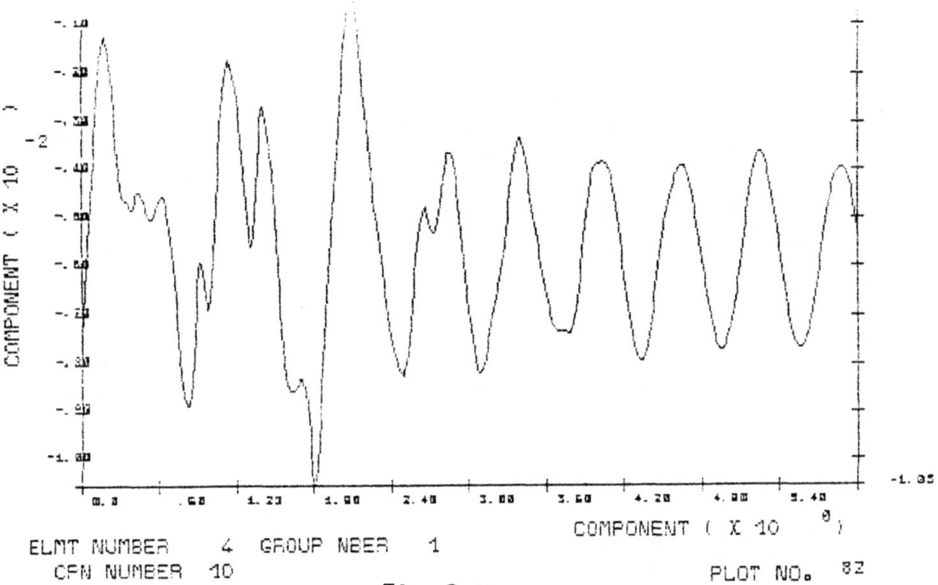

Fig.3a ε_{yy}

CALCUL ELASTOPL. - P.P. BARRAGE EN TERRE PHASE2

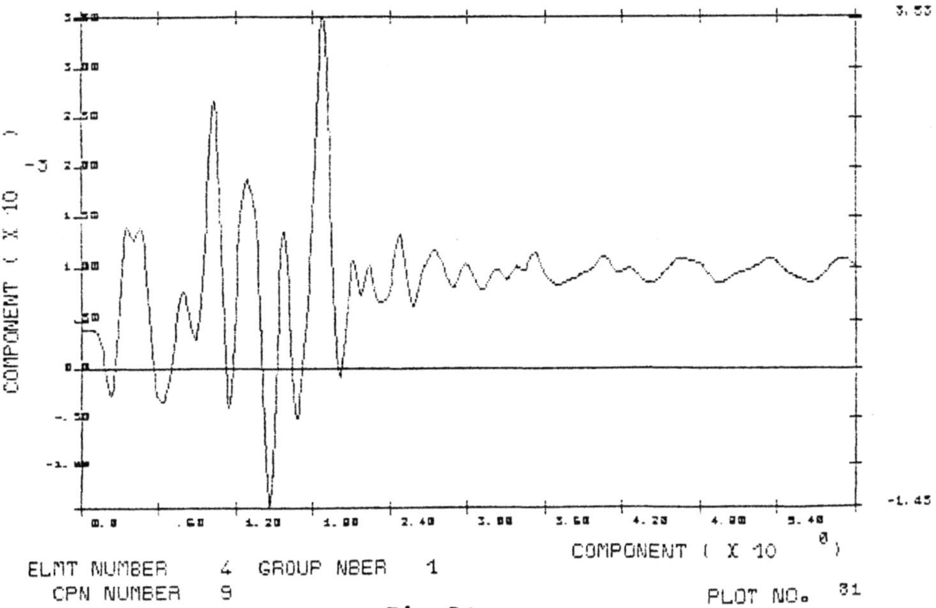

Fig.3b ε_{xx}

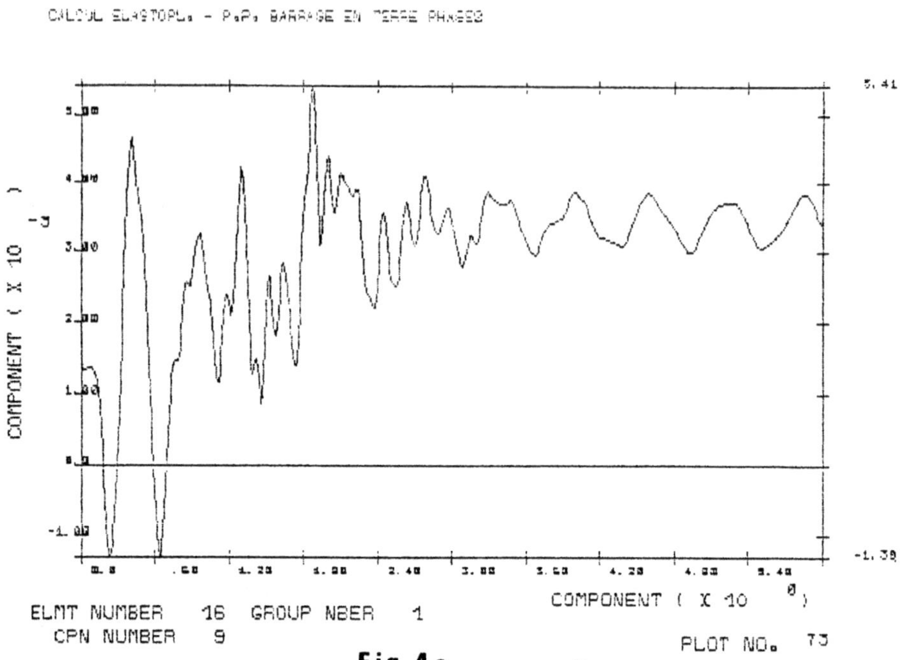

Fig.4a ε_{yy}

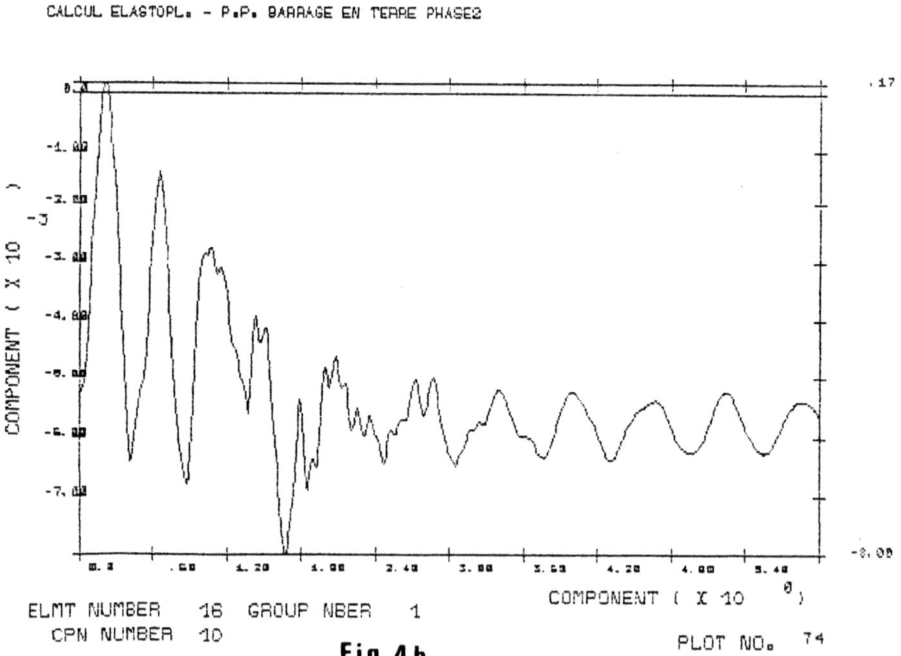

Fig.4b ε_{xx}

Part VI: Reinforced Concrete

Some Aspects of Modelling Reinforced Concrete Structures by Finite Elements

G. MEHLHORN, D. DINGES*, M. KEUSER*, W. KOLMAR*
Gesamthochschule Kassel, Universität, West Germany
*Technical University of Darmstadt, West Germany

Introduction

In addition to ultimate strength criteria and supposed safety factors, the behaviour of a reinforced concrete structure at service loads should be one of the design criteria of the structure. Especially the crack widths control and the deflections at the service loads have to be checked to assure that they will be within reasonable limiting values, dictated by functional requirements of the reinforced concrete structures. If instability problems occur, nonlinear geometric effects shall be accounted for when establishing the equilibrium conditions between external and internal forces.

Therefore the prediction of the overall behaviour of the structure under all load steps is of special interest to ensure a satisfactory designing of the structure. Because reinforced concrete is an inhomogeneous material and the deformation behaviour of a reinforced concrete structure is extremely nonlinear even under working loads, calculation methods based on the classical theory of elasticity are not sufficient to predict the realistic behaviour of a reinforced concrete structure. Of all calculation methods that were developped to consider the deviation of the behaviour of reinforced concrete structures from the linear elastic range the Finite Element Method (FEM) is one of the most powerful tools. In this contribution the FEM is presumed to be known as well as the problems interacting to material modelling of reinforced concrete applying the FEM in the analysis of reinforced concrete structures.

The purpose of this contribution is only to discuss a selection of some special aspects of modelling reinforced concrete using the FEM, as transfer of shear forces across cracks, bond modelling, material modelling of concrete

Finite Element Methods for Nonlinear Problems
Europe-US Symposium, Trondheim, Norway 1985
Editors: Bergan, Bathe, Wunderlich

and layered element approaches.

Transfer of Shear Forces Across Cracks in Reinforced Concrete Structures

In the Finite Element Analysis of reinforced concrete structures the modelling of shear transfer across the cracks is essentially important. Because of the rough crack surfaces, the dowel stiffness and the axial stiffness of the reinforcement bars intersecting the cracks, stresses can still be transferred parallel and perpendicular to the cracks. To model the influence of concrete cracking in nonlinear finite element computer programs two possibilities have been developped, the smeared and the discrete crack approach. If the local behaviour of single cracks is of interest, the discrete crack model has to be used. If the global behaviour of a structure has to be investigated, the smeared crack approach is the more effective method.

The behaviour of a discrete crack can be described by a simplified model (Fig. 1).

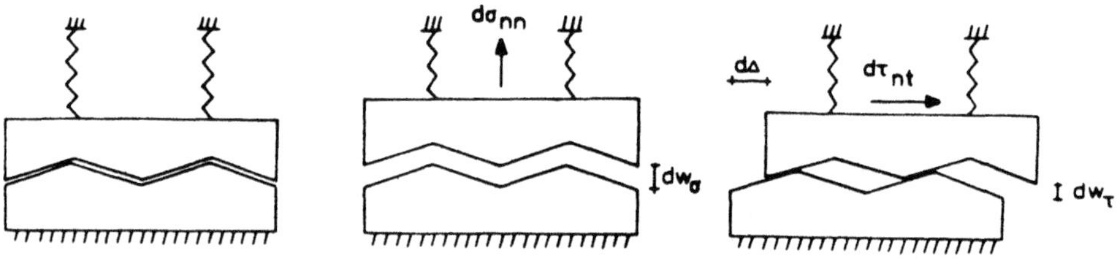

Fig. 1 Simplified model of a discrete crack

Under external normal tensile stresses σ_{nn} perpendicular to the crack, only the crack width w will change. If the crack surfaces are loaded by external shear stresses τ_{nt}, the crack width w will increase according to increasing relative tangential displacements Δ in the crack plane.

We can describe this model by the flexibility matrix [33]:

$$\begin{bmatrix} dw \\ d\Delta \end{bmatrix} = \begin{bmatrix} F_{nn} & F_{nt} \\ F_{tn} & F_{tt} \end{bmatrix} \cdot \begin{bmatrix} d\sigma_{nn} \\ d\tau_{nt} \end{bmatrix} \tag{1}$$

In this equation $F_{tn} = 0$ because $d\sigma_{nn}$ will not change $d\Delta$.

The stiffness matrix $\underset{\sim}{B}^{cr}$ follows by inversion of the flexibility matrix:

$$\begin{bmatrix} d\sigma_{nn} \\ d\tau_{nt} \end{bmatrix} = \begin{bmatrix} B_{nn} & B_{nt} \\ B_{tn} & B_{tt} \end{bmatrix} \cdot \begin{bmatrix} dw \\ d\Delta \end{bmatrix} \tag{2}$$

Herein

$$B_{nn} = \frac{1}{F_{nn}} \quad ; \quad B_{tn} = 0$$

$$B_{nt} = - \frac{F_{nt}}{F_{nn} \cdot F_{tt}} \quad ; \quad B_{tt} = \frac{1}{F_{tt}}$$

The matrix $\underset{\sim}{B}^{cr}$ is an unsymmetric one because increasing of $d\sigma_{nn}$ only leads to increasing of the crack width, but increasing of $d\tau_{nt}$ leads to changing of the crack width and the tangential displacements of the crack surfaces.

For incremental load steps we symmetrize the stiffness matrix:

$$\begin{bmatrix} d\sigma_{nn} \\ d\tau_{nt} \end{bmatrix} = \begin{bmatrix} B_{nn} & B_{nt} \\ B_{nt} & B_{tt} + (B_{tn} - B_{nt}) \cdot \frac{dw}{d\Delta} \end{bmatrix} \cdot \begin{bmatrix} dw \\ d\Delta \end{bmatrix}$$

With $\overline{B}_{nn} = B_{nn}$; $\overline{B}_{nt} = B_{nt}$

$$\overline{B}_{tt} = B_{tt} + (B_{tn} - B_{nt}) \cdot \frac{dw}{d\Delta}$$

we obtain:

$$\begin{bmatrix} d\sigma_{nn} \\ d\tau_{nt} \end{bmatrix} = \begin{bmatrix} \overline{B}_{nn} & \overline{B}_{nt} \\ \overline{B}_{nt} & \overline{B}_{tt} \end{bmatrix} \begin{bmatrix} dw \\ d\Delta \end{bmatrix} \tag{3}$$

The term $\overline{B}_{tt}$ depends on dw and dΔ. Therefore an iteration procedure is necessary to solve equation (3).

The contact element [1] to [3] with two double nodes i - k and l - m to connect the two crack surfaces can be used (Fig. 2).

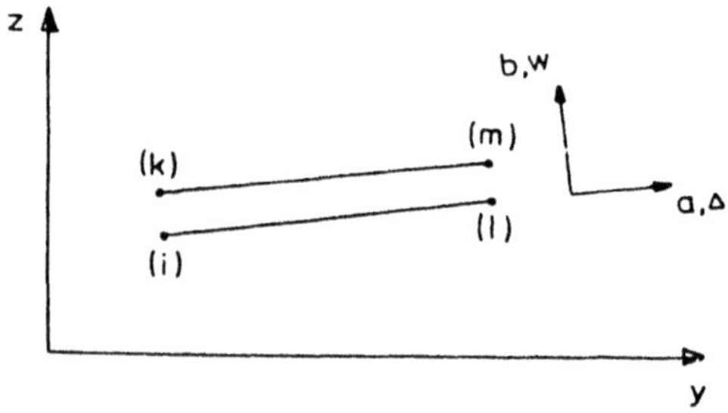

Fig. 2 Contact element

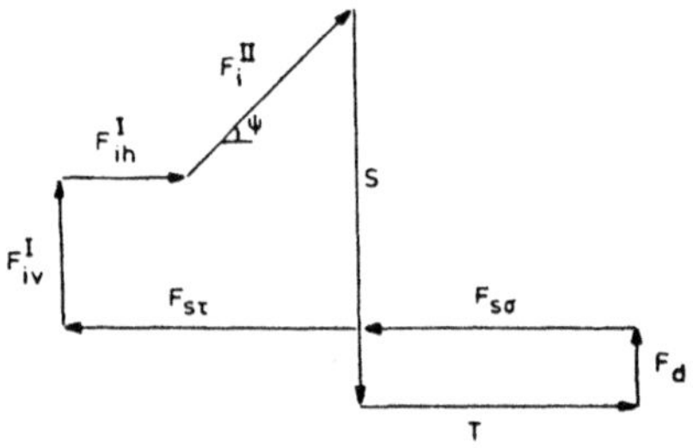

Fig. 3 Equilibrium of the forces in a reinforced crack with the direction of the reinforcement perpendicular to the crack plane

Assuming a linear displacement variation between the two double nodes

$$\begin{bmatrix} w \\ \Delta \end{bmatrix} = \begin{bmatrix} 1 & x & 0 & 0 \\ 0 & 0 & 1 & x \end{bmatrix} \cdot \begin{bmatrix} \alpha_1 \\ \alpha_2 \\ \alpha_3 \\ \alpha_4 \end{bmatrix} \tag{4}$$

we obtain the element stiffness matrix in local coordinates:

$$\begin{bmatrix} R^{w}_{ik} \\ R^{w}_{lm} \\ R^{\Delta}_{ik} \\ R^{\Delta}_{lm} \end{bmatrix} = \frac{L}{6} \cdot \begin{bmatrix} 2\,\overline{B}_{nn} & \overline{B}_{nn} & 2\,\overline{B}_{nt} & \overline{B}_{nt} \\ \overline{B}_{nn} & 2\,\overline{B}_{nn} & \overline{B}_{nt} & 2\,\overline{B}_{nt} \\ 2\,\overline{B}_{nt} & \overline{B}_{nt} & 2\,\overline{B}_{tt} & \overline{B}_{tt} \\ \overline{B}_{nt} & 2\,\overline{B}_{nt} & \overline{B}_{tt} & 2\,\overline{B}_{tt} \end{bmatrix} \begin{bmatrix} w_{ik} \\ w_{lm} \\ \Delta_{ik} \\ \Delta_{lm} \end{bmatrix}$$

or

$$\underset{\sim}{R} = [\underset{\sim}{k}^*] \cdot \underset{\sim}{d} \qquad (5)$$

The stiffness matrix [k] related to the global coordinates has to be computed by matrix transformations.

Walraven [4] developped an analytical model to describe the behaviour of discrete unreinforced and reinforced cracks under shear stresses. This model is based on experimental test results. Walraven showed that the transfer of stresses across a reinforced crack can be described by a polygon of forces (Fig. 3).

The shear force S must be in equilibrium with the dowel force F_d and the vertical components of the force F_i related to aggregate interlock which is subdivided by Walraven into two parts: type I and type II.

"Aggregate Interlock Type I" represents the global roughness of the crack surfaces and "Aggregate Interlock Type II" describes the shear transfer due to the local disturbances of the concrete around the reinforcement bars. F_i^{I} and F_i^{II} are the forces resulting from aggregate interlock types I and II. The subscripts v and h denote the vertical and horizontal components.

The tension force T must be in equilibrium with the force in the reinforcement bars $F_{s\sigma}$ and the horizontal components of the forces $F_{s\tau}$ due to aggregate interlock types I and II:

$$F_d + F_{iv}^{I} + F_i^{II} \cdot \sin\psi = S \tag{6}$$

$$F_{s\sigma} = T \tag{7}$$

$$F_{s\tau} = F_{ih}^{I} + F_i^{II} \cdot \cos\psi \tag{8}$$

The equations (6) to (8) are valid for an inclination of the reinforcement to the crack surface of $\theta = 90°$.

The forces can be determined as a function of the crack width w, the relative tangential displacement Δ of the crack surfaces, the concrete compressive strength f_c', the diameter ϕ of the reinforcement bars, the inclination θ of the direction of the reinforcement bars to the crack surface and the reinforcement ratio ρ.

Walraven [4] gives the relation between w and Δ for concretes used mostly ($20 < f_c' < 40$ N/mm²) as

$$\Delta = 1{,}4 \cdot w^{1{,}2} \tag{9}$$

This formula is independent of the ratio of the reinforcement. The inclination ψ of the concrete compressive struts follows from equation (9) by differentiation:

$$\frac{d\Delta}{dw} = 1{,}68 \cdot w^{0{,}2}$$

and hence it follows:

$$\psi = \arctan\,(0{,}595 \cdot w^{-0{,}2}) \tag{10}$$

To determine the unknown forces $F_{s\tau}$ and F_i^{II} iteratively, we calculate at first the crack width w_σ from the tension force T and the given stiffness B_{nn} of the reinforcement. Then the forces F_d, F_{iv}^{I} and F_{ih}^{I} as well as the angle ψ are known, too. We obtain now the unknown forces $F_{s\tau}$ and F_i^{II} from the polygon of forces (Fig. 3). The force $F_{s\tau}$ will be considered as additional nodal force on both sides of the crack surfaces and the iteration procedure will continue.

All the normal and shear forces are related to the same area, wherefore the components of the forces may be expressed by the stresses and we obtain the shear stiffness from equation (11) for an arbitrary inclination θ of the direction of the reinforcement to the crack surface (Fig. 4).

$$\frac{d\tau_{nt}}{d\Delta} = \frac{d\sigma_d}{d\Delta} + \frac{d\sigma^I_{iv}}{d\Delta} + 1{,}68 \cdot w^{0{,}2} \cdot \left(\frac{d\sigma_{s\tau}}{d\Delta} \sin\theta - \frac{d\sigma^I_{ih}}{d\Delta}\right) + 0{,}2\, w^{-1} \cdot$$

$$(\sigma_{s\tau} \cdot \sin\theta - \sigma^I_{ih})$$

The terms σ_d, σ^I_{iv}, σ^I_{ih}, $\frac{d\sigma_d}{d\Delta}$, $\frac{d\sigma^I_{iv}}{d\Delta}$ and $\frac{d\sigma^I_{ih}}{d\Delta}$ are given by equations (12) to (17):

$$\sigma_d = 14{,}37\, \rho \cdot w^{0{,}432} \cdot (w + 0{,}2)^{-1} \cdot \phi^{0{,}25} \cdot f_c'^{\,0{,}38} \qquad (12)$$

$$\sigma^I_{iv} = -\frac{f_c'}{30} + 2{,}52 \cdot w^{0{,}4} + (0{,}328 \cdot w^{0{,}493} - 0{,}28\, w^{1{,}2})\, f_c' \qquad (13)$$

$$\sigma^I_{ih} = -\frac{f_c'}{20} + 1{,}89 \cdot w^{0{,}57} + (0{,}267 \cdot w^{0{,}648} - 0{,}21\, w^{1{,}2})\, f_c' \qquad (14)$$

$$\frac{d\sigma_d}{d\Delta} = 0{,}735 \cdot \rho \cdot \phi^{-0{,}25} \cdot f_c' \, \frac{0{,}38\, w^{-0{,}736} - 6{,}6\, w^{0{,}232}}{(w + 0{,}2)^2} \qquad (15)$$

$$\frac{d\sigma^I_{iv}}{d\Delta} = 0{,}6 \cdot w^{-0{,}8} + (0{,}096 \cdot w^{-0{,}707} - 0{,}2) \cdot f_c' \qquad (16)$$

$$\frac{d\sigma^I_{ih}}{d\Delta} = 0{,}641 \cdot w^{-0{,}63} + (0{,}103 \cdot w^{-0{,}552} - 0{,}15)\, f_c' \qquad (17)$$

Using the smeared crack approach the shear deformation of a cracked reinforced concrete element consists of the mean value of the relative displacements Δ_m of the two crack surfaces divided by the mean crack spacing a_m and of the shear strain γ of the uncracked concrete between the cracks (Fig. 5).

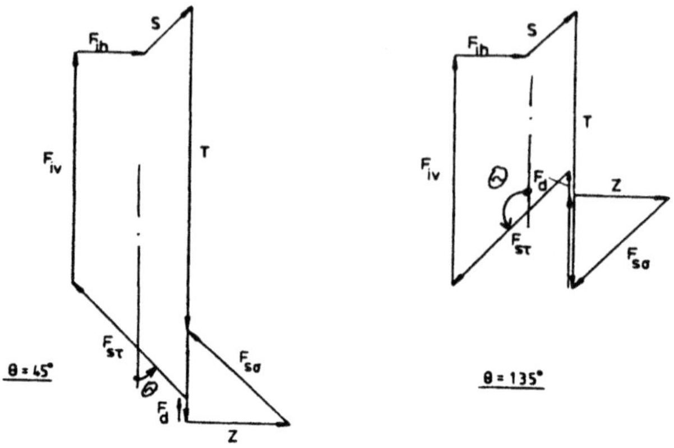

Fig. 4 Force polygon in a reinforced crack with arbitrary direction of the reinforcement intersecting the crack

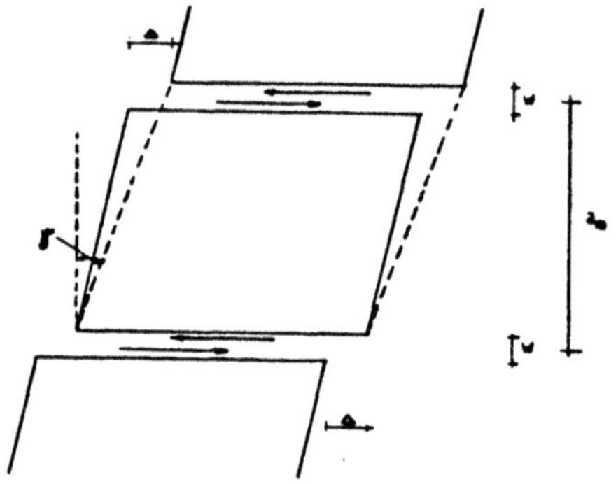

Fig. 5 Shear strain of a cracked reinforced concrete element

The reduction of the shear modulus can be determined by

$$\alpha = \frac{\tau_{nt}/G_c}{\Delta_m/a_m + \tau_{nt}/G_c} \tag{18}$$

The contribution of the cracks to the shear deformations can be determined by the discrete crack model described above.

A variation of different parameters influencing the shear reduction factor α showed that this factor is mainly influenced by the ratio of the reinforcement of the bars bridging the crack [5], [6]. For higher reinforcement ratios the decrease of the shear modulus is less steep than for smaller ones.

The comparison of our results to other shear reduction models used in different investigations, for example those of Al-Mahaidi [7] or Cedolin/Dei Poli [8] showed that these formulations cannot reproduce the physical behaviour very well, although the Cedolin/Dei Poli's solution gives in many cases good results, if the influence of the reinforcement ratio is not important. Therefore we developped a more fitting formulation to calculate the shear reduction factor α depending on the mean tensile strain perpendicular to the crack ε_m and on the reinforcement ratio ρ of the bars bridging the crack:

$$\alpha = \frac{- \ln \dfrac{\varepsilon_m\ [^\circ/_{\circ\circ}]}{c_1}}{c_2} \qquad (19)$$

with

$$c_1 = 7 + 5 \cdot \frac{\rho - 0{,}005}{0{,}015}$$

$$c_2 = 10 - 2{,}5 \cdot \frac{\rho - 0{,}005}{0{,}015}$$

for $\rho \leqq 0{,}02 \mathrel{\hat{=}} 2\ \%$

For higher reinforcement ratios than 2 % the τ-Δ relationship changes only slightly. Therefore formula (19) still may be used for reinforcement ratios higher than 2 %, but than the ratio ρ may be considered as $\rho = 0{,}02$.

As an example, the beam A1 from the tests of Bresler/Scordelis [9] has been analyzed. Thereby different models to reduce the shear stiffness have been investigated. Fig. 6 shows the beam with its reinforcement and the used mesh for the FE-Analysis. The results are shown in fig. 7. For a constant reduction of the shear stiffness the best agreement with the test results was obtained for values of α between 0,10 and 0,20. Very good results gave the model of Cedolin/Dei Poli [8] and the reduction of α according to equation (19) including the influence of the reinforcement ratio, too.

The formulation of Al-Mahaidi [7] gives too low ultimate loads due to the very rapid reduction of α after cracking.

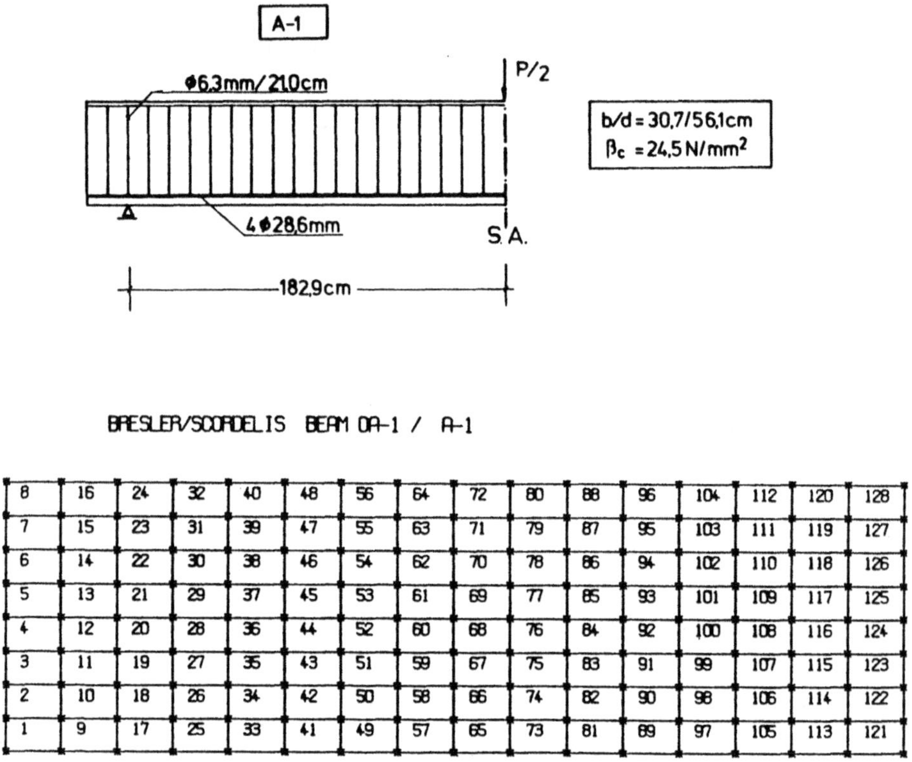

Fig. 6 Test beam of Bresler/Scordelis [9] and finite element mesh

The comparison of shear stiffness formulations for reinforced concrete elements shows that a realistic description of this influence is necessary if significant shear stresses must be transferred across the cracks. The results demonstrate that a constant reduction of the shear modulus is only a rough simplification of the real behaviour and only gives satisfactory results if rather low values of α are chosen. Better agreement with the real behaviour can be obtained with models reducing the shear stiffness depending on the mean tensile strain perpendicular to the cracks and

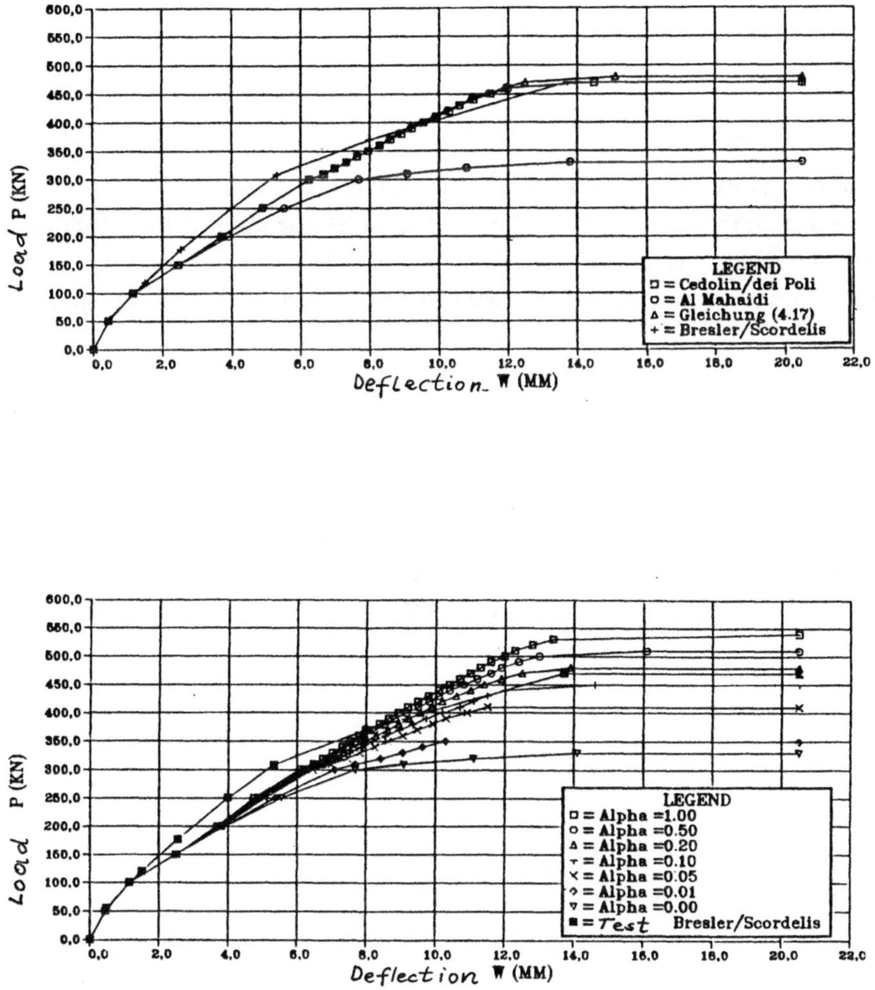

Fig. 6a Midspan deflections for the different shear stiffness reduction models

depending on the reinforcement ratio of the bars intersecting the cracks.

Bond between concrete and reinforcement

The contact element (Fig. 2) is very well applicable to model bond between concrete and reinforcement [1] to [3]. Using an isoparametric formulation it allows the modelling of curved geometry (Fig. 8) and leads to significant advantages in the consideration of nonlinear bond stress-slip relations.

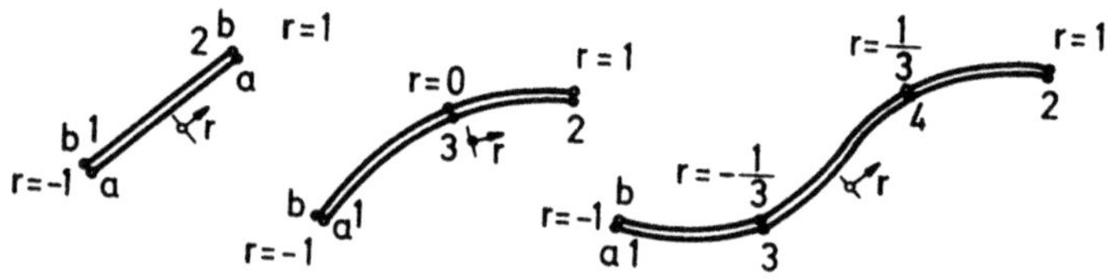

Fig. 7 Isoparametric contact elements

The quality of a bond element depends mainly on two basic assumptions: the displacement function and the bond stress-slip relation.

The displacement function is one of the most important assumptions that are made to derive the stiffness matrix of an element on the basis of the deformation method. For example, Gallagher [10] compares plane stress elements with different displacement functions assuming linear elastic material behaviour. Generally, elements with higher order functions give better results than those with lower ones. The reason therefore is that they can approximate the exact solution better. Low order elements behave stiffer because of their higher energy dissipation.

The results are also valid for contact elements on principle. In [1] contact elements with different displacement functions and bond-link elements [11] are compared to each other. It has been shown that the bond-link element behaves too stiff. This is caused by the constant displacements in the element. This element approximates the exact solution by a stair function and therefore its energy dissipation is too high, especially concerning anti-symmetric portions of the displacements which can be modelled only poorly. If, for example, an antisymmetric, linear slip curve is modelled, only using two degrees of freedom, the energy dissipation of the bond-link element is three times too large compared to the theoretical solution. With the same degrees of freedom, a contact element with a linear displacement function gives an exact solution.

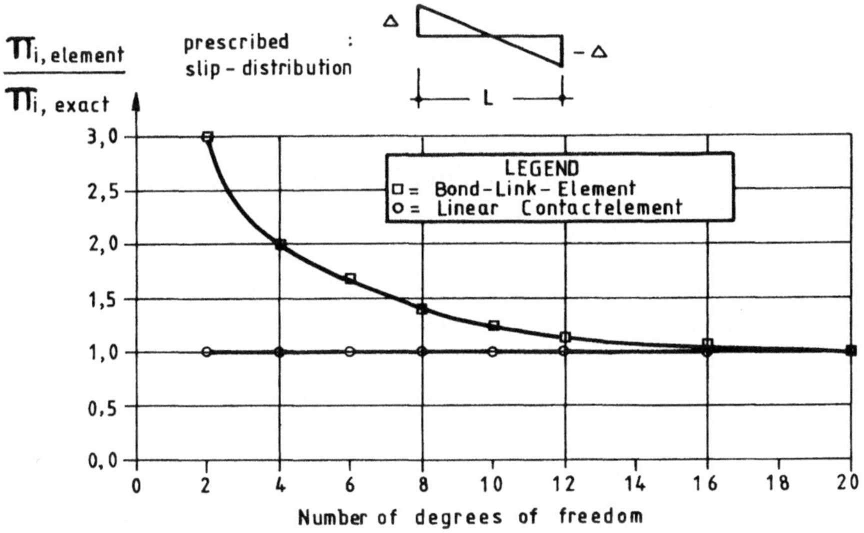

Fig. 8 Energy dissipation of contact elements

For higher order displacement functions the difference between the theoretical solution and the FE approximation decreases.

For practical use elements with linear or especially with quadratic displacement functions proved to be most suitable. On the one side, these elements show a sufficient or even good behaviour concerning their energy dissipation which can be seen as a measure for the quality of the approximation of the exact solution. Elements with cubic or higher order displacement functions only give little improvement. On the other side, the idealization of the structure usually requires a certain number of elements so that an increase of the order of the displacement function would require addiitonal degrees of freedom. The results obtained by theoretical considerations, assuming linear elastic bond behaviour could be confirmed by linear and materially nonlinear FE-calculations. When nonlinear behaviour is concerned the advantages of linear and quadratic elements increase. For example, the realistic approximation of the slip curve between two cracks requires 10 bond-link elements at least, whereas equivalent results can be obtained with 4 linear or 3 quadratic elements. Higher order elements do not bring a significant improvement.

The bond stress-slip relation is the constitutive equation of a contact element and corresponds to the material stress-strain relations of a 2-d or 3-d element. The large scale of different experimental results and analytical models which have been published in the past shows that the bond behaviour is influenced by many factors. Besides, the definition of slip is very different and often unclear.

The stress-slip relation of a contact element describes first of all the behaviour of the contact interface. The parameters which influence bond can be devided into two groups. Things like tensile strength of concrete, the position of the bars during casting of the concrete and the related rib area are independent of the stress state and remain constant during the calculation. Especially Rehm [12] and Eligehausen et. al. [13] have investigated this field. Nilson [14] found out in his experimental research that the bond stress-slip relation is not constant along the bar surface but it depends on the local stress state in the bond zone. Therefore a second group of influence parameters on bond behaviour can be defined which characterize effects like transversal pressure, internal cracking of concrete and the destruction of the concrete before the ribs. As these influences depend on the stress state, they cannot be prescribed before the start of a FE-calculation but they must be updated in every load step of a nonlinear analysis.

In his tests, Dörr [15] found out that bond depends strongly on transversal pressure. More generally it can be said, that there is an interaction between normal and tangential stresses in the contact interface. As this relation is not orthogonal, its consideration leads to an unsymmetric element stiffness matrix. As the majority of the available FE-programs are not capable to handle unsymmetric matrices, in this case an alternative procedure can be applied by adding the unsymmetric portions of the stiffness matrix to the nodal force vector.

In the last years, an analytical bond model for FE-algorithms has been developped which is suited for research as well as for special problems of engineering practice. In this bond model four levels of discretisation of a structure are defined, using the modelling of the reinforcement as the parameter to chose the level for an analysis. Fig. 9 shows the basic bond stress-slip relation which is, for example, suited for structures like test specimen with a fine and detailed element mesh.

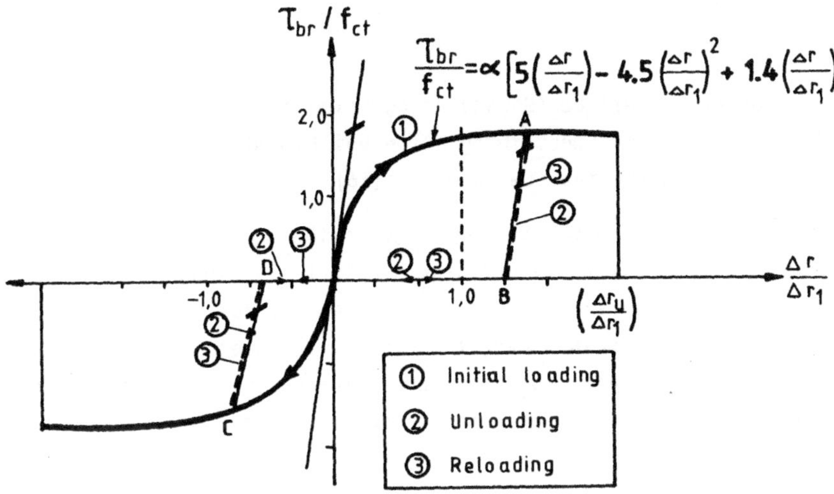

Fig. 9 Normalized bond stress-slip relation

In this bond model all the influence parameters mentioned before are included by the factor α. Actually, the model is suited for calculations without time effects like impact, cyclic or long-time loading. Loading, unloading and reloading are included. This is necessary for monotonically increasing loads, too, for example at crack openings which cause local stress redistributions. In the future, it is planned to develop the model also for time-dependent calculations.

Efficient formulation of the layered degenerated shell element

The "Degenerated Shell Element" has shown that it is very useful in dealing with both, linear and geometrically nonlinear shell problems [16], [17]. Additionally, nonlinear elastic material problems can be handled, if the shell is devided into layers over its thickness [18] to [22]. Therefore the only modification which actually has to be made is to introduce a numerical integration over the shell thickness. For the element stiffness matrix this leads to

$$\underline{K} = \int_v \underline{B}^{*T}_{(\zeta)} \cdot \underline{D}^*_{(\zeta)} \cdot \underline{B}^*_{(\zeta)} \, dv$$

$$= \sum_{NG} WG \, (\sum_{NT} WT \, \underline{B}^{*T}_{(\zeta)} \cdot \underline{D}^*_{(\zeta)} \cdot \underline{B}^*_{(\zeta)} \cdot \det \underline{J}^*) \qquad (20)$$

with

NG, WG : Number of surface (Gauss) integration points and weights
NT, WT : Number of thickness integration points and weights
* indicates the matrices used in [16]

In order to simplify matters we will only regard the geometrically linear case. For the geometrically nonlinear case see [22]. The inner loop in equation (20) has to be carried out at each Gauss integration point. Assuming 6-10 integration points over the element thickness, the formation of the element stiffness matrix becomes very time-consuming. Subsequently we will show that the summation over the thickness can be reformulated to obtain

$$\sum_{NT} WT \, (\underline{B}^{*T}_{(\zeta)} \cdot \underline{D}^*_{(\zeta)} \cdot \underline{B}^*_{(\zeta)} \cdot \det \underline{J}^*) = \underline{B}^T \, (\sum_{NT} WT \cdot \underline{D}_{(\zeta)}) \, \underline{B} \cdot \det \underline{J} \qquad (21)$$

The constitutive matrix has to be formed again for each layer. However, the B-matrix no longer depends on ζ and thus the triple product has to be executed only once at each Gauss point!

Coordinates and displacements

Global and local coordinate systems are defined in Fig. 10:

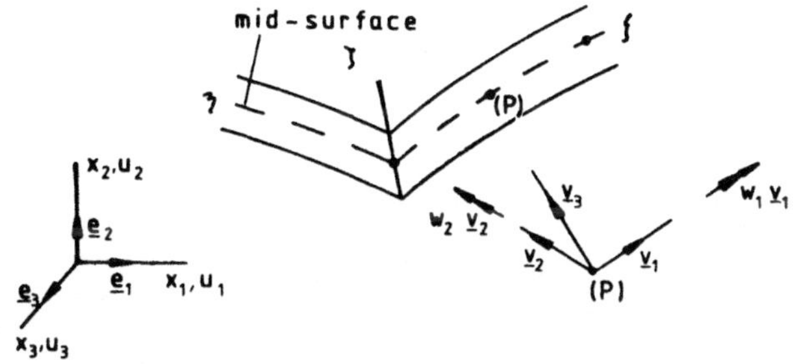

Fig. 10 Global and local coordinate systems

Coordinates and displacements are interpolated in the usual way

$$\underline{x}_{(\xi,\eta,\zeta)} = \underline{\tilde{x}}_{(\xi,\eta)} + \frac{\zeta\, t}{2}\, \underline{v}_{3\,(\xi,\eta)}$$

$$= \sum_{p=1}^{K} N_{p\,(\xi,\eta)} \cdot \underline{x}_{(p)} + \frac{\zeta\, t}{2} \sum_{p=1}^{K} N_{p\,(\xi,\eta)} \cdot \underline{v}_{3\,(p)} \tag{22}$$

where

$$\underline{x} = [x_1\ x_2\ x_3]^T \tag{23}$$

$$\underline{v}_{i(p)} = [v_{i1}\ v_{i2}\ v_{i3}]^T_{(p)} \tag{24}$$

$$N_{p\,(\xi,\eta)} = N_p \tag{25}$$

are the global coordinates, the local base vectors and the usual shape functions for node (p), respectively. The above assumption of a constant shell thickness is not necessary.

Further

$$\underline{u} = \sum_{p=1}^{K} N_p \cdot (\underline{u}_{(p)} + \zeta\, \underline{a}_{(p)} \cdot \underline{w}_{(p)}) \tag{26}$$

where

$$\underline{u} = [u_1\ u_2\ u_3]^T \tag{27}$$

$$\underline{a}_{(p)} = \begin{bmatrix} a_{11} & a_{12} \\ a_{21} & a_{22} \\ a_{31} & a_{32} \end{bmatrix}_{(p)} \quad \text{and} \quad \begin{aligned} a_{j1} &= -\frac{t}{2}\ v_{j2} \\ a_{j2} &= +\frac{t}{2}\ v_{j1} \end{aligned} \tag{28}$$

are the global displacements and direction cosines.

$$\underline{w}_{(p)} = [w_1\ w_2]^T_{(p)} \tag{29}$$

are the "rotational"degrees of freedom which are always referred to the local system $\underline{v}_{i(p)}$.

Jacobian Matrix

For a 3-dimensional case the Jacobian matrix is defined as

$$\underline{J}^* = \begin{bmatrix} x_{1,\xi} & x_{2,\xi} & x_{3,\xi} \\ x_{1,\eta} & x_{2,\eta} & x_{3,\eta} \\ x_{1,\zeta} & x_{2,\zeta} & x_{3,\zeta} \end{bmatrix} \tag{30}$$

We now introduce the assumption that

$$t << r_1\ , \qquad t << r_2 \tag{31}$$

where r_α are the principle radii of curvature of the mid-surface [23]. Thus the derivatives $\underline{x}_{,\xi}$ and $\underline{x}_{,\eta}$ do not change along ζ yielding

$$J_{1j} = x_{j,\xi} \approx \tilde{x}_{j,\xi} = \sum_{p=1}^{K} N_{p,\xi} \cdot x_{j(p)}$$

$$J_{2j} = x_{j,\eta} \approx \tilde{x}_{j,\eta} = \sum_{p=1}^{K} N_{p,\eta} \cdot x_{j(p)} \qquad (32)$$

$$J_{3j} = x_{j,\zeta} = \frac{t}{2} v_{j3} = \frac{t}{2} \sum_{p=1}^{K} N_p \cdot v_{j3(p)}$$

Note that now the coefficients of the Jacobian matrix are independent of ζ!

Strains and Stresses

For the geometrically linear 3-D case, Green-Lagrange strains are defined as

$$\gamma_{ij} = \frac{1}{2} (u_{i,j} + u_{j,i})$$

$$\gamma_{ij} = \gamma_{ji} \qquad (33)$$

and

$$\underline{\gamma} = [\gamma_{11} \ \gamma_{22} \ \gamma_{33} \ \gamma_{23} \ \gamma_{13} \ \gamma_{12}]^T$$

The corresponding stresses are the components of the 2nd Piola-Kirchhoff stress tensor with

$$s_{ij} = s_{ji} \qquad (34)$$

In the following we will require the relations

$$\underline{s} = [s_{11} \ s_{22} \ s_{33} \ s_{23} \ s_{13} \ s_{12}]^T$$

$$\underline{\sigma}_\gamma = [s_{11} \ s_{22} \ s_{33} \ 2\,s_{23} \ 2\,s_{13} \ 2\,s_{12}]^T \qquad (35)$$

$$\underline{\sigma} = [\underline{\sigma}_\gamma \quad \zeta\underline{\sigma}_\gamma]^T$$

Transformations between local and global coordinates are given by:

$$\underline{\bar{s}} = \underline{T} \cdot \underline{s}$$

$$\underline{\bar{\gamma}} = \underline{T} \cdot \underline{\gamma}$$

$$\underline{\sigma}_\gamma = \underline{T}^T \, \underline{\bar{\sigma}}_\gamma \tag{36}$$

Constitutive equations

Referring to local coordinates the stress normal to the mid-surface is neglected [16], [23]:

$$\bar{s}_{33} = 0 \tag{37}$$

In the special case of a linear isotropic material we have in local coordinates:

$$\begin{bmatrix} \bar{s}_{11} \\ \bar{s}_{22} \\ \bar{s}_{33} \\ 2\,\bar{s}_{23} \\ 2\,\bar{s}_{13} \\ 2\,\bar{s}_{12} \end{bmatrix} = \begin{bmatrix} 1 & 0 & 0 & 0 & 0 & 0 \\ & 1 & 0 & 0 & 0 & 0 \\ & & 0 & 0 & 0 & 0 \\ & & & 2(1-\nu) & 0 & 0 \\ & \text{sym} & & & 2(1-\nu) & 0 \\ & & & & & 2(1-\nu) \end{bmatrix} \cdot \begin{bmatrix} \bar{\gamma}_{11} \\ \bar{\gamma}_{22} \\ \bar{\gamma}_{33} \\ \bar{\gamma}_{23} \\ \bar{\gamma}_{13} \\ \bar{\gamma}_{12} \end{bmatrix} \tag{38}$$

or

$$\underline{\bar{\sigma}}_\gamma = \underline{D}^* \cdot \underline{\bar{\gamma}}$$

Strain Displacement-Matrix

The displacement derivatives with respect to global coordinates are

$$u_{i,j} = J^{-1}_{j1} \; u_{i,\xi} + J^{-1}_{j2} \; u_{i,\eta} + J^{-1}_{j3} \; u_{i,\zeta} \tag{39}$$

$$N_{p,j} = J^{-1}_{j1} N_{p,\xi} + J^{-1}_{j2} N_{p,\eta} = N_{,j(p)} \tag{40}$$

Substituting from (40) and (26) into (39) we obtain

$$u_{i,j} = \sum_{p=1}^{K} N_{p,j} \cdot u_{i(p)} + \sum_{p=1}^{K} J^{-1}_{j3} N_p \, a_{i\alpha(p)} \, w_{\alpha(p)} \tag{41}$$

$$+ \zeta \sum_{p=1}^{K} N_{p,j} \cdot a_{i\alpha(p)} \cdot w_{\alpha(p)}$$

The strain-displacement-relations (33) can now be written in the following way

$$\underline{\gamma} = \sum_{p=1}^{K} \underline{B}^*_{(p)} \cdot \underline{U}_{(p)}$$

$$= \sum_{p=1}^{K} (\underline{B}_{o(p)} + \zeta \, \underline{B}_{1(p)}) \cdot \underline{U}_{(p)} = \underline{B}^* \cdot \underline{U} \tag{42}$$

with

$$U_{(p)} = [u_1 \; u_2 \; u_3 \; w_1 \; w_2]^T_{(p)}$$

Note that neither $\underline{B}_{o(p)}$ nor $\underline{B}_{1(p)}$ depend on ζ! Matrices $\underline{B}_{o(p)}$ and $\underline{B}_{1(p)}$ are listed in [22].

Modification of constitutive equations

The constitutive relations referred to global coordinates are

$$\underline{\sigma}_\gamma = \underline{D}_{\gamma(\zeta)} \cdot \underline{\gamma}_{(\zeta)} \tag{43}$$

Defining

$$\underline{\alpha} = \sum_{p=1}^{K} \underline{B}_{o(p)} \underline{U}_{(p)} \; ; \quad \beta = \sum_{p=1}^{K} \underline{B}_{1(p)} \underline{U}_{(p)} \tag{44}$$

with

$$\begin{aligned} \underline{\alpha} &= [\alpha_{11} \; \alpha_{22} \; \alpha_{33} \; \alpha_{23} \; \alpha_{13} \; \alpha_{12}]^T \\ \underline{\beta} &= [\beta_{11} \; \beta_{22} \; \beta_{33} \; \beta_{23} \; \beta_{13} \; \beta_{12}]^T \end{aligned} \tag{45}$$

we can write

$$\underline{\gamma} = \underline{\alpha} + \zeta \, \underline{\beta} \tag{46}$$

Finally defining

$$\underline{\varepsilon} = [\underline{\alpha} \quad \underline{\beta}]^T \tag{47}$$

We obtain for the modified stress-strain-relations

$$\begin{bmatrix} \underline{\sigma}_\gamma \\ \zeta \, \underline{\sigma}_\gamma \end{bmatrix} = \begin{bmatrix} \underline{D}_\gamma & \zeta \, \underline{D}_\gamma \\ \zeta \, \underline{D}_\gamma & \zeta^2 \, \underline{D}_\gamma \end{bmatrix} \cdot \begin{bmatrix} \underline{\alpha} \\ \underline{\beta} \end{bmatrix} \tag{48}$$

or

$$\underline{\sigma} = \underline{D}_{(\zeta)} \cdot \underline{\varepsilon}$$

Principal of virtual displacements

The state of equilibrium is expressed by the principal of virtual displacements:

$$\delta \, W_{ext} = \int_V s_{ij} \cdot u_{j,i} \, dv = \int_V \delta \, \underline{\gamma}^T \, \underline{\sigma}_\gamma \, dv \tag{49}$$

It can easily be shown that

$$\delta\underline{\gamma}^T \cdot \underline{\sigma}_\gamma = \delta\underline{\varepsilon}^T \cdot \underline{\sigma} \tag{50}$$

Substitution from the above deduced relations yields for an arbitrary displacement field

$$\underline{K} \cdot \underline{U} = \underline{R} \tag{51}$$

The element stiffness matrix $\underline{K}$ can now be expressed in the following way

$$\begin{aligned} K &= \int_V \underline{B}^{*T}_{(\zeta)} \cdot \underline{D}^*_{\gamma(\zeta)} \; \underline{B}^*_{(\zeta)} \, dv \\ &= \int_V [\underline{B}_0 \; \underline{B}_1]^T \cdot \underline{D}_{(\zeta)} \cdot [\underline{B}_0 \; \underline{B}_1] \, dv \\ &= \int_V \underline{B}^T \cdot \underline{D}_{(\zeta)} \cdot \underline{B} \, dv \\ &= \sum_{NG} WG \cdot \underline{B}^T \left(\sum_{NT} WT \cdot \underline{D}_{(\zeta)}\right) \underline{B} \det \underline{J} \end{aligned} \tag{52}$$

As desired, the total dependence on ζ is transferred to the D-matrix.

Some experiences concerning the development of concrete models for plates and shells

The essential features that have to be incorporated into a material model for reinforced concrete are described in numerous publications. Here we intend to make some remarks about our experience when developing models for plates and shells.

Plane state of stress

In classical shell theories a state of plane stress is assumed (Kirchhoff/ Love). A contribution of shear energy is neglected. Modern computers make it possible to deal with highly nonlinear materials by dividing the shell's cross-section into layers [20, 22]. Employing the degenerated shell element which includes contributions of shear energy very good results are obtained for linear elastic material problems. For materially nonlinear problems the question arises how the lateral shear stress-strain relations can be formulated. There are different approaches - for different material models [21, 24 ... 27]. In our approach we assume $\overline{s}_{33} = 0$ as usual and uncoupling between $\overline{s}_{\alpha\beta}$ and $\overline{\gamma}_{\alpha 3}$.

For lateral shear we use the same relations as those used for the elastic behaviour of the in-plane shear, $\overline{\sigma}_{12} = \alpha_s \cdot G_o \cdot \overline{\varepsilon}_{12}$, i. e. the shear retention across cracks is taken into consideration.

Furthermore, the lateral shear stresses $\overline{s}_{\alpha 3}$ are not included in the yield criterion [25], (see also [24] and [27] for different approaches).

Comparing results where thin plates were analyzed no differences could be realized [32]. The essential advantage of the above treatment is, however, that all the routines once developped and tested for in-plane loaded structures can directly be used for shell-type structures!

Choice of material model

Looking at the great amount of publications on (reinforced) concrete models, nearly most of them are based on one of the following theories [28, 29]

Nonlinear elasticity:
- Isotropic models
- Orthotropic models
- Octahedral models

Theory of plasticity:
- Elastic-perfect plasticity models
- Elastic-hardening plasticity models

Although we realize and appreciate the criticism of the models mentioned above [31] we asked ourselves:

Concerning the experimental data of concrete being available, are they reliable enough that, for example, endochronic theory will really provide better results?

Does the "ordinary" user of a finite element program know how to handle all the input parameters required for a universal model?

We confined ourselves to short-term proportional loading and concentrated on the orthotropic and plasticity-based models.

Modularity to maintain flexibility

Orthotropic and plasticity-based models seem to be completely different. Provided a tension cut-off is used for the plasticity-based models, the major difference reduces to how the compression behaviour is handled. Thus why should the tensile behaviour be "recomputed" for a new model? We chose the subsequent modules to describe the computation of a stress increment due to an increment in strains. For iteration (i) within a load step we carry out:

IN - Provide linear and nonlinear parameters and quantities computed by iteration (i-1)

DEPS - Compute $\Delta \underline{\varepsilon}^{(i)}$ $(\Delta\, \underline{u}^{(i)})$ and $\underline{\varepsilon}^{(i)} = \underline{\varepsilon}^{(i-1)} + \Delta \underline{\varepsilon}^{(i)}$

TENS - Compute stress increment $\Delta \underline{\sigma}^{(i)}_{lin}$ due to linear material behaviour and tensile failure. Compute $\underline{\sigma}^{(i)} = \underline{\sigma}^{(i-1)} + \Delta \underline{\sigma}^{(i)}_{lin}$

CRIT - Decide whether the compression module has to be executed or not.

COMPR - Compute $\Delta \underline{\sigma}^{(i)} = \underline{D}_T^{(i-1)} \cdot \Delta \underline{\varepsilon}^{(i)}$ by considering additionally compressive behaviour and $\underline{\sigma}^{(i)} = \underline{\sigma}^{(i-1)} + \Delta \underline{\sigma}^{(i)}$

OUT - Store all quantities required for next iteration

Thus identical routines concerning module TENS are employed when analyzing:

a) in-plane loaded plates using the orthotropic, elastic-perfectly plastic or elastic-hardening plastic model

b) layered plates/shells using one of the above mentioned models!

Full details of the required models are given in [32].

Comparison of different material models

Some of the results analyzing various structures are reported in chapter 7 of [32]. Rather close agreement of the "different" models is obvious. The major reason for this behaviour is due to the identical treatment of the tensile behaviour. The results for the orthotropic model, however, were obtained using a modified Newton-Raphson iteration whereas for the plasticity-based models convergence could only be achieved by employing a pure Newton-Raphson iteration. All quantities of iteration (i-1) - except of course displacements - refer to the last accepted state of equilibrium [17], [32]. Referring to the last iteration within a load step will lead to completely wrong results.

References

1. Keuser, M.: Verbundmodelle für nichtlineare Finite-Element-Berechnungen von Stahlbetonkonstruktionen. VDI-Verlag, Düsseldorf, 1985.

2. Mehlhorn, G.; Keuser, M.: Isoparametric contact elements for analysis of reinforced concrete structures. Proceedings of the seminar on finite element analysis of reinforced concrete structures, Tokyo, 1985.

3. Mehlhorn, G.; Kollegger, J.; Keuser, M.; Kolmar, W.: Nonlinear contact problems - a finite element approach implemented in ADINA. Computers & Structures,Vol 21, pp. 69-81, 1985.

4. Walraven, J. C.: Aggregate interlock: a theoretical and experimental analysis. Delft University Press, 1980.

5. Kolmar, W.: Beschreibung der Kraftübertragung über Risse in nichtlinearen Finite-Element-Berechnungen von Stahlbetontragwerken. Dissertation D17, Darmstadt, 1985.

6. Kolmar, W.; Mehlhorn, G.: Comparison of shear stiffness formulations for cracked reinforced concrete elements. Proceeding of the International Conference on Computer Aided Analysis and Design of Concrete Structures. Vol. 1, pp. 133-147, Pineridge Press, Swansea, 1984.

7. Al-Mahaidi, R. S. H.: Nonlinear finite element analysis of reinforced concrete deep members. Report 79-1, Department of Structural Engineering, Cornell University, 1979.

8. Cedolin, L.; Dei Poli, S.: Finite element studies of shear critical R/C beams. Journal of the Engineering Mechanics Division, Proceedings of the ASCE, Vol. 103, pp. 395-410, 1977.

9. Bresler, B.; Scordelis, A. C.: Shear Strength of Reinforced Concrete Beams. Journal of the American Concrete Institute, Vol. 60, January 1963, pp. 51-74.

10. Gallagher, R. H.: Finite-Element-Analysis. Springer-Verlag Berlin, Heidelberg, New York, 1976.

11. Ngo, D.; Scordelis, A.: Finite Element Analysis of Reinforced Concrete Beams. ACI-Journal, March 1967, pp. 152-163.

12. Rehm, G.: Über die Grundlagen des Verbundes zwischen Stahl und Beton. Deutscher Ausschuß für Stahlbeton, Heft 128, Berlin, 1961.

13. Eligehausen, R.; Popov, E.; Bertero, V.: Local Bond Stress-Slip Relationship of Deformed Bars under Generalized Exitations. Rep. No. UCB/EERC-83/23, Earthquake Engineering Center, University of California, Berkeley (Cal.), 1983.

14. Nilson, A.: Nonlinaer Analysis of Reinforced Concrete by the Finite Element Method. ACI-Journal, Sept. 1968, pp. 757-766.

15. Dörr, K.: Ein Beitrag zur Berechnung Stahlbetonscheiben unter besonderer Berücksichtigung des Verbundverhaltens, Dissertation D17, TH Darmstadt, 1980.

16. Ramm, E.: A Plate/Shell Element for Large Deflections and Rotations. In "Formulations and Computational Algorithms in Finite Element Analysis", MIT Press, Cambridge, Mass., 1977.

17. Bathe, K.-J.: Finite Element Procedures in Engineering Analysis. Prentice Hall, Englewood Cliffs, New Jersey, 1982.

18. Hand, F. R.; Pecknold, D. A.; Schnobrich, W. C.: A layered finite element nonlinear analysis of reinforced concrete plates and shells. Technical report 72-2011, University of Illinois, Urbana, 1972.

19. Link, J.; Mehlhorn, G.; Schäfer, H.: Berechnung von dünnen Stahlbetonplatten bei Berücksichtigung eines wirklichkeitsnahen Werkstoffverhaltens. 1. Zwischenbericht, Heft Nr. 1 der Forschungsberichte aus dem Institut für Massivbau der Technischen Hochschule Darmstadt, 1972.

20. Chan, E.: Nonlinear Geometric, Material and Time Dependent Analysis of Reinforced Concrete Shells with Edge Beams. Dept. of Civil Engineering, University of California, Berkeley (Cal.), 1982.

21. Sättele, J. M.: Ein finites Elementkonzept zur Berechnung von Platten und Schalen bei stofflicher und geometrischer Nichtlinearität. Institut für Baustatik, Universität Stuttgart, 1980.

22. Dinges, D.; Keuser, M.; Kolmar, W.; Mehlhorn, G.: Untersuchung verschiedener Elementsteifigkeitsmatrizen auf ihre Eignung zur Berechnung von Stahlbetonkonstruktionen. Zweiter Zwischenbericht zum Forschungsvorhaben, Institut für Massivbau, TH Darmstadt, 1982.

23. Timoshenko, S. P.; Woinowsky-Krieger, S.: Theory of Plates and Shells. Mc Graw-Hill, 1959.

24. Kompfner, T. A.: Ein finites Elementkonzept für die geometrisch und physikalisch nichtlineare Berechnung von Stahlbetonschalen.

25. Figueiras, J. A.: Ultimate Load Analysis of Anisotropic and Reinforced Concrete Plates and Shells. Ph. D. Thesis, Dept. of Civil Engineering, University College of Swansea, 1983.

26. Owen, D. R. J.; Hinton, E.: Finite Element in Plasticity - Theory and Practice. Pineridge Press, Swansea, 1980.

27. Rahman, H. H. A.: Computational Models for the Nonlinear Analysis of Reinforced Concrete Flexural Slab Systems. Ph. D. Thesis, University of Wales, Swansea, 1982.

28. Chen, W.-F.: Plasticity in Reinforced Concrete. Mc Graw-Hill, New York, 1982.

29. Shareff, S. S.; Buyukozturk, O.: Constitutive Modeling of Concrete in Finite Element Analysis. Research Report No. R83-16, MIT, Cambridge, Mass. 1983.

30. Dinges, D.; Keuser, M.; Kolmar, W.; Mehlhorn, G.: Untersuchung verschiedener Elementsteifigkeitsmatrizen auf ihre Eignung zur Berechnung von Stahlbetonkonstruktionen. Dritter Zwischenbericht, GhKassel/TH Darmstadt, 1983.

31. Bazant, Z. P.: Comment on Orthotropic Models for Concrete and Geomaterials. Journal of the Engineering Mechanics Division, ASCE, Vol.109, No. EM3, 1983.

32. Dinges, D.; Keuser, M.; Kolmar, W.; Mehlhorn, G.: Untersuchung verschiedener Elementsteifigkeitsmatrizen auf ihre Eignung zur Berechnung von Stahlbetonkonstruktionen. Vierter Zwischenbericht, GhKassel/TH Darmstadt, 1984.

33. Bazant, Z. P.; Gambarova, P.: "Rough Cracks in Reinforced Concrete", Reprint 3579, ASCE-Convention, Boston, April 1979.

Development and Application of a Gradient-Dependent Fracture Criterion for Finite-Element Analysis of Reinforced-Concrete Surface Structures

H. A. MANG, J. EBERHARDSTEINER, H. WALTER
Technical University of Vienna, Austria

Summary

The modulus of rupture of concrete is not a material property in the classical sense of this term. Likewise, biaxial fracture stresses of concrete are not pure material properties. Criticism in the literature of a fixed failure envelope representing the geometric locus of all biaxial states of fracture stresses was the motivation for the development of a new fracture criterion for the analysis of reinforced-concrete surface structures by the Finite Element Method. This criterion is based on the well-known fact that critical values of physical quantities at a point of a given body frequently also depend on the gradients of these quantities at this point. For the limiting case of vanishing stress gradients, peak stresses from Kupfer's biaxial tests are taken as the biaxial tensile strength. For non-vanishing stress gradients, a magnifying factor for Kupfer's biaxial tensile strength at the considered point is determined with the help of a hypothesis of comparison. The influence of the new fracture criterion on the structural response of a hypar groined vault subjected to dead and snow load is discussed in the numerical investigation.

Introduction

For ultimate load analysis of surface structures made of reinforced concrete (RC), Rankine's strength criterion is almost exclusively used as fracture criterion. According to this criterion, fracture of concrete occurs if $|\sigma_1| \geq |\sigma_{1u}|$ where σ_1 is the algebraically larger principal stress and σ_{1u} is the corresponding peak stress in the $\sigma_1 - \varepsilon_1$ diagram obtained from a biaxial test characterized by monotonically increasing, uniformly distributed principal stresses such that $\alpha_1 = \sigma_2/\sigma_1$ remains constant (proportional loading). For $1 \leq \alpha_2 = 1/\alpha_1 \leq \infty$ (biaxial-tension domain) and $-\infty \leq \alpha_2 \leq \alpha_{2L}$ (part of tension-compression domain) with $\alpha_{2L} = f(f_{cu}/f_{tu})$, where f_{cu} is the prism strength and f_{tu} is the uniaxial (direct) tensile strength of concrete, it is assumed that fracture occurs by cracking of concrete normal to the direction of σ_1. Cracked concrete retains its capacity

Finite Element Methods for Nonlinear Problems
Europe-US Symposium, Trondheim, Norway 1985
Editors: Bergan, Bathe, Wunderlich

to carry internal forces in the direction parallel to the cracks so long as $f_{cu} < \sigma_2 < f_{tu}$. Because of tension stiffening, activated by bond-slip between the reinforcement and the surrounding concrete, the latter also retains part of its strength normal to the cracks. For $\alpha_{2L} < \alpha_L \leq 1$ (part of tension-compression domain, biaxial-compression domain), it is assumed that concrete fails by crushing, thereby losing all of its strength.

Criticism of a fixed failure envelope in the $\sigma_1 - \sigma_2$ plane, representing the geometric locus of all biaxial states of fracture stresses $\sigma_{1u}(\alpha_1)$, $\sigma_{2u}(\alpha_1)$, was raised first by Bažant and Cedolin [1], [2], [3] in the context of blunt crack propagation in Finite Element (FE) analysis. Restricting attention to linear elasticity, for a sharp crack, representing the limiting case of a blunt crack band, the stress gradients at the tip of the crack tend to infinity. Obviously, Rankine's strength criterion, based on fixed values of $\sigma_{1u}(\alpha_1)$, $\sigma_{2u}(\alpha_1)$, is not suitable for situations characterized by large stress gradients tending to infinity in the limit of finite discretizations.

By adjusting the uniaxial tensile strength of concrete to the element size, Bažant and Cedolin [1] obtained objective results (i.e., results converging to the exact solutions as the FE grids are refined) for the problem of a Mode I crack in an infinite solid and a finite strip, respectively, subjected at infinity to uniaxial stresses normal to the cracks. The same applies to the approach of adopting the energy release rate as the controlling parameter for crack extension [1].

Recently, Bažant [4] investigated the size effect in fracture criteria for concrete. He introduced a so-called "size reduced-strength" representing a characteristic of the entire structure rather than of the material. Although it is doubtful whether, in general, such a global characteristic exists, Bažant's criticism of fixed values of fracture stresses is justified.

The purpose of the present paper is to report on the development and application of a new fracture criterion for concrete for ultimate load analysis of thin RC plates and shells. It is based on the well-known fact that critical values of physical quantities at a point of a given body frequently also depend on the gradients of these quantities at the considered point. In the context of ultimate load analysis of RC surface structures this means that the biaxial tensile strength of concrete is assumed to depend on the stress gradients. Thus, it is not a material property in the classical sense of this term.

Ultimate Load Analysis of RC Surface Structures by the FEM

This Section is a brief summary of Section 2 of an article by Mang and Eberhardsteiner [5].

The new fracture criterion for concrete was incorporated into a computer program for ultimate load analysis of RC surface structures by the FEM. Mechanical fundamentals and analysis algorithms of this program are described in detail by Floegl [6]. The kinematics of deformation is approximated by means of the Kirchhoff-Love assumptions. A popular shell theory based on these assumptions is the one by Koiter [7]. It is used in the mentioned computer program. Consideration of geometric nonlinearity is restricted to taking account of moderately large rotations. As was shown by Mang et al. [8], this theory is sufficient for ultimate load analysis of RC surface structures.

Curved, triangular, C^1-conforming, thin-shell elements [9] are used as the discretization tool. The elements are subdivided into sufficiently many thin layers such that approximately a plane state of stress may be assumed to exist in each layer. The state of stress is determined at seven integration points of each layer.

The constitutive equations employed for the intact concrete were proposed by Liu, Nilson and Slate [10]. These equations are based on biaxial tests conducted by Kupfer [11]. The tests were carried out for uniformly distributed normal stresses. This constitutive model is a so-called "equivalent uniaxial model". Advantages and disadvantages of this constitutive model are discussed, e.g., by Chen and Ting [12]. For surface structures analyzed for monotonically increasing, proportional loads, the chosen constitutive model is adequate.

Within the framework of the new fracture criterion, which will be presented in the next Section, the state of concrete is checked at all integration points of all layers of each finite element. If this criterion signals cracking of concrete at a certain integration point, it is assumed that a band of equidistant parallel cracks normal to σ_1, distributed over the tributary domain of this point, will open. This assumption is the basis for the conventional stress-controlled "smeared" technique of modelling the cracked concrete.

In the mentioned computer program for ultimate load analysis of thin RC shells by the FEM, tension stiffening is considered on the basis of a constitutive model proposed by Floegl and Mang [13].

New Fracture Criterion for Concrete

This Section is a revised and abridged version of Section 3 of [5].

Determination of magnifying factor for Kupfer's biaxial tensile strength

It is well known from pertinent tests that the bending tensile strength (modulus of rupture) of concrete beams, f_{btu}, increases with increasing stress gradient. By analogy, it is concluded that the biaxial tensile strength of concrete at a point of a surface structure increases with increasing stress gradients at this point. For the limiting case of vanishing stress gradients, the peak stress $\sigma_{1u}(\alpha_1)$ $(\sigma_{1u} \geqq \sigma_{2u})$ in the $\sigma_1 - \varepsilon_1$ diagram obtained from Kupfer's biaxial tests is taken as the biaxial tensile strength at the considered point. For non-vanishing stress gradients, a magnifying factor k for σ_{1u} is determined by means of a suitable hypothesis. Thus, $\sigma_{1u}(\alpha_1)$ represents a lower bound to the true biaxial tensile strength $\sigma^*_{1u}(\alpha_1) = k\sigma_{1u}$ where $k \geqq 1$. The new criterion for failure of concrete by cracking is $\sigma_1 \geqq k\sigma_{1u} \geqq \sigma_{1u}$. For $-0.1 \leqq \alpha_2 \leqq 1$ (part of tension-compression domain, biaxial-compression domain) σ_{1u} is not increased. For this range of values for α_2, with the exception of the very small subdomain $\alpha_{2L} \cong -0.02 \leqq \alpha_2 \leqq -0.1$, concrete fails by crushing. Reasons for setting $k = 1$ in case of crushing of concrete are (a) lack of experimentally corroborated knowledge about the dependence of concrete failure by crushing on the stress gradient and (b) the fact that global failure of thin RC surface structures is usually triggered by cracking rather than crushing of concrete [5].

The condition for determination of k is [5]:

$$\frac{U^*_{0R}}{U_{0R}}\Bigg|_{(r,\ell,e)} = \frac{U^{(b)}_{0R}\,(\frac{\ell}{3}<x<\frac{2\ell}{3},\ z=\frac{h}{2})}{U^{(t)}_{0R}} \,. \tag{1}$$

The left side of this equation refers to the considered integration point in the FE model of the investigated surface structure. The symbol $|(r,\ell,e)$ identifies this point as the integration point r of concrete layer ℓ of finite element e. The right side of equation (1) is related to a fictitious beam with rectangular cross-section of yet to be determined depth h, subjected, for $\ell/3 < x < 2\ell/3$, to pure bending (Fig.1(a)), and to uniform axial tension (Fig.1(c)), respectively. The argument of $U_{0R}^{(b)}$ refers to the center part of the tensile fiber of the beam shown in Fig.1(a).

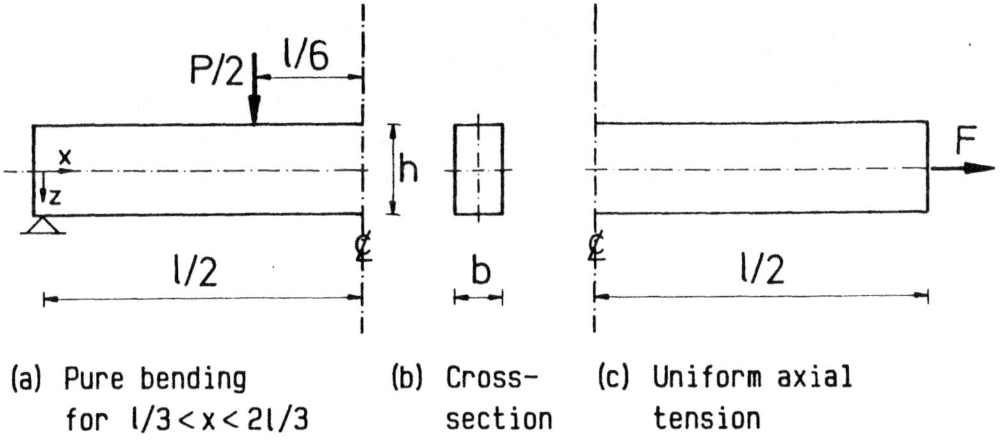

Fig.1. Fictitious concrete beam [5]

U_{0R}^{*} is the true strain-energy density released at the considered integration point when concrete fails by cracking. U_{0R} is the strain-energy density which would be released at this point when concrete fails by cracking, if the stress gradients were zero and if $\alpha_1 = \sigma_{2u}/\sigma_{1u} = \sigma_{2u}^{*}/\sigma_{1u}^{*}$, noting that Kupfer's tests were carried out for proportional loading. $U_{0R}^{(b)}$ ($U_{0R}^{(t)}$) is the strain-energy density released at incipient cracking of the fictitious beam of Fig.1(a) (Fig.1(c)), assuming that cracks start opening when the peak stress in the stress-strain diagram is reached. Equation (1) relates the *unknown* influence of the stress gradients on the biaxial tensile strength of concrete at the considered point of the surface structure to the *known* influence of the stress gradient on the uniaxial tensile strength of concrete for a fictitious beam of yet to be determined depth, subjected to pure bending [5].

As was shown in [5], equation (1) leads to

$$\frac{\sigma_{1u}^{*}}{\sigma_{1u}} = k = \frac{f_{btu}(h)}{f_{tu}} \tag{2}$$

expressing that the ratio $\sigma_{1u}^{*}/\sigma_{1u}$ at the considered point of the surface structure is the same as the ratio f_{btu}/f_{tu} for a concrete beam of yet to be determined depth h.

Fig.2 shows a comparison of a typical stress-strain diagram obtained from one of Kupfer's biaxial tests, characterized by uniform stress distributions, with a corresponding stress-strain diagram based on nonuniform stress distributions. $E_C^{(0)}/(1-\nu_C\alpha_1)$ and $E_C^{(0)*}/(1-\nu_C\alpha_1)$ are effective initial tangent moduli; $E_C^{(0)}$ and $E_C^{(0)*}$ are initial tangent moduli; ν_C is Poisson's ratio. σ_{1u}^* and $E_C^{(0)*}$ are unknown quantities which would have to be determined from suitable tests for nonuniform loading. If results from such tests were available, theoretical considerations for determination of k would be superfluous [5].

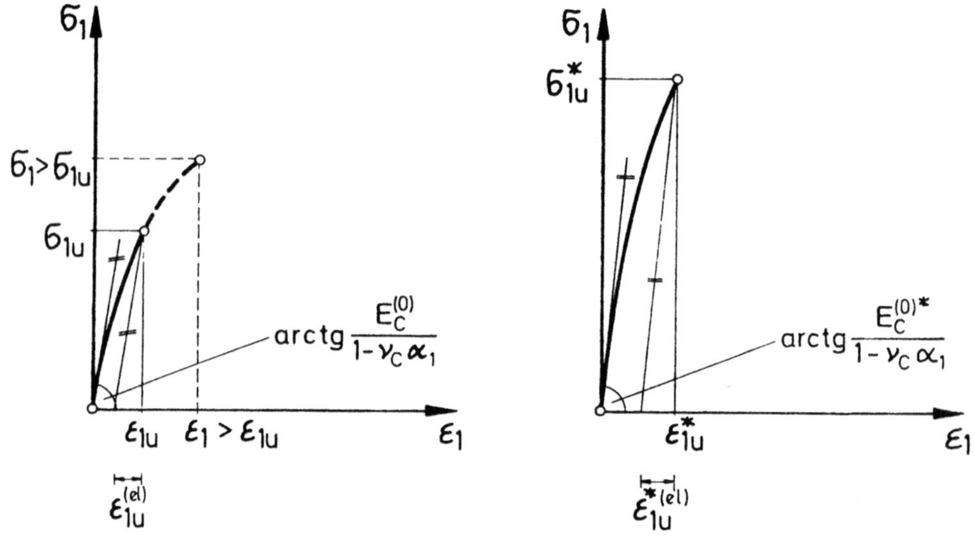

(a) $\sigma_1-\varepsilon_1$ diagram for uniform loading, based on Kupfer's biaxial tests

(b) corresponding $\sigma_1-\varepsilon_1$ diagram for nonuniform loading

Fig.2. Magnification of Kupfer's biaxial tensile strength σ_{1u} [5]

Determination of k from tests for beams

If h was known, k(h) could be determined from modulus-of-rupture and direct-tensile-strength tests. For this purpose, equation (2) is rewritten as [5]

$$k = \frac{f_{btu}}{f_{btu}^{(10)}} \cdot \frac{f_{btu}^{(10)}}{f_{tu}} \tag{3}$$

where $f_{btu}^{(10)}$ is the bending tensile strength of a 10 cm deep beam.

According Hellmann [14], $f_{btu}^{(10)}$ and f_{tu} can be expressed in terms of the cube strength of concrete, f'_{cu}, as follows:

$$f_{btu}^{(10)} = c_{bt}(f'_{cu})^{\frac{2}{3}} , \qquad f_{tu} = c_t(f'_{cu})^{\frac{2}{3}} . \tag{4}$$

Evaluating a large number of test results from the literature, Hellmann reported the following average values of c_{bt} and c_t: $c_{bt} = 0.98$ and $c_t = 0.52$. Thus,

$$k = 1.88 \frac{f_{btu}}{f_{btu}^{(10)}} . \tag{5}$$

Mayer [15] evaluated a large number of pertinent test results from the literature. The $f_{btu}/f_{btu}^{(10)}$ – h diagram shown in Fig.3(a) is based on this evaluation. The diagram shows that the bending tensile strength of concrete increases with decreasing depth of the beam. Consequently, f_{btu} increases with increasing stress gradient. According to equation (5), k(h) is obtained by multiplying the ordinates of the mentioned diagram by 1.88. Fig.3(b) shows the k – h diagram. The dashed part of the diagram is an extrapolation from the part of the curve which is based on experimental results. In general, the former is of no practical relevance.

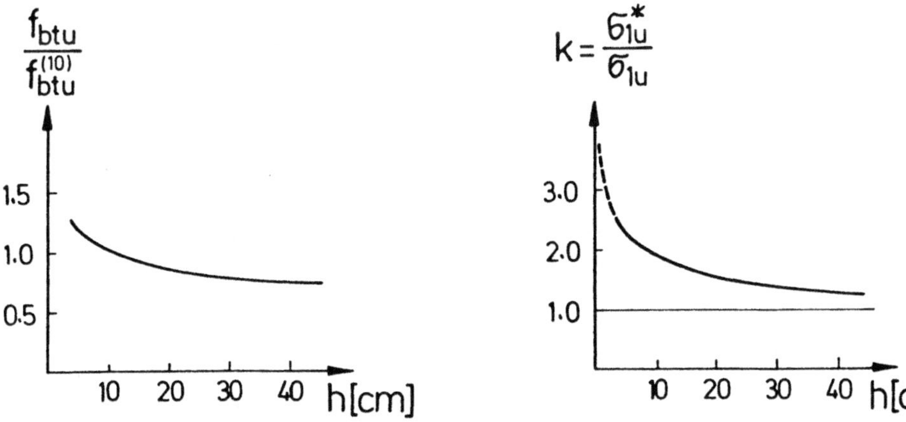

(a) Dependence of bending tensile strength of concrete on depth of beam [15]

(b) Dependence of biaxial tensile strength of concrete on depth of fictitious beam

Fig.3. Determination of magnifying factor k [5]

Determination of h from a Hypothesis of Comparison

The depth of the beam of Fig.1 is determined from the following hypothesis of comparison, valid for uncracked concrete [5]:

$$\left.\frac{|\nabla\hat{U}_0|}{\hat{U}_0}\right|_{(r,\ell,e)} = \left.\frac{\dfrac{\partial U_0^{(b)}}{\partial z}}{U_0^{(b)}}\right|_{\frac{\ell}{3}<x<\frac{2\ell}{3},\quad z=\frac{h}{2}} . \tag{6}$$

where $\hat{U}_0|_{(r,\ell,e)}$ is the elastic part of the strain-energy density at the considered integration point, $|\nabla\hat{U}_0||_{(r,\ell,e)}$ is the absolute value of the gradient vector $\nabla\hat{U}_0$ at this point,

$$U_0^{(b)}\left(\frac{\ell}{3}<x<\frac{2\ell}{3},\ z=\frac{h}{2}\right) = \frac{1}{2E_C^{(0)*}}\left(\frac{P\ell}{bh^2}\right)^2 \tag{7}$$

is the strain-energy density at a point in the center part of the tensile fiber of the beam of Fig.1(a) and

$$\frac{\partial U_0^{(b)}}{\partial z}\left(\frac{\ell}{3}<x<\frac{2\ell}{3},\ z=\frac{h}{2}\right) = \frac{2}{E_C^{(0)*}h}\left(\frac{P\ell}{bh^2}\right)^2 \tag{8}$$

is the gradient of $U_0^{(b)}$ at the mentioned point. Since Hooke's law holds for the fictitious bending beam, it appears logical to use the elastic part of U_0 and of ∇U_0 for the comparison expressed by equation (6).

With the help of the relations (7) and (8), h is obtained from equation (6) as

$$h = 4\left.\frac{\hat{U}_0}{|\nabla\hat{U}_0|}\right|_{(r,\ell,e)} . \tag{9}$$

Details concerning computation of $\hat{U}_0$ and $|\nabla\hat{U}_0|$ are given in [5].

Numerical Investigation

This Section contains a study of the influence of the new fracture criterion on the structural response of a hypar groined vault subjected to dead and snow load. Fig.4(a) shows the shell structure. It was previously analyzed by Schaper [16]. The side lengths 2ℓ of the square in plan view are 30.48 m. The height h_E of the shell at the mid-points of the four edges is 11.43 m above the plane spanned by the supports. The height h_E of the shell is 9.60 m above the mentioned plane. The shell has a uniform thickness t of 0.1016 m.

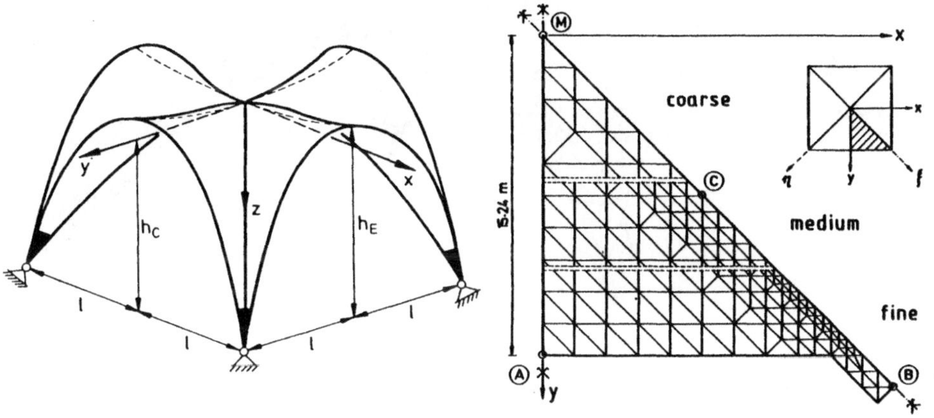

(a) Hypar groined vault (b) Mapping of FE grids on x-y plane

Fig.4. Hypar groined vault

It is beyond the scope of the present paper to discuss structural deficiencies of the described hypar groined vault. For a practical design, appropriate stiffening of the groins would have to be provided, such that shear forces and vertical components of axial forces could be carried for all possible load cases. Moreover, such a design would have to be more specific about the supports. As was shown by Eberhardsteiner and Mang [17], considerable strainless deformations will occur in the given hypar groined vault unless the shell is stiffened adequately in the support region. The shaded area in Fig.4(a) symbolizes such a stiffening.

For dead and snow load it is sufficient to consider only one octant of the hypar groined vault. Fig.4(b) shows the mapping of the FE grids (coarse grid: FE grid 1, medium grid: FE grid 2, fine grid: FE grid 3) of the considered shell octant on the x-y plane. The medium and the fine grid represent successive consistent refinements of the groin region characterized by large stress gradients. Fig.5 shows the discretization of the whole hypar groined vault (FE grid 3).

In explicit form, the middle surface Ω of the hypar shown in Fig.6 is given as

$$z = \frac{x^2}{a} - \frac{y^2}{b} \tag{10}$$

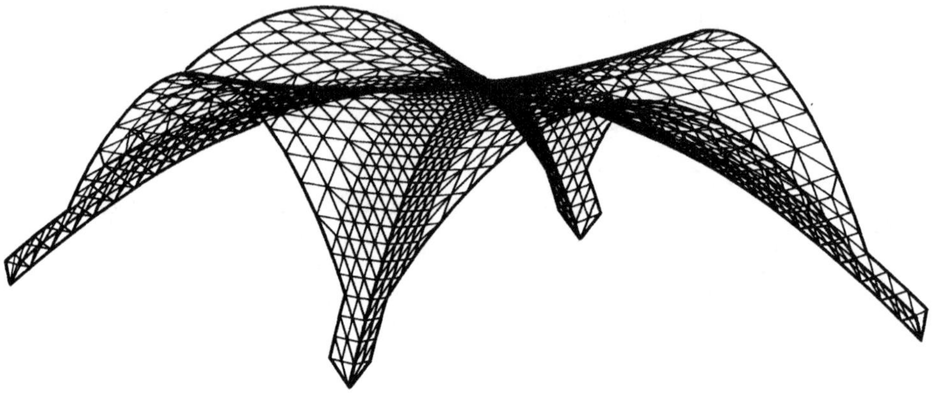

Fig.5. Discretization of hypar groined vault (FE grid 2)

where $a = \ell^2/h_E = 20.23$ m and $b = \ell^2/(h_E-h_C) = 126.92$ m. The previously mentioned computer program is restricted to orthogonal parameter lines [6]. Parameter lines $\alpha = x =$ const., $\beta = y =$ const., representing a logical choice for a parametric representation of the middle surface of the hypar, are non-orthogonal one to another. Employing the concept of "discrete orthogonalization of parameter lines" at the integration points of the finite elements [17], it is not necessary to extend the available computer program to skew curvilinear co-ordinates. Such an extension would require a substantial amount of recoding.

Fig.6 refers to the description of Ω by a parameter form, given as [17]

$$x = \alpha + c\frac{\beta^2}{b}, \quad y = \beta, \quad z = \frac{1}{a}\left(\alpha + c\frac{\beta^2}{b}\right)^2 - \frac{\beta^2}{b} = \frac{x^2}{a} - \frac{y^2}{b} \tag{11}$$

where c is a constant to be determined from the orthogonality criterion for parameter lines, applied to point $P(x_0,y_0,z_0)$. This criterion is given as

$$F(x_0,y_0,z_0,c) = \left(\frac{\partial x}{\partial\alpha}\frac{\partial x}{\partial\beta} + \frac{\partial y}{\partial\alpha}\frac{\partial y}{\partial\beta} + \frac{\partial z}{\partial\alpha}\frac{\partial z}{\partial\beta}\right)\Big|_{(x_0,y_0,z_0,c)} = 0, \tag{12}$$

resulting in

$$c = \frac{2\frac{x_0}{a}}{\left(2\frac{x_0}{a}\right)^2 + 1} . \tag{13}$$

Fig.6 shows that at points on Ω, except at point P, parameter lines $\alpha =$ const., in general, intersect parameter lines $\beta = y =$ const. non-orthogonally.

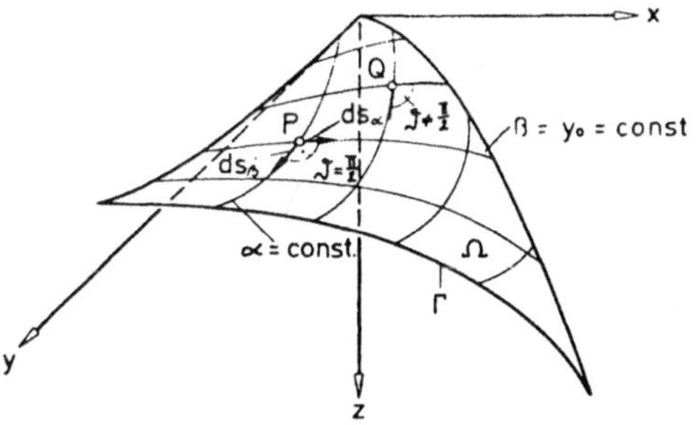

Fig.6. Orthogonalization of parameter line α = const. at point $P(x_0,y_0,z_0)$ on Ω [17]

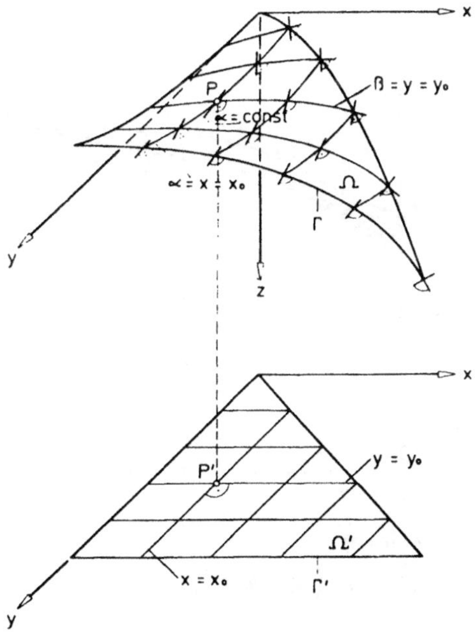

Fig.7. Discrete orthogonalization of parameter lines α = const. on Ω [17]

Fig.7 refers to discrete orthogonalization of parameter lines α = const. on the middle surface of the considered octant of the hyperbolic paraboloid groined vault. It is characterized by applying equation (12) to all integration points of the finite elements.

Before presentation of selected results of ultimate load analysis of the hypar groined vault subjected to dead and snow load, typical results of a convergence study restricted to consideration of dead load and to linear elasticity are shown (Fig.8). For this mode of analysis, only $E_C^{(0)}$, ν and γ (specific weight of concrete) are needed. Numerical values of these quantities are given in Table 1. Fig.8(a) shows plots of the vertical displacement w of the parameter line $\alpha = x = 0$ for the three mentioned FE grids. Very good convergence of the solution is found. The reason for the partial deviation of Schaper's results, obtained with the help of SAP IV [18], from the authors results is the sensitivity of the displacements on the stiffness of elastic springs used by Schaper to model the supports of the shell [16].

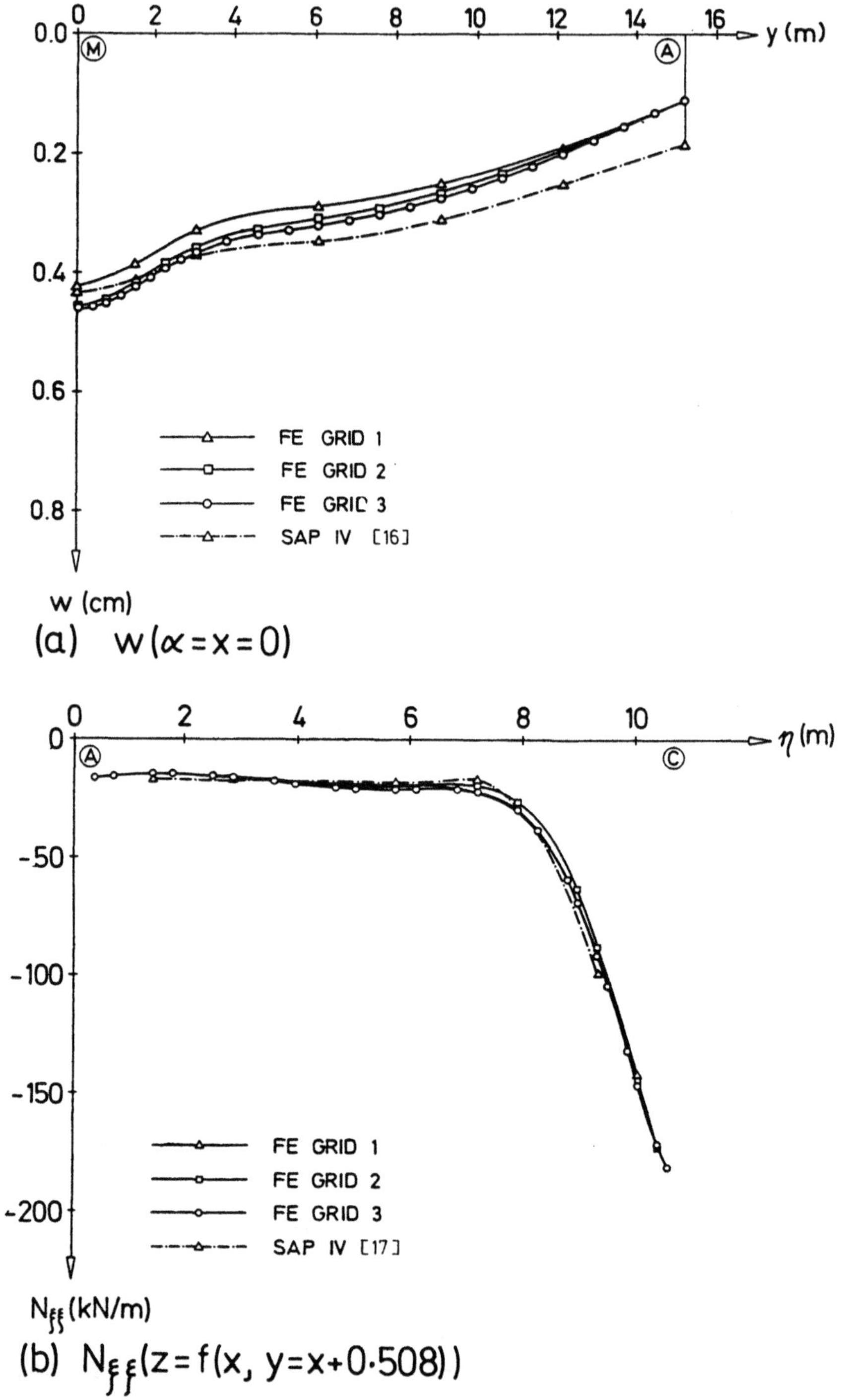

Fig.8. Typical results of convergence study based on linear elastic analysis

Fig.8(b) illustrates that the convergence of the solution for the membrane forces $N_{\xi\xi}$ along $z = f(x, y = x + 0.508)$ is very good.

Table 1 Material properties of concrete and steel

Concrete			Steel		
$E_C^{(0)}$	3.00	[MN/cm²]	$E_S^{(0)}$	20.6	[MN/cm²]
ν_C	0.20	-	$E_S^{(h)}$	2.06	[MN/cm²]
γ	28.83	[kN/m³]	σ_{Sy}	42.0	[kN/cm²]
f_{cu}	-3.00	[kN/cm²]	σ_{Su}	50.0	[kN/cm²]
f_{tu}	0.191	[kN/cm²]			
α_{2L}	-0.0226	-			

Ultimate load analysis of the hypar groined vault is based on the load combination $g + \chi s$ where (apart from the groin region where the thickness of the shell is larger than 0.1016 m [16]) $g = 2.929$ kN/m² is the dead load per unit area of the middle surface of the shell, $s = 1.464$ kN/m² is the reference snow load per unit area of the horizontal projection of the middle surface of the shell and χ is a dimensionless factor by which s is amplified. The value of the load factor for which the analysis signals collapse of the hypar groined vault is denoted as χ_{cr}.

Additional material data needed for ultimate load analysis of the shell are compiled in Table 1. Quantities in Table 1 which were not defined previously are τ_{Bu} (bond strength), $E_S^{(0)}$ (modulus of elasticity of steel), $E_S^{(h)}$ (hardening modulus of steel), σ_{Sy} (yield stress of steel), and σ_{Su} (ultimate stress of steel). The quantities $E_S^{(0)}$, $E_S^{(h)}$, σ_{Sy} and σ_{Su} define a bilinear stress-strain diagram for steel. Information on two orthogonal nets of reinforcement (one near the outer and one near the inner surface of the shell) is given in [16]. The reinforcement nets are treated as layers of "smeared reinforcement". The finite elements are divided into five concrete layers.

Fig.9 shows the dependence of χ_{cr} on the FE grids for $k \geq 1$ (new fracture criterion) as well as for $k = 1$ (conventional fracture criterion). The plots are based on values for χ_{cr} for n = 1,4, and 16 where n = 4 and 16, respectively, is the number of finite elements into which a typical element of FE grid 1 (n=1), coinciding at one of its edges with the groin, must be subdivided in order to obtain a typical element of FE grid 2 and 3, respectively, coinciding at one of its edges with the groin. This mode of interpreting the convergence speed of the results for χ_{cr} reflects the local character of the mesh refinement which is a consequence of the restriction of large stress gradients to the vicinity of the groin. Because of these gradients the present problem is a tough test for the developed computer program for ultimate load analysis of RC surface structures [6]. Fig.9 shows that the difference of corresponding results for χ_{cr} obtained for k ≧ 1 and k = 1, respectively, increases with increasing n. In view of the large stress gradients in the groin region, this result is not astonishing. Since the new fracture criterion considers the increase of the biaxial tensile strength of concrete with increasing stress gradients, it might be expected that $\chi_{cr}(k=1)$ must be smaller than $\chi_{cr}(k \geqq 1)$. For the given problem, however, $\chi_{cr}(k=1) \geqq \chi_{cr}(k \geqq 1)$.

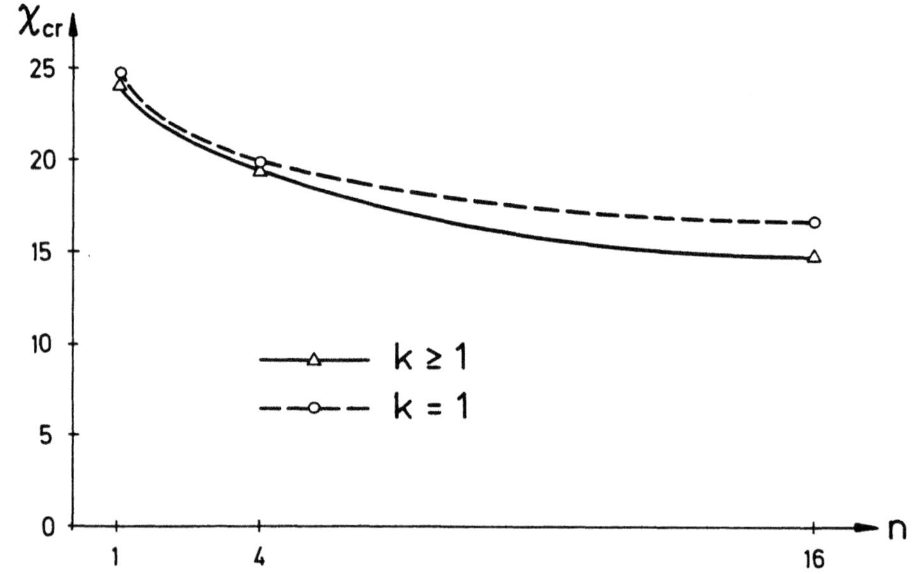

Fig.9. Convergence study: dependence of χ_{cr} on the FE grids for k ≧ 1 as well as for k = 1

Since the collapse of a structure is a global phenomenon, it is impossible to "extrapolate" from the character of a locally applied fracture criterion to the global behavior of a structure. (In this context it may be of interest that ultimate load analysis of a hyperbolic cooling tower [5] has given the "expected" result, namely, $\chi_{cr}(k \geqq 1) \geqq \chi(k=1)$).

Fig.10 shows plots of the vertical displacement of the crown of the shell (point M), w_M, versus the load function χ for k ≧ 1 and k = 1. The plots refer to FE grid 3. According to this figure and to similar figures published by Schaper [16], the structural response of the considered hypar groined vault can be characterized as brittle.

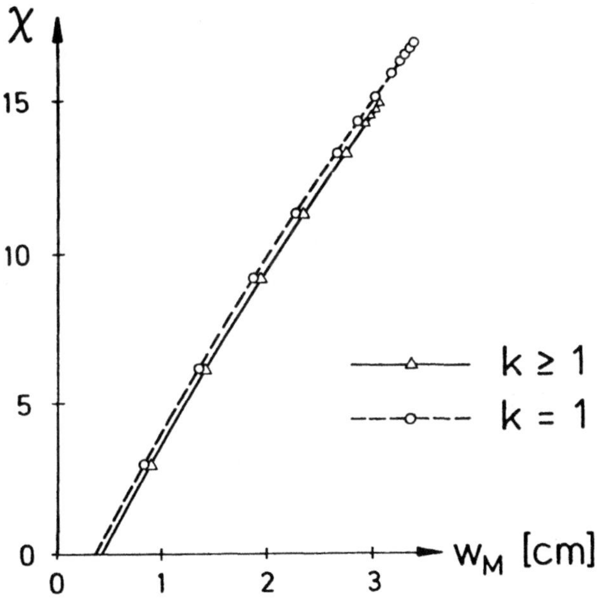

Fig.10. Load-displacement diagrams for the crown of the shell (FE grid 3)

Fig.11 shows plots of the vertical displacement w of the groin for $k \geq 1$ and $k = 1$ at $\chi = 3.0$ and 14.6, respectively. The plots refer to FE grid 3. Near the support, the response of the shell is slightly stiffer for $k \geq 1$ than it is for $k = 1$. The opposite is true for the crown of the shell.

Fig.12 shows plots of "smeared cracks" in two different concrete layers, at $\chi = 14.0$ for $k \geq 1$. The plots refer to FE grid 3. The direction of the dashes indicates the direction of the cracks. The lengths of the dashes are proportional to the tributary areas of the respective integration points. The total number of "cracked integration points" is 1532. (The total number of integration points is 12705.) For $k = 1$, the total number of "cracked integration points" is 2503. The additional cracks are restricted to the groin region, characterized by a cluster of cracks. Therefore, plots of "smeared cracks" for $k = 1$ of the form of the plots of Fig.12 would not permit illustrating the difference of the crack distribution based on $k \geq 1$ and $k = 1$, respectively. This is why such plots have been omitted in the paper.

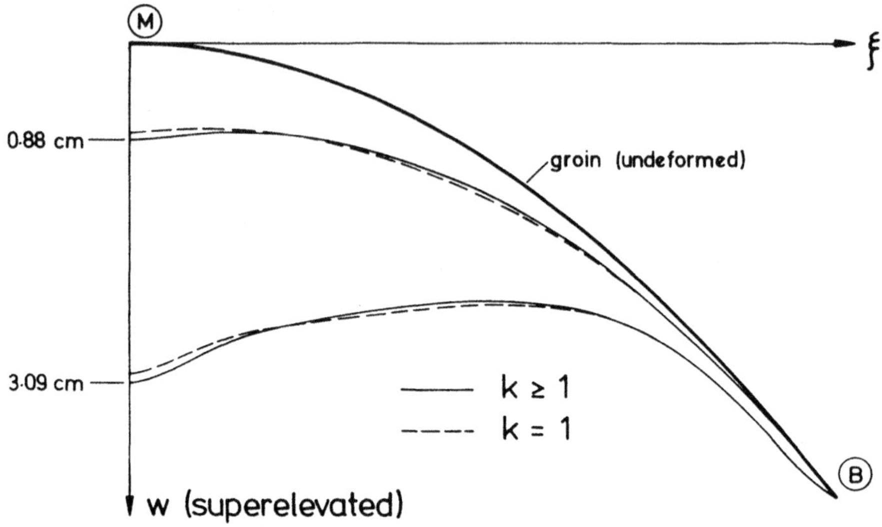

Fig.11. Vertical displacement of the groin (FE grid 3)

Conclusions

Based on results from ultimate load analysis of a hypar groined vault presented in this paper and of a hyperbolic cooling tower reported in [5], it appears that the classical strength criterion is acceptable for ultimate load analysis of these RC surface structures. This criterion ignores the strength-increasing effect of the stress gradients. Consideration of this influence is the nucleus of the new fracture criterion.

The ultimate load obtained for the hypar groined vault on the basis of the new fracture criterion is smaller than the one resulting from the conventional strength criterion. This is in contradiction to the conclusion drawn from ultimate load analysis of the hyperbolic cooling tower, according to which slightly conservative results should be expected when using the conventional strength criterion [5]. Since the collapse of a structure is a global phenomenon, it is impossible to draw general conclusions for the collapse of a surface structure from the nature of a fracture criterion applied at individual integration points.

For the hypar groined vault, the difference of corresponding results for the load factor χ_{cr} associated with collapse of the shell increases as the grid

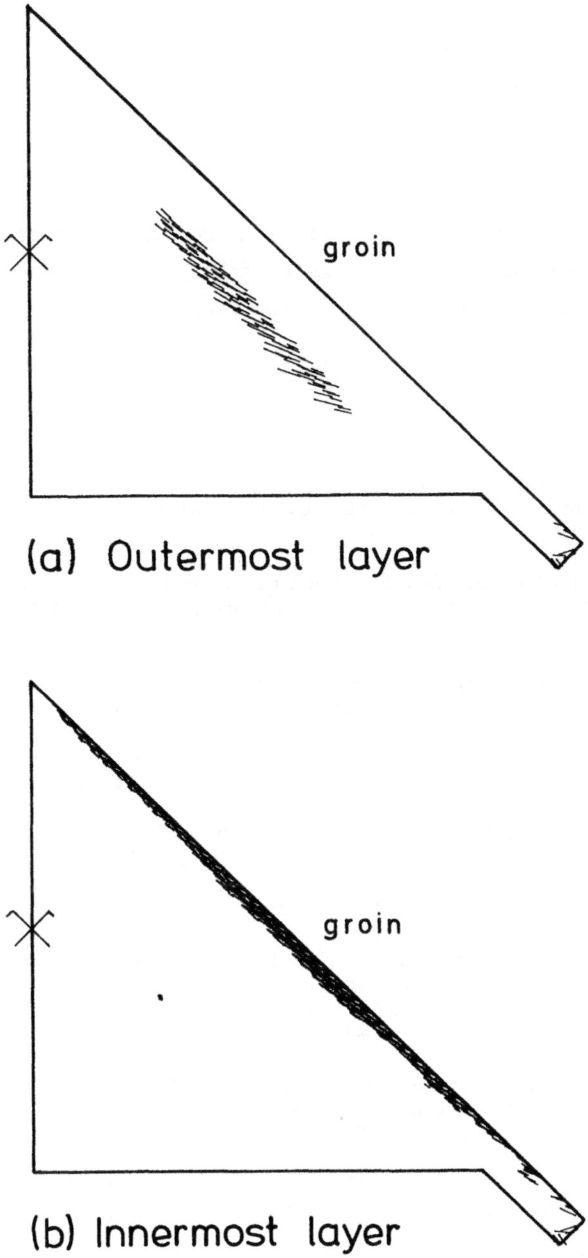

Fig.12. "Smeared cracks" in two different layers, at χ = 14.0 for k ≥ 1 (FE grid 3)

is refined. Additional grid refinements would be necessary to prove the objectivity of the results. This would require additional computer resources. At present, they are not available at the authors' academic institution.

Acknowledgments

Financial support of the Fonds zur Förderung der wissenschaftlichen Forschung (Austrian Science Foundation) and the Hochschuljubiläumsstiftung der Stadt Wien (University Jubilee Foundation of the City of Vienna) is gratefully acknowledged.

References

1. Bažant, Z.P.; Cedolin, L.: Blunt crack band propagation in finite element analysis. Journal of the Engineering Mechanics Division, ASCE, 105 (1979) 297-315.

2. Bažant, Z.P.; Cedolin, L.: Fracture mechanics of reinforced concrete. Journal of the Engineering Mechanics Division, ASCE, 106 (1980) 1287-1306.

3. Bažant, Z.P.; Cedolin, L.: Finite element modelling of crack band propagation. Journal of Structural Engineering, ASCE, 109 (1983) 69-92.

4. Bažant, Z.P.: Size effect in blunt fracture: concrete, rock, metal. Journal of Engineering Mechanics, ASCE, 110 (1984) 518-535.

5. Mang, H.A.; Eberhardsteiner, J.: Collapse analysis of thin RC shells on the basis of a new fracture criterion. Proceedings of the US-Japan Seminar on Finite Element Analysis of Reinforced Concrete Structures, Japan Concrete Institute, Tokyo, 1985.

6. Floegl, H.: Traglastermittlung dünner Stahlbetonschalen mittels der Methode der Finiten Elemente unter Berücksichtigung wirklichkeitsnahen Werkstoffverhaltens sowie geometrischer Nichtlinearität, Dissertation, Technische Universität Wien, Wien, 1981.

7. Koiter, W.T.: General equations of elastic stability for thin shells. Proceedings of the Symposium on the Theory of Shells to Honor Lloyd Hamilton Donell. University of Houston, Houston, 1967.

8. Mang, H.A.; Floegl, H.; Trappel, F.; Walter, H.: Wind-Loaded reinforced concrete cooling towers: buckling or ultimate load? Engineering Structures 5 (1983) 163-180.

9. Thomas, G.R.; Gallagher, R.H.: A triangular thin shell finite element: linear analysis. NASA Contractor Report 2482, Washington D.C., 1975.

10. Liu, T.C.A.; Nilson, A.H.; Slate, F.O.: Biaxial stress-strain relations for concrete. Journal of the Structural Division, ASCE, 98 (1972) 1025-1034.

11. Kupfer, H.: Das Verhalten des Betons unter mehrachsiger Kurzzeitbelastung unter besonderer Berücksichtigung der zweiachsigen Beanspruchung. Deutscher Ausschuß für Stahlbeton, Vol.229, Berlin: Wilhelm Ernst 1973.

12. Chen, W.F.; Ting, E.C.: Constitutive models for concrete structures. Journal of the Engineering Mechanics Division, ASCE, 106 (1980) 1-19.

13. Floegl, H.; Mang, H.A.: On tension stiffening in cracked reinforced concrete slabs and shells considering geometric and physical nonlinearity. Ingenieur-Archiv 51 (1981) 215-242.

14. Hellmann, H.G.: Besichtigungen zwischen Zug- und Druckfestigkeit des Betons. beton 19 (1969) 68-70.

15. Mayer, H.: Die Berechnung der Durchbiegung von Stahlbetonbauteilen. Deutscher Ausschuß für Stahlbeton, Vol.194, Berlin: Wilhelm Ernst 1967.

16. Schaper, G.: Comparative linear and nonlinear analysis of hyperbolic paraboloid groined vaults. Report No. US-SESM 80-12, Dept. of Civil Engineering, University of California, Berkeley, California, 1980.

17. Eberhardsteiner, J.; Mang, H.A.: Discrete orthogonalization of parameter lines on curved surfaces for finite element analysis of thin shells. International Journal for Numerical Methods in Engineering 21 (1985) 837-851.

18. Bathe, K.-J.; Wilson, E.L.; Peterson, F.E.: SAP IV - a structural analysis program for static and dynamic response of linear systems. Report No. EERC 73-11, Dept. of Civil Engineering, University of California, Berkeley, California, 1974.

19. Dodds, R.H.; Darwin, D.; Leibengood, L.O.: Stress controlled smeared cracking in RC beams. Journal of Structural Engineering, ASCE, 110 (1984) 1959-1976.

A Finite Element Technique for Analysis of Reinforced and Prestressed Concrete Structures

F. DAMJANIC
University of Split, Yugoslavia

Summary

A finite element technique for the two-dimensional and axisymmetric analysis of reinforced and prestressed concrete structures under both time-dependent static and transient thermal loading conditions is described. The material model is computationally simple but capable to simulate the dominant nonlinear behaviour of concrete and reinforcement. The model includes the simulations of tensile cracking (smeared crack concept), nonlinear multiaxial compressive behaviour of concrete and yielding of reinforcement. The solution technique described involves concurrently solving an uncoupled set of equations (the transient heat transfer equations and the incremental equilibrium equations) within a time interval to provide both the temperature and displacement increments. One numerical example illustrates the applicability of the method.

Introduction

A large number of numerical models and techniques for the analysis and design of reinforced concrete structures has been developed in the last two decades, and the literature abounds with examples of their implementation and application (See for example the state-of-the-art Proceedings ICC'84/1/). It is, however, to realise that the efficiency of numerical analysis is closely linked to the availability of reliable constitutive properties of concrete. Namely, although reinforced concrete is probably the most commonly used of all structural materials, concrete constitutive properties have not been completely identified, and there is no, at the present, generally accepted material law available to model its behaviour. The complexities of concrete behaviour under different conditions and existing experimental evidence are such that it is extremely difficult, if not impossible, to establish a numerical model which can be adequate for every real situation. Certain idealizations and simplifications have to be employed, if the model is to be used in the numerical ana-

Finite Element Methods for Nonlinear Problems
Europe-US Symposium, Trondheim, Norway 1985
Editors: Bergan, Bathe, Wunderlich

lysis of practical engineering problems. In fact, the final choice of a model depends on the application envisaged. As the result, numerous numerical models and approaches with considerable simplifying assumptions have been proposed as well as a large number of very sophisticated models /1,2/. However, in many cases the accuracy of these elaborated models far outweighs the accuracy of the available material properties, and they often require extensive computations. In addition, the complexity of practical engineering problems makes numerical analysis more expensive.

A finite element technique for the analysis of reinforced and prestressed concrete structures described in this work is computationally simple but capable to simulate the dominant nonlinear concrete and steel behaviour. The basic technique has been earlier developed and tested /3-5/, further improved /6/ and applied to the solution of practical engineering problems /7-10/.

The assumptions adopted in the material modelling are based on the known phenomenological observations and include the simulations of concrete cracking, tension stiffening (or tensile strain-softening), nonlinear multiaxial compressive behaviour of concrete and yielding of reinforcement. The constitutive relationship are incorporated in the general nonlinear finite element solution technique /11/, and utilizing this numerical formulation the BETHEV44 computer code for both the static ultimate load and transient thermal stress analysis of planar and axisymmetric structural concrete problems is developed.

In this work the finite element formulation for the transient thermal stress analysis is exclusively described. The problem of both time dependent static and thermal loading of concrete structures demands the inclusion of time effects in the analysis. Therefore, the technique involves concurrently solving an uncoupled set of equations within a time interval to provide both the temperature and displacement increments. A two-level time stepping scheme is employed to predict temperature changes occurring within a time interval and, in fact, an elasto-viscoplastic material model is employed, using an explicit forward-difference scheme incorporating an equilibrium iteration procedure, to evaluate the thermal stress field at particular time stations. The use of a fully coupled theory can be avoided in a majority of practical structural concrete problems, particularly in such cases, where both temperature and stress changes occur sufficiently slow, and where the structural deformation process has a negligible effect on the temperature field. However, the staggered approach adopted allows the material properties to be temperature dependent.

The results of one numerical example illustrate the applicability of the method and computer code.

Material modelling

In this section the constitutive model for the concrete is described in a form suitable for numerical computation. For reasons of space, only essential expressions are presented here. Detailed formulations can be found elsewhere /12/.

Compressive behaviour of concrete

The response of concrete under a multiaxial compressive stress state is assumed to be linear elastic until the yield criterion is reached. This criterion is formulated in terms of the first two stress invariants (the mean normal stress, I_1, and the shear stress invariant J_2) as

$$F(\underline{\sigma}) = f(\underline{\sigma}) - \sigma_o = [\beta(3J_2) + \alpha I_1]^{1/2} - \sigma_o = 0 \qquad (1)$$

where α and β are material parameters and σ_o is the equivalent effective stress taken as the ultimate compressive stress f_c from a uniaxial test. The parameters α and β are determined from the uniaxial and biaxial experimental results of Kupfer et al. /13/ giving $\alpha = 0.355\,\sigma_o$ and $\beta = 1.355$. The expression (1) is compared in Fig. 1 with the experimental results from Ref. /13/ in biaxial stress space. When the effective stress, defined by the yield function, reaches the ultimate stress f'_c, a viscoplastic response is assumed. For associated viscoplasticity the viscoplastic strain rate /11/ is given by

$$\dot{\underline{\varepsilon}}^{vp} = \gamma_c \langle \Phi \left[(f(\underline{\sigma}) - \sigma_o)/\sigma_o \right] \rangle (\delta f(\underline{\sigma})/\delta \underline{\sigma}) \qquad (2)$$

in which γ_c is a fluidity parameter and $\langle\Phi\rangle$ is the flow function taken as non-zero for $f(\underline{\sigma}) > \sigma_o$ only. The viscoplastic strain increment, according to an Euler forward-difference representation, is then defined as

$$\Delta \underline{\varepsilon}_n^{vp} = \dot{\underline{\varepsilon}}_n^{vp} \, \Delta t_n \qquad (3)$$

in which Δt_n is a time interval (In the static ultimate load analysis the time intervals represent only a fictitious time). The appropriate time step length can be selected satisfying the stability criterion given in Ref. /14/. The compressive yielding is permitted up to a limiting strain at which the concrete is assumed to crush. This crushing condition is developed by simply converting the yield criterion (1) described in terms of stresses directly into strains.

$$\beta(3 \; {}_{\varepsilon}J_2) + \alpha \; {}_{\varepsilon}I_1 = \varepsilon_{cu}^2 \qquad (4)$$

where ${}_{\varepsilon}I_1$ and ${}_{\varepsilon}J_2$ are strain invariants and ε_{cu} is an ultimate total strain extrapolated from uniaxial test results. When ε_{cu} reaches the value specified as the ultimate strain, the material is assumed to lose all its characteristics of strength and rigidity.

Tensile behaviour of concrete

The response of concrete under tensile stresses is also assumed to be linear elastic until the cracking condition, defined by a maximum tensile stress criterion (See Fig. 1), is reached. Cracks are assumed to form in planes perpendicular to the direction of maximum principal tensile stress (σ_1, σ_2, σ_3) as soon as this stress (or stresses) reaches the specified concrete tensile strength f_t'. The elasticity modulus and Poisson's ratio are then reduced to zero in the direction perpendicular to the cracked plane, and a reduced shear modulus, G*, (based on the strain component normal to the crack (Fig. 3)) is employed. For example, the stress-strain relationship for concrete cracked in the 1-pri-

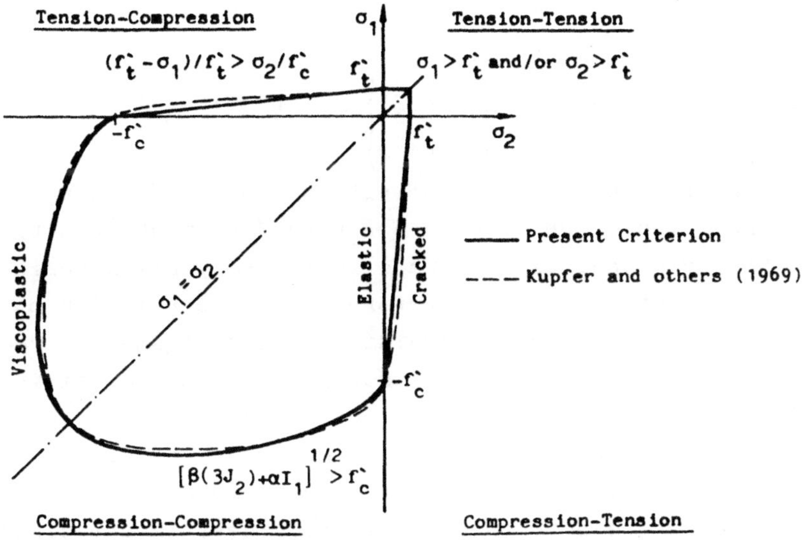

Fig.1. Compressive yield and tensile cracking criterion for concrete

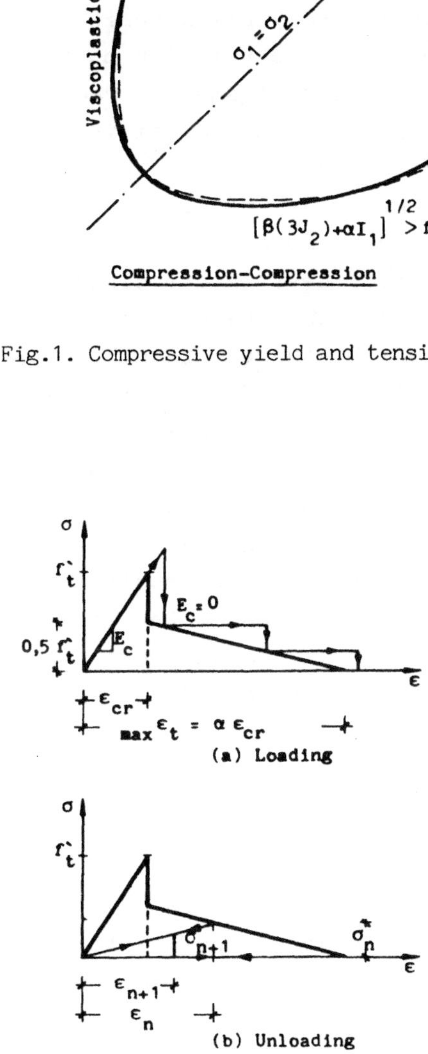

Fig.2. Tension stiffening model for concrete

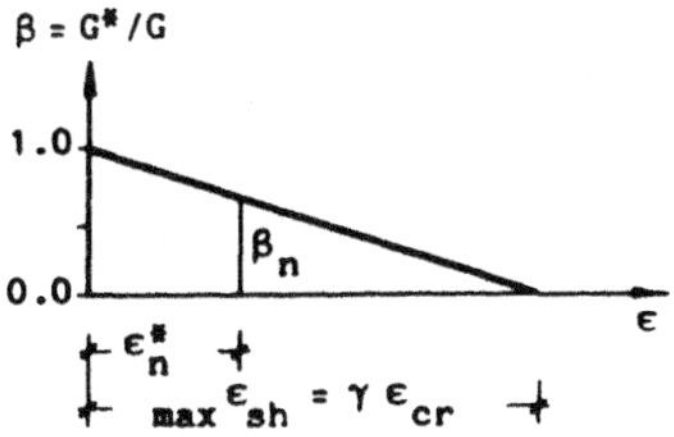

Fig.3. Reduced shear stiffness for cracked concrete

ncipal direction for plane stress situations takes the form

$$\begin{Bmatrix} \sigma_1^* \\ \sigma_2^* \\ \tau_{12}^* \end{Bmatrix} = \begin{bmatrix} 0 & 0 & 0 \\ 0 & cE & 0 \\ 0 & 0 & G^* \end{bmatrix} \begin{Bmatrix} \varepsilon_1^* \\ \varepsilon_2^* \\ \gamma_{12}^* \end{Bmatrix} = \underline{D}^* \underline{\varepsilon}^* \qquad (5)$$

where c=1 for cracking in the 1-direction and c=0 for cracking in both planes. Thus, the cracked concrete "becames" anisotropic, with the principal directions of anisotropy being defined by the crack directions, and therefore the above relation must be transformed to the global coordinate system. A smeared representation for cracked concrete, more convenient for finite element formulation, is adopted, which implies that cracks are not discrete but distributed across a region of the finite element. The post-cracking tensile behaviour is modelled by an adopted procedure, where both the tensile stiffness in the direction perpendicular to the cracked plane and the cracked shear stiffness are assumed to be functions of the current tensile strain component normal to this plane (Figs. 2 and 3). This approach takes into account aggregate interlock and dowel action, but full bond is assumed at the steel-concrete interface. The post-cracking model is adjusted by the tension stiffening (α) and shear retention (γ) parameters. The role of the parameters is examined comparing the numerical results with the results from structural tests /6/. It is shown that the parameter values significantly depend on the type of structural fracture expected. From the series of numerical experiments it has been found that the adequate parameter values are /6/:

Flexural fracture: α = 20 - 25, γ = 20 - 25
Shear fracture: α = 5 - 10, γ = 10 - 15

Compressive and tensile behaviour of steel

The steel reinforcing bars or prestressing cables are modelled by discrete bar or membrane elements lie within,or at the surface of, the "concrete" elements as shown in Fig. 4. The steel behaviour is therefore idealized using a uniaxial elasto-visco-plastic model resisting only the axial force in the reinforcing bar (or plane) direction. The corresponding numerical formulation can be found in Ref. /11/.

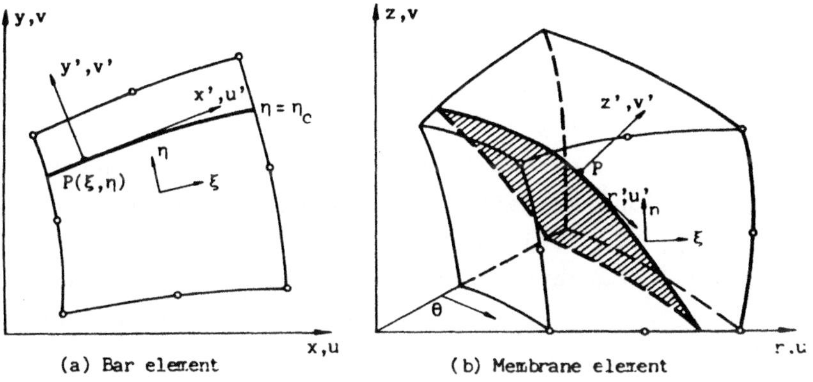

Fig.4. Bar and membrane elements within concrete elements

Finite element solution procedure

The 8-node Serendipity isoparametric elements and reduced integration rule are exclusively employed in the present analysis. The isoparametric elements are used to discretize structural concrete while special elements, embedded in the isoparametric elements, as shown in Fig. 5, are employed to simulate reinforcement /15/. The progress of cracked zones, as well as the compressive behaviour of concrete is analysed and monitored for each element integration point.

The solution technique described in this section involves concurrently solving an uncoupled set of equations (the transient heat transfer equations and the incremental equilibrium equations) within a time interval to provide both the temperature and displacement increments. However, the same finite element mesh and the same algorithm for the solution of the nonlinear problems, based on an incremental and iterative modified Newton-Raphson scheme /11/ is used.

Transient thermal analysis

The temperature field $\underline{\mathbf{T}}_n$ existing in the solid at any instant of time t_n must satisfy the following equation system

$$\underline{\mathbf{K}}(\underline{\mathbf{T}}_n)\ \underline{\mathbf{T}}_n + \underline{\mathbf{C}}(\underline{\mathbf{T}}_n)\ \underline{\mathbf{T}}_n + \underline{\mathbf{f}}(\underline{\mathbf{T}}_n, t) = 0 \tag{6}$$

in which $\underline{\mathbf{K}}$ and $\underline{\mathbf{C}}$ are, respectively, the thermal conductivity and heat capacity matrices, which may be temperature dependent, and $\underline{\mathbf{f}}$ represents the thermal loading which may vary with respect to both temperature and time. Employing the usual two-level scheme for the time discretisation of (6) results in the solution algorithm for the variation of temperature within a time interval $\Delta t_n = t_{n+1} - t_n$. This algorithm is fully described (Also, its stability and accuracy is extensively discussed) in Refs. /16,17/, and therefore, there is no need to be presented here.

Thermal stress determination

As the thermal transient solution proceeds with the temperature field, i.e. the temperature increment $\Delta\underline{\mathbf{T}}_n = \underline{\mathbf{T}}_{n+1} - \underline{\mathbf{T}}_n$, being evaluated at each time interval Δt_n, the corresponding thermal strain increment, $\Delta\underline{\boldsymbol{\varepsilon}}_n^t$, can be computed in the standard manner. Thus, for example, for plane stress situations,

$$\Delta\underline{\boldsymbol{\varepsilon}}_n^t = \begin{Bmatrix} \Delta\varepsilon_x^t \\ \Delta\varepsilon_y^t \\ \Delta\gamma_{xy}^t \end{Bmatrix} = \begin{Bmatrix} \alpha\,\Delta T_n \\ \alpha\,\Delta T_n \\ 0 \end{Bmatrix} \tag{7}$$

in which $\alpha_t = \alpha_t(T_n)$ is the thermal expansion coefficient.Expressing the total strain in terms of the displacement increment $\Delta\underline{\mathbf{d}}_n$, and the viscoplastic component according to (3), the stress increment is given by

$$\Delta\underline{\boldsymbol{\sigma}}_n = \underline{\mathbf{D}}_n\,(\underline{\mathbf{B}}\,\Delta\underline{\mathbf{d}}_n - \dot{\underline{\boldsymbol{\varepsilon}}}_n^{vp}\,\Delta t_n) - \Delta\underline{\boldsymbol{\sigma}}_n^t \tag{8}$$

in which $\underline{\mathbf{B}}$ is the usual strain-displacement matrix and $\underline{\mathbf{D}}_n$ is the elasticity matrix modified to account for tensile cracking behaviour, as shown for a plane stress situation in (5). The term $\Delta\underline{\boldsymbol{\sigma}}_n^t$ represents the initial stress system associated with the temperature change occurring during the time interval Δt_n. So that

$$\Delta\underline{\boldsymbol{\sigma}}_n^t = \underline{\mathbf{D}}_n\,\Delta\underline{\boldsymbol{\varepsilon}}_n^t \tag{9}$$

The incremental form of the equilibrium equations is

$$\int_\Omega \underline{\mathbf{B}}^T\,\Delta\underline{\boldsymbol{\sigma}}_n\,d\Omega = \Delta\underline{\mathbf{f}}_n \tag{10}$$

in which $\Delta \underline{\mathbf{f}}_n$ represents the change in equivalent nodal loads resulting from body forces and tractions occurring during the time interval Δt_n. Substituting from (8) in (10) results in the following expression for the displacement increment

$$\Delta \underline{\mathbf{d}}_n = [{}_T\underline{\mathbf{K}}_n]^{-1} \quad \Delta \underline{\mathbf{V}}_n \tag{11}$$

where the current stiffness matrix ${}_T\underline{\mathbf{K}}_n$ (including the contribution of reinforcement, $\underline{\mathbf{K}}_n^r$ /15/) is

$$ {}_T\underline{\mathbf{K}}_n = \int_\Omega \underline{\mathbf{B}}^T \underline{\mathbf{D}}_n \underline{\mathbf{B}} \, d\Omega + \underline{\mathbf{K}}_n^r \tag{12}$$

and the incremental "pseudo-loads", $\Delta \underline{\mathbf{V}}_n$, are given by

$$\Delta \underline{\mathbf{V}}_n = \Delta \underline{\mathbf{f}}_n + \Delta \underline{\mathbf{f}}_n^t + \Delta \underline{\mathbf{f}}_n^{vp} + \Delta \underline{\boldsymbol{\Phi}}_n \tag{13}$$

in which

$$\Delta \underline{\mathbf{f}}_n^t = \int_\Omega \underline{\mathbf{B}}^T \Delta \underline{\boldsymbol{\sigma}}_n^t \, d\Omega \tag{14}$$

$$\Delta \underline{\mathbf{f}}_n^{vp} = \int_\Omega \underline{\mathbf{B}}^T \underline{\mathbf{D}}_n \underline{\mathbf{B}} \, \dot{\underline{\boldsymbol{\varepsilon}}}_n^{vp} \Delta t_n \, d\Omega \tag{15}$$

$$\Delta \underline{\boldsymbol{\Phi}}_n = \int_\Omega \underline{\mathbf{B}}^T (\underline{\boldsymbol{\sigma}}_n - \underline{\boldsymbol{\sigma}}_n^t) \, d\Omega + \underline{\mathbf{f}}_n + \underline{\mathbf{f}}_n^t \tag{16}$$

where $\Delta \underline{\mathbf{f}}_n$, $\Delta \underline{\mathbf{f}}_n^t$ and $\Delta \underline{\mathbf{f}}_n^{vp}$ represent the change of equivalent nodal forces due to mechanical and thermal loads, and viscoplastic flow respectively. The vector $\Delta \underline{\boldsymbol{\Phi}}_n$ represents the residual or out--of-balance forces, in which $\underline{\mathbf{f}}_n$ and $\underline{\mathbf{f}}_n^t$ are the total mechanical and thermal loads exsisting at the time station t_n. With all incremental values known, quantities can be updated to give the solution at time t_{n+1} as follows

$$\underline{\mathbf{d}}_{n+1} = \underline{\mathbf{d}}_n + \Delta \underline{\mathbf{d}}_n , \qquad \underline{\boldsymbol{\sigma}}_{n+1} = \underline{\boldsymbol{\sigma}}_n + \Delta \underline{\boldsymbol{\sigma}}_n \tag{17}$$

The above time stepping procedure can be performed concurrently with the thermal process as many time intervals as required.

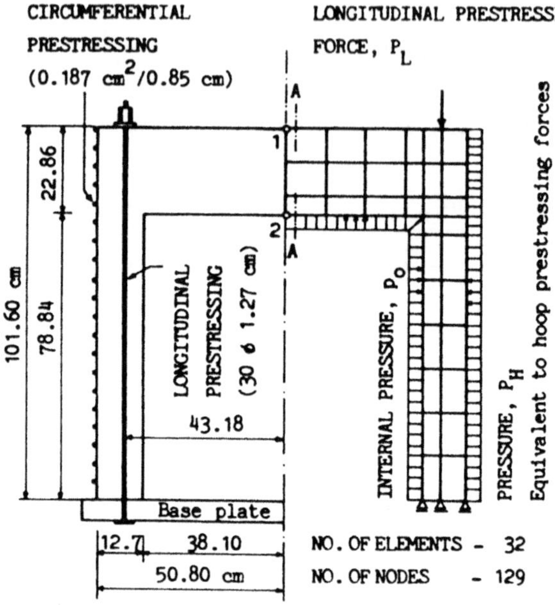

Fig.5. Prestressed concrete vessel showing problem details and finite element idealisation

Table 1. Problem parameters

Mechanical parameters		Thermal parameters	
Elasticity modulus, E	2965kN/m²	Heat conduct., K	0.014W/cm °C
Poisson's ratio, ν	0.15	Heat capacity, C	0.00054Wh/cm³ °C
Compres. strength, f'_c	4.758kN/m²	Radiation param., α_c	0.0021W/cm² °C
Tensile strength, f'_t	0.345kN/m²	Amb. temp., T_a	20°C
Fluidity param., γ_c	0.00025/hr	Init. temp., T_o	20°C
Tens. stiff. param., α	20		
Shear retent. param., γ	25	**Prestressing**	
Ult. comp. strain, ε_{cu}	0.004		
Therm. exp. param., α_t	0.00001	Longitud. force, P_L	3363.000kN
		Hoop prestress. force per cable, P_h	18.577kN
		Equivalent pressure due to hoop prestressing, p_h	0.352kN/cm²

Numerical example

The applicability of the technique described is illustrated by the analysis of a prestressed concrete vessel under time dependent interval pressure and thermal loading. The structure con-

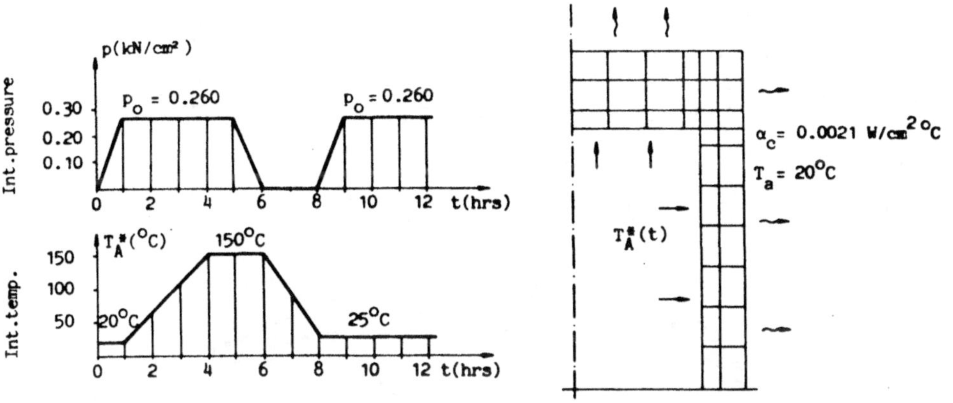

(a) Internal pressure and temperature variations with time

(b) Thermal boundary conditions

Fig.6. Time-dependent pressure and thermal loading

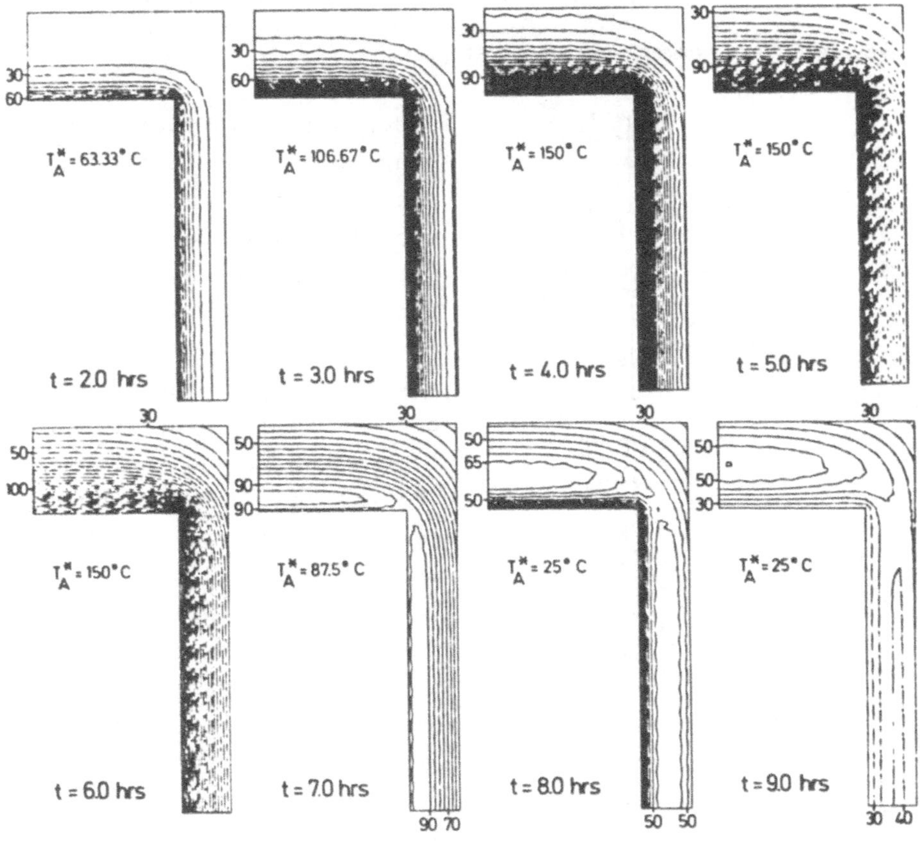

Fig.7. Isotherms for a one-half of the P.C. vessel at some representative time stations (Isotherms at 5°C intervals)

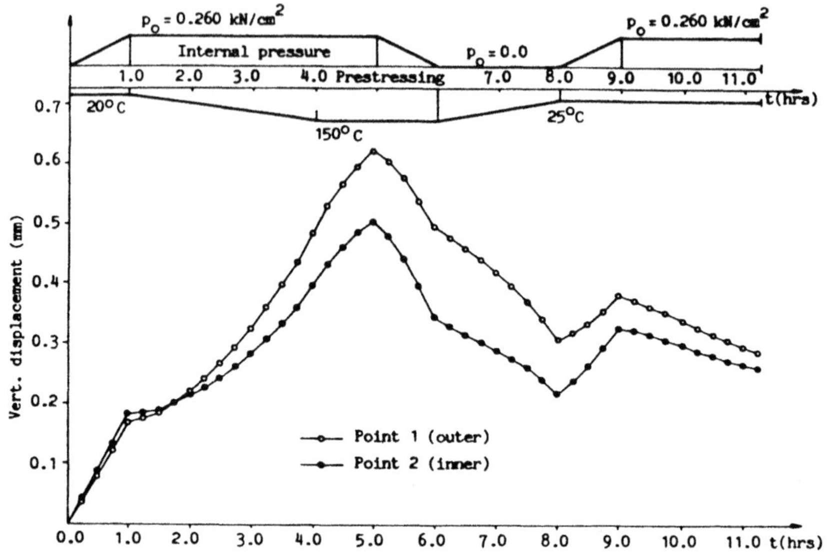

Fig.8. Vertical displacement variation at points 1 & 2

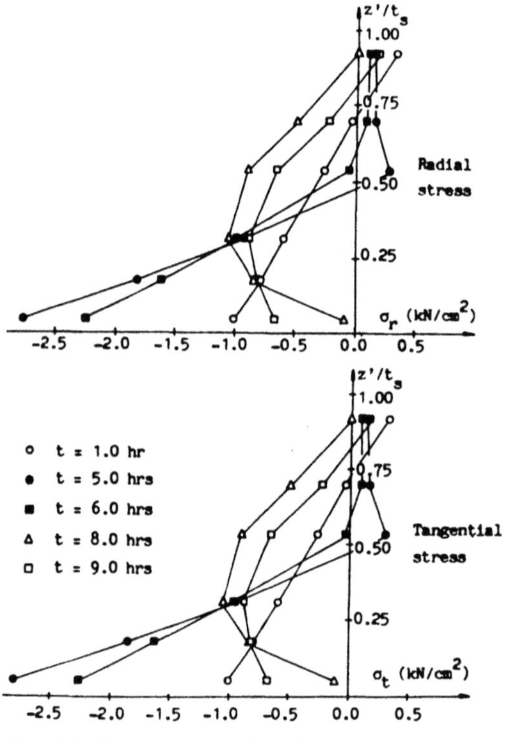

Fig.9. Stress distribution across section AA at some representative time station

sidered in this example corresponds to one of two small scale pressure vessel models, experimentally tested in /18/. Details of the vessel, together with the location of the prestressing system, are given in Fig.5. The material properties and prestressing forces are listed in Table 1. Several authors have numerically analysed this problem under the conditions of prestress and increasing internal pressure /5,15,19,20/, and a close correspondence has been achieved with the experimental results. The conclusions of the previous analyses indicate that failure is primarely attribu-

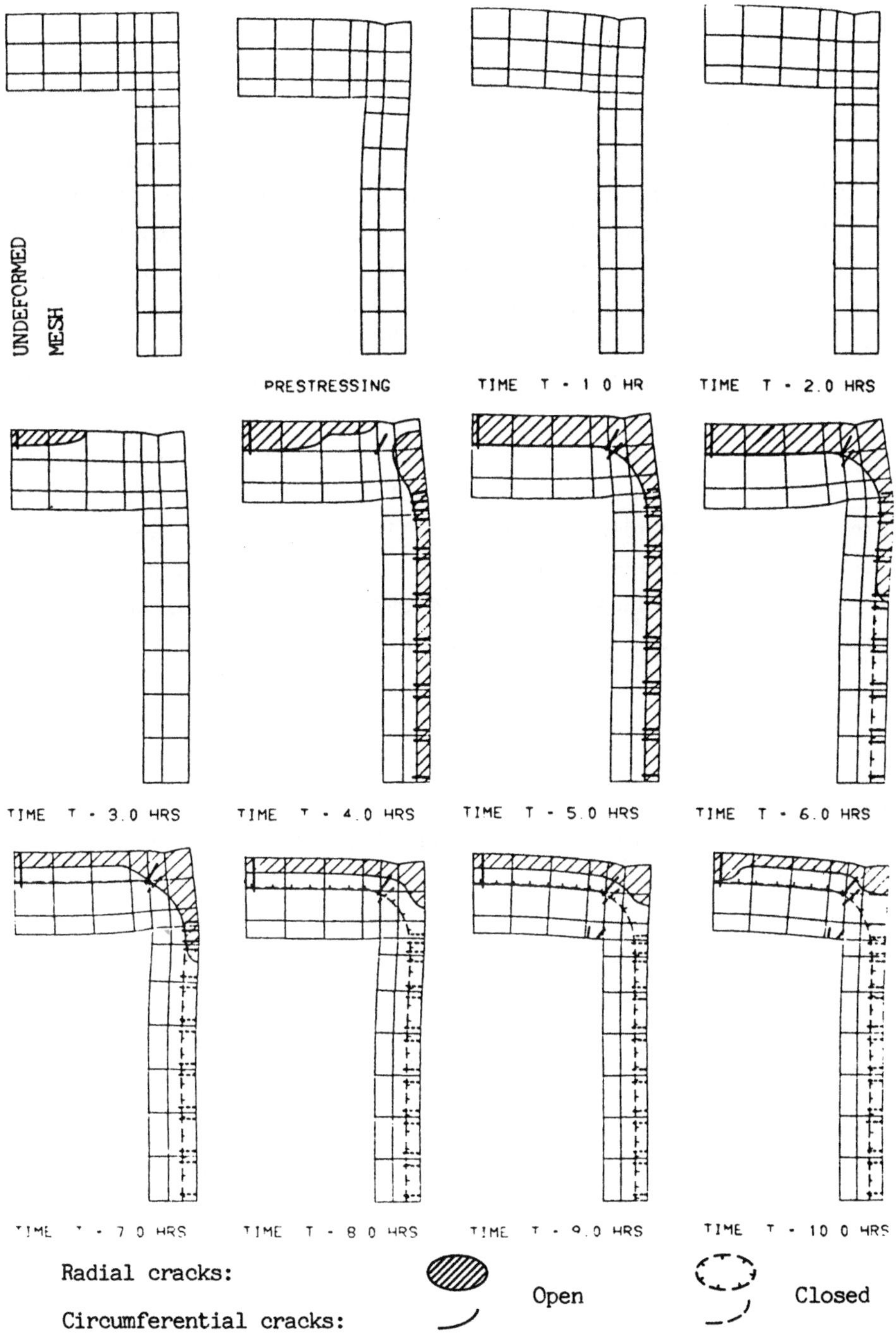

Fig.10. Deformation and cracking history due to the time-dependent loading of Fig.6 (Displacement magnified by 150)

table to tensile cracking.

Here, the vessel is analysed using the fictitious pressure and temperature cicles indicated in Fig.6. The boundary conditions assumed for thermal transient analysis are also included in Fig. 6. The Euler-Backward time stepping scheme is employed for temperature analysis, while the fully explicit approach is considered for stress analysis with a time step length $\Delta t = 0.25$ hr. Isotherms for a one-half of the vessel at some representative time stations are shown in Fig. 7, and Fig. 8 shows the displacement response of the internal and external surfaces with time. Fig.9 shows the solution of stress across the section AA at various time stations. The deformed vessel profiles and the development of zones of tensile cracking are illustrated in Fig. 10. This relatively complex stress behaviour illustrates that use of a full thermal transient solution may be necessary in order to obtain a realistic numerical response; even for cases in which the temperature changes are not very severe.

Conclusion

Finite element technique has been presented for the nonlinear analysis of reinforced and prestressed concrete structures under both time-dependent static and transient loading conditions. The numerical example presented illustrates the applicability of the method.

References

1. Damjanić, F.; Hinton, E.; Owen, D.R.J.; Bićanić, N.; Simović, V. (eds.) Computer-aided analysis and design of concrete structures. Swansea: Pineridge Press 1984.

2. Rots, J.G.; Nauta, P.; Kusters, G.M.A.; Blaauwendraad, J.: Smeared crack approach and fracture localization in concrete. Delft: Heron 30 (1985).

3. Damjanić, F.: Reinforced concrete failure prediction under both static and transient conditions. University of Wales, Swansea: Ph.D. Thesis C/Ph/71/83, 1983.

4. Owen, D.R.J.; Damjanić, F.: A finite element formulation for transient thermal analysis of nonlinear material problems. Trans. 7th Int. Conf. SMiRT (Chicago) H (1983) 169-176.

5. Owen, D.R.J.; Figueiras, J.A.; Damjanić, F.: Finite element analysis of reinforced and prestressed concrete structures including thermal loading.

Comp. Meths. Appl. Mechs. Engng. 41 (1983) 323-366.

6. Damjanić, F.; Owen, D.R.J.: Practical considerations for modelling of post-cracking concrete behaviour for finite element analysis of reinforced concrete structures. In: Damjanić, F. et al. (eds.) Computer--aided analysis and design of concrete structures. Swansea: Pineridge Press (1984) 693-706.

7. Damjanić, F.: Post-construction finite element analysis of a high pressure hydraulic reinforced concrete pipe (In Croatian). Civil Engineering Institute, Zagreb, Yugoslavia 1984.

8. Damjanić, F.: Finite element analysis of a buried concrete pipe under internal pressure. In: Middleton, J.; Pande, G.N.(eds.) Proc. Int. Conf. NUMETA'85 (Swansea). Rotterdam: Balkema (1985) 679-686.

9. Damjanić, F.; Radnić, J.: A finite element analysis of reinforced concrete water pressure pipe. Trans. 8th Int. Conf. SMiRT (Brussels) H (1985). In press.

10. Damjanić, F.: A finite element formulation for analysis of reinforced concrete structures. Proc. Int. Conf. FEICOM'85, Bombay (1985). In press.

11. Owen, D.R.J.; Hinton, E.: Finite elements in plasticity-Theory and practice. Swansea: Pineridge Press 1980.

12. Damjanić, F.: Practical finite element technique for analysis of reinforced concrete structures. Finite element modelling of concrete structures. Swansea: Pineridge Press. In press.

13. Kupfer, H.; Hilsdorf, H.K.; Rüsh, H.: Behaviour of concrete under biaxial stresses. ACI. J. 66 (1969) 656-666.

14. Owen, D.R.J.; Damjanić, F.: Viscoplastic analysis of solids-Stability considerations. In: Hinton, E. et al.(eds.) Recent advances in nonlinear computational mechanics, Ch.8. Swansea: Pineridge Press(1982) 225-253.

15. Phillips, D.V.; Zienkiewicz, O.C.: Finite element nonlinear analysis of concrete structures. Proc. Inst. Civ. Engrs. 61 (1976) 59-88.

16. Damjanić, F.; Owen, D.R.J.: Practical considerations for thermal transient finite element analysis using isoparametric elements. Nucl. Engrg. Design 69 (1982) 109-126.

17. Owen, D.R.J.; Damjanić, F.: Reduced numerical integration in thermal transient finite element analysis. Comp. Structures 17 (1983) 261-276.

18. Sozen, A.M.; Paul, S.L.: Structural behaviour of a small-scale prestressed concrete reactor vessel. Nucl. Engrg. Design 8 (1968) 403-414.

19. Marchertas, A.H.; Fistedis, S.H.; Bažant, Z.P.; Belytschko, T.B.: Analysis and application of prestressed concrete reactor vessels for LMFBR containment. Nucl. Engrg. Design 49 (1978) 155-173.

20. Bathe, K.J.; Ramaswamy, S.: On three-dimensional nonlinear analysis of concrete structures. Nucl. Engrg. Design 52 (1979) 385-409.

Nonlinear Analysis of Concrete and Masonry Structures

M. A. CRISFIELD, J. WILLS
Transport and Road Research Laboratory, Crowthorne, U. K.

Summary

The paper describes the application of non-linear finite element methods to the collapse analysis of concrete beam-and-slab bridges. In addition, a masonry arch bridge is analysed and the soil-structure interaction is investigated.

Beam-and-slab bridges: the idealisation

Three idealisations are possible for beam-and-slab bridges: (1) folded-plates (2) plate with discrete eccentric stiffeners and (3) plate with smeared eccentric stiffeners. At present, the folded-plated option is too expensive especially for multi-beam configurations. The most obvious solution is to introduce the "eccentric stiffener formulation" proposed by Gustafsson and Wright [1] (Fig. 1). Using this approach, the beams are added to the plates without the introduction of any additional variables. In particular, the axial displacement in the beam at depth z (Fig. 1(b)) is given by: $u = \bar{u} + z\theta$ where $\bar{u}$ is the axial displacement in the plate and θ the corresponding rotation variable. For many plate elements, the rotation θ takes a quadratic variation along the beam (ie. with x in Fig.1(b)). Consequently, u will also take a quadratic variation and so should $\bar{u}$ if the formulation is not to be overstiff. These considerations apply to basic r.c. slabs as well as beams-and-slabs because cracking effectively introduces eccentricity. As a result, the adopted bending element [2] has been supplemented by a quadratic membrane element and the final configuration involves five degrees-of-freedom at the corner nodes and two at the mid-sides. A layered technique is used (Fig.1(a)). In particular, the beam is divided into a number of

Finite Element Methods for Nonlinear Problems
Europe-US Symposium, Trondheim, Norway 1985
Editors: Bergan, Bathe, Wunderlich

"bits" (say 6 concrete and 3 for the reinforcement or prestressing) at the centroids of which, the material responses are "lumped". The reinforcement in the slab is treated in a similar manner but the through-thickness response of the concrete is integrated with the aid of Gaussian integration.

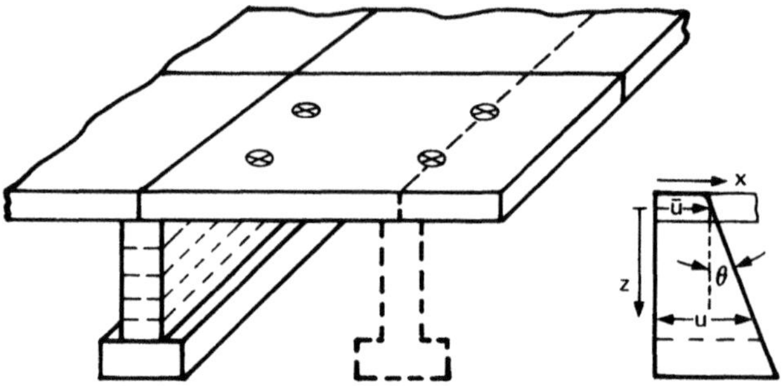

(a) Beam options (b) Displacements

Fig. 1 Beam-and-slab modelling

There are two obvious choices for the positioning of the beams. They can be placed either at the plate-element junctions or, alternatively, along a row of Gauss points (Fig.1(a)). In the latter case, each "bit" of a beam is treated in a very similar fashion to the smeared reinforcement in the slab. Consequently, a beam could be treated very economically if it were placed along a row of Gauss points. However, this would be geometrically restrictive. Hence, a "smeared beam option" has been provided in which the beams may be placed anywhere with their properties being automatically proportioned and "lumped" to the surrounding Gauss points. This action is performed by a pre-processor which will either handle discrete beams or patches of reinforcement.

Concrete Modelling and Nonlinear solution strategy

The concrete modelling is fairly typical of conventional smeared procedures [3] and is further detailed in [4]. The uniaxial stress-strain relationships take the form indicated in Fig.2 in which the tensile softening is introduced to account for the "tension stiffening" [3]. This phenomenon produces a stiffer response than a purely brittle failure because the concrete, adjacent to a crack, carries tension on account of the shear-transfer from the reinforcement. The degree of the resulting tensile-softening is purely empirical and, for the present analyses, the strain at point B (Fig. 2(a)) is set to fifteen times the strain at point A.

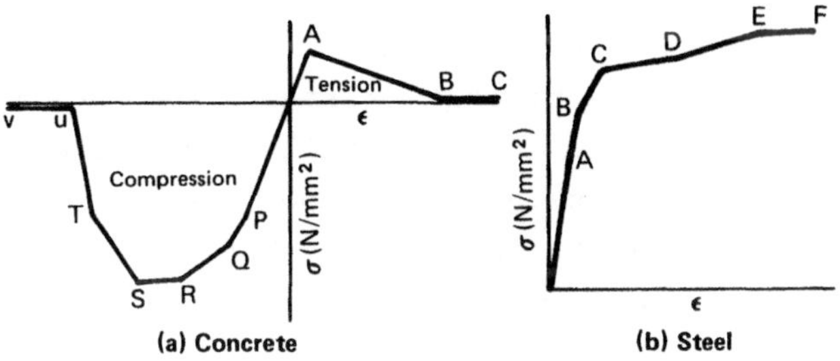

Fig. 2 Uniaxial stress/strain curves

The tensile softening is related to a fixed crack direction while the compressive hardening and eventual softening is related to a concrete plasticity model [4] After cracking, the in-plane tangential shear stiffness is reduced by a "shear retention factor" [3] of 0.1. It is accepted that the adopted model has a number of limitations [4] but the object of the paper is to describe the application of current methods to real engineering structures.

In general, the loading is applied in stages i.e. beam self-weight and prestressing, slab self-weight, "H.A. distributed live loading" [5] and, finally, "H.B. abnormal vehicle loading" [5] . To this end, the computer program applies a load vector of $\underset{\sim}{q}_1 + \lambda \underset{\sim}{q}_2$ where, for any stage, the vectors $\underset{\sim}{q}_1$ and $\underset{\sim}{q}_2$ are fixed while λ is a variable loading scalar. To apply the final abnormal vehicle (H.B. loading), the user inputs the first load increment factor $\Delta\lambda_o$ and, from that stage, the program automatically adjusts the increments with a view to maintaining a constant number of iterations (say 5) at any increment. These iterations are applied until both a displacement and a scaled force-norm criterion are satisfied [6] . The modified Newton-Raphson method is usually adopted with a new tangent stiffness matrix being formed and factorised at each increment. Secant-accelerations may be added and line-searches are always included [6,7] .

Once the current-stiffness parameter [8] has reduced to about 0.4, the program automatically switches to the "spherical arc-length method" [9] which should allow the solution to be traced beyond the maximum load. Within each increment, the incremental displacement norm $\Delta \underset{\sim}{p}^T \Delta \underset{\sim}{p}$ is fixed instead of the normal loading parameter, $\Delta\lambda$. Load reversals are automatically initiated whenever the determinant of the tangent stiffness matrix changes sign. Difficulties can be experienced in the presence of bifurcations or "material instabilities" [4,7]. In times of trouble, a new tangent stiffness matrix may be formed within an increment [7]. If this fails, the increment is automatically cut. This technique allows the computer program to be left without manual intervention for long periods when the work load on the computer is slack.

The AASHO bridge

Figure 3 shows a prestressed-beam and insitu-slab bridge that was tested to destruction at Ottawa, Illinois in 1961. The dimensions and material properties were extracted from Refs. [10,11]. Collapse was induced with the aid of the overload-vehicle that is shown in the figure, which was driven thirty times across the bridge for each of a number of increasing load levels. In order to simulate this effect numerically, the vehicle was placed at the position (Fig. 3) at which it would induce the maximum bending moment in the bridge when the latter was treated as a beam.

Figure 3 shows that a first local maximum (point A) was obtained at a load level that was only two-thirds of the computed maximum load. The falling curves following this and the next two local maxima were associated with "crack localisations" [4,7] in the vicinity of the rear axles. At point E, the tangent stiffness matrix gave a negative pivot and the load was therefore automatically reduced. This process led to the "spurious" equilibrium point F at which level the determinant was positive so that the load was automatically reversed until it regained a position very close to the original point E. A negative pivot was again encountered and, this time, the solution dropped along the "correct" equilibrium path. The numerical difficulties would appear to relate to alternative equilibrium paths that may be associated with strain localisation [7]. Nonetheless, the computer program overcame the problem and the computation was finally stopped when a row of prestressing cables reached

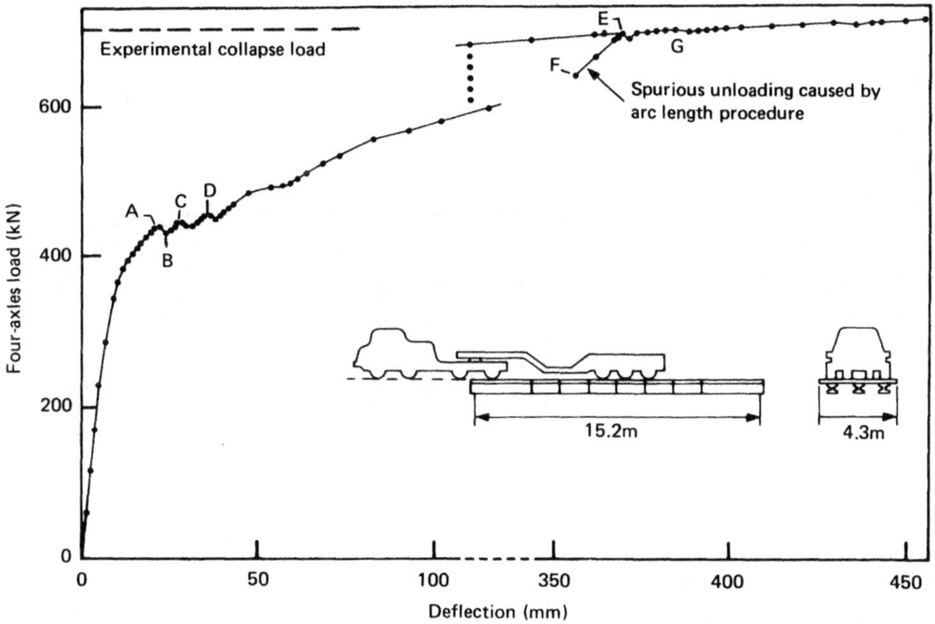

Fig. 3 Load/deflection response for beam-and-slab bridge with 90% initial prestress (to allow for creep and shrinkage)

their maximum strain. This occurred at a load level that was very close to the experimental collapse which was also caused by rupturing of the cables.

<u>TRRL half-scale bridge</u>

Figure 4 shows a single-span half-scale bridge deck that was recently tested to destruction at the Transport and Road Research Laboratory [12] . The first preliminary finite element analysis used a coarse 4x6 mesh with the beams being smeared to the plate Gauss points while the second mesh involved 9x10 plate elements coupled with discrete beams. With reference to Fig.2, the following properties were adopted:

	A	B	C	P,Q	R	S	T	
stress:	4.0	0.01	0.01	30	43	43.2	0.	beam
$\epsilon \times 10^6$	107	1605	300000	800	1800	2150	6500	concrete
stress:	3.5	0.01	0.01	27.	35.	35.2	0.0	slab
$\epsilon \times 10^6$	130	1950	400000	1000	1750	2500	9000	concrete

	A	B	C	D	
stress:	466	566	666		reinforcing
strain	0.002251	0.01	0.02		steel
stress:	1442	1638	1704	1819	prestressing
strain	0.0074	0.0095	0.012	0.050	steel

The stress/strain curves for the concrete were deduced from early cube-strength tests in conjunction with data in Ref.13. The theoretical and experimental solutions are compared in Fig. 5. There are kinks in the curves at a load-factor of 1.0 because up to that load level the standard H.B. abnormal vehicle loading was supplemented by H.A. uniformly-distributed and knife-edge loadings [5] while beyond that load-factor only the H.B. loading was increased. The stiffer response provided by idealisation A partly relates to the coarser-mesh and partly to the lower number of integration points (3 as against 5) that were used through the depth of the slab. Compressive damage was therefore recorded later. The computed collapse for both idealisations involved compressive softening in the region Z of Fig. 4. For idealisation A, a limit-point was successfully passed and the load was dropping steeply before the solution procedure failed. However, for idealisation B, the numerical procedure failed to converge following the detec-

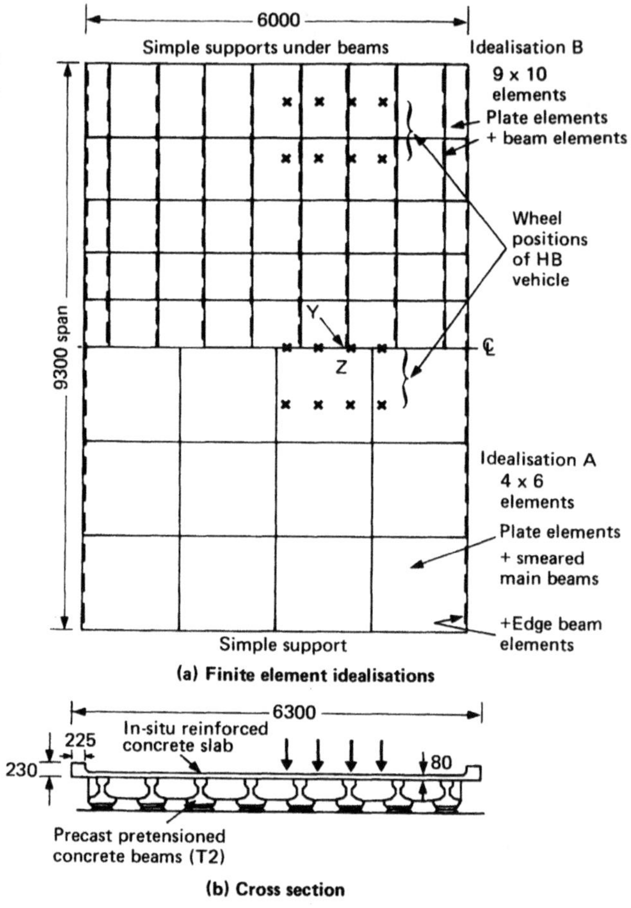

Fig. 4 TRRL half-scale bridge
(all dimensions in mm)

tion of a negative pivot.

The computed load/deflection relationship for idealisation B was in close agreement with the experimental results up to a factor of about 3.1. However, at about this load level, a horizontal "shear-like" failure occurred in the webs of some of the more heavily loaded beams. These failures appeared to have been associated with transverse bending in the beams which were restrained at the ends by rigid diaphragms. Such a mode of failure could only be investigated numerically with the aid of a non-linear folded-plated analysis.

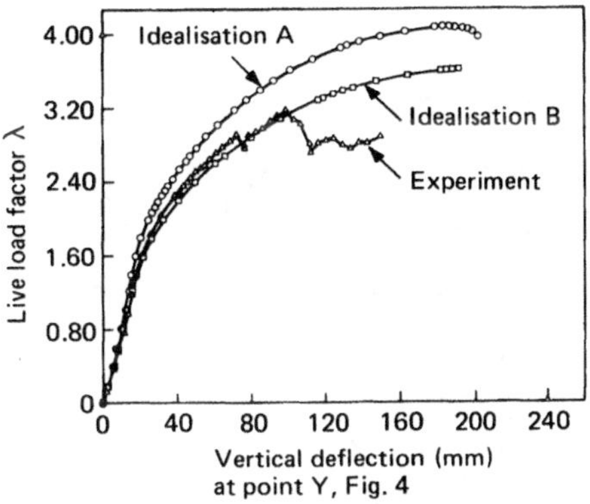

Fig. 5 Load/deflection relationship for TRRL half-scale bridge

Masonry arches

Reference [14] contains a review of previous work on the analysis of masonry arch-bridges which includes reference to earlier finite element work by Sawko and Towler [15]. It also gives details of a masonry arch bridge (Fig. 6)that was tested to virtual destruction at Girvan, Scotland in 1984 [16] . The arch-ring consisted of red sand-stone voussoirs connected by weak limestone mortar joints. In Figure 6, the finite element mesh is not designed to coincide with the joints but rather to provide a "converged solution" when the ring is treated as a continuum. A "one-dimensional approach" [14] was initially adopted with the effects of the fill being replaced by dead weights. Each element had three degrees-of-freedom (two translations and one rotation) at the end nodes and a single membrane displacement variable at each mid-side node. The formulation involved a set of locally shallow arch elements with transformations to global co-ordinates. Moderately-large-deflection theory was adopted.

The weak mortar joints allow the arch to be treated as a no-tension material in the circumferential direction and, in addition, reduce the effective E value so that it is about one-

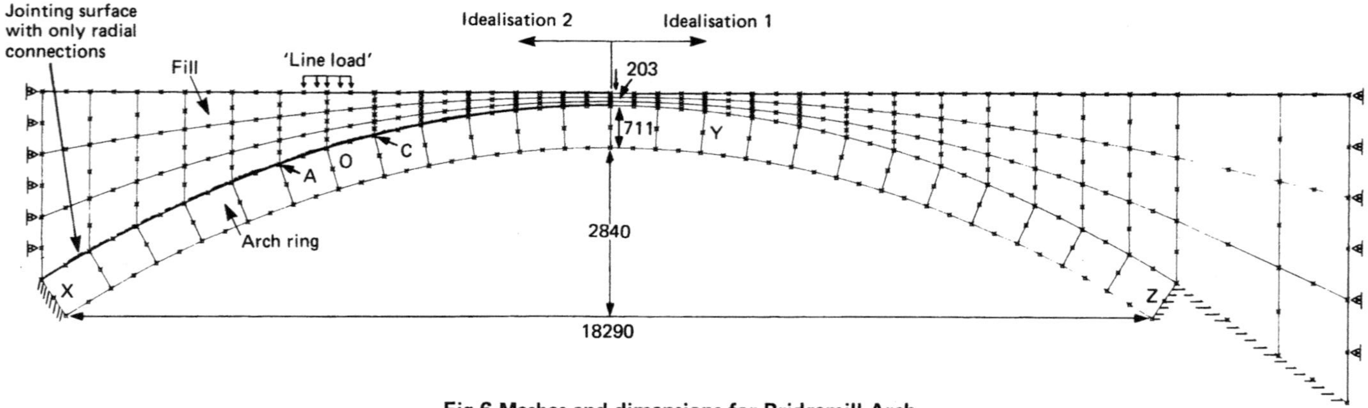

Fig.6 Meshes and dimensions for Bridgemill Arch
(dimensions in mm)

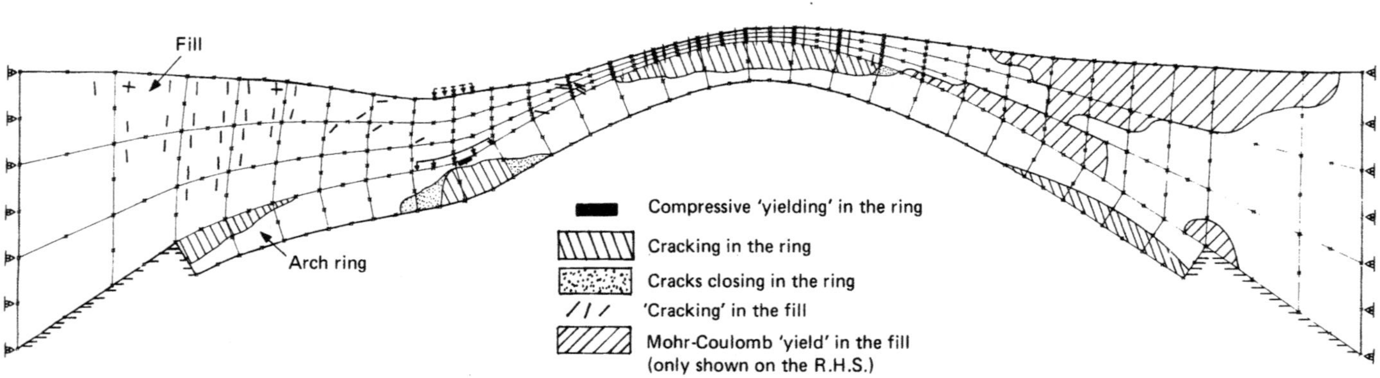

Fig. 8 Computed deformations (exaggerated), cracks and 'yielded states' in Bridgemill Arch

third [16] of the E value of the stone on its own ($15000N/mm^2$) - [16]. In addition, the effective compressive strength will be less than that of the stone ($44N/mm^2$ - [16]). A value of $15N/mm^2$ has been used for the present studies. For the 1-D analysis, the material nonlinearity was accounted for by numerically integrating through the depth of the arch-ring [14]. These integrations were performed at the two-point Gauss stations along the length of the element. Numerical experiments [14] indicated that the Gaussian technique was more accurate than a constant thickness centroidal rule for the integrations through the depth and that at least eight integration points were required.

The live loading involved a "line load" across the 8300mm width of the bridge. This loading was applied at the quarter-span position through a concrete beam of 750mm width. For the 1-D analyses, the loading was simulated by applying a uniformly-distributed load between points A and C in Fig. 6. In this way, it was hoped to allow for the distribution through the "fill and surfacing". The resulting load/deflection relationships are compared with the experimental values in Fig. 7 where the plotted deflection is the vertical deflection at point 0 in Fig. 6. As anticipated, when the E value of the stone was used for the ring material, the computed response was too stiff. However, when the E value was reduced to $4000N/mm^2$, reasonable correlation was obtained with the experimental results up to a load of about 2100kN but the computed collapse load was about 25 per cent too low. In contrast, a "mechanism" solution gave a very reasonable collapse load of

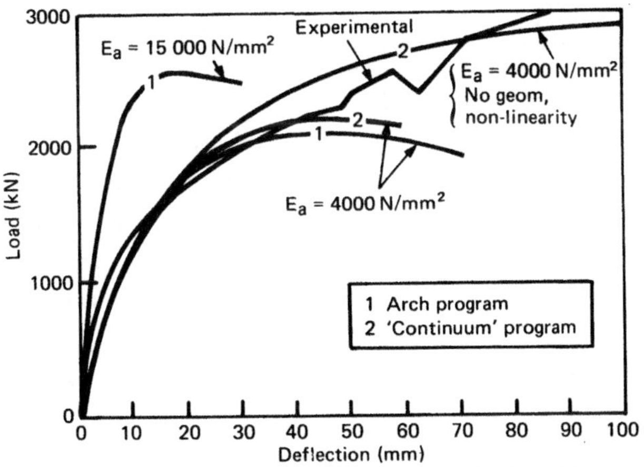

Fig. 7 Load/deflection curves for Bridgemill Arch (No fill)

between 2880 and 3200 kN depending on the assumptions made regarding the distribution through the fill. This "mechanism" or "plastic" solution [14,17] assumes an infinite E value and an infinite compressive strength and postulates the formation of four "hinges" at approximately sections XOYZ in Figure 6. The finite element computer program can effectively reproduce this solution when the E value and compressive strengths are set to very high values [14]. However, with a more realistic E value, the cracks do not fully propagate through the arch (Fig. 8) and the collapse load is reduced. This reduction in load is much less pronounced when the effects of geometric non-linearity are omitted (Fig. 7) and, in these circumstances, the cracks penetrate further through the ring. The arch-ring might appear to be too deep for geometric effects to be significant. However, in the later loading stages, much of the ring is cracked so that the effective section is significantly reduced. The better correlation, that is produced when geometric nonlinearity is omitted, is probably fortuitous because the stiffening effects of the fill and spandrel walls have been omitted.

With a view to the later modelling of the fill, the arch was reanalysed with a two-dimensional continuum program. Initial attempts adopted a concrete model with a very low tensile strength and E value. However, cracks were produced in non-radial directions under self-weight and adversely affected the response. The real stone arch is much weaker in the circumferential direction in which the mortar joints dominate the response. Consequently, the stresses and strains at the Gauss points were resolved into "tangential" and "normal" directions. Only the stiffness in this tangential direction was degraded on account of cracking or compressive yielding. To save computer time and for consistency with the adopted material model, a single element division was used through the depth of the arch and plane-sections were forced to remain plane. To this end, the hierarchical mid-depth variables were set to zero. In addition, a special 2x8 Gaussian integration procedure was introduced with the eight points covering the depth of the ring. As expected, the resulting formulation gave a close correlation with the "one-dimensional" results (Fig. 7).

The previous results have only allowed for the self-weight effects of the fill. The solutions in Fig. 7 would indicate that the neglected lateral stiffening of the fill (and, possibly, the stiffening of the spandrel walls) has contributed about a quarter of the strength of the arch. For deeper arches, a larger contribution could be expected and, for semi-circular arches, these "second-order" effects must become "first-order" [14]. Consequently, attempts are being made to model the soil/structure interaction. Early studies using non-linear spring elements for the soil indicated that the lateral soil pressure could, indeed, provide a quarter of the strength of the arch [14].

Present studies are directed towards a more direct (and expensive) modelling of the fill. To this end, two-dimensional continuum elements have been added to the arch-ring (Figs. 6 and 8). While conditions of plane-stress were applied in the arch-ring, plane-strain was assumed to act in the fill-elements. A Mohr-Coulomb yield criterion with a no-tension cut-off was applied and an associated flow-rule was used. For each of the solutions shown in Fig. 9, a low cohesion of 0.01 N/mm^2 was combined with a Poisson's ratio of 0.2 and an angle-of-friction of 37 degrees. The E value and compressive strength of the arch-ring were maintained at 4000 and 15 N/mm^2 respectively. The "line loading" should strictly have been applied above the fill. However, when such a loading was introduced, the numerical results indicated a local failure in the fill. The stiffening effects of the road-surfacing would need to be modelled if such a local failure were to be avoided. In the meantime, the loading was, as before, applied directly to the arch-ring and was uniformly distributed between points A and C in Fig. 6.

The initial solutions involved the idealisation 1 shown on the right-hand-side of Fig. 8. When the E value of the fill was set to 40N/mm^2, the computed collapse load was very close to the experimental value. However, the load/deflection response was too stiff (Fig. 9) and was very similar to the response computed from the one-dimensional model with non-linear "soil springs" [14]. The computed deformations, cracks and "plastic states" at collapse are shown in Fig. 8 where the deflections have been multiplied by a factor of 12.5 before being added to the co-ordinates. The "yielded fill" is only shown on the right-hand-side of the figure to avoid confusion with the "tensile cut-off cracks" on the left-hand-side. It is the lateral compressive resistance on the right-hand-side where the arch is moving into the soil that produces the additional strength. A better correlation in the shape of the load/deflection curve was obtained by reducing the E value of the fill to 7.5N/mm^2 (Fig. 9). However, this value would appear rather low and an alternative "device" for reducing the stiffness was sought. To this end, free circumferential slip was allowed between the arch-ring and the fill (see idealisation 2 of Fig. 6). As anticipated, both the stiffness and final strengths were significantly reduced (Fig. 9). If slip did occur at the ring/fill interface it would not be frictionless and allowance should be made for frictional slip.

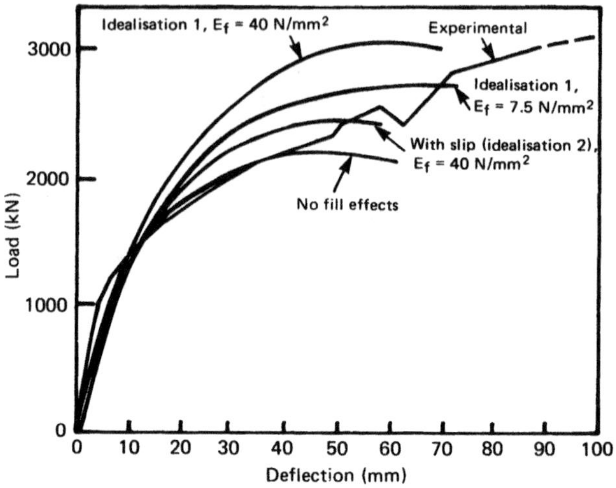

Fig. 9 Load/deflection curves for Bridgemill Arch (with fill)

Numerically, both smaller increments and many more iterations were required once the fill-elements were added. In addition, the line-search algorithm was much more heavily used. The numerical difficulties were surprisingly most severe in the early increments of live loading, following the previous application of self-weight.

Conclusions and suggestions

Finite element beam-and-slab models can be successfully used to assess the strength of concrete bridges provided a form of flexural, but not necessarily ductile, failure is involved. The "smeared stiffener" technique provides an economic model for preliminary analyses. In order to investigate non-flexural failure modes, a folded-plated model may be required. This would be prohibitively expensive for multi-beam structures. However, it should be possible to use a partitioned mixed-idealisation.

Finite element methods can be used to help assess the strength of masonry arches. Further work should include allowance for a frictional rather than a frictionless slip. The effects of the road-surfacing should also be modelled as, in general, should movement at the abutments. The "continuum model" for the arch-ring is computationally convenient but has a number of limitations. In particular, one might anticipate, at the local level, an eventual stiffening in the response following the squeezing of the mortar. Either the continuum stress/strain curve could be altered or, at a greater cost, special mortar-joint elements could be introduced. On the numerical side, attention should be paid to the poor convergence characteristics that were encountered once the fill-elements were added.

Acknowledgements

The work described in this paper forms part of the programme of the Transport and Road Research Laboratory and is published by permission of the Director.

References

1. Gustafsson, W.C; Wright, R.N.; Analysis of skewed composite girder bridges. J. Struct. Div., Proc. ASCE, 94 (1966) 315-317.

2. Crisfield, M.A.; A four-noded thin-plate bending element using shear constraints; a modified version of Lyons' element. Comp. Meth. in Appl. Mech. & Engng., 38 (1983) 93-120.

3. ASCE. Finite element analysis of reinforced concrete; state-of-the-art report. Struct. Div. Committee on Conc. & Masonry Structs. New York: ASCE 1982.

4. Crisfield, M.A.: Difficulties with current numerical models for reinforced-concrete and some tentative solutions. in: Computer Aided Analysis & Design of Concrete Structures. Part 1, ed. C.Taylor et al., Swansea: Pineridge Press (1984) 331-358.

5. British Standards Institute: BS5400: Part 2: London 1978.

6. Crisfield, M.A.: Accelerated solution techniques and concrete cracking. Comp. Meth. in Appl. Mech. & Engng. 33 (1982) 585-607.

7. Crisfield, M.A.: Overcoming limit points with material softening and strain localisation. in: Numerical Solutions for Non-linear Problems, Vol. 2, ed. C.Taylor et al. Swansea: Pineridge Press (1984), 244-277.

8. Bergan, P.G.; Horrigmoe, G.; Krakeland, K.; Soreide, T.H.: Solution techniques for non-linear finite element problems. Int. J. for Num. Meth. in Engng. (1980) 1667-1696.

9. Crisfield, M.A.: A fast incremental/iterative procedure that handles snap-through. Computers & Structs. 13 (1981) 261-290.

10. Highway Research Board. The AASHO road test. Reports 2 & 4, Special Reports 61B & 61D, Nat. Academy of Sci. - Nat. Res. Council Reports 951 & 953, Washington, 1962.

11. Petersen, S.W.; Kostem, C.N.: The inelastic analysis of beam-slab highway bridge structures, Lehigh Univ., Fritz Eng. Lab. Report 378B.5 (1975).

12. Withey, K.: Tests on half-scale concrete bridges: Results from a prestressed beam and insitu slab deck. TRRL Working Paper, in preparation.

13. Regan, P.E.; Yu, C.W.: Limit state design of structural concrete. London: Chatto & Windus 1973.

14. Crisfield, M.A.: Finite element and mechanism methods for the analysis of masonry and brickwork arches. TRRL Research Report 19, Transport and Road Research Laboratory, Crowthorne, Berks., England. 1985.

15. Sawko, F.; Towler, K.: Structural behaviour of brickwork arches. in: Proc. 7th Int. Conf. on Loadbearing brickwork, ed. H.W. West, Brit. Ceramic Soc. London 1981.

16. Hendry, A.W; Davies, S.R; Royles, R.: Test on a stone masonry arch at Bridgemill, Girvan; Univ. of Edinburgh, Contractor's Report to the Transport and Road Research Lab. 1984.

17. Heyman, J.: The estimation of the strength of masonry arches. Proc. Instn. Civ. Engrs., Part 2, 69 (1980) 921-937.

Finite Element Analysis of the Non Linear Response of Concrete Dams Subjected to Internal Loads

E. OÑATE, J. OLIVER, G. BUGEDA
Universidad Politécnica de Cataluña, Barcelona, Spain

Summary

The paper presents a finite element formulation for the modelling of cracking in concrete dams under severe internal forces due to water expansion of mortar. Details of the constitutive model for plain concrete used are given in the paper together with some aspects of the finite element solution process. Also a joint element introduced for the modelling of construction joints is briefly presented. Finally, numerical results for the analysis of a gravity dam for which experimental results are available are given.

Introduction

It is well known that many existing concrete dams, operating in practice, show extensive cracking or microcracking zones. A situation of this kind is unusual in most typical reinforced and pre-stressed concrete structures where the amount of permissible cracking is strictly restricted at design level. Thus, in such structures appearance of cracking over certain prescribed limits is understood as structural failure, and severe security and repair actions are undertaken.

However, in concrete dams existance of a certain amount of cracking or microcracking is almost unavoidable, and the structural analyst should be able to predict the global stability of the dam in the presence of a high percentage of cracking zones.

Cracking in concrete dams can be originated by many reasons like:

a) A wrong structural design of the dam.
b) Errors in the construction phase of the dam.
c) A wrong evaluation of the material properties of concrete.

In any case, appearance of cracking in concrete dams causes the entrance of water into the dam body. This is usually prevented by an adequate distribu-

Finite Element Methods for Nonlinear Problems
Europe-US Symposium, Trondheim, Norway 1985
Editors: Bergan, Bathe, Wunderlich

tion of drains which avoid water to progress inside the dam. On the other hand, unexpected cracking can reduce significantly the security coefficient of the dam which in some cases, can come down to dangerous figures.

Classical methods for dam analysis have encountered serious difficulties to take into account the effect of cracking in the structural response of the dam. However, recent numerical models and computational techniques developed for the analysis of concrete structures [1] are a clear step forward for the prediction of cracking in dams even at early design stages.

One of the most serious problems related with cracking in concrete dams is the combined effect of cracking and water action on the material properties of concrete. Here, expansion of mortar in presence of water plays a paramount role in the amount and propagation of cracks in the dam. A typical situation of this nature is the following: a) Cracking onsets in the water face of the dam due to the internal forces originated by the expansion of concrete in contact with water. b) As waters comes into the dam body new zones of concrete expand. This causes new internal forces, the opening of new cracks and also closing of previously opened ones. c) The combined effect of opening and closing of cracks + entrance of water + mortar expansion continues until water reaches a drainage zone and no further concrete regions are allow to come into contact with water. This situation obviously depends on the amount of mortar expansion which, in most cases, is of little importance. However, unexpected high values of mortar expansion can reproduce the process described earlier and cause severe cracking damage in the dam which, if not adequately prevented, can put in danger the global stability of the structure.

Expansion of concrete in presence of water is a well known effect which causes continuosly increasing positive straining in concrete up to a (usually) maximum stabilized value. The effect of concrete expansion can be interpreted as similar (and opposite) to shrinkage, with the magnitude of the maximum straining depending on the material properties of mortar. Typical values of such straining are of the order of 0.3 ‰ which means that, even in normal situations, mortar expansions effects should not be ignored in practical computations.

In this paper a numerical model for the analysis of concrete dams subjected to internal forces due to water expansion of mortar is presented. However, other internal forces due to effects like shrinkage , creep or temperature could be treated in a similar manner.

The paper is divided into four parts. In the first, a detailed discussion of the constitutive model used for concrete is presented. The second part deals with the obtention and solution of the discretised systems of equilibriums equations. In the third part, a joint element included for the modelling of construction joints is briefly presented. Finally, in the last part of the paper the efficiency of the formulation is checked in the two dimensional analysis of a concrete dam for which some experimental results are available.

Constitutive model for concrete.

We present here a model for the elastic-cracking behaviour of plain concrete based in a "smeared approach". The model presented can be considered an extension of the work of the Borst and Nauta [2] , [3], de Borst [4] , Rahman [5] , amongst others [6] - [8] , for the modelling of cracks in reinforced concrete structures. Special emphasis is given in this paper to the treatment of the development of compression stresses in the cracks, and to the modelling of the different stress paths due to the alternance of increasing and decreasing strains caused by the effect of internal forces. Details of the model are presented next:

a) **Linear behaviour.**

The constitutive equation for concrete is based in the standard one-dimensional stress-strain curve of Fig. 1a with the following properties:

- **Compression zone:** Compression stresses in dams have small or moderate values, i.e. the non linear part of the classical one-dimensional compression curve and plasticity effects can be neglected. Thus, a linear elastic behaviour with constant Young's modulus E has been taken.
- **Tension zone:** Concrete is assumed to have a linear elastic behaviour until the maximum tensile stress f'_t is reached. The sudden drop in tensile stress after crack formation has modelled using an artificial "tension stiffenning" line (there is no reinforcing in the structure) to improve the numerical stability of the solution algorithm.

The two dimensional constitutive model for the elastic response of concrete under moderate compression stresses is directly obtained in a simplified form

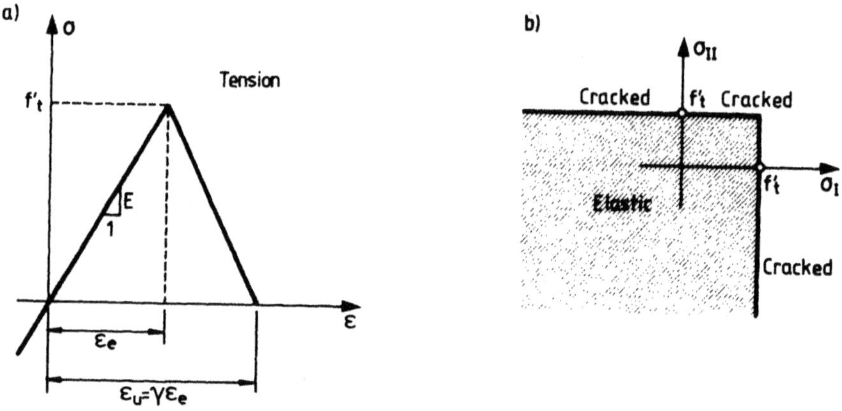

Fig.1 One dimensional (a) and two dimensional (b) constitutive curves for plain concrete under moderate compression stress.

from the one dimensional case (see Fig. 1b). The stress-strain relationship can be written as

$$\underset{\sim}{\sigma} = \underset{\sim}{D}\,\underset{\sim}{\varepsilon} \tag{1}$$

where for 2-D problems

$$\underset{\sim}{\sigma} = [\,\sigma_x,\ \sigma_y,\ \tau_{xy}\,]^T \tag{2}$$

$$\underset{\sim}{\varepsilon} = [\,\varepsilon_x,\ \varepsilon_y,\ \gamma_{xy}\,]^T$$

and $\underset{\sim}{D}$ is the standard 2D elasticity matrix [9] which is a function of the elastic Young modulus E and the Poisson's ratio ν .

In the following we will explain the methodology used for establishing the onset of the first and second crack at a point, and the development of stre sses at the cracks.

Onset of the first crack at a point.

The first crack appears orthogonal to the direction of maximum principal stress when this reaches the tensile stress f'_t. The stress field at the crack is then calculated as follows:

Step 1 : A local orthogonal coordinate set x' y' is established such that axe y' coincide with the direction of the crack (Fig. 2a-1).

Step 2 : Local strains at the crack are computed as

$$\underset{\sim}{\varepsilon}' = \underset{\sim}{T}\,\underset{\sim}{\varepsilon} \tag{3}$$

where

$$\underset{\sim}{\varepsilon}' = [\,\varepsilon_{x'}\,,\ \varepsilon_{y'}\,,\ \gamma_{x'y'}\,]^T \tag{4}$$

and

$$\underset{\sim}{T} = \begin{bmatrix} \cos^2\alpha & \sin^2\alpha & \sin\alpha\ \cos\alpha \\ \sin^2\alpha & \cos^2\alpha & -\sin\alpha\ \cos\alpha \\ -2\mathrm{sen}\,\alpha\cos\alpha & 2\sin\alpha\ \cos\alpha & \cos^2\alpha - \mathrm{sen}^2\alpha \end{bmatrix} \tag{5}$$

is the transformation matrix with α being the angle between axes x' and x (see Fig. 2a)

Step 3 : The stress $\sigma_{x'}$, orthogonal to the crack direction, is computed from the corresponding strain $\varepsilon_{x'}$ using the one dimensional stress-strain constitutive curve (see Fig. 2a-2). This implies an "uncoupled" behaviour between the two orthogonal stresses at the crack. Finally, stresses $\sigma_{y'}$ and $\tau_{x'y'}$ are calculated as

$$\sigma_{y'} = E^* \varepsilon_{y'}\ ,\ \tau_{x'y'} = \beta\, G\, \gamma_{x'y'}\ ,\ G = \frac{E}{2(1+\nu)} \tag{6}$$

where β is a shear retention factor.

In eq(6) $E^* = E$ for plane stress problems. However, for plane strain situa tions the existance of a non zero normal transverse stress σ_z implies that an equivalent Young's modulus $E^* = \frac{E}{1-\nu 2}$ should be taken.

Step 4 : Global stresses are calculated from local ones as

$$\underset{\sim}{\sigma} = \underset{\sim}{T}^T\,\underset{\sim}{\sigma}' \quad \text{with} \quad \underset{\sim}{\sigma}' = [\,\sigma_{x'}\,,\ \sigma_{y'}\,,\ \tau_{x'y'}\,]^T \tag{7}$$

<u>Onset of the second crack at a point.</u>

A second principal stress can reach the tensile limit at a cracked point

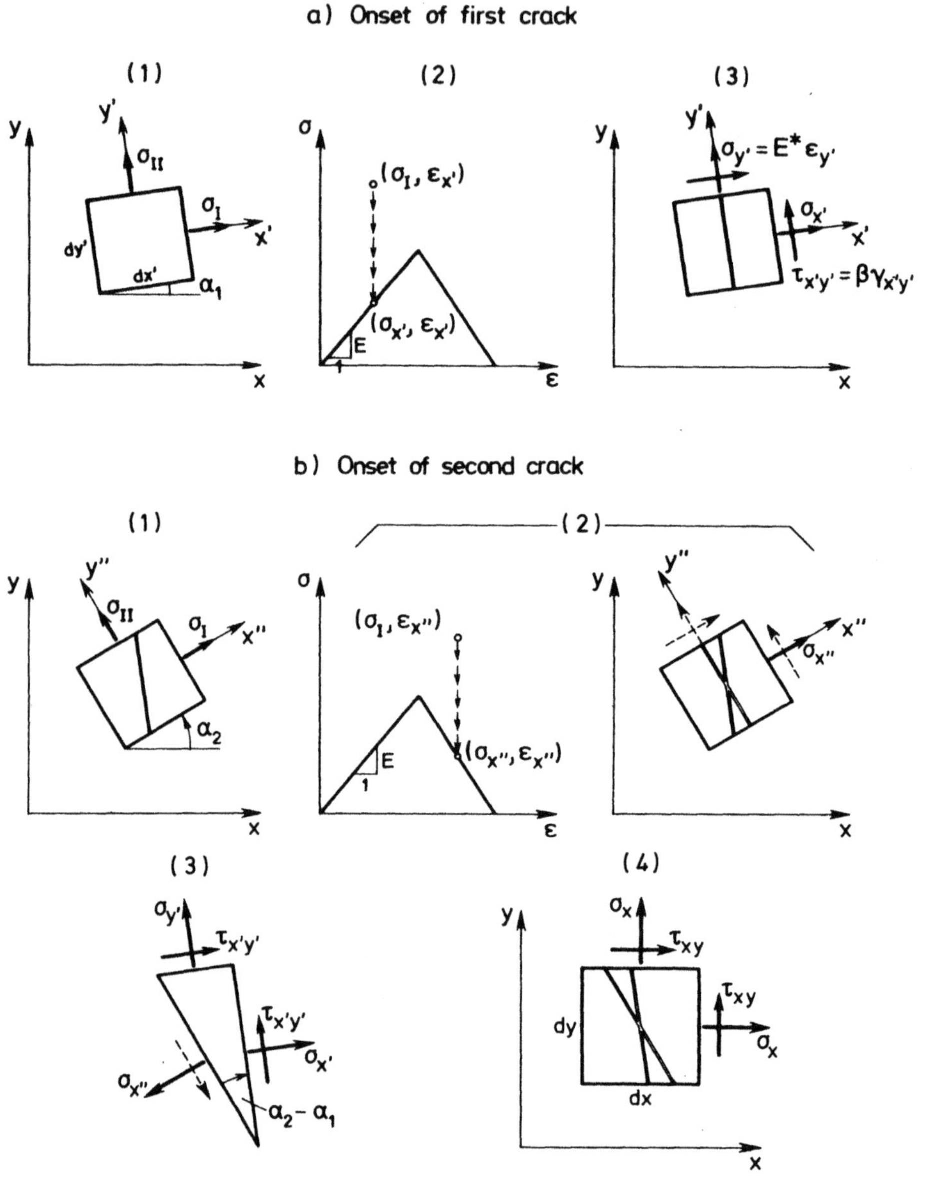

Fig. 2 Onset of first and second cracks at a point

due to the change in principal stresses after the stress relaxation during the onset of the first crack, or alternatively, to the aplication of a new load increment. In any of these cases two possibilities are encountered:

a) The angle between the first and the second (possible) crack is less than a prescribed treshold value (the value of 60° has been taken in the examples analysed). In this case the second crack is not allowed and the corresponding principal stress orthogonal to the crack is set equal to f'_t. Finally, step b-3 described below is performed.

b) If the angle between the first and the second crack is greater than the treshold value, a second crack is allowed in the orthogonal direction to the maximum principal stress and the following steps are performed:

Step b-1. Local strains $\underset{\sim}{\varepsilon}''$ and stresses $\underset{\sim}{\sigma}''$ are calculated in the local system of the second crack (see Fig. 2b-1).

Step b-2. The stress normal to the second crack $\sigma_{x''}$ is relaxed to the one dimensional constitutive curve according to the value of $\varepsilon_{x''}$.

Step b-3. Equilibrium in the double cracked region (Fig.2b-3) is imposed by evaluating the stress normal to the first crack, $\sigma_{y'}$, using - the standard Mohr's expression

$$\sigma_{y'} = (2\sigma_{x'} - \sigma_{x'}(1 + \cos 2(\alpha_2 - \alpha_1)) - 2\tau_{x'y'}\,\mathrm{sen}2(\alpha_2 - \alpha_1)) / (1 - \cos 2(\alpha_2 - \alpha_1)) \tag{8}$$

Step b-4. Finally, the global stresses in the double cracked point are computed using eq.(7).

<u>Previously cracked points. Stress paths.</u>

The change in internal forces within the structure can easily produce changes in the sign of the crack stress. We will define here a "closed crack" as that for which the stress in its corresponding orthogonal direction has passed from a possitive (tensile) value to a negative (compression) one. Obviously, the term "closed crack" has not any physical meaning from the point of view of non-permeability to water. However, it helps to clearly define the sign change in the crack stress.

The model to simulate the stress paths at a crack is shown in Fig. 3. The main features of such model are briefly described below:

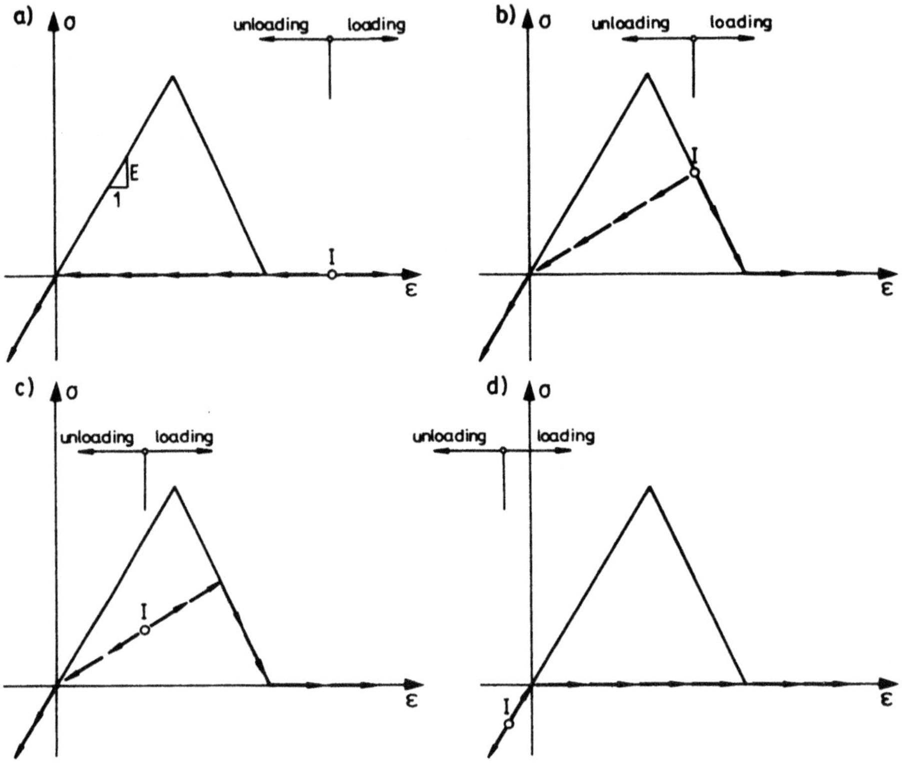

Fig.3 Stress paths at a cracked point I.

- Fig. 3a. The initial stress point at the crack is on the horizontal region of the one dimensional stress-strain curve (the ultimate tensile strain ε_u has been reached): Unloading and further loading takes place at zero stress level for possitive deformation and following the compression line for negative deformation values.

- Fig. 3b. The initial stress point is in the descending branch of the tension stiffening curve: Then, unloading for possitive stress values (opened crack) follows the line which conects the stress point and the origin, and the compression line for negative ones (closed crack). Further loading follows the tension-stiffening line.

- Fig. 3c. The initial stress point is in an intermediate position along the unloading line, and the crack is still opened: Then, further loading or unloading follows the paths described in the previous case.

Fig. 3d. The initial stress point is in the compression branch (closed crack): Further loading or unloading follows the compression line if the crack remains closed. For positive straining (new opening of crack), loading follows the line of zero stress.

The above alternatives, though they are quite obvious in most cases, present a number of possibilities which should be carefully studied at the programming stage. In order to clarify concepts, we will summarize the steps to be followed for existing single and double cracked situations.

1. Single cracked point (Fig. 2a).

Step 1 : Local strains $\underset{\sim}{\varepsilon}'$ are computed from eq. (3).

Step 2 : The stress normal to the crack, $\sigma_{x'}$, is computed following the loading and unloading processes described earlier.

Step 3 : Stresses $\sigma_{y'}$ and $\tau_{x'y'}$ are computed using eq (6) and the new principal stresses obtained.

Step 4 : The onset of a second crack is checked following the steps described in previous section.

2. Double cracked point (Fig. 2b).

Step 1 : The local strains in each cracked direction are calculated using eq. (3).

Step 2 : The two stresses $\sigma_{x'}$ and $\sigma_{x''}$, orthogonal to each crack, respectively are computed following the loading and unloading processes described earlier.

Step 3 : The shear stress $\tau_{x'y'}$ along the first cracked direction is computed using eq. (6).

Step 4 : The stress $\sigma_{y'}$, along the direction of the first crack is obtained using eq. (8).

Step 5 : The global strains $\underset{\sim}{\sigma}$ in the double cracked point are computed using eq. (7) and the local strains $\underset{\sim}{\sigma}'$.

Discretised equilibrium equations. Non linear solution algorithm.

The system of discretised equilibrium equations are obtained via the virtual work principle and standard finite element procedures [9] . The resulting non linear system has been solved using a Newton-Raphson algorithm combined with an incremental process for the application of the internal - forces due to concrete expansion. Thus, for the ith iteration of the mth load increment corresponding to internal and external forces at time t, the displacement increments are obtained solving the following matrix equation:

$$\underset{\sim}{K}_T\,({}^{t}\underset{\sim}{a}_m^{i})\quad \Delta\,{}^{t}\underset{\sim}{a}_m^{i} \;=\; -\;\underset{\sim}{\Psi}\,({}^{t}\underset{\sim}{a}_m^{i}) \tag{9}$$

where $\underset{\sim}{K}_T$ is the tangent matrix and

$$\underset{\sim}{\Psi}\,({}^{t}\underset{\sim}{a}_m^{i}) = \int_{\Omega} \underset{\sim}{B}^T\,\underset{\sim}{\sigma}\,(\lambda_m\,{}^{t}\underset{\sim}{\varepsilon}_o,\ {}^{t}\underset{\sim}{a}_m^{i})\,d\Omega \;-\; \lambda_m\,\underset{\sim}{R} \tag{10}$$

is the residual force vector. In (10) $\underset{\sim}{B}$ is the standard strain matrix ${}^{t}\underset{\sim}{\varepsilon}_o$ the expansion strains at time t, ${}^{t}\underset{\sim}{a}_m^{i}$ the displacement vector for - the ith iteration of the mth increment of load at time t, λ_m the load increment coefficient and $\underset{\sim}{R}$ the vector of external loads (assumed to remain constant in time).

It can be proved that the tangent matrix can be written in the form [9]

$$\underset{\sim}{K}_T = \int_{\Omega} \underset{\sim}{B}^T\,\underset{\sim}{D}^{*}\,\underset{\sim}{B}\;d\Omega \tag{11}$$

where the tangent constitutive matrix $\underset{\sim}{D}^{*}$ is given by [3]

$$\underset{\sim}{D}^{*} = \underset{\sim}{D} - \underset{\sim}{D}\,\underset{\sim}{M}\,(\underset{\sim}{D}^{cr} + \underset{\sim}{M}^T\,\underset{\sim}{D}\,\underset{\sim}{M})^{-1}\,\underset{\sim}{M}^T\,\underset{\sim}{D} \tag{11}$$

for the cracked points and

$$\underset{\sim}{D}^{*} = \underset{\sim}{D} \tag{12}$$

for the non cracked points. In above $\underset{\sim}{D}$ is the standard linear elasticity - constitutive matrix of eq. (1) and $\underset{\sim}{D}^{cr}$ is a constitutive matrix which relates the stress increments with the strains due to the opening and closing of cracks only. For each crack, matrix $\underset{\sim}{D}^{cr}$ is obtained as [3]

$$\underset{\sim}{D}^{cr} = \begin{bmatrix} \dfrac{E^* \; E_s}{E^* - E_s} & 0 \\ 0 & \dfrac{\beta}{1-\nu} \; \dfrac{E}{2(1+\nu)} \end{bmatrix} \tag{13}$$

where E_s is the slope of the unloading branch, as defined in Fig. (4). In eq. (11), $\underset{\sim}{M}$ is a transformation matrix relating the three global strains with the two local strains at a crack, $\varepsilon_{x'}$, $\gamma_{x'y'}$. For each crack matrix M is given by

$$\underset{\sim}{M} = \begin{bmatrix} \cos\alpha & -\,\mathrm{sen}\,\alpha\,\cos\alpha \\ \mathrm{sen}^2\alpha & \mathrm{sen}\,\alpha\,\cos\alpha \\ 2\,\mathrm{sen}\,\alpha\,\cos\alpha & \cos^2\alpha - \mathrm{sen}^2\alpha \end{bmatrix} \tag{14}$$

where α has been defined in Fig. (3).

It is worth noting that plasticity or non linear elastic effects in concrete can be easily incorporated in this formulation simply substituting matrix $\underset{\sim}{D}$ of eq. (11) by the corresponding elastoplastic or the non linear elastic tangent constitutive matrix.

Once the displacement increments for time t have been obtained, the internal strains at time $t + \Delta t$ are explicitely computed from the appropiate experimental curve and the incremental process is restarted. Convergence of the solution process can be improved using any of the non linear equation solution algorithms currently available [5] , [10] ,[11] .

Use of joint elements.

Joint elements are very useful to model the effect of discontinuities in rock masses or concrete structures and different alternative joint formulation for two and three dimensional problems have been recently proposed [12], [13] . In dam problems joint elements can be used to reproduce the effect of construction or drainage joints. In this section we will briefly present the main features of the one dimensional joint element chosen to be coupled with the concrete formulation previously presented.

The joint element used here is an extension of the one presented by Carol et al [12] for the analysis of rock masses. The geometry of the joint element is shown in Fig 4 . The displacement field at the joint is characterised by the values of the relative displacements between the nodes of the upper and lower sides of the joint and sampled at the three internal points, 1', 2' and 3' of the joint midline (see Fig. 4a).

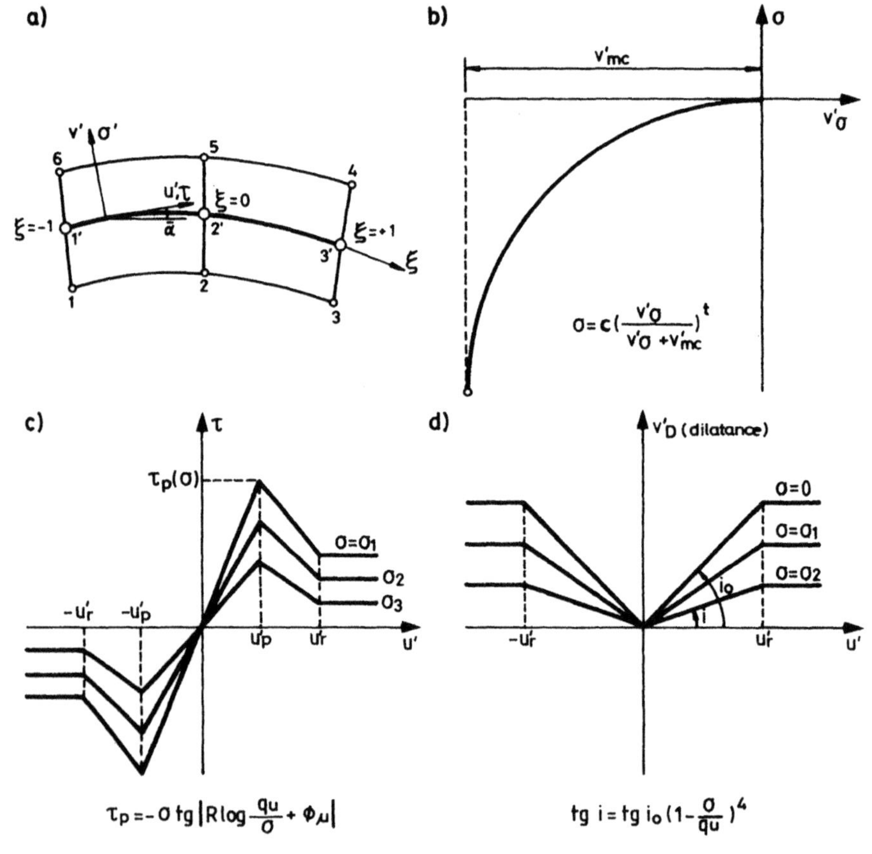

Fig.4 a) Joint element, b-d) Constitutive equations for the joint.

It is easy to show that the tangent stiffness matrix of the joint element can be written as [12]

$$\underset{\sim}{K}_T^J = \int_l \bar{\underset{\sim}{B}}^T \underset{\sim}{D}_t \bar{\underset{\sim}{B}} \, ds \tag{15}$$

where l_j is the lenght of the joint midline and $\bar{B}$ is the "relative displacement" matrix which relates the two relative displacements at a point of the joint mid line and the twelve global displacements of the six nodes of the two joint faces. Matrix $\bar{B}$ can be written as

$$\bar{B} = \left[B_1 \ , B_2 \ , B_3 \ , -B_3 \ , -B_2 \ , -B_1 \right] \tag{16}$$

with

$$B_i = \begin{bmatrix} -N_i \cos \bar{\alpha} & -N_i \operatorname{sen} \bar{\alpha} \\ N_i \operatorname{sen} \bar{\alpha} & -N_i \cos \bar{\alpha} \end{bmatrix} \tag{17}$$

where $\bar{\alpha}$ is defined in Fig. 4a and N_i is the standard one dimensional shape function of the three noded element of nodes 1' , 2' , 3' (see Fig. 4a). On the other hand, the constitutive tangent matrix D_t , can be obtained from the three basic constitutive joint equations relating: normal compression stress σ -relative normal displacement v_σ (Fig. 4b), tangential stress τ -tangential relative displacement v'(Fig. 4c), and the dilatancy normal displacement v'_D -tangential relative displacement u' (Fig. 4d), respectively. The ten parameters associated with the three constitutive equations must be obtained experimentaly. The constitutive equations are completed with the relationship between the total relative normal displacement, v', and displacements v'_σ and v'_D given simply by

$$v' = v'_\sigma + v'_D$$

The explicit form of matrix D_t of eq. (15) can be found in ref. [12].

Example. Analysis of a gravity dam under concrete expansion effects.

The crack model presented in the paper has been applied to the analysis of a real gravity dam in which unexpected high concrete expansion strains were found after construction. The geometry of the dam allows a two dimensional study. Details of the simplified geometry of the transverse section of the dam, boundary conditions and finite element mesh of standard isoparametric eight node elements used are shown in Fig.5a-b. A 3 x 3 Gaussian rule for the numerical evaluation of the element integrals has been used. Zone II in Fig. 5a models the effect of piers for which plane stress assumptions have been taken. On the other hand, zone I represents the dam body for which usual plane strain conditions are assumed. The material properties of concrete are the following:

- Young's modulus, $E = 0{,}2 \times 10^6$ Kg/cm^2
- Poisson's ratio, $\nu = 0{,}2$
- Maximum tensile stress, $f'_t = 30$ Kg/cm^2
- Tension stiffening coefficient, $\gamma = 1{,}05$
- Density = 0.24×10^{-2} Kg/cm^3
- Thickness zone I (plane strain) = 17,5 mts.
- Thickness zone II (plane stress) = 4 mts.
- Shear retention factor, $\beta = 0.5$

In Fig. 5a the zones with different expansion curves for concrete are shown. The expansion strain versus time curves for each zone obtained from simple one dimensional tests, are plotted in Fig. 5c (Note the unusual high values of the maximum straining of 1 and 1,2 ‰ in zones 2 and 3, respectivily). Line S-S' in Fig. 5a , represents a drainage joint establishing the limit for the entrance of water within the dam. However, no joint elements were used in this study.

The analysis has been performed in three consecutive steps under the following loading conditions:

- Step I - Self weight
- Step II - (I) + hydrostatic pressure
- Step III - (I) + (II) + transient internal forces due to concrete water expansion. Step III initiates the effects of transient expansion forces in the dam body. At zero time only the integrating points adjacent to the water face of dam are allow to expand causing new cracking into the dam body. Water is assumed to reach instantaneously the cracked points and new internal forces are consequently activated.

The curves of vertical and horizontal displacements versus time of two dam points have been plotted in Fig. 5d where experimental results available have also been plotted for comparison. The time scale used in the numerical study has been adequately adjusted using the experimental results. Good agreement has been obtained between experimental and numerical values even for the relatively coarse finite element mesh used, which shows the efficiency of the numerical model proposed.

Results for the compression and traction stress distributions at various times are shown in Fig. 6 . It is worth noting that as cracking progresses

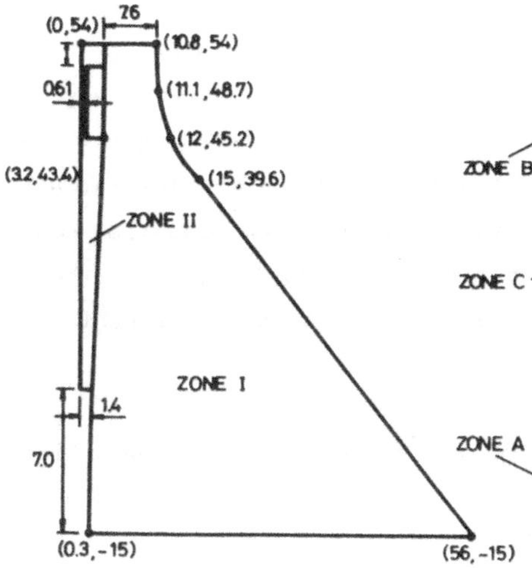

a) Dam geometry (in mts.)

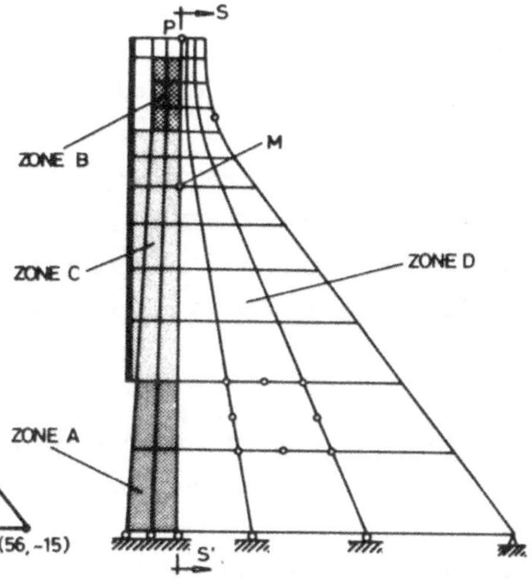

b) Finite element mesh

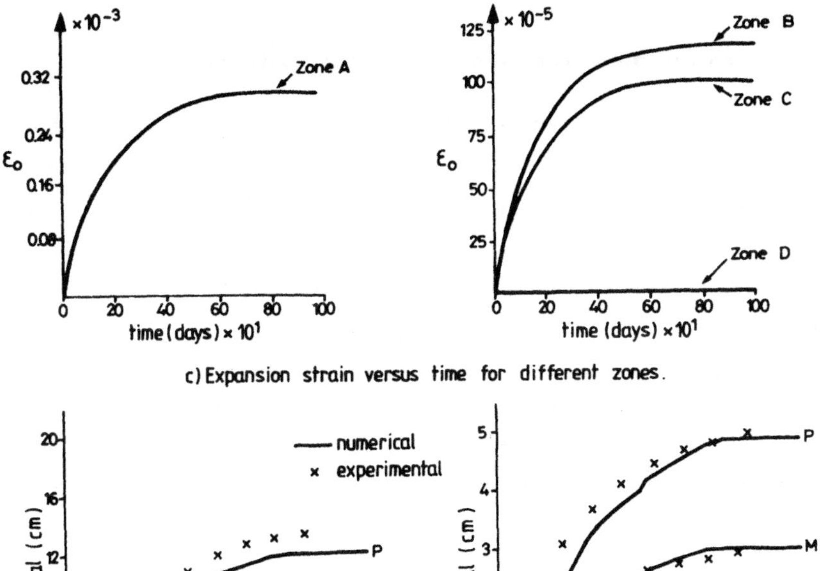

c) Expansion strain versus time for different zones.

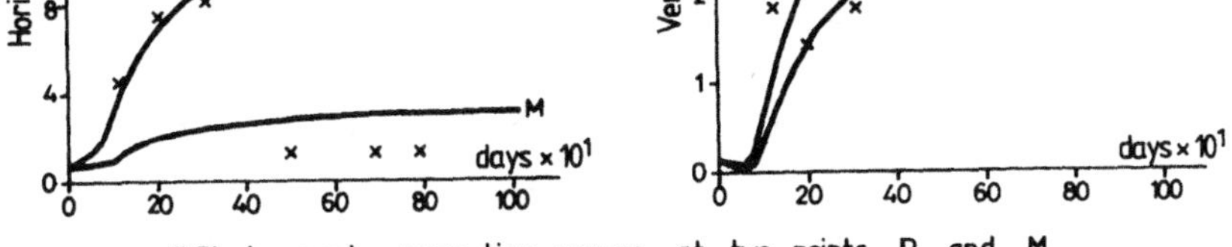

d) Displacement versus time curves at two points P and M.

Fig. 5

into the dam the stress distribution suffers severe changes from the initial compression field (see Fig. 6) to a situation in which high compression - stresses develop at the zones closed to the up stream and downstream boundaries, whereas most of the upper part of the dam body is under tensile stresses. The final structural behaviour of the dam can thus be interpreted to be equivalent to that of a "hollow" dam in which the structural load carrying capacity is carried out by the lateral compression zones only, i.e. the internal tractioned region does not contribute to the structural strenght of the dam.

In Fig. 7 , the evolution of cracking for different times is shown. Distinction between opened cracks (under tension or zero stress) and closed cracks (under compression stress) is made in the figure. Note the fast evolution - of the cracked region and the closing of cracks in the upstream compression zone for the final stages of the deformation.

The following conclussions can be extracted from this analysis:

1) Expansion effects create an extensive amount of cracking in the dam body.

2) Expansion effects cause important changes in the stress distribution in the dam. Initialy compressed zones change to tractioned ones (cracked) - in which compression stresses can again be developed (closed cracks).

3) Displacements and stresses converge to a steady state equilibrated configuration which guarantees the stability of the dam even under severe cracking conditions. The structural response of the cracked dam is then equivalent to that of a hollow dam and only the upstream and downstream compression zones contribute to the global stability of the structure.

Final conclussions.

The use of a smeared crack approach for the finite element analysis of concrete dams under internal forces has been presented. The model predicts well - the opening and closing of cracks in plain concrete produced by the sign - changes in strains due to the internal loading. The model also seems adequate for the evaluation of cracking damage in concrete dams subjected to internal forces due to water expansion of concrete. The example analysed shows - the extent of such a damage in certain situations and the importance of a - good prediction of the internal cracking and stress redistribution in the -

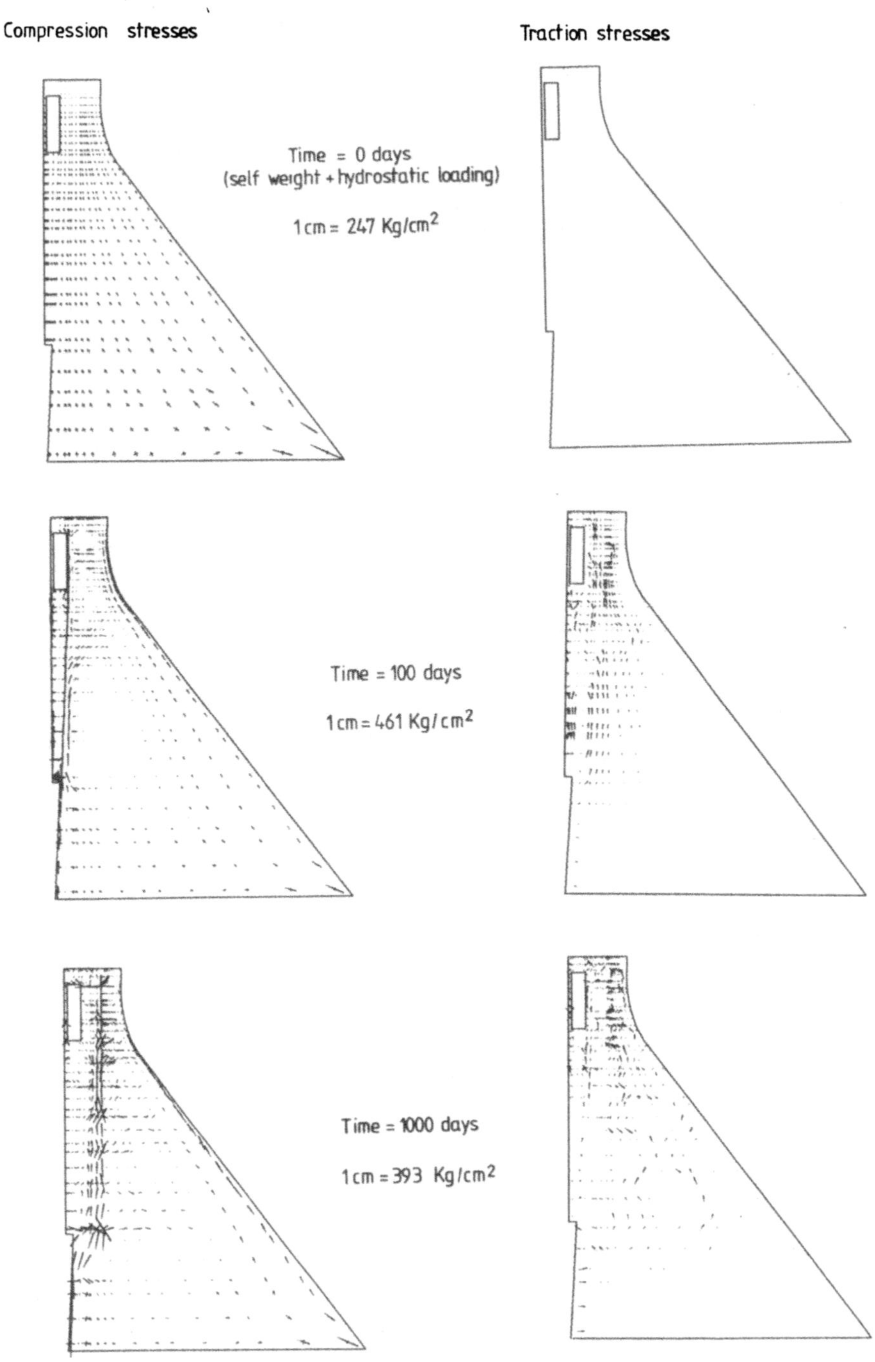

Fig.6 - Principal stresses at various times.

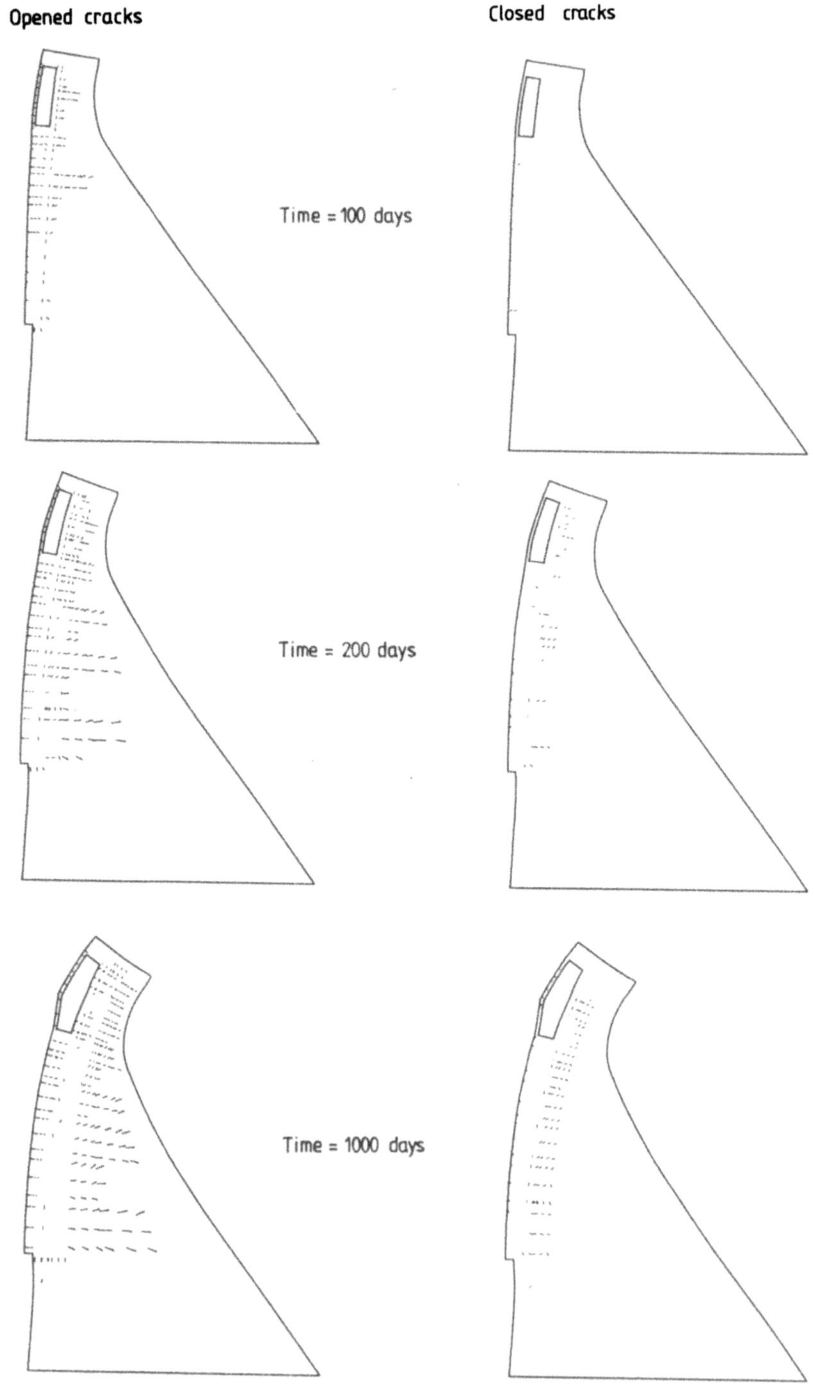

Fig.7 - Development of cracks and deformed shapes of dam.

dam to guarantee structural stability. Further developments of the numerical model, including the use of the concrete joint element presented in the paper to simulate internal joints in cracked dams are currently under study.

References.

1. Damjanic, F., Hinton, E., Owen, D.R.J., Biçanic,N.,Simonic,V. **Proc. International Conference on Computer Aided Analysis and Design for Concrete Structures**, Pineridge Press Swansea, 1984.

2. de Borst, R. and Nauta, P.; Smeared crack analysis of reinforced concrete beams and slabs failing in shear. Proc. Int. Conf. Computer Aided Analysis and Design for Concrete Structures. Edited by F. Damjanic et al . Pineridge Press, Swansea, 1984.

3. de Borst, R. and Nauta, P.; Non orthogonal cracks in a smeared finite element model. Internal report. Institute TNO for Building Materials and Structures, Delft, The Netherlands, 1984.

4. de Borst, R. Application of advanced solution techniques to concrete cracking and non associated plasticity. **Numerical Methods in Non Linear Problems.** (Eds. C. Taylor, E. Hinton, D.R.J. Owen and E. Oñate). Pineridge Press, Swansea, Vol. 2, 314-325, 1984.

5. Rahman, H.H.A.; Computational models for the non linear analysis of reinforced concrete for flexural systems. **Ph.D. Thesis.** Univ. College of Swansea, C/Ph/66/82, U.K., 1982.

6. Lin, C.S.; Non linear analysis of reinforced concrete slabs and shells. **SESM Report** n. 73-1, Univ. of California, Berkeley, USA, 1973.

7. Rashid, Y.R.; Analysis of prestressed concrete pressure vessels. **Nucl. Eng. Design.** 7, 334-344, 1968.

8. Suidan, M. and Schnobrich, W.C. Finite element analysis of reinforced concrete. **ASCE. J. Struct. Div.** 99, 2109-2122, 1973.

9. Zienkiewicz, O.C. **The finite element method.** Mc. Graw Hill, 1979.

10. Crisfield, M.A.; An arc-length method including line-searchers and accelerations. **Int. J. Num. Engng.** 19, 1269-1289, 1983.

11. Crisfield, M.A.; Difficulties of current numerical models for reinforced concrete and some tentative solutions. Proc. **Int. Conf. Computer Aided Analysis and Design of Concrete Structures.** (Eds. F. Damjanic et al). Pineridge Press, Swansea, Vol. 1, 331-358, 1984.

12. Carol, I., Alonso, E.E.; A new joint element for the analysis of fractured rock. **International Congress on Rock Mechanics,** F. 147-151, Melbourne, Australia, 1983.

13. Goodman, R.E. and John, C.St.; Finite element analysis for discontinous rocks. **Numerical Methods in Geotechnical Engineering.** Ed. by C.S. Desar and U.T. Christian, Mc. Graw Hill, 1977.

Part VII: Contact Problems

On the Influence of Contact Boundary Conditions on the Stability of Shell Structures Under Follower Loads

F. G. RAMMERSTORFER, W. AULI, F. D. FISCHER*
Technical University of Vienna, Austria
*University for Mining and Metallurgy, Leoben, Austria

Summary

After some general considerations regarding stability of conservatively versus non-conservatively and statically versus dynamically loaded structures the influence of boundary conditions on the stability of wind loaded and liquid filled earthquake loaded cylindrical cantilever shells is studied in the paper. Special attention is drawn to the contact problem which arises if the shell base (including a bottom plate) may partially uplift under the action of axial tensile membrane forces. Such problems appear in the analysis of liquid storage tanks.

1. INDIVIDUAL STABILITY CRITERIA

This chapter describes very briefly the different kinds of stability loss of elastic structures under static or dynamic loading.

1.1 Static conservative loading

In the case of conservative loading, i.e. either deformation-independent or follower forces with a global potential belonging to the force field /1/, the tangent stiffness matrix ${}^{m}K$ of the problem discretized by the finite element method is symmetric. The static stability criterion /2/ for static loading leads to

$$ {}^{m^*}K\,\delta u = 0 \quad \text{with } \delta u \neq 0 \qquad (1) $$

(m^* critical equilibrium configuration)
The critical load multiplier can be approximated by the solution of the linearized eigenvalue problem /3/

Finite Element Methods for Nonlinear Problems
Europe-US Symposium, Trondheim, Norway 1985
Editors: Bergan, Bathe, Wunderlich

$$[{}^{o}K_c + {}^{m}\eta K_{NL}\ ({}^{m}\lambda)]\ \delta u=0 \qquad (2)$$

with

$$^{m}K = {}^{o}K_c + {}^{m}K_{NL} \qquad (3)$$

(${}^{o}K_c$ is the constant initial stiffness matrix for the unloaded structure).

The approximative critical values of the load multiplier λ are

$$^{m}\hat{\lambda}^*_i = {}^{m}\eta_i\ {}^{m}\lambda \approx \lambda^*_i \qquad (4)$$

If, as assumed, ${}^{m}K$ and hence ${}^{o}K_c$ as well as ${}^{m}K_{NL}$ are symmetric only real eigenvalues of (2) may appear /4/.

Furthermore, also the eigenfrequencies ω_i, being the solutions of the eigenvalue problem

$$({}^{m}K - {}^{m}\omega_i^2\ M)\ \Phi_i = 0 \qquad (5)$$

cannot be complex /4/ and, hence, flutter instability can be excluded.

1.2 Static non-conservative loading

If the static load is a follower load, i.e. a deformation dependent loading, and if several criteria are fulfilled - see chapter 2 - so that ${}^{m}K$ is non-symmetric then flutter instability may occur and the kinetic criterion for stability loss must be applied /2/.

The following conditions are met in the critical equilibrium state m*:

a) The lowest real eigenvalue of eqn. (5) becomes zero, i.e.

$$^{m*}\omega_1^2 = 0 \qquad (6)$$

leading to divergence (buckling) instability (as it is valid for conservatively loaded structures, too) or

b) two real eigenvalues approach each other if ${}^{m}\lambda$ approaches ${}^{m*}\lambda = \lambda^*_1$:

$$m \rightarrow m^*\ ({}^{m}\lambda \rightarrow \lambda^*) \Rightarrow {}^{m}\omega_i^2 \rightarrow {}^{m}\omega_{i+1}^2 \quad \text{with} \quad {}^{m*}\omega_i^2 = {}^{m*}\omega_{i+1}^2 \qquad (7)$$

and remain conjugate complex for $\lambda > \lambda^*_1$:

$$^{m}\lambda > \lambda^* \Rightarrow \mathrm{Re}({}^{m}\omega_i^2) = \mathrm{Re}({}^{m}\omega_{i+1}^2) > 0 \wedge \mathrm{Im}({}^{m}\omega_i^2) = -\mathrm{Im}({}^{m}\omega_{i+1}^2) \qquad (8)$$

indicating flutter instability, i.e. the amplitude of the periodic motion following any perturbation grows with time;

- self-excited vibrations appear.

If non-conservatively loaded structures lose their stability due to divergence, then the application of the static criterion eqns. (1) to (4), leads to the same buckling load λ^* as the kinetic criterion, eqn. (6), does. However, in general, it needs to be proven that flutter instability is excluded if the static criterion is applied.

1.3 Dynamic loading

Bolotin /5/, by considering the solutions of the equations of motion, describes several excitation-frequency domains which contain frequencies leading to parameter resonance. Bolotin's analytical approach is applied to simple structures under periodic conservative and non-conservative loading. - Some examples analyzed in more detail in /6/ will be given in Chapter 3.

Additionally, parametric combination resonance may also lead to vibrations with amplitudes growing with time if certain combinations of eigenmodes are excited.

Another aspect of consideration of the stability of dynamically loaded structures is to ask which amplitude of the load multiplier Λ, with $\lambda(t) = \Lambda f(t)$, $f(t)$ being a prescribed function of time, (e.g. a step function, a prerecorded function), will bring about a sudden change of the response time history of the structure. This critcal value $\Lambda^*(f)$ is called dynamic buckling load /7/. Examples will be given in Chapter 3.

2. THE INCREMENTAL EQUILIBRIUM EQUATION FOR DEFORMATION DEPENDENT LOADING

In /1,3/ a detailed derivation of the individual contributions to ${}^{m}K = {}^{o}K_c + {}^{m}K_u + {}^{m}K_g$ (${}^{m}K_u$ is the initial displacement matrix, ${}^{m}K_g$ is the geometric or initial stress matrix) is given for elastic structures under deformation-independent loading. In /1/

the fourth contribution to ${}^{m}K$, namely the load stiffness matrix ${}^{m}K_L$, and the actual load vector are obtained for follower loading from the virtual work principle so that the linearized incremental equilibrium equation has the following form :

$$ {}^{m}K\Delta u = ({}^{o}K_c + {}^{m}K_u + {}^{m}K_g - {}^{m}K_L) \ \Delta u = {}^{m+1}R - {}^{m}F \qquad (9) $$

In order to show the influence of the deformation dependence of the load on the tangent stiffness matrix ${}^{m}K$ and on the load vector ${}^{m}R$ a very brief summary of the derivation of the incremental equilibrium equation with notations used in /1,6/ is given:
The principle of virtual work

$$ \delta W = \delta W_{(a)} + \delta W_{(i)} = 0 \qquad (10) $$

for states of equilibrium shows a significant difference between deformation dependent and deformation independent loading in the $\delta W_{(a)}$ - term (work of the external forces). For simplicity in the following considerations only proportional pressure loading acting normal to the deformed surface is applied, i.e.

$$ {}^{m}p = {}^{m}\lambda\,{}^{m}f\,{}^{m}n \qquad (11) $$

${}^{m}f$ describes the reference pressure field, ${}^{m}n$ is the actual normal vector of the loaded surface. In /1/ a distinction is made between a space attached deformation independent reference pressure field

$$ {}^{m}f = f({}^{m}x) \qquad (12) $$

${}^{m}x$ being the coordinate of the considered point of the deformed surface, and a body attached pressure field

$$ {}^{m}f = f({}^{o}x) \qquad (13) $$

${}^{o}x$ being the coordinates of the unloaded surface.

The virtual work of the pressure forces in the state m+1 is

$$ {}^{m+1}\delta W_{(a)} = \int_{{}^{m+1}A} {}^{m+1}\lambda\,{}^{m+1}f\,{}^{m+1}n^{T}\delta u \ d({}^{m+1}a) \qquad (14) $$

If the normal vector is expressed in terms of the coordinates

$$ {}^{m+1}x = {}^{m}x + \Delta u \qquad (15) $$

and after some linearization /1,6/ (neglection of products $\frac{\partial \Delta u_i}{\partial r}\,\frac{\partial \Delta u_j}{\partial s}$; r, s being the local coordinates) the following expression is obtained

$$^{m+1}\delta W_{(a)} = e_{ijk}\,^{m+1}\lambda \iint_{rs} {}^{m+1}f \frac{\partial^m x_j}{\partial r} \frac{\partial^m x_k}{\partial s} \delta u_i \, dr ds \; +$$

$$e_{ijk}\,^{m+1}\lambda \iint_{rs} {}^{m+1}f \left(\frac{\partial \Delta u_j}{\partial r} \frac{\partial ^m x_k}{\partial s} + \frac{\partial ^m x_j}{\partial r} \frac{\partial \Delta u_k}{\partial s} \right) \delta u_i \, dr ds \qquad (16)$$

with the permutation operator

$$e_{ijk} = \begin{cases} 0 \text{ for } i=j \text{ or } j=k \text{ or } i=k \\ 1 \text{ for } (i,j,k)=(1,2,3) \text{ or } (2,3,1) \text{ or } (3,1,2) \\ -1 \text{ for } (i,j,k)=(1,3,2) \text{ or } (2,1,3) \text{ or } (3,2,1) \end{cases} \qquad (17)$$

After the introduction of the finite element interpolation functions the first term on the right hand side of eqn. (16) contributes to the load vector ^{m+1}R, while the second term leads to the so-called load stiffness matrix ^mK_L. If K_L is non-symmetric and, hence, ^mK is non-symmetric, too, a non-conservative problem is treated.

As shown by Schweizerhof /1/ K_L can be devided into boundary terms and interior terms and the following sufficient conditions for symmetry of ^mK_L(and, consequently, of ^mK) can be obtained from /1/ :

a) The boundary terms have to be zero, which is the case either if $^mf \equiv 0$ along the boundary for all states or if at least two displacement components are restricted or prescribed at the boundaries.

b) In the case of body attached reference pressure fields the additional condition $f(^0x)=$ const. (i.e. no variation of the pressure field in the interior) has to be met.

The incremental equilibrium equation (9) can accordingly be applied iteratively to ensure equilibrium. Furthermore, it can be extended to dynamic problems by adding the inertia and damping forces /6/.

3. SOME TYPICAL PROBLEMS

With the NISA-80 program /8/ after implementing - as described in /6/ - the follower load capabilities for normal and tangential surface loads of 2-D, 2-D-degenerated (beam), 3-D and shell elements, the Wilson and Newmark time integration schemes and a special eigenvalue solver /9/ for non-symmetric linear eigenvalue problems, a tool has been developed which allows the analysis of the classes of stability problems discussed in Chapter 1.
This will be demonstrated by the following examples.

3.1 The beam column

The simple problem of an end loaded beam column is suitable to check the usefulness of the algorithms implemented into NISA-80 and to demonstrate some different kinds of stability loss.
The following data apply to the examples :
$E=2.06*10^7 N/cm^2$, $A=100\ cm^2$, $J=1000\ cm^4$, $l=1000\ cm$, $\rho = 0.008\ kg/cm^3$

3.1.1 Conservative loading

A conservative compression load acting centrically and parallel to the undeformed axis leads, of course, to buckling if this load is quasi-statically increased to P_E , the Euler load. However, if this load is applied periodically

$$P(t)=P_o+P_t \cos \Omega t \tag{18}$$

then parameter resonance will appear for several intervals of Ω . After the introduction of abbreviations according to /5/

$$\alpha = \frac{4P_o l^2}{\pi^2 EJ}, \quad \beta = \frac{4P_t l^2}{\pi^2 EJ}, \quad \lambda = \frac{\rho A l^4 \Omega^2}{4\pi^4 EJ} \tag{19}$$

the instability domains are shown in Fig.1. This solution is restricted to load levels $P_o + P_t \ll P_E$.
If a frequency of excitation Ω is chosen which, according to /5/, would not lead to parameter resonance and if the load is

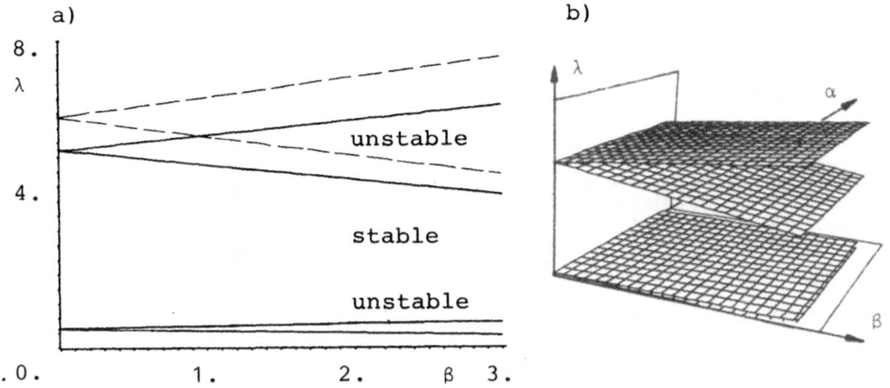

Fig. 1: Instability domains for conservatively loaded beam column
a) $\alpha = 0$, dotted line: Bolotin's solution /5/
solid line: more accurate evaluation of Bolotin's derivation - see /6/
b) Instability domains for varying α

applied with increasing amplitude

$$P(t) = P_t(t)\cos\Omega t \qquad (20)$$

with $P_t(t)$ slowly increasing with time, we will find that, even if we apply a certain imperfection, the load amplitude $P_t(t)$ can be increased above the Euler load P_E before the appearance of considerable transversal deflections, which is often called "dynamic buckling" - see Fig. 2. This example shows clearly that the appearance of vanishing and even negative eigenvalues ω^2 is <u>not</u> sufficient to enforce "dynamic buckling" in the aforementioned sense, if the time interval with $\omega^2 < 0$ is not sufficiently large. The results of this investigation will also be important with regard to the stability analysis of earthquake loaded liquid filled cylindrical shells as described in Chapter 5. It should be mentioned that the results depend to a relatively high degree on the kind of perturbation applied to bring the column to buckling.

3.1.2 Follower end load

If the load is a statically applied follower load (tangential end load) we are dealing with Beck's problem /10/ which leads to flutter instability. However, if the loaded end is elastically

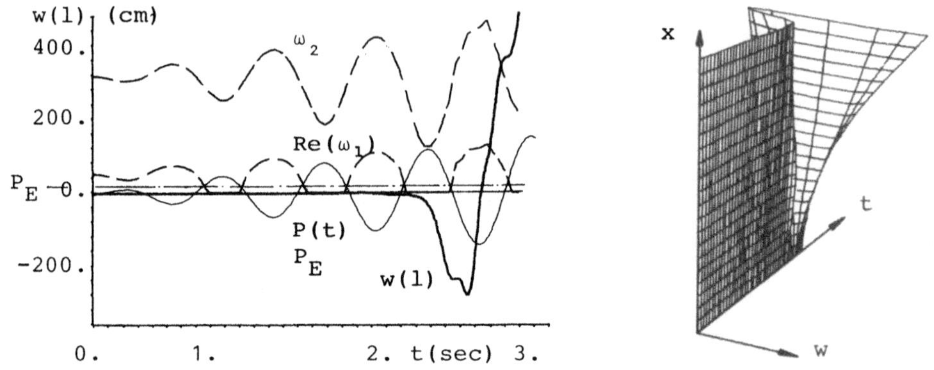

Fig. 2: Periodically conservatively end loaded beam column: time history of
——— the load
——— the transversal end deflection
– – – – the two lowest eigenfrequencies

supported an increasing support stiffness will result in a change from flutter to divergence instability /2/ despite the fact that the stiffness matrix remains non-symmetric as long as the support stiffness has a finite value.

The typical behaviour of Beck's column as calculated with the modified NISA-80 program is shown in Fig. 3. The results shown in Fig. 3 are obtained by including large displacements effects. The eigenvalues (Fig. 3a) were calculated by solving the eigenvalue problem stated by eqn. (5), however, with ${}^{m}K$ being non-symmetric, using an algorithm derived from Dirschmid's solver /9/. The time integration to compute the flutter motion (Fig. 3b and 3c) was performed by the use of the Newmark method.
It should be mentioned that, even though the unsymmetry of the ${}^{m}K$ matrix of the problem under consideration is very weak, the eigenvalues in terms of natural frequencies differ considerably as compared to the related conservative problem.

Let us now consider a follower load with periodically varying intensity in analogy to the conservative problem described above. The consideration of parameter resonance based on Bolotin's approach /5/ gives instability domains for the frequency of

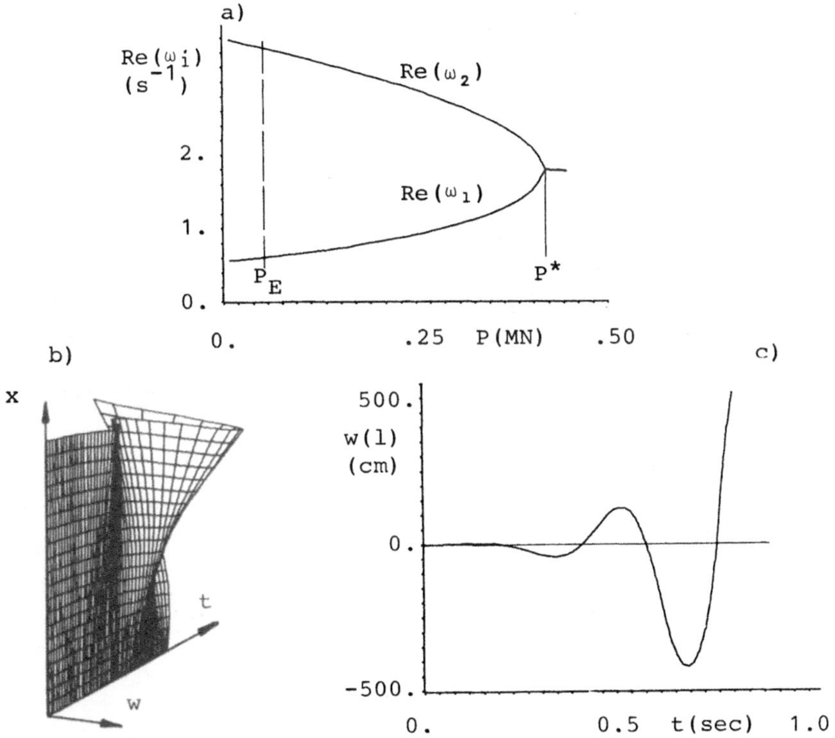

Fig. 3:Results of Beck's problem - nonlinear analysis
a) lowest two eigenvalues indicating flutter instability at $P = P^*$
b) and c) flutter motion for $P > P^*$ ($P=1.2\ P^*$)

excitation as shown in Fig. 4. Here, again, these results are restricted to loads $P_o + P_t \ll P^*$.

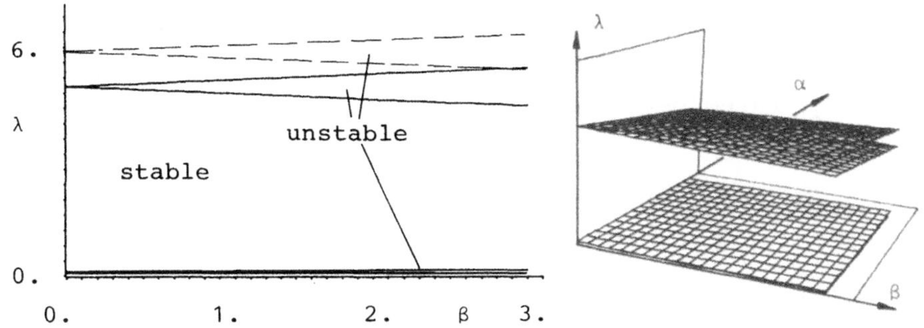

Fig. 4:Instability domains for tangentially loaded beam column

The consideration of dynamic buckling by observing the time history of the transverse deflection is demonstrated in Fig. 5. A follower load amplitude increasing linearly with time results in a sudden increase of the transverse deflection starting from the very small perturbation motion at a load level lying above the flutter load P*(Fig. 5.a).

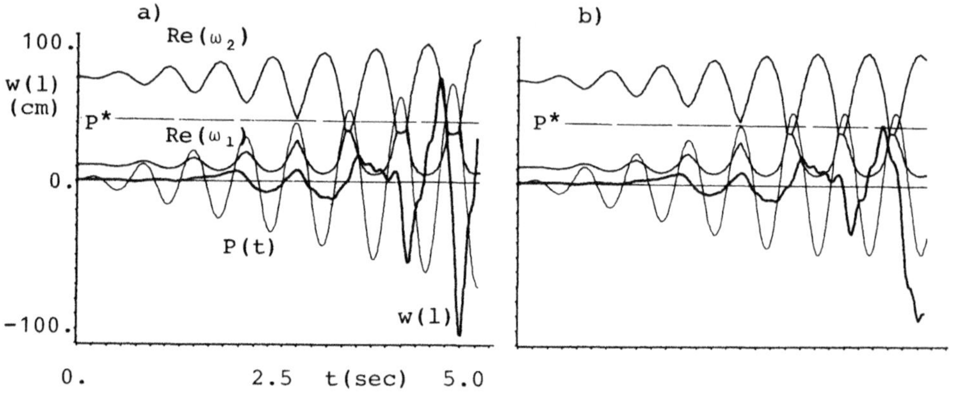

Fig. 5:Periodically tangentially and loaded beam column:
time history of
——— the load
━━━ the transversal end deflection
——— Re (ω_1) and Re (ω_2)

The consideration of the eigenfrequencies shows that some periods before the "dynamic buckling" becomes evident the two lowest eigenfrequencies become conjugate complex. One can conclude that several time intervals of fulfilled flutter conditions (i.e. $Re(\omega_1) = Re(\omega_2)$), are necessary for the formation of considerable flutter motions.

This conclusion is confirmed by the investigation documented in Fig. 5b. One can recognize the formation of flutter vibrations during the repeated periods of load application with P_t slightly larger than P*.

3.2 The circular arch

Another problem, the circular arch under a parabolically distributed pressure load, is treated in order to study principal features of the behaviour of wind loaded cylindrical shells as discussed in Chapter 4.

The circular arch shown in Fig. 6 should serve as a very coarse substitute of the wall portion belonging to the stagnation zone of the wind loaded shell treated in Chapter 4.

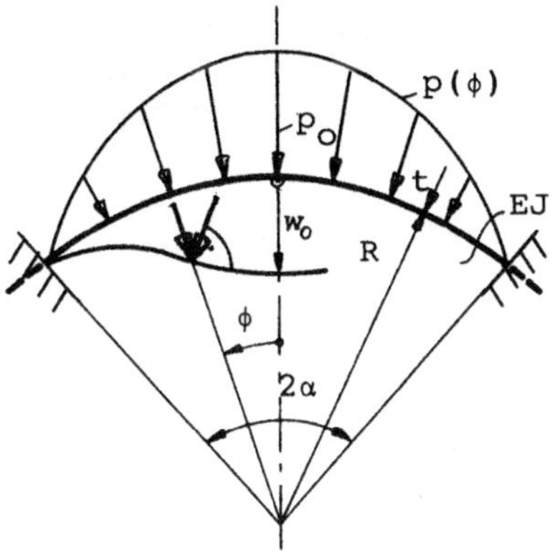

$p_{Oref}=2.778.10^{-4}$
$t = 1$ cm
$E = 2.06\times10^{7}$ N/cm^2
$A = 1$ cm^2
$J = 0.833$ cm^4
$\alpha = 40^{\circ}$
$R = 20$ m

Fig. 6:The circular arch

The arch is discretized by 12 four-noded beam elements and the geometrically nonlinear analysis is carried out both under conservative deformation-independent body-attached pressure loads and under non-conservative follower pressure loads. In both cases the pressure intensity is given by

$$ {}^{m}p(\phi) = {}^{m}\lambda p_o (1-(\frac{\phi}{\alpha})^2) \qquad (21)$$

The load displacement paths as well as eigenvalue curves for both buckling loads according to eqn. (4) and the eigenfrequencies according to eqn. (5) are shown in Fig. 7.
Despite the fact that the pressure load vanishes at the boundaries and, hence, no boundary terms appear in the load stiffness matrix, the non-uniformity of $p(\phi)$ leads to unsymmetries in ${}^{m}K$. However, these unsymmetries are so small compared to the order of magnitude of the matrix elements that their influence is negligible. Therefore, the response of the structure (Fig. 7) clearly shows that divergence buckling and not flutter instability can be predicted also for the wind loaded tank. However, the buckling load and even more the prebuckling deformation differ between the conservative load case and the follower load case. The same trends will be observed later in the tank analysis.

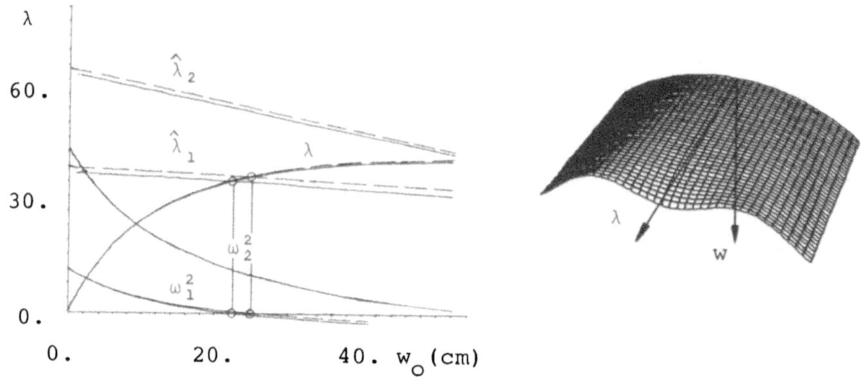

Fig. 7: Response of the statically loaded circular arch
— — — — conservative radial pressure load
———— nonconservative normal pressure load

The question whether gusty wind leads to significantly different buckling loads compared to quasi-static wind pressure application should also be treated in a qualitative manner by the use of the arch-substitute model under follower pressure loads varying in time in order to simulate a gusty wind with increasing intensity.

The frequency of the wind force is chosen to be the half of the fundamental frequency of free bending vibrations of the unloaded arch. Figure 8a shows the behaviour of the perfect arch. The bifurcation point ($\omega_1^2=0$) is completely ignored and a "dynamic snap through" is recognized at a load level at which ω_2^2 vanishes. However, even if a very small antisymmetric imperfection is applied, for example a small pertubation moment M_ϕ at the top point of the arch, the dynamic buckling appears at the load level at which ω_1^2 vanishes; see figure 8b. This structure is loaded with a frequency which is rather low as compared to the natural frequency and the dynamic effects are, consequently, not that pronounced. From these results one might conclude that in addition to the follower load effects also the dynamic effects may be neglected in the analyses of the buckling of wind loaded tanks, which has already been pointed out in /1/.

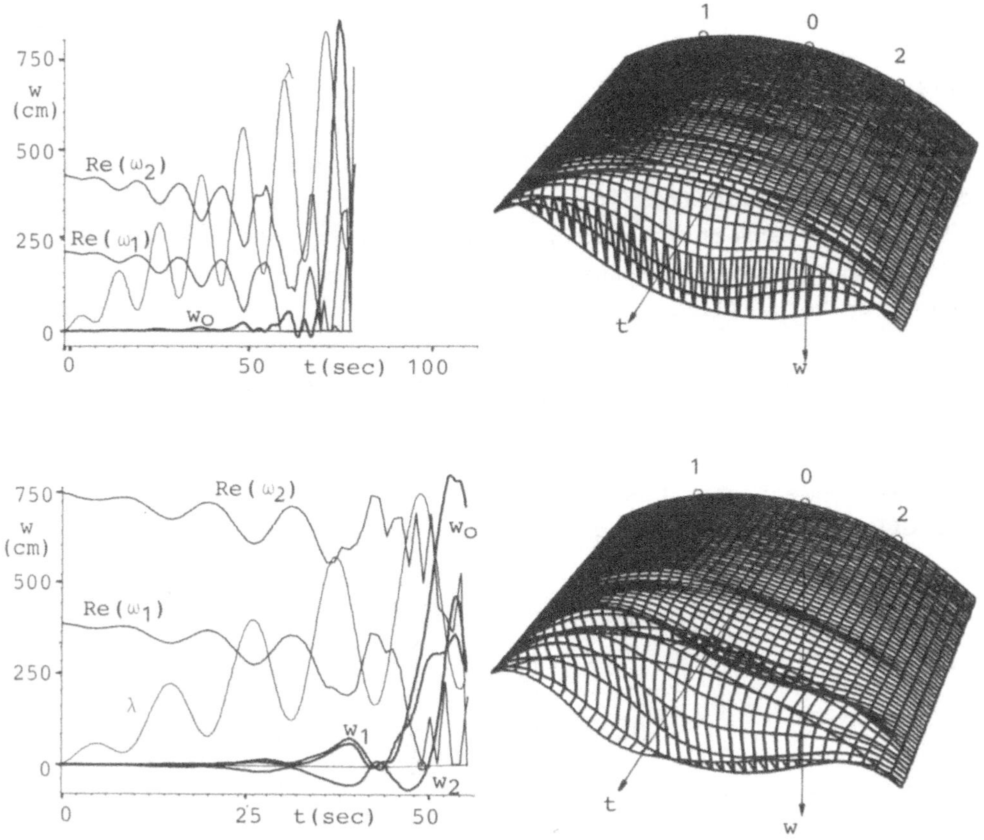

Fig. 8: Response of the dynamically loaded circular arch - substituting a circular cylindrical shell under gusty wind load.
a) Perfect case, b) imperfect case

4. THE WIND LOADED CYLINDRICAL SHELL

The stability of wind loaded cylindrical shells plays an important role in liquid storage tank design and is treated in numerous papers, as for example /1,11,12/. The influence of the follower load character is studied for fixed base tanks in Ref. /1/. As shown among others by Sobel /13/ the influence of the boundary conditions on the stability of pressure loaded cylindrical shells is considerable. Hence, it is important to take the possible partial uplifting of unanchored tanks into account if their stability is evaluated /14/.

In /14/ a shell-spring model is developed which allows the analysis of unanchored tanks using a model consisting of the tank wall, i.e. the cylindrical shell with a wall thickness tapering from bottom to top, resting on a non-linear elastic foundation. The non-linear effective spring characteristic of the elastic edge support can be obtained either analytically or numerically as shown in Ref. /15/.

The wind loaded tank given in Fig. 9 is modelled by the NISA-80 degenerated shell elements /8/ modified in order to take follower loads into account /6/ and non-linear spring elements with a load displacement character corresponding to the uplift-resistance effect of the bottom plate of the tank /14/.

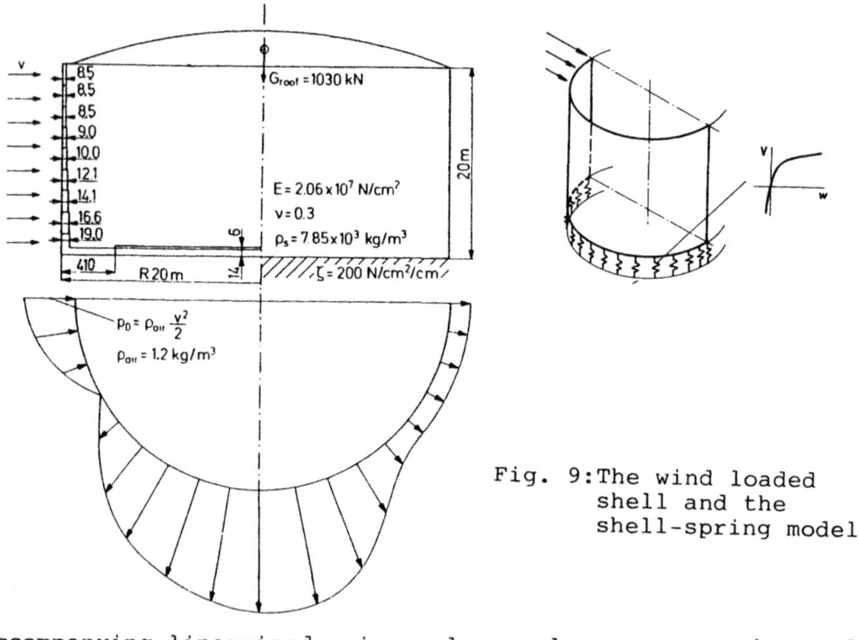

Fig. 9: The wind loaded shell and the shell-spring model

Accompanying linearized eigenvalue analyses can now be performed either based on /14/ :

$$[\,{}^{o}K_c + {}^{m}\hat{\eta}\,({}^{m}K_u + {}^{m}K_g - {}^{m}K_L + {}^{m}(\Delta K_s))]\ \delta\hat{u} = 0 \qquad (22)$$

or

$$[{}^{o}K_c + {}^{m}(\Delta K_s) + {}^{m}\hat{\hat{\eta}}({}^{m}K_u + {}^{m}K_g - {}^{m}K_L)]\,\delta\hat{\hat{u}} = 0 \qquad (23)$$

where ${}^{m}(\Delta K_S)$ means the change of the contributions of the contact springs to the elastic stiffnes matrix. The approximations of the critical load multiplier ${}^{m}\hat{\lambda}^{*} = {}^{m}\hat{\eta}\;{}^{m}\lambda$ and ${}^{m}\hat{\hat{\lambda}}^{*} = {}^{m}\hat{\hat{\eta}}\;{}^{m}\lambda$ approach the same value if the buckling state is reached. The advantage of (23) in comparison with (22) is that it is easily to be implemented in finite element programs which allow accompanying eigenvalue analyses for geometrically nonlinear structures by implementing the stiffness matrix of the contact springs as materially nonlinear only contributions. However, the formulation (23) underestimates the influence of uplifting while (22) overestimates this influence.

First of all, the tank model is loaded with quasi-statically increased deformation-independent wind pressure forces. After incrementing the wind velocity up to 150 km/h an eigenvalue analysis according to eqn. (23) is performed.

Fig. 10a shows the fundamental buckling mode. The lowest eigenvalue is ${}^{m}\hat{\eta}_1 = 4.53$. Of course, this value has to be considerably decreased due to the high imperfection sensibility of wind loaded cylindrical shells /12/.
In order to show the significant influence of the partial uplifting of the shell on the stability limit the same tank as investigated so far is also considered in an anchored configuration. The accompanying eigenvalue analysis also performed at v = 150 km/h leads in this case to an eigenvalue of ${}^{m}\hat{\eta}_1 = 10.38$. The corresponding buckling mode is shown in Fig.10 b. It differs significantly from the one obtained for the unanchored tank but it is in good qualitative agreement with the buckling mode shown in an earlier paper /12/.
The estimation of the stability limits shows that the possibility of partial uplifting of the tank base leads to a drastic reduction of the safety of the tank regarding the stability under wind loading.
Taking the follower load character of the wind forces into account leads to a change in the stiffness matrix due to ${}^{m}K_L$ and in

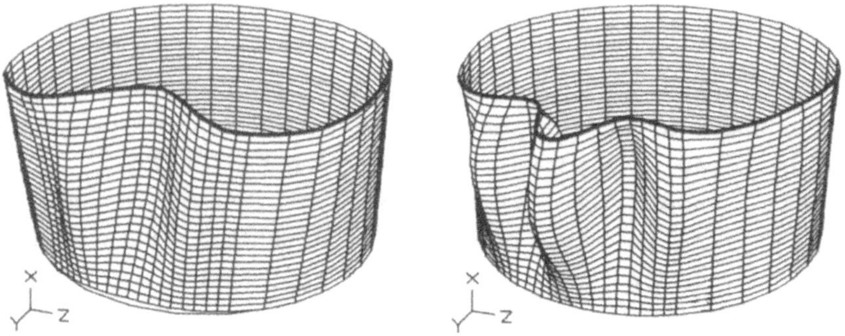

Fig.10:Fundamental buckling mode of the tank
a) unanchored b) anchored

the load vector ${}^{m}R$.
The stiffness matrix ${}^{m}K$ becomes slightly non-symmetric. The influence of the follower forces is, regarding deformations up to the design wind speed of v = 15o km/h, completely negligible and, regarding the buckling loads, very small, too.

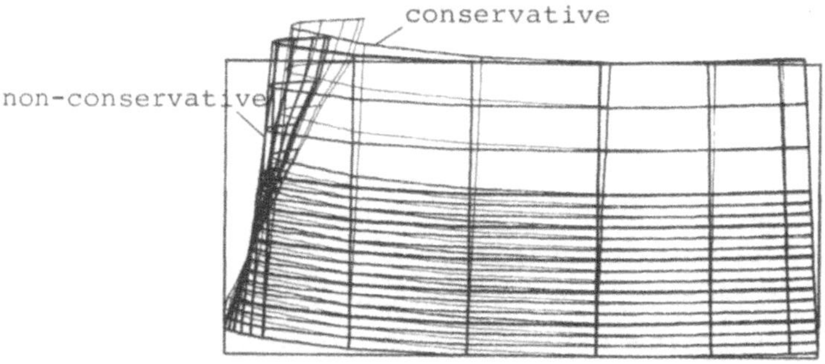

Fig.11:Deformations of the wind loaded tank (v = 250 km/h)

Fig. 11 shows the difference in deformations between the conservatively loaded and the follower pressure loaded tank which becomes only noticeable at an unrealistic wind level of v=250 km/h. Hence, in combination with the results obtained for the arch problem (Chapter 3) one can conclude that the boundary conditions of wind loaded tanks (anchored versus unanchored) have a strong influence on the stability loss, the follower force character and dynamic effects due to gusty wind pressure are, however , not so pronounced.

5. THE EARTHQUAKE LOADED LIQUID FILLED TANK

If an earthquake excites a liquid filled storage tank, dynamically activated pressure loads act on the tank wall and on the bottom plate in addition to the hydrostatic pressure.
The calculation of this pressure distribution is, for example, described in /16/. This pressure causes an overturning moment and, as a consequence, axial membrane forces at the tank base which may lead - for tensile forces - to partial uplift or - for compression forces - to buckling.
Since this kind of buckling is mainly caused by the reaction forces for which the influence of the follower force character of the fluid pressure is very weak the analysis can be based on deformation independent conservative loading. However, the boundary conditions have a strong influence on the distribution of the axial membrane forces and, hence, on the stability limit which is expressed for example in terms of maximum ground acceleration of an assumed earthquake.

The shell spring model discussed in Chapter 4 is also applicable in this case. The effective spring characteristic must include the fluid pressure on the bottom plate.

When the same tank as considered above under wind load is excited by an earthquake corresponding to the USAEC-response spectrum the maximum pressure distribution appearing during the earthquake can be estimated as being

$$p(x,\Phi)=P(x)\ \cos(\Phi) \qquad (24)$$

with P(x) as shown in Fig. 12.
The justification of the application of the response spectrum method for this nonlinear problem is based on the fact that all nonlinear effects (uplifting) have only a small influence on the fundamental natural frequency of the system in the relevant

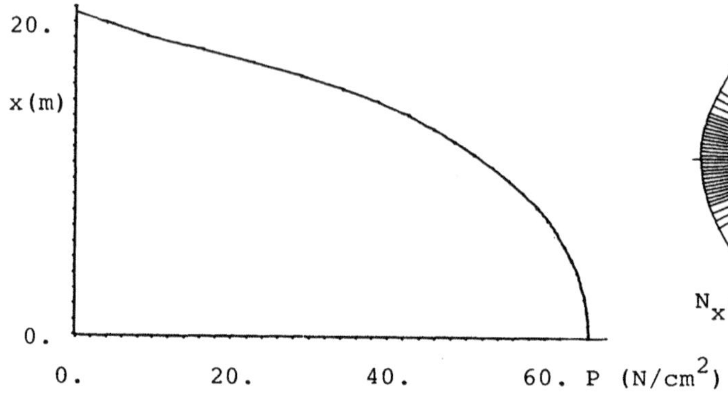

Fig.12: Maximum dynamically activated pressure distribution due to earthquake loading

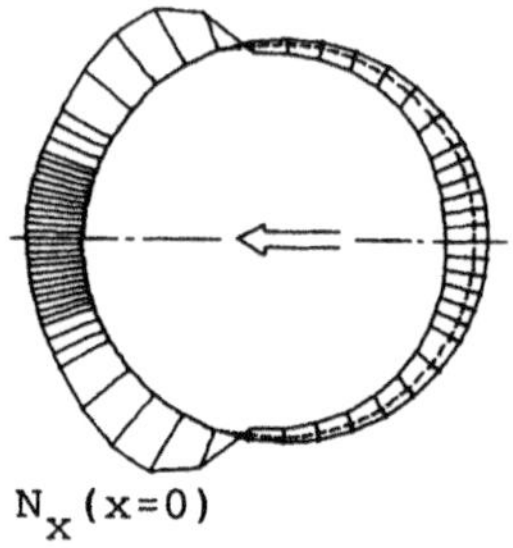

Fig 13: Axial membrane force distribution at the shell base

deformation regime. The nonlinear prebuckling analysis results in a considerable base uplift even if the maximum ground acceleration of the earthquake has an order of magnitude of observed earthquakes. The corresponding axial membrane forces in the shell at the base, $N_x(x = 0, \phi)$ are shown in Fig. 13. This distribution differs to a high degree from the cosine-distribution predicted for anchored tanks /16/. It shows two maxima, both lying excentrically with respect to the earthquake direction. This is the reason why the rather unexpected but in reality already observed /18/ excentrical local buckling pattern appears, as shown in Fig. 14.

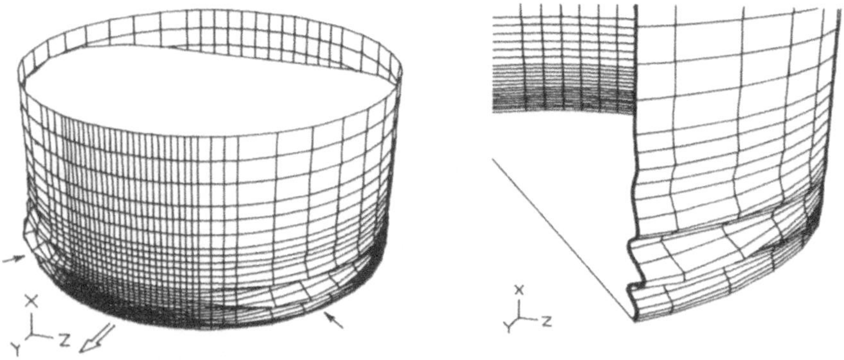

Fig.14: Fundamental buckling mode of the earthquake loaded liquid filled tank

The critical load multiplier obtained from the solution of the eigenvalue problem (23) solved for the load situation discussed above leads to a maximum value of the critical axial membrane stress distribution which is very close to the classical buckling stress of axially loaded perfect cylinders.

Despite the fact that axially compressed cylinders are very sensitive to imperfection, experimental investigations on earthquake loaded liquid filled shells /17/ have shown that these axial membrane forces can really appear during the earthquake without buckling of the shell. The explanation of this surprising result is partially given in /16/ and lies additionally in the transient character of the earthquake loading which - taking the results of the periodically loaded column (Chapter 3.1) into account - may be the reason of an underestimation of the buckling load in a quasi-static analysis.

6. CONCLUSIONS

It has been shown that different mechanisms of stability loss such as static or dynamic buckling as well as flutter instability, can be treated by the finite element method. In cases where flutter instability cannot be excluded a non-symmetric dynamic eigenvalue problem has to be solved in order to detect critical loads even if the load is applied statically.

However, in many cases a quasi-static conservative, i.e. symmetrized, approach leads to reasonably good results regarding the estimation of the stability limit, even if dynamic follower loads are acting. As examples for such structures wind and earthquake loaded unanchored liquid storage tanks have been investigated.

7. ACKNOWLEDGEMENTS

Parts of the work presented in this paper are sponsored by the FWF (Fonds zur Förderung der wissenschaftlichen Forschung Vienna, Austria) under the Project No. S30-04.

REFERENCES

1 Schweizerhof, K.: Nichlineare Berechnung von Tragwerken unter verformungsabhaengiger Belastung mit finiten Elementen, Rep. Inst. f. Baustatik, Univ. Stuttgart: 1982

2 Rammerstorfer, F.,G.,Jump Phenomena Associated With the Stability of Geometrical Nonlinear Structures. Recent Advances in Non-linear Computational Mechanics (Eds. E.Hinton, D.R. Owen, C. Taylor). Pineridge Press, Swansea : 1982.

3 Ramm, E.: Geometrisch nichtlineare Elastostatik und Finite Elemente, Habilitationsschrift, Univ. Stuttgart: 1975.

4 Becker, J.; Dreyer, H.J.; Haake, W.; Nabert, R. : Numerische Mathematik fuer Ingenieure. B. G. Teubner. Stuttgart : 1977.

5 Bolotin, W.W. : Kinetische Stabilitaet elastischer Systeme. VEB Deutscher Verlag der Wissenschaften, Berlin : 1961.

6 Auli, W.: Statische und dynamische Stabilitaet elastischer Leichtbaukonstruktionen unter konservativer und nichtkonservativer Belastung. Dissertation. ILFB, Technical Univ. Vienna: 1985 (to appear).

7 Ganapathi, M.; Varadan, T.,K.: Dynamic Buckling of orthotropic Shallow Spherical Shells, Computers and Structures 15 (1982), 517-520

8 Ramm, E.; Stegmueller, H.: Program NISA80 - Input Description, Rep. Inst. f. Baustatik, Univ. Stuttgart: 1984.

9 Dirschmid, H.J.: Contributi al problema parziale della ricerca degli autovalori, Rep. Univ. Degli Studi Di Firenze: 1979.

10 Beck , M.: Die Knicklast des einseitig eingespannten tangential gedrueckten Stabes. ZAMP 3 (1952) , 225-228.

11 Kundurpi, P.S.; Gopalacharyulu, S.; Johns, D.J.: Stability of Cantilever Shells Under Wind Loads, J.Eng. Mech.Div., Am.Soc.Civ.Eng. 101 (1975), 517-530.

12 Brendel, B.; Ramm, E.; Fischer ,F.D.; Rammerstorfer, F.G.: Linear and Nonlinear Stability Analyses of Thin Cylindrical Shells Under Wind Loads. J. Struct.Mech. 9 (1981), 91-113.

13 Sobel, L.H.: Effects of Boundary Conditions on the Stability of Cylinders Subject to Lateral and Axial Pressures. AIAA Journal 2 (1964), 1437-1440.

14 Rammerstorfer, F.G.; Auli, W.; Fischer, F.D.: Uplifting and Stability of Wind Loaded Vertical Cylindrical Shells. Engineering Computations (to appear).

15 Fischer, F.D.; Rammerstorfer, F.G.; Auli W.: Strength and Stability of Earthquake Loaded Liquid Filled Tanks, Transactions 8th SMiRT, Vol. B,(1985), BK2/1*, 475-486

16 Fischer D.F.; Rammerstorfer, F.G.: The Stability of Liquid-Filled Cylindrical Shells Under Dynamic Loading. Buckling of Shells (Ed. E. Ramm). Springer, Berlin, Heidelberg,New York: 1982.

17 Shih, C.F.; Babcock, C.D.: Buckling of Oil Storage Tanks in SPPL Tank Farm During the 1979 Imperial Valley Earthquake. ASME-84-PVP-74: 1984.

18 Shibata,H.; Akyama,H.: Seismic Capacity Testing of a Thin Wall 500 Ton Cylindrical Tank, MITI-NRC SIE Meeting, Palo Alto, Calif.: 1984.

Nonlinear Analysis of Shells of Revolution Including Contact Conditions

W. WUNDERLICH, H. CRAMER, W. REDANZ
Ruhr-University Bochum, West Germany

Summary

The paper deals with a special method for shells of revolution with unilateral contact constraints. It is based on a problem-oriented approach with a Fourier expansion of loads and variables in circumferential direction. The properties of the resulting ring elements are obtained by numerical integration to any desired degree of accuracy. To retain the advantages of the Fourier expansion also in the nonlinear range all nonlinear terms are handled as pseudo-loads. For convergence improvement of the initial load method on the system level an orthogonalisation of the solution base is employed. Its capabilities with respect to its stability and accuracy are demonstrated by inclusion of the highly nonlinear unilateral contact constraints without friction. On the boundaries of the shell the constraints are also evaluated by Fourier expansion at discrete points of contact or release. As a numerical example a cylinder under horizontal loading and release conditions on the foundation is treated.

1. Introduction

The numerical evaluation of the load carrying capacity of shells of revolution leads in general to a considerable effort, especially if in addition to nonaxisymmetric loads the behaviour is nonlinear, or in cases where contact conditions on the boundaries have to be taken into account. Specific formulations and computer programs which take advantage of the special geometry should be employed to decrease time consuming calculations and high cost. This is of special importance for buckling and ultimate load investigations.

For linear problems the Fourier decomposition of loads and variables along the circumferential direction is a well known procedure to split the full-size problems into a number of smaller ones. Due to the orthogonality of the trigonometric functions

Finite Element Methods for Nonlinear Problems
Europe-US Symposium, Trondheim, Norway 1985
Editors: Bergan, Bathe, Wunderlich

for this case a set of equations for the different harmonics can be obtained [1]. This semi-analytical approach is very efficient and, for regular axisymmetric geometry, it has been found to be more advantageous than standard two-dimensional discretisation procedures. Unfortunately, the formulation for shells of revolution in the nonlinear range leads to a coupling of all harmonics and to fully populated system equations which would also be obtained by an usual two-dimensional discretisation. To reduce the high effort connected with this approximation, a special approach has been proposed [2-4] which tries to retain the advantages of the Fourier expansion also in the nonlinear range by handling the nonlinear terms of the coefficient matrix as pseudo-loads on the right hand side iteratively. The equations then remain uncoupled with small bandwith and, in addition, during the iteration procedure the system matrix has to be formed only once.

On the other hand, this iteration procedure shows convergence only in certain cases. Numerical difficulties are encountered e.g. for regions with nonlinear hardening or with highly interdependent harmonics. For this reason, a modified iteration has been applied in the geometrically and material nonlinear regime [4]. It may be classified as a special type of conjugate gradient method with a problemoriented preconditioning to perform the solution of the algebraic equations. This approach is based on the successive approximation of the vector of the incremental nodal unknowns by a linear combination of orthogonal base vectors. These are taken to be orthogonal with respect to the current tangential stiffness matrix. The explicit knowledge of the latter is, however, not required in the actual solution procedure because iterations are performed exclusively on vectors. In this way the efficiency of the semi-analytical approach can be utilized in the nonlinear range as well and the above-mentioned difficulties no longer occur.

In this paper, this type of iteration procedure using orthogonal base vectors is applied to shells of revolution with contact constraints. The iteration with the orthogonalisation procedure is investigated for boundaries in contact or in release, which lead to a highly nonlinear behaviour in addition to the nonlinear characteristics of the thin shell itself. It is intended to show that with this orthogonalisation of the base vectors the approach using Fourier-decomposition can also be transferred to shells of revolution with contact constraints.

2. Method of solutions via ring elements

The basic equations and way of solution to analyse shells of revolution in the linear and nonlinear range have been described in detail in [1-4], so only a summary of the essential features is outlined here. The nonlinear differential equations are formulated such that the forces and displacements appearing in the natural boundary conditions are chosen as unknowns. This leads to first-order derivatives with respect to the meridional coordinate and to eight first order differential equations with four displacement quantitities and four corresponding stress resultants as unknowns. The linearized incremental equations may then be written in the abbreviated form:

$$\frac{\partial}{\partial s}\underline{z}(\Theta,s) = \underline{A}(s,\frac{\partial}{\partial\Theta},\overset{o}{\underline{z}},\underline{E}_T)\,\underline{z}(\Theta,s) + \bar{\underline{z}}(\Theta,s) \tag{1}$$

In equ. (1) s, Θ, $\underline{A}$, $\underline{z}$, $\overset{o}{\underline{z}}$, $\underline{E}_T$ denote the meridional and circumferential coordinate, the matrix containing the geometric shell parameters, the incremental state vector, the vector of the present configuration and the current tangent moduli. The variables in $\underline{z}(\Theta,s)$ and the load vector $\bar{\underline{z}}$ are decomposed by the Fourier expansion:

$$\underline{z}(\Theta,s) = \overset{o}{\underline{z}}(s) + \sum_{n=1}^{N} \underline{\Psi}_n(\Theta)\,\overset{(n)}{\underline{z}}(s) \tag{2}$$

where $\underline{\Psi}_n$ is a diagonal matrix containing trigonometric functions. Separating equ. (1) into linear and nonlinear parts leads to

$$\frac{d}{ds}\overset{(n)}{\underline{z}}(s) = (\overset{(n)}{\underline{A}}_L + \overset{(n)}{\underline{A}}_{GP}(\overset{o}{\underline{z}},\underline{E}_T))\,\overset{(n)}{\underline{z}}(s) + \overset{(\underline{n})}{\bar{\underline{z}}}(s) \tag{3}$$

in which $\underline{A}_L$ is a block-diagonal matrix of uncoupled blocks with respect to the Fourier harmonics, while $\underline{A}_{GP}$ is fully populated, implying relations between all Fourier terms. Application of a discretisation procedure at this stage would lead to a fully populated tangential stiffness matrix $\underline{k}_T(\overset{\circ}{\underline{v}})$ and thus to a full coupling of all Fourier terms of the nodal degrees of freedom $\overset{(n)}{\underline{v}}$ and the associated load vector $\overset{(n)}{\underline{p}}$. For the global structure this would lead to the system:

$$\underline{K}_T(\overset{\circ}{\underline{V}})\ \underline{V} = \bar{\underline{P}}\ . \tag{4}$$

To avoid this coupling the "inital-load" method is used and the nonlinear influences are treated as pseudo-loads by

$$\underline{z}_{GP} = \underline{A}_{GP}(\overset{\circ}{\underline{z}},\underline{E}_T)\underline{z}\ . \tag{5}$$

Equ. (3) can then be written in the form

$$\frac{d}{ds}\overset{(n)}{\underline{z}}(s) = \overset{(n)}{\underline{A}_L}\ \overset{(n)}{\underline{z}}(s) + \overset{(n)}{\bar{\underline{z}}}(s) + \overset{(n)}{\underline{z}_{GP}}(s,\overset{\circ}{\underline{z}},\underline{E}_T,\underline{z})\ . \tag{6}$$

This equation is now again uncoupled with respect to the various Fourier harmonics and has to be solved iteratively. This property is of great advantage compared with a full twodimensional discretisation. Here, ring-elements may be used for each harmonic. Through the numerical integration of equ. (6) in the initial state the corresponding transfer matrix can be evaluated to a given relative accuracy limit. Its transformation to the stiffness matrix of a ring-element eventually leads to a system of equations for each increment and harmonic for the whole structure :

$$\underline{K}_L\ \underline{V} = \bar{\underline{P}} + \underline{P}_{GP}(\overset{\circ}{\underline{V}},\underline{V})\ . \tag{7}$$

3. Nonlinear boundary conditions

In addition to geometrical and material nonlinearities the load carrying capacity of shells of revolution under nonaxisymmetric loading may be strongly affected by nonlinear support conditions. Thus, in realistic calculations nonlinear boundary conditions have to be taken into account. Two types of unilateral conditions are treated in this paper. They describe the boundary conditions for tensionless rigid or elastic support.

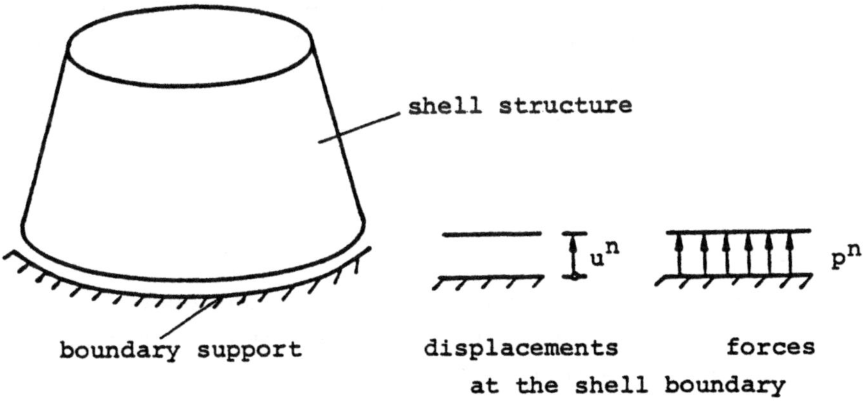

Fig. 1: Boundary displacements and tractions

a) Shell on rigid foundation

For the shell boundary on a rigid foundation the boundary condition takes the form (see Fig. 1):

$$\text{if } u^n > 0, \quad \text{then} \quad p^n = 0 \quad \text{on} \quad S^r \tag{8a}$$

$$\text{if } u^n = 0, \quad \text{then} \quad p^n \leq 0 \quad \text{on} \quad S^c. \tag{8b}$$

In (8a), (8b) S^r and S^c denote those parts of the boundary surface, where release or contact occurs, respectively.

b) Shell on elastic support

Another type of unilateral bounday condition is described by the constraints:

$$\text{if } u^n > 0, \quad \text{then} \quad p^n = 0 \quad \text{on} \quad S^r \tag{9a}$$

$$\text{if } u^n \leq 0, \quad \text{then} \quad p^n = \frac{u^n}{f} \quad \text{on} \quad S^c \tag{9b}$$

where f denotes the flexibility of the support.

4. Spatial discretisation of the contact surfaces

The method used to discretise the problem in the context of the finite element method employs ring elements using the stiffness method described above, except in the simulation of the contact surface. To obtain a discrete representation of the latter let us first consider that part of the boundary at which the shell is in contact with the surface S^c. The term

$$\delta \int_{S^c} (\underline{p}^T \underline{u} - \frac{1}{2} \underline{p}^T \underline{f} \, \underline{p}) \, ds^c \qquad (10)$$

represents the incremental virtual work of the boundary forces and displacements and is added to the virtual work formulation which forms the basis of the standard displacement method. To include the contact conditions displacements and contact forces are varied independently as the basic unknowns of the problem [5,6]. For rigid support this approach is equivalent to the common Lagrange multiplier method [7].

Equ. (10) can be considered as the combination of the kinematic and equilibrium equations

$$\int_{S^c} \delta \underline{p}^T (\underline{u} - \underline{f} \, \underline{p}) \, ds^c = 0 \qquad (10a)$$

$$\int_{S^c} \delta \underline{u}^T (\underline{p} - \underline{p}_{shell}) \, ds^c = 0 \qquad (10b)$$

in the contact surface. It incorporates condition (9a) and with f = 0 also equation (8a).

For the finite element discretisation the circumferential contact line is divided into equidistant parts, with the displacements $\underline{u}$ and the contact forces $\underline{p}$ being assumed to be a step function. Then the integration can be expressed by a summation.

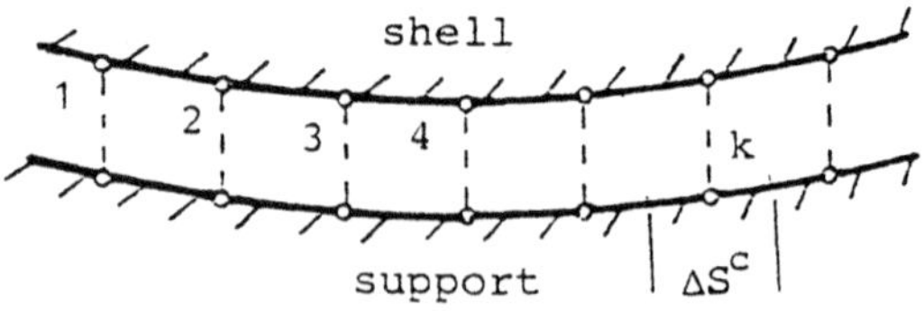

Fig. 2: Schematic representation of the discretisation and integration points along the boundary surface

Thus, expression (10) becomes

$$\sum_{i=1}^{3} \sum_{k=1}^{N^c} \delta(p_k^i \; u_k^i - \frac{1}{2} \, p_k^i \; f^{(i)} \; p_k^i) \Delta s^c \tag{11}$$

where i = 1,2,3 denote the three spatial directions and N^c is the total number of discrete contact-elements or integration points.

By assembling the global system matrix, the boundary forces p^i are included in the vector of nodal unknowns, Thus, for one integration point equ. (11) leads to the contact element matrix

$$\begin{aligned} \underline{c}_k \; \underline{z}_k &= \Delta s^c \begin{bmatrix} 0 & 1 \\ 1 & -f^{(i)} \end{bmatrix} \begin{bmatrix} u^i \\ p^i \end{bmatrix}_k \\ &= \begin{bmatrix} \underline{0} & \underline{g}^T \\ \underline{g} & -\underline{f} \end{bmatrix} \begin{bmatrix} \underline{u} \\ \underline{p} \end{bmatrix}_k \end{aligned} \tag{12}$$

It represents the mixed finite element matrix of an elastic spring. The structure of the resulting global incremental equations can be arranged in the matrix form typical for mixed finite element methods:

$$\underbrace{\begin{bmatrix} \underline{K}_T & \underline{G}^T \\ \underline{G} & -\underline{F} \end{bmatrix}}_{\underline{C}} \underbrace{\begin{bmatrix} \underline{V} \\ \underline{P} \end{bmatrix}}_{\underline{Z}} = \underbrace{\begin{bmatrix} \bar{\underline{P}} \\ \underline{0} \end{bmatrix}}_{\bar{\underline{Z}}} \tag{13}$$

where $\underline{G}$, $\underline{G}^T$ and $\underline{F}$ represent the assemblage of the element matrices $\underline{g}_k$, $\underline{g}_k^T$ and $\underline{f}_k$. The vector $\underline{P}$ includes the unknown contact pressures $\underline{p}_k$.

In the incremental loading process the regions with contact or release and the number of active contact conditions, or elements, as well as the size and structure of the coefficient matrix will normally change. The resulting difficulties can be avoided, if the contact pressures $\underline{p}_k$ for all possible contact nodes are included in the global system of equations, so that the total number of discrete degrees fo freedom is fixed. This is particularly important in this approach because the discrete degrees of freedom $\underline{u}_k$, $\underline{p}_k$ appearing in (13) have to be transformed into their discrete harmonic counterparts. This requires a special treatment of the nodal pressures associated with released boundary points.

The constraint conditions (8b) and (9b) have to be included in the global system of equations for each released discrete boundary point. One common way to do this is to replace all elements of the corresponding row except the diagonal element by zero. This uncouples the unknown nodal pressures from the other degrees of freedom and results in

$$b_k^{(i)} \; p_k^i = 0 \;, \qquad \text{or} \tag{14}$$

$$p_k^i = 0 \quad \text{for} \quad b_k^{(i)} \neq 0 \;, \tag{15}$$

where $b_k^{(i)}$ represents the diagonal element.

The elements of the corresponding columns may be arbitrarily chosen because they are multiplied by $p_k^i = 0$.

An equivalent procedure may be performed on the element level by adding a special element for each released boundary point of the type

$$\underline{c}_k^r \; \underline{z}_k = \Delta s^r \begin{bmatrix} 0 & a^{(i)} \\ 0 & b^{(i)} \end{bmatrix}_k \begin{bmatrix} u^i \\ p^i \end{bmatrix}_k \tag{16}$$

where $a^{(i)}$ and $b^{(i)}$ may take arbitrary values except $b^{(i)} = 0$. Hence, it is seen that with $a^{(i)} \neq 0$ a nonsymmetric element matrix arises from (16). $a^{(i)}$ and $b^{(i)}$ represent no mechanical properties. Their choice is important only from the numerical point of view, particularly in connection with interative solution procedures. This is more detailed using a simplified one-dimensional model in chapter 5.

To retain the approach with ring elements using Fourier decomposition in the circumferential direction the discrete values $\underline{u}_k$, $\underline{p}_k$ have to be developed into the associated degrees of freedom equivalent to a Fourier analysis:

$$\underline{z}_k = \begin{bmatrix} \underline{u} \\ \underline{p} \end{bmatrix}_k = \underline{T}_k \; \underline{z} = \underline{T}_k \left[\overset{(0)}{\underline{z}}, \overset{(1)}{\underline{z}}, \; . \; . \; . \; . \; , \overset{(N)}{\underline{z}} \right], \tag{17a}$$

$$\sum_{k=1}^{N^c} \delta \underline{z}_k^T \underline{c}_k^c \underline{z}_k + \sum_{k=1}^{N^r} \delta \underline{z}_k^T \underline{c}_k^r \underline{z}_k = \delta \underline{z}^T \{ \sum_{k=1}^{N^c} \underline{T}_k^T \underline{c}_k^c \underline{T}_k + \sum_{k=1}^{N^r} \underline{T}_k^T \underline{c}_k^r \underline{T}_k \} \underline{z}^T, \tag{17b}$$

where $\underline{z}^T = \left[\overset{(0)}{\underline{z}}{}^T, \overset{(1)}{\underline{z}}{}^T, \ldots, \overset{(N)}{\underline{z}}{}^T \right]$ is a vector containing the N+1 Fourier terms of the nodal degrees of freedom. One obtains a full coupling of all Fourier harmonics even if the behaviour of the shell is linear. To maintain the uncoupling of the system matrix on the left hand side the element matrices for contact and release are separated into two parts. The first is added to the left hand side whereas the second is added to the right hand side of the system of equations. On the left hand side a matrix representing an elastic spring with a constant flexibility f^* in the circumferential direction is used. To approximate the real behaviour at the boundary surface in an average manner a mean flexibility f^* with $f < f^* < \infty$ has to be chosen:

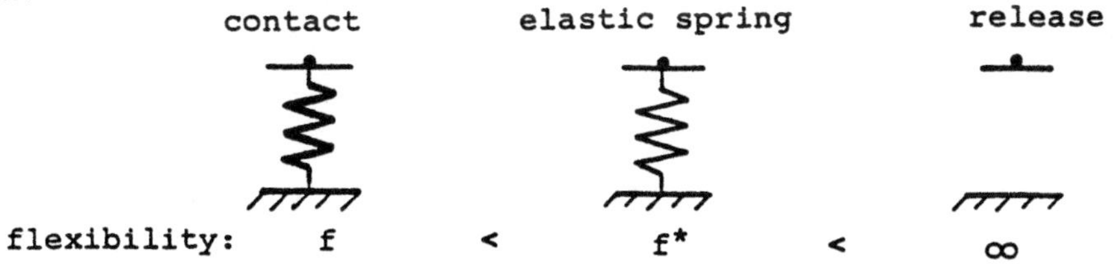

Fig. 3: Model of support conditions at discrete points

The element description is separated as follows:

For contact:

$$\begin{bmatrix} \underline{0} & \underline{I} \\ \underline{I} & -\underline{f} \end{bmatrix} \begin{bmatrix} \underline{u} \\ \underline{p} \end{bmatrix} = \underbrace{\begin{bmatrix} \underline{0} & \underline{I} \\ \underline{I} & -\underline{f}^* \end{bmatrix} \begin{bmatrix} \underline{u} \\ \underline{p} \end{bmatrix}}_{\text{left hand side}} + \underbrace{\begin{bmatrix} \underline{0} & \underline{0} \\ \underline{0} & \underline{f}^* - \underline{f} \end{bmatrix} \begin{bmatrix} \underline{u} \\ \underline{p} \end{bmatrix}}_{\text{right hand side}} \tag{18}$$

For release:

$$\begin{bmatrix} \underline{O} & \underline{a} \\ \underline{O} & \underline{b} \end{bmatrix} \begin{bmatrix} \underline{u} \\ \underline{p} \end{bmatrix}_k = \underbrace{\begin{bmatrix} \underline{O} & \underline{I} \\ \underline{I} & -\underline{f}^* \end{bmatrix} \begin{bmatrix} \underline{u} \\ \underline{p} \end{bmatrix}_k}_{\text{left hand side}} + \underbrace{\begin{bmatrix} \underline{O} & \underline{a}-\underline{I} \\ -\underline{I} & \underline{b}+\underline{f}^* \end{bmatrix} \begin{bmatrix} \underline{u} \\ \underline{p} \end{bmatrix}_k}_{\text{rightbhand side}} . \tag{19}$$

This results in the mixed finite element matrix equation system with symmetric left hand side

$$\begin{bmatrix} \underline{K}_L & \underline{L}^T \\ \underline{L} & -\underline{F} \end{bmatrix} \begin{bmatrix} \underline{V} \\ \underline{P} \end{bmatrix} = \begin{bmatrix} \underline{\bar{P}} \\ \underline{O} \end{bmatrix} + \begin{bmatrix} \underline{O} & -\Delta\underline{R} \\ -\Delta\underline{L} & \Delta\underline{F} \end{bmatrix} \begin{bmatrix} \underline{V} \\ \underline{P} \end{bmatrix} \tag{20}$$

$$\text{or with} \quad \underline{K}_T\ \underline{V} = \underline{K}_L\ \underline{V} - \underline{P}_{GP}\ (\overset{\circ}{\underline{V}},\underline{V}) \tag{21}$$

into

$$\begin{bmatrix} \underline{K}_L & \underline{L}^T \\ \underline{L} & -\underline{F}^* \end{bmatrix} \begin{bmatrix} \underline{V} \\ \underline{P} \end{bmatrix} = \begin{bmatrix} \underline{\bar{P}} \\ \underline{O} \end{bmatrix} + \begin{bmatrix} \underline{P}_{GP} - \Delta\underline{R}\ \underline{V} \\ -\Delta\underline{L}\ \underline{V} + \Delta\underline{F}\ \underline{P} \end{bmatrix} \tag{22}$$

$$\underline{C}_L \qquad \underline{Z} \quad = \quad \underline{\bar{Z}} \quad + \quad \underline{Z}_N(\overset{\circ}{\underline{Z}},\underline{Z})$$

in which the Fourier harmonics on the left hand side are still uncoupled.

The system of equations (20) has to be solved iteratively as the pseudo-load and initial displacement vector $\underline{Z}_N(\overset{\circ}{\underline{Z}},\underline{Z})$ also depends on the unknowns. This may be done straightforwardly by successive iterations in the manner of the initial load method:

$$\underline{Z}_j = \underline{C}_L^{-1} \left[\underline{\bar{Z}} + \underline{Z}_N(\overset{\circ}{\underline{Z}},\ \underline{Z}_{j-1}) \right] . \tag{23}$$

The initial load and displacement vector $\underline{Z}_N$ depends on the solution vector $\underline{Z}_{j-1}$ of the (j-1)-th iteration step. The coefficient matrix of $\underline{C}_L$ has to be triangularized only once. All other operations can be performed on the element level. The iteration procedure itself will be described in chapter 6.

5. Considerations concerning convergence in contact problems

So far, the parameters f^*, a and b appearing in the formulation of the element matrices in (18) and (19) have not been specified in more detail. As mentioned above they do not represent real mechanical properties. However, they are important for the convergence of iterative solution procedures. To give an insight into their significance for the iteration process we consider the following simplified model.

Let us assume that the behaviour at one contact point can be represented by one discrete contact-element and a spring stiffness C_s that characterises the stiffness of the remaining structural system.

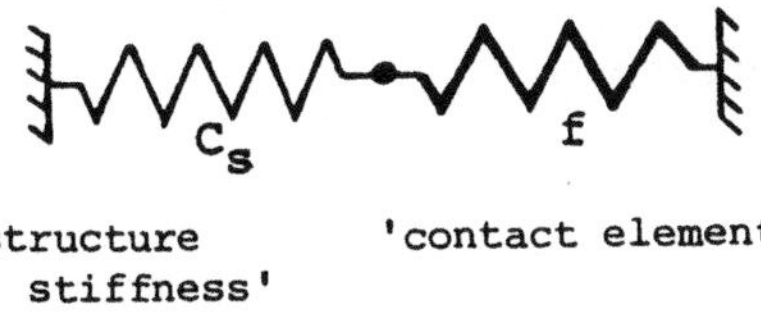

Fig. 4: Simplified structural model for a contact point

To get an estimate for the influence of f^* we consider the above successive iteration scheme for this simple model. It takes the form

$$\underbrace{\begin{bmatrix} C_s & 1 \\ 1 & -f^* \end{bmatrix}}_{\underline{A}} \underbrace{\begin{bmatrix} u \\ p \end{bmatrix}_j}_{\underline{x}_j} = \underbrace{\begin{bmatrix} \bar{P} \\ 0 \end{bmatrix}}_{\underline{r}} + \underbrace{\begin{bmatrix} 0 & 0 \\ 0 & -(f^* - f) \end{bmatrix}}_{\underline{B}} \underbrace{\begin{bmatrix} u \\ p \end{bmatrix}_{j-1}}_{\underline{x}_{j-1}} . \tag{24}$$

In this case the convergence may be assessed by considering the corresponding eigenvalue problem

$$\lambda \, \underline{A} \, \underline{x} = \underline{B} \, \underline{x} . \tag{25}$$

Its eigenvalue given by

$$\lambda = \frac{f^* - f}{\frac{1}{C_s} + f^*} \tag{26}$$

is always less than one for $f^* - f \geq 0$. This implies that the iteration procedure is stable. Small values of $f^* - f$ thus have a beneficial influence on the convergence behaviour.

Using the same consideration for a released boundary point one obtains the scheme

$$\begin{bmatrix} C_s & 1 \\ 1 & -f^* \end{bmatrix} \begin{bmatrix} u \\ p \end{bmatrix}_j = \begin{bmatrix} \bar{P} \\ 0 \end{bmatrix} + \begin{bmatrix} 0 & 1-a \\ 1 & -(b+f^*) \end{bmatrix} \begin{bmatrix} u \\ p \end{bmatrix}_{j-1} \tag{27}$$

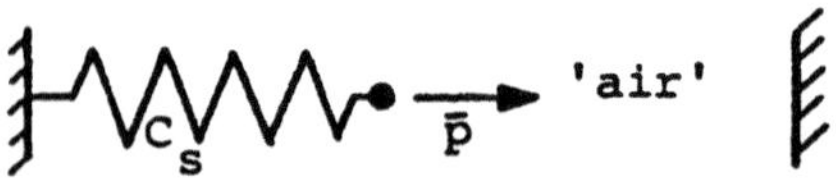

Fig. 5 : Simplified model for a released boundary point

An eigenvalue analysis of this problem generally shows that the eigenvalues λ are real and the $|\lambda|$ is less than one only for $a = 1$. In this case the eigenvalue is given by

$$\lambda = \frac{\frac{1}{C_s} + (b + f^*)}{\frac{1}{C_s} + f^*} \; . \tag{28}$$

The scheme (27) is stable for positive values of C_s in the range of $-2f^* < b < 0$.

Cases of particular interest are

$$\text{I)} \quad b = -f^* \;, \quad \lambda = \frac{\frac{1}{C_s}}{\frac{1}{C_s} + f^*} \tag{29}$$

and

$$\text{II)} \quad b = -2f^* \;, \quad \lambda = \frac{\frac{1}{C_s} - f^*}{\frac{1}{C_s} + f^*} \tag{30}$$

For $b = -f^*$ the best convergence characteristics are obtained for small values of f^*. This is contrary to the condition for best convergence at a contact point. Thus, to obtain equally good convergence behaviour in all regions a suitable choice of f^* related to $\frac{1}{C_s}$ is essential. A choice of $f^* \approx \frac{1}{C_s}$ is advantageous. This is particularly true in the case (30) where $b=-2f^*$. In summary, the following choice of a, b and f^* seems to be beneficial for the convergence behaviour of the iterative solution procedure:

$$a = 1 \; , \quad f^* \approx \frac{1}{C_s} \; , \qquad (31)$$
$$-2f^* < b < -f^* \; .$$

Therefore an appropriate a priori estimate of the average shell structure stiffness over the whole loading range and a corresponding choice of the flexibility parameter f^* may accelerate the convergence of the procedure. This is also the case if the modified iteration procedure described below is used.

6. Iterative solution procedure

An important drawback of the usual successive iteration procedure (23) is its slow convergence near limit points and its likely divergence when stiffening in some sections of the structure occurs.

To avoid these difficulties a modified iteration procedure similar to that described in [4] for the standard stiffness method was developed to obtain the solution of (13). It is based on the approximation of the generalized virtual work expression

$$\delta \underline{Z}^T (\underline{C}_T \, \underline{Z} - \bar{\underline{Z}}) = 0, \qquad (32)$$

which follows from (13), and uses a linear combination of trial vectors $\underline{y}_j$ to approximate $\underline{Z}$ in the sense of weighted residuals. This leads to

$$\underline{Z} = \sum_{j=1}^{M} \alpha_j \, \underline{y}_j \; , \qquad (33)$$

where α_j are appropriate weighting factors. Written more compactly as

$$\underline{Z} = \underline{Y} \, \underline{\alpha} \; , \qquad (34)$$

$\underline{Y} = [\, \underline{y}_1, \underline{y}_2, \ldots, \underline{y}_M \,]$ is the matrix containing the global trial vectors which are used as a reduced basis for the approximation of $\underline{Z}$. The matrix $\underline{\alpha}^T = [\, \alpha_1, \alpha_2, \ldots, \alpha_M \,]$ contains the associated weighting factors. The accuracy of the solution depends on the number and the choice of the base vectors $\underline{y}_j$. The number of these base vectors is increased by one in each iteration step by using a modification of expression (23):

$$\underline{y}_j = \underline{C}_L^{-1} \left[\bar{\underline{Z}} + \underline{Z}_N (\overset{\circ}{\underline{Z}}, \underline{z}_{j-1}) \right] \tag{35}$$

where $\underline{z}_{j-1}$ now is the solution of equation (32) for j-1 base vectors. As already mentioned, this implies an expansion of the basis in each iteration step. At each step the vector $\underline{z}_{j-1}$ is found by solving

$$\underline{Y}^T \underline{C}_T \underline{Y} \underline{\alpha} = \underline{Y}^T \bar{\underline{Z}} \tag{36}$$

for α. Note that $\underline{Y}^T$ contains all j-1 base vectors. The products $\underline{C}_T \underline{y}_i$ may be calculated on the element level, and the tangential system matrix does not appear explicitly. This is seen from

$$\underline{C}_T \underline{y}_i = \underline{C}_L \underline{y}_i - \underline{Z}_N (\overset{\circ}{\underline{Z}}, \underline{y}_i) . \tag{37}$$

The resulting system of equations

$$\underline{W} \underline{\alpha} = \underline{r}$$

$$\text{with} \quad \underline{W} = \underline{Y}^T \underline{C}_T \underline{Y} , \quad \text{and} \quad \underline{r} = \underline{Y}^T \bar{\underline{Z}} \tag{38}$$

may be solved efficiently starting from the triangularized matrix of the previous iteration step [8]. The so calculated lower and upper triangular matrices $\underline{L}$ and $\underline{R}$, $\underline{W} = \underline{L}\underline{R}$, can be used to transform the base vectors into

$$\underline{X}^{*T} = \underline{L}^{-1} \underline{Y}^T , \tag{39a}$$

$$\underline{X} \underline{\delta} = \underline{Y} \underline{R}^{-1} \underline{\delta} , \tag{39b}$$

$$\underline{\delta} = \underline{R} \underline{\alpha} , \tag{39c}$$

where $\underline{W} = \underline{L} \underline{R}$, $\underline{X} = [\underline{x}_1, \underline{x}_2, \ldots, \underline{x}_j]$,

$\underline{X}^* = [\underline{x}_1^*, \underline{x}_2^*, \ldots, \underline{x}_j^*]$, and $\underline{\delta}^T = [\delta_1, \delta_2, \ldots, \delta_j]$.

Since $\underline{X}^*$ and $\underline{C}_T\underline{X}$ are orthogonal, the unknown weighting factors δ may now be determined by the expression

$$\underline{I}\ \underline{\delta} = \underline{X}^{*T}\ \bar{\underline{Z}}\ . \tag{40}$$

The approximate solution for the j-th iteration step is given by

$$\underline{Z}_j = \sum_{i=1}^{j} \underline{x}_i\ \delta_i$$

$$= \underline{Z}_{j-1} + \underline{x}_j\ \delta_j . \tag{41}$$

and $$\delta_j = \underline{x}_j^{*T}\ \bar{\underline{Z}}\ . \tag{42}$$

The iteration process may be terminated if the convergence criterion

$$\|\underline{Z}_j - Z_{j-1}\|_2 / \|\underline{Z}_j\|_2 < \varepsilon\ , \tag{43}$$

where $\|\ldots\|_2$ denotes the Euclidean norm, is satisfied.

The starting vector can be chosen to be $\underline{Z}_o = \underline{0}$ or better, the approximate solution of the previous load step. In general, global convergence is achieved after a maximum number of steps, which is equal to the total number of degrees of freedom. In the case of positive definite matrices $\underline{C}_L$ and $\underline{C}_T$, the method described can be associated with the conjugate gradient method [4,8,9], where $\underline{C}_L$ is used for preconditioning. Due to the orthogonality of the used trigonometric functions $\underline{C}_L$ is sparsely populated so that preconditioning is not time consuming. On the other hand it is essential for the convergence of the iteration procedure.

7. Numerical example

In this chapter numerical results of a quasistatic analysis of a liquid filled cylindrical shell under horizontal base exitation are presented. Similar problems were treated in [10,11] and the geometric and loading parameters chosen are the same as those in [11]. Also the pressure distribution along the meridian was held fixed (see Fig. 6) and only the amplitudes were increased incrementally in proportion to the factor λ. In this fashion an estimate of the maximum load-carrying capacity of the cylinder was obtained.

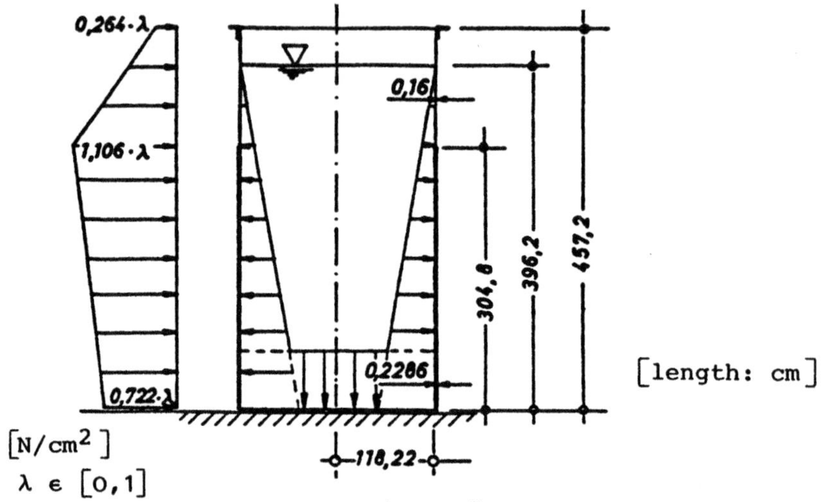

Material properties: $E = 6{,}86\text{x}10^6 \ \text{N/cm}^2$

$\nu = 0{,}3$, $\rho_L(\text{water}) = 1{,}0\text{x}10^{-5} \ \text{Ns}^2/\text{cm}^4$

Fig. 6: Model tank under consideration

It was assumed that the pseudo-static pressure has circumferential distribution in the first Fourier harmonic, and the hydrostatic load-component is axisymmetric. For simplicity, no bottom plate was included in the model. The corresponding loads were simulated by equivalent ring loads at the circumferential boundary line. In the analysis this line was modelled with 13 Fourier harmonics (n=0,1,2,...,12). It follows that the contact and release conditions may be enforced at 24 discrete points around the circumference. In the axial direction the shell was discretised with 19 ring elements of varying lengths (see Fig. 10). In all directions the lower end of the shell is free to uplift. As a result the structural stiffness of the shell will be somewhat underestimated. Fig. 7 shows the load-deflection behaviour of the shell for various points around the circumference at ground-level. It is seen that at a load level of approximately 50 percent of the maximum applied load the shell begins to uplift at Point A (see. Fig. 7). Upon further in-

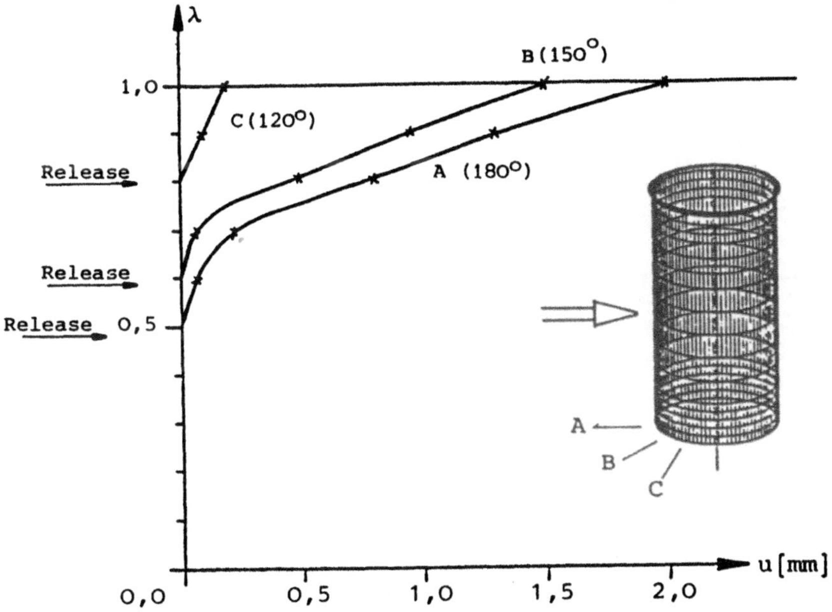

Fig. 7: Load-Deflections curves for various points at ground level

crease of λ a zone of release spreads around the circumference until at $\lambda = 1$ only half of the circumference is still in contact with the ground. Naturally this is accompanied by a fundamental change in the load-carrying behaviour of the shell, as can be seen in Figs. 8 and 9. Before the shell loses contact at point A it shows a considerable stiffening behaviour. This is evident from Fig. 8 in which the load parameter λ versus lateral displacements is plotted at various points around the circumference at a distance of about 2.5 R above the bottom.

Fig. 9 a/b shows the distribution of the axial and circumferential membrane forces at $\lambda = 0.4$. It can be observed that prior to uplift the circumferential shell response is essentially governed by the first Fourier harmonic, and thus directly corresponds to the distribution of the external pressure. Significant nonlinear influences are present only in the axial direction. The state of stresses and displacements changes, however, significantly once contact is lost over a certain region. Generally speaking it results in a gradual but quite noticeable decrease of the shell's overall stiffness. This is seen from the

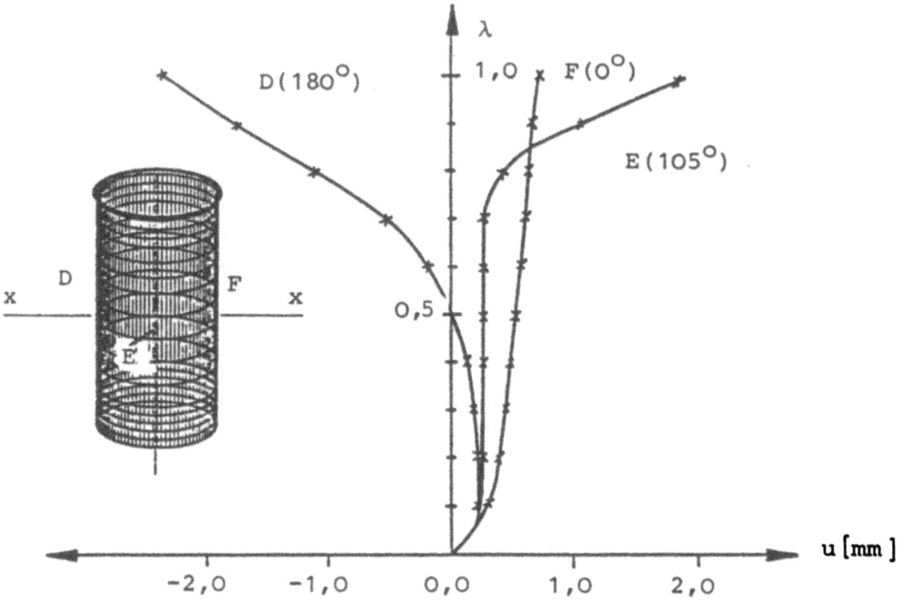

Fig. 8 : Load deflection relationship for lateral displacements in x - x

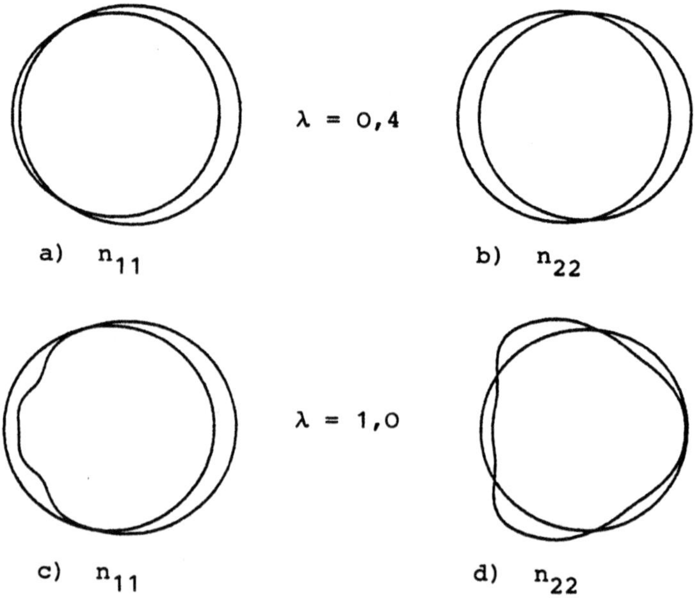

Fig. 9: Membrane Forces in x-x before and after release

decreasing slopes of the curves in Fig. 8 for values of λ larger than about 0.6. However, it also results in a marked redistribution of stresses in the circumferential direction as may be seen from the lower part of Fig. 9 where the same stress resultants have been plotted as in Figs. 9, a, b but at a higher value of λ. In particular, the circumferential distribution of the meridional membrane force n_{22} changes considerably. This is to be expected, because in the release area tensile stresses are found to occur which are counteracted by additional compressive stresses in the transition region of the remaining contact zone. In those regions of the contact zone which are at a greater distance from the release area, no significant stress-redistributions have been found.

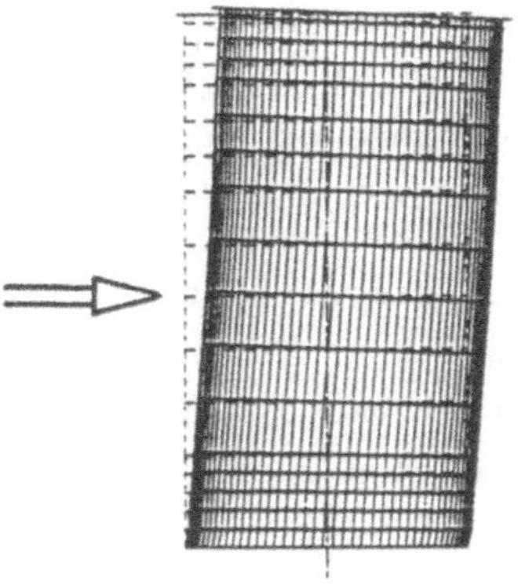

a) λ = 0,4

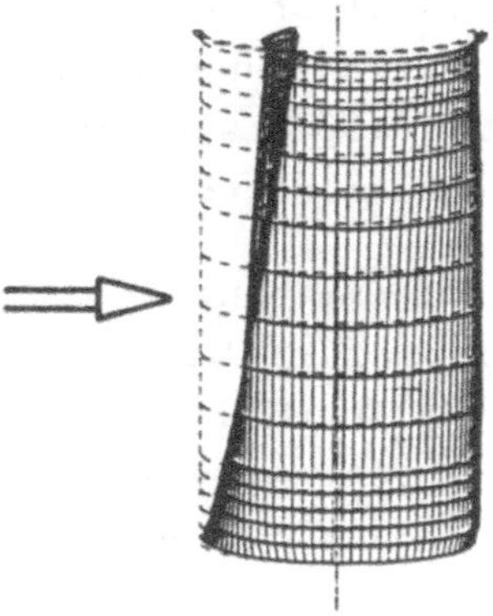

b) λ = 1,0

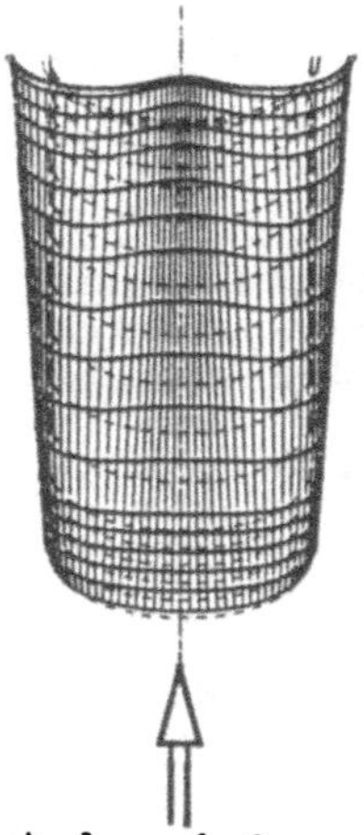

c) λ = 1,0

Fig. 10: Deformed shapes of the tank at various load levels.

Fig. 10 a shows a plot of the deformed shell viewed from the side at a value of $\lambda = 0.4$, that is, before release has occured at point A. Figs. 10b and c show lateral and front views of the deformation shell at $\lambda = 1$. Note that the uplift regions are visible at the lower left end of the shell. To improve the visibility of the vertical displacements somewhat different scaling factors have been chosen in preparing the plots of Figs. 10a and b/c so that both are not directly comparable.

The above analysis was based on a numerical model with approximately 980 degrees of freedom. The convergence of the iteration procedure was very satisfactory. During the first loading steps an average of 4 to 5 iterations were needed whereas at higher load levels, when significant stress redistributions occur, a total of 20 to 30 iteration steps were found to be sufficient. From a numerical standpoint, therefore, the method presented appears to be efficient and capable of handling complex problems of this type.

From a more mechanical point of view, the discretisation chosen would appear to be somewhat too coarse. This is the probable reason why in this analysis no local failure mode with a relatively short wavelenght was observed at the most highly compressed region of the shell bottom.

References

[1] Wunderlich, W.: "On the analysis of shells of revolution by transfer matrices". (in German), Ing.-Archiv, Vol. 36 (1967), 262-279.

[2] Wunderlich, W., Rensch, H.J., Obrecht, H.: "Analysis of elastic-plastic buckling and imperfection-sensitivity of shells of revolution". In: "Buckling of Shells", Ramm, E., (ed.). Springer Verlag, Berlin (1982), 137-174.

[3] Rensch, H.J.: "Elastic-plastic buckling and imperfection sensitivity of torispherical shells". Techn.-wiss. Mitteilungen Nr. 82-13, Institut für konstruktiven Ingenieurbau, Ruhr-Universität Bochum (1982).

[4] Wunderlich, W., Cramer, H., Obrecht, H.: Application of ring elements in the nonlinear analysis of shells of revolution under nonaxisymmetric loading". FENOMECH 1984, 3rd Int. Conf. on Finite Elements in Nonlinear Mechanics, Stuttgart 1984.

[5] Francavilla , A., Zienkiewicz, O.C.: "A note on numerical computation of elastic contact problems". Int. Num. Meth. Engng. Vol. 9, 913-924, 1975.

[6] Hughes, T.R.J., Taylor, R.L., Sackmann, J.L., Curnier, A., Kanoknukulchai, W.: "A finite element method for a class of contact-impact problems". Comp. Meth. Appl. Mech. Engng. Vol. 8, 249, 1986.

[7] Wriggers, P., Nour-Omid, B.: "Solution methods for contact problems", Report No. UCB/SESM-84/09, University of California, Berkeley, 1984.

[8] Faddejew, D.K., Faddejewa, W.N.: "Numerische Methoden der linearen Algebra". R. Oldenbourg Verlag, 1970.

[9] Ortega, J.M., Rheinboldt, W.C.: "Iterative solution of nonlinear equations in several variables". Academic Press, New York (1970).

[10] Niwa, A.: "Seismic behaviour of tall liquid storage tanks", Earthquake Engng. Research Center Report VCB/EERC-78/04, Univ. of Berkeley, 1978.

[11] Fischer, D.F., Rammerstorfer, G.F.: "The stability of liquid-filled cylindrical shells under dynamic loading". In: "Buckling of Shells", Ramm, E. (ed.), Springer Verlag, Berlin (1982), 569-597.

Finite Element Postbuckling Analyis of Shells with Nonlinear Contact Constraints

E. STEIN, W. WAGNER, P. WRIGGERS
University of Hannover, West Germany

Summary

In this paper a class of nonlinear problems is discussed where stability as well as postbuckling behaviour is coupled with contact constraints. The resulting finite element formulation is applied to structural problems of beams and shells undergoing finite elastic deflections and rotations. For the examination of the postbuckling behaviour the arc-length method is used. Furthermore, the contact conditions are introduced via a perturbed Lagrangian formulation. The performance of the element formulation and the algorithms is demonstrated by numerical examples.

1. Introduction

In engineering applications often stability or postbuckling behaviour of structural members like thin shells and beams have to be investigated. This can only be done by considering the geometrically nonlinear behaviour. Finite deflections and rotations have to be included for the postbuckling range. In some engineering applications the structures will undergo contact. This adds another type of nonlinearity to the formulation due to the inequality constraints. The coupling of these types of nonlinearities and the related algorithms are the main aspects of the paper. The contents of the paper may be outlined as follows.

In section 2, we develop an axisymmetric shell element as a special case of a shell theory including large deflections and moderate rotations. A Mindlin type of shell theory is employed which is derived using direct tensor notation, Wagner /1/. The resulting element formulation is based on linear shape functions for the three independent variables using only one point numerical integration for all terms. In the linear case it reduces to an element similar to that given by Zienkiewicz et. al. /2/. With this element formulation also geometrically nonlinear beams may be treated. Elements of this type have been used in a linearized version by e.g. Stein,Wriggers /3/, Wriggers,Klee /4/.

Finite Element Methods for Nonlinear Problems
Europe-US Symposium, Trondheim, Norway 1985
Editors: Bergan, Bathe, Wunderlich

To include finite rotations, an updated Lagrangian procedure is used which then restricts the application of the element only to small strains. This is valid for most of the engineering shell problems. Although there exist theories for elastic axisymmetric shells including finite rotations and strains, Reissner /5/, Noor /6/ the foregoing formulation has been used because of its simplicity. In addition, incremental algorithms have to be employed for tracing the postbuckling behaviour of structures. These algorithms allow in most applications only small load steps near limit points. Therefore, for this class of problems the theory presented by Reissner /5/ exhibits no advantage over a simple updating scheme.

In section 3 we give a summary of the contact formulation. The constraint conditions for the frictionless contact between deformable bodies are introduced via a perturbed Lagrangian procedure for the case of finite deformations, see Wriggers,Simo/7/. The formulation leads to different contact elements including the Lagrangian multiplier and the penalty approach, see Hallquist /8/. To overcome numerical problems associated with the Lagrangian multiplier approach a partitioning method will be used to solve the nonlinear set of equations.

Section 4 is concerned with the algorithmic part. Here, methods which are used for the uncoupled problems are combined. The arc-length method, see Riks /9/, Wempner /10/, Wessels /11/, for tracing limit points is employed together with active set strategies, see e.g. Luenberger /12/ for the determination of contact regions in the context of an overall Newton iteration. A main object is the developement of a consistent Hessian matrix for the global Newton iteration which up to now is only possible for a fixed set of active constraints.

In section 5, the performance of the algorithms and the element formulation is demonstrated for thin and moderate thick beam and shell applications by numerical examples.

2. Derivation of an axisymmetric nonlinear shell element

2.1 Geometrically nonlinear shell theory

In this chapter we describe the main ideas and equations of a geometrically nonlinear shell theory including transverse shear. For a detailed description we refer to e.g. Wagner /1/. In Figure 1 , the initial (i.c.) and the current configuration (c.c.) of the shell with the essential notations are shown. Quantities in the i.c. are marked by an upper circle o.

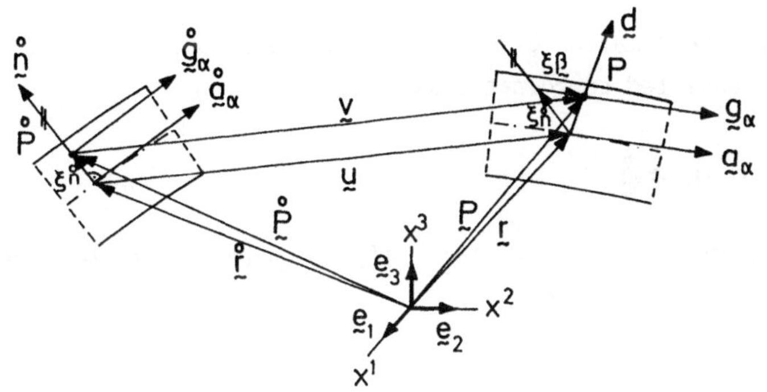

Fig. 1 Geometry of a thin shell in the initial and in the current configuration

A point $\mathring{P}$ is described in the shell space by

$$\mathring{\underset{\sim}{P}} = \mathring{\underset{\sim}{r}} + \xi \mathring{\underset{\sim}{n}} \quad . \tag{1}$$

Furthermore, we can introduce the base vectors of the middle surface

$$\mathring{\underset{\sim}{a}}_\alpha = \mathring{\underset{\sim}{r}}_{,\alpha} \qquad ; \quad n = \frac{\mathring{\underset{\sim}{a}}_1 \times \mathring{\underset{\sim}{a}}_2}{|\mathring{\underset{\sim}{a}}_1 \times \mathring{\underset{\sim}{a}}_2|} \tag{2}$$

and in the shell space

$$\mathring{\underset{\sim}{g}}_\alpha = \mathring{\underset{\sim}{P}}_{,\alpha} = \mathring{\underset{\sim}{a}}_\alpha + \xi \mathring{\underset{\sim}{n}}_{,\alpha} \quad ; \quad \mathring{\underset{\sim}{g}}_3 = \mathring{\underset{\sim}{n}} \quad . \tag{3}$$

With the curvature tensor

$$\mathring{\underset{\sim}{B}} = - \mathring{\underset{\sim}{n}}_{,\alpha} \otimes \mathring{\underset{\sim}{a}}^\alpha \tag{4}$$

we get the relation between the base vectors

$$\mathring{\underset{\sim}{g}}_i = (\mathring{\underset{\sim}{1}} - \xi \mathring{\underset{\sim}{B}}) \, \mathring{\underset{\sim}{a}}_i = \mathring{\underset{\sim}{Z}} \mathring{\underset{\sim}{a}}_i \, . \tag{5}$$

In equation (5) we introduce the shifter tensor $\underset{\sim}{Z}$ and the unit tensor

$$\overset{\circ}{\underset{\sim}{1}} = \overset{\circ}{\underset{\sim}{a}}_i \otimes \overset{\circ}{\underset{\sim}{a}}{}^i = \overset{\circ}{\underset{\sim}{a}}_\alpha \otimes \overset{\circ}{\underset{\sim}{a}}{}^\alpha + \overset{\circ}{\underset{\sim}{n}} \otimes \overset{\circ}{\underset{\sim}{n}} . \tag{6}$$

The material deformation gradient $\underset{\sim}{F}$ in shell space is derived for the introduced convected coordinates as

$$\underset{\sim}{F} = \underset{\sim}{g}_i \otimes \overset{\circ}{\underset{\sim}{g}}{}^i . \tag{7}$$

According to Pietraszkiewicz /13/,/14/, the material defomationgradient of the middle surface is given by

$$\underset{\sim}{G} = \underset{\sim}{F}(\xi = 0) = \underset{\sim}{a}_i \otimes \overset{\circ}{\underset{\sim}{a}}{}^i = \underset{\sim}{a}_\alpha \otimes \overset{\circ}{\underset{\sim}{a}}{}^\alpha + \underset{\sim}{d} \otimes \overset{\circ}{\underset{\sim}{n}} \tag{8}$$

with

$$\underset{\sim}{F} = \underset{\sim}{Z}\,\underset{\sim}{G}\,\overset{\circ}{\underset{\sim}{Z}}{}^{-1} . \tag{9}$$

The Green strain tensor in shell space

$$\underset{\sim}{E} = \frac{1}{2}(\underset{\sim}{F}^T \underset{\sim}{F} - \underset{\sim}{1}) \tag{10}$$

can be represented in the form, see /13/

$$\underset{\sim}{E} = \frac{1}{2} \overset{\circ}{\underset{\sim}{Z}}{}^{T-1} (\overset{1}{\underset{\sim}{E}} + \xi \overset{2}{\underset{\sim}{E}} + \xi^2 \overset{3}{\underset{\sim}{E}}) \overset{\circ}{\underset{\sim}{Z}}{}^{-1} . \tag{11}$$

$\overset{1}{\underset{\sim}{E}}$, $\overset{2}{\underset{\sim}{E}}$ and $\overset{3}{\underset{\sim}{E}}$ are strain tensors defined on the middle surface of the shell.

Next, we express the strains by displacements. According to Fig. 1 , we get the displacement vector

$$\underset{\sim}{v} = \underset{\sim}{u} + \xi \underset{\sim}{\beta} \tag{12}$$

with

$$\underset{\sim}{\beta} = \underset{\sim}{d} - \overset{\circ}{\underset{\sim}{n}} . \tag{13}$$

The displacement gradient can be introduced as

$$\underset{\sim}{H} = \underset{\sim}{u}_{,\alpha} \otimes \overset{\circ}{\underset{\sim}{a}}{}^{\alpha} \tag{14}$$

and similar we define

$$\underset{\sim}{V} = \underset{\sim}{\beta}_{,\alpha} \otimes \overset{\circ}{\underset{\sim}{a}}{}^{\alpha} \, . \tag{15}$$

We can use an orthorgonal projection tensor to consider only the normal components of the displacement gradient

$$(\underset{\sim}{u}_{,\alpha} \cdot \underset{\sim}{n})\, \underset{\sim}{n} \otimes \overset{\circ}{\underset{\sim}{a}}{}^{\alpha} = \overset{\circ}{\underset{\sim}{P}} \underset{\sim}{H} \qquad \overset{\circ}{\underset{\sim}{P}} = \overset{\circ}{\underset{\sim}{n}} \otimes \overset{\circ}{\underset{\sim}{n}} \, . \tag{16}$$

In the case of a "moderate rotation theory", see Koiter /16/, Pietraszkiewicz /13/, Simmonds, Danielson /17/, a separation of the strains into membrane- (m), shear- (s), bending- (b) and nonlinear- (nl) terms leads to the following expressions

$$\underset{\sim}{E}^{m} = \frac{1}{2}(\overset{\circ}{\underset{\sim}{1}}{}^{T} \underset{\sim}{H} + \underset{\sim}{H}^{T} \overset{\circ}{\underset{\sim}{1}}) \qquad \underset{\sim}{E}^{nl} = \frac{1}{2}(\underset{\sim}{H}^{T} \overset{\circ}{\underset{\sim}{P}}{}^{T} \overset{\circ}{\underset{\sim}{P}} \underset{\sim}{H})$$

$$\underset{\sim}{E}^{s} = (\underset{\sim}{\beta} + \underset{\sim}{H}^{T} \overset{\circ}{\underset{\sim}{n}}) \otimes \overset{\circ}{\underset{\sim}{n}} \tag{17}$$

$$\underset{\sim}{E}^{b} = -\underset{\sim}{H}^{T} \overset{\circ}{\underset{\sim}{B}} + \overset{\circ}{\underset{\sim}{1}}{}^{T} \underset{\sim}{V}$$

see e.g. Wagner /1/.

Due to the moderate rotation theory, it can be shown that $\underset{\sim}{\beta}$ has only tangential components.

The numerical treatment is based on the weak formulation in shell space. For the description of finite rotations in conjunction with the applied "moderate rotation theory" an updated Lagrangian formulation has to be used.

In the current configuration, see Bathe et. al. /18/

$$\delta\pi(\underset{\sim}{v}) = \int_{^{1}B} {}^{2}_{1}\underset{\sim}{S} \cdot \delta^{2}_{1}\underset{\sim}{E}\, dv - \int_{^{0}\partial B_{\sigma}} {}^{2}_{0}\bar{\underset{\sim}{t}} \cdot \delta \underset{\sim}{v}\, da \tag{18}$$

holds. $\underset{\sim}{S}$ denotes the 2nd Piola-Kirchhoff-tensor and $\bar{\underset{\sim}{t}}$ is the vector of the monogenetic surface loads.

Restricting ourselves to small strains and with respect to

$$\underset{\sim}{S}^{(SMS)} = \overset{\circ}{\underset{\sim}{Z}} \underset{\sim}{S} \overset{\circ}{\underset{\sim}{Z}}{}^{T} \tag{19}$$

we can formulate the principle of virtual work in terms of the shell middle surface

$$\delta\pi(\underset{\sim}{u},\underset{\sim}{\beta}) = \int_{{}^{1}\mathcal{M}} [\underset{\sim}{N}\cdot(\delta\underset{\sim}{E}^{m}+\delta\underset{\sim}{E}^{nl})+\underset{\sim}{Q}\cdot\delta\underset{\sim}{E}^{s}+\underset{\sim}{M}\cdot\delta\underset{\sim}{E}^{b}]\,da + \int_{{}^{0}\mathcal{M}_{\sigma}} [\bar{\underset{\sim}{f}}\cdot\delta\underset{\sim}{u}+\bar{\underset{\sim}{m}}\cdot\delta\underset{\sim}{\beta}]\,d\mathring{a}\,. \tag{20}$$

Assuming a homogeneous and isotropic elastic material , we introduce the linear hyperelastic material law

$$\underset{\sim}{S} = \underset{\sim}{\mathbb{C}}[\underset{\sim}{E}] \tag{21}$$

with

$$\underset{\sim}{\mathbb{C}} = 2\mu\,\mathring{\underset{\sim}{\mathbb{I}}} + \lambda\,\mathring{\underset{\sim}{1}}\otimes\mathring{\underset{\sim}{1}}\,. \tag{22}$$

This leads in the case of shear elastic shells to constitutive equations for the commonly used stress and couple resultant tensors

$$\begin{aligned} \underset{\sim}{N} &= \frac{Ch}{1-\nu^2}\,\underset{\sim}{\mathbb{H}}[\underset{\sim}{E}^{m}+\underset{\sim}{E}^{nl}] \\ \underset{\sim}{Q} &= XGh\,\underset{\sim}{\mathbb{K}}[\underset{\sim}{E}^{s}] \\ \underset{\sim}{M} &= \frac{Ch^3}{12(1-\nu^2)}\,\underset{\sim}{\mathbb{H}}[\underset{\sim}{E}^{b}] \end{aligned} \tag{23}$$

with the constitutive tensors

$$\begin{aligned} \underset{\sim}{\mathbb{H}} &= (1-\nu)\,\mathring{\underset{\sim}{a}}_{\alpha}\otimes\mathring{\underset{\sim}{a}}_{\beta}\otimes\mathring{\underset{\sim}{a}}^{\alpha}\otimes\mathring{\underset{\sim}{a}}^{\beta} + \nu\,\mathring{\underset{\sim}{a}}_{\alpha}\otimes\mathring{\underset{\sim}{a}}^{\alpha}\otimes\mathring{\underset{\sim}{a}}_{\beta}\otimes\mathring{\underset{\sim}{a}}^{\beta} \\ \underset{\sim}{\mathbb{K}} &= \mathring{\underset{\sim}{a}}_{\alpha}\otimes\mathring{\underset{\sim}{n}}\otimes\mathring{\underset{\sim}{a}}^{\alpha}\otimes\mathring{\underset{\sim}{n}}\,. \end{aligned} \tag{24}$$

Within the finite element analysis, the solution of the nonlinear equation (20) is obtained by an iterative Newton scheme . For this purpose, a sequence of linearized problems has to be solved which is stated below

$$\begin{aligned} \delta\pi(\underset{\sim}{u},\underset{\sim}{\beta}) &+ \int_{{}^{1}\mathcal{M}} \Big\{ \frac{Ch}{1-\nu^2}(\delta\underset{\sim}{E}^{m}+\delta\underset{\sim}{E}^{nl})\cdot\underset{\sim}{\mathbb{H}}[\Delta\underset{\sim}{E}^{m}+\Delta\underset{\sim}{E}^{nl}] \\ &+ \frac{Ch^3}{12(1-\nu^2)}\delta\underset{\sim}{E}^{b}\cdot\underset{\sim}{\mathbb{H}}[\Delta\underset{\sim}{E}^{b}] + XGh\,\delta\underset{\sim}{E}^{s}\cdot\underset{\sim}{\mathbb{K}}[\Delta\underset{\sim}{E}^{s}]\Big\}\,da \\ &+ \int_{{}^{1}\mathcal{M}} \mathring{\underset{\sim}{P}}\delta\underset{\sim}{H}\cdot\underset{\sim}{N}\mathring{\underset{\sim}{P}}\Delta\underset{\sim}{H}\,da = 0\,. \end{aligned} \tag{25}$$

For example, the variation $\delta \underset{\sim}{E}^{b}$ and the linearization $\Delta \underset{\sim}{E}^{b}$ of the strain-tensors are defined as

$$\delta \underset{\sim}{E}^{b} = -\delta \underset{\sim}{H}^{T} \overset{\circ}{\underset{\sim}{B}} + \overset{\circ}{\underset{\sim}{1}}{}^{T} \delta \underset{\sim}{v} \qquad \Delta \underset{\sim}{E}^{b} = -\Delta \underset{\sim}{H}^{T} \overset{\circ}{\underset{\sim}{B}} + \overset{\circ}{\underset{\sim}{1}}{}^{T} \Delta \underset{\sim}{v} \quad . \tag{26}$$

The iteration is terminated when a norm of the residual $\delta\pi$ remains under a prescribed tolerance.

2.2 Geometrically nonlinear axisymmetric shell element

Starting from the strain tensor defined in equation (17) and using the conditions for axisymmetry

$$u_{\Phi} = 0 , \quad \beta_{\Phi} = 0 , \quad (\dots)_{,\Phi} = 0 \tag{27}$$

the physical strains may be expressed as follows, notation see Figure 2

$$\begin{aligned} E^{m}_{ss} &= u_{,s} & E^{m}_{\Phi\Phi} &= \frac{1}{r}(u \sin\Theta - w \cos\Theta) \\ E^{b}_{ss} &= -\beta_{s,s} & E^{b}_{\Phi\Phi} &= -\frac{1}{r}\beta \sin\Theta - \frac{\cos\Theta}{r^2}(u \sin\Theta - w \cos\Theta) \\ E^{s}_{s} &= \frac{1}{2}(W_{,s} - \beta_s) & E^{nl}_{ss} &= \frac{1}{2} W^2_{,s} . \end{aligned} \tag{28}$$

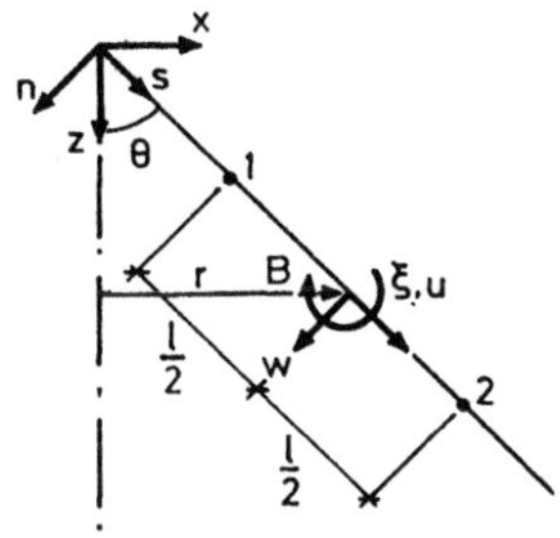

Fig. 2. Notation for axisymmetric shell element

The $\underline{B}$ - matrices belonging to these strain measures can be found in appendix 1. The linear strains are almost identical with those used by Zienkiewicz /2/ exept the second term in $E^{b}_{\phi\phi}$ which results from the consequently derived tensorial strain measures. The influence of this term is neglectable, Wagner /1/ .

The appropriate elasticity matrix $\underline{D}$ is defined by

$$\underline{D} = \begin{bmatrix} \underline{D}^m & \underline{0} & \underline{0} \\ \underline{0} & \underline{D}^b & \underline{0} \\ \underline{0} & \underline{0} & \underline{D}^s \end{bmatrix} \tag{29}$$

where the matrices $\underline{D}^m$, $\underline{D}^b$ and the scalar D^s stand for membrane-, bending- and shear- rigidities, respectively. For isotropic materials these are

$$\underline{D}^m = \frac{C\,h}{1-\nu^2} \begin{bmatrix} 1 & \nu \\ \nu & 1 \end{bmatrix} \quad \underline{D}^b = \frac{C\,h^3}{12(1-\nu^2)} \begin{bmatrix} 1 & \nu \\ \nu & 1 \end{bmatrix} \quad D^s = X G h \,. \tag{30}$$

Due to the shear elastic shell theory ,only first derivatives exist in the strain measures. Therefore, only C° - continuity is required for the shape functions. In the case of a straight element, linear shape functions are possible and the element formulation becomes extremely simple. Within the numerical integration of the element matrices we have to introduce reduced or selective integration schemes. Otherwise the well known effect of shear locking occurs, see e.g. Hughes, Taylor, Kanoknukulchai /19/ . Reduced or selective integrations and mixed methods are closely related, see Malkus, Hughes /20/ . Another way to motivate reduced integration schemes is the interpretation of the shear stiffness matrix as a penalty term for a Kirchhoff-type shell theory, Zienkiewicz /2/ . For the sucessful application of the element only a one point integration is necessary, see Zienkiewicz /2/. Especially for pure membrane situations, we observe better results if two point integration is used for the membrane and bending terms, see Wagner /1/. However, the improvements are not so significant. Therefore, in the case of nonlinear problems we choose a one point integration because of its simplicity.

In the element formulation a Timoshenko beam element is included which is obtained by slightly modifications of the $\underline{B}$- and $\underline{D}$-matrices , Wagner /1/ .

The numerical finite element analysis is based on equation (25). Using the expressions given in appendix 1, an explicit version of (25) follows for axisymmetric shells

$$\delta \underline{v}^T \{ \int_{{}^1\mathcal{M}} [(\underline{B}^m + \underline{B}^{nl})^T \underline{D}^m (\underline{B}^m + \underline{B}^{nl}) + \underline{B}^{bT} \underline{D}^b \underline{B}^b + \underline{B}^{sT} D^s \underline{B}^s] \, da \, \Delta \underline{v} + \int_{{}^1\mathcal{M}} \underline{G}^T \hat{N} \underline{G} \, da \, \Delta \underline{v} + \underline{R} \} = \delta \underline{v}^T \{ \underline{K}_T \Delta \underline{v} + \underline{R} \} = 0 \, . \tag{31}$$

3. Contact formulations for axisymmetric shells

In the context of finite element techniques for contact problems several aspects have to be considered. These are the enforcement of the contact conditions, the incorporation of the constraint equations into the weak form and the proper algorithm.

In the case of thin shells and beams exact solutions of simple contact problems show that the contact is only pointwise , see e.g. Kalnins, Updike /21/. Therefore, an enforcement of the contact conditions on a nodal basis is the appropriate choice. Our formulation is based on a kinematic model which was suggested by Hallquist /8/ for the penalty method.

Two different types of strategies are commonly used to incorporate contact constraints into the global finite element equations: the Lagrangian multiplier and the penalty method. Since these methods have different drawbacks and advantages we will look at both approaches. Thus, we add the contact conditions via a perturbed Lagrangian formulation to the global equations which contains as limit cases the penalty and the Lagrangian multiplier method. The mathematical foundation of the perturbed Lagrangian method may be found in Oden /22/.

The third aspect in finite element analysis of contact problems is associated with the algorithmic part which will be addressed in section 4.

3.1 Discrete contact conditions

We will restrict ourselves to frictionless contact of the shell elements described in section 2. In a pointwise formulation of the contact constraints the penetration g of a node into the surface of the shell element (Fig. 3) is indicated by the inequality

$$g = (\underset{\sim}{x}_s - \underset{\sim}{x}_1) \cdot \underset{\sim}{n} > 0 \tag{32}$$

In equation (32) $\underset{\sim}{x}$ denotes the current position vector of a node i.

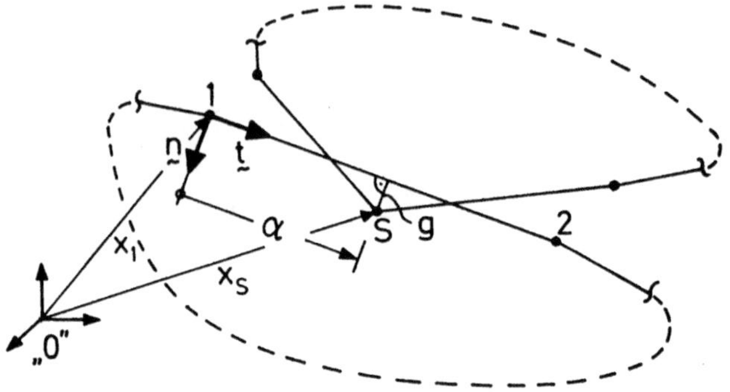

Fig. 3. Contact of a node with a shell element

3.2 Perturbed Lagrangian formulation of the contact tangent

Due to the perturbed formulation, Oden /22/, the contribution of a single contact node (Fig. 3) to the total potential energy of the bodies being in contact is given by

$$\pi_c(\underset{\sim}{v},\lambda) = \lambda\, g(\underset{\sim}{v}) - \frac{1}{2\varepsilon}\lambda^2 \quad ; \quad \varepsilon > 0 \tag{33}$$

Here λ denotes the Lagrangian multiplier associated with the nodal contact force. For finite element computations the weak form is obtained by the variation of (33). With the variation of the normal vector, see Wriggers, Simo /7/,

$$\delta\underset{\sim}{n} = -\frac{1}{\|x_2 - x_1\|}(\underset{\sim}{t}\otimes\underset{\sim}{n})(\delta\underset{\sim}{v}_2 - \delta\underset{\sim}{v}_1)$$

the variation of g yields

$$\delta g = \left\{(\delta\underset{\sim}{v}_s - \delta\underset{\sim}{v}_1)\cdot\underset{\sim}{n} - \frac{(\underset{\sim}{x}_s - \underset{\sim}{x}_1)}{\|\underset{\sim}{x}_2 - \underset{\sim}{x}_1\|}\cdot(\underset{\sim}{t}\otimes\underset{\sim}{n})(\delta\underset{\sim}{v}_2 - \delta\underset{\sim}{v}_1)\right\} . \tag{34}$$

Introducing the surface coordinate α as shown in Fig. 3 leads to another expression for δg

$$\delta g = \left\{\delta\underset{\sim}{v}_s - (1-\alpha)\,\delta\underset{\sim}{v}_1 - \alpha\,\delta\underset{\sim}{v}_2\right\}\cdot\underset{\sim}{n} \tag{35}$$

which is better suited for computational purposes. Furthermore, we can see from this expression immediately that only forces normal to the surface 1-2 are transmitted. This reflects the frictionless contact.

With equation (35) the weak form for one contact segment may be stated as follows

$$\delta\pi_c = \langle \delta \underline{v}_c^T, \delta\lambda \rangle \begin{Bmatrix} \lambda \underline{N}_s \\ -\frac{1}{\varepsilon}\lambda + g \end{Bmatrix} \tag{36}$$

where in equation (36) the definitions

$$\underline{N}_s^T = \langle \underline{n}^T, -(1-\alpha)\underline{n}^T, -\alpha\underline{n}^T \rangle \quad ; \quad \underline{v}_c^T = \langle \underline{v}_s^T, \underline{v}_1^T, \underline{v}_2^T \rangle$$

have been used.

A consistent linearization of the weak form, equation (36), has been given by Wriggers, Simo /7/. The matrixform of the resulting contact tangent yields

$$\langle \delta \underline{v}_c^T, \delta\lambda \rangle \left(\begin{bmatrix} \underline{K}(\lambda,g) & \underline{N}_s \\ \underline{N}_s^T & -\frac{1}{\varepsilon} \end{bmatrix} \begin{Bmatrix} \Delta \underline{v} \\ \Delta\lambda \end{Bmatrix} = - \begin{Bmatrix} \lambda \underline{N}_s \\ -\frac{1}{\varepsilon}\lambda + g \end{Bmatrix} \right) \tag{37}$$

with

$$\underline{K}(\lambda,g) = -\frac{\lambda}{\|\underline{x}_2 - \underline{x}_1\|} \left\{ \underline{N}\,\underline{T}_s^T + \underline{T}_s \underline{N}^T + \frac{g}{\|\underline{x}_2 - \underline{x}_1\|} \underline{N}\,\underline{N}^T \right\}$$

and

$$\underline{T}_s^T = \langle \underline{t}^T, -(1-\alpha)\underline{t}^T, -\alpha\underline{t}^T \rangle \quad ; \quad \underline{N}^T = \langle 0, -\underline{n}^T, \underline{n}^T \rangle .$$

In the limit $\varepsilon \to \infty$ equation (37) reduces to a pure Lagrangian formulation. Solving, on the other hand, equation (36) for λ yields the approach given by Simo et. al. /23/ for the penalty method. For completeness the associated matrix formulations are specified in appendix 2. In the case of linearized kinematics the term $\underline{K}$ vanishes in equation (37). Then the classical linear Lagrangian multiplier formulation see Hughes et. al /24/ or the penalty formulation Hallquist /8/ may be recovered.

4. Algorithms for geometrically nonlinear shell problems including contact constraints

4.1 Arc-Length-Methods

The sensitivity of shells, especially cylindrical or flat shells, against imperfections is well known. Therefore it is necessary to investigate the postcritical range. The usually employed incremental algorithms with load control or displacement control fail at limit points. To overcome this problem the arc-length-method is introduced where loads and displacements have to be changed simultaneously within an iteration. Variants of the method were originally introduced by Riks /9/, Wempner /1o/ and Wessels /11/. Up to now many authors have applied these procedures successfully to different engineering problems in the original or modified versions, see e.g. Crisfield /25/ , Ramm /26/ .

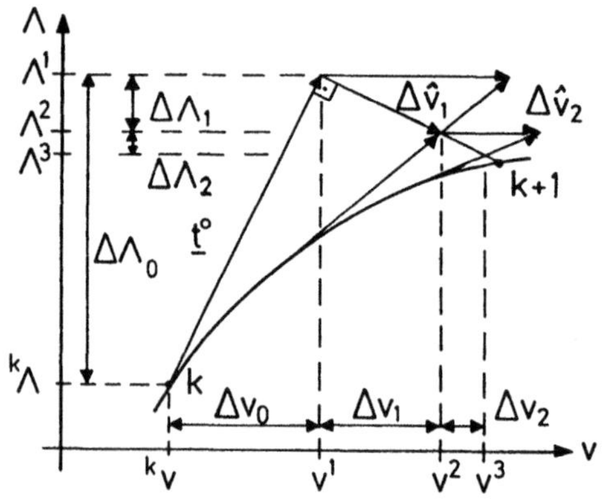

Fig. 4. Notation used for the arc-length method

Basically, in arc-length-methods an additional constraint equation is introduced to restrict a vector (arc-length) in the load displacement space to a prescribed value . Here, we adopt a commonly used scheme, see e.g. Ramm /26/, which imposes the linear constraint equation

$$(\Delta \underline{v}_0 , \Delta \Lambda_0)^T (\underline{v}^i - {}^k\underline{v} - \Delta \underline{v}_0 \, , \, \Lambda^i - {}^k\Lambda - \Delta \Lambda_0) = 0 \qquad (38)$$

on the weak form (20). Due to the matrix notation in appendix 1 equation

(20) is given by

$$\underline{F}^{i} - \Lambda^{i}\,\Delta\underline{P} = 0\,. \tag{39}$$

Λ^{i} denotes the load parameter which is treated like an unknown and is determined with the aid of equation (38). Fig. 4 shows the notation employed in equation (38) and (39). Using equation (31) the linearization of the system of equations (38) and (39) yields

$$\begin{bmatrix} \underline{K}_T^{i} & -\Delta\underline{P} \\ \Delta\underline{v}_0^{T} & \Delta\Lambda_0 \end{bmatrix} \begin{Bmatrix} \Delta\underline{v}_i \\ \Delta\Lambda_i \end{Bmatrix} = - \begin{Bmatrix} \underline{F}^{i} - \Lambda^{i}\Delta\underline{P} \\ (\Delta\underline{v}_0, \Delta\Lambda_0)^T(\underline{v}^{i} - {}^{k}\underline{v} - \Delta\underline{v}_0, \Lambda^{i} - {}^{k}\Lambda - \Delta\Lambda_0) \end{Bmatrix} \tag{40}$$

where the coefficient matrix is unsymmetric. This system can be solved by a partitioning method which then works with symmetric matrices only. The algorithm is given in section 4.3.

4.2 Contact algorithms

Since contact conditions impose inequality constraints on the functional associated with equation (20) a nonlinear programming problem has to be solved. For this purpose, from the optimization theory different methods are available. Here, we will use a type of an active set strategy together with Newtons method. The main problem in this approach is how to change the working set of active constraints. This effects the asymptotic convergence properties of the algorithm which then may not be of second order as usual in the application of the standard Newton method. So far, no effective algorithms exist which determine the right active set and exhibit second order convergence properties. However, if the appropriate active set is known, then due to the consistent contact tangent, the algorithm converges quadratically.

An active set of the inequality constraints is determined by a check of equation (32) during the iteration. If a node penetrates into a surface then the constraint (32) is activated for this node and after that treated like an equality constraint. On the other hand a constraint equation for a node is removed from the working set if in the penalty approach the penetration becomes a gap ($g < 0$) or if in the Lagrangian multiplier method the multiplier indicates tension. Such a method is relatively simple , effective and converges in most applications. It should be noted that the procedure stated above exhibits no global convergence properties. If no convergence is achieved within the contact iteration the basic but slower active set strategy , see e.g. Luenberger /12/ has to be used.

4.3 Combined algorithm

In sections 4.1 and 4.2 two independent algorithms have been described. Both algorithms are based on the Newton method. For the calculation of contact problems in the postbuckling range the contact algorithm is combined with the arc-lenght method. The final version of the algorithm is stated below for an updated Lagrangian scheme (notation see appendix 2).

1. Set initial values $\underline{v}_0 = \underline{0}$

2. For k=1,....,n - load-increments do

   Update coordinates and stresses ${}^{k}\underline{x} = {}^{k-1}\underline{x} + {}^{k-1}\underline{v}$

   Apply load increment $\Delta \underline{P}_K$

3. For i=1,....,m until convergence do

   Check contact: $g_s < 0$ inactive constraint

   $g_s \geq 0$ active constraint, add contact contribution $\underline{C}_s$ to tangent stiffness $\underline{K}_T$

   Solve:

   Penalty method: $\underline{\Gamma}^i = \varepsilon \underline{g}^i$

   Lagrangian multiplier method: $\underline{\Gamma}^i = \underline{\lambda}^i$

$$\underline{R}^i = \underline{F}^i - \Lambda^i \Delta \underline{P} + \underline{N}^i \underline{\Gamma}^i$$

$$(\underline{K}_T + \varepsilon \underline{C}\underline{C}^T)^i \Delta \underline{v}_i^{I} = \Delta \underline{P} \qquad \left[\begin{array}{c|c} \underline{K}_T + \underline{K}(\lambda, g) & \underline{N} \\ \hline \underline{N}^T & \underline{0} \end{array}\right]^i \left[\begin{array}{c} \Delta \underline{v}^{I} \\ \hline \Delta \underline{\lambda}^{I} \end{array}\right]^i = \left[\begin{array}{c} \Delta \underline{P} \\ \hline \underline{0} \end{array}\right]$$

$$(\underline{K}_T + \varepsilon \underline{C}\underline{C}^T)^i \Delta \underline{v}_i^{II} = -\underline{R}^i \qquad \left[\begin{array}{c|c} \underline{K}_T + \underline{K}(\lambda, g) & \underline{N} \\ \hline \underline{N}^T & 0 \end{array}\right]^i \left[\begin{array}{c} \Delta \underline{v}^{II} \\ \hline \Delta \underline{\lambda}^{II} \end{array}\right]^i = -\left[\begin{array}{c} \underline{R}^i \\ \hline g^i \end{array}\right]$$

Calculate the incremental load scaling factor:

$$\Delta\Lambda_i = - \frac{\Delta \underline{v}_0^T \Delta \underline{v}_i^{II}}{\Delta \underline{v}_0^T \Delta \underline{v}_i^I + \Delta\Lambda_0}$$

Calculate incremental displacements (and Lagrangian multipliers):

$$\Delta \underline{v}_i = \Delta\Lambda_i \, \Delta \underline{v}_i^I + \Delta \underline{v}_i^{II} \qquad \Delta \underline{\lambda}_i = \Delta\Lambda_i \, \Delta \underline{\lambda}_i^I + \Delta \underline{\lambda}_i^{II}$$

Update displacements, load scaling factor (and Lagrangian multipliers):

$$\underline{v}^{i+1} = \underline{v}^i + \Delta \underline{v}_i \qquad \underline{\lambda}^{i+1} = \underline{\lambda}^i + \Delta \underline{\lambda}_i$$

$$\Lambda^{i+1} = \Lambda^i + \Delta\Lambda_i$$

Check for convergence:

$\|\underline{R}^{i+1}\| \le tol$ terminate go to 2.
else go to 3.

As can be seen from the algorithm, the penalty approach is much simpler to implement. However, a large value for the penalty parameter has to be selected to fulfil the contact constraints as good as possible. This may lead to a poor conditioning of the coefficient matrix which even can get worse near limit points where diagonal terms in the tangent matrix tend to zero. Therefore, the appropriate choice of the penalty parameter ε is not obvious but may be obtained from error estimates, see Wriggers, Nour-Omid /28/. In some problems where the determinant of $\underline{K}$ approaches zero ε has to be changed during the iteration. On the other hand the Lagrangian multiplier approach as stated in the algorithm fulfils the contact conditions exactly but leads to zero diagonal terms in the tangent matrix and has an increased number of unknowns.

5. Examples

In this chapter some examples are presented to discuss the behaviour of the element formulation within the updated Lagrangian procedure and the performance of the combined algorithms.

5.1 Contact of a spherical shell

In the first example we consider an elastic shell which is in contact with a rigid plate. Since the ratio of radius to thickness is 30 the shell is moderately thick and the influence of transverse shear has to be taken into account. The geometry of the shell is shown in Fig. 5 which also represents the load deflection curve. 21 elements were sufficient to discretize the problem. The Lagrange multiplier formulation has been chosen for the contact calculations. For small load increments (3-5) iterations were used to achieve convergence whereas for larger load increments (6-8) iteration were needed. The results are in good agreement with an analytic solution given by Updike, Kalnins /21/.

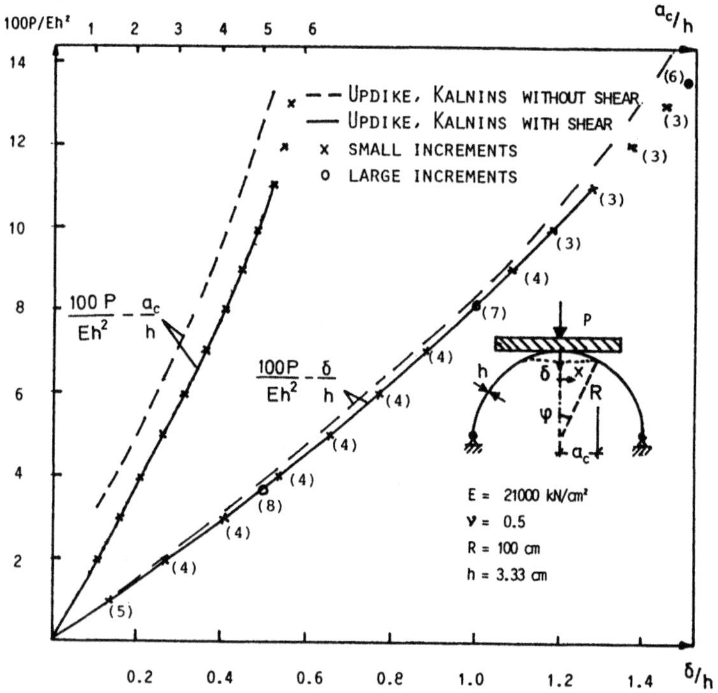

Fig. 5. Contact between a spherical shell and a rigid plate

The distribution of the contact stresses for different angles φ is exhibited in Fig. 6. Despite of the simplicity of the axisymmetric shell element the contact stresses agree very good with the stresses obtained from the analytic solution.

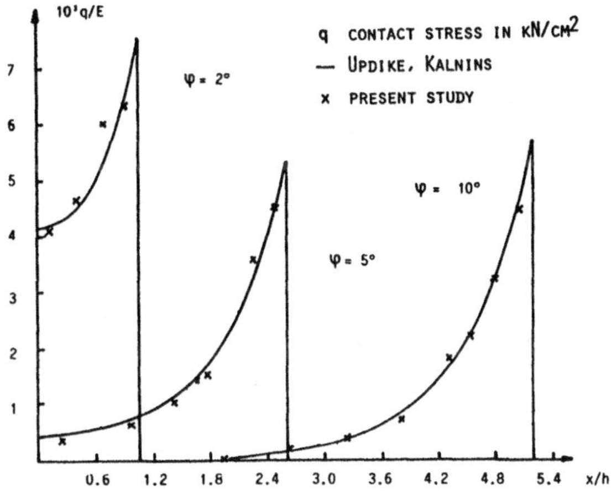

Fig. 6. Contact stresses for different angles φ = 2.5°, 5°, 10°

5.2 Snap-through of a flat axisymmetric shell with contact

This example shows the influence of contact constraints over the snap-through behaviour of flat shells. The geometry of the system is desribed in Fig. 7. The rigid contact surface has been chosen in such a way that contact only appears in the postcritical range. The problem has been modelled by 20 axisymmetric shell elements. Not more than 5 iterations were necessary to achieve convergence within the Lagrangian multiplier formulation for a given load increment of $\Delta P=1.0$ kN/cm^2 and an initial value for the load scaling factor of $\Delta\Lambda=0.2$. Fig. 8 shows a plot of the load ordinate versus the displacement of node 17 (see Fig. 7). The lowest curve is associated with the unconstraint case. According to the distance between the rigid surface and the shell the other branches of the load deflection curve occur. It may be observed that the stiffening of the system by the rigid support does not effect the snap-through phenomenon in the postbuckling range.

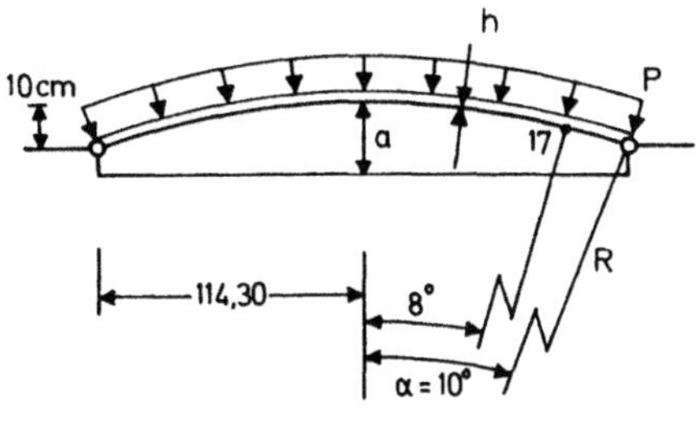

E = 21000. KN/cm²
ν = 0.3
R = 658.23 cm
α = 10°
a = 10,15,20 cm
h = 1.0 cm

Fig. 7. Flat axisymmetric shell with rigid contact surface

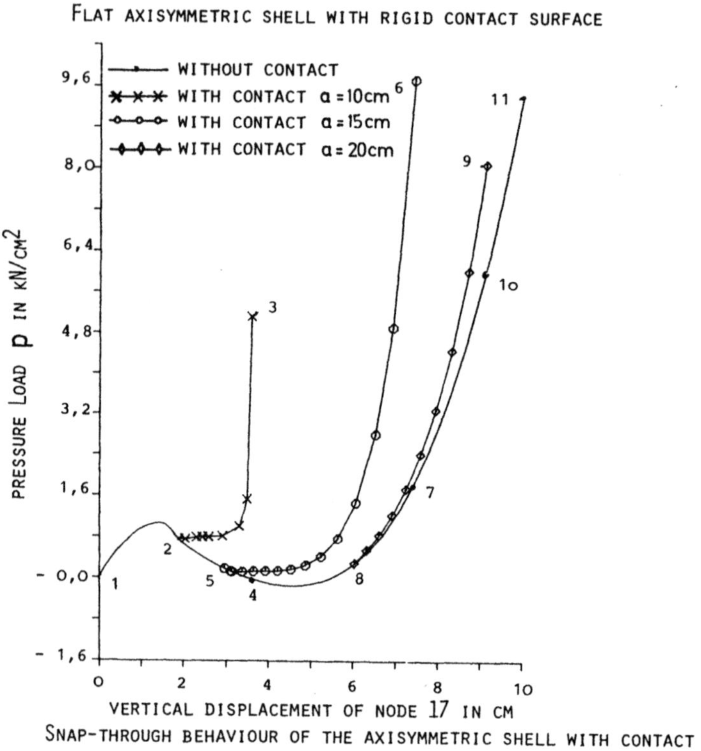

Fig. 8. Snap-through behaviour of the axisymmetric shell with contact

5.3 Circular Arch under point load

As a third example we consider an unsymmetrically supported circular arch under a vertical point load. This example is selected to check the performance of the element together with the updated Lagrangian scheme by a problem of finite deflections and rotations. The axisymmetric shell element is degenerated to a beam element which can model moderate rotations. The finite element mesh consists of 20 equally spaced elements with linear shape functions. Fig. 9 exhibits the load deflection curve for the top displacement of the arch. Since this curve exhibits a displacement reversal in addition to a load reversal the arc-length method is the only possible algorithm to follow the load deflection path. At the end of the displacement reversal part of the load deflection curve the solution deviates from that given by Simo et.al. /23/. This is due to the choice of a coarser finite element mesh.

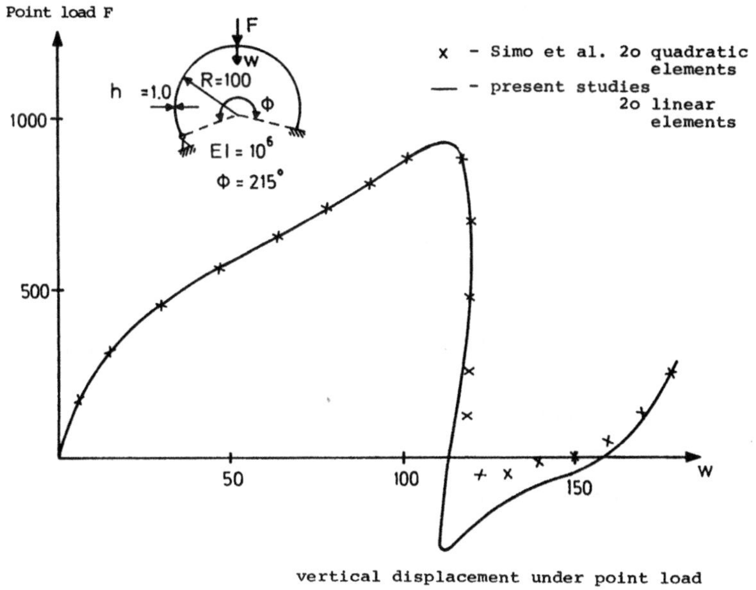

Fig. 9. Circular arch under point load.

5.4 Bending of a spherical shell

The application of the element to an axisymmetric shell problem with finite deflections and rotations is considered as a final example. For this purpose, we examine a spherical shell under point load at the apex. The system is shown in Fig. 10 which also exhibits the deformed shapes for different load levels. The finite element mesh consist of 24 linear shell elements. This example was also treated by Fried /27/ who used the same number of elements but with cubic shape functions. Our solution is in very good agreement with that by Fried which furthermore shows the nice performance of the simple and inexpensive element formulation given in section 2.

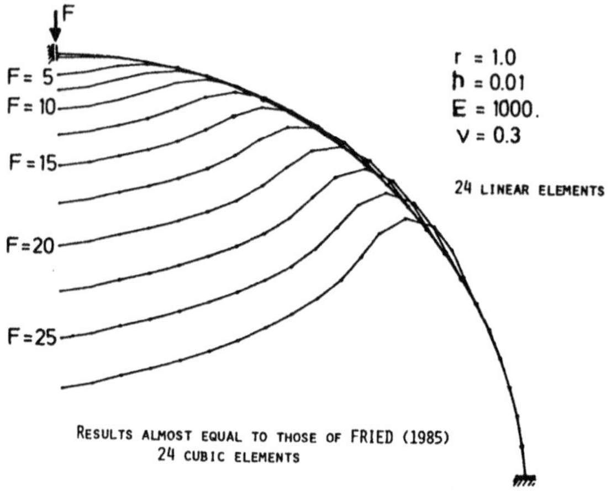

Fig. 10. Spherical shell under vertical point load

5.5 Postbuckling behaviour of two spherical shells being in contact

In the last example we consider the contact of two deformable shells. The upper shell represents a hollow sphere and is modelled by 40 shell elements. The lower shell is a spherical cap which is discretized by 20 elements. The data of both shells are contained in Fig. 11. Here the use of the penalty method for the contact formulation is advantegeous. Since a small penetration occurs due to the penalty method the contact algorithm does not loose contact within the iteration which may happen using the Lagrange multiplier approach. To trace the postbuckling behaviour the arc-length method has been applied. The number of iterations within each load step applying a full Newton scheme varies between 3 iterations when the contact surface does not change and 5 to 7 iterations elsewhere. Near limit points an increasing number of iterations may occur. In this case the arc-length is reduced automatically after 7 iterations to trace the limit point. Fig. 11 exhibits the load deflection curve for the vertical tip displacement of the upper shell. At two different load levels snap through occurs. Considering only the displacement of the lower shell we observed a pure bending behaviour of the upper shell while the lower shell was responsible for the snap-through behaviour. Fig.12 shows deformed shapes at different load steps which are marked in the load deflection curve of Fig.11. Finally a study of different thickness values of the lower shell shows drastically in Fig. 13 the sensivity (pure bending to snapping-through) of shell structures against changing the parameter h/R.

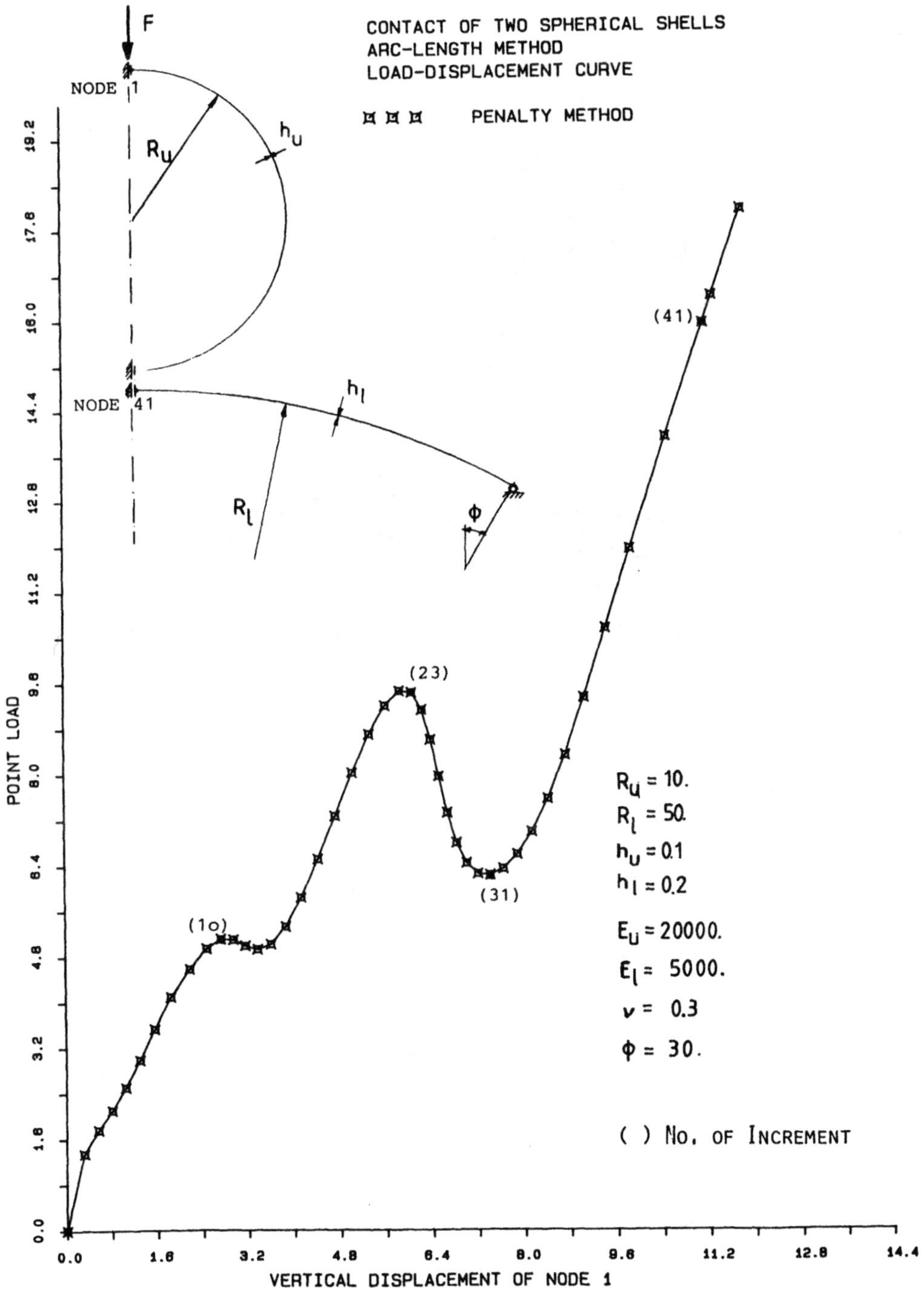

Fig. 11 Load deflection curve for the contact of two spherical shells.

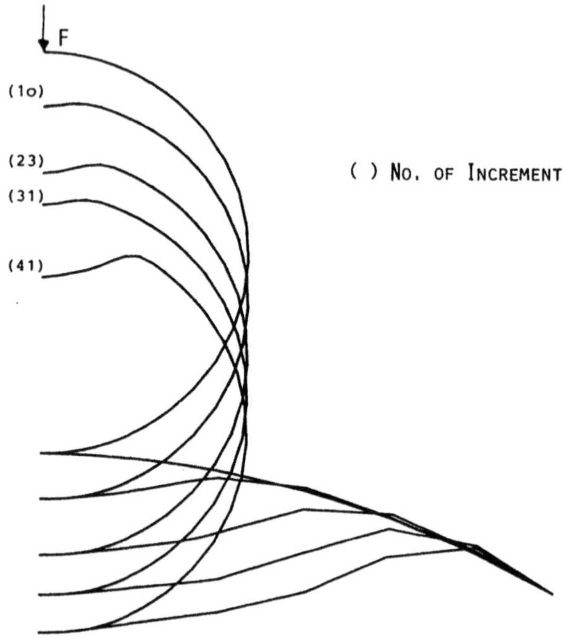

Fig.12 Deformed shapes of the contact between two spherical shells at different load steps

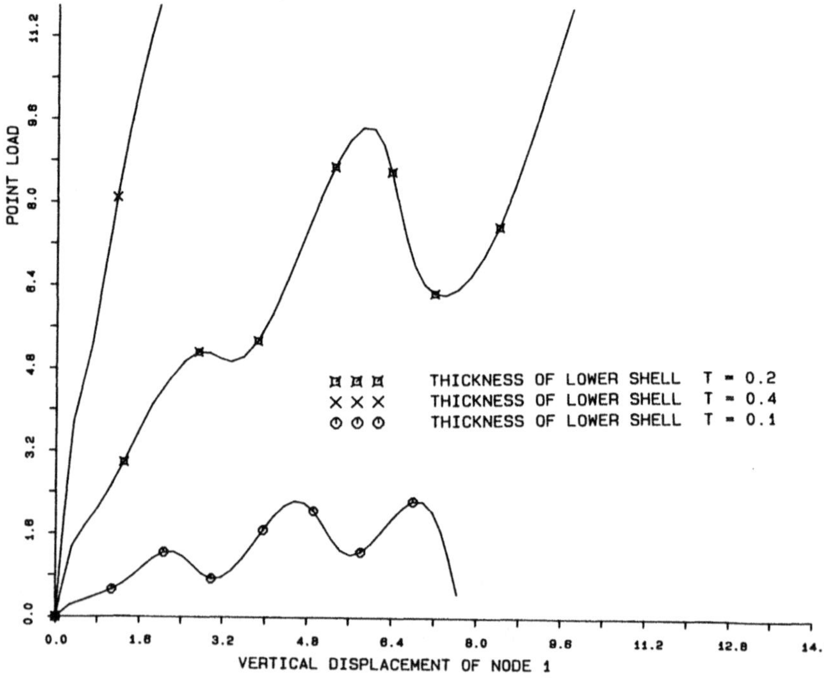

Fig. 13 Load deflection curve for different thicknesses of the lower shell

6. Conclusions

In this paper a simple axisymmetric shell element has been developed which is based on a Mindlin type kinematic and includes the possibility of analysing large displacements and rotations restricted to small strains.
Since the aim of the paper was a treatment of stability problems of several deformable bodies being in contact, a unified contact formulation for the Lagrangian multiplier and the penalty method was given and combined with the well known arc-length method for tracing the snap through behaviour of the shell structures.
Consistent linearizations have been applied to arrive at tangent matrices for the overall Newton algorithm and to ensure quadratical convergence rates.

Appendix 1

$\underline{B}$ - matrices for axisymmetric shell element:

$$\underline{B}^{m(e)} = \begin{bmatrix} N_{i,s} & 0 & 0 \\ \frac{\sin\Theta}{r} N_i & -\frac{\cos\Theta}{r} N_i & 0 \end{bmatrix}$$

$$\underline{B}^{b(e)} = \begin{bmatrix} 0 & 0 & -N_{i,s} \\ -\frac{\sin\Theta\cos\Theta}{r^2} N_i & \frac{\cos^2\Theta}{r^2} N_i & -\frac{\sin\Theta}{r} N_i \end{bmatrix}$$

$$\underline{B}^{s(e)} = \begin{bmatrix} 0 & N_{i,s} & -N_i \end{bmatrix}$$

$$\underline{B}^{nl(e)} = \begin{bmatrix} 0 & w_{,s} N_{i,s} & 0 \\ 0 & 0 & 0 \end{bmatrix} = \underline{G}^{(e)}$$

Appendix 2

Normal force used in the initial stress matrix:

$$\hat{N} = n_{ss}$$

Residual:

$$\underline{R}^{(e)} = \int_{{}^1\mathcal{M}^{(e)}} [(\underline{B}^m + \underline{B}^{nl})^T \underline{D}^m (\underline{B}^m + \tfrac{1}{2}\underline{B}^{nl}) + \underline{B}^{bT} \underline{D}^b \underline{B}^b + \underline{B}^{sT} \underline{D}^s \underline{B}^s] da \, \underline{v}^{(e)}$$

$$- \Lambda \, \Delta \underline{P}^{(e)} = \underline{F}^{(e)} - \Lambda \, \Delta \underline{P}^{(e)}$$

For the penalty approach equation (37) yields

$$\delta \underline{v}_c^T \varepsilon \, (\underline{C}_s \underline{C}_s^T \Delta \underline{v}_c + g \, \underline{N}_s) = 0$$

with

$$\underline{C}_s \underline{C}_s^T = \underline{N}_s \underline{N}_s^T - \frac{g}{\|\underline{x}_2 - \underline{x}_1\|} \left\{ \underline{T}_s \underline{N}^T + \underline{N} \, \underline{T}_s^T + \frac{g}{\|\underline{x}_2 - \underline{x}_1\|} \underline{N} \, \underline{N}^T \right\}$$

Equation (37) is given for the Lagrangian multiplier approach by

$$\langle \delta \underline{v}_c^T , \delta\lambda \rangle \left(\left[\begin{array}{c|c} \underline{K}(\lambda, g) & \underline{N}_s \\ \hline \underline{N}_s^T & 0 \end{array} \right] \left\{ \begin{array}{c} \Delta \underline{v}_c \\ \hline \Delta\lambda \end{array} \right\} + \left\{ \begin{array}{c} \lambda \underline{N}_s \\ \hline g \end{array} \right\} \right) = 0$$

For more than one node being in contact all contributions of the constraint conditions have to be assembled. This leads for the penalty method to the matrix formulation

$$\delta \underline{v}^T \varepsilon \, (\underline{C} \, \underline{C}^T \Delta \underline{v} + \underline{N} \, \underline{g}) = 0$$

and finally for the Lagrangian multiplier method to

$$\langle \delta \underline{v}^T , \delta\underline{\lambda} \rangle \left(\left[\begin{array}{c|c} \underline{K}(\underline{\lambda}, \underline{g}) & \underline{N} \\ \hline \underline{N}^T & \underline{0} \end{array} \right] \left\{ \begin{array}{c} \Delta \underline{v} \\ \hline \Delta\underline{\lambda} \end{array} \right\} + \left\{ \begin{array}{c} \underline{N} \lambda \\ \hline \underline{g} \end{array} \right\} \right) = 0$$

References

1. Wagner, W.: Eine geometrisch nichtlineare Theorie schubelastischer Schalen mit Anwendung auf Finite-Element- Berechnungen von Durchschlag- und Kontaktproblemen. Forschungs-und Seminarberichte aus dem Bereich der Mechanik der Universität Hannover, Nr. F 85/2, 1985.

2. Zienkiewicz, O.C.; Bauer, J.; Morgan, K.: A Simple and Efficient Element for Axisymmetric Shells. Int. J. Num. Meth. Engng. 11 (1977) 1545-1558.

3. Stein, E.; Wriggers, P.: Stability of Rods with Unilateral Constraints, a Finite Element Solution. Comp. & Struct. 19 (1984) 205-211.

4. Wriggers, P.; Klee, K.-D.: Zur Berechnung von Stäben unter nichtkonservativer Belastung mit Hilfe eines inkrementellen Arbeitsprinzipes. ZAMM 64 (1984) T322 - T324.

5. Reissner, E.: On finite Symmetrical Deflections of Thin Shells of Revolution. J. Appl. Mech. Trans. ASME Serie E 36 (1969) 267-270.

6. Noor, A. K.; Peters, J. M.; Andersen, C., M.: Mixed Models and Reduction Techniques for Large-Rotation Nonlinear Problems. Comp. Meth. Appl. Mech. Engng. 44 (1984) 67-89.

7. Wriggers, P., Simo, J.C.: A Note on Tangent Stiffnesses for Fully Nonlinear Contact Problems. to appear

8. Hallquist, J.O.: NIKE2D: An Implicit, Finite- Deformation, Finite-Element Code for Analyzing the Static and Dynamic Response of Two-dimensional Solids. Rept. UCRL- 52678, University of California, Lawrence Livermore National Laboratory, 1979.

9. Riks, E.: The Application of Newtons Method to the Problem of Elastic Stability. J. Appl. Mech. 39 (1972) 1060-1066.

10. Wempner, G.A.: Discrete Approximations Related to Nonlinear Theories of Solids. Int. J. Solids Struct. 7 (1971) 1581-1599.

11. Wessels, M.: Das statische und dynamische Durchschlagproblem der imperfekten flachen Kugelschale bei elastischer rotationssymmetrischer Verformung. Mitt. des Inst. f. Statik der Univ. Hannover, Nr. 23, 1977.

12. Luenberger, D.G.: Linear and nonlinear programming. 2nd. Edition, Addison-Wesley Pub., Reading, Massachusetts,1984.

13. Pietraszkiewicz, W.: Introduction to the Non-linear Theory of Shells. Mitt. aus dem Inst. f. Mechanik, Nr. 10, Ruhr-Universität Bochum, 1977.

14. Pietraszkiewicz, W.: Finite Rotations and Lagrangean Description in the Non-linear Theory of Shells. Polish Academy of Sciences, Institute of fluid-flow machinery, Polish Scientific Publishers, Warszawa-Poznan, 1979.

15. Berg, A.: Beiträge zur geometrisch nichtlinearen Theorie und inkrementellen Finite-Element-Berechnung dünner elastischer Schalen. Forschungs- und Seminarberichte aus dem Bereich der Mechanik der Universität Hannover, Nr. F 83/2, 1983.

16. Koiter, W.T.: On the Nonlinear Theory of thin elastic Shells. Proc. Koninl. Ned. Akad. Wetenshap Serie B69 (1966) 1-54.

17. Simmonds, J.G.; Danielson, D.A.: Nonlinear Shell Theory with Finite Rotations and Stress-Function Vectors. J. Appl. Mech. Trans. ASME Serie E 39 (1972) 1085-1090.

18. Bathe, K.J.; Ramm, E.; Wilson, E.L.: Finite Element Formulations for Large Deformation Dynamic Analysis. Int. J. Num. Meth. Engng. 9 (1975) 353-386.

19. Hughes, T.J.R.; Taylor, R.L.; Kanoknukulchai, W.: A simple and efficient finite element for plate bending. Int. J. Num. Meth. Engng. 11 (1977) 1529-1543.

20. Malkus, D.S.; Hughes, T.R.J.: Mixed finite element methods - reduced and selective integration techniques: a unification of concepts. Comp. Meth. Appl. Mech. Engng. 15 (1978) 63-81.

21. Updike, D.P.; Kalnins, A.: Contact Pressure between an Elastic Spherical Shell and a Rigid Plate. J. Appl. Mech. Trans. ASME Serie E 39 (1972) 1110-1114.

22. Oden, J.T.: Exterior Penalty Methods for Contact Problems in Elasticity . in: Nonlinear Finite Element Analysis in Structural Mechanics, ed. Wunderlich, Stein, Bathe, Springer - Verlag,Berlin,Heidelberg,New-York, 1981.

23. Simo, J.C.; Wriggers, P.; Schweizerhof, K.; Taylor, R. L.: Postbuckling Analysis involving Inelasticity and Unilateral Constraints. Report No. UCB/SESM 84/15, University of California, Berkeley, 1984.

24. Hughes, T.R.J.; Taylor, R.L.; Sackman, J.L.; Curnier, A.; Kanoknukulchai, W.: A Finite Element Method for a Class of Contact - Impact Problems. Comp. Meth. Appl. Mech. Engng. 8 (1976) 249-276.

25. Crisfield, M.A.: A fast and incremental/iterative solution that handles "snap-through". Comp. & Struct. 13 (1981) 55-62.

26. Ramm, E.: Strategies for Tracing the Nonlinear Response Near Limit Points. in: Nonlinear Finite Element Analysis in Structural Mechanics, ed. Wunderlich, Stein, Bathe, Springer Verlag Berlin-Heidelberg-New-York 1981 .

27. Fried, I.: Nonlinear Finite Element Analysis of the Thin Elastic Shell of Revolution. Comp. Meth. Appl. Mech. 48 (1985) 283-299.

28. Wriggers, P.; Nour-Omid, B.: Solution Methods for Contact Problems. Report No. UCB/SESM 84/09, University of California, Berkeley, 1984.

Augmented Lagrangian Techniques for Solving Frictionless Contact Problems in Finite Elasticity

R. GLOWINSKI*, M. VIDRASCU*, P. LE TALLEC
*INRIA, Le Chesnay Cedex, France
LCPC, Paris, France

1. Summary

In nonlinear situations, the consistency between the choice of the mathematical formulation of the problem and the choice of the numerical algorithm for its solution appears to be critical. Finding such compatible choices for different classes of nonlinear problems has been for us a major direction of research in the past few years. To illustrate this work, we will discuss the numerical solution of frictionless contact problems in Finite Elasticity, by augmented Lagrangian techniques.

In the contact-free problem, such techniques write the original equilibrium equations as a saddle-point problem to be solved by a Uzawa type algorithm (GLOWINSKI-LE TALLEC [1], [2]). In the present study, the formulation is obtained either by first treating the contact constraints by a penalty method and then using the techniques of [1] to write the penalized problem under a saddle-point form, or by treating directly all the constraints of the problem by augmented Lagrangian techniques. These formulations, once approximated by compatible mixed finite elements, can easily be solved by different new variants of the Uzawa algorithm. These variants are obtained by considering formally saddle-point problems as dual evolution problems to be solved by Alternating Directions Methods. The presence of the penalized contact terms nevertheless complicates the matricial system to be solved at each step, which may become nonlinear but stays monotone and therefore can be solved by a preconditioned conjugate gradient method.

2. Basic formulations of the problem

2.1. Initial variational formulation. The problem consists in determining the final equilibrium position $\underset{\sim}{x}+\underset{\sim}{u}(x)$ of any particle $\underset{\sim}{x}$ of an elastic body which occupies a known domain Ω in its stress free reference configuration, which is subjected to a given distribution of external loads, and whose motion is restricted by the presence of rigid bodies in its neighborhood.

Finite Element Methods for Nonlinear Problems
Europe-US Symposium, Trondheim, Norway 1985
Editors: Bergan, Bathe, Wunderlich

For handling the large deformations which are involved, we label any particle $\underset{\sim}{x}$ by its position in the reference configuration Ω (Lagrange coordinates) and we relate both $\underset{\sim}{x}$ and its displacement $\underset{\sim}{u}(\underset{\sim}{x})$ to a fixed rectangular Cartesian coordinate system. The body is subjected to body forces of intensity $\underset{\sim}{f}$ per unit volume of Ω and to surface tractions of intensity $\underset{\sim}{g}$ per unit area applied on a portion Γ_2 of the boundary of Ω. In addition, the displacement takes on prescribed values $\underset{\sim}{u}_1$ on the complementary part Γ_1 of Γ_2 in Γ, and the body is incompressible, which implies that $\underset{\sim}{u}$ preserves the volumes and therefore satisfies

$$\det(\underset{\sim}{Id} + \nabla \underset{\sim}{u}) = 1 \quad \text{in} \quad \Omega .$$

Finally, the presence of Q rigid bodies restricting the displacements is taken into account through the unilateral constraints

$$\underset{\sim}{a}_j(\underset{\sim}{x}) . \underset{\sim}{u}(\underset{\sim}{x}) - b_j(\underset{\sim}{x}) \leq 0, \ \forall \underset{\sim}{x} \in \Gamma_2^j, \ \forall \ 1 \leq j \leq Q, \qquad (2.1)$$

with $\underset{\sim}{a}_j(\underset{\sim}{x}) \in \mathbb{R}^3$, $b_j(\underset{\sim}{x}) \in \mathbb{R}$ and $\Gamma_2^j \subset \Gamma_2$ given. Implicitly, this supposes that the contact between the elastic body and the piecewise linear rigid mating parts is perfect and frictionless.

For simplicity, we will restrict ourselves to Mooney-Rivlin materials subjected to dead loading ($\underset{\sim}{f}$ and $\underset{\sim}{g}$ independent of $\underset{\sim}{u}$) although the proposed technique obviously applies to any hyperelastic material under any type of loading. In this framework, the mechanical problem described above is classically written as

<u>Minimize the potential energy</u> J(.) <u>over the set</u> K <u>of kinematically admissible displacement fields</u> (2.2)

with

$$J(\underset{\sim}{v}) = \int \{C_1(|\nabla \underset{\sim}{v} + \underset{\sim}{Id}|^2 - 3) + C_2(|\text{adj}(\nabla \underset{\sim}{v} + \underset{\sim}{Id})|^2 - 3)\} \, d\Omega$$

$$- \int_\Omega \underset{\sim}{f} . \underset{\sim}{v} \, d\Omega$$

$$- \int_{\Gamma_2} \underset{\sim}{g} . \underset{\sim}{v} \, d\Gamma, \quad (\text{adj } \underset{\sim}{H} = \text{transpose of cofactor matrix}) \qquad (2.3)$$

$$K^{-} = \{\underset{\sim}{v}, \underset{\sim}{v} = \underset{\sim}{u}_1 \text{ on } \Gamma_1, \det(\underset{\sim}{Id}+\nabla \underset{\sim}{v}) = 1 \text{ in } \Omega, \quad \underset{\sim}{a}_j(\underset{\sim}{x}).\underset{\sim}{v}(\underset{\sim}{x}) - b_j(\underset{\sim}{x}) \leq 0, \forall \underset{\sim}{x} \in \Gamma_2^j, \forall\, 1 \leq j \leq Q\} \tag{2.4}$$

2.2. Penalty formulation. Treating the contact constraints (2.1) by a penalty method transforms (2.2) into the new problem

<u>Minimize the penalized energy</u> $J_\varepsilon(.)$ <u>over the set</u> K <u>of extended kinematically admissible displacement fields</u>, (2.5)

with ε a given small number and

$$J_\varepsilon(\underset{\sim}{v}) = J(\underset{\sim}{v}) + \frac{1}{3\varepsilon} \sum_{j=1}^{Q} \int_{\Gamma_2^j} [\mathrm{Sup}(0, \underset{\sim}{a}_j.\underset{\sim}{v}-b_j)]^3 \, d\Gamma \,, \tag{2.6}$$

$$K = \{\underset{\sim}{v}, \underset{\sim}{v} = \underset{\sim}{u}_1 \text{ on } \Gamma_1, \det(\nabla\underset{\sim}{v}+\underset{\sim}{Id}) = 1 \text{ in } \Omega\} \tag{2.7}$$

Here, we use a cubic penalty term for the unilateral constraints in order to get a smoother transition when a contact first appears. The addition of these cubic terms would correspond at the physical level to the replacement of the rigid mating parts by very stiff nonlinear springs.

2.3. Augmented Lagrangian formulation of the penalized problem
As in [1], we decompose the problem (2.5) by introducing the new variable $\underset{\sim}{F}$ on which we impose the constraints

$$\underset{\sim}{F} = \underset{\sim}{Id} + \nabla\underset{\sim}{u}, \quad \det \underset{\sim}{F} = 1 \,.$$

Then, (2.5) splits into two parts : one related to the generic variable $\underset{\sim}{v}$, involving the penalty terms, the external forces and the term $C_1(|\underset{\sim}{Id}+\nabla\underset{\sim}{v}|^2-3)$, the other one related to a new generic variable $\underset{\sim}{H}$ and involving the incompressibility constraint. The variables $\underset{\sim}{v}$ and $\underset{\sim}{H}$ satisfy the constraint

$$\underset{\sim}{H} = \underset{\sim}{Id} + \nabla\underset{\sim}{v},$$

to be treated by an augmented Lagrangian technique. More precisely, by adding to the penalized energy J_ε both a penalty and a duality term associated to the constraint $\underset{\sim}{H} = \underset{\sim}{Id} + \nabla\underset{\sim}{v}$, we introduce the augmented Lagrangian

$$\mathcal{L}_R(\underset{\sim}{v},\underset{\sim}{H},\underset{\sim}{\mu}) = \int_\Omega \{C_1(|\underset{\sim}{Id}+\nabla\underset{\sim}{v}|^2-3)+C_2(|\mathrm{adj}\ \underset{\sim}{H}|^2-3)\}\,d\Omega\ -$$
$$-\int_\Omega \underset{\sim}{f}.\underset{\sim}{v}\,d\Omega - \int_{\Gamma_2} \underset{\sim}{g}.\underset{\sim}{v}\,d\Gamma + \frac{1}{3\varepsilon}\sum_{j=1}^{Q}$$
$$\int_{\Gamma_2^j} (\mathrm{Sup}(0,\underset{\sim}{a}_j.\underset{\sim}{v}-b_j))^3 d\Gamma$$
$$+\int_\Omega \{\frac{R}{2}|\underset{\sim}{Id}+\nabla\underset{\sim}{v}-\underset{\sim}{H}|^2-\underset{\sim}{\mu}.(\underset{\sim}{Id}+\nabla\underset{\sim}{v}-\underset{\sim}{H})\}p(\underset{\sim}{x})\,d\Omega\ , \qquad (2.8)$$

where R is an arbitrary positive constant and $p(\underset{\sim}{x})$ an arbitrary positive function, which, for example, can be taken equal to the first elastic constant $C_1(\underset{\sim}{x})$. To (2.8), we then associate the Lagrangian problem

<u>Find</u> $\{\underset{\sim}{u},\underset{\sim}{F},\underset{\sim}{\lambda}\}\in\{X+\underset{\sim}{u}_1\}\times Y\times Z$ <u>such that</u>

$$\frac{\partial\mathcal{L}_R}{\partial\underset{\sim}{v}}(\underset{\sim}{u},\underset{\sim}{F},\underset{\sim}{\lambda}).\underset{\sim}{v} = 0\ ,\quad \forall\underset{\sim}{v}\in X\ ,$$
$$\mathcal{L}_R(\underset{\sim}{u},\underset{\sim}{F},\underset{\sim}{\lambda}) \le \mathcal{L}_R(\underset{\sim}{u},\underset{\sim}{H},\underset{\sim}{\lambda}),\quad \forall\underset{\sim}{H}\in Y, \qquad (2.9)$$
$$\frac{\partial\mathcal{L}_R}{\partial\underset{\sim}{\mu}}(\underset{\sim}{u},\underset{\sim}{F},\underset{\sim}{\lambda}).\underset{\sim}{\mu} = 0,\quad \forall\underset{\sim}{\mu}\in Z,\ ,$$

with

$$X = \{\underset{\sim}{v},\underset{\sim}{v} = 0 \text{ on } \Gamma_1\}\ , \qquad (2.10)$$
$$Z = [L^2(\Omega)]^{N\times N}\ , \qquad (2.11)$$
$$Y = \{\underset{\sim}{H}\in Z,\ \det\underset{\sim}{H} = 1\}\ , \qquad (2.12)$$
$$\frac{\partial\mathcal{L}_R}{\partial\underset{\sim}{v}}(\underset{\sim}{u},\underset{\sim}{F},\underset{\sim}{\lambda}).\underset{\sim}{v} = \int_\Omega \{C_1(\underset{\sim}{Id}+\nabla\underset{\sim}{u})+Rp(\underset{\sim}{Id}+\nabla\underset{\sim}{u}-\underset{\sim}{F})-p\underset{\sim}{\lambda}\}.\nabla\underset{\sim}{v}\,d\Omega$$
$$-\int_\Omega \underset{\sim}{f}.\underset{\sim}{v}\,d\Omega - \int_{\Gamma_2}\underset{\sim}{g}.\underset{\sim}{v}\,d\Gamma + \frac{1}{\varepsilon}\sum_{j=1}^{Q}\int_{\Gamma_2^j}[\mathrm{Sup}(0,\underset{\sim}{a}_j.\underset{\sim}{u}-b_j)]^2 \qquad (2.13)$$
$$\underset{\sim}{a}_j.\underset{\sim}{v}\,d\Gamma\ ,$$
$$\frac{\partial\mathcal{L}_R}{\partial\underset{\sim}{\mu}}(\underset{\sim}{u},\underset{\sim}{F},\underset{\sim}{\lambda}).\underset{\sim}{\mu} = -\int_\Omega p\,\underset{\sim}{\mu}.(\underset{\sim}{Id}+\nabla\underset{\sim}{u}-\underset{\sim}{F})\,d\Omega\ . \qquad (2.14)$$

Exactly as proved in [1] for the contact free problem, for R sufficiently large, <u>the Lagrangian problem</u> (2.9) <u>and the Euler-Lagrange equations associated to the penalized minimization</u>

problem (2.5) are equivalent, the solutions of (2.9) being independent of the choice of R or of $p(\underset{\sim}{x})$. The fundamental interest of (2.9) compared to (2.5) is that it admits a compatible simple numerical algorithm for its solution, which treats independently each leg of (2.9).

2.4. Formal interpretation of (2.9). Let us introduce

$$F(\underset{\sim}{H}) = \mathcal{L}_o(0,\underset{\sim}{H},0) \quad \text{if} \quad \underset{\sim}{H} \in Y ,$$
$$= +\infty \quad \text{if not}$$

$$G(\underset{\sim}{v}) = \mathcal{L}_o(\underset{\sim}{v},0,0)$$

$$<\underset{\sim}{H},\underset{\sim}{\lambda}> = \int_\Omega p\, \underset{\sim}{H}.\underset{\sim}{\lambda}\, d\Omega \quad , \forall \underset{\sim}{H}, \underset{\sim}{G} \in Z ,$$

$$B : X \to Z , \underset{\sim}{v} \to B\underset{\sim}{v} = \nabla \underset{\sim}{v} .$$

Under these notations, the augmented Lagrangian defined in (2.8) can be written

$$\mathcal{L}_R(\underset{\sim}{v},\underset{\sim}{H},\underset{\sim}{\mu}) = F(\underset{\sim}{H})+G(\underset{\sim}{v})+ \frac{R}{2}<B\underset{\sim}{v}-\underset{\sim}{H},B\underset{\sim}{v}-\underset{\sim}{H}>-<\underset{\sim}{\mu},B\underset{\sim}{v}-\underset{\sim}{H}> .$$

Let us now pretent that F and G are convex, differentiable with invertible gradients F' and G', respectively on X and Z. Then the solution of (2.9) would satisfy

$$F'(B\underset{\sim}{v})+ p\underset{\sim}{\lambda} = G'(\underset{\sim}{v})- B^T p\underset{\sim}{\lambda} = 0,$$

that is $B\, G'^{-1}(B^T p\lambda) - F'^{-1}(-p\underset{\sim}{\lambda}) = B\underset{\sim}{v} - B\underset{\sim}{v} = 0$.

Formally, the solution of (2.9) amounts therefore to the computation of the steady state solution of the evolution equation

$$\frac{d\lambda}{dt} + BG'^{-1}(B^T p\lambda) - F'^{-1}(-p\underset{\sim}{\lambda}) = 0 . \tag{2.15}$$

2.5. Augmented Lagrangian formulation of the original problem. The penalty formulation (2.5) is only an approximation of the original problem (2.2) ; to be accurate this approximation requires ε to be small. On the other hand, numerical algorithms for the solution of (2.5) will converge less rapidly if ε is small. One way to solve this dilemna is to treat the contact constraints (2.1) themselves by augmented Lagrangian techniques.

For that purpose, adding to the original energy J(.) both a penalty and a duality term for the constraint $\underset{\sim}{H} = \underset{\sim\sim}{Id} + \nabla\underset{\sim}{u}$ and for the constraints $\underset{\sim}{a}_j.\underset{\sim}{v}-b_j = h_j$, we introduce the augmented Lagrangian

$$\mathcal{L}_R^C(\underset{\sim}{v},(\underset{\sim}{H},h_j),(\underset{\sim}{\mu},\mu_j)) = \int_\Omega \{C_1(|\nabla\underset{\sim}{v}+\underset{\sim\sim}{Id}|^2-3)+C_2(|\mathrm{adj}\underset{\sim}{H}|^2-3)$$

$$- \underset{\sim}{f}\cdot\underset{\sim}{v}\}d\Omega - \int_{\Gamma_2} \underset{\sim}{g}.\underset{\sim}{v}d\Gamma + \int_\Omega \{\frac{R}{2}|\underset{\sim\sim}{Id}+\nabla\underset{\sim}{v}-\underset{\sim}{H}|^2 - (\underset{\sim\sim}{Id}+\nabla\underset{\sim}{v}-\underset{\sim}{H})\}p\, d\Omega$$

$$+ \sum_{j=1}^{Q} \int_{\Gamma_2^j} \{R_j(\underset{\sim}{a}_j.\underset{\sim}{v}-b_j-h_j)^2 - \mu_j(\underset{\sim}{a}_j.\underset{\sim}{v}-b_j-h_j)\}\, d\Gamma\ , \tag{2.16}$$

with R, R_j, and p arbitrary. To (2.16), we finally associate the problem

<u>Find</u> $\{\underset{\sim}{u},(\underset{\sim}{F},f_j),(\underset{\sim}{\lambda},\lambda_j)\} \in X \times Y_C \times Z_C$ <u>such that</u>

$$\frac{\partial\mathcal{L}_R^C}{\partial\underset{\sim}{v}}(\underset{\sim}{u},(\underset{\sim}{F},f_j),(\underset{\sim}{\lambda},\lambda_j)),\underset{\sim}{v} = 0,\ \forall\underset{\sim}{v} \in X,$$

$$\mathcal{L}_R^C(\underset{\sim}{u},(\underset{\sim}{F},\underset{\sim}{f}_j),(\underset{\sim}{\lambda},\lambda_j)) \le \mathcal{L}_R^C(\underset{\sim}{u},(\underset{\sim}{H},\underset{\sim}{h}_j),(\underset{\sim}{\lambda},\lambda_j)),$$

$$\forall\{\underset{\sim}{H},h_j\} \in Y_C \tag{2.17}$$

$$\frac{\partial\mathcal{L}_R^C}{\partial\underset{\sim}{\mu}}(\underset{\sim}{u},(\underset{\sim}{F},\underset{\sim}{f}_j),(\underset{\sim}{\lambda},\lambda_j)).\underset{\sim}{\mu} + \sum_{j=1}^{Q} \frac{\partial\mathcal{L}_R}{\partial\mu_j}(\underset{\sim}{u},(\underset{\sim}{F},f_j),(\underset{\sim}{\lambda},\lambda_j)).\mu_j = 0,$$

$$\forall\{\underset{\sim}{\mu},\mu_j\} \in Z_C$$

with $Z_C = Z \times \prod_j L^2(\Gamma_2^j)$ and

$$Y_C = \{\{\underset{\sim}{H},h_j\} \in Z,\ \det\underset{\sim}{H} = 1,\ h_j \le 0\}\ .$$

Problem (2.17) is equivalent to the original problem (2.2). Moreover, the multipliers λ_j represent the contact forces at equilibrium, which means that friction can be introduced in (2.17) by acting properly on the local variables h_j (gap between the body and the rigid support) and λ_j (contact forces).

Problem (2.17) has the same structure as (2.9) ; in particular, a dual evolution problem (2.15) can also be associated to (2.17) if we replace $\mathcal{L}_R$ by $\mathcal{L}_R^C$, Y by Y_C and Z by Z_C. In the following, to simplify the presentation, we will only discuss the numerical solution of (2.9), but obviously the algorithms to be proposed will apply identically to (2.17).

3. Numerical algorithms

3.1. The basic algorithm. The basic algorithm which is naturally associated to the augmented Lagrangian formulation (2.9) is an Uzawa algorithm defined as follows

$$\text{Let } \{\underset{\sim}{\lambda}^o, \underset{\sim}{F}^{-1}\} \text{ be given in } Z \times Y \quad ; \tag{3.1}$$

then, for $n \geq 0$, $\underset{\sim}{F}^{n-1}$ and $\underset{\sim}{\lambda}^n$ being known, we compute $\{\underset{\sim}{u}^n, \underset{\sim}{F}^n\}$ in $X \times Y$ by relaxation, i.e. by setting

$$\underset{\sim}{F}_o^n = \underset{\sim}{F}^{n-1} \tag{3.2}$$

and by computing $\underset{\sim}{u}_k^n$ and $\underset{\sim}{F}_k^n$ for $1 \leq k \leq m$ by solving sequentially

$$\frac{\partial \mathcal{L}_R}{\partial \underset{\sim}{v}}(\underset{\sim}{u}_k^n, \underset{\sim}{F}_{k-1}^n, \underset{\sim}{\lambda}^n) \cdot \underset{\sim}{v} = 0, \ \forall \underset{\sim}{v} \in X, \underset{\sim}{u}_k^n \in X + \underset{\sim}{u}_1, \tag{3.3}$$

$$\mathcal{L}_R(\underset{\sim}{u}_k^n, \underset{\sim}{F}_k^n, \underset{\sim}{\lambda}^n) \leq \mathcal{L}_R(\underset{\sim}{u}_k^n, \underset{\sim}{H}, \underset{\sim}{\lambda}^n), \ \forall \underset{\sim}{H} \in Y, \ \underset{\sim}{F}_k^n \in Y \ ; \tag{3.4}$$

once $\{\underset{\sim}{u}^n, \underset{\sim}{F}^n\} = \{\underset{\sim}{u}_m^n, \underset{\sim}{F}_m^n\}$ is known, $\underset{\sim}{\lambda}^n$ is updated by

$$\underset{\sim}{\lambda}^{n+1} = \underset{\sim}{\lambda}^n - R(\nabla \underset{\sim}{u}^n - \underset{\sim}{F}^n) \ . \tag{3.5}$$

The practical implementation of this algorithm on the computer requires

(i) the replacement of the sets X, Y and Z by finite element approximations,

(ii) more details on the solution of steps (3.3) and (3.4).

Compatible finite element approximations of the sets X, Y and Z are proposed in [1], [2] or [3], and the convergence of the finite element solution of (2.9) towards the continuous solution of (2.2) is studied in LE TALLEC [4]. The numerical treatment of step (3.4) is completely described in [1] (plane and axisymmetric problems) or in [2] (three-dimensional case); in fact (3.4) reduces to the solution in parallel of one-dimensional algebraic equations to be solved by a Newton's method. As for step (3.3), its numerical treatment will be discussed in Section 4, but before we introduce a new variant of Algorithm (3.1)-(3.5).

3.2. Review of several Alternate Directions Schemes. Let us consider the evolution problem

$$\frac{du}{dt} + A_1(u) + A_2(u) = 0, \ u(0) = u_0 , \tag{3.6}$$

with A_1 and A_2 two operators from the Hilbert space Z into itself. Denoting by DT a given time step and by u^n an approximation of u(nDT), Problem (3.6) can be solved numerically by one of the following Alternate Directions Schemes

Douglas-Rachford (unconditionnally stable)([5])

Set $u^o = u_o$, then, for $n \geq 0$ and u^n known, determine u^{n+1} by solving sequentially

$$\frac{\hat{u}^{n+1} - u^n}{DT} + A_1(\hat{u}^{n+1}) + A_2(u^n) = 0,$$

$$\frac{u^{n+1} - u^n}{DT} + A_1(\hat{u}^{n+1}) + A_2(u^{n+1}) = 0.$$

θ-scheme-(stiff A stable for certain values of θ)([6]) .

Set $u^o = u_o$, then, for $n \geq 0$ and u^n known, determine u^{n+1} by solving sequentially

$$(u^{n+\theta} - u^n)/\theta DT + A_1(u^{n+\theta}) + A_2(u^n) = 0 ,$$

$$(u^{n+1-\theta} - u^{n+\theta})/(1-2\theta)DT + A_1(u^{n+\theta}) + A_2(u^{n+1-\theta}) = 0,$$

$$(u^{n+1} - u^{n+1-\theta})/\theta DT + A_1(u^{n+1}) + A_2(u^{n+1-\theta}) = 0.$$

3.3. Application to the numerical solution of (2.9). As seen in (2.15), formally the solution of (2.9) amounts to the calculation of the steady state solution of (2.15). To compute this steady state, one may think of solving the whole evolution problem (2.15) by one of the Alternate Direction Schemes that have been just proposed.

First, if we use the Douglas-Rachford scheme on (2.15) with $A_1(\lambda) = B(G'^{-1}(B^T p\lambda))$ and $A_2(\lambda) = -F'^{-1}(-p\lambda)$, we recover exactly the basic algorithm (3.1)-(3.5) with R = DT and m = 1. Indeed, setting $F^n = F'^{-1}(-p\lambda^n)$ and $u^n = G'^{-1}(B^T p\,\hat{\lambda}^n)$, the Douglas-Rachford scheme becomes

$$\left(\frac{1}{p} B^{-T} G'(u^{n+1}) - \lambda^n\right)/DT - F^n + Bu^{n+1} = 0,$$

$$\left(-\frac{1}{p} F'(F^{n+1}) - \lambda^n\right)/DT - F^{n+1} + B\,u^{n+1} = 0,$$

$$\lambda^{n+1} = -\frac{1}{p} F'(F^{n+1}) = \lambda^n - DT(Bu^{n+1} - F^{n+1}),$$

in which we immediately recognize (3.3)-(3.5).

Now, if we apply the θ-scheme to (2.15) with $A_1(\lambda) = -F'^{-1}(-p\lambda)$, $A_2(\lambda) = B(G'^{-1}(B^T p\lambda))$, and DT = R, we obtain formally the following algorithm :

$$u^n = G'^{-1}(B^T p\,\lambda^n),$$

$$\left(-\frac{1}{p} F'(F^{n+\theta}) - \lambda^n\right)/\theta R - F^{n+\theta} + Bu^n = 0,$$

$$\lambda^{n+\theta} = -\frac{1}{p} F'(F^{n+\theta}) = \lambda^n - \theta R(Bu^n - F^{n+\theta}),$$

$$\left(\frac{1}{p} B^{-T} G'(u^{n+1-\theta}) - \lambda^{n+\theta}\right)/(1-2\theta)R - F^{n+\theta} + Bu^{n+1-\theta} = 0,$$

$$\lambda^{n+1-\theta} = \frac{1}{p} B^{-T} G'(u^{n+1-\theta}) = \lambda^n - (1-2\theta)R(Bu^{n+1-\theta} - F^{n+\theta}),$$

$$\left(-\frac{1}{p} F'(F^{n+1}) - \lambda^{n+1-\theta}\right)/\theta DT - F^{n+1} + Bu^{n+1-\theta} = 0,$$

$$\lambda^{n+1} = \lambda^{n+1-\theta} - \theta R(Bu^{n+1-\theta} - F^{n+1}).$$

Written in terms of the augmented Lagrangian (2.8), this algorithm takes on a practical form, called ALG4, which can be used for the numerical solution of (2.9) although it was derived formally in a very restrictive framework. This new algorithm ALG4, which is as simple to implement as is (3.1)-(3.5), and which sometimes converges much faster, is

Let $\underset{\sim}{\lambda}^{o} \in Z$ be given. Then, for $n \geq 0$, and $\underset{\sim}{\lambda}^{n}$ known, solve sequentially

$$\frac{\partial \mathcal{L}_o}{\partial \underset{\sim}{v}}(\underset{\sim}{u}^n, 0, \underset{\sim}{\lambda}^n).\underset{\sim}{v} = 0, \ \forall \underset{\sim}{v} \in X, \ \underset{\sim}{u}^n \in X + \underset{\sim}{u}_1 \ ; \tag{3.7}$$

$$\mathcal{L}_{\theta R}(\underset{\sim}{u}^n, \underset{\sim}{F}^{n+\theta}, \underset{\sim}{\lambda}^n) \leq \mathcal{L}_{\theta R}(\underset{\sim}{u}^n, \underset{\sim}{H}, \underset{\sim}{\lambda}^n), \ \forall \underset{\sim}{H} \in Y, \ \underset{\sim}{F}^{n+\theta} \in Y \ ; \tag{3.8}$$

$$\underset{\sim}{\lambda}^{n+\theta} = \underset{\sim}{\lambda}^n - \theta R(\nabla \underset{\sim}{u}^n - \underset{\sim}{F}^{n+\theta}), \tag{3.9}$$

$$\frac{\partial \mathcal{L}}{\partial \underset{\sim}{v}}(1-2\theta)R(\underset{\sim}{u}^{n+1-\theta}, \underset{\sim}{F}^{n+\theta}, \underset{\sim}{\lambda}^{n+\theta}).\underset{\sim}{v} = 0 \ , \quad \forall \underset{\sim}{v} \in X, \ \underset{\sim}{u}^{n+1-\theta} \in X + \underset{\sim}{u}_1 \ ; \tag{3.10}$$

$$\underset{\sim}{\lambda}^{n+1-\theta} = \underset{\sim}{\lambda}^{n+\theta} - (1-2\theta)R(\nabla \underset{\sim}{u}^{n+1-\theta} - \underset{\sim}{F}^{n+\theta}),$$

$$\mathcal{L}_{\theta R}(\underset{\sim}{u}^{n+1-\theta}, \underset{\sim}{F}^{n+1}, \underset{\sim}{\lambda}^{n+1-\theta}) \leq \mathcal{L}_{\theta R}(\underset{\sim}{u}^{n+1-\theta}, \underset{\sim}{H}, \underset{\sim}{\lambda}^{n+1-\theta}) \ , \quad \forall \underset{\sim}{H} \in Y, \ \underset{\sim}{F}^{n+1} \in Y \ ; \tag{3.12}$$

$$\underset{\sim}{\lambda}^{n+1} = \underset{\sim}{\lambda}^{n+1-\theta} - \theta R(\nabla \underset{\sim}{u}^{n+1-\theta} - \underset{\sim}{F}^{n+1}) \tag{3.13}$$

4. The problem in displacements

Once X,Y and Z are replaced by finite element approximations X_h, Y_h and Z_h, the displacement problem (3.3) (or (3.7) or (3.10)) becomes

$$\frac{\partial \mathcal{L}_R^h}{\partial \underset{\sim}{v}}(\underset{\sim}{u}_k^n, \underset{\sim}{F}_{k-1}^n, \underset{\sim}{\lambda}^n).\underset{\sim}{v} = 0, \ \forall \underset{\sim}{v} \in X_h, \underset{\sim}{u}_k^n \in X_h + \underset{\sim}{u}_1 \ , \tag{4.1}$$

with $\mathcal{L}_R^h$ given by

$$\begin{aligned} \mathcal{L}_R^h(\underset{\sim}{v}, \underset{\sim}{H}, \underset{\sim}{\mu}) = & \int_\Omega \{C_1(|\underset{\sim}{Id} + \nabla \underset{\sim}{v}|^2 - 3) + C_2(|\mathrm{adj}\ \underset{\sim}{H}|^2 - 3) - \underset{\sim}{f}.\underset{\sim}{v}\}\, d\Omega \\ & - \int_{\Gamma_2} \underset{\sim}{g}.\underset{\sim}{v}\, d\Gamma + \frac{1}{3\varepsilon} \sum_{j=1}^{Q} \int_{\Gamma_2^j} \left(\mathrm{Sup}\left(0, \underset{\sim}{a}_j.\underset{\sim}{v} - b_j\right)\right)^3 d\Gamma \\ & + \int_\Omega \{\tfrac{R}{2}|\underset{\sim}{Id} + \nabla_h \underset{\sim}{v} - \underset{\sim}{H}|^2 - \underset{\sim}{\mu}.(\underset{\sim}{Id} + \nabla_h \underset{\sim}{v} - \underset{\sim}{H})\} p(\underset{\sim}{x})\, d\Omega \ , \end{aligned} \tag{4.2}$$

and $\nabla_h \underset{\sim}{v}$ the L^2 projection of $\nabla \underset{\sim}{v}$ on Z_h. As a function of $\underset{\sim}{v}$, $\mathcal{L}_R^h$ is the sum of a positive definite quadratic part and of a convex function. Therefore, (4.1), which is highly nonlinear, is nevertheless associated to a symmetric strongly monoton operator and a Polak-Ribière conjugate gradient method with preconditioning appears to be perfectly well adapted for its numerical solution. Introducing a preconditioning operator A from X_h onto itself, this method, when applied to (4.1), defines the following algorithm (GLOWINSKI [7, p.205]).

Initialization

* set u^o = previously computed solution in displacements ($= u^{\tilde{n}}_{k-1}$) ;
* solve $\langle Ag^o, v\rangle = \frac{\partial \mathcal{L}^h_R}{\partial v}(u^o,.,.)$, $\forall v \in X_h$, $g^o \in X_h$;
* set $z^o = g^o$;

Loop for i = 0 until satisfied

* solve $\mathcal{L}^h_R(u^i - \rho_i\, z^i,.,.) \le \mathcal{L}^h_R(u^i - \rho\, z^i,.,.), \forall \rho > 0, \rho_i > 0$;
* set $u^{i+1} = u^i - \rho_i\, z^i$;
* solve $\langle Ag^{i+1}, v\rangle =_h \frac{\partial \mathcal{L}^h_R}{\partial v}(u^{i+1},.,.).v$, $\forall v \in X_h, g^{i+1} \in X_h$;
* compute $\beta_i = (\frac{\partial \mathcal{L}^h_R}{\partial v}(u^{i+1},.,.).(g^{i+1}-g^i))/(\frac{\partial \mathcal{L}^h_R}{\partial v}(u^i,.,.).g^i)$;
* set $z^{i+1} = g^i + \beta_i\, z^i$.

The above conjugate gradient algorithm is performed at each resolution of (3.3), that is once per iteration of the Uzawa algorithm (3.1)-(3.5) (or twice per iteration of ALG4). If A is correctly chosen and if its associated matrix has been computed and factorized once for all, this algorithm is not very costly because a reasonable estimate of the solution is known (mainly the solution u^n_{k-1} at the previous call) and because no matrix construction or factorization is involved. This is particularly important in three-dimensional computations where, in addition, one can modify the preconditioning operator A in order to minimize the occupied memory.

The choice of the preconditioning operator is obviously very important. Our present strategy consists in choosing a block diagonal incomplete Cholesky factorization of the operator

$$\langle A_o u, v\rangle = \int_\Omega \{2\, C_1 \nabla u.\nabla v + Rp\, \nabla_h u.\nabla_h v\}\, d\Omega$$

which corresponds to the quadratic part of the function $\mathcal{L}_R^h(v,.,.)$. This strategy is simple general and robust but other choices are possible.

In the case of the augmented Lagrangian formulation (2.17), the problem in displacements gets simpler and better conditioned, and thus the above algorithm always converges fast and is far less sensitive to the choice of a preconditioning operator.

5. Numerical results

We present below the three-dimensional numerical study of a beam, with a T section, made of a Mooney-Rivlin material with $C_1 = .5$ and $C_2 = .05$, whose geometry is indicated in Fig. 1 (length=30, heighth=1., width=8), which is fixed on both ends, and whose upper surface is subjected to a uniform dead loading of density .05 per unit area. Two cases are considered depending whether there is a horizontal plane limiting the displacement (Fig. 3, case with contact) or not (Fig. 2). This horizontal plane, if it exists, is located at 3 unit length below the original position of the beam.

For symmetry reasons, only one fourth of the beam is considered, and is approximated by 198 8 nodes isoparametric cubic finite elements. The numerical results were obtained after 20 iterations of algorithm (3.1)-(3.5), acting on the Lagrangian formulation (2.9) and treating the contact constraints by penalization with $\varepsilon = 10^{-5}$.

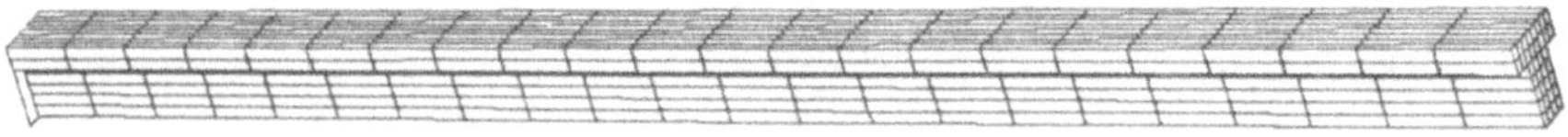

Fig.1 : Original configuration of the beam

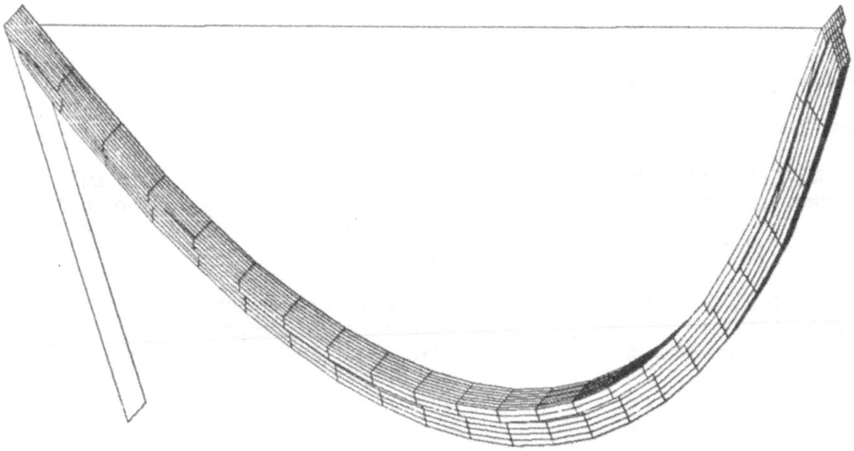

Fig.2 : Deformed configuration of the beam:case without contact.

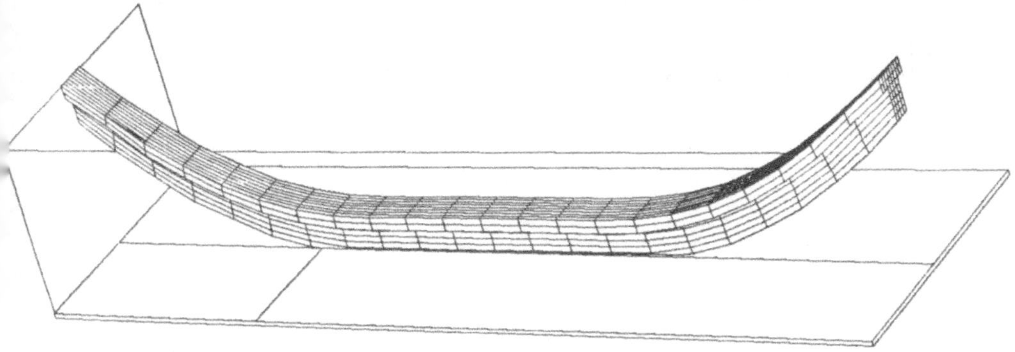

Fig.3 : Deformed configuration of the beam:case with contact.

REFERENCES

1 GLOWINSKI-LE TALLEC : Numerical solution of equilibrium problems in incompressible finite elasticity by augmented Lagrangian methods I. Two dimensional and axisymmetric problems, SIAM J. of Appl. Math. 42 (1982) pp. 400-429.

2 GLOWINSKI-LE TALLEC : Numerical solution of equilibrium problems in incompressible finite elasticity by augmented Lagrangian methods II. Three-dimensional problems, SIAM J. of Appl. Math., 44 (1984) 710-733.

3 LE TALLEC-VIDRASCU : Une méthode numérique pour les problèmes d'équilibre de corps hyperélastiques compressibles en grandes déformations. Numer. Math. 43 (1984) 199-224.

4 LE TALLEC : Existence and Approximation results for nonlinear mixed problems : Application to incompressible finite elasticity, Numer. Math. 38 (1982) 365-382.

5 DOUGLAS-RACHFORD : On the numerical solution of heat conduction problems in two or three space variables. Trans. Amer. Math. Soc. 82 (1956) 421-439.

6 BEALE-MAJDA : Rates of convergence for viscous splitting of the Navier-Stokes equations, Math. Comp. 37 (1981) 243-260.

7 GLOWINSKI : Numerical methods for nonlinear variational problems, New-York, Springer-Verlag (1984).

Part VIII:
Fluids

Entropy-Stable Finite Element Methods for Compressible Fluids: Application to High Mach Number Flows with Shocks

T. J. R. HUGHES, M. MALLET, L. P. FRANCA
Stanford University, USA

Summary
A brief overview of a new finite element method for the compressible Euler and Navier-Stokes equations is presented. The equations are written in terms of entropy variables. When combined with an appropriate Petrov-Galerkin formulation, the discrete solution automatically satisfies the entropy production inequality. A discontinuity capturing term is included in the Petrov-Galerkin weighting function enabling non-oscillatory yet crisp shock profiles to be obtained. Numerical examples at high Mach numbers illustrate the good behavior of the method.

The Compressible Euler and Navier-Stokes Equations

In computational fluid dynamics, the traditional starting point for discretization of the compressible Euler and Navier-Stokes equations has been the form of the equations written in terms of so-called *conservation variables*:

$$\boldsymbol{U}_{,t} + \boldsymbol{F}_{i,i} = \boldsymbol{F}^{v}_{i,i} + \boldsymbol{F}^{h}_{i,i} + \boldsymbol{\mathcal{F}} \tag{1}$$

where, in three dimensions,

$$\boldsymbol{U} = \begin{pmatrix} U_1 \\ U_2 \\ U_3 \\ U_4 \\ U_5 \end{pmatrix} = \rho \begin{pmatrix} 1 \\ u_1 \\ u_2 \\ u_3 \\ e \end{pmatrix} \quad \text{(conservative variables)} \tag{2}$$

$$\boldsymbol{F}_i = u_i \boldsymbol{U} + p \begin{pmatrix} 0 \\ \delta_{1i} \\ \delta_{2i} \\ \delta_{3i} \\ u_i \end{pmatrix} \quad \text{(Euler flux)} \tag{3}$$

$$\boldsymbol{F}^{v}_i = \begin{pmatrix} 0 \\ \tau_{1i} \\ \tau_{2i} \\ \tau_{3i} \\ \tau_{ij} u_j \end{pmatrix} \quad \text{(viscous flux)} \tag{4}$$

Finite Element Methods for Nonlinear Problems
Europe-US Symposium, Trondheim, Norway 1985
Editors: Bergan, Bathe, Wunderlich

$$\boldsymbol{F}_i^h = \begin{pmatrix} 0 \\ 0 \\ 0 \\ 0 \\ 0 \\ -q_i \end{pmatrix} \quad \text{(heat flux)} \tag{5}$$

$$\boldsymbol{\mathcal{F}} = \rho \begin{pmatrix} 0 \\ b_1 \\ b_2 \\ b_3 \\ b_i u_i + r \end{pmatrix} \quad \text{(source vector)} \tag{6}$$

The notation is as follows: ρ is the density; u_i is the velocity; e is the total energy density; p is the thermodynamic pressure; δ_{ij} is the Kronecker delta (i.e., $\delta_{ij} = 1$ if $i = j$, otherwise $\delta_{ij} = 0$); τ_{ij} is the viscous stress; q_i is the heat flux vector; b_i is the body force per unit mass; r is the heat supply per unit mass; an inferior comma denotes partial differentiation (e.g. $\boldsymbol{U}_{,t} = \partial \boldsymbol{U}/\partial t$, the partial derivative with respect to time); and repeated indices indicate summation (e.g. $\boldsymbol{F}_{i,i} = \partial \boldsymbol{F}_i/\partial x_i = \partial \boldsymbol{F}_1/\partial x_1 + \partial \boldsymbol{F}_2/\partial x_2 + \partial \boldsymbol{F}_3/\partial x_3$, in which x_i denotes the i-th Cartesian coordinate). The total energy density may be expressed as

$$e = \iota + \frac{u^2}{2} \tag{7}$$

where ι is the internal energy density and

$$\frac{u^2}{2} = \frac{u_1^2 + u_2^2 + u_3^2}{2} \tag{8}$$

is the kinetic energy density. To (1) we append the following constitutive equations:

$$\iota = c_v \theta \tag{9}$$

$$p = (\gamma - 1)\rho\iota \tag{10}$$

$$\tau_{ij} = \lambda u_{k,k}\delta_{ij} + \mu(u_{i,j} + u_{j,i}) \tag{11}$$

$$q_i = -\kappa\theta_{,i} \tag{12}$$

where c_v is the specific heat at constant volume; θ is the absolute temperature; $\gamma = c_p/c_v$, in which c_p is the specific heat at constant pressure; λ and μ are the viscosity coefficients; and κ is the conductivity. The specific heats are assumed to be positive constants. Equation (1) represents conservation of mass, balance of momenta and balance of energy.

Consider a semidiscrete Galerkin weighed residual formulation and, for simplicity, ignore boundary condition terms. The variational statement is: find a function $\boldsymbol{U}$ such that for all weighting functions $\boldsymbol{W}$,

$$0 = \int_{\Omega} \boldsymbol{W}^T (\boldsymbol{U}_{,t} + \boldsymbol{F}_{i,i} - \boldsymbol{F}^v_{i,i} - \boldsymbol{F}^h_{i,i} - \boldsymbol{\mathcal{F}}) d\Omega \tag{13}$$

We assume $\boldsymbol{W}$ and $\boldsymbol{U}$ are expanded in terms of typical C^0 finite element interpolation functions. This makes (13) well-defined in the sense of distributions. Classical stability estimates are developed by replacing $\boldsymbol{W}$ in (13) with the solution $\boldsymbol{U}$. In the present circumstances this does *not* even make dimensional sense, viz.

$$0 = \int_{\Omega} \boldsymbol{U}^T (\boldsymbol{U}_{,t} + ...) d\Omega$$

$$= \frac{1}{2} \frac{d}{dt} \int_{\Omega} \rho^2 \underbrace{(1 + u^2 + e^2)}_{???} d\Omega + ...$$

This suggests the inappropriateness of (13) as a basis for numerical approximation and has led us to seek an alternative formulation.

Entropy Variables

This section is based upon Hughes, Franca and Mallet [4] which extends earlier work of Harten [3] and Tadmor [7]. The recent work of Dutt [2] also considers stable boundary conditions. The flux terms in (1) can be written as

$$\boldsymbol{F}_{i,i} = \boldsymbol{F}_{i,U} \boldsymbol{U}_{,i} = \boldsymbol{A}_i \boldsymbol{U}_{,i} \tag{14}$$

$$\boldsymbol{F}^v_i = \boldsymbol{K}^v_{ij} \boldsymbol{U}_{,j} \tag{15}$$

$$\boldsymbol{F}^h_i = \boldsymbol{K}^h_{ij} \boldsymbol{U}_{,j} \tag{16}$$

where

$$\boldsymbol{A}_i = \boldsymbol{A}_i(\boldsymbol{U}) \tag{17}$$

$$\boldsymbol{K}^v_{ij} = \boldsymbol{K}^v_{ij}(\boldsymbol{U}) \tag{18}$$

$$\boldsymbol{K}^h_{ij} = \boldsymbol{K}^h_{ij}(\boldsymbol{U}) \tag{19}$$

With these, (1) becomes

$$\boldsymbol{U}_{,t} + \boldsymbol{A}_i \boldsymbol{U}_{,i} = (\boldsymbol{K}_{ij} \boldsymbol{U}_{,j})_{,i} + \boldsymbol{\mathcal{F}} \tag{20}$$

where

$$\boldsymbol{K}_{ij} = \boldsymbol{K}^{v}_{ij} + \boldsymbol{K}^{h}_{ij} \tag{21}$$

The thermodynamic entropy density, η , is given by

$$\eta = c_v s \tag{22}$$

where

$$s = \ln\left(\frac{p}{p_o}\left(\frac{\rho}{\rho_o}\right)^{-\gamma}\right) \tag{23}$$

The (physical) *entropy variables*, $\boldsymbol{V}$, are defined by

$$\boldsymbol{V}^T = H_{,\boldsymbol{U}} \tag{24}$$

where

$$H = -\rho s \tag{25}$$

Consider a change of variables:

$$\boldsymbol{U} = \boldsymbol{U}(\boldsymbol{V}) \tag{26}$$

The new variables satisfy

$$\boldsymbol{A}_0 \boldsymbol{V}_{,t} + \widetilde{\boldsymbol{A}}_i \boldsymbol{V}_{,i} = (\widetilde{\boldsymbol{K}}_{ij} \boldsymbol{V}_{,j})_{,i} + \boldsymbol{\mathcal{F}} \tag{27}$$

where

$$\boldsymbol{A}_0 = \boldsymbol{U}_{,\boldsymbol{V}} \tag{28}$$

$$\widetilde{\boldsymbol{A}}_i = \boldsymbol{A}_i \boldsymbol{A}_0 \tag{29}$$

$$\widetilde{\boldsymbol{K}}_{ij} = \boldsymbol{K}_{ij} \boldsymbol{A}_0 \tag{30}$$

The coefficient matrices possess the following properties:

(i) $\boldsymbol{A}_0$ is symmetric and positive definite and the $\widetilde{\boldsymbol{A}}_i$'s are symmetric.

(ii) The matrix

$$\widetilde{\boldsymbol{K}} = \begin{pmatrix} \widetilde{\boldsymbol{K}}_{11} & \widetilde{\boldsymbol{K}}_{12} & \widetilde{\boldsymbol{K}}_{13} \\ \widetilde{\boldsymbol{K}}_{21} & \widetilde{\boldsymbol{K}}_{22} & \widetilde{\boldsymbol{K}}_{23} \\ \widetilde{\boldsymbol{K}}_{31} & \widetilde{\boldsymbol{K}}_{32} & \widetilde{\boldsymbol{K}}_{33} \end{pmatrix} \tag{31}$$

is symmetric and positive semi-definite.

The present definition of $\boldsymbol{V}$ enables us to derive the following result:

$$0 = \boldsymbol{V}^T(\boldsymbol{A}_0\boldsymbol{V}_{,t} + \widetilde{\boldsymbol{A}}_i\boldsymbol{V}_{,i} - (\widetilde{\boldsymbol{K}}_{ij}\boldsymbol{V}_{,j})_{,i} - \boldsymbol{\mathcal{F}})$$

$$= H_{,t} + (H u_i)_{,i} + \boldsymbol{V}_{,i}^T\widetilde{\boldsymbol{K}}_{ij}\boldsymbol{V}_{,j} - (\boldsymbol{V}^T\widetilde{\boldsymbol{K}}_{ij}\boldsymbol{V}_{,j})_{,i} - \boldsymbol{V}^T\boldsymbol{\mathcal{F}}$$

$$= \frac{1}{c_v}(-(\rho\eta)_{,t} - (\rho\eta u_i)_{,i} + c_v\boldsymbol{V}_{,i}^T\widetilde{\boldsymbol{K}}_{ij}\boldsymbol{V}_{,j} - (\frac{q_i}{\theta})_{,i} + \frac{\rho r}{\theta}) \tag{32}$$

which leads to the local form of the *Clasius-Duhem inequality,* or *second law of thermodynamics,* namely

$$(\rho\eta)_{,t} + (\rho\eta u_i)_{,i} + (\frac{q_i}{\theta})_{,i} - \frac{\rho r}{\theta} \geq 0 \tag{33}$$

Remark: Harten [3] has proposed a family of generalized entropy functions for the compressible Euler and Navier-Stokes equations neglecting heat conduction. Equation (25) represents a particular member of this family. Line 2 of equation (32) can be obtained for all members of Harten's family, but not line 3.

Entropy Production

A Galerkin formulation of (27) , in which $\boldsymbol{W}$ and $\boldsymbol{V}$ are expanded in terms of C^0 finite element interpolations, can be written as (cf. (13)):

$$0 = \int_\Omega \boldsymbol{W}^T(\boldsymbol{A}_0\boldsymbol{V}_{,t} + \widetilde{\boldsymbol{A}}_i\boldsymbol{V}_{,i} - (\widetilde{\boldsymbol{K}}_{ij}\boldsymbol{V}_{,j})_{,i} - \boldsymbol{\mathcal{F}})d\Omega \tag{34}$$

Selecting $\boldsymbol{W} = \boldsymbol{V}$ in (34) and proceeding as in (32) enables us to derive a global statement of the Clausius-Duhem inequality in terms of the discrete solution:

$$\int_\Omega ((\rho\eta)_{,t} + (\rho\eta u_i)_{,i} + (\frac{q_i}{\theta})_{,i} - \frac{\rho r}{\theta})d\Omega = c_v\int_\Omega \boldsymbol{V}_{,i}^T\widetilde{\boldsymbol{K}}_{ij}\boldsymbol{V}_{,j}d\Omega \geq 0 \tag{35}$$

This is an important result: *the discrete solution always satisfies the Clausius-Duhem inequality.*

Another interpretation of (35) is that the rate-of-growth of a convex function of the discrete solution, namely $H = -\rho s$, is appropriately bounded from above. *Thus, the fundamental stability property possessed by solutions of the Navier-Stokes equations is automatically inherited by discrete solutions.* We refer to this property as being *entropy-stable.*

It is well known that the Galerkin finite element method is ineffective when flows contain shocks or sharp layers. It may be noted that, *for the case of the Euler equations, entropy will always be conserved by a Galerkin finite element solution*, that is

$$0 = \int_{\Omega} \boldsymbol{V}^T(\boldsymbol{A}_0\boldsymbol{V}_{,t} + \widetilde{\boldsymbol{A}}_i\boldsymbol{V}_{,i})d\Omega = \frac{1}{c_v}\int_{\Omega}((\rho\eta)_{,t} + (\rho\eta u_i)_{,i})d\Omega \tag{36}$$

The entropy production associated with exact discontinuous solutions cannot be represented by a Galerkin formulation with C^0 functions. A Petrov-Galerkin formulation, in which $\boldsymbol{W}$ is altered to produce appropriate entropy production in the discrete solution, can be effectively used to rectify the situation.

Petrov-Galerkin Formulation

The Petrov-Galerkin formulation that we currently favor has the following form:

$$0 = \sum_{e=1}^{n_{el}}\int_{\Omega^e} \widetilde{\boldsymbol{W}}^T(\boldsymbol{A}_0\boldsymbol{V}_{,t} + \widetilde{\boldsymbol{A}}_i\boldsymbol{V}_{,i} - (\widetilde{\boldsymbol{K}}_{ij}\boldsymbol{V}_{,j})_{,i} - \boldsymbol{\mathcal{F}})d\Omega\ +$$

$$+ \text{ viscous and heat flux continuity and boundary terms} \tag{37}$$

where Ω^e is the domain of element e, n_{el} is the number of elements, and

$$\widetilde{\boldsymbol{W}} = \boldsymbol{W} + \boldsymbol{T}_i\boldsymbol{W}_{,i} \tag{38}$$

$$\boldsymbol{T}_i = \boldsymbol{T}_i^{(1)} + \boldsymbol{T}_i^{(2)} \tag{39}$$

in which $\boldsymbol{T}_i^{(1)}$ is a generalized streamline matrix and $\boldsymbol{T}_i^{(2)}$ is a discontinuity capturing matrix. A detailed description of these matrices for the Euler and Navier-Stokes equations is beyond the scope of this paper. To illustrate the concepts involved we shall consider the *linear, scalar, advection-diffusion equation*:

$$\phi_{,t} + \boldsymbol{a}\cdot\nabla\phi = \nabla\cdot(\boldsymbol{\kappa}\nabla\phi) + f \tag{40}$$

where ϕ is the dependent variable, $\boldsymbol{a}$ is the given velocity vector field, $\boldsymbol{\kappa}$ is the diffusivity tensor and f is the body source term. A Petrov-Galerkin formulation for (40) , analogous to (37) , may be written as follows:

$$0 = \sum_{e=1}^{n_{el}}\int_{\Omega^e} \widetilde{w}(\phi_{,t} + \boldsymbol{a}\cdot\nabla\phi - \nabla\cdot\boldsymbol{\kappa}\nabla\phi - f)d\Omega\ +$$

$$+ \text{ diffusive flux continuity and boundary terms} \tag{41}$$

where

$$\tilde{w} = w + \tau_1 \boldsymbol{a} \cdot \nabla w + \tau_2 \boldsymbol{a}_{\|} \cdot \nabla w \tag{42}$$

in which τ_1 and τ_2 are parameters which control the magnitudes of the streamline and discontinuity-capturing terms, respectively. The role of the streamline term has been discussed at length elsewhere (see, e.g., Brooks and Hughes [1] and Johnson et al. [6]). The vector $\boldsymbol{a}_{\|}$ denotes the projection of $\boldsymbol{a}$ onto $\nabla\phi$, that is,

$$\begin{cases} \boldsymbol{a}_{\|} = \dfrac{(\boldsymbol{a} \cdot \nabla\phi)}{|\nabla\phi|^2} \nabla\phi, & \text{if } \nabla\phi \neq 0 \\ \quad = 0, & \text{if } \nabla\phi = 0 \end{cases} \tag{43}$$

Note that $\boldsymbol{a}_{\|}$ is a homogeneous function of degree zero in $\nabla\phi$. Let

$$\boldsymbol{a}_{\perp} = \boldsymbol{a} - \boldsymbol{a}_{\|} \tag{44}$$

A simple, but important, result is that

$$\boldsymbol{a}_{\|} \cdot \nabla\phi = \boldsymbol{a} \cdot \nabla\phi \tag{45}$$

The presence of the discontinuity capturing term results in improved stability. This may be seen by restricting our attention to the interaction of $\tilde{w}$ with $\boldsymbol{a} \cdot \nabla\phi$, viz.

$$\tilde{w}(\ldots + \boldsymbol{a} \cdot \nabla\phi + \ldots) = \ldots + w\boldsymbol{a} \cdot \nabla\phi + \nabla w \cdot \underbrace{\tau_1 \boldsymbol{a}\boldsymbol{a}^T}_{\text{streamline operator}} \nabla\phi +$$

$$+ \nabla w \cdot \underbrace{\tau_2 \boldsymbol{a}_{\|}\boldsymbol{a}_{\|}^T}_{\text{discontinuity-capturing operator}} \nabla\phi + \ldots \tag{46}$$

The last term engenders control over gradients in the direction $\nabla\phi$ and significantly increases the robustness of the streamline-upwind/Petrov-Galerkin method in the presence of sharp layers without degrading convergence rate (see Hughes, Mallet and Mizukami [5] for numerical examples and further details). The price that one pays for this improvement is that the new method – streamline-upwind/Petrov-Galerkin plus discontinuity capturing – is always ***nonlinear*** in that $\boldsymbol{a}_{\|}$ depends upon $\nabla\phi$. This is of no concern when the original equations are nonlinear, such as for the compressible Euler and Navier-Stokes equations.

Numerical Examples

The calculations presented in this section are for the compressible Euler equations. An explicit time-stepping algorithm was employed although, with the exception of the elevon cove, all the results obtained were steady. Four-node bilinear quadrilateral elements were used throughout. The results for the first two problems were obtained with the entropy variables derived from equation (25) . The last three problems employed variables derived from the following generalized entropy function advocated by Harten [3]:

$$H = -\rho \frac{2\gamma}{\gamma - 1} exp(\frac{s}{2\gamma}) \tag{47}$$

Figure 1 shows results for an oblique shock at Mach 2. A reflected shock at Mach 2.9 is illustrated in Figure 2. A Mach 8 bow shock calculation is shown in figure 3. In each of these cases, excellent agreement with theoretical results was obtained.

Results for a model of the space shuttle elevon cove are presented in Figure 4. The Mach number is 7 and a number of interesting flow features are apparent, namely, an expansion at the throat inlet followed by an oblique shock, recirculation in the entry of the throat due to discontinous initial velocity data, and a shock-tube effect throughout the throat.

Results for a NACA0012 airfoil are presented in Figure 5. Despite the coarseness of the mesh employed, excellent results are obtained.

References

1. Brooks, A.N.; Hughes, T.J.R.: Streamline Upwind / Petrov Galerkin Formulations for Convection Dominated Flows with Particular Emphasis on the Incompressible Navier-Stokes Equations. Computer Methods in Applied Mechanics and Engineering 32 (1982) 199-259.

2. Dutt, P.K.: Stable Boundary Conditions and Difference Schemes for Navier-Stokes type Equations. Ph.D. Thesis, University of California, Los Angeles, 1985.

3. Harten, A.: On the Symmetric Form of Systems of Conservation Laws with Entropy. Journal of Computational Physics 49 (1983) 151-164.

4. Hughes, T.J.R.; Franca, L.P.; Mallet, M.: Symmetric Forms of the Compressible Euler and Navier-Stokes Equations and the Second Law of Thermodynamics. Computer Methods in Applied Mechanics and Engineering, in press.

5. Hughes, T.J.R.; Mallet, M.; Mizukami, A.: Beyond SUPG. Computer Methods in Applied Mechanics and Engineering, in press

6. Johnson, C.; Navert, U.; Pitkaranta: Finite Element Methods for Linear Hyperbolic Problems. Computer Methods in Applied Mechanics and Engineering 45 (1985) 285-312.

7. Tadmor, E.: Skew-selfadjoint forms for Systems of Conservation Laws. Journal of Mathematical Analysis and Applications 103 (1984) 428-442.

1a.

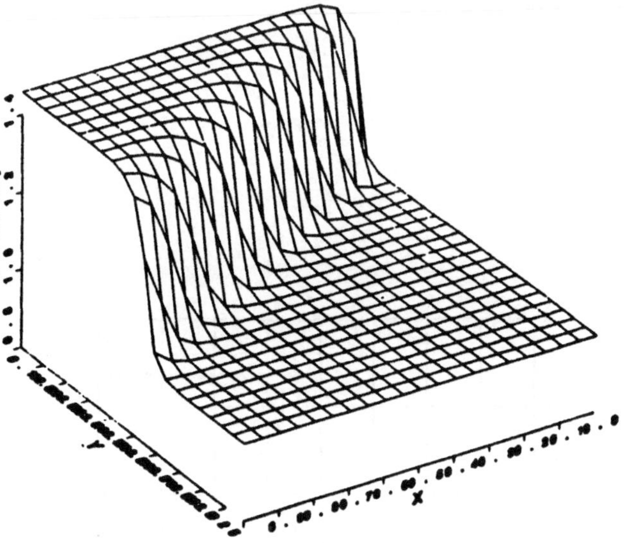

1b.

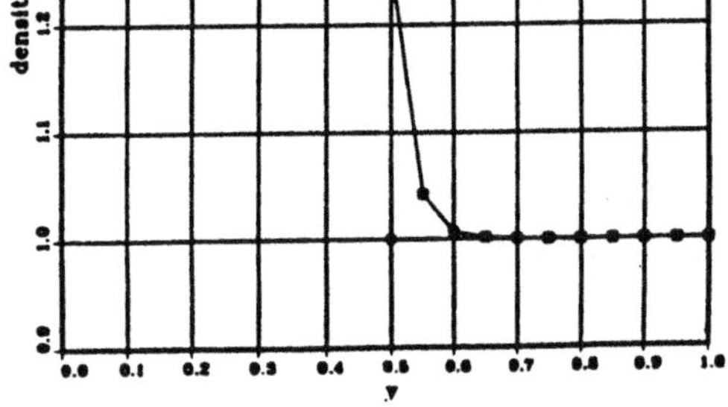

Figure 1 : Mach 2 oblique shock on a 10 degree wedge; (a) density elevation; (b) density profile compared with exact solution at x = 0.9.

2a.

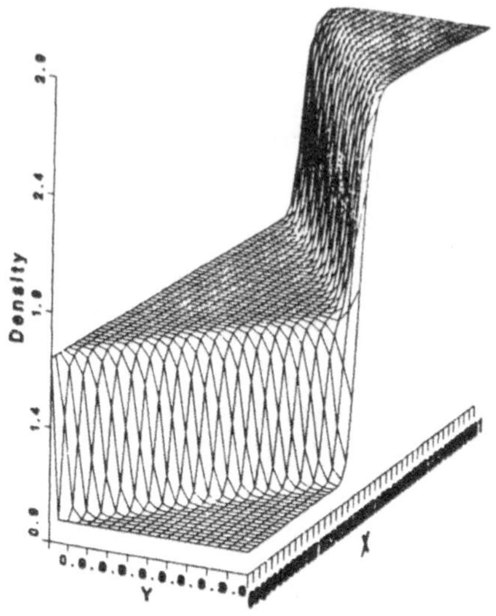

2b.

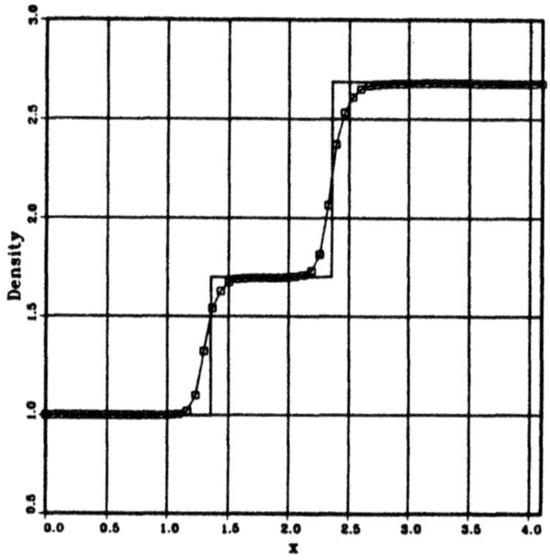

Figure 2 : Mach 2.9 reflected shock; (a) density elevation; (b) density profile compared with exact solution at y= 0.25.

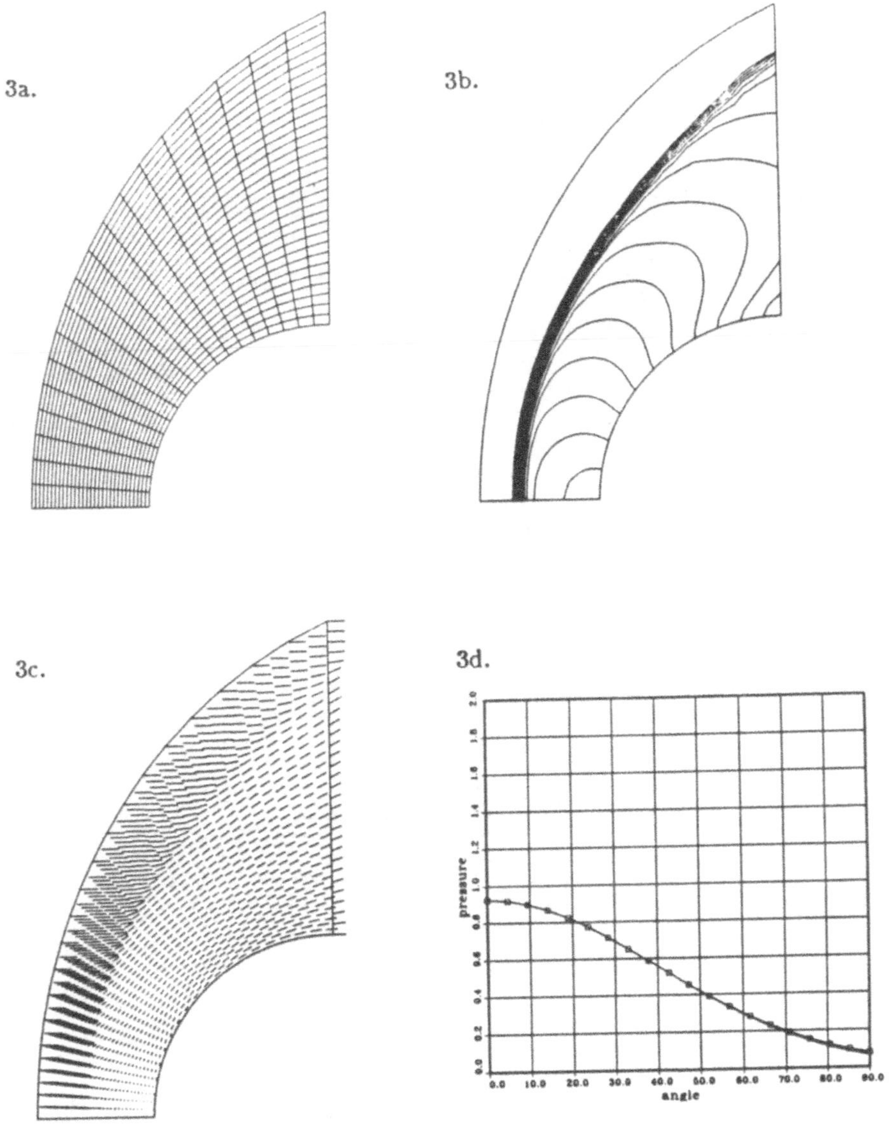

Figure 3 : Mach 8 bow shock; (a) mesh; (b) **Mach number contours**; (c) velocity vectors; (d) wall pressure profile compared with exact solution.

4a.

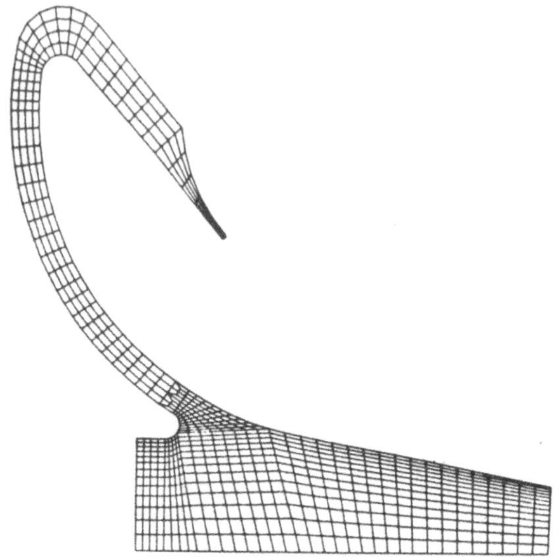

4b.

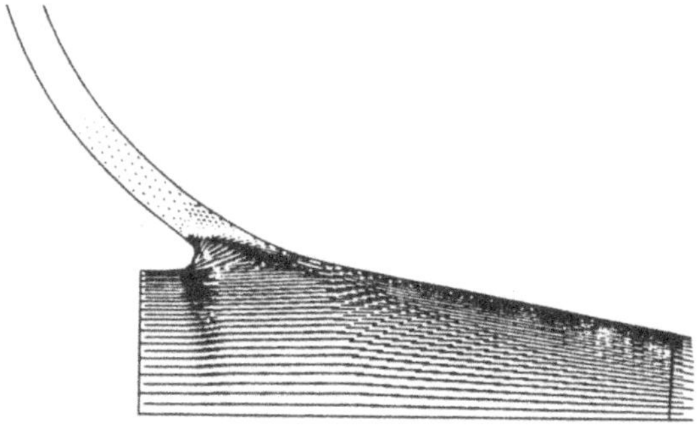

4c.

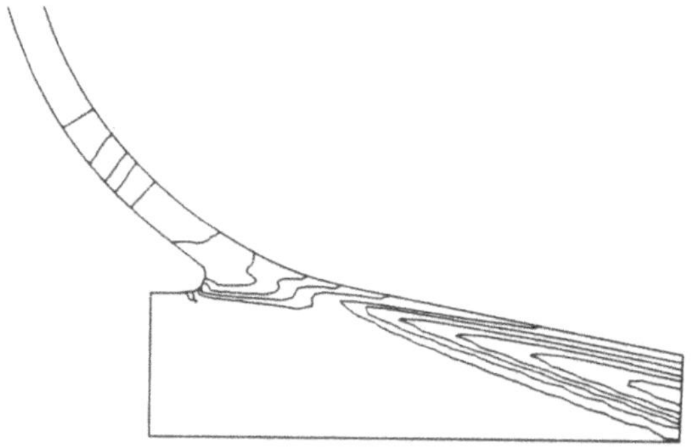

Figure 4 : Mach 7 elevon cove calculation; (a) mesh; (b) velocity vectors; (c) density contours.

5a.

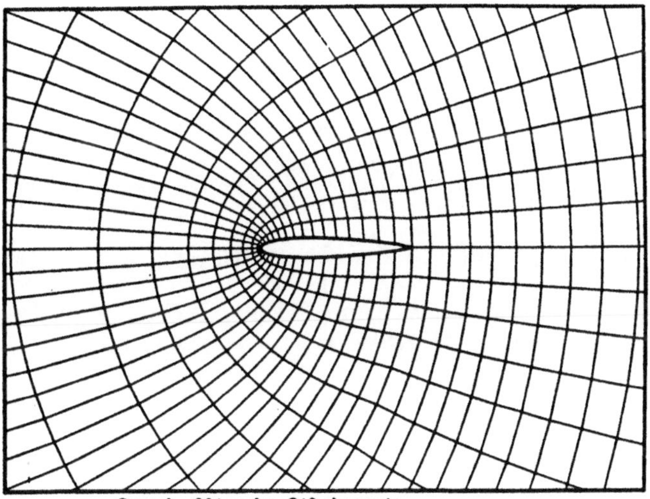

C mesh , 904 nodes ,840 elements

5b.

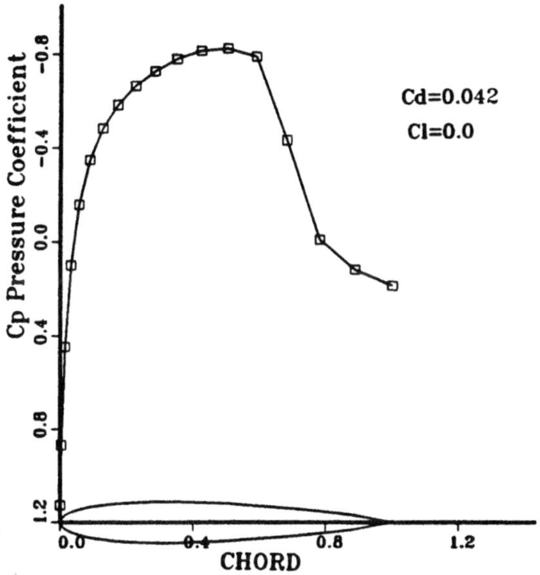

Figure 5 : NACA0012 airfoil at zero angle of attack, Mach number = 0.85; (a) mesh; (b) pressure coefficient.

Analysis on a Taylor Weak-Statement Algorithm for CFD Applications

A. J. BAKER, J. W. KIM
University of Tennessee, Knoxville, USA

Summary

One view of finite element analysis in computational fluid dynamics (CFD) is as a formal procedure for establishing suitable trial and test functions for the high speed flow problem class. A Taylor series weak-statement formulational procedure is developed herein that contains a family of parameters eligible for constraint according to norm specifications. The derived family of Taylor weak-statements is shown to contain, as special cases, a wide variety of independently-derived algorithms published over the past decade. A theoretical analysis is completed that identifies an optimal test space for the problem class.

Introduction

Much research has been conducted over the past decade on derivation of finite element weak-statement algorithms applicable to high speed flow problems. This CFD problem class is characterized by a large Reynolds or Peclet number, with the resulting equation system approximating (over a large part of the solution domain) the inviscid hyperbolic conservation law system termed the Euler equations. The earliest developments include those of Wahlbin [1], Dendy [2] and Raymond and Garder [3] for linear scalar first-order hyperbolics. The family of Petrov-Galerkin algorithms followed shortly, cf., Heinrich, et.al. [4], Hughes et.al. [5], Morton and Parrott [6], as well as the penalty-Galerkin algorithm of Baker and Soliman [7] and the characteristic-Galerkin method, cf. Morton [8]. In every instance, for a given space of trial functions, a test space was identified that yielded an optimum error estimate in some norm. Upon dissection, one of the additional terms so generated invariably corresponds to a second spatial derivative that yields an "artificial viscosity" term with the action of smoothing the inherent short wave length dispersion error.

Recently, Donea [9] published the "Taylor-Galerkin" weak-statement formulation, wherein the weak-statement is formed on a Taylor series reexpressed in terms of the flux vectors of a first order hyperbolic conservation law (system). The astute interchanging of weak-statement and Taylor series immediately added perspective to

Finite Element Methods for Nonlinear Problems
Europe-US Symposium, Trondheim, Norway 1985
Editors: Bergan, Bathe, Wunderlich

independently derived algorithm statements for CFD problem classes ranging from transonic potential flow to high speed Navier-Stokes solutions, cf., Baker [10]. Realizing the potential richness of this formulational procedure, a project is now underway to examine the details of a Taylor weak-statement algorithm for first-order hyperbolics. This paper derives and presents the algorithm construction for a model equation, and identifies constraint statements for optimal error estimates in familiar norms.

Taylor Weak-Statement Formulation

Consider the homogeneous hyperbolic conservation law system,

$$L(\mathbf{q}) = \frac{\partial \mathbf{q}}{\partial t} + \frac{\partial \mathbf{f}(\mathbf{q})}{\partial \mathbf{x}} \equiv \mathbf{q}_t + \mathbf{f}_{\mathbf{x}} = 0 \tag{1}$$

where boldface indicates a vector-valued function, $\mathbf{x}$ spans the n-dimensional domain R^n and contains the elements x_j, $1 \leq j \leq n$, and the solution domain is $\Omega \equiv R^n \times [t_o, t)$. The elements of $\mathbf{q}(\mathbf{x},t)$ are the dependent variables, and $\mathbf{f}(\mathbf{q})$ is the flux vector. For example, for the Euler equations,

$$\mathbf{q} \equiv \begin{Bmatrix} \rho \\ \rho \mathbf{u} \\ \rho e \end{Bmatrix} \qquad \mathbf{f} \equiv \begin{Bmatrix} \rho \mathbf{u} \\ \mathbf{u}\,\rho\mathbf{u} + p\delta \\ \mathbf{u}\,\rho e + \mathbf{u}p \end{Bmatrix} \tag{2}$$

and ρ is density, $\rho\mathbf{u}$ is the momentum vector, ρe is specific internal energy, p is pressure and δ is the Kronecker delta. For a polytropic perfect gas, $p = (\gamma-1)(\rho e - \tfrac{1}{2}\rho\mathbf{uu})$. For the Euler equations $\mathbf{f}$ is a homogeneous function of degree one in $\mathbf{q}$ with a similarity transformation that simultaneously diagonalizes the Jacobian $[J]$, i.e.,

$$\mathbf{f} = [J]\mathbf{q} = [S][\Lambda][S]^{-1}\mathbf{q} \tag{3}$$

In one-dimension, for example, the elements of the diagonal matrix $[\Lambda]$ are $\lambda_i = \{u, u+a_o, u-a_o\}$ where a_o is the sound speed.

The elements of construction of a weak-statement discrete approximate solution to eqn. 1 involve identification of a finite dimensional subspace $S^h \subset H^1$ from which the trial functions $\mathbf{q}^h(\mathbf{x},t)$ may be extracted, and a subspace $V^h \subset H^1_0$ of test functions $\mathbf{v}^h(\mathbf{x})$ which are arbitrary. The straight-forward extension of the Galerkin methods developed for elliptic boundary value problems utilize $\mathbf{q}^h$ and $\mathbf{v}^h$ to define the Galerkin weak-statement

$$\int_{R^n} v^h L(q^h) \equiv 0 \qquad \text{for all } v^h \in V^h \tag{4}$$

The usual assumption for $q^h(\cdot)$ is that space and time are separable, hence eqn. 4 yields an ordinary differential equation system for the values of $q^h(x,t)$ at the nodes $\{X_j\}$ of the discretization $\cup R^n_h$ of R^n. Thus, the Galerkin weak-statement eqn. 4 provides the data needed to form a single term Taylor series for the temporal evolution of the nodal variable array $\{Q(t)\}$.

Donea [9] established the astute artifice of interchanging the order of the weak-statement and Taylor series, yielding the "Taylor-Galerkin" algorithm. Since eqn. 1 defines an evolutionary problem statement, there must exist the Taylor series

$$\hat{f} = q_{n+1} - q_n - \Delta t\, q_t\Big|_{n+\theta} - 1/2\, \Delta t^2\, q_{tt}\Big|_{n+\theta} - 1/6\, \Delta t^3\, q_{ttt}\Big|_{n+\theta} - \dots = 0 \tag{5}$$

which is a valid form, to within the specific truncation error term, for Θ taking the values (0, ½, 1) and $t_{n+1} = t_n + \Delta t$. Equation 1 permits restatement of the derivatives expressed in eqn. 5 as,

$$q_t = -f_x \tag{6}$$

$$q_{tt} = -f_{xt} = -f_{tx} = -(f_q q_t)_x = +(f_q f_x)_x$$
$$\equiv -\alpha(f_q q_t)_x + (1-\alpha)(f_q f_x)_x \tag{7}$$

$$q_{ttt} = -f_{xtt} = \beta(f_q f_x)_{xt} \tag{8}$$

where $0 \leq \alpha \leq 1$ denotes the partition for evaluating the equivalent expressions for q_{tt}, and $\beta = (0,1)$ for retention of the term q_{ttt}. Substituting, eqn. 5 takes the form

$$\hat{f} = q_{n+1} - q_n + \Delta t\, f_x\Big|_{n+\theta} + \frac{\Delta t^2}{2}(\alpha f_q q_t - (1-\alpha) f_q f_x)_x\Big|_{n+\theta} - \frac{\Delta t^3}{6}\,\beta (f_q f_x)_{xt}\Big|_{n+\theta} - \dots \tag{9}$$

Substitution of the trial function q^h (hence also f^h) into eqn. 9 thus yields the discrete approximate Taylor series $\hat{f}^h = \hat{f}(u^h)$. The Taylor Galerkin weak-statement then directly produces the final linear algebra statement

$$F(\mathbf{Q}, \Theta, \Delta t, \alpha, \beta) = \int_{R^n} \mathbf{v}^h \hat{\mathbf{f}}^h = 0 \tag{10}$$

It is easy to verify that the first three terms in the Taylor-Galerkin weak-statement eqn. 10, using eqn. 9, reproduce exactly the Galerkin weak-statement algorithm eqn. 4. Further, Donea and colleagues have reported extensively on eqn. 10 for $\alpha = 0$, cf., [11-12]. Conversely, Baker [10] reports a rearrangement of eqn. 10 that yields a Taylor series for definition of the test functions $\mathbf{v}^h$ in terms of the solution Courant number and the trial functions $\mathbf{q}^h$. For example, for the one-dimensional continuity equation, eqns. 1-2, and for an assumed constant advection velocity $\mathbf{u} = a\hat{i}$, eqn. 9 becomes

$$\hat{f} = \rho_{n+1} - \rho_n + \Big[\Delta t(a\rho)_x + \frac{\Delta t^2}{2}(\alpha a \rho_t + (\alpha - 1)a \cdot a \rho_x)_x - \frac{\Delta t^3}{6}(\beta a \cdot a \rho_x)_{xt}\Big]_{n+\theta} \tag{11}$$

Using a Green-Gauss theorem to project $(\cdot)_x$ onto $\mathbf{v}^h$, the Taylor Galerkin weak-statement eqn. 10 becomes

$$\begin{aligned} F(\mathbf{RHO}, \Theta, \Delta t, \alpha, \beta) &= \int_{R^1} v^h \hat{f}^h \, dx \\ &= \int_{R^1} \Big[v^h(\rho^h_{n+1} - \rho^h_n) + v^h \Delta t\, a(\rho^h_x)_{n+\theta} \\ &\quad - v^h_x a \frac{\Delta t^2}{2}(\alpha \rho^h_t + (\alpha-1)a\rho^h_x)_{n+\theta} \\ &\quad + v^h_x a^2 \frac{\Delta t^3}{6} \beta(\rho^h_{xt})_{n+\theta} \Big] dx + \oint_{\partial R} v^h(\cdot) \cdot \hat{\mathbf{n}}\, d\sigma \end{aligned} \tag{12}$$

where $(\cdot)_{n+\theta} \equiv \Theta(\cdot)_{n+1} + (1-\Theta)(\cdot)_n$, and the last integral permits direct application of flux boundary conditions. The terms in eqn. 12 not contained in the Galerkin weak-statement have the common multiplier $a\Delta t$, which can be replaced by the Courant number definition $c \equiv a\,\Delta t/\Delta x = a\,\Delta t/h$, which is a scalar constant for a uniform discretization of R^1. Making this substitution, defining $(\rho^h_{xt})_{n+\theta} \equiv (\rho^h_{n+1} - \rho^h_n)_x\, \Delta t^{-1}$ and $(\rho^h_t)_{n+\theta} \equiv (\rho^h_{n+1} - \rho^h_n)/\Delta t$ for $\Theta = 1/2$, and combining like terms yields

$$\begin{aligned} F(\mathbf{RHO}, \Theta{=}½, \Delta t, \alpha, \beta) = \int_{R^1} \Big[&(v^h + chv^h_x(-\alpha + \beta ch \tfrac{\partial}{\partial x}))(\rho^h_{n+1} - \rho^h_n) \\ &+ \frac{\Delta t}{2}(v^h + ch\, v^h_x(1-\alpha))(a\rho^h_{n+1} + a\rho^h_n)_x \Big] dx \end{aligned} \tag{13}$$

where $\partial/\partial x \equiv (\cdot)_x$. Therefore, in comparison to the "classical" Galerkin weak-statement, eqn. 4, wherein $v^h = \rho^h$, the Taylor Galerkin weak-statement has defined the elements of the test space as

$$v^h \Rightarrow \{\mathbf{v}^h\} = \begin{Bmatrix} v^h + chv_x^h(-\alpha + ch\beta \frac{\partial}{\partial x}) \\ \\ v^h + chv_x^h(1-\alpha) \end{Bmatrix} \tag{14}$$

Thus, the two elements of $\{\mathbf{v}^h\}$ are distinct for the $\partial/\partial t$ and $a\partial/\partial x$ terms of the model equation and are dependent upon α and β.

A Taylor Weak-Statement Algorithm

Equations 10 and 14 at first view provide a tantalizing insight into a construction methodology for a test space $\mathbf{v}^h$ that might be optimal in some norm. For example, the second term in eqn. 14b is a diffusion term with artificial diffusion coefficient $\nu \equiv (1-\alpha)ch$. For $\beta \equiv 0$, and except for the negative sign on α in eqn. 14a, eqn. 14 defines the test space of the algorithms of both Dendy [2] and Raymond and Garder [3] for $\alpha \equiv ½$. Similarly, to within the sign, the algorithm of Wahlbin [1] is recovered after elimination of the factor of h on v_x^h, while for $\alpha \neq ½$, the penalty Galerkin algorithm of Baker and Soliman [7] is recovered in functional form. Certainly, a wide variety of finite difference algorithms may also fit, as well as Petrov-Galerkin and perhaps characteristic Galerkin algorithms.

These observations prompt the following hypothesis: for the one-dimensional unsteady advection equation,

$$L(u) = u_t + f(u)_x \equiv u_t + au_x = 0 \tag{15}$$

and for an arbitrary trial space $u^h \varepsilon S^h \subset H^1$, establish a Taylor weak-statement to determine a test space $v^h \varepsilon V^h$ that is optimal in a suitable norm. The Taylor series expansion for eqn. 15 for arbitrary Θ is,

$$\begin{aligned} \frac{1}{\Delta t}(u^{n+1} - u^n) &= (1-\Theta)u_t^n + \Theta u_t^{n+1} \\ &+ \frac{\Delta t}{2}\left\{(1-\alpha)u_{tt}^n - \Theta u_{tt}^{n+1}\right\} \\ &+ \frac{\Delta t^2}{6}\left\{(1-\Theta)u_{ttt}^n + \Theta u_{ttt}^{n+1}\right\} + \ldots \end{aligned} \tag{16}$$

where superscript n denotes the time level of evaluation. Replace the derivatives in eqn. 16 using eqn. 15 as follows,

$$u_t = -f_x = -(au)_x$$

$$u_{tt} = -\alpha(au_t)_x + \beta(af_x)_x$$

$$u_{ttt} = \gamma(a^2u_{tx})_x - \mu(a^2f_{xx})_x \tag{17}$$

where the parameter set $\{\alpha, \beta, \gamma, \mu\}$ is arbitrary, ie., no a **priori** constraint relates α and β or γ and μ. Substituting eqn. 17 into eqn. 16 yields the replacement for the Taylor series in eqn. 9 as,

$$\begin{aligned}\hat{f}(u) = &\left\{1 + \alpha\frac{\Delta t}{2}\frac{\partial a^{n-\theta}}{\partial x} - \gamma\frac{\Delta t^2}{6}\frac{\partial}{\partial x}(a^2)^{n+\theta}\frac{\partial}{\partial x}\right\}\left(\frac{u^{n+1}-u^n}{\Delta t}\right)\\ &+ \frac{\partial}{\partial x}\left\{(1-\theta)f^n + \theta f^{n+1}\right\}\\ &- \beta\frac{\Delta t}{2}\frac{\partial}{\partial x}\left\{(1-\theta)(a^2)^n\frac{\partial u^n}{\partial x} - \theta(a^2)^{n+1}\frac{\partial u^{n+1}}{\partial x}\right\}\\ &+ \mu\frac{\Delta t^2}{6}\frac{\partial}{\partial x}\left\{(1-\theta)(a^3)^n\frac{\partial^2 u^n}{\partial x^2} + \theta(a^3)^{n+1}\frac{\partial^2 u^{n+1}}{\partial x^2}\right\}\end{aligned} \tag{18}$$

where,

$$a^{n-\theta} \equiv (1-\theta)a^n - \theta a^{n+1}$$

$$(a^2)^{n+\theta} \equiv (1-\theta)(a^2)^n + \theta(a^2)^{n+1} \tag{19}$$

and using the following linearizing assumptions

$$(1-\theta)(au_t)^n - \theta(au_t)^{n+1} \equiv \left\{(1-\theta)a^n - \theta a^{n+1}\right\}\left\{\frac{u^{n+1}-u^n}{\Delta t}\right\}$$

$$(1-\theta)(a^2u_{tx})^n + \theta(a^2u_{tx})^{n+1} \equiv \left\{(1-\theta)(a^2)^n + \theta(a^2)^{n+1}\right\}\left\{\frac{u^{n+1}-u^n}{\Delta t}\right\}_x$$

$$(1-\theta)(a^2(au_x)_x)^n + \theta(a^2(au_x)_x)^{n+1} \equiv (1-\theta)(a^3)^n u^n_{xx} + \theta(a^3)^{n+1}u^{n+1}_{xx} \tag{20}$$

The weak-statement, eqn. 14, for the Taylor expansion eqn. 18, upon definition of a suitable space of trial functions u^h, and for an arbitrary space of test functions v^h, becomes

$$\{F\} \equiv \int_{R^1} v^h\hat{f}(u^h) = \int_{R^1}\left[v^h + ch\left(\theta + \frac{\alpha}{2}(1-2\theta)\right)v^h\frac{\partial}{\partial x} + c^2h^2\left(\frac{\gamma}{6} - \frac{\beta\theta}{2}\right)v^h_x\frac{\partial}{\partial x} + c^3h^3\frac{\mu\theta}{6}v^h_x\frac{\partial^2}{\partial x^2}\right]\left(u^{n+1} - u^n\right)$$

$$+ \left[chv^h \frac{\partial}{\partial x} + c^2h^2 \frac{\beta(1-2\Theta)}{2} v_x^h \frac{\partial}{\partial x} - c^3h^3 \frac{\mu(1-2\Theta)}{6} v_x^h \frac{\partial^2}{\partial x^2} \right] u^n \Bigg] dx$$

$$+ \oint_{\partial R} v^h(\cdot) \cdot \hat{n} d\sigma \equiv \mathbf{0} \tag{21}$$

In eqn. 21, $c \Rightarrow c^h \equiv a^h \Delta t/\Delta x = a^h \Delta t/h$ is the Courant number, the superscript on both u^h and c^h has been omitted for clarity, and the surface integral results from use of the Green-Gauss theorem. Thus, the terms in square brackets in eqn. 21 define the Taylor weak-statement test function as a power series in Courant number, the Greek-letter parameters, and derivatives of v^h as a generalization of eqn. 14.

The requirement is to evaluate eqn. 21 for a specific choice of trial and test spaces. For this purpose assume,

$$u^h(x,t) \equiv \sum_{j=1} \psi_j(x) U_j(t)$$

$$= \underset{e}{\cup} \{ N_1(\zeta) \}^T \{ U(t) \}_e \tag{22}$$

where $\psi_j(x)$ are the familiar linear hat (chapeau) functions, with $\psi_j(x_i) = \delta_{ij}$, or conversely, $\{N_1(\zeta)\}$ is the linear finite element basis spanning the generic finite element domain R_e^1. Further, assume that $V^h \subset S^h$ such that $v^h(x) \Rightarrow \{N_1(\zeta)\}$ also. Finally, for simplicity assume that the advection velocity is a uniform constant, hence $c^h \Rightarrow c$, the constant Courant number. For these definitions, eqn. 21 can be evaluated by assembly over the elements of ∂R_e, ie.,

$$\{F(U_j)\} = \underset{e}{S} \{ F(\{U\}_e) \}_e \tag{23}$$

and

$$\begin{aligned} \{F(\cdot)\}_e = & \Big([M]_e + ch(\Theta + \tfrac{\alpha}{2}(1-2\Theta)) [C]_e \\ & + (ch)^2 \left(\tfrac{\gamma}{6} - \tfrac{\beta\Theta}{2}\right) [D]_e \Big) \left(\{U\}_e^{n+1} - \{U\}_e^n \right) \\ & + (ch)^3 \tfrac{\mu\Theta}{6} [D]_e \left(\{U_x\}_e^{n+1} - \{U_x\}_e^n \right) \\ & + \left(c[C]_e + c^2h \frac{\beta(1-2\Theta)}{2} [D]_e \right) \{U\}_e^n \\ & - \left(c^3h^2 \frac{\mu(1-2\Theta)}{6} [D]_e \right) \{U_x\}_e^n \end{aligned} \tag{24}$$

In eqn. 24, $[M]_e$ is the mass matrix, $[C]_e$ is the convection matrix and $[D]_e$ is the diffusion matrix, which for $\{N_1\}$ are the familiar forms

$$[M]_e \equiv \int_{R_e} \{N\}\{N\}^T dx = \frac{h}{6}\begin{bmatrix} 2 & 1 \\ 1 & 2 \end{bmatrix}$$

$$[C]_e \equiv \int_{R_e} \{N\}\{N\}_x^T dx = \frac{1}{2}\begin{bmatrix} -1 & 1 \\ -1 & 1 \end{bmatrix}$$

$$[D]_e \equiv \int_{R_e} \{N\}_x\{N\}_x^T dx = \frac{1}{h}\begin{bmatrix} 1 & -1 \\ -1 & 1 \end{bmatrix} \qquad (25)$$

For $\mu \neq 0$, the form of eqn. 24 has required introduction of the additional nodal degrees of freedom $\{U_x\}$, since the $\{N_1\}$ basis cannot support two derivatives. The matrix multiplication by $[D]_e$, eqn. 25, is achieved by the definitions

$$\{U_x\}_e \equiv \begin{cases} \dfrac{1}{h}\begin{Bmatrix} U_{j-1} - U_{j-2} \\ U_j - U_{j-1} \end{Bmatrix}_e & a \geq 0 \\ -\dfrac{1}{h}\begin{Bmatrix} U_{j-1} - U_j \\ U_j - U_{j+1} \end{Bmatrix}_e & a < 0 \end{cases} \qquad (26)$$

for the element domain R_e containing vertex nodes x_{j-1} and x_j. Thus, eqn. 26 defines an element of flux vector "upwinding" with a corresponding widening of the (finite difference) stencil of eqn. 24.

Accuracy, Stability and Convergence

Equation 21 defines the Taylor weak-statement algorithm, while eqns. 23-26 give the specific form for the projection of both u^h and v^h onto the linear finite element basis. For the additional assumption of a uniform discretization, the assembled form of eqns. 23-26 over the generic element pair yields a finite difference recursion relation form. Noting that,

$$\underset{e}{S}[M]_e = h(1 + \frac{1}{6}\,\delta^2)$$

$$\underset{e}{S}[C]_e = h\,\Delta_o$$

$$\underset{e}{S}[D]_e = -h\,\delta^2 \qquad (27)$$

and for

$$\begin{aligned}\delta^2 U_j &\equiv U_{j-1} - 2U_j + U_{j+1}\\ \Delta_o U_j &\equiv (U_{j+1} - U_{j-1})/2\\ \Delta_- U_j &\equiv U_j - U_{j-1}\\ \Delta_+ U_j &\equiv U_{j+1} - U_j\end{aligned} \tag{28}$$

the Taylor weak-statement algorithm implementation on a uniform grid and for $\{N_1\}$ can be compactly written as

$$\left[1 + \frac{1}{6}\delta^2 + (c\theta + A)\Delta_o - cB\,\delta^2 - c^2E\,\delta^2\Delta_-\right](U^{n+1} - U^n)_j$$
$$+ \left[c\,\Delta_o - cP\,\delta^2 + c^2Q\,\delta^2\Delta_-\right]U_j^n = 0 \tag{29}$$

assuming $a \geq 0$ and for the definitions

$$\begin{aligned}A &\equiv c\,\alpha(1-2\Theta)/2\\ B &\equiv c(\gamma/6 - \beta\Theta/2)\\ E &\equiv c\,\mu\Theta/6\\ P &\equiv c\beta(1-2\Theta)/2\\ Q &\equiv c\mu(1-2\Theta)/6\end{aligned} \tag{30}$$

Thus, the Greek parameter set introduced in eqns. 17-18 is replaced by the set $\{A, B, E, P, Q\}$ which is eligible for optimization according to a specified norm. The present purpose is served by a Fourier modal analysis. Expanding the discrete solution $u^h(j\Delta x, n\Delta t)$ as

$$U_j^n \equiv e^{i\omega(j\Delta x - \bar{\gamma}n\Delta t)} \tag{31}$$

where $\omega \equiv 2\pi/\lambda$ and λ is the Fourier mode wavelength, and letting $\bar{\gamma} \equiv \bar{\beta} + i\bar{\delta}$ where $i \equiv \sqrt{-1}$, upon substitution of eqn. 31 into eqn. 29 and proceeding through the solution details yields,

$$\bar{\beta} = a\left[\frac{1 - \beta_2 m^2 + \gamma_2 m^4 - \delta_2 m^6 + \ldots}{1 - \beta m^2 + \gamma m^4 - \delta m^6 + \ldots}\right]$$
$$= a\left[1 - (\beta_2 - \beta)m^2 + \left[(\gamma_2 - \gamma) - \beta(\beta_2 - \beta)\right]m^4\right.$$
$$\left. - \left[(\delta_2 - \delta) - \gamma(\beta_2 - \beta) - \beta(\gamma_2 - \gamma) + \beta^2(\beta_2 - \beta)\right] m^6 + O(m^8)\right]$$

$$\bar{\delta} = am\left[\frac{-\alpha_1 + \beta_1 m^2 - \gamma_1 m^4 + \ldots}{1 - \beta m^2 + \gamma m^4 - \ldots}\right]$$

$$= am\Big[-\alpha_1 + (\beta_1 - \beta\alpha_1)m^2 - \Big\{(\gamma_1 - \gamma\alpha_1) - \beta(\beta_1 - \beta\alpha_1)\Big\}m^4 + O(m^6)\Big] \quad (32)$$

where $m \equiv \omega\Delta x$ and the various coefficients α, α_1, ..., μ_2 are polynomials in A, B, ..., Q and Courant number c.

The algorithm order-of-accuracy is controlled by the vanishing of coefficients of m^P in eqn. 32. Only $\alpha_1 = 0$ is independent of the definition $\mu > 0$, eqn. 20, which yields $P = -(A+c\Theta)$. We consider two cases:

Case 1 ($\mu \equiv 0$): In this instance, only the parameters A and B remain to be determined since $P = -(A+c\Theta)$. In eqn. 32a, requiring the coefficient of m^2 to vanish yields $B = 0$, and $P = (15)^{-1/2}$ causes the coefficient of m^4 to vanish. Thus, the constraint of optimal spatial order of accuracy yields

$$\bar{\beta} = a\left[1 + O(m^6)\right]$$
$$\bar{\delta} = \left[\frac{-a}{12\sqrt{15}}m^3 + O(m^5)\right] \quad (33)$$

for $P = -(A+c\Theta) = (15)^{-1/2}$ and $B = 0$, and for $E \equiv 0 \equiv Q$ in eqn. 29. For a given integration rule specifying Θ, A is related to the integration step size via

$$A_{opt} = -(15^{-1/2} + c\,\Theta) = -\left(15^{-1/2} + \frac{a\Theta\Delta t}{\Delta x}\right) \quad (34)$$

Except for the recognition of eqn. 34, these definitions for A, B and P yield identically the Raymond-Garder [3] algorithm, as well as the essential aspects of the Galerkin algorithms of Wahlbin [1] and Dendy [2]. The penalty Galerkin algorithm of Baker and Soliman [7] is also related, having been based on a modification to the Raymond-Garder formulation. Note that eqn. 34 is incompatible with the Taylor series definition of A, eqn. 20, for $\Theta = ½$ which is admissible for the cited algorithms.

Case 2 ($\mu \neq 0$): This definition adds E and Q to the constraint set for optimization, recall eqn. 30. The elimination of coefficients in eqn. 32 to $O(m^6)$ yields

$$\bar{\beta} = a[1 + O(m^6)]$$
$$\bar{\delta} = a[\tilde{\mu} m^5 + O(m^7)] \qquad (35)$$

for
$$\tilde{A} \equiv A + c\Theta$$
$$B = (15\tilde{A} + 2)/30c = -Q$$
$$E = -(5\tilde{A} + 1)/30c^2$$
$$P = -\tilde{A}$$
$$\tilde{\mu} = -(\tilde{A} + ½)/180 \qquad (36)$$

and $\tilde{A}$ is arbitrary. The principal difference between eqn. 33 and eqn. 35 is the order reduction of the artificial diffusion term to m^5, plus a smaller multiplier dependent upon $\tilde{A}$. The solution eqn. 36 permits replacement of B, E, P and Q and clearing the Courant number multipliers in the Taylor weak-statement, eqn. 29, yielding the final form

$$\left[1 + \frac{1}{6}\delta^2 + \tilde{A}\Delta_o - \left(\frac{15\tilde{A}+2}{30}\right)\delta^2 + \left(\frac{5\tilde{A}+1}{30}\right)\delta^2\Delta_-\right](U^{n+1}-U^n)_j$$
$$+ \left[c\Delta_o + c\tilde{A}\delta^2 - c\left(\frac{15\tilde{A}+2}{30}\right)\delta^2\Delta_-\right]U_j^n = 0 \qquad (37)$$

The Courant number and integration implicitness defined by the development leading to eqn. 29 remains specified within $\tilde{A} \equiv A + c\Theta$.

The fact that $\tilde{A}$ remains arbitrary admits an additional constraint specification dependent upon the problem definition. For example, a monotone constraint might be employed for solutions with sharp fronts, e.g., shocks, contact discontinuities. Alternatively, for smooth solutions the elimination of the $O(m^6)$ term in eqn. 32a yields,

$$\frac{1}{180}\left(\tilde{A}^2 + \tilde{A} + \frac{26}{140}\right) = 0 \qquad (38)$$

with solutions $\tilde{A}_o = (-0.24645, -0.75355)$. Since stability requires $\tilde{\mu} < 0$, eqn. 36, the smaller value for $\tilde{A}_o$ is not admissible, hence

$$\tilde{\mu}_o = -(-0.24645 + ½)/180 \approx -1/720 \qquad (39)$$

Assuming for notational simplicity that $\tilde{A}_{opt} \equiv -1/4$, the corresponding optimal parameters for the Taylor weak-statement algorithm, eqn. 29 are

$$\begin{aligned} A_o &= -(1/4 + c\Theta) \\ B_o &= -7/120c = -Q_o \\ E_o &= 1/120c^2 \\ P_o &= 1/4 \end{aligned} \tag{40}$$

Thus, comparing the two cases, inclusion of the terms in μ(eqn. 30) has uniformly improved the Taylor weak-statement algorithm order of accuracy by $O(m^2)$ and has reduced the artificial viscosity coefficient from $a(\Delta x)^3/12\sqrt{15}$ to $a(\Delta x)^5/720$. This has accrued at the expense of increasing the bandwidth of eqn. 29 by one node support, yet still employs only the simplest linear basis for u^h and v^h. For the additional order constraint, and defining $\nu \equiv -A_{opt}$ (since $A_{opt} < 0$), the order-optimized Taylor weak-statement eqn. 37 assumes the compact form ($a > 0$)

$$\left[1 - \nu\Delta_o + \left(\frac{1+5\nu}{10}\right)\delta^2 + \left(\frac{1-5\nu}{30}\right)\delta^2\Delta_-\right](U^{n+1}-U^n)$$
$$+ \left[c\Delta_o - c\nu\delta^2 - c\left(\frac{2-15\nu}{30}\right)\delta^2\Delta_-\right]U_j^n = 0 \tag{41}$$

A comparison can now be made among many Galerkin (and finite difference) algorithms for the discussed model problem within the framework of the Taylor weak-statement eqn. 29. Table 1 presents this comparison for terms in eqn. 29 combined as noted in the column headers. For these data, (sgn a) $\equiv$ sign of the advection velocity, which has been appended throughout as required (for consistency). Further, the listings under the header "1/6-cB" are added for completeness, even though $\Theta = 0$ was defined for all but the penalty-Galerkin algorithm (developed for a non-linear equation system). Recall also that the "1/6" stems from the mass matrix assembly, see eqn. 27, a term that is absent in all the comparison finite difference algorithms.

As noted in the introduction, every algorithm construction for the inviscid model problem creates an artificial diffusion term, which in the notation of the Taylor weak-statement is $-cP\delta^2$. Therefore, cP is the "artificial viscosity" coefficient and, viewing the appropriate column in Table 1, the functional form for the dozen listed algorithms is essentially invariant (since ν, ν_d, k', etc. all have an implied connection to Courant number). For the model problem, this is the only added term in the donor cell upwind algorithm (Courant-Isaacson-Rees), which is well verified to exhibit excessive artificial diffusion. Lax-Wendroff refines the diffusion coefficient to $c^2/2$, which is the functional form of every other algorithm including characteristic-Galerkin.

The next most common added term is 1/6 - cB, which appears in the Jacobian for implicit algorithms ($\Theta > 0$) as a balancing term for -cP (and the mass matrix

TABLE I

Comparison of Various Algorithms Within Taylor Weak - Statment, Eqn. 29

Coefficients in Equation 29

METHOD	Eqn/ Ref	θ	$c\theta + A$	$\frac{1}{6} - cB$	$-c^2E$	$-cP$	c^2Q	COMMENTS
Taylor Galerkin	(30)	all	$c\theta + \frac{\alpha c}{2}(1-2\theta)$	$\frac{1}{6} - c^2(\frac{\gamma}{6} - \frac{\beta\theta}{2})$	$-c^3(\frac{\mu\theta}{6})$	$-c^2\,\frac{\beta(1-2\theta)}{2}$	$c^3\,\frac{\mu(1-2\theta)}{6}$	$\theta \equiv 0 \equiv \alpha \equiv \gamma,\ \beta \equiv 1 \equiv \mu$, ref. [9].
Taylor Weak-Statement	(37)	all	$\tilde{A}$	$\frac{1}{6} - \frac{\pm 15\tilde{A}+2}{30}$	$\frac{5\tilde{A}\pm 1}{30}$	$c\tilde{A}$	$-c\,\frac{\pm 15\tilde{A}+2}{30}$	+ if $a > 0$, - if $a < 0$, $\tilde{A} \equiv c\theta + A$ arbitrary
Optimal Taylor Weak-Statement	(41)	all	$-(\mathrm{sgn}\ a)\nu$	$\frac{1+5\nu}{10}$	$(\mathrm{sgn}\ a)\frac{1-5\nu}{30}$	$-c(\mathrm{sgn}\ a)\nu$	$-c\,\frac{2-15\nu}{30}$	$\nu = 1/4 \equiv -\tilde{A}_{opt}$
Wahlbin	[1]	0	$-(\mathrm{sgn}\ a)$	$\frac{1}{6} - c(\mathrm{sgn}\ a)\theta$	0	$-c(\mathrm{sgn}\ a)$	0	
Dendy	[2]	0	$-(\mathrm{sgn}\ a)k'$	$\frac{1}{6} - c(\mathrm{sgn}\ a)k'\theta$	0	$-c(\mathrm{sgn}\ a)k'$	0	$\mathrm{Re}\,[\phi, k\phi_x] \le C_1\lVert\phi\rVert^2,\ k = k'\Delta x,\ C_1 > 0$
Raymond-Garder	[3]	0	$-(\mathrm{sgn}\ a)\nu$	$\frac{1}{6} - c(\mathrm{sgn}\ a)\nu\theta$	0	$-c(\mathrm{sgn}\ a)\nu$	0	$\nu = 1/\sqrt{15}$
penalty-Galerkin	[7]	1/2	$-\nu_i$	$\frac{1}{6} - c(\mathrm{sgn}\ a)\nu_d\theta$	0	$-c(\mathrm{sgn}\ a)\,\nu_d$	0	$\nu_i, \nu_d \approx C_1(1/\sqrt{15}),\ 0 \le C_1 \le 2$ for Euler eqns.
Euler-Petrov-Galerkin	[6]	0	0	$\frac{1}{6} - \frac{\nu}{6}$	0	$-c(\mathrm{sgn}\ a)\nu/2$	0	$0 \le \nu \le c < 1$
Crank-Nicolson-Petrov-Galerkin	[6]	1/2	$\frac{c}{2} - (\mathrm{sgn}\ a)\nu/2$	$\frac{1}{6} - \frac{\nu}{2}\left\{(\mathrm{sgn}\ a)c - \frac{1}{6}\right\}$	0	$-c(\mathrm{sgn}\ a)\nu/2$	0	$0 \le \nu \le c$
Euler-Characteristic Galerkin	[8]	0	0	0	0	$-c^2/2$	$c^3/6$	$\lvert c\rvert \le 1$
Donor-Cell Upwind	[14]	0	0	0	0	$-c(\mathrm{sgn}\ a)1/2$	0	Courant - Isaacson - Rees
Lax-Wendroff	[15]	0	0	0	0	$-c^2/2$	0	$\lvert c\rvert \le 1$
Warming-Beam Explicit	[16]	0	0	0	0	$-c^2/2$	$-c(1-c)/2$	$\lvert c\rvert \le 1,\ a > 0$

assembly). All the Galerkin algorithms share the functional form including Petrov-Galerkin but excluding characteristic-Galerkin. In distinction, the characteristic and Taylor Galerkin algorithms share identical c^3-dependence in the flux-limiting term c^2Q, while the Taylor weak-statement and Warming-Beam have both a linear and quadratic dependence. Finally, the entire family of Galerkin algorithms, including Crank-Nicolson-Petrov-Galerkin but excluding characteristic-Galerkin, share the identical functional form for the term $c\Theta + A$, which is absent in all cited finite difference algorithms.

Summary and Conclusions

A Taylor weak-statement finite element algorithm for a first order hyperbolic conservation law (system) has been derived. The development framework is a weak-statement for a Taylor series completed with an arbitrary mix of derivative replacements using the conservation law statement, recall eqns. 17-18. The resulting weak-statement defines the space of test functions as a power series in Courant number for an arbitrary test function $v^h(x)$, eqn. 21.

The Taylor weak-statement algorithm has been implemented for identical test and trial spaces using the simplest linear basis and a definition of the nodal flux degrees of freedom, eqn. 26. A Fourier stability analysis removes all but one arbitrary coefficient $(\tilde{A})$ for arbitrary integration implicitness Θ, eqn. 37. This algorithm is uniformly fifth-order accurate and stable for all Θ and $\tilde{A} < 0$. An increase to sixth-order accuracy yields $\tilde{A}_{opt} \equiv -1/4$, whereupon the artificial viscosity coefficient is $O(10^{-3})$. The resulting Taylor algorithm is functionally compared to ten independently derived Galerkin and finite difference algorithm constructions and found to contain most essential functional aspects of each. Numerical experiments are required, and are now being conducted, to firmly quantize relative performance aspects of the Taylor weak-statement algorithm for the range of these comparisons.

References

1. Wahlbin, L.B., "A Dissipative Galerkin Method Applied to Some Quasi-linear Hyperbolic Equations," **R.A.I.R.O.**, 8, (1974), 109-117.

2. Dendy, J.E., "Two Methods of Galerkin Type Achieving Optimum L^2 Rates of Convergence for First-order Hyperbolics," **SIAM J. Numerical Analysis**, 11, (1974), 637-653.

3. Raymond, W.H., and Garder, A., "Selective Damping in a Galerkin Method for Solving Wave Problems with Variable Grids," **Monthly Weather Review**, 104, (1976), 1583-1590.

4. Heinrich, J.C., Huyakorn, P.S., Zienkiewicz, O.C., and Mitchell, A.R., "An 'Upwind' Finite Element Scheme for Two-Dimensional Convective Transport Equation," **Int. J. Num. Mtd. Engr.**, 11, (1977), 134-143.

5. Hughes, T.J.R., and Brooks, A.N., "A Multidimensional Upwind Scheme With No Crosswind Diffusion," in T.J.R. Hughes (ed), **Finite Element Methods for Convection Dominated Flows**, ASME, AMD-34, New York, (1979), 19-35.

6. Morton, K.W., and Parrott, A.K., "Generalized Galerkin Methods for First Order Hyperbolic Equations," **J. Comp. Phys.**, 36, (1980), 249-270.

7. Baker, A.J., and Soliman, M.O., "A Finite Element Algorithm For Computational Fluid Dynamics," **AIAA J.**, 21, (1983), 816-827.

8. Morton, K.W., "Characteristic Galerkin Methods for Hyperbolic Problems," **Proc. 5th GAMM Conf. Num Mtd. Fluid Mechanics**, Rome, (1983).

9. Donea, J., "A Taylor-Galerkin Method For Convective Transport Problems," **Int. J. Num. Mtd. Engr.**, 20, (1984), 101-119.

10. Baker, A.J., "On a Penalty Finite Element Algorithm for High Speed Flow," **J. Comp. Mtd. Appl. Mech. and Engr.**, 49, 1985, to appear.

11. Donea, J., Giuliani, S., Laval, H., and Quartapelle, L., "Time-Accurate Solution of Advection-Diffusion Problems By Finite Elements," **J. Comp. Mtd. Appl. Mech. and Engr.**, 45, (1984), 123-145.

12. Selmin, V., Donea, J., and Quartapelle, L., "Taylor-Galerkin Method for Non-Linear Hyperbolic Equations," in C. Taylor, et.al.(ed), **Proc. Int. Conf. Num. Mtd. for Transient and Coupled Problems**, Pineridge Press, Swansea, (1984).

13. Selmin, V., Donea, J., and Quartapelle, L., "Finite Element Method for Nonlinear Hyperbolic Problems," **J. Comp. Mtd. Appl. Mech. and Engr.**, (1985), to appear.

14. Courant, R., Isaacson, E., and Rees, M., "On the Solution of Nonlinear Hyperbolic Differential Equations," **Comm. Pure Appl. Math.**, 5, (1952), 243-255.

15. Lax, P.D., and Wendroff, B., "Systems of Conservation Laws," **Comm. Pure Appl. Math.**, 13, (1960), 217-237.

16. Warming, R.F., and Beam, R.M., "Upwind 2nd order difference schemes and applications in Unsteady Aerodynamics Flows," **Proc. AIAA 2nd Comp. Fluid Dynamics Conference**, Hartford, Connecticut, (1975), 17-28.

An Optimal 3D Finite Element for Incompressible Media

M. BERCOVIER
The Hebrew University of Jerusalem, Israel

1. INTRODUCTION

FEM approximation of incompressible materials and incompressible viscous flows are usually based on the same family of elements. Basis functions must be suitable for both problems. In preceding studies we pointed out that second order (quadratic) elements must be of Lagrange type in order to be consistent. This is slightly frustrating since it means that 3 dimensional problems must be approximated by means of the 27 node brick, dealing with a 81 by 81 element matrix (not counting the eventual pressure degrees of freedom). Standard serendipity elements such as the 20 node (sometimes one takes the 21 node elements) include all the basic quadratic functions and are nearly optimal and far more economic. We are thus faced with a challenge: can we construct a quadratic element for incompressible fields similar to a serendipity element? The answer is yes and we shall give here an element based on the 21 node serendipity one to what we add 6 additional vector basis functions. Using the local incompressibility constraints we end up with a 66 degrees of freedom element the cost of which compares well to the 60 degrees of freedom of the classic 20 node element. A numerical example illustrates the effectiveness of this new element, and we conclude by mentioning current developments. We limit ourselves to the simple case of Stokes equations.

2. REDUCED CONSTRAINTS FOR STOKES EQUATIONS

We consider the following problem: given a (smooth and bounded) domain Ω in R^3, minimizes

$$E(\mathbf{u}) = \mu\int_\Omega \sum_{ij} [1/2\,(\mathbf{u}_{i,j} + \mathbf{u}_{j,i})]^2 d\Omega - \int_\Omega \mathbf{f}\,\mathbf{u}\, d\Omega \qquad (1)$$

among all $\mathbf{u}$ such that $\mathbf{u}$ is null on the boundary $\partial\Omega$ (Dirichlet problem) and subject to the constraint:

$$\operatorname{div} \mathbf{u} = 0 \qquad (2)$$

Finite Element Methods for Nonlinear Problems
Europe-US Symposium, Trondheim, Norway 1985
Editors: Bergan, Bathe, Wunderlich

We thus limit ourselves to the simplest case possible. The corresponding mixed formulation is:

Find a saddle point to

$$L(\mathbf{u},p) = E(\mathbf{u}) + \int_\Omega p \operatorname{div} \mathbf{u} \, d\Omega \tag{3}$$

Existence of unicity of (3) is based in the case of Stokes equations on Ladyzhenskaya's lemma:

$$\sup_{\mathbf{v} \in (H_0^1(\Omega))^3} \frac{(\int p_\Omega \operatorname{div} \mathbf{v} \, d\Omega)}{||\mathbf{v}||_{H_0^1}} \geq K \, ||p||_{L^2(\Omega)/R} \tag{4}$$

We refer to the book of Temam [13] for details. Brezzi [5] (and in an other form Babuska [1]) has shown that (4) is actually an abstract necessary and sufficient condition for consistent approximations of mixed variational problems. Thus if V is our trial space for u, and W the one for the pressure p, these spaces must so be chosen that:

$$\sup_{\mathbf{v} \varepsilon V} \frac{| \int p \operatorname{div} \mathbf{v} \, d\Omega |}{|| \mathbf{v} ||_V} \geq K \, ||p||_W \tag{5}$$

If (5) is satisfied then FEM solution of (3) always exists and error estimates to the continuous solution can be given. This mixed formulation can then be replaced by the corresponding consistent penalty formulation, see Bercovier [2] for the theory. (5) is what we called the Brezzi-Babuska condition for mixed finite elements. More details can be found in Bercovier et all [4] and in Engelman et all [7].

If we consider quadratic approximations for **u** at the element level, we shall take p in the space of first order polynomials, that is we impose at element level the constraints

$$\int_K p \operatorname{div} \mathbf{u} \, dK = 0 \; ; \; p = \{1, x_1, x_2, x_3\} \tag{6}$$

This is weaker than the original constraint (2). Let us rewrite (6) as:

$$\int_K \operatorname{div} \mathbf{u} \, dK = 0 \qquad \text{(p constant)} \tag{7.1}$$

$$\int_K x_i \operatorname{div} \mathbf{u} \, dK = 0 \qquad (x_i = 1, 2, 3, \text{ first moments}) \tag{7.2}$$

If our element has a node at the center (such as for the 21 or the 27 node elements) we can modify the basis functions in such a way as to eliminate the degrees of freedom at the center and satisfy (7.2) for all remaining basis functions. Such constructions are given by Griffiths [9]. Thus the main difficulty comes from the mean constraint (7.1). It is this constraint that is not satisfyed neither by the 8 node brick nor by the 20 node elements. This leads to the presence of "pressure modes" (see Malkus et all [11] and Sani et all [12]).

Fortin [8] has shown how to add degrees of freedom to the 8 node brick in order to build basis functions that will satisfy (7.1). The resulting approximation is of first order only. We shall apply his method to quadratic elements and give a non trivial numerical example.

3. THE SERENDIPITY $S_2 - P_1$ ELEMENTS

Following Fortin we want to add the degree of freedom **u . n** at each mid face of a Q element (Figure 1.). It can be shown that such an element will satisfy condition (7.1). What we obtain are new basis functions that are not standard in that they are vector basis functions and not the classic "scalar" ones. Dealing with "isoparametric" geometries is not simple. In order to expose our ideas in the clearest possible way we shall limit ourselves in the present case to simple rectangular bricks and work only on the reference element. The general case is given in Bercovier-Goldberger [3].

We start by using the standard 20 node serendipity shape functions. Set $t_0 = tt_i$, $s_0 = ss_i$, $r_0 = rr_i$ $(r_i, s_i, t_i = \pm 1)$ then at the corner nodes:

$$\phi_i = 1/8(1 + r_0)(1 + s_0)(1 + t_0)(r_0 + s_0 + t_0 - 2) \qquad (8)$$
$$i = 1, 2, 3, 4, 5, 6, 7, 8.$$

at the midside nodes:

$$\phi_i = 1/4(1 - r^2)(1 + s_0)(1 + t_0) \text{ for } i = 9, 11, 17, 19$$

$$\phi_i = 1/4(1 - s^2)(1 + r_0)(1 + t_0) \text{ for } i = 13, 14, 15, 16 \qquad (9)$$

$$\phi_i = 1/4(1 - t^2)(1 + r_0)(1 + s_0) \text{ for } i = 10, 12, 18, 20$$

To that we had the central (bubble) shape functions:

$$\phi_{21} = (1 - r^2)(1 - s^2)(1 - t^2) \qquad (10)$$

and modify accordingly the first 20 shape functions

$$\phi_i \leftarrow \phi_i + 1/4\ \phi_{21} \qquad \text{for } i = 1, 2, \ldots, 8$$

$$\phi_i \leftarrow \phi_i - 1/4\ \phi_{21} \qquad \text{for } i = 9, 10, \ldots, 20$$

so that they have a null value at node 21.

Up to now we have the standard Q21 serendipity + central node element. We now add directional degrees of freedom at mid faces (nodes: 22, 23, 24, 25, 26, 27). Denote by $\mathbf{e}_k$ the unit vector in direction k = 1(r), 2(s) and 3(t). Then:

$$\tilde{\phi}_{24} = +1/2\ r(1+r)(1-s^2)(1-t^2)\ \mathbf{e}_1$$
$$\tilde{\phi}_{26} = -1/2\ r(1-r)(1-s^2)(1-t^2)\ \mathbf{e}_1 \tag{11}$$

$$\tilde{\phi}_{22} = -1/2\ s(1-s)(1-r^2)(1-t^2)\ \mathbf{e}_2$$
$$\tilde{\phi}_{27} = +1/2\ s(1+s)(1-r^2)(1-t^2)\ \mathbf{e}_2 \tag{12}$$

$$\tilde{\phi}_{25} = -1/2\ t(1-t)(1-r^2)(1-s^2)\ \mathbf{e}_3$$
$$\tilde{\phi}_{23} = +1/2\ t(1+t)(1-r^2)(1-s^2)\ \mathbf{e}_3 \tag{13}$$

Consider now the 63 basis functions given by:

$$\tilde{\phi}_{i,k} = \mathbf{e}_k\,\phi_i \qquad i = 1, \ldots, 21\ ;\ k = 1, 2, 3$$

we have to modify them according to the introduction of (11), (12) and (13). For the faces r = ±1, by (11) we have:

$$\tilde{\phi}_{i,1} = (\phi_i + 1/4\ \phi_{26})\ \mathbf{e}_1 \qquad \text{for } i = 1, 4, 5, 8 \tag{14}$$

$$\tilde{\phi}_{i,1} = (\phi_i - 1/2\ \phi_{26})\ \mathbf{e}_1 \qquad \text{for } i = 12, 13, 16, 20 \tag{15}$$

$$\tilde{\phi}_{i,1} = (\phi_i + 1/4\ \phi_{24})\ \mathbf{e}_1 \qquad \text{for } i = 2, 3, 6, 7 \tag{16}$$

$$\tilde{\phi}_{i,1} = (\phi_i - 1/2\ \phi_{24})\ \mathbf{e}_1 \qquad \text{for } i = 10, 14, 15, 18 \tag{17}$$

Remain unchanged:

$$\tilde{\phi}_{i,1} = \phi_i \; e_1 \quad \text{for } i = 9,11,17,19,21$$

We proceed in the same manner for faces s = ± 1 and faces t = ±1. Hence in each direction we have a total of 23 degrees freedom.

We can show then that the Brezzi-Babuska condition is satisfied independently of the element size.

The 3 degrees of freedom are now eliminated following an idea due to Hecht [10]. By integration by part we have on the each element:

$$\int_K x_i \, \text{div} \, (\phi_{21} \, \mathbf{e}_j) \, dK = - \, \delta_{ij} \int_K \phi_{21} dK \quad dK \; ; i = 1,2,3 \; ; j = 1,2,3 \quad (18)$$

where δ_{ij} is Kronecker's symbol.

We thus modify the 66 basis functions (all except the 3 at the center node) in such a way that

$$\int_K \text{div} \, (\tilde{\phi}_{i,k}) x_l \quad dx = 0 \text{ for } i = 1, ...,20; \; k = 1,2,3 \; ; \; l = 1,2,3$$

$$\int_K \text{div} \, (\tilde{\phi}_i) x_l \quad dx = 0 \text{ for } i = 22, 23, 24, 25, 26, 27; \; l = 1, 2, 3$$

These functions are:

$$\tilde{\tilde{\phi}}_{i,k} = \tilde{\phi}_{i,k} + \sum_l \left(\int_K x_l \, \partial_k \phi_{ik} \, dK \, / \int_K \phi_{21} \, dK \right) \phi_{21} \, \mathbf{e}_l \quad (19)$$

$$i = 1, ..., 20 \; ; \; k = 1, 2, 3 \; ; \quad \phi_{ik} \text{ component k of } \tilde{\phi}_{i,k}$$

The same modifications applies to the functions $\tilde{\phi}_i$ i = 22, ..., 27.

NOTE:

We now have vector basis functions, at each node these basis functions are different for each direction. We cannot call our element an isoparametric element, any more. It is a new approach were the geometric definition of the element and the actual basis functions have little in common!

4. PRESSURE AND PENALIZATION

We are now left with only one constraint per element (7.1). Hecht in the above mentioned paper has shown how to further modify the new basis functions in order to satisfy (7.1). This construction is delicate and we prefer to keep (7.1) as a constraint. This means that a Lagrange multiplier p is defined element per element, with only one degree of freedom per element. In an other publication we shall show how to use such a pressure for time dependent problems. In the present computations we have implemented our penalty method. We include the stiffness matrix derived element by element from:

$$1/\varepsilon \, \left(\int_K \operatorname{div} \mathbf{u} \, dK\right)^2 \quad ; \qquad \varepsilon \ \text{"small"} \tag{20}$$

For nearly parallepipedic element (20) can be replaced by the one point formula:

$$1/\varepsilon \, [\, \operatorname{meas}(K) \operatorname{div} \mathbf{u}\,(21)]^2 \tag{21}$$

a cheap and efficient penalty matrix.

Numerical stability is even less impaired by a small ε here than in our reduced constarint/penalty 27 node element since we have one quarter of the constraints only to impose. We can see that (20) or (21) is in fact for each element a rank one correction of the global stiffness matrix.

5. NUMERICAL EXAMPLE

This new finite element was implemented into Engelman's Fluid Dynamics Code FIDAP [6] and a 3D problem was run using it, comparaison then could be made with a computation using the classic 27 node element. In both cases the penalty approach was choosen. The domain is a unit cube with a non uniform mesh (cf. Figure 2.). The upper plate (z = 1) is moving at unit velocity in direction Ox. Only Stokes equations were considered here. On bottom and side walls we set a zero velocity, the penalty parameter was given by $\varepsilon = 10^{-6}$. Graphic output for velocities and pressure from both runs were identical. To illustrate the difference (of order 10^{-3}) we give along the line x = .5, y = .5, the velocity in direction Ox (Table (1)).

The first spectacular advantage of our new element is the improved time solution due to the reduction of the number of degrees of freedom. Our mesh consisted of 125 elements. The 27 node elements gave rise to a total of 2187 degrees of freedom on a half bandwidth of 335 (maximum 549). We next considered the 26 + 6 node element (that is including the three degrees of freedom at node 21). We had a problem of size 1587 degress of freedom by a mean bandwidth of 256 (maximum 409). Finally our reduced element gave rise to a problem of size

1212 degrees of freedom by a mean bandwidth of 230 (maximum 334). Table (2) give the runtime statistics provided by FIDAP. Runs were made on The Hebrew University CDC Cyber 855. A cubic domain is of course the optimal shape for our results (i.e. bandwidth and degrees of freedom reduction), thus the factor 3.5 we obtained. But we can still expect a factor 2 to 3 on general domains.

6. CONCLUSION

An optimal quadratic 3D finite element for incompressible media has been introduced. It seems to be to that kind of problems equivalent to what the 21 node brick is for thick shells, the cheapest among the really robust finite elements. We are currently extending its use to Navier-Stokes equations (steady state and time dependent problems). We are also implementing this approach in rubber type materials. Finally we think that it may be worth looking at this method of vector basis functions for the definition of thick shells where lo king could be avoided.

ACKNOWLEDGMENT

Inclusion of this element into FIDAP was done by Y. Ben-Dor. Runs were performed by N. Goldberger. We also thank M. Engelman for the right to use FIDAP.

REFERENCES

[1] Babuska I., Error Bounds for the Finite Element Method. Numer. Math., Vol. 16, 322 - 333, 1971.

[2] Bercovier M., Perturbation of Mixed Variational Problems. Application to Mixed Finite Element Methods. R.A.I.R.O. (An. Numerique), Vol. 12, 211 - 236, 1978.

[3] Bercovier M., Goldberger N., to appear.

[4] Bercovier M., Hasbani Y., Gilon Y., Bathe K.J., On a Finite Element Procedure for Non Linear Incompressible Elasticity. In Hybrid and Mixed Finite Element Methods, S.N. Atluri, R.H. Gallagher and O.C. Zienkiewicz editors, J. Wiley and Sons, New York, 1983.

[5] Brezzi F., On the existence Uniqueness and Approximation of Saddle Point Problems Arising from Lagrangian Multipliers. R.A.I.R.O. (An. Numerique), Vol. 8 , 129 - 151, 1974.

[6] Engelman M., FIDAP, A Fluid Dynamics Analysis Program, User/Theoretical Manual. FDI, Evanston, Illinois.

[7] Engelman M., Sani R.L., Gresho P.M., Bercovier M., Consistent Versus Reduced Integration Penalty Methods for Incompressible Media Using Several Old and New Elements. Int. J. Num. Methods Fluids, Vol. 2, 25 -42, 1982.

[8] Fortin M., Old and New Finite Elements for Incompressible Flows. Int. J. Num. Methods Fluids, Vol. 1, 347 -364, 1981.

[9] Griffiths D.F., An Approximately Divergence-Free 9 Node Velocity Element (with Variations) for Incompressible Flows. Int. J. Num. Methods Fluids, Vol. 1, 323 - 347, 1981.

[10] Hecht F., Use of Divergence Free Basis in Finite Element Methods. In Analysis of Laminar Flow Over a Backward Facing Step, K. Morgan, J. Perriaux and F. Thomasset ed., vieweg, Braunschweig, 1984.

[11] Malkus D.S. and Fuller Jr. E.R., An Isoparametric Finite Element Model for Large Strain Elastostatics. Journal of Res. Nat. Bur. Stand., Vol. 86, 79 - 109, 1981.

[12] Sani R., Gresho P., Lee R., Griffiths D. and Engelman M., The Cause and Cure (?) of the Spurious Pressure Generated by Certain FEM Solutions of the Incompressible Navier-Stokes Equations. Part 1 and 2, Int. J. Num. Methods Fluids, Vol. 1, 17 - 43 and 171 - 204, 1981.

[13] Temam R., Navier-Stokes Equations. North Holland, Amsterdam , 1977.

u Velocity 27 Node	u Velocity New Element	z coordinate
0.	0.	0.
- 0.0960	- 0.0954	0.180
- 0.1750	- 0.1744	0.361
- 0.2150	- 0.2143	0.487
- 0.2130	- 0.2140	0.614
- 0.1410	- 0.1420	0.702
+ 0.0360	+ 0.0313	0.790
+ 0.2360	+ 0.2310	0.852
+ 0.5100	+ 0.5093	0.914
+ 0.7390	+ 0.7400	0.957

Table (1).

	27 Node	21 + 6 node	66 degrees of freedom element
Element Generation	23.83	29.85	31.44
LDL^T Decomposition and solution	759.1	329.3	204.5

Table (2).

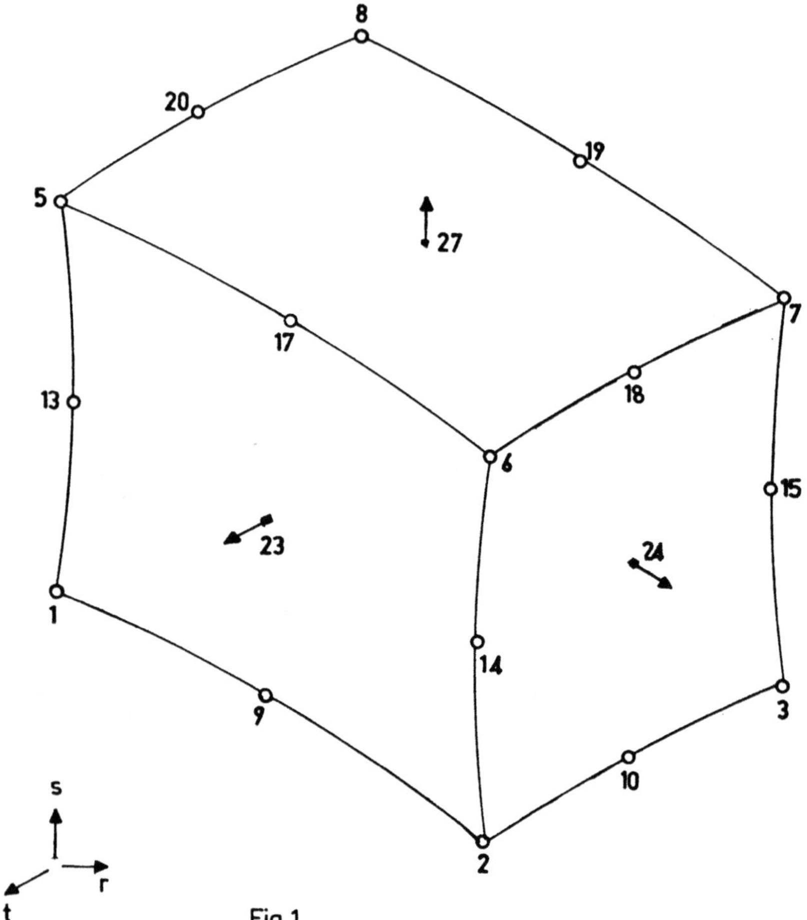

Fig. 1

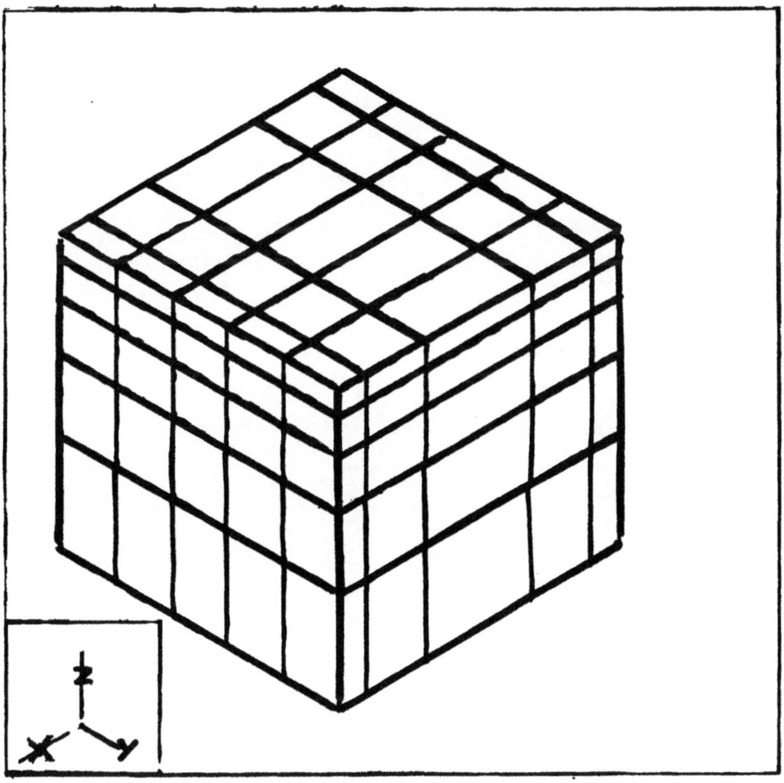

FIGURE 2

A Spectral Element Method for Solution of the Two- and Three-Dimensional Time-Dependent Incompressible Navier-Stokes Equations

G. E. KARNIADAKIS, E. T. BULLISTER, A. T. PATERA
Massachusetts Institute of Technology, Cambridge, USA

Abstract

The isoparametric spectral element method is a high-order technique for solution of the time-dependent incompressible Navier-Stokes equations that combines the accuracy of spectral methods with the geometric flexibility of finite element schemes. In the spectral element discretization, the domain is broken up into curvilinear brick elements, and the independent and dependent variables represented as high-order Lagrangian interpolants through Chebyshev collocation points. The nonlinear terms in the governing equations are then treated with explicit collocation, while the pressure and diffusion contributions are handled implicitly with standard variational projection operators. The implicit equations are inverted using either static condensation or conjugate gradient iteration, both techniques being amenable to a high degree of vectorization and parallelism. Results are presented for moderate Reynolds number flow in pipe expansions, flow past cylinders, and flow over three-dimensional roughness elements.

1. Introduction

Over the past several decades, numerous schemes have been proposed for the numerical solution of the Navier-Stokes equations for incompressible fluid flow. These techniques fall into roughly three categories as regards spatial discretization: finite difference [1], finite element [2], and spectral [3]. Although the differences between these techniques might have initially been quite clear, more recently there is an increasing trend towards construction of algorithms which span these traditional distinctions.

A particularly attractive approach to this confluence of numerical algorithms is the combination of finite element and spectral concepts. At the root of both of these methods is the principle of weighted-residuals, the differences arising primarily in the choice of expansion functions and associated spatial projection operators. In particular,

Finite Element Methods for Nonlinear Problems
Europe-US Symposium, Trondheim, Norway 1985
Editors: Bergan, Bathe, Wunderlich

whereas finite element techniques are typically based on fixed (low-) order interpolants, convergence being obtained in the limit of increased domain sub-division, spectral methods are based on global interpolants, convergence resulting from increase in the order of the representation. A successful synthesis of these concepts would lead to techniques with rapid (exponential) convergence ("spectral"), that were sufficiently flexible to handle general geometric configurations ("finite element").

There have, indeed, been several methods proposed for the solution of (second-order) elliptic problems using high-order expansions local to finite (macro-) elements, termed p- or combined h-p type finite elements [4], global elements [5], and spectral elements [6,7], respectively. All of these methods are similar in that they use variational projection operators in conjunction with local polynomial expansions, for which exponential convergence to smooth solutions can be proven by techniques similar to those used for low-order (or h-type) finite element schemes [8]. In addition to these theoretical results, there is ample numerical evidence [6,7] indicating that these methods do, in fact, combine geometric flexibility with high-order convergence for this class of problems.

There are several difficulties associated with extension of these high-order finite element techniques to the solution of real fluid flow problems. First, the equations of incompressible flow are not simple second-order elliptic (or parabolic) operators, but rather have significant (even "dominant") nonlinear hyperbolic contributions for large Reynolds numbers. Second, the vector Navier-Stokes equations are coupled and constrained (through incompressibility), resulting in greatly increased mathematical and computational complexity over the case of uncoupled, unconstrained scalars. Lastly, high-order techniques achieve rapid convergence only at the expense of increased coupling between nodes. For real problems in several space dimensions, the issue of efficiency is critical if such high-order techniques are to prove useful relative to their low-order counterparts.

The isoparametric spectral element technique [6,7,9], addresses these issues related to application of high-order finite-element methods to moderate Reynolds numbers flows. Work to date using the spectral element method has been limited to two-dimensional simulations of flow and heat transfer in Cartesian geometries, in which the boundary conditions on velocity are either imposed at solid walls, or taken to be spatially periodic [6,7,9,10]. It has been shown in numerous linear [6,7] and nonlinear [11] model hyperbolic and elliptic problems, as well as for the full Stokes problem [7], that the method does, indeed, give exponential convergence in space for problems with sufficiently smooth solutions. Furthermore, the use of static condensation [7] and fast solvers [12] results in operation counts (per degree-of-freedom) commensurate with those for low-order techniques, indicating that the spectral increase in accuracy reflects a real increase in computational efficiency. In summary, the spectral element method appears to be a viable approach for solution of incompressible ("smooth") flows in complex geometry, in particular at larger Reynolds numbers where accuracy requirements as regards numerical diffusion, dispersion, and resolution are increasingly stringent.

In this paper, we consider extensions of the spectral element technique to axisymmetric geometries, general inflow/outflow boundary conditions, and full three-dimensional configurations. The basic numerical discretizations are described, and examples given for flow in pipe expansions, flow past a cylinder (the von Karman vortex street), and flow past three-dimensional roughness elements.

2. Governing Equations

We consider the problem of unsteady flow and forced convection in a general domain D, with boundary ∂D composed of solid walls, ∂D_S, inflow/outflow boundaries, $\partial D_I \cup \partial D_O$, and periodic surfaces, ∂D_P. The governing equations for the flow are the incompressible Navier-Stokes equations,

$$\vec{v}_t = \vec{v} \times \vec{\omega} - \nabla \Pi + R^{-1} \nabla^2 \vec{v} \qquad \text{in D}, \qquad (1a)$$

$$\nabla \cdot \vec{v} = 0 \qquad \text{in } D, \qquad (1b)$$

where $\vec{v}(\vec{x},t)$ ($= u\hat{x} + v\hat{y} + w\hat{z}$) is the velocity, $\vec{\omega} = \nabla x \vec{v}$ is the vorticity, and $\Pi = p + {}^1/_2 |\vec{v}|^2$ is the dynamic pressure. Here R is the Reynolds number, $R = UL/\nu$, where ν is the kinematic viscosity of the fluid, and U and L are units of velocity and length, respectively. The appropriate boundary conditions for $\vec{v}$ on ∂D will be discussed in the context of the individual applications presented in the later sections.

The associated equation for forced convection heat transfer is given by

$$T_t + \nabla \cdot (\vec{v}T) = (RPr)^{-1} \nabla^2 T \qquad \text{in } D, \qquad (2)$$

where $T(\vec{x},t)$ is the temperature, Pr is the Prandtl number, $Pr = \nu/\alpha$, and α is the thermal diffusivity of the fluid. Boundary conditions on the temperature can either be Dirichlet, Neumann (flux), mixed (heat transfer coefficient), or periodicity [9].

3. Temporal Discretization

To discretize the Navier-Stokes equations (1) in time, we use a semi-implicit fractional-step method [7,13], with third-order Adams-Bashforth for the nonlinear convective terms, and an implicit (Crank-Nicolson) method for the viscous terms. The motivation for an explicit nonlinear/-implicit diffusion time-stepping scheme is that, for time-dependent flows, the disparity between the accuracy and stability restriction for the convective terms is typically not sufficient to warrant a Newton iteration, whereas the (linear) viscous terms are both stiffer and more easily inverted.

The semi-discrete equations for $\vec{v}^n(\vec{x}) = \vec{v}(\vec{x}, n\Delta t)$ are then,

$$\tilde{\vec{v}}^{n+1} - \vec{v}^n = \Delta t \sum_{q=0}^{2} \beta_q (\vec{v} x \vec{\omega})^{n-q} \qquad \text{in } D, \qquad (3)$$

followed by the (split) pressure step,

$$\hat{\tilde{\vec{v}}}^{n+1} - \tilde{\vec{v}}^{n+1} = -\nabla \Pi \qquad \text{in } D, \qquad (4a)$$

$$\nabla^2 \Pi = \nabla \cdot \hat{\tilde{v}}^{n+1} \qquad \text{in D}, \qquad (4b)$$

$$\nabla \Pi \cdot \hat{n} = \begin{cases} \hat{\tilde{v}}^{n+1} \cdot \hat{n} \text{ on } \partial D_S \\ \lambda \text{ on } \partial D_I \cup \partial D_O \end{cases}, \qquad (4c)$$

and finally the viscous contributions,

$$\tilde{v}^{n+1} - \hat{\tilde{v}}^{n+1} = {}^{\Delta t}/_{2} R^{-1} \nabla^2 (\tilde{v}^{n+1} + \tilde{v}^{n}) \qquad \text{in D}, \qquad (5)$$

with the full boundary conditions imposed on the viscous step, (5). Here $\beta_0 = 23/12$, $\beta_1 = -16/12$, and $\beta_2 = 5/12$ are the Adams-Bashforth coefficients. Also, λ is given by $\lambda = \int\int \hat{\tilde{v}} \cdot d\vec{S} / \int\int dS$ (integration being carried out over $\partial D_O \cup \partial D_I$), corresponding to the minimum adjustment required to the boundary velocity after the nonlinear step to insure solvability of the pressure equation, (4b).

Although the flow is not incompressible at the end of the full step (more precisely, there are divergence errors in thin boundary layers near solid surfaces due to the inviscid pressure boundary condition (4c)), the Stokes part of the scheme given above has been shown to be $O(\Delta t)$ in the velocities on numerous (real and model) test problems [7]. There are larger ($O(\Delta t^{1/2})$) errors in velocity gradients and the pressure near solid boundaries, however, as will be shown in the later sections, this does not preclude accurate predictions of drag and transport (typically dominated by spatial errors). Although of low order, the semi-implicit/-splitting scheme has the important advantages of being extremely simple to implement, and of reducing all the implicit operations to uncoupled Helmholtz equations. Note that the spectral element spatial discretization is not restricted to the particular time-stepping scheme given here, and higher-order methods have been implemented [6].

The time-stepping scheme used for the energy equation (2) is essentially the same as that for the velocity, with third-order Adams-Bashforth for the convective terms, and Crank-Nicolson for the conduction terms,

$$\hat{T}^{n+1} - T^{n} = -\Delta t \sum_{q=0}^{2} \beta_q \nabla \cdot (\vec{v}T)^{n-q} \qquad \text{in D}, \qquad (6)$$

$$T^{n+1} - \hat{T}^{n+1} = {}^{\Delta t}/_{2}(RPr)^{-1}\nabla^2(T^{n+1}+T^n) \qquad \text{in } D, \qquad (7)$$

respectively. All boundary conditions are imposed on the conduction step, (7).

4. Spatial Discretization

The isoparametric spectral element discretization proceeds by first breaking up the computational domain into general (curvilinear) brick elements, and expanding the geometry, velocity, pressure, and temperature in tensor product bases in terms of the local coordinates, $\vec{r}$ ($=r\hat{r} + s\hat{s} + t\hat{t}$). In element k ($[-1,1]x[-1,1]x[-1,1]$), we therefore have

$$\begin{aligned}[\vec{x},\vec{v},\Pi,T]^k_N(\vec{r}) &= \sum_{i=0}^{N}\sum_{j=0}^{N}\sum_{m=0}^{N} [\vec{x},\vec{v},\Pi,T]^k_{ijm} h_i(r)h_j(s)h_m(t) \\ &= \sum_i [\vec{x},\vec{v},\Pi,T]^k_i g_i(\vec{r}) \qquad ,\end{aligned} \qquad (8)$$

where the $h_i(z)$ are Nth-order Lagrangian interpolants through the Gauss-Lobatto Chebyshev collocation points [6]. Local variables are related to the global decomposition by global node numbers, thus insuring C^0-continuity of the basis, as required for the weak formulation to be presented below. The interpolation (8) can be easily shown to result in exponential convergence for infinitely smooth functions as $N \Rightarrow \infty$ [3]. The choice of collocation points is important as regards the accuracy of the numerical quadratures and the collocation approximations [7].

It is clear from the semi-discrete equations in (3-7) that to solve the Navier-Stokes equations we must be able to accurately represent and invert both (nonlinear) hyperbolic and elliptic operators. Considering first the convective (hyperbolic) contributions, (3) and (6), we use a mixed collocation / Galerkin approach,

$$\nabla\cdot(\vec{v}T) \Rightarrow \quad \Sigma'_k \iiint g_i g_j \nabla_{jp}\cdot(\vec{v}^k_p T^k_p)\,|J^k|\,d\vec{r} \quad , \qquad (9)$$

(summation over repeated indices assumed), where Σ'_k denotes direct stiffness summation [8], ∇_{jp} is the discrete gradient operator (i.e., the gradient at point j due to node p), and J is the Jacobian of the global to local transformation defined by (8). In some sense, (9) corresponds to collocation of the derivative terms, followed by a Galerkin approximation for the resulting convective flux. This maintains many of the desirable features of a full Galerkin formulation (e.g., conservation [9]), while eliminating the triple products which can be prohibitively expensive for high-order methods [3]. It has been shown [6,11] that the high-order approximation (9) exhibits very little numerical dispersion and diffusion, an important property for high-Reynolds number simulations.

For the pressure and diffusion contributions, (4-5,7), we use the standard (symmetric, negative-definite) variational formulation for second-order elliptic equations [8],

$$\nabla^2 T - \gamma^2 T \;\Rightarrow\; \Sigma'_k \iiint \{-\nabla_{pi} g_p g_q \nabla_{qj} - \gamma^2 g_i g_j\} T_j |J^k| d\vec{r}, \qquad (10)$$

where ∇_{ij} and J are defined as in (9). Solution of general Helmholtz equations by (10) has been demonstrated numerically [7] to give exponential convergence to smooth solutions as $N \Rightarrow \infty$, even for general geometries with relatively deformed macro-elements. The variational formulation (10) is readily extended to the case of cylindrical co-ordinates (as is (9)), although some care must be taken to re-formulate the equations so as to insure exponential convergence (in appropriate radially-weighted norms) in the presence of the pole singularity [14].

The key to the computational efficiency of both the hyperbolic and elliptic contributions, (9) and (10), respectively, is the sum factorization that follows from the tensor product bases, (8). For instance, to perform sums of the form $\iiint g_i g_p T_p d\vec{r}$ in three space dimensions, we write

$$\iiint g_i g_p T_p d\vec{r} = (\int h_i h_p dz\; (\int h_j h_q dz\; (\int h_k h_r dz\; T_{pqr}\;))), \qquad (11)$$

which gives an operation count of $O(N^4)$, rather than the more obvious (and prohibitive) $O(N^6)$. As regards the pressure and (implicit) diffusion terms, we must not only evalu ate, but also invert, operators of the form (10) or (11). In two space dimensions, a direct static condensation solver [8] is used, as discussed in detail in [7,12]. In three space dimensions, static condensation is no longer a viable approach due to the relatively large number of elemental boundary points, and we use instead a conjugate gradient iterative technique [15-18], guaranteed to converge given the symmetric negative-definite form of the elliptic contributions. In the context of our time-dependent algorithms, we have found that, even with no pre-conditioner, relatively few iterations are required per time step to achieve acceptable convergence, giving operation counts (per element per time step) of roughly $O(N^4)$. Furthermore, the conjugate gradient algorithm for spectral elements leads to optimal utilization of modern machine architecture, with natural coarse-grained parallelism, minimal inter-elemental communication, insignificant storage, and minimum vector lengths of $O(N^2)$.

Armed with the hyperbolic and elliptic discrete operators and associated inversion procedures described above, solution of the time-discretized Navier-Stokes equations (3-5), and energy equations (6-7), readily follows [6,7,9]. Slight modifications are required in the pressure step, (4), to insure that the divergence is zero in an appropriate weak sense [7], however the basic approximations employed are analogous to (9-10) above, and it is believed that the accuracy and efficiency results quoted for the individual operators do, indeed, extend to the full Navier-Stokes equations.

5. Flow Simulations

As our first example, we demonstrate the cylindrical geometry capability by considering axisymmetric flow in a pipe expansion with expansion ratio (inlet to outlet) of $d_1/d_2 = 1/2$. The exact geometry and an associated spectral element mesh are shown in Fig. 1. At the inlet we specify parabolic flow, on the pipe walls we require no slip (zero velocity), and at outflow we impose a (fully-developed) zero Neumann condition.

In Fig. 2 we plot the steady streamline pattern (obtained by time integration until a steady-state is achieved) at a Reynolds number of 200. Here the Reynolds number is defined as $R = U_1 d_1/\nu$, where d_1 is the inlet pipe diameter, and U_1 is the inlet centerline velocity. In Fig. 3 the non-dimensional re-attachment length, L_r/d_1, is plotted as a function of Reynolds number over the range $0 < R < 400$. Our results are seen to be in very good agreement with the available experimental data [19].

As our next example we consider the classical problem of two-dimensional flow past an isolated cylinder of diameter d, for Reynolds numbers $R = U_\infty d/\nu$ in the range $0 < R < 200$. We also solve the associated heat transfer forced convection problem, taking the incoming stream to be at temperature $T = 0$, and the cylinder walls to be at temperature $T = 1$. A typical spectral element mesh used is shown in Fig. 4, where it is seen that very high resolution is achieved near the cylinder, as well as moderately high resolution in the wake downstream. The boundary conditions on the computational mesh are taken to be potential flow imposed at the upstream boundary and side "walls", no-slip required on the cylinder walls, and outflow Neumann conditions at the downstream boundaries. At larger Reynolds numbers these boundary conditions are not entirely satisfactory, as the splitting scheme creates non-physical boundary layers at the artificial "walls" of the domain.

For sufficiently low Reynolds numbers ($R < 40$), the flow past a cylinder achieves a steady state. However, for $R > 40$ the stable state is a limit cycle, corresponding to the well-known von Karman vortex street. In Figs. 5a and 5b we show results for $R = 100$ in the form of instantaneous streamlines and isotherms ($Pr=2$), respectively, at a particular time during the cycle. From the (approximately material) isotherms, a clear picture of the Karman street appears. Fig. 6 shows the time behavior of the streamwise velocity at a particular point in the flow domain at $R=200$, from which it can be seen that the Strouhal number of the calculated flow is $St = fd/U_\infty = 0.204$, in reasonable agreement with the experimental value of $St = .19$ [20]. Note also the significant higher harmonic content in the signal. In Fig. 7 we plot the detailed streamlines near the cylinder at one point in the cycle at R=200, showing the complex structure of the oscillatory separation point. Our results appear to bo very similar to those obtained numerically using low-order finite element methods [21].

From the numerical data we can calculate the time-averaged drag coefficient, $C_D = F_D/(^1/_2 \rho U_\infty^2 d)$, and the time,space-averaged Nusselt number, Nu (equivalent here to the average flux over the cylinder surface). In Fig. 8 these two transport measures are plotted as a function of Reynolds number over the range $0 < R < 200$, where it is seen that there is excellent agreement with experiment and available correlations [22-24]. The heat transfer calculation is, in fact, a more severe test of the numerical solution, as it reflects both the boundary layer structure and the wake region of the flow. The computational resources needed to achieve the high accuracy demonstrated here are in fact quite "modest", the calculations requiring roughly 100 seconds on a CRAY-XMP per shedding cycle for a 3,150-node mesh.

As our last example, we consider the full *three-dimensional* flow over a roughness element in a channel. More precisely, the analysis is for pressure-driven flow ($d\bar{p}/dx$ imposed) in an infinite array of cubical disturbances, one periodicity unit of which is shown in Fig. 9. The domain is discretized by 17 identical spectral elements, each with resolution up to 9x9x9. This flow, in addition to being a vehicle for understanding three-dimensional separation, is a good model for circuit board and electronic component cooling configurations, for generic (large) roughness elements, and for extended surfaces in heat exchange devices. The boundary conditions are taken to be periodicity on all "vertical" faces ($\hat{x}$ and $\hat{z}$), with no-slip on all faces of the cube ("chip"), and no-slip on the channel walls ($\hat{y}$ faces). The temperature is taken to be unity on the cube and bottom channel wall, and zero on the top channel wall. We define the Reynolds number of the flow as $R = {}^{d\bar{p}}/_{dx} h^3/2\rho\nu^2$, where h is the half-channel width.

At R = 450 we find that the flow reaches a stable steady-state for large times. In Fig. 10 we show the projection of the velocity vectors and associated temperature fills (for Pr= 1) on a $\hat{z}$-plane halfway through the cube, and on a $\hat{y}$-plane halfway up the cube. These pictures demonstrate the complex three-dimensional separation phenomena, the multiple shear layers, and the effect of the separated flow on the resulting temperature distribution. In Fig. 11 we plot the pressure contours on the cube surface, showing the high pressure (dark pattern) at impingement, and the low pressure (light patterns) on the downstream side of the sharp corner. A slight adverse pressure gradient can also be seen on the top face of the protuberance. Note also the relatively high pressure at the chip base, corresponding to the flow curvature at that point. Lastly, in Fig. 12, we show a surface map of the the Nusselt number (flux) on the cube surface, in which the effect of the flow structure on the resulting thermal boundary layers can be clearly seen. Note, in particular, how the layers thicken and then thin again as the flow turns down at the downstream edge of the cube. At Reynolds numbers slightly higher than those reported here we find that the flow becomes unsteady, an indication that the transition process in this three-dimensional channel flow has begun.

Acknowledgements

This work was supported by the NSF under Grant MEA-8212469, the ONR under Contract N00014-85-K-0208, and by NASA under Grant NAG1-574. Computations were performed on the BCS CRAY-XMP, the NASA-Ames CRAY-XMP, and a DEC VAX 750 supplied in part by a DEC External Research Program Grant.

References

1. Roache, P.J,"Computational Fluid Dynamics," Hermosa Publishers, Albuquerque, 1982.
2. Thomasset,F.,"Implementation of Finite Element Methods for Navier-Stokes Equations,"Springer,1981.
3. Gottlieb,D.O. & Orszag,S.A.,"Numerical Analysis of Spectral Methods,"SIAM,Philadelphia,1977.
4. Babuska,I. & Dorr,M.R., Numer. Math., **37** (1981), 257.
5. Delves,L.M. & Phillips,C., J. Inst. Math. Appl., **25** (1980), 177.
6. Patera,A.T.,J. Comput. Phys., **54** (1984), 468.
7. Korczak,K.Z. & Patera,A.T.,J. Comput. Phys., to appear.
8. Strang,G. & Fix,G.J.,"An Analysis of the Finite Element Method,"Prentice-Hall,Englewood Cliffs, 1973.

9. Ghaddar,N.K.,Karniadakis,G.E., & Patera,A.T., Numerical Heat Transfer, to appear.
10. Ghaddar,N.K.,Korczak,K.Z.,Mikic,B.B., & Patera,A.T., submitted to J. Fluid. Mech.
11. Basdevant,C.,Deville,M.,Haldenwang,P.,Lacroix,J.M., Orlandi,P.,Ouazzini,J., & Patera,A.T., Computers and Fluids, to appear.
12. Patera,A.T., submitted to J. Comput. Phys.
13. Orszag,S.A. & Kells,L.C., J. Fluid Mech., **96** (1980), 159.
14. Karniadakis,G.E., Ph.D. Thesis, Dept. of Mech. Engr., M.I.T.
15. Hestenes,M.R. & Stiefel,E., Nat. Bur. Standards J. Res., **49** (1952), 409.
16. Kershaw,D.S., J. Comput. Phys., **26** (1978), 43.
17. Glowinski,R.,Periaux,J.,& Pironneau,O., in "Preconditioning Methods: Theory and Applications" (ed: D.J. Evans), Gordon & Breach, New York, 1983.
18. Bullister,E.T., Ph.D. Thesis, Dept. of Mech. Engr., M.I.T.
19. Macagno, E.O, Hung, T.K., J. Fluid Mech, **28** (1967), 43.
20. Berger, E., Wille, R., Ann. Rev. Fluid Mech., **4** (1972), 313.
21. Gresho,P.M.,Chan,S.T.,Lee,R.L., & Upson,C.D., Int. J. Num. Meth. Fluids, **4** (1984), 619.
22. Jordan, S., Fromm, J., Phys. Fluids, **15** (1972), 371.
23. Fand, R.M., Int. J. Heat Mass Transfer, **8** (1965), 995.
24. Eckert, E.R.G., Drake, R.M., "Analysis of Heat and Mass Transfer," McGraw Hill, New York, 1972.

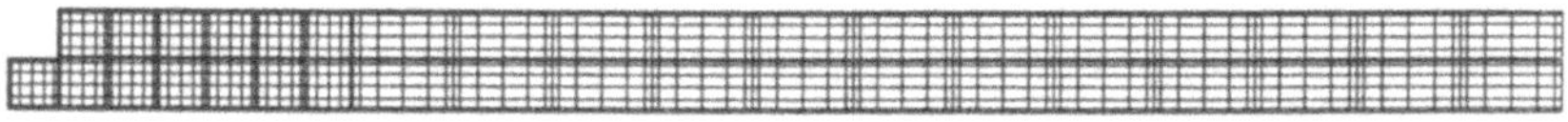

Fig. 1. Typical spectral element mesh used for the pipe expansion calculations ($d_2/d_1 = 2$).

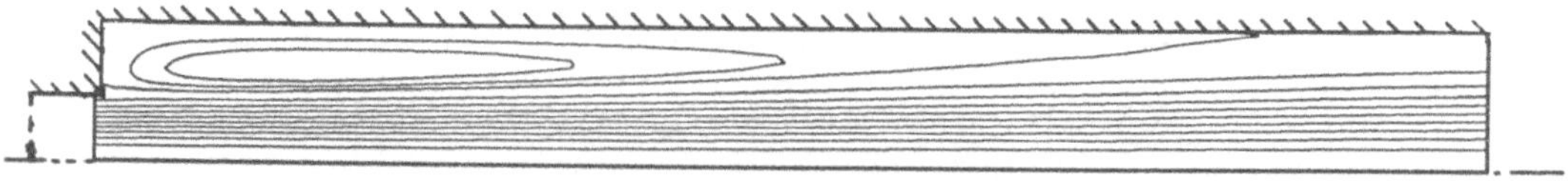

Fig. 2. Steady streamlines for R = 200. The non-dimensional re-attachment length is $L_r/d_1 = 8.45$.

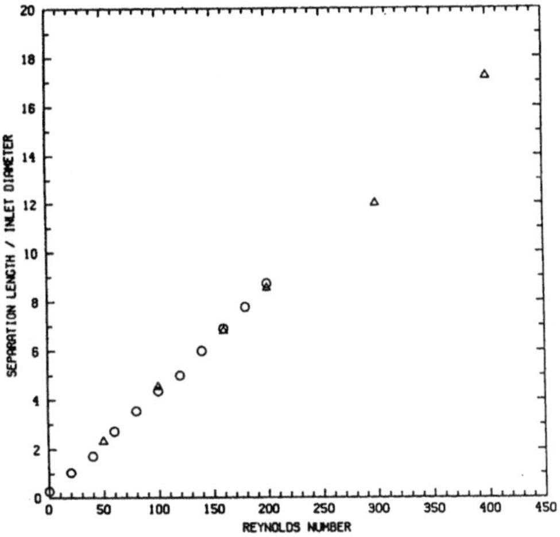

Fig. 3. L_r/d_1 versus Reynolds number; △ - our calculation, ○ - reference [19].

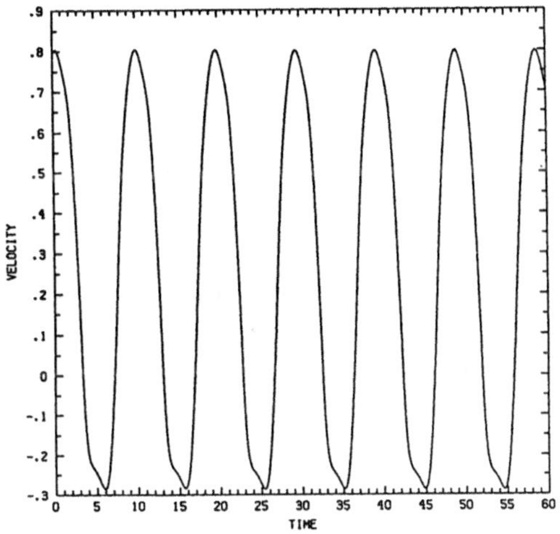

Fig. 6. Streamwise velocity as a function of time for R = 200.

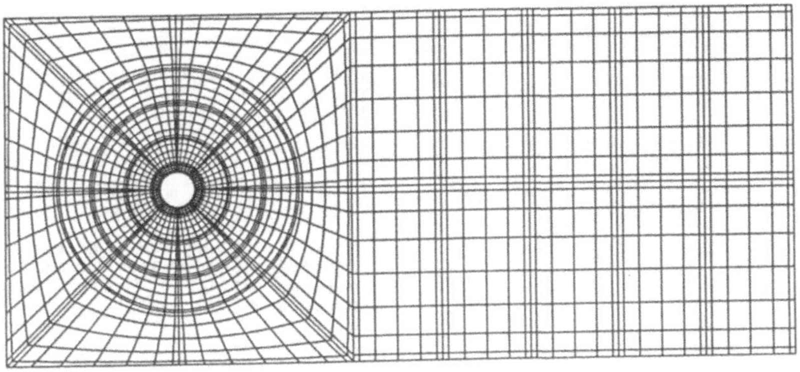

Fig. 4. Typical spectral element mesh used for the flow past a cylinder calculations.

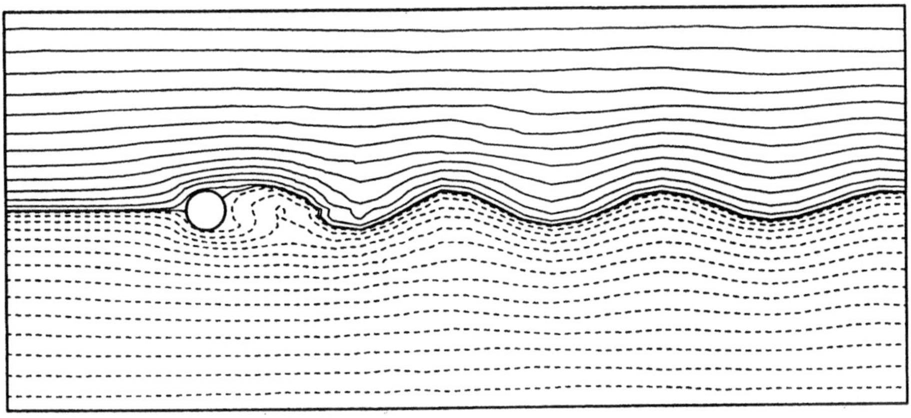

Fig. 5a. Instantaneous streamlines at R = 100.

Fig. 5b. Instantaneous isotherms at R = 100, Pr = 2.

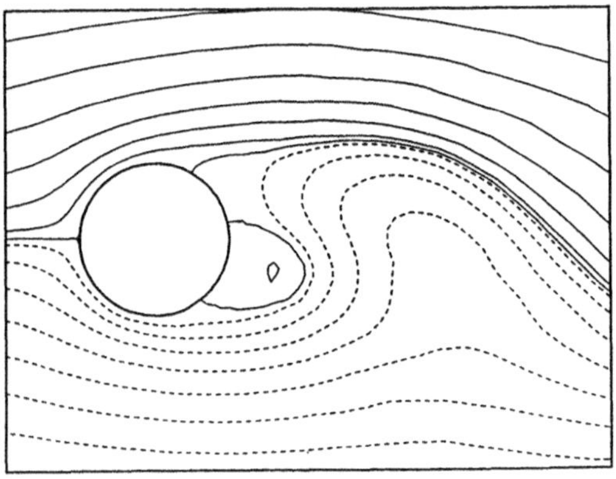

Fig. 7. Instantaneous streamlines near the cylinder for R = 200.

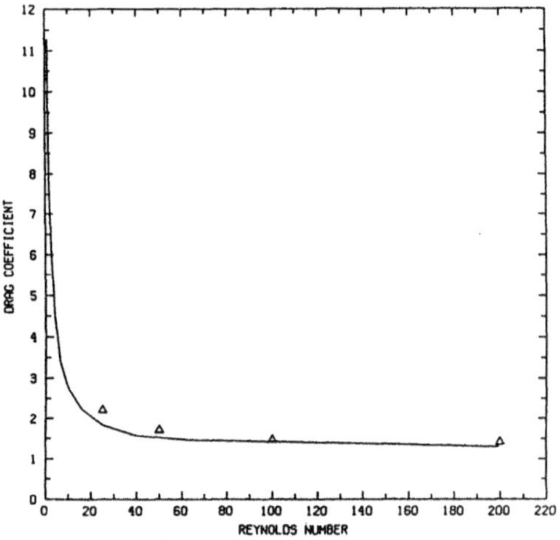

Fig. 8a. Drag coefficient as a function of Reynolds number; Δ - our calculation, solid line - reference [22].

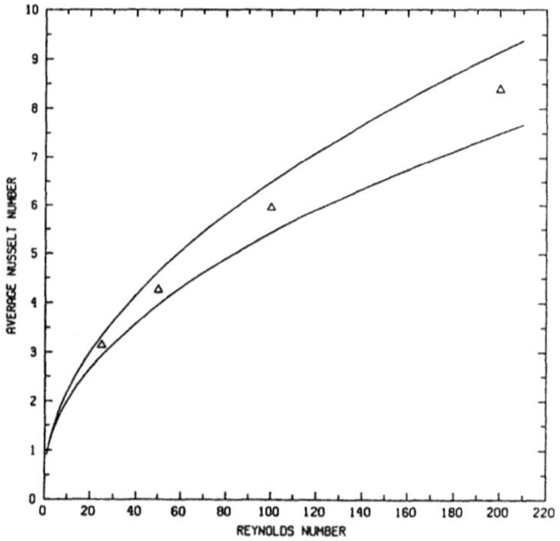

Fig. 8b. Nusselt number as a function of Reynolds number for Pr = 1; Δ - our calculation, upper curve - reference [23], lower curve - reference [24].

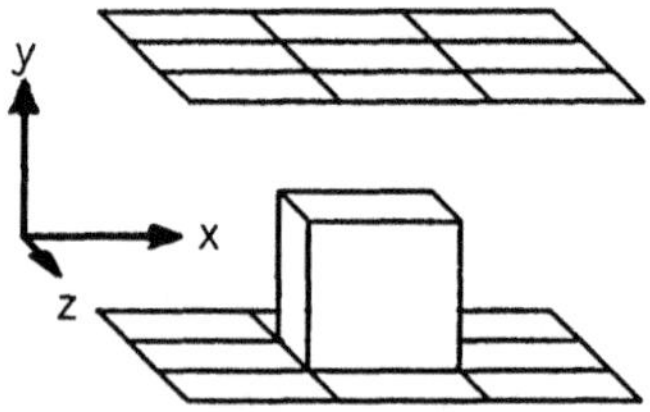

Fig. 9. Periodicity unit of array of three-dimensional roughness elements.

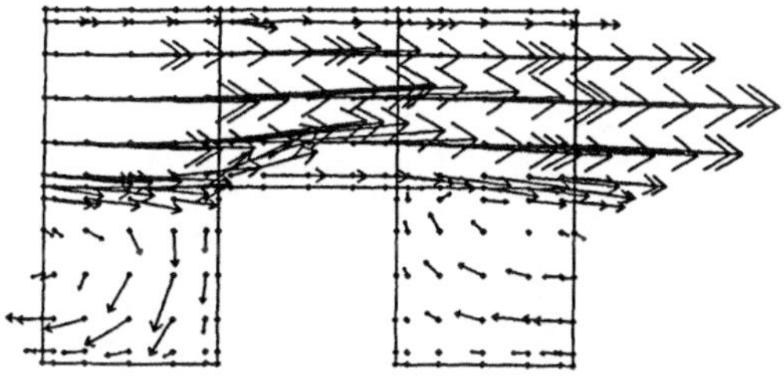

Fig. 10a. Velocity vectors on $\hat{z}$-plane halfway through the chip.

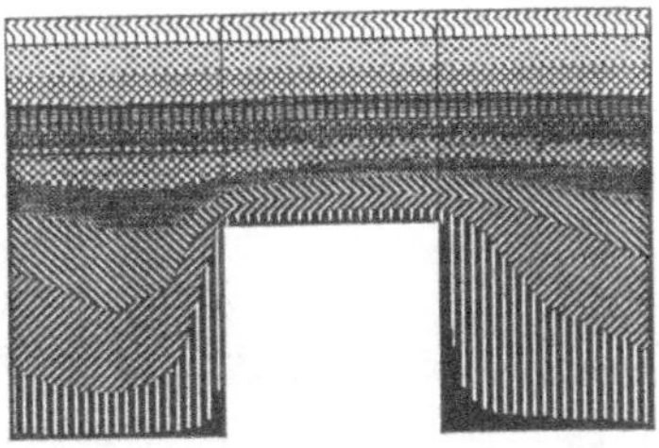

Fig. 10b. Isotherms for the $\hat{z}$-plane halfway through the chip (see legend this page).

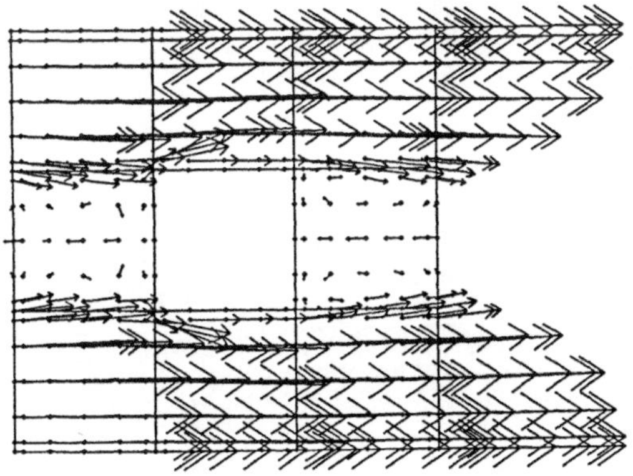

Fig. 10c. Velocity vectors on $\hat{y}$-plane halfway up the chip.

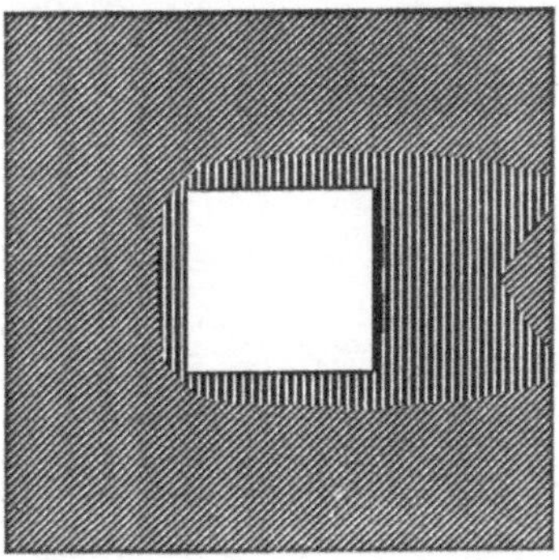

Fig. 10d. Isotherms for the $\hat{y}$-plane halfway up the chip (see legend this page).

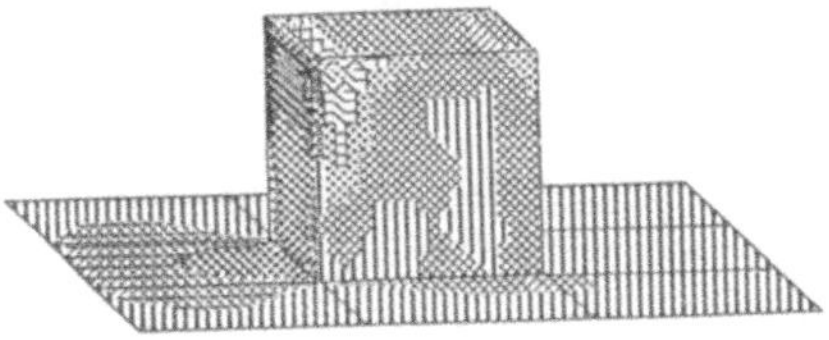

Fig. 11. Pressure contours on the surface of the cube and bottom channel wall (see legend this page).

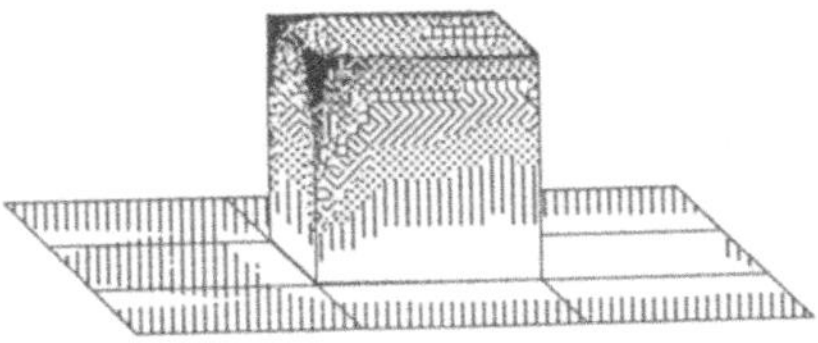

Fig. 12. Nusselt number (flux) contours on the surface of the cube and bottom channel wall (see legend this page).

Legend:

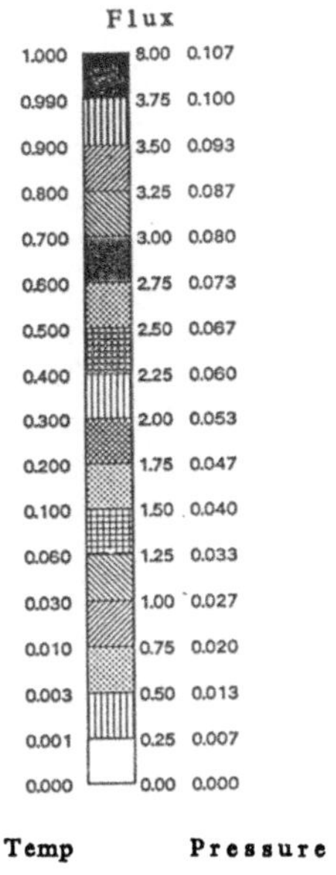

C. A. Brebbia, J. C. F. Telles, L. C. Wrobel

Boundary Element Techniques

Theory and Applications in Engineering

1984. 284 figures. XIV, 464 pages.
ISBN 3-540-12484-5

Contents: Approximate Methods. - Potential Problems. - Interpolation Functions. - Diffusion Problems. - Elastostatics. - Boundary Integral Formulation for Inelastic Problems. - Elastoplasticity. - Other Nonlinear Material Problems. - Plate Bending. - Wave Propagation Problems. - Vibrations. - Further Applications in Fluid Mechanics. - Coupling of Boundary Elements with Other Methods. - Computer Program for Two-Dimensional Elastostatics. - Appendix A: Numerical Integration Formulas. - Appendix B: Semi-Infinite Fundamental Solutions. - Appendix C: Some Particular Expressions for Two-Dimensional Inelastic Problems. - Subject Index.

Finite Element Systems

A Handbook

Editor: **C. A. Brebbia**

3. revised edition. 1985. 456 figures.
XXVI, 767 pages. A Computational Mechanics Centre Publication. ISBN 3-540-15116-8

Contents: Introduction. - Current Finite Element Systems.

Springer-Verlag
Berlin Heidelberg
New York Tokyo

Springer

GPSR Compliance

The European Union's (EU) General Product Safety Regulation (GPSR) is a set of rules that requires consumer products to be safe and our obligations to ensure this.

If you have any concerns about our products, you can contact us on ProductSafety@springernature.com

In case Publisher is established outside the EU, the EU authorized representative is:

Springer Nature Customer Service Center GmbH
Europaplatz 3
69115 Heidelberg, Germany

Batch number: 10844661

Printed by Printforce, the Netherlands